Handbook of Drug Metabolism

Handbook of Drug Metabolism

edited by

Thomas F. Woolf
Parke-Davis Pharmaceutical Research
Ann Arbor, Michigan

CRC Press is an imprint of the
Taylor & Francis Group, an **informa** business

CRC Press
Taylor & Francis Group
6000 Broken Sound Parkway NW, Suite 300
Boca Raton, FL 33487-2742

First issued in paperback 2019

ISBN-13: 978-0-8247-0229-8 (hbk)
ISBN-13: 978-0-367-39985-6 (pbk)

Visit the Taylor & Francis Web site at
http://www.taylorandfrancis.com

and the CRC Press Web site at
http://www.crcpress.com

Preface

A scientist starting a career in the field of drug metabolism research will soon realize that general and comprehensive textbooks on the subject are limited or several years outdated. It has now been well over 25 years since Bert La Du, George Mandel, and E. Leong Way edited their classic book, *Fundamentals of Drug Metabolism and Drug Disposition*, and 22 years since Bernard Testa and Peter Jenner published their well-conceived and written text, *Drug Metabolism: Chemical and Biochemical Aspects.* The more recent comprehensive book on the subject by Testa and Jenner, *Concepts in Drug Metabolism, Part B*, is almost 20 years old. In the early 1980s, as a young drug metabolism investigator, I frequently encountered questions and issues dealing with a specific aspect or a new development in metabolism. In such instances, an updated general reference text would have been an invaluable resource, and as I gained more professional experience and increased my interactions with colleagues in both academic and industrial metabolism groups, it was apparent that many of my colleagues also felt the same need for an updated book dealing with drug metabolism. In many ways, this book is the result of discussions with colleagues and mentors such as William Trager from the University of Washington, Neal Castagnoli, Jr., of Virginia Polytechnic Institute, and Peter G. Welling of Parke-Davis Pharmaceutical Research.

This textbook was designed to provide the reader with information on basic concepts in metabolism, as well as specialized topics such as drug–drug interactions and enzyme regulation. Basically, the book is set up like a graduate course in drug metabolism. It is intended to be not only a useful reference for industrial, academic, and government drug metabolism scientists, but also a teaching tool for those pursuing a career in drug metabolism.

The first few chapters are designed to give the reader an overview of drug metabolism from a historical, kinetic, and chemical context. The chapter by Philip C. Smith should give the reader a feel for the importance of pharmacokinetics in interpreting drug metabolism. Important topics such as clearance, volume of distribution, sequential metabolism, and nonlinear kinetics are included. Neal Castagnoli, Jr., and colleagues prepared a chapter describing some of the chemistry involved in metabolism and the concepts of Phase 1 and Phase 2 metabolites. Especially interesting in this well-written chapter are topics dealing with tertiary amine metabolism and reactive metabolite chemistry.

Felix A. de la Iglesia and colleagues have put together a detailed chapter on hepatic architecture to give the scientist a feel for the complexity and unique properties of this organ system. The book then shifts to enzymology as Paul R. Ortiz de Montellano describes the cytochrome P450 oxidative system. Here the mechanism by which cytochrome

P450 oxidizes substrates is presented in detail along with information on some side reactions of this enzyme system. Allen E. Rettie and Michael B. Fisher describe non-P450 oxygenases capable of carrying out xenobiotic oxidation reactions including flavin-containing monooxygenase (FMO), aldeyde oxidase, and xanthine oxidase. The cytochrome P450 oxidative system is the topic of Brian Burchell's chapter.

Ann K. Daly has prepared an excellent chapter dealing with pharmacogenetics, a topic of great importance as we try to deal with intersubject variabilities in drug response. She covers not only cytochrome P450 polymorphisms but also *N*-acetyltransferases, as well as glutathione conjugation enzymes. F. Peter Guengerich describes the molecular and biochemical aspects of enzyme inhibition. Wayne Hooper next gives an overview of metabolic-based drug interactions, a topic of great importance to the clinician as well as to regulatory agencies. Magnus Ingelman-Sundberg and Martin J. J. Ronis provide a state-of-the-art chapter on cytochrome P450 regulation. Lung metabolism, particularly important for some environmental toxins, is covered by Garold S. Yost.

A. David Rodrigues introduces the reader to the integration of biotechnology to drug metabolism through use of heterologously expressed human drug metabolizing enzymes. This chapter describes how expressed enzymes are used by industrial scientists and some of the precautions of which the scientist needs to be aware when interpreting the results. Thomas Friedberg and colleagues have put together a chapter presenting detailed information on expressed human drug-metabolizing enzymes that comes from a molecular biological perspective. Also mentioned are various efforts with transgenic animals for metabolism experiments. Steven A. Wrighton and his fellow contributors provide a detailed description of perhaps the most useful and accepted in vitro model for metabolism, namely that employing the liver subcellular microsomal fraction. Michael W. Sinz discusses hepatocytes in in vitro metabolism, a crucial topic in the area of xenobiotic biotransformations. Marián Kukan describes the use of liver perfusion experiments in drug discovery and discusses applications with respect to first-pass metabolism and in vitro/in vivo correlations. Munir Pirmohamed and B. Kevin Park consider microsomal experiments and couple them with cytotoxicity studies; thus, their chapter provides practical applications of in vitro methodologies for studying toxicities. John R. Cashman discusses FMO enzymology and its role in drug metabolism, while Shiyin Yee and Wesley W. Day bring to the reader an appreciation for the Caco-2 cell models in drug transport and metabolism process. I. D. Wilson and colleagues elucidate the role of nuclear magnetic resonance spectroscopy as it pertains to drug metabolism.

Philip B. Inskeep and Wesley W. Day have gone to great lengths to describe the practical and applied aspects of drug metabolism as applied to the drug discovery and development paradigm. This chapter, along with the preceding in vitro chapters, illustrates the characterization of the metabolic fate of new chemical entities. In the final chapter, William F. Pool highlights the regulations, guidelines and objectives of clinical metabolism studies.

The *Handbook of Drug Metabolism* is an effort to update the literature on our understanding of drug metabolism concepts and applications by bringing together collaborators from academic and industrial laboratories. While this book provides the reader with many of the important concepts and applications, it is not exhaustive. Topics such as the specific Phase 2 enzymes, extrahepatic metabolism, homology and pharmacophore modeling of cytochromes P450, species differences, and information management are areas that are intended to be topics of subsequent volumes of this handbook.

Thomas F. Woolf

Contents

Contributors

Brian Burchell, Ph.D. Department of Molecular and Cellular Pathology, Ninewells Hospital and Medical School, University of Dundee, Dundee, United Kingdom

John R. Cashman, Ph.D. Human BioMolecular Research Institute, San Diego, California

Kay Castagnoli Department of Chemistry, Virginia Tech, Blacksburg, Virginia

Neal Castagnoli, Jr., Ph.D. Department of Chemistry, Virginia Tech, Blacksburg, Virginia

Ann K. Daly, Ph.D. Department of Pharmacological Sciences, University of Newcastle upon Tyne Medical School, Newcastle upon Tyne, United Kingdom

Wesley W. Day, Ph.D., D.A.B.T. Department of Clinical Research, Pfizer Inc., Groton, Connecticut

Felix A. de la Iglesia, M.D. Department of Pathology and Experimental Toxicology, Parke-Davis Pharmaceutical Research Division, Warner-Lambert Company, Ann Arbor, Michigan

Sean Ekins, Ph.D.* Department of Drug Disposition, Lilly Research Laboratories, Eli Lilly & Company, Indianapolis, Indiana

Michael B. Fisher, Ph.D. Department of Medicinal Chemistry, University of Washington, Seattle, Washington

* *Current affiliation*: Central Research Laboratories, Pfizer Inc., Groton, Connecticut

Thomas Friedberg, Ph.D. Biomedical Research Centre, Ninewells Hospital and Medical School, University of Dundee, Dundee, United Kingdom

F. Peter Guengerich, Ph.D. Department of Biochemistry and Center in Molecular Toxicology, Vanderbilt University School of Medicine, Nashville, Tennessee

Jeffrey R. Haskins, Ph.D. Department of Pathology and Experimental Toxicology, Parke-Davis Pharmaceutical Research Division, Warner-Lambert Company, Ann Arbor, Michigan

C. J. Henderson, Ph.D. Biomedical Research Centre, Ninewells Hospital and Medical School, University of Dundee, Dundee, United Kingdom

Wayne D. Hooper, Ph.D. Department of Medicine, The University of Queensland at Royal Brisbane Hospital, Brisbane, Queensland, Australia

Magnus Ingelman-Sundberg, Ph.D. Division of Molecular Toxicology, Institute of Environmental Medicine, Karolinska Institute, Stockholm, Sweden

Philip B. Inskeep, Ph.D. Animal Health Safety and Metabolism, Pfizer Inc., Groton, Connecticut

Marián Kukan, Ph.D. Laboratory of Perfused Organs, Institute of Preventive and Clinical Medicine, Bratislava, Slovakia

J. C. Lindon, Ph.D. Department of Chemistry, Imperial College School of Medicine, London, United Kingdom

Stéphane Mabic, Ph.D. Department of Chemistry, Virginia Tech, Blacksburg, Virginia

Jukka Mäenpää, Ph.D.* Department of Drug Disposition, Lilly Research Laboratories, Eli Lilly & Company, Indianapolis, Indiana

J. K. Nicholson, Ph.D. Department of Chemistry, Imperial College School of Medicine, London, United Kingdom

Paul R. Ortiz de Montellano, Ph.D. Department of Pharmaceutical Chemistry, University of California–San Francisco, San Francisco, California

B. Kevin Park, Ph.D. Department of Pharmacology and Therapeutics, The University of Liverpool, Liverpool, United Kingdom

Munir Pirmohamed, Ph.D. Department of Pharmacology and Therapeutics, The University of Liverpool, Liverpool, United Kingdom

* *Current affiliation*: Leiras Oy, Helsinki, Finland

William F. Pool, Ph.D. Department of Pharmacokinetics, Dynamics, and Metabolism, Parke-Davis Pharmaceutical Research Division, Warner-Lambert Company, Ann Arbor, Michigan

M. P. Pritchard, Ph.D. Biomedical Research Centre, Ninewells Hospital and Medical School, University of Dundee, Dundee, United Kingdom

Allan E. Rettie, Ph.D. Department of Medicinal Chemistry, University of Washington, Seattle, Washington

Donald G. Robertson, Ph.D. Department of Pathology and Experimental Toxicology, Parke-Davis Pharmaceutical Research Division, Warner-Lambert Company, Ann Arbor, Michigan

A. David Rodrigues, Ph.D. Department of Drug Metabolism, Merck Research Laboratories, West Point, Pennsylvania

Martin J. J. Ronis, Ph.D. Department of Pediatrics, University of Arkansas for Medical Sciences, Arkansas Children's Hospital Research Institute, Little Rock, Arkansas

Michael W. Sinz, Ph.D. Department of Pharmacokinetics, Dynamics, and Metabolism, Parke-Davis Pharmaceutical Research Division, Warner-Lambert Company, Ann Arbor, Michigan

Philip C. Smith, Ph.D. Division of Pharmaceutics, School of Pharmacy, University of North Carolina at Chapel Hill, Chapel Hill, North Carolina

I. D. Wilson, Ph.D. Safety of Medicines Department, Zeneca Pharmaceuticals, Macclesfield, Cheshire, United Kingdom

C. R. Wolf, Ph.D. Biomedical Research Centre, Ninewells Hospital and Medical School, University of Dundee, Dundee, United Kingdom

Steven A. Wrighton, Ph.D. Department of Drug Disposition, Lilly Research Laboratories, Eli Lilly & Company, Indianapolis, Indiana

Shiyin Yee, Ph.D. Department of Drug Metabolism, Pfizer Central Research, Groton, Connecticut

Garold S. Yost, Ph.D. Department of Pharmacology and Toxicology, University of Utah, Salt Lake City, Utah

Handbook of Drug Metabolism

1
Pharmacokinetics of Drug Metabolites

Philip C. Smith
School of Pharmacy, University of North Carolina at Chapel Hill, Chapel Hill, North Carolina

I. INTRODUCTION

Drug metabolites and their disposition in vivo are well recognized by scientists, clinicians, and regulatory agencies to be important when evaluating a new drug entity. In the past two decades increased attention has been placed on drug metabolism for several reasons. First, the number of drugs with active metabolites, by design (i.e., prodrugs) or by chance, has increased (1,2), as exemplified by the present interest in the potential contributions of morphine-6-glucuronide toward the analgesic activity of the age-old drug, morphine (3). In addition, with the advent of methods to establish the metabolic genotype and characterize the phenotype of individual patients (4,5) and the identification of specific isoforms of enzymes of metabolism, there is an increased appreciation of how elimination of a drug by metabolism can influence drug bioavailability and clearance and, ultimately, affect its efficacy and toxicity. These rapidly evolving methods may soon permit cost-effective individual optimization of drug therapy based on a subject's metabolic capability (5), just as renal creatinine clearance has been used for years to assess renal function and permits individualized dose adjustment for drugs cleared by the kidney (6). Finally, the appreciation of the role of bioactivation in the potential toxicity of drugs and other xenobiotics (7) requires that metabolites continue to be evaluated and scrutinized for possible contributions to adverse effects observed in vivo. Although the importance of drug metabolism is seldom questioned, the interpretation and use of pharmacokinetic data on the disposition of metabolites is neither well understood nor fully implemented by some investigators. The objective of this chapter is to provide a basis for the interpretation and use of metabolite pharmacokinetic data from preclinical and clinical investigations.

Several previous authors have reviewed methods and theory for the analysis of metabolite pharmacokinetics, with literature based on simple models, as early as 1963 by Cummings and Martin (8). More recently, thorough theoretical analyses and reviews have been published, notably by Houston (9,10), Pang (11), and Weiss (12). The topic of metabolite kinetics is now also found in some commonly employed textbooks on pharmacokinetics (13), although the topic is usually not presented or taught in a first, introductory course on pharmacokinetics. This chapter is not intended to present all aspects of basic pharmacokinetics that may be necessary for a through understanding of metabolite kinetics; therefore, we recommend that some readers consult other sources (13–16) if an introduction to basic pharmacokinetic principles is needed. This chapter will also not attempt to present or discuss all possible permutations of metabolite pharmacokinetics, but will make an effort to present and distinguish what can be assessed in humans and animals in vivo given commonly available experimental methods, which may occasionally be augmented by in vitro studies.

II. METABOLITE KINETICS FOLLOWING A SINGLE INTRAVENOUS DOSE OF PARENT DRUG

A. General Considerations in Metabolite Disposition

Much of the theory presented here will be based on primary metabolites, as shown in Scheme 1, that are formed directly from the parent drug or xenobiotic for which the initial dose is known. In contrast, secondary or sequential metabolites, as indicated in Scheme 2, are formed from one or more primary metabolites. The theory and resultant equations for the analysis of sequential metabolite kinetics are often more complex (see Sec. V; 11). Because most metabolites of interest are often primary ones, this review will focus on these, unless otherwise noted. Scheme 1 is the simplest model for one metabolite that can be measured in vivo with other elimination pathways for the parent drug, either by metabolism or excretion (e.g., biliary or renal), represented as a combined first-order elimination term k_{other}. Pharmacokinetic models will be presented here for conceptual reasons, but where model-independent or ''noncompartmental'' methods are appropriate, their applications will be discussed.

Here A is the amount of drug or xenobiotic administered, $A(\text{m})$ is the amount of a particular metabolite present in the body with time. When sequential metabolism is occurring metabolites are distinguished with a subscript; $A(\text{m}_1)$ is the amount of primary metabolite present with time, $A(\text{m}_2)$ is the amount of secondary, or sequential, metabolite formed with time. $A(\text{m})_{\text{elim}}$ is the amount of the primary metabolite of interest that is excreted (e.g., biliary or renal) or further metabolized. The parameter k_f is the first-order

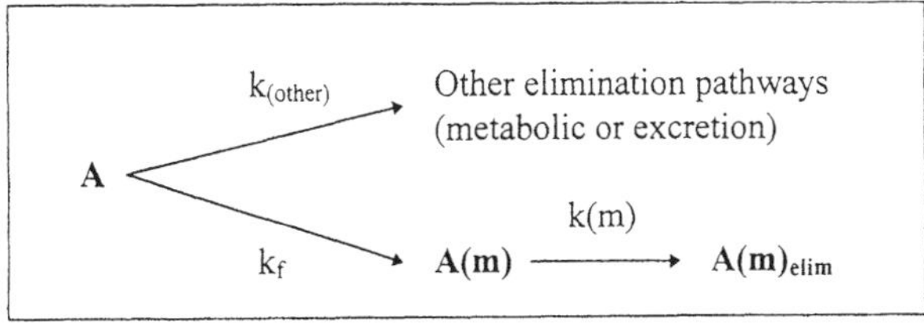

Scheme 1 Drug metabolism to a primary metabolite followed by urinary excretion with parallel elimination pathways.

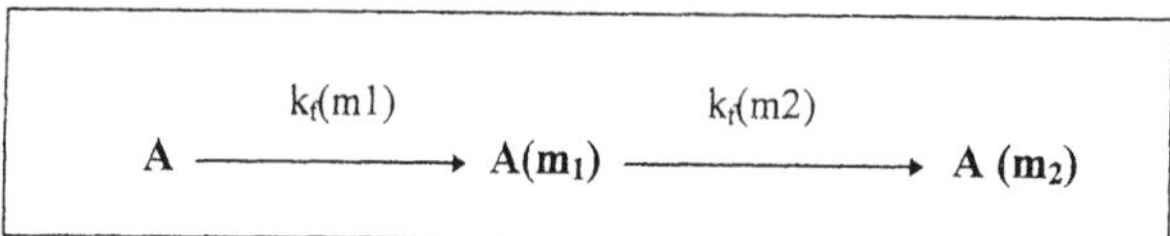

Scheme 2 Sequential metabolism to a secondary metabolite m_2, from a primary metabolite m_1.

formation rate constant for the metabolite and k_{other} represents a first-order rate constant for the sum of formation of other metabolites and elimination by other pathways. The constant $k(m)$ is the elimination rate constant for the metabolite, whereas the sum of k_f and k_{other} is k, the total first-order rate constant for the overall elimination of the parent drug. With this simple model, and derivations from this, the following assumptions will be employed unless otherwise noted:

1. The elimination and distribution processes are first order and thus linear; that is, they are not influenced by the concentration of drug or metabolite in the body. For example, saturation of enzyme and transport systems, cosubstrate depletion, saturable plasma protein, or tissue binding do not occur.
2. All drug metabolism represents irreversible elimination of the parent drug, thus, there is no reversible metabolism, enterohepatic recycling, or bladder resorption.
3. For simplicity, a one-compartment model will be used that assumes rapid distribution of parent and metabolite within the body.
4. There is no metabolism that results in metabolite being eliminated without first being presented to the systemic circulation.

From Scheme 1, the following equation is used to describe the rate of change in the amount of metabolite in the body at any time, which is equal to the rate of formation less the rate of elimination,

$$\frac{dA(m)}{dt} = k_f \cdot A - k(m) \cdot A(m) \tag{1}$$

This rate of input (i.e., formation) and output (i.e., elimination) is analogous to the form of the equation for first-order drug absorption and elimination (17). The amounts of metabolite and parent drug present in the body on initial intravenous bolus dosing of the drug are zero and the administered dose (D), respectively. The disposition of parent drug can be described with an exponential term, as shown in Eq. (2),

$$A = D \cdot e^{-k \cdot t} \tag{2}$$

Substitution of A into Eq. (1) permits solving for $A(m)$ as a function of time (17),

$$A(m) = \frac{k_f \cdot D}{k(m) - k}[e^{-k \cdot t} - e^{-k(m) \cdot t}] \tag{3}$$

Because the amount of metabolite is often unknown, but instead metabolite concentration $C(m)$ is measured in plasma, which can be expressed by dividing both sides of Eq. (3) by the volume of distribution of the metabolite $V(m)$ as follows:

$$C(\mathrm{m}) = \frac{k_\mathrm{f} \cdot D}{V(\mathrm{m}) \cdot (k(\mathrm{m}) - k)} [e^{-k \cdot t} - e^{-k(\mathrm{m}) \cdot t}] \quad (4)$$

Equations (3) and (4) describe the amount and concentration, respectively, of a primary metabolite in the body over time after an intravenous bolus dose of the parent drug. Immediately after dosing there is no metabolite present, and the amount of metabolite will then reach a maximum when the rate of formation equals the rate of elimination of the metabolite. This peak occurs when $t_{\mathrm{m,peak}} = \ln[k/k(\mathrm{m})]/[k - k(\mathrm{m})]$ (17). Here k and $k(\mathrm{m})$ determine the shape of the drug and metabolite concentration versus time profiles, whereas k_f influences the fraction of the dose that is metabolized, thus affecting the magnitude of the metabolite concentration. It is apparent that the relative magnitude or ratio of the two rate constants for the elimination of parent drug and metabolite determines the overall profile of the metabolite relative to that of the parent drug, with two limiting cases described in the following two sections.

B. Formation Rate-Limited Metabolism

In the first example, if $k(\mathrm{m}) \gg k$, then the metabolite is eliminated by either excretion or further sequential metabolism much more rapidly than the rate at which the parent drug is eliminated. Since $k = k_\mathrm{f} + k_{\mathrm{other}}$, it also follows that $k(\mathrm{m}) \gg k_\mathrm{f}$. Under this condition, defined as *formation rate-limited* (FRL) metabolism, the exponential term describing metabolite elimination in Eqs. (3) and (4) $e^{-k(\mathrm{m}) \cdot t}$ declines rapidly to zero relative to the exponential term describing parent drug elimination $e^{-k \cdot t}$, and the term in the denominator $(k(\mathrm{m}) - k)$ approaches the value $k(\mathrm{m})$. Thus, shortly after an intravenous bolus dose of parent drug, Eq. (3) simplifies to,

$$A(\mathrm{m}) = \frac{k_\mathrm{f} \cdot D}{k(\mathrm{m})} [e^{-k \cdot t}] \quad (5)$$

Equation (4) can be simplified similarly and, if one then takes the natural log (ln) of both sides of Eq. (5), then the amount of metabolite in the body can be described by a linear relation relative to time,

$$\ln A(\mathrm{m}) = \ln\left(\frac{k_\mathrm{f} \cdot D}{k(\mathrm{m})}\right) - k \cdot t \quad (6)$$

A similar relation to Eq. (6) can be derived from Eq. (4) using concentrations rather than amounts when FRL metabolism applies. This log-linear relation, common to first-order systems, indicates that when $k(\mathrm{m}) \gg k$ the terminal half-life measured for the metabolite amount or concentration versus time curves represent that of the parent drug, not that of the metabolite. This is shown in Figs. 1A and 2. Moreover, as $k(\mathrm{m})$ increases the $t_{\mathrm{m,peak}}$ for the metabolite approaches zero. In this case, the metabolite will reach peak concentrations very quickly after a bolus dose of the parent drug.

Equation (5) can also be rearranged to indicate that the rate of elimination of metabolite $k(\mathrm{m})A(\mathrm{m})$ approximates its rate of formation from the parent drug, $k_\mathrm{f}D\,[e^{-kt}] = k_\mathrm{f}A$, where A is the amount of parent drug in the body at any time after the dose,

$$k(\mathrm{m})A(\mathrm{m}) = k_\mathrm{f} \cdot D \cdot e^{-kt} \quad (7)$$

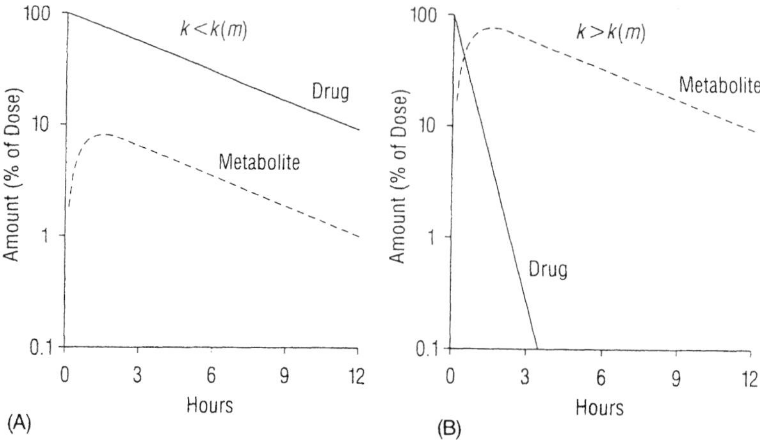

Fig. 1 Drug and metabolite profiles simulated for the two common cases from Scheme 1. (A) Formation rate-limited metabolism where $k(\mathrm{m}) > k$ (k_{f}, k_{other} and $k(\mathrm{m})$ are 0.2, 0, and 2, respectively). (B) Elimination rate-limited metabolism where $k(\mathrm{m}) < k$ (k_{f}, k_{other}, and $k(\mathrm{m})$ are 2, 0, and 0.2, respectively. Shown are amounts expressed as percentage of the dose; if converted to concentrations of parent drug and metabolite the relative ratios of the curves may change as determined by V and $V(\mathrm{m})$, respectively. (From Ref. 13.)

Thus, with FRL metabolism an apparent equilibrium exists between the formation and elimination of metabolite such that the ratio of metabolite to parent drug is approximately constant soon after a dose of the parent drug. Because the term De^{-kt} describes the amount of parent drug in the body at any time, shortly after an intravenous bolus dose of the parent drug, the ratio of amount of metabolite to drug is,

$$A(\mathrm{m}) = \frac{k_{\mathrm{f}}}{k(\mathrm{m})} A \tag{8}$$

Because the volume of distributions of parent V and metabolite $V(\mathrm{m})$ are assumed constant, Eq. (8) can be rewritten by multiplying by volume terms to provide concentrations and clearance terms where, $CL_{\mathrm{f}} = k_{\mathrm{f}} V$ and $CL(\mathrm{m}) = k(\mathrm{m})V(\mathrm{m})$, and the units of clearance are volume/time,

$$C(\mathrm{m}) = \frac{k_{\mathrm{f}} \cdot V}{k(\mathrm{m}) \cdot V(\mathrm{m})} C = \frac{CL_{\mathrm{f}} \cdot C}{CL(\mathrm{m})} \tag{9}$$

Although the metabolite concentration in the body $C(\mathrm{m})$ at any time after a dose of parent drug can be determined by assay of plasma samples, if $k(\mathrm{m}) \gg k$, it is not possible to estimate $k(\mathrm{m})$ unambiguously. An estimate of $k(\mathrm{m})$ can be determined only by obtaining metabolite plasma concentration versus time data following an intravenous dose of the metabolite itself. These relations do, however, indicate that the concentration ratio of metabolite versus parent will essentially be constant over time (see Fig. 1A). This ratio will be useful when relations of concentration and clearance are discussed later.

With FRL metabolism the observed apparent half-life of metabolite from concentration versus time curves is related to a first-order elimination rate constant of the parent

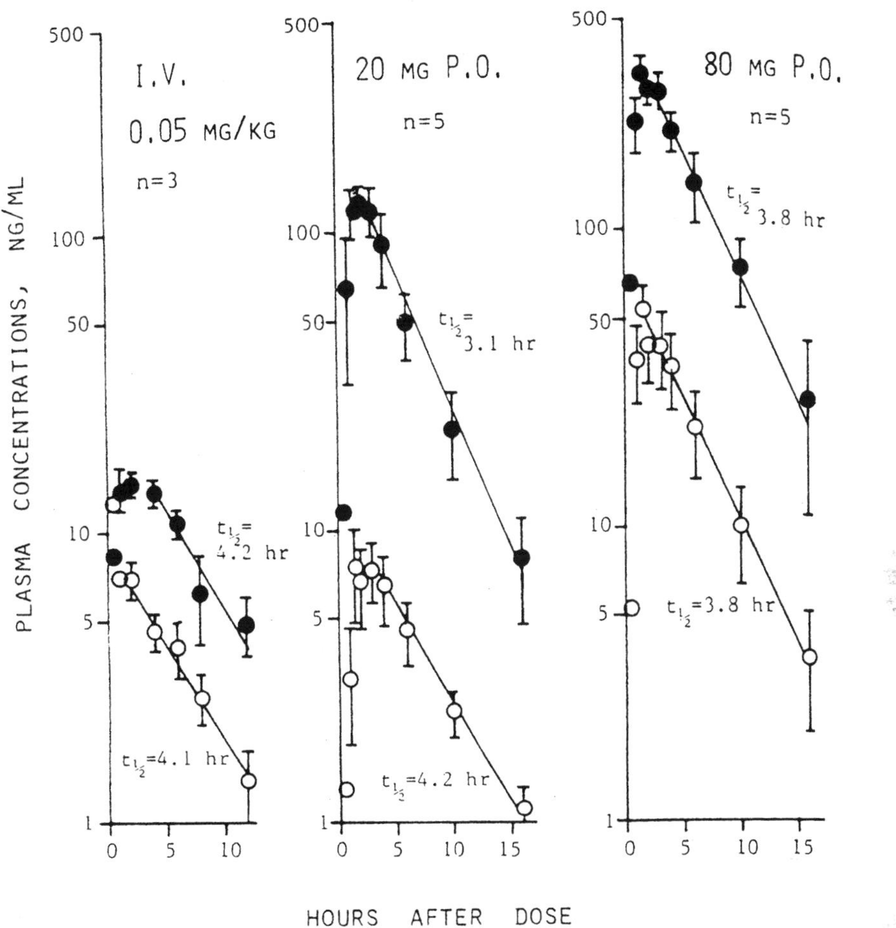

Fig. 2 Plasma levels of naphthoxylacetic acid (NLA; ●), and propranolol (○; mean ± SEM) after single intravenous doses of propranolol in normal human subjects. NLA exhibits FRL metabolism with parallel half-lives and rapid attainment of peak levels. The higher ratios of NLA/propranolol at equilibrium are due to a much lower ratio of $CL(m)/CL$, however, $V(m)/V$ must be lower still [see Eq. (15)] to provide $k(m)/k > 1$, characteristic of FRL metabolism. (From Ref. 18.)

drug $t_{1/2} = \ln 2/k$, and the elimination half-life of the parent drug is longer than that of the metabolite if the metabolite were dosed independently. Because the metabolite is eliminated much faster than it is formed, its true half-life is not apparent and the concentration versus time profile of the metabolite follows that of the parent drug as shown in Figs. 1A and 2 where the log of the amount or plasma concentration are plotted on the ordinate. For naphthoxylacetic acid, a metabolite of propranolol (see Fig. 2), its plasma profile parallels that of propranolol whether the parent drug is given intravenously or orally (18). Even if the parent drug displayed more complex disposition characteristics with an initial distribution phase noted after an IV bolus dose (i.e., a two-compartment model) or perhaps secondary absorption peaks owing to enterohepatic recycling, one would still expect to see a parallel profile for a metabolite subject to FRL metabolism. The observation of parallel metabolite and parent profiles after dosing the parent drug can also occur in cases of reversible metabolism, thus this possibility should also be considered (see Sec. VIII).

C. Elimination Rate-Limited Metabolism

The second limiting case is when $k(\mathrm{m}) \ll k$; that is, the elimination half-life of the metabolite is much longer than that of the parent drug. Here the metabolite is eliminated by either excretion or sequential metabolism, with a first-order rate constant that is much smaller than the rate constant for elimination of the parent drug ($k = k_\mathrm{f} + k_\mathrm{other}$), and this situation is defined as *elimination rate-limited* (ERL) metabolism. There is no requirement for the relative magnitude of $k(\mathrm{m})$ and k_f. Because of the differences between $k(\mathrm{m})$ and k, the exponential term $e^{-k \cdot t}$ in Eqs. (3) and (4) declines rapidly relative to the exponential term $e^{-k(\mathrm{m}) \cdot t}$, and the denominator, $(k(\mathrm{m}) - k)$, approaches the value of $-k$. Thus, when $k(\mathrm{m}) \ll k$, after most of the parent drug has been eliminated, Eq. (3) which describes an IV bolus of parent drug simplifies to,

$$A(\mathrm{m}) = \frac{k_\mathrm{f} \cdot D}{k}[e^{-k(\mathrm{m}) \cdot t}] \tag{10}$$

With the case of ERL, Eq. (4) also simplifies to express metabolite concentration versus time after an intravenous dose of the parent drug. This simplification indicates that with ERL metabolism, a log-linear plot of amount or concentration of metabolite versus time would have a terminal slope reflecting the true elimination half-life for the metabolite [i.e., $t_{1/2} = \ln 2/k(\mathrm{m})$]. This is shown in Fig. 1B and an example of this type of metabolite profile is exemplified by the disposition of *N*-desalkylhalazepam, a metabolite of halazepam shown in Fig. 3 (19). Under the condition of ERL metabolism, the elimination half-life of the metabolite is unambiguously and clearly resolved from that of the parent drug. Prodrugs are generally designed to follow ERL metabolism where the prodrug is rapidly metabolized to the active moiety that persists in the body for a much longer time than the parent prodrug (e.g., aspirin forming salicylate or mycophenolate mofetil forming mycophenolic acid). Considerations of possible accumulation of metabolite under ERL metabolism after prolonged dosing of parent drug will be addressed in section IV.

D. Rates of Metabolite Elimination Approximately Equal to Rates of Elimination of Parent Drug

The foregoing conditions of FRL and ERL metabolism permit simplification of the equations describing the disposition of the metabolite. However, when $k(\mathrm{m})$ is close to the value of k, log-linear plots of metabolite concentration versus time do not in theory become apparently linear in the terminal phase of a concentration versus time profile because neither exponential term of Eqs. (3) or (4) will become negligible and drop out as time progresses. Error in the analysis of plasma concentrations will also contribute to inability to discern a value for $k(\mathrm{m})$ from such a plot. Under these conditions, the use of log-linear plots to estimate $k(\mathrm{m})$, and subsequently, the elimination half-life of the metabolite, will lead to an underestimate of $k(\mathrm{m})$ (10). Therefore, caution should be used when interpreting metabolite elimination rates and half-life data when clear distinction of $k(\mathrm{m})$ from k cannot be made. In practice, when $k(\mathrm{m}) \cong k$ it is possible that noise in the data may make the

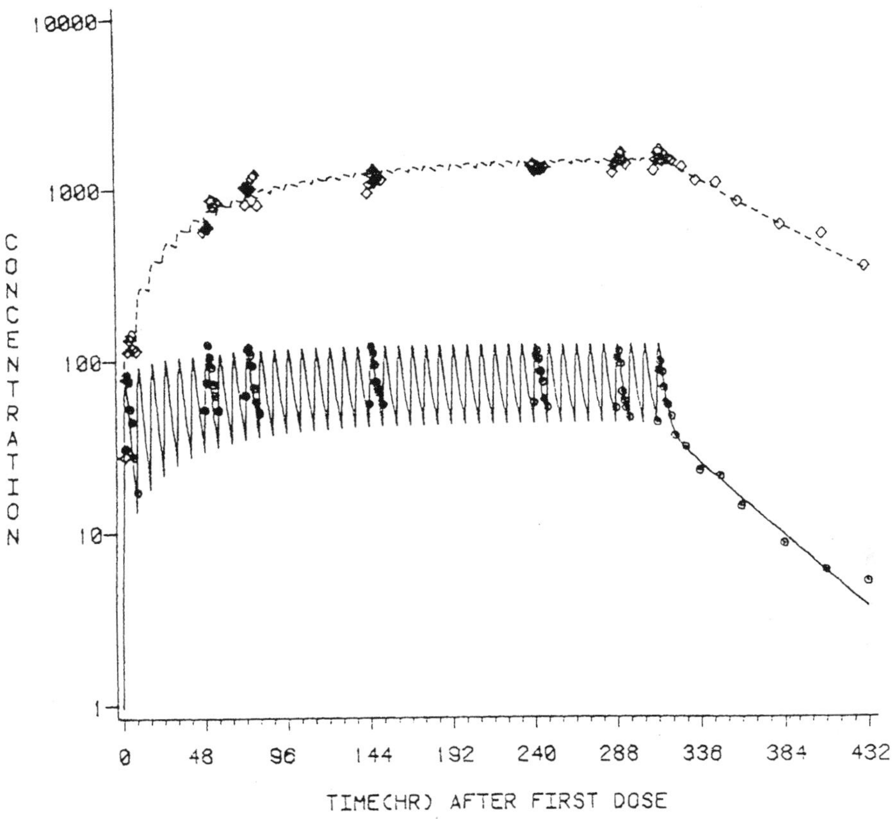

Fig. 3 Plasma levels (ng/mL) of halazepam (⊙) and *N*-desalkylhalazepam (◇) after 40 mg halazepam every 8 h for 14 days showing characteristics of elimination rate-limited metabolism. The estimated elimination half-lives for halazepam and *N*-desalkylhalazepam are 35 and 58 h, respectively. Notable for the longer half-life of the metabolite are a longer time to achieve steady state than the parent drug and smaller fluctuations between doses, as seen in the simulated fit of the data. (From Ref. 19.)

terminal phase of the log concentration versus time plots appear reasonably log-linear. A discussion of how to analyze data when it appears that $k(\mathrm{m}) \cong k$ is presented in the earlier review by Houston (10).

E. Clearance and Volume of Distribution for Metabolites

Most of the foregoing discussion dealt with amounts of drug and metabolite in the body and methods to simply distinguish FRL and ERL metabolism and also to estimate $k(\mathrm{m})$ and half-life for a metabolite with ERL metabolism. However, in most instances, amounts are not known because concentrations of metabolite are determined in plasma or blood over time after dosing the parent drug. Because metabolites are seldom administered to humans, volume of distribution of the metabolite cannot be determined to allow unambiguous conversion of observed metabolite concentrations to amounts in the body. It may be possible, given availability of metabolite, to administer metabolite to animals to determine relevant pharmacokinetic parameters; however, extrapolation of pharmacokinetic values from animals to humans is complex and problematic. If the preformed metabolite can be administered to humans, relevant pharmacokinetic parameters can be determined, as commonly employed for the parent drug (14–17). However, one needs to be considerate of the possibility that preformed metabolite dosed exogenously into the systemic circulation may behave differently than metabolite formed within specific tissues of the body such as the liver or kidney. Much of the discussion here will focus on basic clearance concepts that are applicable to metabolites, given limited knowledge of their disposition in humans, and may provide insight into the disposition of the metabolite if some assumptions are made.

The mass balance relation in Eq. (1) can be modified by multiplying $k(\mathrm{m})$ by $V(\mathrm{m})$ and then dividing $A(\mathrm{m})$ by $V(\mathrm{m})$, which provide the values of metabolite clearance $CL(\mathrm{m})$ and $C(\mathrm{m})$, respectively. Similarly, multiplying and dividing the other terms in Eq. (1) by the volume of distribution of the parent drug V, provides CL_f (k_f V) and C, respectively, where CL_f is the fractional clearance of parent drug to form the metabolite and C is the concentration of the parent. CL_f can also be expressed as the product $f_\mathrm{m}CL$, where f_m is the fraction ($f_\mathrm{m} = k_\mathrm{f}/k = CL_\mathrm{f}/CL$) of systematically available dose of parent drug that is converted irreversibly to the metabolite of interest. For the purposes of the discussion here, it will be assumed that any metabolite formed is systemically available and not subject to sequential metabolism or excretion without being presented to the systemic circulation. With this assumption and the foregoing substitutions, Eq. (1) can be rewritten as,

$$\frac{\mathrm{d}A(\mathrm{m})}{\mathrm{d}t} = CL_\mathrm{f} \cdot C - CL(\mathrm{m}) \cdot C(\mathrm{m}) \tag{11}$$

It is useful to consider the integration of Eq. (11) relative to time from zero to infinity after an intravenous bolus of the parent drug. Since metabolite amounts in the body at times zero and infinity are zero, and the terms of CL_f and $CL(\mathrm{m})$ are assumed to be constant, the integral of concentration versus time is the area under the concentration versus time curve (AUC). The following relation is obtained,

$$CL_\mathrm{f} \cdot \mathrm{AUC} = CL(\mathrm{m}) \cdot \mathrm{AUC(m)} \tag{12}$$

Where AUC(m) and AUC are the area under the plasma concentration versus time curve for the metabolite and parent drug, respectively. Since the product of a clearance and an AUC term is an amount, $CL_\mathrm{f} \cdot \mathrm{AUC}$ equals the amount of metabolite formed from the parent that reaches the systemic circulation, which for an intravenous dose equals $f_\mathrm{m}D$. The value of CL_f or f_m is usually not unambiguously known, but may occasionally be estimated with some assumptions, (e.g., no sequential metabolism and all metabolite formed is excreted in the urine). Because CL_f is defined as $f_\mathrm{m} \cdot \mathrm{CL}$, this can be substituted into Eq. (12) and then rearranged to provide,

$$\frac{f_\mathrm{m} \cdot CL}{CL(\mathrm{m})} = \frac{\mathrm{AUC(m)}}{\mathrm{AUC}} \tag{13}$$

This relation indicates that the relative AUCs of the metabolite versus parent drug will be dictated by the elimination clearances of metabolite and parent drug and the magnitude of the fraction of the dose that is directed toward the particular metabolite. For example, the AUC of morphine-3-glucuronide (M3G) is much greater than the AUC of morphine (ratio about 7.8). Since the value of f_m cannot exceed unity, and a collection of urine long enough to estimate total recovery indicated that M3G averages 44% of the IV dose, then it is apparent that the ratio of $CL/CL(m)$ must be much greater than 1 (about 17). It was reported that the half-life of M3G and morphine were 3.9 and 1.7 h (20), respectively, thus M3G follows ERL metabolism which is consistent with the much lower clearance of the metabolite contributing to its slow rate of elimination. With relative measures of clearance available from Eq. (13), one can also estimate the relative magnitude of the volume of distribution between metabolite and parent drug. Morphine being basic has a fairly large volume of distribution, estimated to be 4 L/kg (20,21). In this case of ERL metabolism for which the relative values of k and k(m) can be determined, one can substitute the relation, $CL = kV$ into Eq. (13) and rearrange the equation to estimate relative values for the volumes of distribution,

$$\frac{V}{V(\mathrm{m})} = \frac{1}{f_\mathrm{m}}\left[\frac{k(\mathrm{m}) \cdot \mathrm{AUC(m)}}{k \cdot \mathrm{AUC}}\right] \tag{14}$$

Using the data presented in Fig. 4 and associated data (20), Eq. (14) provides an estimate of V(m) for M3G of about 0.5 L/kg, which is roughly one-seventh the value for V of morphine in adults. M3G has not been administered to humans; however, following an infusion of a diamorphine (a prodrug of morphine) to infants the V(m) of M3G was esti-

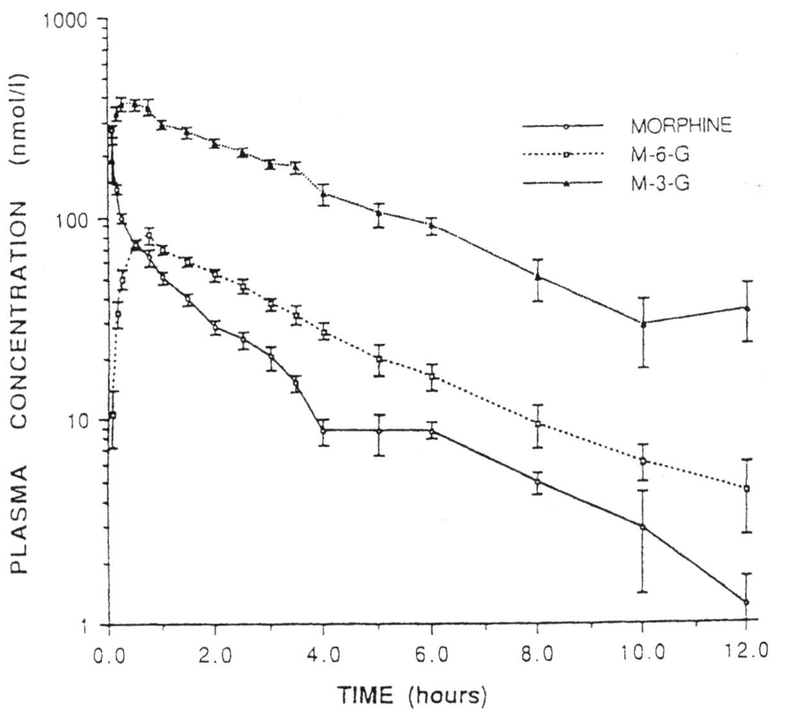

Fig. 4 Plasma levels (mean ± SEM) for morphine, morphine-6-glucuronide (M6G), and morphine-3-glucuronide (M3G) in humans after a 5-mg–intravenous bolus. The plot and associated data indicate that M3G has FRL metabolism as its average half-life is more than twofold longer than that of morphine. (From Ref. 20.)

mated to be 0.55 L/kg. Also when its active analgetic isomer M6G, was given to humans, it was determined to have a small volume of distribution of only 0.3 L/kg (22). This example shows that the much larger AUC for M3G relative to morphine is due to a smaller clearance for the metabolite. The high peak M3G concentration is likely due to the rapid formation of the metabolite, which has a smaller volume of distribution because of its lower partitioning into tissues relative to the much more lipophilic parent drug, morphine.

The elimination rate constant, which is a parameter dependent on clearance and inversely dependent on the volume of distribution (i.e., $k = CL/V$), is lower for M3G owing to $CL(\mathrm{m})$ being substantially smaller than CL. This relation is summarized by the following,

$$\frac{k(\mathrm{m})}{k} = \frac{CL(\mathrm{m}) \cdot V}{CL \cdot V(\mathrm{m})} \tag{15}$$

In cases where metabolism is FRL, the value of $k(\mathrm{m})$ cannot be estimated; thus, the relative volumes of distribution cannot be determined using Eq. (14), even when f_{m} is known. However, Eq. (13) is quite useful in estimating the important parameter $CL(\mathrm{m})$, which can be used to predict average concentrations of the metabolite on chronic administration, as will be discussed later. From Eq. (13) and the example of naphthoxylacetic acid/propranolol shown in Fig. 2 where AUC(m)/AUC is much greater than 1, it is apparent that the clearance of naphthoxylacetic acid is much smaller than that of its parent drug [$CL(\mathrm{m})/CL \ll 1$] because the value of f_{m} cannot exceed unity, propranolol forms other known metabolites and only 14% of the dose was excreted in urine as naphthoxylacetic acid. When Eq. (15) is then considered and since $k(\mathrm{m})/k$ must exceed the value of 1 for FRL metabolism, it is apparent that the ratio of $V/V(\mathrm{m})$ must be large to compensate for the small ratio of $CL(\mathrm{m})/CL$ for naphthoxylacetic acid/propranolol. Thus, $V(\mathrm{m})$ must be much smaller than V, which is also confirmed by the high concentrations of naphthoxyacetic acid relative to that of propranol shown in Fig. 2 soon after the dose.

Consideration of the two primary pharmacokinetic parameters, clearance and volume, in Eq. (15) also provides an understanding of why FRL is more common than ERL metabolism [i.e., $k(\mathrm{m})/k$ is greater than 1 for most metabolites]. Most metabolites are more polar than the parent drug owing to oxidation, hydrolysis, or conjugation; thus, they often distribute less extensively in the body [$V(\mathrm{m}) < V$]. Exceptions to this may be metabolic products from methylation or acetylation that may be similar or more lipophilic than the respective parent drug. Most metabolites also have higher clearances than the parent drug, owing to susceptibility to further phase II metabolism, enhanced biliary or renal secretion once a polar or charged functional group is added by biotransformation (e.g., oxidation to a carboxylic acid, conjugation with glucuronic acid, glycine, glutathione, or sulfate), or reduced protein binding, which may increase renal filtration clearance and increase clearance of metabolites with low extraction ratios. Together, these effects of a smaller volume and higher clearance result in FRL metabolism being the most commonly observed behavior for metabolites.

In ERL metabolism, volume of distribution of a metabolite can be estimated using Eq. (14) if the parent drug can be administered as an intravenous dose to humans. With FRL metabolism, $k(\mathrm{m})$ cannot often be unambiguously determined from plasma metabolite concentration versus time data in humans; therefore, $V(\mathrm{m})$ cannot be easily determined. However, volume of distribution for a metabolite in humans may be extrapolated from the values of $V(\mathrm{m})$ obtained in animals after intravenous dosing of the metabolite, if such data are available. Because volume is a parameter that, to great extent, is dependent on physiocochemical properties of a compound and binding to tissues, this parameter when corrected for differences in plasma protein binding tends to be more amenable to interspecies scaling than is clearance (23,24). With a prediction of $V(\text{m})$ in humans based on interspecies scaling, Eq. (15) may be used to estimate $k(\text{m})$ for cases of FRL metabolism if f_{m} or $CL(\text{m})$ are known.

Volume of distribution for a metabolite at steady state [$V(\text{m})_{\text{ss}}$] can also be estimated from mean residence time (MRT) measurements as discussed in the following section.

F. Mean Residence Time for Metabolites After an Intravenous Dose of Parent Drug

Mean residence time (MRT) in the body is a measure of an average time that a molecule spends in the body after a dose and is a pharmacokinetic parameter that can be employed to describe metabolite disposition. MRT is considered a noncompartmental parameter based on statistical moment theory (25); however, its use does assume that processes of metabolite formation and clearance are first-order and linear (i.e., not dose- or time-dependent), the metabolite is formed irreversibly, and the metabolite is eliminated only from the sampling compartment (i.e., no peripheral tissues eliminate the metabolite by excretion or further metabolism). Hepatic and renal clearance are generally considered as part of the sampling compartment. There are more complex methods to estimate MRT that may accommodate reversible metabolism (26), although they are seldom employed or reported. MRT(m) is a time-average parameter that is dependent on the disposition of the metabolite once formed. Thus, MRT(m) is of value in evaluating whether elimination and distribution of the metabolite have changed when the shape of the plasma metabolite concentration versus time curve is altered in response to changes in the disposition of the parent drug. Also, the relation $V(\text{m})_{\text{ss}} = CL(\text{m})\text{MRT}(\text{m})_{\text{m,iv}}$, is useful to determine $V(\text{m})_{\text{ss}}$, if $CL(\text{m})$ can be determined (27).

When an intravenous bolus dose of preformed metabolite is administered, the $\text{MRT}(\text{m})_{\text{m,iv}}$ is calculated as,

$$\text{MRT}(\text{m})_{\text{m,iv}} = \frac{\text{AUMC}(\text{m})_{\text{m,iv}}}{\text{AUC}(\text{m})_{\text{m,iv}}} \tag{16}$$

Where AUMC is the area under the first moment of the concentration versus time curve from the time of dosing, then estimated to infinity (15,25–28). The subscripts indicate the compound and route administered. A similar relation describes the MRT of the parent drug if given as an intravenous bolus. $\text{MRT}(\text{m})_{\text{m,iv}}$ measured after a rapid IV bolus of the metabolite reflects a mean time in the body for elimination and distribution of the metabolite. Measures of AUMC can be subject to substantially more error than AUC, primarily owing to the need to extrapolate a larger portion of the first moment from the last sampling time to infinity (15,25). When the metabolite is formed after IV dosing of the parent drug, the measured mean residence time reflects not only the mean time of metabolite in the body, but also the time required for its formation from the parent drug. Therefore, the MRT(m) is corrected for this contribution by subtracting the MRT of the parent drug,

$$\text{MRT}(\text{m})_{\text{m,iv}} = \frac{\text{AUMC}(\text{m})_{\text{p,iv}}}{\text{AUC}(\text{m})_{\text{p,iv}}} - \frac{\text{AUMC}_{\text{p,iv}}}{\text{AUC}_{\text{p,iv}}} \tag{17}$$

Here, the ratio $AUMC(m)_{p,iv}/AUC(m)_{p,iv}$ is sometimes referred to as the mean body residence time for the metabolite [MBRT(m)], which reflects formation, distribution, and elimination processes, whereas the $MRT(m)_{m,iv}$ may be referred to as the mean disposition residence time, [MDRT(m)], reflecting only elimination and distribution (27). From Eq. (17), $MRT(m)_{m,iv}$ can be determined unambiguously from the plasma concentration versus time profiles of metabolite and parent drug after an intravenous dose of the parent drug, without the need for an intravenous dose of the metabolite (27,28). In a later section, $MRT(m)_{m,iv}$ will be derived following extravascular dosing of the parent drug.

III. METABOLITE KINETICS AFTER A SINGLE EXTRAVASCULAR DOSE OF PARENT DRUG

When a drug is not administered by an intravenous route, the rate of drug absorption from the site of extravascular administration [e.g., the gastrointestinal (GI) tract for peroral, or the muscle for intramuscular], adds additional complexity to understanding the disposition of the metabolite. There are several confounding factors to be considered, the most obvious being both the extent of availability F of the parent drug and its rate of absorption k_a, as defined by a rate constant. Here, for simplicity, a first-order rate of absorption will be employed, although drug inputs that approximate zero-order process are also commonly found, especially with sustained- or controlled-release dosage forms. Additional considerations when extravascular administration is employed is estimating the fraction of the dose transformed into metabolite during absorption, which is referred to as ''first-pass metabolite formation,'' and what fraction of metabolite that is formed during the absorption process reaches the plasma sampling site. These issues will be discussed later. It is instructive to first consider Scheme 3, which represents drugs with little or no first-pass metabolite formation, as commonly found for drugs with high oral availability. Again, for simplicity, it will be assumed that all metabolite formed after absorption of the parent compound reaches the systemic circulation before irreversible elimination [i.e., the availability of metabolite when formed in vivo F(m) is complete].

A. Metabolite Disposition After Extravascular Drug Administration with Limited Metabolite Formation During Absorption

Scheme 3 represents a drug for which negligible metabolite forms during the absorption process. This scenario may be expected for drugs with high oral availability, low hepatic

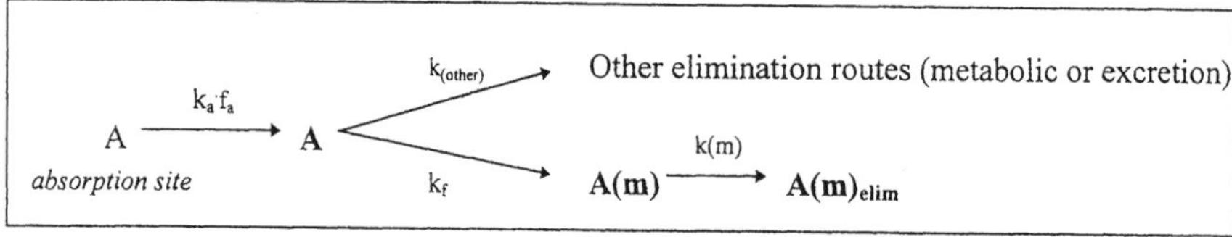

Scheme 3 Absorption of drug, its metabolism to a primary with no first-pass metabolism to the primary metabolite and parallel elimination pathways. The primary metabolite is eliminated by excretion or further metabolism.

and gut wall metabolic clearance or drug administered by other extravascular administration routes where little or negligible metabolite may be formed during absorption (e.g., intramuscular administration). However, this scheme does not necessarily require high availability for the drug, but does assume that drug not reaching the systemic circulation is not due to biotransformation to the metabolite of interest [e.g., low availability may be due to poor dissolution, low GI membrane permeability, degradation at the site of absorption, or the formation of other metabolites]. In this case $F = f_a$ if no other first-pass metabolites are formed. Scheme 3 contains an absorption step that was not present in Scheme 1. This catenary process with rate k_a, drug elimination (k) and metabolite elimination ($k(\mathrm{m})$) in series will influence metabolite disposition depending on the step that is rate-limiting.

For a drug with first-order absorption and elimination, the following biexponential equation describes the disposition of the parent drug (17),

$$C = \frac{k_a \cdot F \cdot D}{V(k_a - k)}[e^{-k \cdot t} - e^{-kat}] \tag{18}$$

This equation is mathematically analogous to Eqs. (3) and (4), and if multiplied by V provides amounts rather than concentrations. If Eq. (18) is substituted into Eq. (1) and solved for the concentration of metabolite over time, one obtains a relation with three exponentials because there are two steps before the metabolite reaching the systemic circulation and one step influencing its elimination,

$$C(\mathrm{m}) = C_1 e^{-k \cdot t} + C_2 e^{-k(\mathrm{m}) \cdot t} + C_3 e^{-k \cdot at} \tag{19}$$

Here the constants C_1, C_2, and C_3 are complex terms derived from the model in Scheme 3, which include the bioavailability of the parent drug F and dose D in their respective numerators, as well as the volume of distribution of the metabolite in the denominator. The values for these constants in Eq. (19) may be estimated by computer fitting of experimental data. Although the values of these constants have limited inherent utility themselves, the fitting process does provide a description of the time-dependent metabolite profile. Through the application of the superposition principle (15), the descriptive equation may be employed to estimate metabolite profiles at steady state after chronic dosing or following irregular multiple dosing regimens of the parent drug.

Because Scheme 3 does not have any reversible processes, it is a catenary chain with simple constants for each exponential term in Eq. (19) that are easily conceptualized from the schematic model. These exponential terms indicate that any one of the processes of absorption, parent drug elimination, or metabolite elimination may be rate-determining. The rate constant associated with the rate-limiting step (i.e., the slowest step) corresponds to the slope observed for the log-linear concentration versus time curve of the metabolite in plasma, assuming that one of the three rate constants is distinctly smaller. For drugs with FRL metabolism, either k or k_a will correspond to the terminal slope (and subsequently the observed half-life), and knowledge of the rates of absorption and elimination of the parent drug may be employed to discern which process or step may control the apparent terminal half-life of the metabolite concentration in plasma. For ERL metabolism the elimination half-life of the metabolite is, by definition, longer than the elimination half-life of the parent drug. Therefore, if absorption is rate-limiting, then the absorption rate may govern not only the observed terminal half-life of the parent drug, but also may dictate the observed apparent terminal half-life of the metabolite. This is shown in Fig. 5 where the rate of absorption of morphine is decreased after administration of a slow-release buccal formulation, such that the elimination rate for morphine cannot be distinguished from that of its metabolites (20). Without administration of an IV dose or an immediate-release tablet of morphine (see Fig. 4), one could not discern from Fig. 5 whether M3G follows ERL or FRL metabolism or whether the terminal half-life of Fig. 5 is due to absorption.

B. Metabolite Disposition After Extravascular Drug Administration with Metabolite Formation During the Absorption Process

In numerous cases, as occurs for drugs with significant first-pass hepatic or gut wall metabolism, metabolite formed during absorption must be considered in characterizing the disposition of metabolite. Scheme 4 presents a model with one primary metabolite and other elimination pathways for the parent drug after absorption. Also introduced here is consideration of the availability of the metabolite once formed in vivo [F(m)].

Here, the availability term F is the product of the fraction of dose of parent drug absorbed f_a, and the fraction of dose reaching the liver that escapes biotransformation to metabolites (first-pass metabolism by the liver or GI tract) during first-pass absorption F_H. The term f_a includes drug not reaching the systemic circulation because of poor solubility, degradation, or low GI membrane permeability, and is often estimated by application of a mass balance approach after administration of radiolabeled drug. For example, if there is little radioactivity recovered in feces after an IV dose, then summation of the fraction of radiolabel in urine, tissues, and expired breath after an oral dose of radiolabeled material would provide an estimate of f_a. The term $(1 - F_H)$ represents the fraction of drug absorbed that forms metabolites during the first-pass, which is also commonly defined as the extraction ratio across the organ E. Also introduced here, is consideration of the availability of the metabolite F(m), which is that fraction of metabolite formed that is subsequently systemically available (e.g., not subject to sequential metabolism or biliary excretion). In practice, F(m) can be determined unambiguously only after administration of the preformed metabolite by both the IV and extravascular routes, which is seldom possible in humans, although it may be feasible in animals. If other metabolites are also formed during absorption, then a fraction of the extraction ratio would represent the primary metabolite of interest (12). First-pass formation of metabolite by the gut wall is not considered separately here because the source of metabolite measured in the systemic circulation (either from liver or GI wall) cannot easily be distinguished in human studies.

Assuming that there is no gut wall metabolism and that the fraction of metabolite formed during the first-pass of drug through the liver is the same as subsequent passes (i.e., no saturable first-pass metabolism), then the amount of metabolite formed and presented to the systemic circulation is equal to the product of the AUC(m) and CL(m) (10),

$$f_a \cdot f_m \cdot F(\mathrm{m}) \cdot D_{po} = CL(\mathrm{m}) \cdot \mathrm{AUC(m)}_{p,po} \tag{20}$$

Where f_m is the fraction of dose that is converted to the metabolite of interest and the subscripts indicate that the parent drug was administered orally. A similar relation exists for an IV dose of parent drug; however, f_a would be equal to unity. When the relation of Eq. (20) is applied to both oral and IV doses of the parent drug (equal doses are used

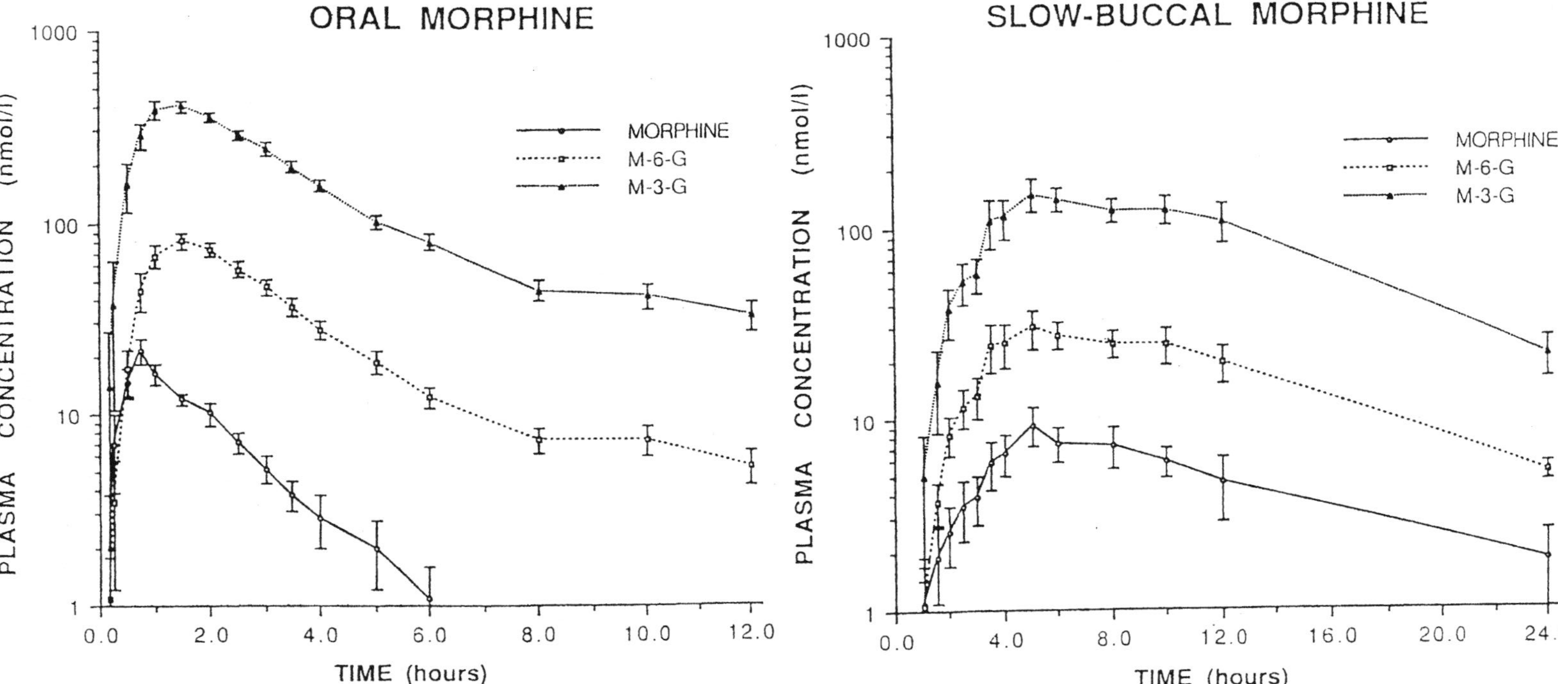

Fig. 5 Comparison of immediate release oral (11.7 mg) versus slow-release buccal (14.2 mg) administration of morphine in humans on the profile of morphine, morphine-6-glucuronide (M6G), and morphine-3-glucuronide (M3G). The common terminal half-life observed for parent drug and both metabolites shown for the slow-release buccal formulation are longer than obtained after the immediate-release dosage form, suggestive of absorption being rate-limiting. (From Ref. 20.)

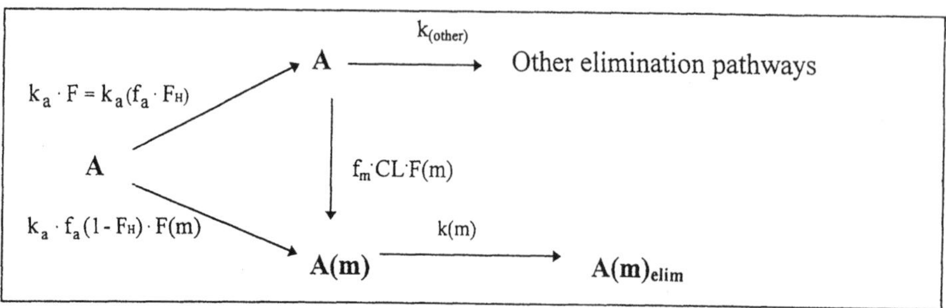

Scheme 4 Absorption of drug, metabolism to a primary metabolite with first-pass formation of metabolite followed by excretion or metabolism.

here thus doses cancel), the fraction of an oral dose of parent drug that is absorbed f_a can be estimated by measuring the metabolite exposure AUC(m) (10),

$$f_a = \frac{\mathrm{AUC(m)}_{\mathrm{p,po}}}{\mathrm{AUC(m)}_{\mathrm{p,iv}}} \tag{21}$$

With the assumption that first-pass loss of absorbed drug is due to only hepatic elimination, then F_H can be estimated from the relation, $F = f_a F_H$, where $F = \mathrm{AUC_{po}}/\mathrm{AUC_{iv}}$ if doses of parent drug are equal by both oral and IV routes. If the value of f_a, as determined with Eq. (21), is greater than 1 this would suggest that GI wall metabolism is occurring. When the data of naphthoxylacetic acid/propranolol in Fig. 2 is analyzed in this manner, f_a was estimated to be 0.98, which indicates that the absorption of propranolol is essentially complete and much of the formation of this metabolite is hepatic.

C. Estimating Fraction of Metabolite Formed and Formation Clearance

Commonly, values for f_m are reported for drugs administered to humans based on collection of metabolite in urine after a dose of the parent drug. This assumes that all metabolite formed reaches the systemic circulation [i.e., $F(m) = 1$] and is then excreted into urine, or that identification of sequential metabolites is accurate and they are also efficiently excreted in urine. If metabolite can be administered independently, then additional calculations of f_m can be employed. Dosing preformed metabolite IV provides $\mathrm{AUC(m)}_{\mathrm{m,iv}}$, which can be compared with metabolite exposure from an IV dose of parent drug (29),

$$f_m \cdot F(\mathrm{m}) = \frac{\mathrm{AUC(m)}_{\mathrm{p,iv}} \cdot M}{\mathrm{AUC(m)}_{\mathrm{m,iv}} \cdot D} \tag{22}$$

where M and D are molar doses of metabolite and parent, respectively. This relation assumes that systemic clearance of the metabolite is independent of whether metabolite was dosed exogenously or formed from the parent in vivo. The term F(m) is present in Eq. (22) because availability of metabolite formed from IV parent drug may be less than complete. If the preformed metabolite were instead administered orally where it must also pass the liver before reaching the systemic circulation, the F(m) term cancels (10),

$$f_m = \frac{AUC(m)_{p,iv} \cdot M}{AUC(m)_{m,po} \cdot D} \tag{23}$$

The experimental approach of dosing the metabolite orally may be more easily performed because it avoids the preparation and administration of an IV dose. However, use of Eq. (23) must now assume that the metabolite is well absorbed from the intestine [i.e., $f_a(m)$ is unity].

Once f_m is estimated, CL_f is simply the product of f_m and the total clearance of the parent drug $CL_f = f_m \cdot CL$. An alternative approach to determine rate of metabolite formation and cumulative extent of formation is by the application of deconvolution analysis (30,31). With data from an IV dose of the metabolite (i.e., with a known input of the metabolite), the subsequent rate and extent of metabolite formation can be obtained by deconvolution (31) after any known input dose of the parent drug. This approach was applied to determine both metabolite formation rates and f_m for M6G after IV bolus and infusion doses of morphine, as shown in Fig. 6 (22). The values of f_m when estimated by use of Eq. (22) [assuming $F(m) = 1$] and by deconvolution were 12 ± 2 (mean $\pm$ s.d.) and $9 \pm 1\%$ of the morphine dose, respectively, which is within anticipated experimental error (22). The advantage of the deconvolution approach is that it provides metabolite formation rates over time that may be helpful in analyzing the system, and it has few assumptions for its application. A disadvantage of the method is it requires an exogenous

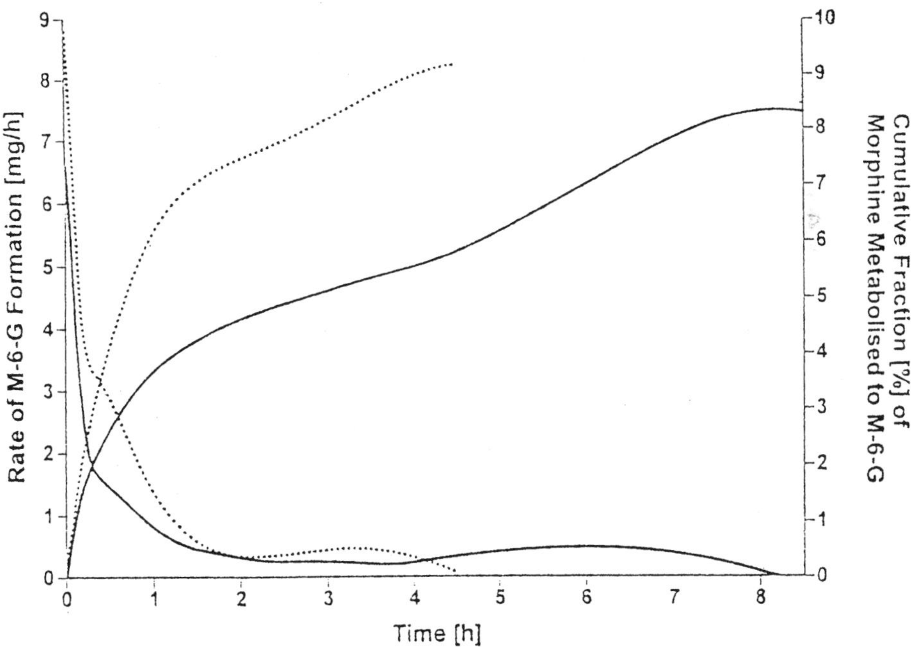

Fig. 6 Deconvolution analysis of morphine-6-glucuronide (M6G)/morphine to determine the rates of M6G formation (left ordinate, decending lines) and fraction of dose metabolized to M6G (right ordinate, ascending lines) after intravenous doses of morphine. Solid lines are data from a 0.13-mg/kg bolus followed by 0.005-mg/kg/h infusion for 8 h. Dotted lines are data from a 0.24-mg/kg bolus followed by 0.0069-mg/kg/h infusion for 4 h. (From Ref. 22.)

dose of the preformed metabolite and assumes metabolite formed in vivo behaves similarly to that dosed as preformed metabolite.

D. Mean Residence Time for Metabolite After an Extravascular Administration of Parent Drug

The $MRT(m)_{m,iv}$ can also be determined after an extravascular dose of the parent drug, as long as the assumptions stated in Section II.F apply. The ability to obtain $MRT(m)_{m,iv}$ without an IV dose of the metabolite when there is no first-pass metabolism has been presented by several authors (27,28,32), and modifications to accommodate first-pass metabolite formation have been considered (33,34).

If there is no first-pass formation of metabolite during the absorption process, then $MRT(m)_{m,iv}$ can be determined by correction of the mean residence time of metabolite in the body for contributions from the parent drug, which now include its absorption as well as distribution and elimination that were accounted for in Eq. (17),

$$MRT(m)_{m,iv} = \frac{AUMC(m)_{p,po}}{AUC(m)_{p,po}} - \frac{AUMC_{p,po}}{AUC_{p,po}} \tag{24}$$

The ratio $AUMC_{p,po}/AUC_{p,po}$ is defined as the MRT_{po} of the parent drug when given orally (or by any other extravascular route) and is the sum of the MRT_{iv} and the mean absorption time (MAT) of the parent drug. Because the MRT_{po} is simply a ratio determined from observed plasma concentration profile of the parent drug after oral administration, $MRT(m)_{m,iv}$ can be estimated without IV administration of either parent or metabolite. As mentioned in Section II.F, $MRT(m)_{m,iv}$ provides a parameter for the assessment of disposition (distribution and elimination) of the metabolite without consideration of its rate or time course of formation.

If there is first-pass formation of metabolite, then Eq. (24) may introduce significant error in determining $MRT(m)_{m,iv}$ (33,34). To correct for this error, the contribution of first-pass metabolism from an IV dose of the parent drug $MRT_{p,iv}$ = AUMC/AUC and the fraction of the absorbed dose that escapes first-pass metabolism F_H need to be considered. Assuming that all metabolite formed from the first-pass is due to a single metabolite,

$$MRT(m)_{m,iv} = \left(\frac{AUMC(m)_{p,po}}{AUC(m)_{p,po}}\right) - \left(\frac{AUMC_{p,po}}{AUC_{p,po}}\right) + (1 - F_H)\, MRT_{p,iv} \tag{25}$$

It is apparent that when first-pass metabolism is minimal (i.e., F_H is unity), then Eq. (25) collapses to Eq. (24). In addition, when $MRT_{p,iv}$ is small relative to the other terms, as occurs with transient prodrugs such as aspirin, then the last term of Eq. (25) is again insignificant. However, when there is substantial first-pass metabolism, then use of Eq. (24) would provide a value that underestimates $MRT(m)_{m,iv}$. In practice Eq. (24) may provide a value that is negative, indicating that there must be some first-pass metabolism occurring (34).

IV. METABOLITE DISPOSITION AFTER CHRONIC ADMINISTRATION OF PARENT DRUG

Long-term administration of drugs, either by extravascular routes, multiple infusions, or continuous infusions, is commonly employed in ambulatory and hospital settings. In these cases, active metabolites, toxic metabolites, or the effect of the metabolite on the disposition of the parent drug should be considered. Here one needs to consider the accumulation of both the parent drug and metabolite to steady-state concentrations and the relation between the steady-state characteristics of the parent relative to that of the metabolite because these may differ from those observed after only a single dose of drug. The critical pharmacokinetic descriptors for the disposition of the metabolite on chronic drug administration are the time to achieve steady state; peak, trough, and average concentrations at steady state; the accumulation at steady state relative to a single dose of parent drug; and metabolite clearance. Volume of distribution is a parameter that has little contribution to achieving steady-state levels, but does affect the swings from peak to trough concentrations at steady state owing to its contribution in determining the half-life of parent drug and metabolite.

A. Time to Achieve Steady State for a Metabolite

Following prolonged extravascular administration or IV infusion for a sufficiently long time, parent drug and metabolite will eventually achieve steady-state concentrations where the input of parent drug and metabolite (i.e., formation) into the systemic circulation on average equal their respective rates of elimination. The time to achieve such a steady state is dependent on the half-life of the parent or metabolite and, if first-order elimination processes occur, generally it takes about 3.3 half-lives to reach 90% of the ultimate steady-state level and after 5 half-lives a compound would achieve about 97% of the theoretical steady-state level. In the case of a primary metabolite formed from the parent drug, it will be the slowest, rate-limiting process that governs when a metabolite reaches steady-state levels in the body. Therefore, for metabolites with FRL metabolism, the elimination half-life of the parent compound will control the time to achieve steady-state levels of the metabolite, whereas for ERL metabolism, the longer half-life of the metabolite will dictate the time needed for the metabolite to reach steady-state levels. ERL metabolism is shown in Fig. 7, in which the two metabolites of morphine take as much as 30 h to accumulate to steady-state concentrations after bolus injection and continuous IV infusion of diamorphine (a prodrug of morphine) to infants, even though morphine rapidly attained steady state (21). In this example the mean elimination half-life of the metabolites in 19 infants were estimated by the rate of attainment to steady state (15) to be 11.1 h and 18.2 h for M3G and M6G, respectively (21), whereas these metabolites have reported half-lives estimated to be only several hours in adults (20,22).

After discontinuation of chronic or repetitive parent drug administration, the rate of decline in metabolite concentration will again be dependent on whether FRL or ERL metabolism is operative. In FRL metabolism, the rate of decline for the concentrations of the metabolite will simply follow that of the parent drug. In contrast, for ERL metabolism, levels of metabolite will persist, based on the longer half-life of the metabolite, as shown in Fig. 8 for the metabolites of nitroglycerin (35).

B. Metabolite Concentrations at Steady State After Continuous Administration of Parent Drug

At steady state for the metabolite, the rate of formation equals the rate of elimination of the metabolite [i.e., $\Delta A(\mathrm{m})/\Delta t = 0$]. Considering the fraction of metabolite formed from the parent f_{m} and the systemic bioavailability of the metabolite once formed $F(\mathrm{m})$, from

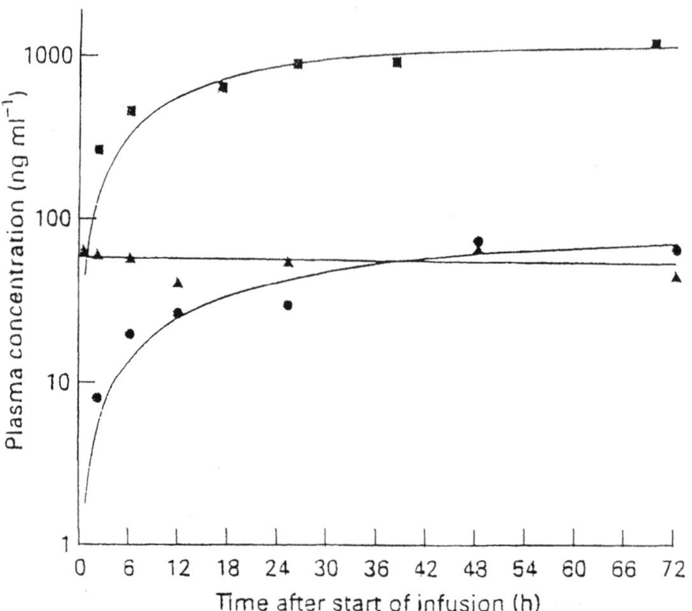

Fig. 7 Disposition of morpine (▲), with the formation of morphine-3-glucuronide (M3G ■), and morphine-6-glucuronide (M6G ●) in a representative infant receiving diamorphine (prodrug of morphine) intravenous bolus followed by an infusion for 72 h. The rate of attainment of steady-state for the metabolites is determined by the slowest rate constant that provides estimated half-lives of 11 and 18 h for M3G and M6G, respectively (From Ref. 21.)

Eq. (11) the following relation is defined at steady state where the left term is the formation rate of metabolite that reaches the systemic circulation and the right term is its rate of elimination (9),

$$f_m \cdot F(m) \cdot CL \cdot C_{ss} = CL(m) \cdot C(m)_{ss} \tag{26}$$

Since at steady-state concentration for the parent drug, $CL \cdot C_{ss}$ is equal to the rate of infusion R_0, this can be substituted into Eq. (26), which on rearrangement provides the average concentration of metabolite at steady-state, $C(m)_{ss,ave}$.

$$C(m)_{ss,ave} = \frac{f_m \cdot \mathrm{F(m)} \cdot R_0}{CL(m)} \tag{27}$$

This equation describes that $C(m)_{ss,ave}$ will be proportional to the infusion rate of the parent drug. However, the difficulties in obtaining estimates of the other terms of the equation without dosing of the metabolite limits the usefulness of Eq. (27). Alternatively, indirect approaches to estimate $C(m)_{ss,ave}$ from AUC data obtained after a single dose of the parent drug will be addressed later.

If instead of IV infusions, drug input is by regular and repetitive extravascular administration, the rate of input R_0 in Eq. (27), is modified to reflect an average drug administered each dosing interval D/τ. There is also a correction for the fraction of extravascular dose absorbed f_a (9),

$$C(m)_{ss,ave} = \frac{f_m \cdot F(m) \cdot f_a \cdot D}{CL(m) \cdot \tau} \tag{28}$$

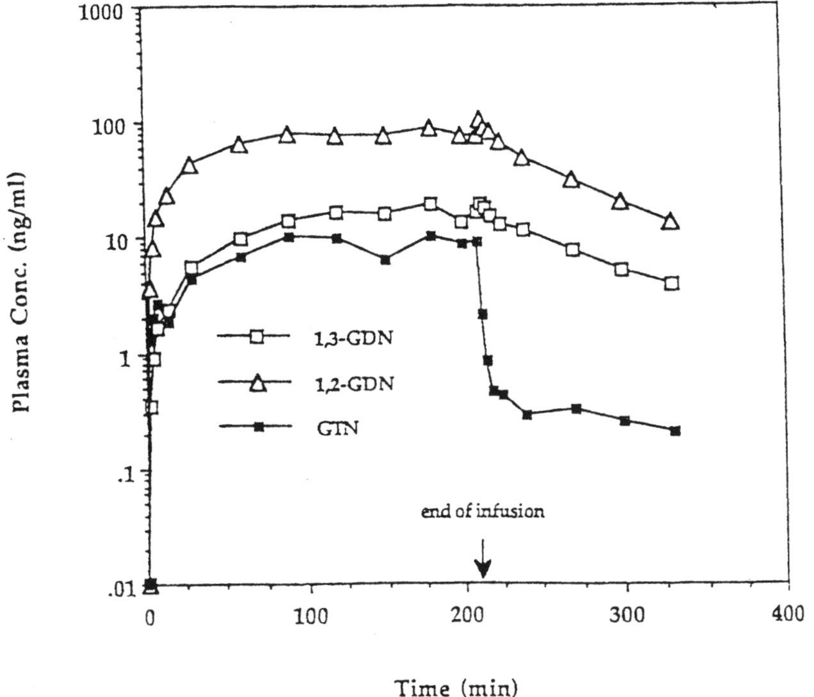

Fig. 8 Plasma levels of glyceral trinitrate (GTN; ■), 1,2-glyceryl dinitrate (1,2-GDN; △) and 1,3-glyceryl dinitrate (1,3-GDN; □) following an IV infusion of 0.070 mg/min GTN to a dog. The half-lives of the metabolites were much longer than the apparent half-life of approximately 4 min for GTN after the end of the infusion, and data suggest that the prolonged activity of GTN may be due to the metabolites. The prolonged residual GTN at very low concentrations is not understood nor is its slower than anticipated rise to steady state. (From Ref. 35.)

Here, the average rate of drug input is $f_a \cdot D/\tau$. The value of $C(\mathrm{m})_{\mathrm{ss,\,ave}}$ obtained reflects the average level at steady state, but gives no information of relative fluctuations from peak to trough metabolite concentrations $C(\mathrm{m})_{\mathrm{ss,max}}/C(\mathrm{m})_{\mathrm{ss,min}}$ at steady state. The extent of this fluctuation for the metabolite levels will depend on whether FRL or ERL metabolism occurs. Assuming that absorption is not the rate-limiting process, for FRL metabolism the metabolite rapidly equilibrates with the parent drug, thus the peak to trough metabolite concentration ratio $C(\mathrm{m})_{\mathrm{ss,max}}/C(\mathrm{m})_{\mathrm{ss,min}}$ will be similar to $C_{\mathrm{ss,max}}/C_{\mathrm{ss,min}}$ obtained for the parent drug. In contrast, for ERL metabolism, the longer elimination half-life of the metabolite will dampen its concentration swings at steady state such that $C(\mathrm{m})_{\mathrm{ss,max}}/C(\mathrm{m})_{\mathrm{ss,min}} < C(\mathrm{m})_{\mathrm{ss,max}}/C_{\mathrm{ss,min}}$.

C. Estimating Metabolite Levels at Steady State from a Single Dose

1. Superposition Principle to Estimate Metabolite Levels at Steady State

When the metabolite and parent drug follow linear processes that are independent of dose and concentrations (i.e., no saturable processes occur), then prediction of metabolite concentrations after multiple doses or with a change of dose can be estimated if the route of drug administration does not change. Under these conditions the time course of the metabolites as well as their average concentrations can be estimated by the principle of superposition (15). However, if the route of drug administration is different between the single-dose administration and the chronic dosing without a change in bioavailability (e.g., single dose is an intravenous bolus, whereas the chronic administration is an intravenous infusion), the average concentration of the metabolite at steady state can be predicted based on relations of clearance and AUC, as described in the following.

Superposition simply takes the plasma concentration versus time profile after a single dose and assumes that each successive dose would behave similarly, although the magnitude of concentration will vary proportionally with the dose administered. Thus both the parent and metabolite profile can be summed over time. This can be done either by fitting the single-dose data to appropriate mathematical functions as a sum of exponentials or polynomials and then summing this to infinity if the dosing regimen has a constant-dosing interval (15), or alternatively, a more simple method that is easily applied to even irregular-dosing intervals is to use a spreadsheet to sum the concentrations of metabolite from successive doses of the parent drug (15), making certain to have values or extrapolated metabolite concentrations for at least five elimination half-lives of metabolite or parent, whichever is longest.

2. *Single-Dose AUC Values to Predict Average Metabolite Levels at Steady State*

If an estimate of the average concentration of metabolite at steady state is desired, then AUC(m) after a single dose can be employed effectively, as shown by Lane and Levy (36). This approach can also be used when the formulation of parent drug changed between single and chronic administration without altering availability of parent drug or fraction of metabolite formed (e.g., single dose is an oral suspension, whereas chronic administration is an sustained-release capsule). From the relation that the average rate of formation of metabolite is equal to its average rate of elimination at steady state, as provided in Eq. (26), these authors then used the relations of AUC and *CL* in Eq. (12), to derive the following from AUC data collected after a single dose of the parent drug,

$$\frac{C(\mathrm{m})_{\mathrm{ss,ave}}}{C_{\mathrm{ss,ave}}} = \frac{\mathrm{AUC(m)}}{\mathrm{AUC}} \tag{29}$$

This relation assumes linearity (i.e., constant clearance) for parent drug and metabolite relative to changes in dose and time. Equation (29) is applicable to any route of administration, but does require that the route used for the single-dose measurements of AUC be the same as that to be employed for steady-state measurements of concentrations. This requirement usually ensures that availabilities of the parent drug and metabolite formed are the same between single and chronic doses. This provides a means to estimate the average metabolite concentrations at steady state from AUC(m)/AUC ratios from a single dose and knowledge of an estimated $C_{\mathrm{ss,ave}}$ for the parent drug. Although initially derived from a one-compartment model, Eq. (29) was later extended with less restrictions by Weiss (12) using a noncompartmental approach.

When the route of administration changes, which may alter availability of the parent drug, the relation of Eq. (29) must consider this change. If the initial single-dose data were obtained from an IV dose, then the ratio of metabolite to parent drug concentrations at steady state after oral dosing will be affected by the fraction of parent drug systemically available (10,12). With an assumption that availability of the metabolite did not change with a change in the route of parent drug administration and there is not significant GI wall metabolism, the following relation can be used,

$$\frac{C(\mathrm{m})_{\mathrm{ss,av}}}{C_{\mathrm{ss,ave}}} = \frac{\mathrm{AUC(m)_{iv}}}{\mathrm{AUC_{iv}}}\left(\frac{1}{F_{\mathrm{H}}}\right) \tag{30}$$

With regular doses and dosing intervals τ, the AUC is equal to $C_{\mathrm{ss,ave}} \cdot \tau$; therefore, Eq. (29) can be rearranged and written more simply as,

$$C(\mathrm{m})_{\mathrm{ss,ave}} = \frac{\mathrm{AUC(m)}}{\tau} \tag{31}$$

where AUC(m) is determined after a single dose of parent drug. Under these conditions of regular dosing to steady state, data on the disposition of the parent drug is not necessarily required when estimating $C(\mathrm{m})_{\mathrm{ss,ave}}$, although most often the parent drug data are available. When a single intravenous bolus dose of parent drug was used to obtain measurement of AUC(m), but then a continuous infusion of rate R_0 is later employed, one can also derive $C(\mathrm{m})_{\mathrm{ss,ave}}$ without complete knowledge of the disposition of parent drug. Since for continuous infusion, $C_{\mathrm{ss,ave}} = R_0/CL$, and for a single IV bolus dose of parent drug, $\mathrm{AUC} = D/CL$, substitution of these well-known relations into Eq. (29) and rearrangement provides (36),

$$C(\mathrm{m})_{\mathrm{ss}} = \frac{R_0 \cdot \mathrm{AUC(m)}}{D} \tag{32}$$

This relation provides an actual level, because it is from an infusion where metabolite reaches a steady-state level. Only knowledge of the dose of drug employed for the single intravenous dose D and the resultant AUC(m) are needed.

V. SEQUENTIAL METABOLISM

A common scheme in drug metabolism is the concept of sequential metabolism during which a parent drug is converted to one or more primary metabolites that are then further converted to a secondary metabolite (see Scheme 2). This process leads to two metabolites as shown in Fig. 9, where propranolol is first oxidized to 4-hydroxypropranolol, which then undergoes sequential metabolism to 4-hydroxypropranolol glucuronide (37). Indeed, sequential metabolism formed the basis for the common nomenclature of phase I and phase II metabolism, coined by Williams (38). Another commonly observed sequence is that of parallel metabolism leading to a common metabolite as exemplified by the metabolism of dextromethorphan via CYP2D6 and CYP3A4 (39) shown in Scheme 5.

For either of the two examples presented in scheme 5, given that first-order processes apply, the equation that describes the disposition of the secondary metabolite would be complex polyexponential functions, with the number of terms determined by the number of first-order processes leading to the pentultimate metabolite of interest. Such processes may also include an absorption step (not shown in Scheme 5) if the parent compound is administered extravascularly. These processes can be dealt with by fitting the data with a polyexponential function or using other appropriate equations as presented by Eq. (19). A mathematical analysis of parallel and sequential systems has recently been developed

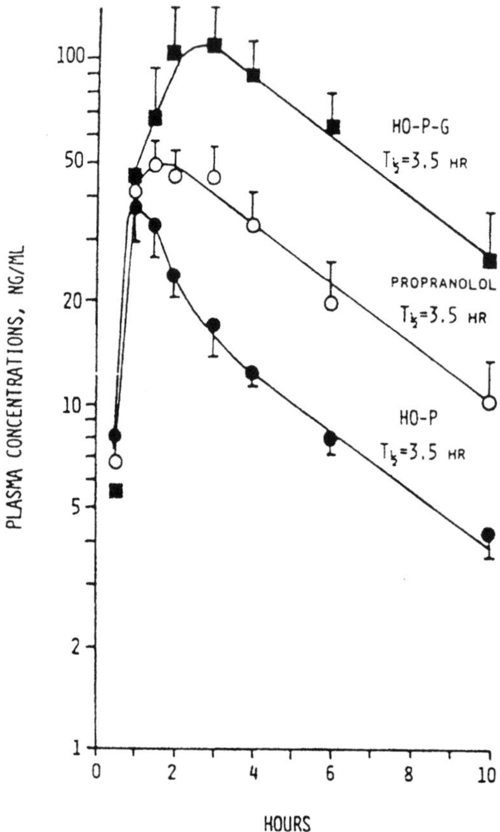

Fig. 9 Disposition of propranol (○), its 4-OH metabolite (●) and sequential formation of its 4-OH-glucuronide metabolite (■) in humans after a single 80-mg–oral dose. Data are mean ± SEM, $N = 6$. The initial rapid increase in 4-OH propranolol is followed by a later peak concentration of 4-OH-glucuronide owing to sequential metabolism. The nonlinearity of the 4-OH metabolite immediately following its peak is likely due to significant formation of this metabolite during absorption, which takes time to distribute to tissues before equilibration with formation of the glucuronide. Based on urinary excretion data most of the 4-OH propranolol is converted to propranolol glucuronide and when Eq. (21) is applied to these two metabolites the values of f_a are substantially greater than 1, suggestive of GI wall metabolism. (From Ref. 37.)

(40). In the more simple, catenary chain represented by Scheme 2, the terminal apparent half-life of the secondary metabolite will be determined by the rate-limiting step (i.e., the step with the smallest rate constant among all steps describing the disposition of the metabolite). It is also possible in sequential metabolism, depending on relative rates and competing pathways, that the primary metabolite (m_1 in Scheme 2) may not be observed at all, either in plasma or in urine.

Sequential metabolism is one of the primary reasons that the systemic availability $F(m)$ of a primary metabolite once formed may be less than complete (41,42), as exemplified by 4-OH propranolol which forms the glucuronide in vivo (see Fig. 9). Other factors influencing the systemic availability of a metabolite may include biliary and renal excretion of metabolites once formed in the liver and kidney, respectively, without access to

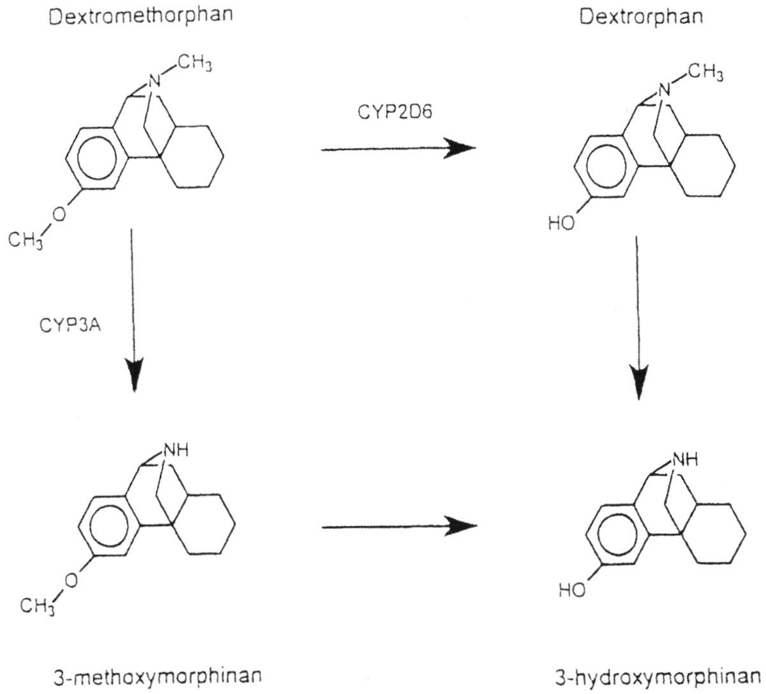

Dextromethorphan

CYP2D6

Dextrorphan

CYP3A

3-methoxymorphinan

3-hydroxymorphinan

Scheme 5 Parallel metabolic pathways of dextromethorphan to a common metabolite. (From Ref. 33.)

the systemic circulation. Further complications arise when one considers the effect of sequential metabolism, intracellular access, and systemic availability of metabolites for evaluating the disposition of metabolite formed in vivo with that of preformed metabolite administered directly into the systemic circulation (36). To estimate actual pharmacokinetic parameters of metabolite, it is usually necessary to administer preformed metabolite, preferably by the intravenous route. However, inherent assumptions with such experiments is that the aforementioned factors are independent of the route of metabolite administration.

VI. RENAL CLEARANCE OF METABOLITES

Many metabolites are eliminated to a significant extent or almost exclusively into the urine. Collection and analysis of urine samples has some advantages, the primary being that it is less invasive than blood sampling. Because urine production volume in a day or during a dosing interval is limited and usually a much smaller volume than V or $V(\mathrm{m})$, urine concentrations of metabolite or parent drug are often a great deal higher than those seen in plasma and thus more easily measured. Urine can also be collected incrementally, usually in intervals of 1 h or more although in humans, although this is often more difficult for small children, infants, and the elderly. For larger animals, catheterization can be performed to obtain continuous urine samples, but for smaller animals, such as the rat, this becomes an invasive procedure that may alter renal function. When incremental urine samples cannot be collected, complete collections over four to five half-lives after a single dose or during one-dosing interval at steady state can be performed. Unfortunately, even though urine collection is noninvasive, urinary excretion data are all too often not included in clinical protocols, because it is considered a secondary measure to plasma drug profiles. However, urinary metabolite excretion data can offer valuable information on the disposition of metabolites, especially when interpreting drug–drug interactions or the basis for altered pharmacokinetics in patient populations.

There are some assumptions that will be made in discussing renal clearance of metabolites that need to be considered. The first is that renal drug metabolism is not occurring (i.e., metabolite measured in plasma is what is filtered or transported into the renal tubule from the plasma, and parent drug does not convert to metabolite in the kidney). If renal drug metabolism does occur, the estimated renal clearance would be considered as an ''apparent clearance'' which is similar to estimating biliary clearance in a cannulated animal when hepatic metabolism of drug occurs without the metabolite being presented to the systemic circulation. Although the concept of drug metabolism in the liver before arrival of metabolite in the bile is well accepted, there appears to be a common, erroneous assumption that drug metabolism is unlikely to occur in the kidney. Another concern that is more easily tested is whether the metabolites are stable once excreted in the urine, for metabolites often reside in the bladder for periods of many hours before voiding. An example of this is the hydrolysis and acyl migration of ester glucuronides which can be quite variable because it is a pH-dependent process (43,44). Finally, it is generally assumed that once drug or metabolite arrives in the urinary bladder, it is irreversibly removed from the systemic circulation. However, studies have shown that resorption from the urinary bladder back into the systemic circulation can be significant for some compounds and this phenomenon should be considered (45). For the purpose of discussion in this section, it will be assumed that these complicating factors are negligible or, if they are occurring, they are reproducible, thus permitting calculations that provide estimates of apparent pharmacokinetic parameters.

A. Determination of Renal Clearance for Metabolites

Renal clearance is one pharmacokinetic parameter that can be determined for most metabolites as well as the parent drug irregardless of the route of administration, dosing regimens, formulations, or availability. The data needed are urinary excretion rates and plasma concentrations. The following general equation applies to calculate renal clearance (CL_R) during a urine collection interval and is expressed for a metabolite (15),

$$CL_R(\mathrm{m}) = \frac{\Delta A(\mathrm{m})_e/\Delta t}{C(\mathrm{m})_{\mathrm{ave}}} \tag{33}$$

Here, $\Delta A(\mathrm{m})_e$ is the amount of metabolite excreted into the urine during the collection interval, Δt is the collection interval, and $C(\mathrm{m})_{\mathrm{ave}}$ is the average plasma metabolite concentration over the interval. Intervals are typically at least 2 or several hours long for studies in humans owing to the difficulty of voiding more frequently. If frequent, short incremental collections of urine are desired, it is a common practice to provide oral fluids, such that adequate (but not excessive) urine production rates are maintained. The necessary collection interval makes the use of incremental renal clearance calculations difficult and less valuable for drugs with very short half-lives, such as less than an hour. The average plasma concentration is usually measured as the concentration at the midpoint of the collection interval or an average of the plasma concentrations at the beginning and end of a collection period.

An alternative calculation for incremental renal clearance is to rearrange Eq. (33) realizing that the product of the average plasma metabolite concentration and the time interval is the incremental AUC(m) for that collection period [i.e., $\Delta\text{AUC(m)} = C(\text{m})_{\text{ave}} \cdot \Delta t$]; thus,

$$CL_{\text{R}}(\text{m}) = \frac{\Delta A(\text{m})_{\text{e}}}{\Delta\text{AUC(m)}} \tag{34}$$

Either of the foregoing incremental renal clearance calculations can be applied to any collection interval, without consideration for the method of drug administration. These can also be applied to chronic multiple dose regimens for which it is common to measure total urinary output over the dosing interval τ.

Incremental renal clearance calculations should result in values that are approximately constant if $CL_{\text{R}}(\text{m})$ is constant across the periods collected (i.e., linear or concentration-independent pharmacokinetics) (13,15–17). If they are not similar, a useful approach is to examine a plot of excretion rate $\Delta A(\text{m})_{\text{e}}/\Delta t$, as a function of metabolite plasma concentration $C(\text{m})_{\text{ave}}$. The slope of such a plot is $CL_{\text{R}}(\text{m})$ and any nonlinearities may become apparent if renal clearance increases or decreases as concentrations of metabolite change over time or parent drug competes for active renal clearance process (i.e., active secretion or reabsorption).

Measuring incremental renal clearance is not routinely performed because it requires multiple urine collection over several intervals which adds the expense of processing and analysis costs, or it may not be feasible for studies with children, small animals, or drugs with very short half-lives. Alternatively, a measure of average renal clearance can be determined following a single dose of parent drug by use of the following a single dose of parent drug by use of the following equation (15),

$$CL_{\text{R}}(\text{m}) = \frac{A_{\text{e}}(\text{m})_{0-\infty}}{\text{AUC(m)}_{0-\infty}} \tag{35}$$

where $A_{\text{e}}(\text{m})$ is the amount of metabolite excreted from initial time of dosing (time zero) to a later time, usually when almost all of the drug and metabolite have been eliminated from the body, four or five elimination half-lives of parent or metabolite (whichever is rate-limiting). AUC(m) is the area under the metabolite plasma concentration versus time curve for the same interval. For studies conducted with multiple doses to steady-state, the interval for collection of urine and measurement of AUC should be the dosing interval τ.

B. Use of Metabolite Excretion to Assess Disposition of Drug and Metabolite

1. *Use of Metabolite Excretion Rate to Assess the Profile of Metabolite in Plasma*

Metabolite concentrations are often more easily measured in urine than in plasma in which concentrations may be below detection limits of the available assay. Assuming that $CL_{\text{R}}(\text{m})$ is constant, the excretion rate into urine should then parallel plasma concentrations of the metabolite as described by Eq. (33) (15). Since a fraction of the total systemic clearance of the metabolite $CL(\mathrm{m})$ may be due to renal excretion $f_e(\mathrm{m}) = CL_R(\mathrm{m})/CL(\mathrm{m})$, the rate of excretion can be expressed as

$$\frac{\Delta A(\mathrm{m})_e}{\Delta t} = f_e(\mathrm{m}) \cdot CL(\mathrm{m}) \cdot C(\mathrm{m})_{\mathrm{ave}} \tag{36}$$

Here, $C(\mathrm{m})_{\mathrm{ave}}$ describes the average course of plasma metabolite concentrations over time, whereas $f_e(\mathrm{m})$ and $CL(\mathrm{m})$ are constant. Thus, urinary excretion data of metabolite when plotted versus time can characterize the profile of metabolite in plasma even when plasma concentrations of metabolite cannot be measured. This profile can be used to estimate the rates of initial increase in metabolite concentrations, the approximate time to peak metabolite levels in plasma, and the apparent elimination half-life of the metabolite.

2. *Use of Metabolite Excretion Rates to Assess the Profile of the Parent Drug*

For drugs exhibiting FRL metabolism, which is the most common situation, urinary excretion rates or fraction of dose excreted in the urine may be used to estimate the plasma profile of the parent drug. With FRL metabolism, the plasma profile of metabolite parallels that of the parent drug, as shown from Eq. (9), which when substituted into Eq. (36) provides,

$$\frac{\Delta A(\mathrm{m})_e}{\Delta t} = f_e(\mathrm{m}) \cdot CL_f \cdot C_{\mathrm{ave}} \tag{37}$$

where C_{ave} is the average plasma concentration of parent drug over time. Thus, urinary excretion rate of the metabolite is directly proportional to the concentration of parent drug, which is expected under the conditions of FRL metabolism. Thus, a plot of metabolite excretion rates versus time provides a description of the parent drug profile and can be used to estimate the duration of the absorption phase, approximate time to peak concentration, and elimination half-life of the parent drug, without the need to assay parent drug in plasma or urine. Also evident from Eq. (37) is that with FRL metabolism when values of $f_e(\mathrm{m})$ and $F(\mathrm{m})$ are 1, then the excretion rate of metabolite equals its rate of formation from the parent drug.

3. *Use of Metabolite Excretion Rates to Assess Formation Clearance of the Metabolite*

When drug concentrations can be measured in plasma and $f_e(\mathrm{m})$ is known, under FRL metabolism, Eq. (37) can be rearranged to estimate formation clearance of the metabolite, CL_f.

$$CL_f = \frac{\Delta A(\mathrm{m})_e/\Delta t}{f_e(\mathrm{m}) \cdot C_{\mathrm{ave}}} \tag{38}$$

Where C_{ave} is the average concentration of drug during the period of urine collection. The use of Eq. (38) assumes that $F(\mathrm{m})$ of the metabolite is unity (i.e., all metabolite formed reaches the systemic circulation), although it does not require $f_e(\mathrm{m})$ to be equal to unity; however, in practice, this relation has been applied to metabolites that are primarily excreted in the urine.

Alternatively, one could utilize amounts of metabolite and drug rather than concentrations using the same assumptions,

$$k_f = \frac{\Delta A(\mathrm{m})_e/\Delta t}{f_e(\mathrm{m}) \cdot A_{ave}} \tag{39}$$

A classic example of the use of this approach is in Levy's studies of salicylate metabolism that did not even require measurement of plasma salicylate concentrations (46). Salicylate has four major metabolites that are rapidly excreted in urine, such that plasma concentrations of these metabolites were difficult to measure before the advent of of modern HPLC. Given that levels of metabolite in plasma were very low and urinary metabolites accounted for almost all of the salicylate dose, Levy estimated the amount of salicylate remaining in the body by using a mass balance approach for which the amount remaining to be excreted at a given time interval represented A_{ave}. By evaluating urinary excretion rates versus A_{ave}, using a linear transformation of the Michaelis–Menton equation, the formation of the salicyluric acid and the phenol glucuronide metabolites were found to be saturable and the Michaelis–Menton constants K_m and V_m were determined, whereas the metabolism to gentisic acid and the acyl glucuronide were nonsaturable with constant k_f values over a large range of amounts of salicylate in the body.

When incremental collections of urine are not obtained, but the urine collection is for a sufficient number of elimination half-lives to approximate a complete collection of metabolite in the urine, then Eq. (37) can be rearranged and the integral of ($C_{ave} \cdot \Delta t$) is the AUC for the parent drug,

$$CL_f = \frac{A(\mathrm{m})_e}{f_e(\mathrm{m}) \cdot \mathrm{AUC}} \tag{40}$$

The use of Eq. (40) is not dependent on rapid elimination of metabolite after formation, thus it is applicable to both conditions of FRL and ERL metabolism as long as urine is collected for a sufficient duration to account for all the metabolite formed after a single dose. The relation is also applicable for the analysis of excretion data after chronic dosing to steady state for which the amounts of metabolite in urine and AUC are measured over the dosing interval τ. This equation as written, does assume that F(m) of the metabolite is unity for the CL_f value to be accurate, although use of an apparent CL_f may be adequate for comparative experiments if F(m) is constant between experiments.

C. Assessment of Renal Metabolism

The use of pharmacokinetic analysis to assess or quantify the extent of renal metabolism in vivo is difficult and is not often employed. Renal and biliary clearance calculations employ the same equations and are both subject to errors if formation of metabolite occurs in the organ with subsequent excretion of metabolite without presentation of all metabolite to the systemic circulation. Biliary clearances of a metabolite can be determined by collection of bile in animals, but rarely in humans, and are often then referred to as *apparent* clearances owing to the understanding that hepatic metabolism is likely occurring. In contrast, seldom are renal clearances of metabolites reported as *apparent* clearances; thus, there is an inherent assumption that none of the metabolite excreted in the urine is formed by the kidney and immediately excreted in the urine. Even though there is little information to determine how often this assumption is valid, given the lower metabolic capability of kidney relative to the liver for most substrates that have been examined, this assumption is usually accepted for renal clearance calculations. Such an assumption would probably be challenged if made for calculating biliary clearance of a metabolite.

If Eqs. (33)–(35) are applied to metabolite excreted in the urine and the value of renal clearance based on blood concentrations was very high, exceeding renal blood flow, then one may suspect that renal metabolism is occurring. This requires the use of clearance values based on blood concentrations, which can be calculated from clearance in plasma and measurement of the blood/plasma concentration ratio, by the relation, $CL \cdot C = CL_B \cdot C_B$, where the subscript B refers to blood. Therefore, ratios of metabolite concentration in blood relative to plasma need to be obtained and are easily determined in vitro. This phenomenon was reviewed by Vree et al. (47) in human studies. It is easy to conceptualize this phenomenon if one obtains extensive excretion of metabolite in urine when a very sensitive assay cannot even measure the metabolite in plasma. If the plasma concentration of metabolite is assigned a value of zero, then CL_R(m) from the foregoing equations is infinity, an unlikely event; therefore, renal metabolism should be considered. It would be more appropriate to conservatively assign the concentration of metabolite in plasma to a value just below the assay quantification limit, rather than zero and then compare the clearance value based on blood concentration to known renal blood flow.

An alternative, though more complex and invasive approach to assess renal metabolism was proposed and evaluated by Riegelman and co-workers (48,49), where parent and radiolabeled metabolite were infused simultaneously to estimate the renal clearance of the metabolite and formation rates. Contributions of hepatic metabolism to metabolite excreted in urine were determined with the assumption that the metabolism was FRL and F(m) of metabolite formed in the liver was unity, (i.e., these metabolites were not excreted in the bile or subject to sequential metabolism). The labeled dose of metabolite provides information supporting renal metabolism because the ratio of labeled to unlabeled metabolite in plasma versus urine will be different if parent drug is converted to metabolite when passing through the kidney. This approach revealed that 60–70% of salicyluric acid metabolite produced from salicylate in a human subject was formed in the kidney (49). With the current availability of liquid chromatography mass spectrometry (LC–MS), this method could be adapted to use stable isotope tracers of metabolites, rather than radiolabeled material.

VII. BILIARY CLEARANCE AND ENTEROHEPATIC RECYCLING OF METABOLITES

Biliary clearance is an excretory route of elimination in many ways similar to renal clearance discussed in the foregoing. For some drugs, xenobiotics, and metabolites, excretion into the bile can represent the major route for elimination. The mechanisms of bile formation, hepatobiliary transport, and excretion processes have been reviewed (50–52). Drugs can be eliminated in the bile by direct excretion, without biotransformation; however, many drugs are excreted after metabolism to more polar, charged metabolites, often by conjugation with sulfate, glucuronic acid, or glutathione. Excretion by the bile can be an irreversible route of elimination for a drug if its metabolites have poor permeability (e.g., glucuronide or sulfate conjugates), the drug molecule is poorly absorbed, or the drug is subject to degradation or complexation in the intestinal contents. In contrast, there are numerous examples of conjugated (phase II) metabolites that are excreted in the bile, cleaved to yield the parent drug, and then the parent drug is reabsorbed into the systemic circulation (i.e., enterohepatic recycling; EHC). The factors influencing EHC can be quite complex as discussed by Pollack and Brouwer (53) using the example of valproic acid disposition in the rat as a model.

A. Measurement of Biliary Clearance

Bile collection can occasionally be obtained from humans, such as in patients treated for biliary obstruction or following a liver transplant. Animal studies of biliary excretion are quite common, often conducted to provide insight into mechanistic processes of drug and metabolite disposition. When bile is collected after administration of the parent drug, *apparent* biliary clearance of metabolite can be determined using Eqs. (33)–(35), in which the excretion rate of bile is substituted for that of urine. The term apparent is usually used because often the drug entering the hepatocyte is first metabolized and then the metabolite is excreted in the bile without subsequent access to the systemic circulation. If a preformed metabolite is administered directly, the same equations can be applied, but may provide much different values for biliary clearance of the metabolite than that determined when parent drug is administered because the metabolite need not be formed in the hepatocyte before excretion in the bile. Because of the potential rate-limiting steps involving active uptake or diffusion into the hepatocyte, metabolism in the hepatocyte and active transport from the hepatocyte into the bile, which may vary when parent drug versus metabolite is administered, the absolute value for metabolite biliary clearance using Eqs. (33)–(35) may not be that informative when the preformed metabolite is administered directly.

More commonly employed when metabolite is formed in the hepatocyte and subsequently excreted in bile with negligible plasma levels is the measurement of excretion rate of metabolite or parent drug in the bile as a function of parent drug concentration in plasma, which provides a measure of apparent biliary clearance,

$$\frac{\Delta A_{\text{bile}}}{\Delta t} = CL_{\text{bile}} \cdot C_{\text{ave}} \tag{41}$$

where ΔA_{bile} can include both parent and metabolite, although often metabolite(s) is(are) dominant, and C_{ave} refers to average plasma concentrations of the parent drug during the collection interval. If only metabolite is present in bile, no metabolite is measurable in plasma and metabolite once formed in the hepatocyte is not released into the systemic circulation, the term CL_{bile} represents an apparent formation clearance ($CL_{\text{f,bile}}$) for the metabolite in the liver. This relation can be rewritten in terms of an AUC, either during an interval or until all drug is eliminated after a single dose, where the subscript *bc* refers the AUC obtained with an exteriorized bile cannula,

$$\Delta A_{\text{bile}} = CL_{\text{bile}} \cdot \Delta \text{AUC}_{\text{bc}} \tag{42}$$

B. Enterohepatic Recycling

Enterohepatic recycling via metabolites is considered as a ''futile cycle'' because of the inefficiency of the metabolic process. The existence of EHC can be proved based on several criteria, as summarized by Duggan and Kwan (54). It can be proved unequivocally in animals studies by (a) linked experiments for which the bile from a donor animal is infused into a recipient; (b) gradual establishment of a portal–systemic plasma concentration gradient after IV dosing of parent drug; and (c) increase systemic clearance by irreversible bile diversion. In contrast, in most human studies, only suggestive evidence of EHC can be obtained. These include observations such as the presence of secondary peaks in plasma concentration versus time curves, often coincident with gallbladder discharge in response to a meal, or the presence of drug-derived material in the feces after administration by routes other than oral or rectal. It should be noted that secondary peaks caused by EHC seen in humans would not initially be expected in the rat, which lacks a gallbladder. However, if reabsorption in the intestine is delayed owing to regional distributions of β-glucuronidase or sulfatase, time-dependent hydrolysis of conjugates or site-specific permeability, secondary peaks may also be seen in the rat (53) and, thus, secondary peaks in humans may be due to reasons other than EHC.

When bile is diverted in animals, the AUC in plasma of parent drug will decrease relative to that obtained without diversion when enterohepatic recycling is operative. CL_{bile} determined with bile diversion using Eq. (43) can then be employed to estimate total biliary exposure or the total amount of drug or metabolite subject to EHC when cycling is operative without bile diversion (48),

$$\Sigma A_{\text{bile}} = CL_{\text{bile}} \cdot \text{AUC} \tag{43}$$

where AUC is the exposure to parent drug in plasma in an animal without biliary diversion, and ΣA_{bile} is the sum of all parent drug and metabolite that may be cycled back to drug when EHC is operative (i.e., it is the cumulative drug exposure to the intestine). This measure of cumulative drug exposure to the intestine due to EHC can result in an amount exceeding the IV dose when EHC is very extensive and efficient. For example, as shown in Table 1, indomethacin exposure in bile of the dog had an estimated 362% of the dose cycled, and the extent of cycling was inversely correlated with observed toxic doses across the species (54).

EHC can be very efficient resulting in futile cycling. When this occurs, the metabolite formed and excreted in the bile does not represent an irreversible loss, instead the EHC process represents a distribution process, with bile–intestine as a peripheral distribution compartment. Thus, the much shorter terminal half-life of valproic acid glucuronide in rats with bile drainage was due to a decrease in the volume of distribution relative to control rats as well as the increase in clearance (53). One measure of the efficiency of EHC was provided by Tse *et al.* (55). From two experiments where bile is collected and AUC for the parent drug in plasma is measured in one treatment after a single intravenous dose, one can calculate the fraction of drug dose excreted in the bile as parent drug and metabolites that are subject to possible cycling, F_b. From this information and the definition that the amount of drug that is reabsorbed from the first cycle is $F_a \cdot F_b \cdot D$, and the assumption that this fraction is constant with all subsequent cycles, the following relation can be derived (55):

$$F_a = \frac{[1 - \text{AUC}_{\text{bc}}/\text{AUC}]}{F_b} \tag{44}$$

where AUC_{bc} for the parent drug is measured with bile cannulation and drainage. The value of F_a obtained is a measure of the fraction of the sum of drug and metabolite excreted in the bile that is then reabsorbed. When using this relation, only metabolites that could be recycled (i.e., are reversible to parent drug in the GI tract) would be included when estimating F_b. When F_b is a significant part of the dose, but AUC_{bc} is equivalent to AUC, it is apparent that bile diversion did not influence the amount of drug reaching the systemic

Table 1 Species Differences in Indomethacin Biliary Exposure and Toxicity[a]

Species	Clearance (mL/min/kg)			Area (μg · min/mL)		Plasma gradient, $C_p^{port}\,dt/C_p^{ven}$	Total exposure, $\Sigma_{bile}^{\%}$	Minimum toxic dosage (mg/kg/day)
	Plasma, $\dot{V}_{Cl,\,p}$	Urine, $\dot{V}_{Cl,\,r}$	Bile, $\dot{V}_{Cl,\,b}$	Venous, $\int_0^\infty C_p^{ven}\,dt$	Portal, $\int_0^\infty C_p^{port}\,dt$			
Dog	8.2	<0.1	13.3	122	310[b]	2.54[b]	362	0.5
Rat	0.32	0.01	0.39	3074	3535	1.15	134	0.75
Monkey	8.3	3.0	2.2	121	121	1.0	26	1.0
Guinea pig	6.25	1.85	1.20	158	181	1.15	21	6.0
Rabbit	3.62	1.09	0.40	278	334	1.20	13	20.0
Man	1.79	0.22	0.16[c]	592	592	1.0[d]	9.5	

Portal/venous concentration ratios greatly exceeded 1 in the dog, and there was a strong inverse correlation between total biliary exposure and minimum toxic dosage.

[a]All disposition data for single intravenous dosage of 1.0 mg/kg except man for which 25 mg total dosage normalized to 1.0 mg/kg.

[b]Based on complete 0- to 2-h portal and systemic plasma profiles; for all other species, mean of more than five measurements at interval specified in text.

[c]Calculated from $f_{bile} = 0.09$ (H. B. Hucker, unpublished).

[d]Assumed.

Source: Ref. 54.

circulation; thus, F_a is zero. In contrast, when a significant part of the dose is excreted in bile and AUC_{bc} is much less than AUC, then recycling is significant and F_a can approach a value of 1.

VIII. REVERSIBLE METABOLISM

Reversible metabolism occurs when a metabolite or biotransformation product and the parent drug undergo interconversion in both directions, as shown in Scheme 6. The scheme can also be written in terms of concentrations and clearance terms, rather than amounts and rate constants, respectively, as shown in Scheme 6B using the common convention in the literature where the drug is considered in compartment **1** and the metabolite is compartment **2**. These compartments are not to be confused with physiological spaces. Here the "metabolite" may be the pharmacologically active species in the case of administration of a prodrug, an active metabolite, or an inactive metabolite.

Although reversible metabolism is less often addressed in reviews of drug metabolite kinetics (10), there are numerous examples occurring across a wide variety of compounds, as noted in a review by Cheng and Jusko (56). These include phase I metabolic pathways for amines, such as imipramine; alcohols, such as corticosteroids and estradiol; lactones; and sulfides/sulfoxides, such as captopril and sulindac. Examples for phase II metabolic pathways include carboxylic acids to their glucuronides, as occurs for ibuprofen; sulfation of phenols, such as estrone; and acetylation of amines, such as procainamide (56). Pharmacokinetic methods for the analysis of reversible metabolism are well described (56–59). More recently, considerations of statistical moment analysis as applied to reversible systems have also been addressed by Cheng (60,61).

Reversible metabolism is often ignored when analyzing the pharmacokinetic data of compounds that undergo metabolite interconversion. The pharmacokinetic parameter estimates so obtained using methods or approaches that do not consider the reversible nature of the system are not true estimates of pharmacokinetic parameters and should be considered apparent parameter values. Indeed, regulatory agencies may not require rigorous evaluation of the true reversible metabolic parameters because the critical parameters of clearance, volume of distribution, and bioavailability can be unambiguously obtained only after direct administration of the preformed metabolite, which is usually not feasible in humans owing to the need to secure an Investigational New Drug Permit (IND) for the IV administration of the metabolite. Moreover, in many cases of reversible metabolism,

A)	**Other pathways**	← k_{other}	A	$\xrightarrow{k_f}$ / $\xleftarrow{k_r}$	A(m)	$\xrightarrow{k(m)}$	$A(m)_{elim}$
B)	**Other pathways**	← CL_{10}	C=A/V	$\xrightarrow{CL_{12}}$ / $\xleftarrow{CL_{21}}$	C(m)=A(m)/V(m)	$\xrightarrow{CL_{20}}$	$A(m)_{elim}$

Scheme 6 Reversible conversion between metabolite and drug with parallel elimination pathways for the parent drug and metabolite.

the metabolite may achieve only low concentrations relative to the parent drug or is inactive, thus a great deal of effort to fully elucidate its pharmacokinetics may be difficult or not justified based on the considerable costs and efforts. There are, however, several common disease states, such as renal or hepatic impairment that may dramatically alter the disposition of the metabolite, thus significantly influencing the disposition of the parent drug, and these should be addressed. When a metabolite is active or of toxicological relevance, or when altered disposition of the metabolite substantially modifies the pharmacokinetic profile of the parent drug, efforts to more fully investigate reversible metabolism are warranted.

A. Determination of Primary Pharmacokinetic Parameters for Reversible Metabolic Systems

The primary pharmacokinetic parameters of clearance, volume, and availability, as well as commonly employed secondary parameters of half-life and mean residence time, will be discussed here. Discussion of less commonly employed parameters can be found in other literature (56–61). However, difficulties in estimating parameters involving reversible metabolism, either because of the need to dose the metabolite directly or of inherent errors in the complex equations employed, generally limit the usefulness of estimating some parameters. Other parameters unique to reversible metabolic systems are descriptors of the reversibility of the process that include the recycling numbers, recycled fraction, and exposure enhancement, which will be discussed in Section VIII.B.

1. Half-Life in Reversible Metabolism Systems

As shown in Fig. 10 for the interconversion of methylprednisone and methylprednisolone which undergo the reversible metabolism of ketone–alcohol common with steroids, after the parent drug and metabolite reach equilibrium, the two compounds decline in parallel when either is administered intravenously (62). Because the terminal half-life reflects a hybrid of the clearance terms for the overall reversible system and the volumes of distribution of parent drug and metabolite, estimating changes in the half-life in response to an altered clearance or volume term is difficult. Although the use of ''sojourn times'' have been proposed as a measure of time a drug or metabolite is in the body before being eliminated or transformed in reversible systems (57), in practice most often an apparent half-life is reported with the understanding that it may be subject to change in response to alterations of the disposition of parent drug or metabolite.

That the terminal half-life of parent and metabolite are parallel when parent is administered could lead to an erroneous assumption that FRL metabolism is operative when, in fact, reversible metabolism is occurring. Confirmation of reversibility can be made by identifying formation of the parent after administration of the metabolite; however, such an experiment may not be feasible in humans. Therefore, in vitro studies with human tissues or animal studies may be necessary to infer reversible metabolism in humans.

2. Clearance Parameters in Reversible Metabolism Systems

For a reversible system as shown in Scheme 6B the following relations can be derived (56,57) for determining clearance values after collection of AUC data from an IV bolus dose of parent drug or metabolite on two separate occasions. These relations assume that

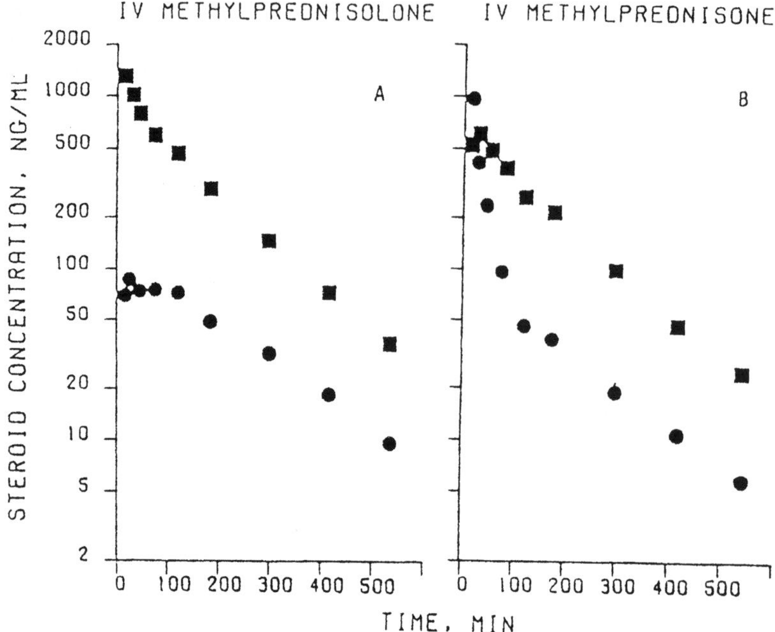

Fig. 10 The reversible metabolism of methylprednisolone (■) and methylprednisone (●) when each is given on separate occasions as a 1.25-mg/kg–intravenous bolus to a rabbit. The parallel profiles, constant concentration ratios at equilibrium and formation of both compounds when either are administered are characteristic of reversible metabolism. (From Ref. 62.)

clearance and distribution processes are linear for both parent drug and metabolite (i.e., independent of concentration).

$$CL_{10} = \frac{\text{Dose}^{p} \cdot \text{AUC}_{m}^{m} - \text{Dose}^{m} \cdot \text{AUC}_{m}^{p}}{\text{AUC}_{p}^{p} \cdot \text{AUC}_{m}^{m} - \text{AUC}_{m}^{p} \cdot \text{AUC}_{m}^{m}} \tag{45}$$

$$CL_{20} = \frac{\text{Dose}^{m} \cdot \text{AUC}_{p}^{p} - \text{Dose}^{p} \cdot \text{AUC}_{p}^{m}}{\text{AUC}_{p}^{p} \cdot \text{AUC}_{m}^{m} - \text{AUC}_{m}^{p} \cdot \text{AUC}_{p}^{m}} \tag{46}$$

$$CL_{21} = \frac{\text{Dose}^{m} \cdot \text{AUC}_{m}^{p}}{\text{AUC}_{p}^{p} \cdot \text{AUC}_{m}^{m} - \text{AUC}_{m}^{p} \cdot \text{AUC}_{p}^{m}} \tag{47}$$

$$CL_{21} = \frac{\text{Dose}^{p} \cdot \text{AUC}_{p}^{m}}{\text{AUC}_{p}^{p} \cdot \text{AUC}_{m}^{m} - \text{AUC}_{m}^{p} \cdot \text{AUC}_{p}^{m}} \tag{48}$$

Here the superscript indicates the dose administered as being from the parent or metabolite, whereas the subscript refers to the compound that is measured in plasma. AUC is the total area under the plasma concentration versus time curve extrapolated to infinity. Similar equations can be derived when the compounds are infused to steady state (59) for which the values needed are infusion rates and steady-state concentrations. These relations are not unique as the model has also been applied to other two-compartment pharmacokinetic systems with elimination occurring from each compartment, such as the reversible distribution and elimination from the maternal–fetal unit (63).

If apparent clearance terms are used for parent drug or metabolite, these will overestimate the true values for CL_{10} and CL_{20}, respectively. The magnitude of the error is a complex relation of all the clearance terms as discussed by Ebling and Jusko (57) and is shown here for CL_{app} of the parent drug,

$$CL_{app} = \frac{D}{\text{AUC}} = CL_{10} + CL_{12}\left[\frac{CL_{20}}{CL_{21} + CL_{20}}\right] \tag{49}$$

Ebling and Jusko (57) defined the term $[CL_{20}/(CL_{21} + CL_{20})]$ an efficiency parameter because it is a fraction that defines the extent of drug clearance by metabolite formation (CL_{12}) resulting in metabolite that does not return or interconvert back to the parent drug (i.e., an irreversible loss). Similar relations have been determined for $CL(\text{m})_{app}$; however, if the metabolite cannot be administered, then only the term CL_{app} will usually be reported.

It is also evident from Eq. (49) that CL_{app} will underestimate the total elimination capacity for the parent drug (i.e., $CL_{10} + CL_{12}$), where CL_{12} is a measure of metabolite formation clearance, CL_f. Therefore, estimating the formation clearance of a metabolite from in vivo studies using, $CL_f = CL_{app} - CL_{10}$, (where CL_{10} is determined from the other clearance pathways) may grossly underestimate total metabolic capability for the particular metabolic pathway in vivo.

3. *Volume of Distribution in Reversible Metabolism Systems*

Volume of distribution at steady state for parent drug and metabolite in reversible metabolic systems are independent, but the equations to calculate the values necessitate consideration of the disposition of both parent drug and metabolite. Indeed, given the structural changes from parent drug to metabolites, as well as potential differences in protein binding and lipophilicity between the parent drug and metabolites, it is reasonable to expect that distribution in the body could be quite different, as previously mentioned for morphine and M6G. In the absence of interconversion, volume of distribution at steady state $[V(\text{m})_{ss,app}]$ is calculated with the following equation for the metabolite,

$$V(\text{m})_{ss,app} = \frac{M \cdot \text{AUMC}_{\text{m}}^{\text{m}}}{(\text{AUC}_{\text{m}}^{\text{m}})^2} \tag{50}$$

where AUMC(m) is the first moment of the plasma concentration versus time curve for the metabolite (17). This equation for apparent V_{ss} is in error if applied to reversible metabolism systems, for $V(\text{m})_{ss,app}$ will overestimate the real $V(\text{m})_{ss}$ because the parent drug reverts back to the metabolite (57). Moreover, $V(\text{m})_{ss,app}$ is not independent of clearance processes as is the true $V(\text{m})_{ss}$, thus changes in clearance terms of either parent or metabolite will modify the value of $V(\text{m})_{ss,app}$ in reversible metabolism systems. The relation between the apparent and real parameter are described for the metabolite as follows (57),

$$V(\text{m})_{ss,App} = V(\text{m}) + V_{ss}\left[\frac{CL_{12} \cdot CL_{21}}{(CL_{12} + CL_{10})^2}\right] \tag{51}$$

The complexities of these relations for volume and clearance combine, such that the following equation for $V(\text{m})_{ss}$ is dependent on dose of metabolite and measured AUC and AUMC data (57),

$$V(\text{m})_{ss} = \frac{M[\text{AUC}_{\text{p}}^{\text{p}2} \cdot \text{AUMC}_{\text{m}}^{\text{m}} - \text{AUC}_{\text{p}}^{\text{m}} \cdot \text{AUC}_{\text{m}}^{\text{p}} \cdot \text{AUMC}_{\text{p}}^{\text{p}}]}{\text{AUC}_{\text{p}}^{\text{p}} \cdot \text{AUC}_{\text{m}}^{\text{m}2} - \text{AUC}_{\text{m}}^{\text{p}} \cdot \text{AUC}_{\text{p}}^{\text{m}2}} \tag{52}$$

Measures of AUMC used for Eqs. (50) and (52) are often criticized for the potential error that may be introduced when extrapolating the first moment curve to infinity. Alternative equations derived for application of data obtained from infusions of metabolite and parent drug to steady-state levels may reduce some of the errors (59). Given the limited ability to dose preformed metabolite to humans and the potential error in determining V_{ss} for parent drug or metabolite within reversible systems, true values for the pharmacokinetic parameters in reversible metabolism systems have been determined in humans for very few compounds. Instead, it is more common to report $V_{ss,app}$ values, with an understanding that such values are inaccurate and subject to change if clearance is altered.

4. *Availability in Reversible Metabolism Systems*

Interconversion of parent and metabolite also complicates the measures of availability, and consideration of this has lead to the development of equations that assess absorption processes independent of clearance (56,64). However, the increased complexity of the relations and the need to dose metabolite independently to assess the values have restricted their application. Thus, apparent availability is often employed, ignoring the contributions of reversible metabolism.

B. Measures of Reversibility in Drug Metabolism

Unique to reversible systems are measures defined as recycled fraction (RF), number of recyclings through the reversible process (R_I) as well as other terms (56,57). RF is a measure of the likelihood of a molecule going back and forth through the reversible system (i.e., being converted in both directions), and is a value between zero and 1 determined from the relative values of clearance,

$$RF = \frac{CL_{12} \cdot CL_{21}}{CL_{11} \cdot CL_{22}} \tag{53}$$

where $CL_{11} = CL_{10} + CL_{12}$ and $CL_{22} = CL_{20} + CL_{21}$.

The number of recyclings, R_I, can exceed 1 and provides a measure of how exposure to parent or metabolite is enhanced by the interconversion process,

$$R_I = \frac{CL_{12} \cdot CL_{21}}{CL_{11} \cdot CL_{22} - CL_{12} \cdot CL_{21}} \tag{54}$$

C. Influence of Altered Metabolite Clearances on the Disposition of Parent Drug

Because metabolite and parent can interconvert, altered irreversible clearance of the metabolite (CL_{20}) can influence the disposition of the parent drug, causing changes in CL_{app} [i.e., Eq. (49)]. Examples of this include the nonsteroidal anti-inflammatory drugs for which renal clearance of labile ester glucuronide metabolite is reduced by renal dysfunction (65). For example, although diflunisal is eliminated almost entirely by metabolism

Table 2 Relative Values of AUC for Diflunisal and Diflunisal Acyl Glucuronide in Rats

Clearance term altered	Percent of normal AUC	
	Diflunisal	Diflunisal acyl glucuronide
CL_{10}	244	244
CL_{12}	145	15
CL_{20}	143	186
CL_{21}	75	147
CL_{23}	93	114

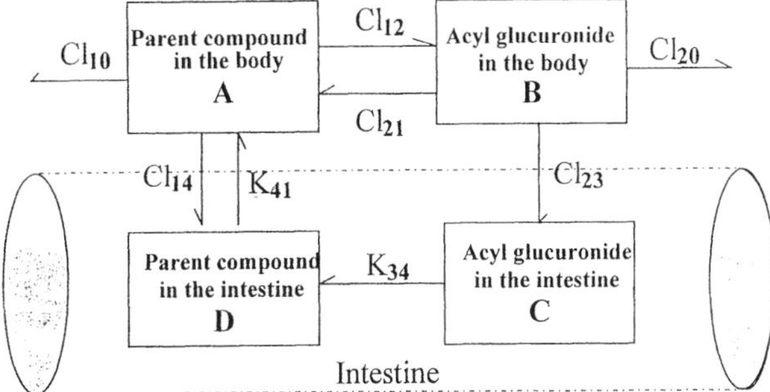

The data were obtained by simulation using the model shown when individual clearance terms in the scheme for reversible metabolism are reduced to 10% of their initial value. The clearance term, CL_{23}, is the biliary clearance of diflunisal acyl glucuronide. Baseline values are 100. Data were obtained by Monte Carlo simulation using initial values obtained in rats from Dickinson and co-workers (66–68).

in the rat, simulation results shown in Table 2 show that when clearances that directly affect diflunisal glucuronide are altered, the AUC of both parent drug and metabolite can occasionally be significantly modified. This has substantial clinical significance, because hepatobiliary or renal disease may alter clearance pathways of the metabolite and cause unanticipated alterations in the disposition of the parent drug even though the parent drug itself is not eliminated directly into bile or into the urine.

IX. NONLINEAR PROCESSES AND METABOLITE DISPOSITION

Metabolite disposition is subject to the same sources of nonlinearity that can occur with parent drug (15). These may include saturable enzymic metabolism, cosubstrate depletion in phase II metabolism, saturable transport for biliary or renal excretion, or saturable protein binding. In summary, metabolite disposition can be affected by nonlinearities involving both the rate of input (i.e., formation), rate of output (i.e., elimination), and distribution. Any nonlinearity in disposition of the parent drug that is the precursor for the metabolite will lead to increased complexities in predicting metabolite disposition; thus the correlation between exposure to parent drug versus exposure to the metabolite will not be predictable as dose is varied, or may vary with time when dosing chronically. In this section, the most common source of nonlinearity will be considered and discussed, assuming that only one clearance parameter is nonlinear at a given time. However, one should realize that multiple nonlinearities may occur simultaneously, especially in animal toxicology studies where drug doses are intentionally pushed to levels much greater than those employed later in humans.

A. Michaelis–Menton Formation Versus Elimination of the Metabolite

The assumption that metabolite formation and elimination occur as first-order processes with constant CL_f and $CL(\mathrm{m})$ generally holds for typical enzyme or active transport systems when concentrations of drug or metabolite, respectively, are below their K_m (Michaelis–Menton constant where the rate is half of the maximum velocity, Vm; (15)) values. If Michaelis–Menton kinetics apply for metabolite formation and elimination, with only a single pathway for formation and elimination of metabolite, Eq. (11) can be rewritten as,

$$\frac{dA(\mathrm{m})}{dt} = \frac{V_{\mathrm{m,f}} \cdot C}{K_{\mathrm{m}} + C} - \frac{V_{\mathrm{m}}(\mathrm{m}) \cdot C(\mathrm{m})}{K_{\mathrm{m}}(\mathrm{m}) + C(\mathrm{m})} \tag{55}$$

It is apparent that $CL(\mathrm{m})$ and $k(\mathrm{m})$ are now functions of $K_m(\mathrm{m})$ and $V_m(\mathrm{m})$,

$$CL(\mathrm{m}) = \frac{V_{\mathrm{m}}(\mathrm{m})}{\mathrm{K_m}(\mathrm{m}) + C(\mathrm{m})} \tag{56}$$

$$k(\mathrm{m}) = \frac{V_{\mathrm{m}}(\mathrm{m})}{[\mathrm{K_m}(\mathrm{m}) + C(\mathrm{m})] - V(\mathrm{m})} \tag{57}$$

and similar relations exist for CL_f and k_f. When $C \ll K_m$ and $C(\mathrm{m}) \ll K_m(\mathrm{m})$, then both right-hand terms of Eq. (55) simplify yielding first-order processes, thus clearance and the rate constant of elimination are essentially constant. However, if the elimination of the metabolite is saturable (i.e., when $C(\mathrm{m}) > K_m(\mathrm{m})$), both $CL(\mathrm{m})$ and $k(\mathrm{m})$ become variable, with $CL(\mathrm{m})$ decreasing as $C(\mathrm{m})$ increases. When saturation of $CL(\mathrm{m})$ occurs, and CL_f is constant (i.e., formation of metabolite is not saturable in the concentration range of drug), then $C(\mathrm{m})/C$ and AUC(m)/AUC will increase as the drug dose increases. Typically, plots of metabolite levels normalized to drug dose should be superimposable, but with saturable elimination the dose-normalized metabolite concentrations will increase with dose level. Even under conditions of FRL metabolism, saturation of metabolite elimination will increase these ratios as dose increases, although the metabolite concentration profile may still appear to be parallel to that of the parent drug profile. However, because the formation clearance (CL_f) of the metabolite did not change with dose, f_m would remain constant. Thus, for a metabolite that is primarily excreted in the urine, the percentage of the dose recovered in the urine as metabolite would be independent of dose.

For metabolites subject of ERL metabolism, saturable metabolite elimination may be noticeable from a profile of metabolite concentration versus time when parent drug is rapidly administered or absorbed. A semilog plot of metabolite concentrations versus time profile may show the classic concave shape at higher concentrations, then become log-linear at lower concentrations (15).

In contrast, if the metabolite formation clearance (CL_f) is saturable, while the elimination of metabolite is not saturable, then $C(\mathrm{m})/C$ and AUC(m)/AUC will decrease as drug dose increases. In this case, f_m will decline as the dose increases, assuming that there are other additional pathways for drug elimination besides formation of the metabolite of interest. With FRL metabolism the apparent half-life of the metabolite will increase if the saturable pathway of metabolism is a significant fraction of the overall elimination of parent drug, because the half-life of parent drug will increase. For ERL metabolism ($k(\mathrm{m}) \ll k$), the profile of the metabolite after escalating single doses of parent drug may not be altered significantly, unless the elimination half-life of the parent drug increased significantly at higher doses, such that value of k decreases to the extent that it approaches that of $k(\mathrm{m})$. In this case of saturable metabolite formation clearance, plotting metabolite concentration versus time profiles normalized to dose would show a decline in the AUC(m) of the profiles as dose increased.

Mean residence time (MRT) concepts can also be applied to metabolites subject to Michaelis–Menton metabolism (26,61,69). Although there has been some debate about using MRT in nonlinear systems, a recent review addressed these concerns and with simulations showed that the values of MRT, V_{ss}, V_m, and K_m could be determined accurately (70). A potential advantage of MRT concepts for evaluating nonlinear systems is that they do not require multiple trials of various compartmental models normally employed when determining Michaelis–Menton parameters (70).

X. CONCLUSION

Even if some of the commonly employed mathematical relations for performing pharmacokinetic analysis of parent drug are also applicable to metabolites, because metabolites are formed in vivo, there are some unique methods applicable to describing metabolite disposition and an effort was made here to identify those methods. Many of the difficulties in analyzing metabolite pharmacokinetics stem from limited information about the rate and extent of their formation in the body (i.e., the input into the body is usually unknown). Thus, many of the relations described in this review are based on derivations that do not require knowledge of the rate of metabolite input, but often make informed estimates of the extent of formation of metabolite from the parent drug or assume that the extent of formation is constant between treatments or linear with dose. In many instances, especially when estimating levels after prolonged administration of the parent drug, the extent of metabolite formed is more important than the rate at which it was formed. When possible, exogenous administration of preformed metabolite by a known input rate circumvents these limitations. However, owing to the potential different behavior of preformed metabolite and that formed at tissue sites within the body (41,42), new assumptions arise that may be difficult to validate. One of the primary advantages of mean residence time approaches for describing metabolite pharmacokinetics is the fewer assumptions made for their application, but a lack of thorough understanding of their theory has limited their use by some investigators. Whatever approach is used, there is much to be gained from a better understanding of the pharmacokinetics of metabolites.

This chapter did not present extensive derivations of some of the mathematical relations provided. Earlier reviews on the pharmacokinetics of metabolites (10–12) or the primary literature cited should be consulted, together with this chapter, for further insight and understanding of the derivations and assumptions of the equations presented. In addition, some basic concepts in pharmacokinetics are needed for the full implementation of some of the relations provided, and these can be found in textbooks on the topic (13–17). Topics such as reversible metabolism and mean residence time concepts are now better understood owing to more recent work, and original literature should be consulted where appropriate. This review should provide a foundation for understanding how the complexities of metabolite formation and elimination may be analyzed. In conjunction with other later chapters in this text, this should assist scientists in the design of in vitro experiments, in vivo animal studies, and clinical trials to characterize the metabolism of drugs and other xenobiotics, such that optimal information can be obtained about the disposition of metabolites in the body with as few assumptions as possible.

ACKNOWLEDGMENTS

The assistance and patience of Dr. Laurene Wang in reading, commenting on, and editing this chapter is greatly appreciated.

REFERENCES

1. A. J. Atkinson and J. M. Strong, Effect of active drug metabolites on plasma level response relationships, *J. Pharmacokinet. Biopharm.* *5*:95–109 (1977).
2. T. A. Sutfin and W. J. Jusko, Compendium of active drug metabolites, *Drug Metabolism and Disposition: Considerations in Clinical Pharmacology*, (G. R. Wilkinson and M. D. Rawlins, eds.), MTP Press, Boston, 1985, pp. 91–159.
3. R. K. Portenoy, H. T. Thaler, C. E. Inturrisi, H. Friedlander-Klar, and K. M. Foley, The metabolite morphine-6-glucuronide contributes to the analgesia produced by morphine infusion in patients with pain and normal renal function, *Clin. Pharmacol. Ther.* *51*:422–431 (1992).
4. J. Brockmoller and I. Roots, Assessment of liver metabolic function, *Clin. Pharmacokinet.* *27*:216–248 (1994).
5. P. B. Watkins, Noninvasive tests of CYP3A enzymes, *Pharmacogenetics* *4*:171–184 (1994).
6. G. R. Matzke and S. P. Millikin, Influence of renal function and dialysis on drug disposition, *Applied Pharmacokinetics*, 3rd ed. (W. E. Evans, J. J. Schentag, W. J. Jusko, and M. V. Relling, eds.), Applied Therapeutics, Vancouver, WA, 1992, pp. 8.1–8.49.
7. S. D. Nelson, Mechanisms of the formation and disposition of reactive metabolites that can cause acute liver injury, *Drug Metab. Rev.* *27*:147–177 (1995).
8. A. J. Cummings and B. K. Martin, Excretion and accrual of drug metabolites, *Nature* *200*: 1296–1297 (1963).
9. J. B. Houston and G. Taylor, Drug metabolite concentration–time profiles: Influence of route of drug administration, *Br. J. Clin. Pharmacol.* *17*:385–394 (1984).
10. J. B. Houston, Drug metabolite kinetics, *Pharmacol. Ther.* *15*:521–552 (1982).
11. K. S. Pang, A review of metabolite kinetics, *J. Pharmacokinet. Biopharm.* *13*:632–662 (1985).
12. M. Weiss, A general model of metabolite kinetics following intravenous and oral administration of the parent drug, *Biopharm. Drug Dispos.* *9*:159–176 (1988).
13. M. Rowland and T. N. Tozer, *Clinical Pharmacokinetics*, Lea & Febiger, Baltimore, 1995.
14. L. Z. Benet, Pharmacokinetics: The dynamics of drug absorption, distribution and elimination, *The Pharmacological Basis of Therapeutics*, 9th ed. (J. G. Hardman, L. E. Limbird, P. B. Molinoff, and R. W. Ruddon, eds.). McGraw Hill, New York, 1996, pp. 3–28.
15. M. Gibaldi and D. Perrier, *Pharmacokinetics*, 2nd ed., Marcel Dekker, New York, 1982.

16. M. Gibaldi, *Biopharmaceutics and Clinical Pharmacokinetics*, 4th ed., Lea & Febiger, Philadelphia, 1991.
17. M. Gibaldi, and D. Perrier, *Pharmacokinetics*, Marcel Dekker, New York, 1975.
18. T. Walle, E. C. Conradi, U. K. Walle, T. C. Fagan, and T. E. Gaffrey, Naphthoxylacetic acid after single and long-term doses of propranolol, *Clin. Pharmacol. Ther. 26*:548–554 (1979).
19. M. Chung, J. M. Hilbert, R. P. Gural, E. Radwanski, S. Symchowicz, and N. Zampaglione, Multiple dose halazepam kinetics, *Clin. Pharmacol. Ther. 35*:838–842 (1984).
20. R. Osborne, S. Joel, D., Trew, and M. Slevin. Morphine and metabolite behavior after different routes of morphine administration: Demonstration of the importance of the active metabolite morphine-6-glucuronide, *Clin. Pharmacol. Ther. 47*:12–19 (1990).
21. D. A. Barrett, D. P. Barker, N. Rutter, M. Pawula, and P. N. Shaw, Morphine, morphine-6-glucuronide and morphine-3-glucuronide pharmacokinetics in newborn infants receiving diamorphine infusions, *Br. J. Clin. Pharmacol. 41*:531–537 (1996).
22. J. Lotsch, A. Stockmann, G. Kobal, K. Brune, R. Waibel, N. Schmidt, and G. Geisslinger, Pharmacokinetics of morphine and its glucuronides after intravenous infusion of morphine and morphine-6-glucuronide in healthy volunteers, *Clin. Pharmacol. Ther. 60*:316–325 (1996).
23. M. Rowland, Physiological pharmacokinetic models and interanimal species scaling, *Pharmacol. Ther. 29*:49–68 (1985).
24. H. Boxenbaum, Interspecies scaling, allometry, physiological time, and the ground plan for pharmacokinetics, *J. Pharmacokinet. Biopharm. 10*:201–227 (1982).
25. S. Riegelman and P. Collier, The application of statistical moment theory to the evaluation of in vivo dissolution time and absorption time, *J. Pharmacokinet. Biopharm. 8*:509–534 (1980).
26. H. Cheng and W. J. Jusko, Mean residence times and distribution volumes for drugs undergoing linear reversible metabolism and tissue distribution and linear or nonlinear elimination from the central compartment, *Pharm. Res. 8*:508–511 (1991).
27. M. Weiss, Drug metabolite kinetics: Noncompartmental analysis, *Br. J. Clin. Pharmacol. 19*: 855–856 (1985).
28. S. A. Veng-Pedersen and W. R. Gillespie, A method for evaluating the mean residence times of metabolites in the body, systemic circulation, and the peripheral tissue not requiring separate I. V. administration of the metabolite, *Biopharm. Drug Dispos. 8*:395–401 (1987).
29. S. A. Kaplan, M. L. Jack, S. Cotler, and K. Alexander. Utilization of area under the curve to elucidate the disposition of an extensively biotransformed drug, *J. Pharmacokinet. Biopharm. 1*:201–215 (1973).
30. D. J. Cutler, Linear system analysis in pharmacokinetics, *J. Pharmacokinet Biopharm. 6*:265–282 (1978).
31. M. D. Karol and S. Goodrich, Metabolite formation pharmacokinetics: Rate and extent determined by deconvolution, *Pharm. Res. 5*:347–351 (1988).
32. Y. Murai, T. Nakagawa, K. Yamaoka, and T. Uno, High-performance liquid chromatographic determination and moment analysis of urinary excretion of flucloxacillin and its metabolites in man, *Int. J. Pharm. 15*:309–320 (1983).
33. M. Weiss, Metabolite residence time: Influence of the first-pass effect, *Br. J. Clin. Pharmacol. 22*:121–122 (1986).
34. K. K. H. Chan and M. Gibaldi, Effects of first-pass metabolism and metabolite mean residence time determination after oral administration of parent drug, *Pharm. Res. 7*:59–63 (1990).
35. F. W. Lee, T. Salmonson, and L. Z. Benet, Pharmacokinetics and pharmacodynamics of nitroglycerin and its dinitrate metabolites in conscious dogs: Intravenous infusion studies, *J. Pharmacokinet. Biopharm. 21*:533–550 (1993).
36. E. A. Lane and R. H. Levy, Prediction of steady-state behavior of metabolite from dosing of parent drug, *J. Pharm. Sci. 69*:610–612 (1980).
37. T. Walle, E. C. Conradi, U. K. Walle, T. C. Fagan, and T. E. Gaffney, 4-Hydroxypropranolol

and its glucuronide after single and long-term doses of propranolol, *Clin. Pharmacol. Ther. 27*:22–31 (1980).

38. R. T. Williams, *Detoxification Mechanisms. The Metabolism of Drugs and Allied Organic Compounds*, John Wiley & Sons, New York, 1947.
39. D. Jones, C. Gorski, B. D. Haehner, E. M. O'Mara, and S. D. Hall, Determination of cytochrome P450 3A4/5 activity in vivo with dextromethorphan *N*-demethylation, *Clin. Pharmacol. Ther. 60*:374–384 (1996).
40. K. S. Pang, Kinetics of sequential metabolism, *Drug Metab. Dispos. 23*:166–177 (1995).
41. K. S. Pang and J. R. Gillette, Sequential first-pass elimination of a metabolite derived from its precursor, *J. Pharmacokinet. Biopharm. 7*:275–290 (1979).
42. R. G. Tirona and K. S. Pang, Sequestered endoplasma reticulum space for sequential metabolism of salicylamide: Coupling of hydroxylation and glucuronidation, *Drug Metab. Dispos. 24*:821–833 (1996).
43. R. A. Upton, J. N. Buskin, R. L. Williams, N. H. G. Holford, and S. Riegelman, Negligible excretion of unchanged ketoprofen, naproxen and probenecid in urine, *J. Pharm. Sci. 69*:1254–1257 (1980).
44. P. C. Smith, J. Hasegawa, P. N. J. Langendijk, and L. Z. Benet, Stability of acyl glucuronides in blood, plasma and urine: Studies with zomepirac, *Drug Metab. Dispos. 13*:110–112 (1985).
45. J. T. Dalton, M. E. Weintjes, and J. L. Au, Effects of bladder reabsorption on pharmacokinetic data analysis, *J. Pharmacokinet. Biopharm. 22*:183–205 (1995).
46. G. Levy, T. Tsuchiya, and L. P. Amsel, Limited capacity for salicyl phenolic glucuronide formation and its effect on the kinetics of salicylate elimination in man, *Clin. Pharmacol. Ther. 13*:258–268 (1972).
47. T. B. Vree, Y. A. Hekster, and P. G. Anderson, Contribution of the human kidney to the metabolic clearance of drugs, *Ann. Pharmacother. 26*:1421–1428 (1992).
48. S. H. Wan and S. Riegelman, Renal contribution to overall metabolism of drugs I: Conversion of benzoic acid to hippuric acid, *J. Pharm. Sci. 61*:1278–1284 (1972).
49. B. V. Lihmann, S. H. Wan, S. Riegelman, and C. Becker, Renal contribution to overall metabolism of drugs IV: Biotransformation of salicylic acid to salicyluric acid in man, *J. Pharm. Sci. 62*:1483–1486 (1973).
50. C. D. Klaassen and J. B. Watkins, Mechanisms of bile formation, hepatic uptake and biliary excretion, *Pharmacol. Rev. 36*:1–67 (1984).
51. R. P. J. Oude Elferink, D. K. F. Meijer, F. Kuipers, P. L. M. Jansen, A. K. Groen, and F. M. M. Groothis, Hepatobiliary secretion of organic compounds: Molecular mechanism of membrane transport, *Biochim. Biophys. Acta 1241*:215–268 (1995).
52. M. Yamazaki, H. Suzuki, and Y. Sugiyama. Recent advances in carrier-mediated hepatic uptake and biliary excretion of xenobiotics, *Pharm. Res. 13*:497–513 (1996).
53. G. M. Pollack and R. L. Brouwer, Physiological and metabolic influences on enterohepatic recirculation: Simulations based upon the disposition of valproic acid in the rat, *J. Pharmacokinet. Biopharm. 19*:189–225 (1991).
54. D. E. Duggan and K. C. Kwan, Enterohepatic recirculation of drugs as a determinant of therapeutic ratio, *Drug Metab. Rev. 9*:21–41 (1979).
55. F. L. S. Tse, F. Ballard, and J. Skinn, Estimating the fraction reabsorbed in drugs undergoing enterohepatic circulation, *J. Pharmacokinet. Biopharm. 10*:455–461 (1982).
56. H. Cheng and W. J. Jusko, Pharmacokinetics of reversible metabolic systems., *Biopharm. Drug Dispos. 14*:721–766 (1993).
57. W. F. Ebling and W. F. Jusko, The determination of essential clearance, volume and residence time parameters of recirculating metabolic systems: The reversible metabolism of methylprednisolone and methylprednisone in rabbits, *J. Pharmacokinet. Biopharm. 14*:557–599 (1986).
58. J. G. Wagner, A. R. DiSanto, W. R. Gillespie, and K. S. Albert, Reversible metabolism and pharmacokinetics: Applications to prednisone and prednisolone, *Res. Commun. Chem. Pathol. Pharmacol. 32*:387–405 (1981).

59. H. Cheng and W. J. Jusko, Constant-rate intravenous infusion methods for estimating steady-state volumes of distribution and mean residence times in the body for drugs undergoing reversible metabolism, *Pharm. Res. 7*:628–632 (1990).
60. H. Cheng, A method for calculating the mean transit times and distribution rate parameters of interconversion metabolites, *Biopharm. Drug Dispos. 14*:635–641 (1993).
61. H. Cheng, Mean residence time of drugs administered non-instantaneously and undergoing linear tissue distribution and reversible metabolism and linear or non-linear elimination from the central compartment, *Biopharm. Drug Dispos. 16*:259–267 (1995).
62. W. F. Ebling, S. J. Szefler, and W. J. Jusko, Methylprednisolone disposition in rabbits. Analysis, prodrug conversion, reversible metabolism. *Drug Metab. Dispos.* 13:296–304 (1985).
63. L. H. Wang, A. M. Rudolph, and L. Z. Benet, Pharmacokinetic studies of the disposition of acetaminophen in the sheep maternal–fetal unit, *J. Pharmacol. Exp. Ther. 238*:198–205 (1986).
64. S. S. Hwang and W. F. Bayne, General method for assessing bioavailability of drugs undergoing reversible metabolism in a linear system, *J. Pharm. Sci. 75*:820–821 (1986).
65. E. M. Faed, Decreased clearance of diflunisal in renal insufficiency—an alternative explanation, *Br. J. Clin. Pharmacol. 10*:185 (1980).
66. J. A. Watt, A. R. King, and R. G. Dickinson, Contrasting systemic stabilities of the acyl and phenolic glucuronides of diflunisal in the rat, *Xenobiotica 21*:403–415 (1995).
67. R. G. Dickinson, A. R. King, and R. K. Verbeeck, Elimination of diflunisal and its acyl glucuronide, phenolic glucuronide and sulfate conjugates in bile-exteriorized and intact rats, *Clin. Exp. Pharmacol. Physiol. 16*:913–924 (1989).
68. J. A. Watt and R. G. Dickinson, Effects of blockage of urine and/or bile flow on diflunisal conjugation and disposition in rats, *Xenobiotica 20*:835–845 (1990).
69. H. Cheng and W. J. Jusko, Mean residence time for drugs showing simultaneous first-order and Michaelis–Menton elimination kinetics, *Pharm. Res. 6*:258–261 (1989).
70. H. Cheng, W. R. Gillespie, and W. J. Jusko, Mean residence time concepts for nonlinear systems, *Biopharm. Drug Dispos. 15*:627–641 (1995).

2

Oxidative Metabolic Bioactivation of Xenobiotics

Stéphane Mabic, Kay Castagnoli, and Neal Castagnoli, Jr.
Virginia Tech, Blacksburg, Virginia

I. INTRODUCTION

Mammals are equipped with a variety of enzyme systems that catalyze the transformation of xenobiotics to form, in general, more polar metabolites that are more readily excreted and that are less likely to have access to and to interact with membrane-bound receptors. In the early 1970s, Williams introduced the concept of phase I and phase II biotransformations, a concept that captured the generally held view that the metabolism of foreign compounds is a detoxication process (1). Phase I or functionalization reactions proceed by oxidative, reductive, and hydrolytic pathways and lead to the introduction of a functional group (OH, SH, NH_2, or CO_2H) and to a modest increase of hydrophilicity (2). Phase II or conjugation reactions modify the newly introduced functional group to form *O*- and *N*-glucuronides, sulfate esters, various α-carboxyamides, and glutathionyl adducts, all with increased polarity relative to the unconjugated molecules (2). The overall sequence is illustrated in Scheme 1 with the biotransformation of Δ^9-tetrahydrocannabinol **1** (3). The parent compound is highly lipophilic; therefore, it tends to partition back into the general circulation following glomerular filtration in the kidney (4). Phase II glucuronidation of the phenylic hydroxyl group generates the more hydrophilic *O*-glucuronide **2**. Alternatively, the allylic methyl group undergoes a series of oxidations to form the corresponding carboxylic acid **3** which, in turn, forms the polar acylglucuronide conjugate **4**. The cited

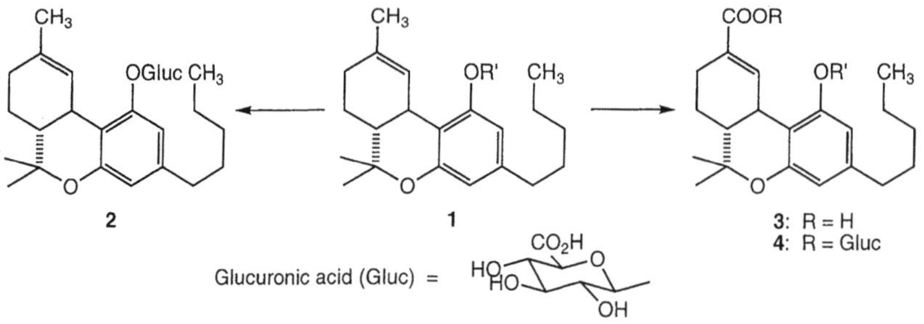

Scheme 1 Biotransformation pathways for the lipophilic drug Δ^9-tetrahydrocannabinol **1**.

literature provides in-depth reviews of these types of biotransformations that the reader may consult for additional details.

Although most biotransformations that xenobiotics undergo lead to polar, less toxic metabolites, a consideration of the structural features and toxic effects of certain compounds suggested that these same biotransformation pathways may also generate chemically reactive species that mediate the toxic effects of the parent compounds. Such toxication reactions were recognized first by the Millers, who discovered that a metabolite, probably the sulfate conjugate **6** (Scheme 2) of the aminoazo dye *N,N*-dimethyl-4-aminoazobenzene (**5**; ''butter yellow,'' a compound that was used during World War II years to color margarine) formed covalent adducts with DNA (5). It soon became apparent that this chemistry was closely linked to the formation of hepatic tumors (6,7). The susceptibility of compounds such as **6** to attack by nucleophilic functionalities present on proteins and nucleic acids is a common theme encountered in a host of metabolic bioactivation reactions. Thus, catalytic pathways that usually protect mammals against the toxicity of lipophilic organic compounds can, when the structural features of the substrate molecules contain latent chemical reactivity, lead to toxic outcomes.

During the past 50 years, a large body of knowledge has accumulated documenting the role of the xenobiotic-metabolizing enzyme systems in the mediation of the toxic effects of drugs, environmental pollutants, and even some endogenous compounds. This chapter will highlight some of the more important of these bioactivation reactions and will focus on enzyme-catalyzed oxidations. The available literature dealing with this topic is extensive (8,9) and should be consulted for additional examples and further details.

II. THE ENZYME SYSTEMS

The cytochrome P450 family of heme-containing proteins is the most important enzyme system in terms of phase I-catalyzed oxidative biotransformations that result in the forma-

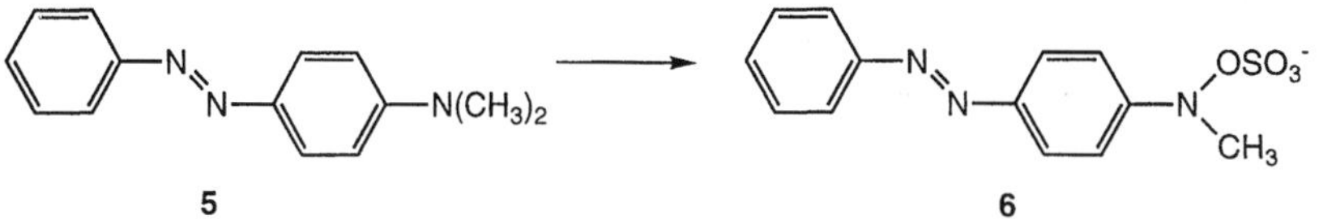

Scheme 2 Possible bioactivation pathway for ''butter yellow'' **5**.

tion of biologically reactive metabolites (10). For comprehensive analyses of specific aspects of the P450 enzymes, the reader is directed to the many excellent books (11–14) and a review (15) on these subjects.

The P450 enzymes are hemoproteins with approximate molecular weights of 50,000 Da. These enzymes catalyze the monooxidation of a wide variety of structurally unrelated compounds, including endogenous steroids and fatty acids and an essentially unlimited number of lipophilic xenobiotics. Those members of this family of enzymes that participate in steroidogenesis are found in the mitochondria and endoplasmic reticulum of steroidogenic organs (16). The drug-metabolizing P450 enzymes are located primarily in the cells' endoplasmic reticulum which, following tissue homogenization, yields the 100,000 $\times$ *g* microsomal fraction. These enzymes are found in high concentrations in the liver, but are present in a variety of other tissues, including lung (17,18), kidney (19,20), gastrointestinal tract (21,22), nasal mucosa (23,24), skin (25), and brain (26–31). The distribution of enzymatic activity is not always uniform. For example, the levels of cytochrome P450 activity are considerably higher in Clara cells than in other types of lung cells (32), which may render these cells more susceptible to pneumotoxins (21).

The evolution of members of this family of enzymes is also evident from a consideration of their molecular composition and diversity (33–35). The possibility of the expression of multiple forms of the P450 enzymes was first suggested by the inducibility of selected metabolic activities (36–38) and, subsequently, by the different substrate selectivities observed in reconstituted enzyme preparations (39,40). More recently, the application of molecular biological techniques has led to the isolation and expression of many cDNAs encoding these enzymes (41,42). As many as 400 genes that encode for the plant and animal P450s have been described (43), including 28 in the human genome (44). In addition to the diversity in enzyme composition, polymorphic variations in humans are also well documented (45–49).

The amino acid sequences of numerous P450 enzymes have been deduced by recombinant DNA techniques. These sequences now form the basis for classifying and naming P450 enzymes (50). In general, P450 enzymes with less than 40% amino acid sequence identity are assigned to different gene families (gene families 1, 2, 3, 4, · · ·). P450 enzymes that are 40–55% identical are assigned to different subfamilies (e.g., 2A, 2B, 2C, 2D, 2E, · · ·). P450 enzymes that are more than 55% identical are classified as members of the same subfamily (e.g., 2A1, 2A2, 2A3, · · ·).

The reactions catalyzed by the P450s involve the four-electron reduction of dioxygen, which is coupled to the two-electron oxidation of a substrate (RH) and the two-electron oxidation of reduced nicotinamide adenine dinucleotide phosphate (NADPH) to form the oxidized product RH(O), $NADH^+$, and water. Enzymatic activity is dependent on a flavoprotein, NADPH–cytochrome P450 reductase, that passes reducing equivalents from NADPH to the cytochrome P450 substrate complex. The cytochrome P450 reaction cycle is shown in Fig. 1 (51–53). In brief, the resting Fe^{III} form of the enzyme first binds substrate (RH) to form a binary complex, Fe^{III}(RH), that undergoes a one-electron reduction. The resulting Fe^{II}(RH) complex binds dioxygen to form a ternary complex, $Fe^{II}(O_2)$(RH), which, following a second one-electron reduction, undergoes O–O bond scission to produce 1 mol of water and the perferryl species $(FeO)^{III}$. Interactions of this highly reactive iron–oxo system with the bound substrate molecule eventually result in oxygen transfer to form the oxidized product RH(O) and to regenerate the resting Fe^{III}(RH) form of the enzyme. The name cytochrome P450 reflects the unusually long wavelength (450 nm vs. 420 nm) of the Soret band observed for the ferrocarbonyl (Fe^{II}–CO) complex of this protein.

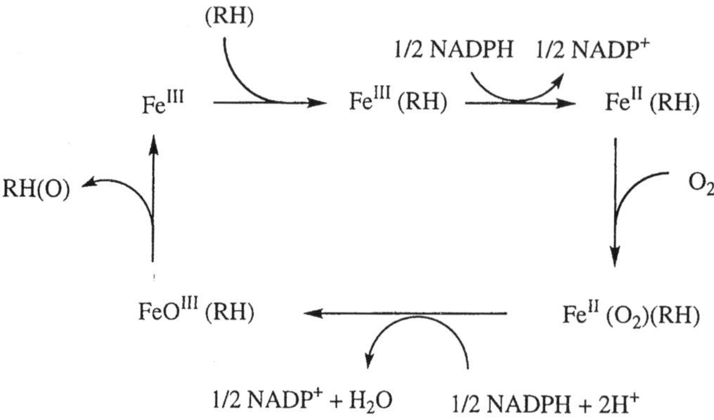

Fig. 1 The cytochrome P450 reaction cycle.

The very broad range of catalytic activity of the P450s is responsible for the complex array of biotransformations associated with these enzymes. Fortunately, the number of these can be reduced to a few types of basic phase I reactions. Often the bioactivation of xenobiotics may involve more than one metabolic step and may require processing of the primary metabolite by a phase II enzyme. As will be illustrated with specific examples, P450-catalyzed epoxidations, particularly those leading to arene oxide formation, *N*-oxidations of aromatic amines and amides, and α-carbon oxidations of nitrosamines are associated with a variety of toxic outcomes, including mutagenesis, carcinogenesis, and cytotoxicity. Detailed catalytic mechanisms to account for these reactions remain to be fully documented (54–57). Recent X-ray crystallographic characterization of bacterial cytochrome P450s (the only members of this superfamily of enzymes to be obtained in crystalline form) is providing important insights into the molecular events involved in the catalytic pathway (58–61).

A second oxidase system that generates biologically reactive (and toxic) metabolites is the flavoprotein monoamine oxidase (MAO; 62–64). Two forms of this enzyme have been characterized in mammalian systems—MAO-A and MAO-B. These enzymes catalyze the α-carbon oxidation of amines **(7)**, a reaction also catalyzed by members of the cytochrome P450s (65), to generate the corresponding iminium metabolites **10**. The reaction pathway is thought to proceed through aminium radical **8** and α-carbon radical **9** intermediates (Scheme 3; 66). Particularly important MAO substrates are the biogenic amines. For example, dopamine **11** (Scheme 4) undergoes a two-electron oxidation to form the corresponding iminium metabolite **12**, as described in general (67). Subsequent hydrolysis of **12** leads to the aldehyde **13**, which is rapidly oxidized to the carboxylic acid DOPAC **(14)**. As discussed later in this chapter, the principal toxicological interest in

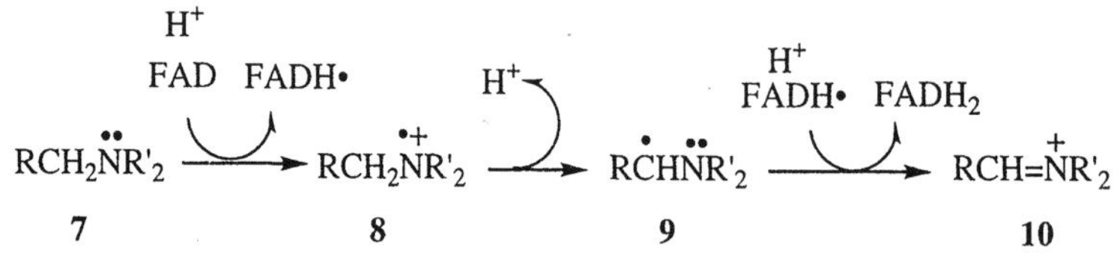

$$\underset{\mathbf{7}}{RCH_2\ddot{N}R'_2} \xrightarrow{H^+,\ FAD \rightarrow FADH\bullet} \underset{\mathbf{8}}{RCH_2\overset{\bullet+}{N}R'_2} \xrightarrow{-H^+} \underset{\mathbf{9}}{R\dot{C}H\ddot{N}R'_2} \xrightarrow{H^+,\ FADH\bullet \rightarrow FADH_2} \underset{\mathbf{10}}{RCH{=}\overset{+}{N}R'_2}$$

Scheme 3 Metabolic scheme for the MAO-catalyzed oxidation of amines.

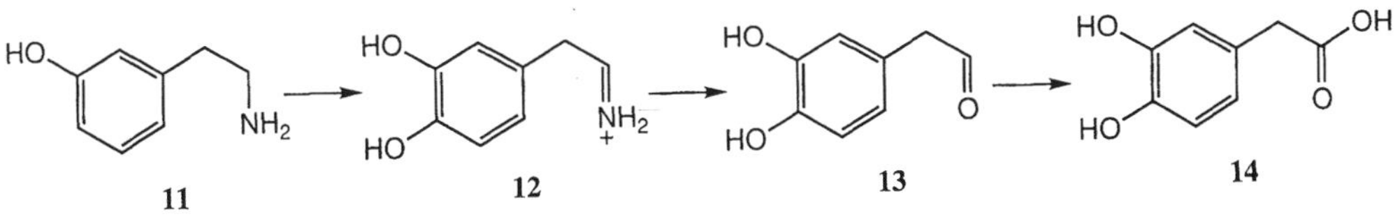

Scheme 4 The metabolic fate of dopamine **11**.

MAO is related to the MAO-B–catalyzed bioactivation of neurotoxic tetrahydropyridine derivatives.

III. SELECTED EXAMPLES OF BIOACTIVATION PATHWAYS

A. Arenes

1. Condensed Polyarenes

Detailed in vitro studies have led to the identification of a variety of chemically reactive metabolites derived from arenes that form covalent adducts with biopolymers, including DNA (68). As shown in Scheme 5, the key step in the bioactivation of arenes **A** is the NADPH-dependent cytochrome P450-catalyzed formation of arene oxides **B** that, because of the associated ring strain, are susceptible to attack by nucleophiles, including those present on biopolymers, which lead to the corresponding adducts **C**. Arene oxides also undergo hydration to form *trans*-dihydrodiols **D**, a reaction catalyzed by the microsomal enzyme hydratase, and reaction with glutathione (GSH) to form the corresponding glutathionyl conjugates **E**, with attack taking place on the more electrophilic allylic carbon atom. The glutathionyl moiety may undergo a 1,2-migration to give, after a series of

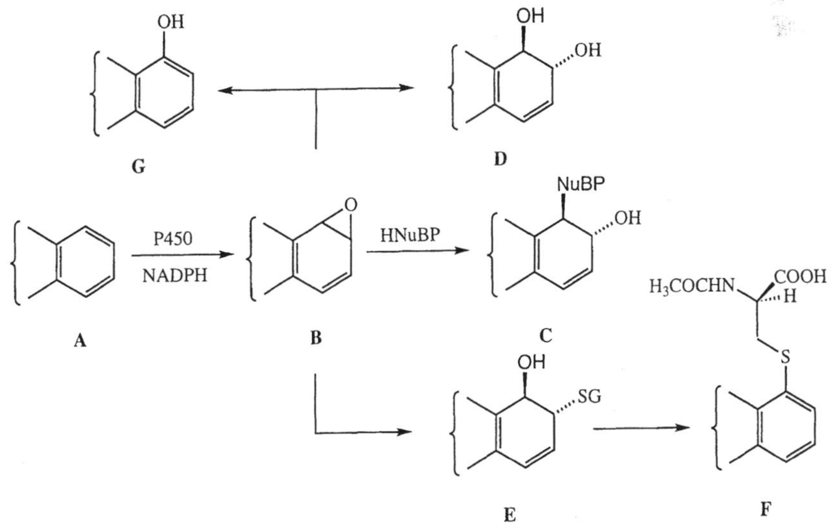

Scheme 5 Cytochrome P450-catalyzed bioactivation of arenes **A** and fate of the resulting arene oxides **B**.

additional reactions, the *N*-acetylcysteinyl or so-called mercapturic acid conjugates **F**, which are excreted in the urine see (**E** → **F**; Scheme 5). A fifth important reaction of arene oxides proceeds through a rearrangement to form the phenolic products **G**. The dihydrodiols and phenolic products may be processed by phase II reactions to form polar conjugates, such as *O*-glucuronides (69–71) and sulfate esters (72–74).

Conclusive evidence that arene oxides are primary reactive intermediates was first established with naphthalene (**15**; Scheme 6) (75). The stereospecific P450-catalyzed oxidation of **15** gives the corresponding (+)-(1*R*,2*S*) epoxide **16** (76) that is converted to the glutathionyl adduct **17**. Compound **17** is known to rearrange, through the episulfonium species **18**, to the (*S*)-1-naphthyl derivative **19** that is processed further to yield the polar mercapturic acid **20** (77–79). Two alternative pathways of the arene oxide **16** involve (a) the microsomal enzyme epoxide hydrase-catalyzed hydration to yield the *trans*-dihydrodiol **21** (80), and (b) the spontaneous isomerization to 1-naphthol **23**, which proceeds by a C-1 to C-2 (NIH) hydride shift (**16** → **22** → **23**) (81,82). This pathway has been examined in considerable detail, with the critical supporting evidence coming from the conversion of naphthalene-2-d_1 (**15-d$_1$**) by **16-d$_1$**, and **22-d$_1$**, to the final product, 1-naphthol-2-d_1 (**23-d$_1$**).

The metabolic fate of condensed polyarenes (also known as polycyclic aromatic hydrocarbons; PAHs) are closely linked to their carcinogenic properties (83,84). The key metabolic step in the bioactivation of these systems is arene oxide formation. The metabolic bioactivation pathway of the potent carcinogen benzo[*a*] pyrene **24**, an extensively studied example of this type of compound, is summarized in Scheme 7. Initial stereoselective oxidation of the 7,8-double bond yields the corresponding (7*R*,8*S*)-oxide **25** (85,86). Intermediate **25** is a good substrate for an epoxide hydratase, which yields the corresponding (*R*,*R*)-*trans*-dihydrodiol **26** (87). This product is oxidized further in a reaction catalyzed by cytochrome P450 to yield a mixture of the diastereomeric 7,8-diol-9,10-epoxides **27** and **28** (88), with isomer **28** in general being the principal product (89). The dihydrodiol epoxides **27** and **28** react readily with DNA (90–92) and are highly mutagenic (93,94) and tumorigenic (95,96). The use of ^{32}P-labeling has made possible the identification and

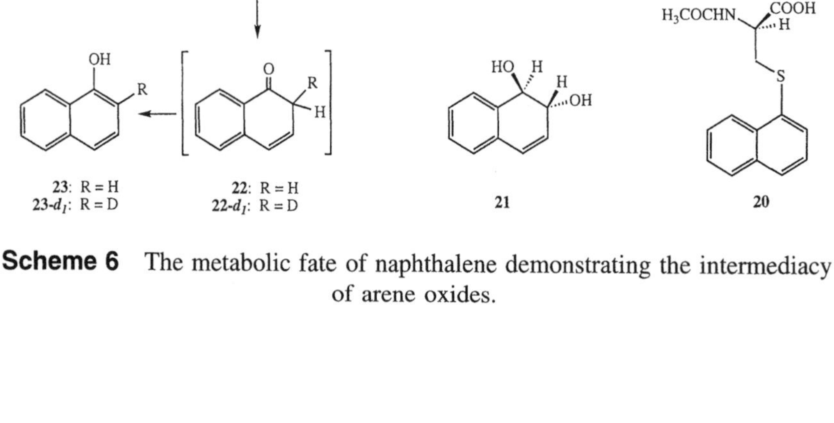

15: R = H
15-d_1: R = D

16: R = H
16-d_1: R = D

17

18

19

NIH Shift

H_3COCHN COOH H

23: R = H
23-d_1: R = D

22: R = H
22-d_1: R = D

21

20

Scheme 6 The metabolic fate of naphthalene demonstrating the intermediacy of arene oxides.

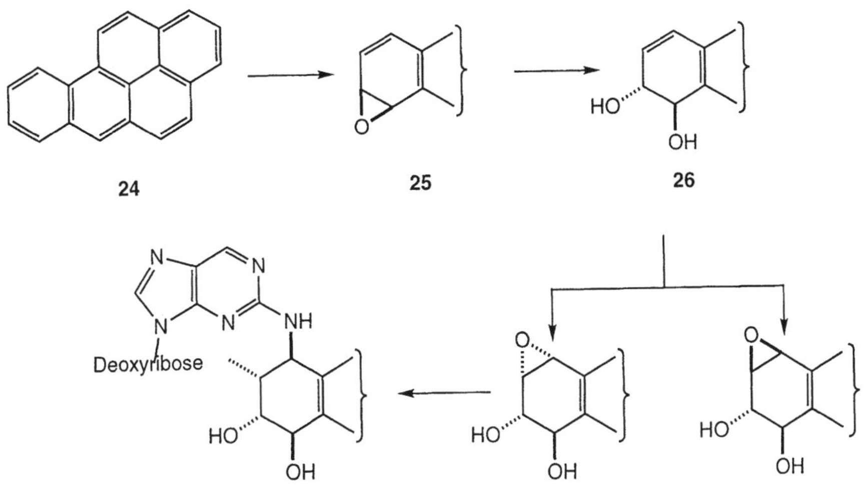

Scheme 7 The bioactivation pathway for benzo[*a*]pyrene **24**.

characterization of a variety of adducts with DNA, and an abundant literature on the structure of these adducts has become available in the last few years (68,97–113). A major product results from attack of the N^2-deoxyguanosine amino group at C-10 of the dihydrodiol oxide **28** to generate the *trans*-adduct shown in structure **29** (114).

Similar metabolic bioactivation pathways have been documented, with a variety of condensed polyarenes including benzo[*c*]phenanthrene **30** (115–119), 7,12-dimethylbenz[*a*]anthracene **31** (120–122), benzo[*g*]chrysene **32** (123–125), and benzofluoranthene **33** (Fig. 2; 101,126). In all of these examples, arene oxide intermediates form adducts with adeninyl and guaninyl groups present on DNA.

2. *Halogenated Arenes*

Polyhalogenated arenes, because of their large-scale industrial use in diverse applications, such as sealants, transformers, and dielectric fluids, have become worldwide environmental contaminants (127–129). Many of these compounds are hepatotoxic (130,131) and carcinogenic (132). Detailed studies have been conducted on bromobenzene **34**, which is used here to illustrate the complex scheme describing the metabolic pathways of this class of compounds (Scheme 8). One bioactivation pathway proceeds through a regioselective 3,4-epoxidation to form the bromoarene oxide **35**. This electrophilic intermediate forms covalent adducts with proteins (133) and DNA (131,134). It also undergoes detoxification by a *trans*-nucleophilic addition reaction with GSH to give the isomeric glutathionyl adducts **36** and **37**, intermediates that ultimately are excreted as the mercapturic acid conjugates **38** and **39**, respectively (110,111). The epoxide **35** also undergoes hydration to yield the *trans*-dihydro **40**, which is oxidized to 4-bromocatechol **41** (135). The arene oxide **35** rearranges through the NIH shift mechanism to form 4-bromophenol **42**, a second source of the bromocatechol **41**. Hanzlik recently reported that bromobenzene also undergoes a P450-mediated hydroxyl insertion reaction to yield 3-bromophenol **43** that is further oxidized to 4- and 3-bromocatechol, **41** and **44**, respectively, and to bromohydroquinone **45** (136). These catechols and hydroquinones may autoxidize to generate the corresponding

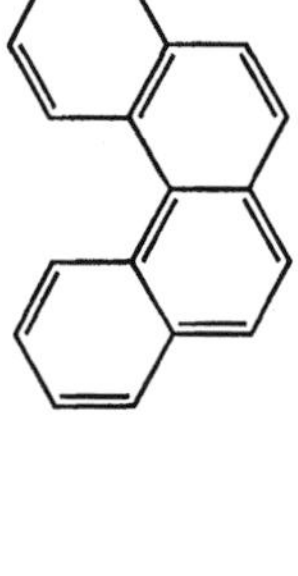

30 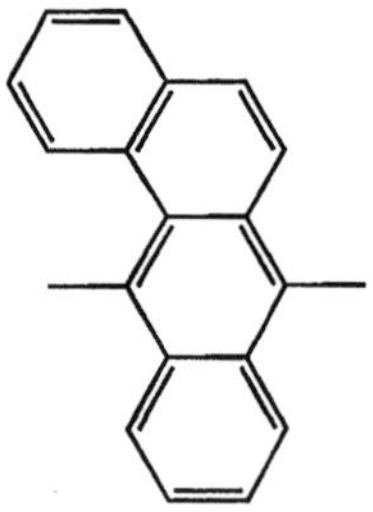 31 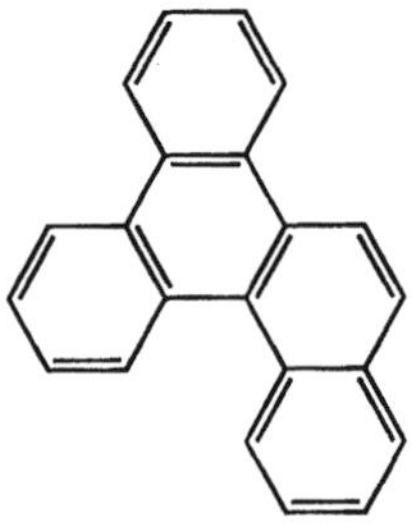 32 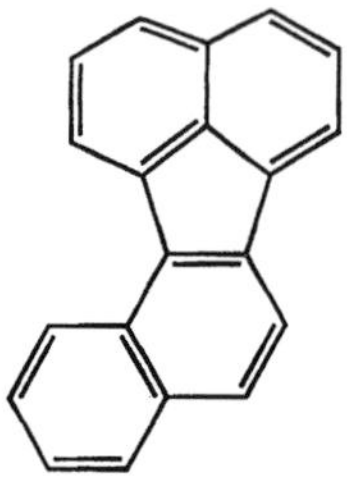 33

Fig. 2 Examples of condensed polyarenes.

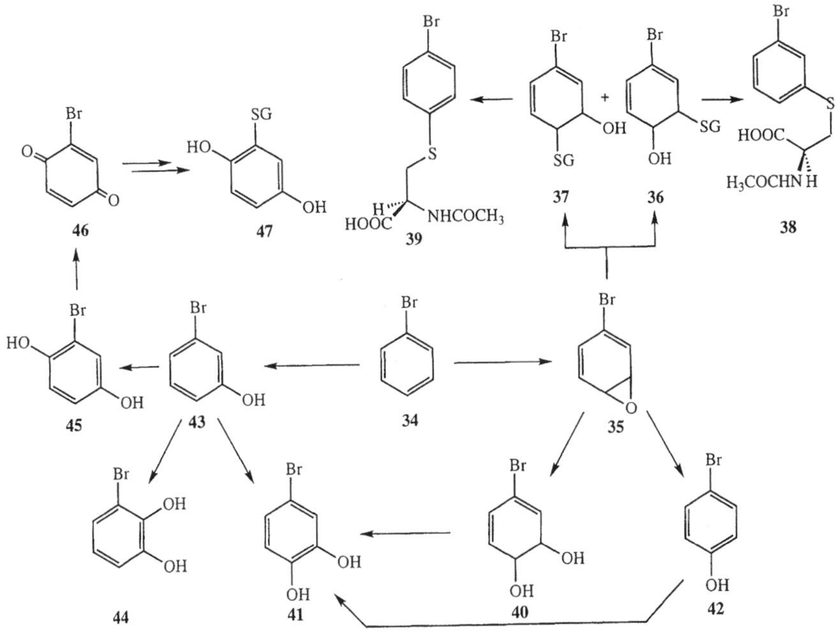

Scheme 8 Metabolic bioactivation pathways for bromobenzene.

o- and *p*-quinones, electrophilic species that also may contribute to the carcinogenicity and hepatotoxicity of bromobenzene through covalent interactions with biomacromolecules (130,137–140). This type of reactivity is illustrated by the glutathionyl derivative **47** which, presumably, is formed by an addition–elimination reaction proceeding through the *p*-quinone **46**, followed by reduction of the resulting adduct.

The pathways described here apply to other polyhalogenated arenes (141–144) with differing ratios of 2,3- and 3,4-epoxide and 3-hydroxy insertion products (121). Quinones and arene oxides are also proposed to be the reactive species mediating the toxicity by reaction with proteins and DNA (145). Because of the complexity of the identification of polychlorobiphenyls and higher members, relatively few thorough studies have been conducted.

B. Phenols, Catechols, and Quinones

The hepatocarcinogenic properties (146,147) of safrole (**48**), a constituent of a variety of essential oils and spices, such as oil of sassafras, basil, and nutmeg (148), have been associated with the alkylation of DNA with the electrophilic sulfate ester **50** (149,150) derived from the cytochrome P450-generated alcohol **49** (Scheme 9) (7). The catechol **51**, formed by a cytochrome P450-catalyzed *O*-dealkylation reaction, however, is the major metabolite detected in rodent urine (151–153). This catechol undergoes further oxidation to the orthoquinone **52**, which spontaneously isomerizes to the quinone methide **53**, a very reactive electrophilic species susceptible to attack by nucleophilic functionalities present on endogenous macromolecules (154,155). Other *para*-alkyl-substituted phenols also form reactive toxic quinone methides (156–159).

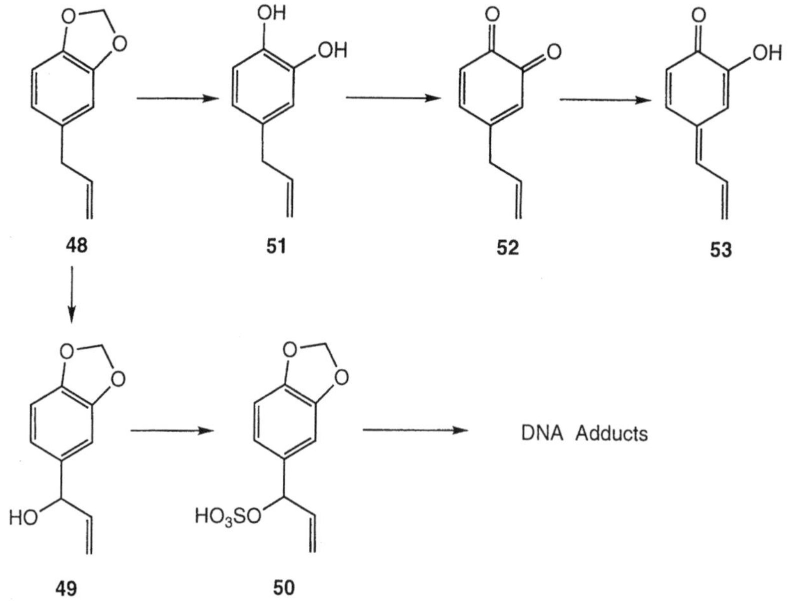

Scheme 9 Metabolic bioactivation pathways for safrole **48**.

Various bioactivation routes, illustrated with estrone (**54**; Scheme 10), have been proposed to account for the adducts formed between DNA and estrogens (160–162) and for the carcinogenic properties of estrogens (163–165). The reactivity of hydroxyarene oxide **55**, which is formed in a reaction catalyzed by a form of cytochrome P450 different from that responsible for the formation of catechol estrogens (166–167), was documented with glutathione (GSH; (168,169)). The role of the epoxide in DNA damage, however, could not be confirmed (170). The main oxidative pathway of **54** (171) leads to the 2,3- and 3,4-catechol derivatives (**56** and **57**, respectively) which are thought to be formed by a cytochrome P450-dependent hydroxyl insertion reaction (172). These catechols undergo further oxidization to the orthoquinones **59** and **60**, both of which react with GSH (173) and alter the structure of DNA (171,174). As illustrated with the 3,4-isomer, quinone methides (such as **61**) also act as bioalkylating agents (173). A third possible toxication pathway for estrone involves a cytochrome P450/P450 reductase-catalyzed redox cyclic process (175) that leads to the release of reactive oxygen species, as shown in Scheme 10. This redox cycling (**56** $\leftrightarrow$ **58** $\leftrightarrow$ **60**) is associated with DNA damage in human breast cancer cells (176).

C. Nitrogen-Containing Compounds

1. *Aromatic Amines and Amides*

The carcinogenic, mutagenic, and cytotoxic properties of many drugs, pesticides, cosmetics, and synthetic intermediates that contain aromatic amino and amido groups are fairly well understood (177–180). Arylamines are bioactivated, through an *N*-hydroxy intermediate, to reactive *N*-*O*-sulfate and *N*-*O*-acetyl metabolites. The reaction sequence is illus-

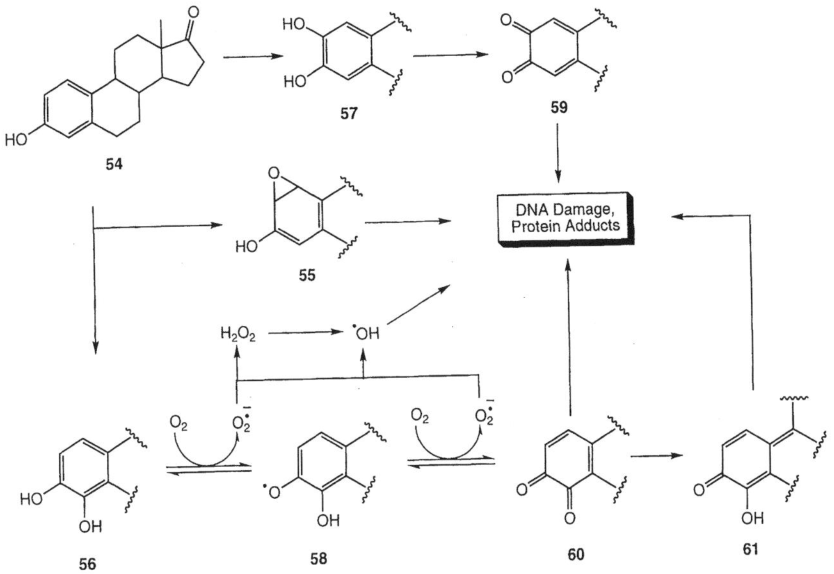

Scheme 10 Metabolic bioactivation pathways for estrone **54**.

trated with three examples: 4-dimethylaminoazobenzene (**5**), acetaminophen (**65**), and the carcinogenic 2-acetylaminofluorene (**72**).

The hepatocarcinogenicity of the azo dye 4-dimethylaminoazobenzene **5** was documented early in rats and involves covalent adducts with DNA (5). Structure–activity studies suggested that the toxicity was mediated by the reactive nitrenium intermediate **64** or its precursor **6** (Scheme 11). The metabolic sequence eventually elucidated involves an initial cytochrome P450-catalyzed *N*-demethylation reaction, leading to the secondary amine **62**, which undergoes *N*-hydroxylation to the hydroxylamine **63**. The formation of **63** appears to be catalyzed by both cytochrome P450 and the microsomal flavin-containing monooxygenases (3). Incubation of **63** with rat hepatic cytosol resulted in the formation of the sulfate conjugate **6**, a compound that reacts with methionine, GSH, and nucleic acids (181,182). The good leaving-group tendency of the sulfonyloxy group is thought to lead to the highly reactive nitrenium **64** that may be the electrophilic species involved in adduct formation.

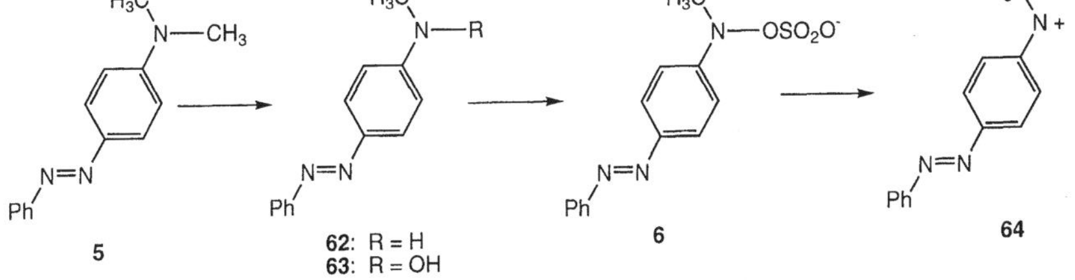

Scheme 11 Metabolic bioactivation pathways for the azo dye 4-dimethylaminoazobenzene **5**.

The bioactivation pathways of acetaminophen (paracetamol) were largely elucidated by Mitchell in the 1970s (183–187) and have since been extensively reviewed (188,189). Although acetaminophen (**65**; Scheme 12) is a safe drug, hepatotoxicity and necrosis (190) can result when exposed to an overdose. The major detoxification modes lead to the glucuronide (**66**) and sulfate (**67**) conjugates (191,192). A key reactive metabolite is the electrophilic *N*-acetyliminoquinone **69** (193). Normally this species is detoxified through formation of the GSH adduct **70** (156,164), but on depletion of GSH, protein adducts are also formed. The bioactivation pathway leading to **69** does not to proceed through an *N*-hydroxy derivative. Instead a two-step cytochrome P450-catalyzed redox reaction proceeding through the phenoxy radical **68** has been proposed (160). Another pathway, however, proceeds by the nitrogen-centered radical **71**, issued from a redox cycle, inducing activity of prostaglandin H synthase and leading to lipid peroxidation (2). Two types of chemical-based mechanisms are, therefore, advanced to explain the hepatotoxic properties of acetaminophen. The first, involves covalent binding of the iminoquinone **69** to hepatic proteins by a Michael-type addition of nucleophilic thiol residues at the C-2/C-6 positions (194,195). Recent immunochemical results support this proposal (196–198). The extent of hepatic injury correlates well with the amount of protein–drug adducts formed in vivo (199). The second mechanism is mediated by an oxidative stress process involving reactive oxygen species and resulting in lipid peroxidation. This pathway is initiated by a redox cycle involving oxidation–reduction of the acetylamino nitrogen atom as shown (200). A third mechanism involving a disturbance of Ca^{+2} homeostasis has been proposed (201). The role of each toxication pathway to produce cell necrosis is still under investigation.

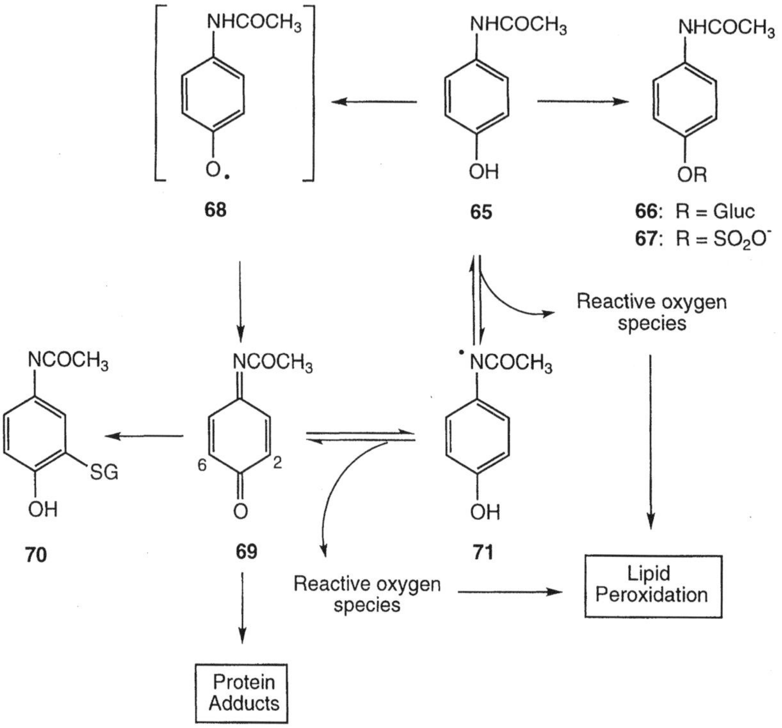

Scheme 12 Metabolic bioactivation pathways for acetaminophen **65**.

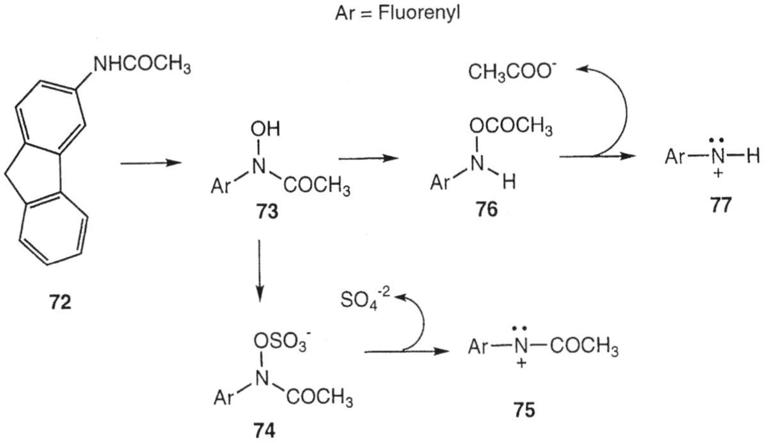

Ar = Fluorenyl

Scheme 13 Metabolic bioactivation pathways for 2-acetylaminofluorene **72**.

Additional details on the bioactivation of acetaminophen will be found in the extensive review literature dealing with this subject (2,188–190,192,202–204).

The aromatic amide 2-acetylaminofluorene (**72**; AAF), binds extensively to DNA (Scheme 13) (205,206). AAF, a potent hepatocarcinogen (207,208) undergoes a variety of metabolic biotransformations leading to reactive species via the primary *N*-hydroxy intermediate metabolite **73** (178,209). In addition to detoxification pathways, such as glucuronidation, the *N*-hydroxy metabolite **73** is converted to the sulfate ester **74** (210,211) that reacts by an S_N1 mechanism, involving loss of the sulfate group, to yield the *N*-acetyl-*N*-arylnitrenium species **75**, a resonance-stabilized species that alkylates DNA (178,212). Intermediate **73** also undergoes an acetyl rearrangement leading to **76** that decomposes to the *N*-arylnitrenium **77** (213).

2. *Cyclic Tertiary Amines*

The neurodegenerative properties of the parkinsonian-inducing nigrostriatal neurotoxin 1-methyl-4-phenyl-1,2,3,6-tetrahydropyridine (MPTP; **78**) are mediated by the pyridinium metabolite **80** (Scheme 14) (64,214). Hepatic cytochrome P450s catalyze both the oxidative *N*-demethylation (215–218) and ring α-carbon oxidation of MPTP (65), the later reaction leading to the 1-methyl-4-phenyl-2,3-dihydropyridinium species $MPDP^+$ (**79**),

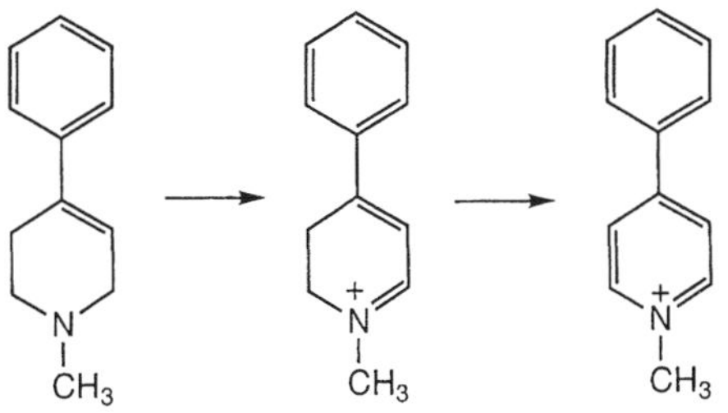

MPTP (**78**) $MPDP^+$ (**79**) MPP^+ (**80**)

Scheme 14 Oxidative metabolism of the parkinsonian-inducing agent MPTP **78**.

which subsequently undergoes autoxidation to the pyridinium metabolite MPP⁺ (**80**). The potent nigrostriatal toxic effects of MPTP, however, are due to its efficient conversion to **80** by brain monoamine oxidase B (MAO-B) catalysis involving the same dihydropyridinium intermediate **79** (219,220).

The detailed molecular events leading to the selective loss of nigrostriatal neurons following exposure to MPTP are still under debate. In addition to the bioactivation step, MPP^+ appears to be actively taken up into nigrostriatal nerve terminals by the dopamine transporter (221,222) and to be localized in the inner mitochondrial membrane by forces derived from the mitochondrial membrane electrochemical potential gradient (223). MPP^+ inhibits complex I of the mitochondrial electron transport chain (224–228), which leads to depletion of ATP (229). The mitochondrial injury resulting from inhibition of complex I may also lead to the production of chemically reactive radical species, such as superoxide radical anion ($O_2^{\cdot -}$), the one-electron reduction product of dioxygen (O_2) (230–237). Recent evidence suggests that the toxicity of MPTP ultimately may rely on the lipid-peroxidizing properties of peroxynitrite ($ONOO^-$), an adduct formed between the free radical nitric oxide (NO) and $O_2^{\cdot -}$ (238,239). Neuronal NO is derived from L-arginine in a reaction catalyzed by neuronal nitric oxide synthase (nNOS) (240–244). The possibility that the nigrostriatal toxicity of MPTP is mediated by $ONOO^-$ has been investigated in rodents and baboons with the aid of the selective nNOS inhibitor 7-nitroindazole (7-NI; **81**; Fig. 3 (245–249). Both mice and baboons were protected by 7-NI from the neurotoxic effects of MPTP. The conclusion that the inhibition of nNOS is the only factor mediating 7-NI's neuroprotection may not be correct, however, because 7-NI, which also inhibits MAO-B, blocks the bioactivation of MPTP to MPP^+ in rodents (250). Furthermore, inhibition of nNOS activity in the common marmoset does not protect against the neurotoxicity of MPTP (251). Studies with nNOS inhibitors that do not alter striatal levels of MPP^+ should help resolve this important issue.

A second class of cyclic tertiary amines of toxicological interest that show carcinogenic (252–254), genotoxic (255), neurotoxic (256), and hepatotoxic (257) properties are the pyrrolizidine alkaloids **82**. Cytochrome P450-mediated ring α-carbon oxidation (Scheme 15), once again at allylic positions, is the principal metabolic bioactivation step (258–261). Two possible intermediary iminium oxidation products, **83** and **84**, may be formed. These products are the protonated forms of the same pyrrolyl metabolite, **85** which, itself, is chemically unstable. Spontaneous cleavage of the R_1OCO–C bond of **85** gives the electrophilic species **86**, which is susceptible to Michael addition to form C-7 adducts **87** (262). Although the initially formed pyrrolyl esters are toxic, compounds that form cross-linked products **89**, presumably through intermediates, such as **88**, belong to the most toxic members of this class of compounds (263).

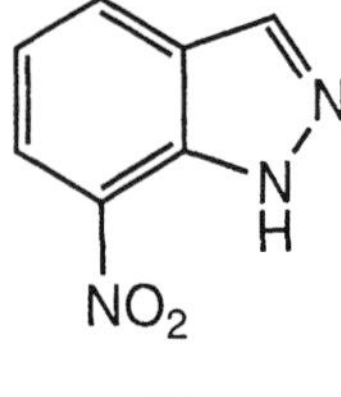

Fig. 3 Structure of 7-nitroindazole **81**.

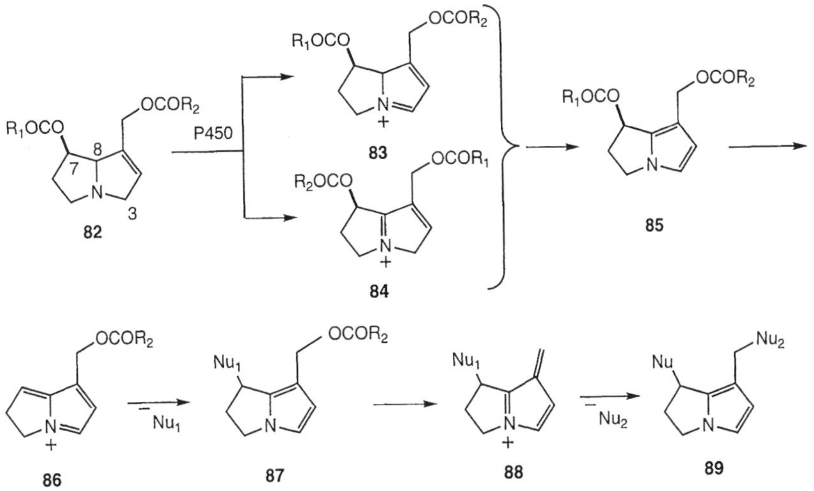

Scheme 15 Metabolic bioactivation pathway of the pyrrolizidine alkaloids.

3. Nitrosamines

Several nitrosamines are potent carcinogens. For example, the tobacco-derived nitrosamine 4-(methylnitrosoamino)-1-(3-pyridinyl)-1-butanone (NNK; **90**; Scheme 16) induces lung tumors in rodents and is believed to be a causative factor in human lung cancer (264). The *N*-nitrosopyrrolidine (**91**; NPYR) induces mainly liver tumors in rats, whereas the closely structurally related *N*-nitrosopiperidine (**92**; NPIP) causes tumors of both the esophagus and the liver (Scheme 17) (265). As detailed in the following, α-hydroxylation is the common bioactivation route for all nitrosamines (266).

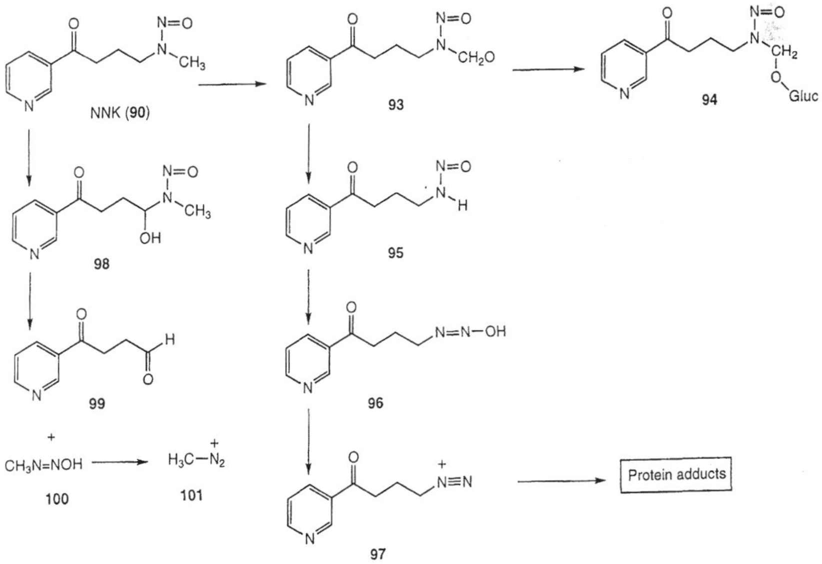

Scheme 16 Metabolic bioactivation of NNK **90**.

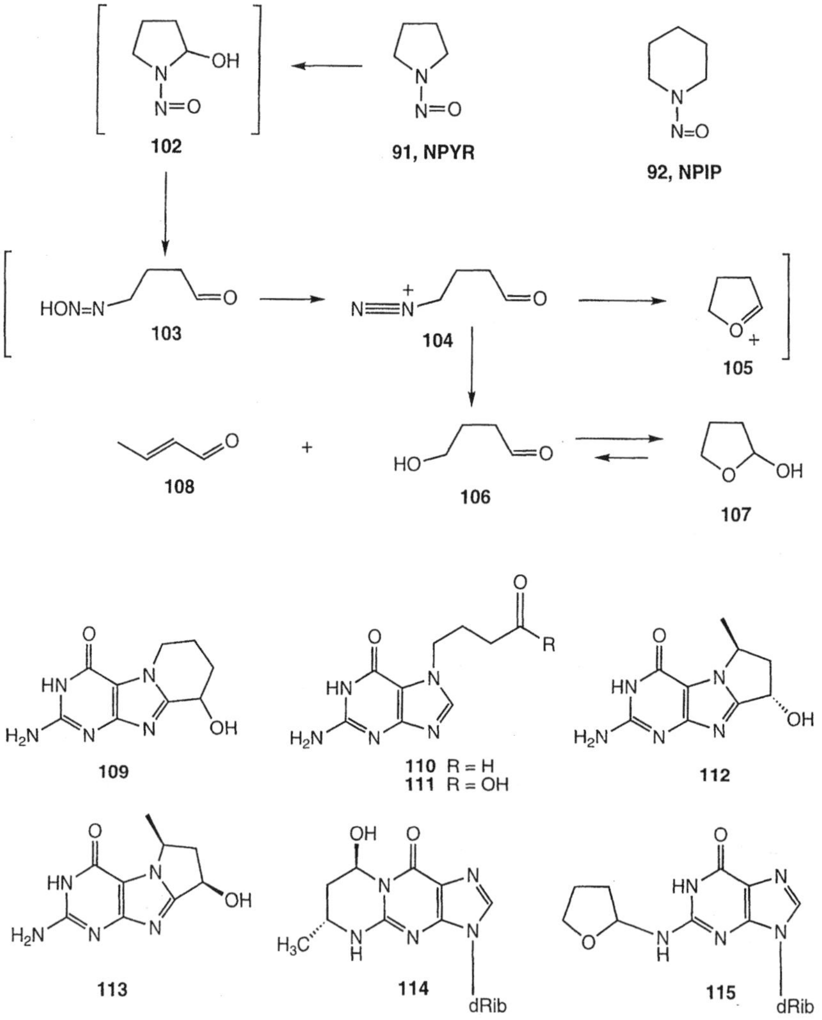

Scheme 17 Metabolic bioactivation of NPYR **91**.

NNK was shown to undergo α-hydroxylation both on the *N*-methyl and α-methylene groups (see Scheme 16). The former oxidation leads to the intermediate carbinol **93** that may be captured as an *O*-glucuronide **94** or that may lose formaldehyde to form the diazotic acid **96** (264). Compound **96** spontaneously converts to the diazonium **97** that will alkylate proteins. The intermediate α-hydroxynitrosamine **98** will cleave to form the ketoaldehyde **99** (267) and methanediazotic acid (**100**), the precursor to a highly reactive methylating agent, the methyldiazonium species **101**.

NPYR **91** undergoes ring opening after ring α-carbon oxidation **102**, leading to intermediate **103**, precursor of the diazonium **104**. Loss of nitrogen from **104** affords, by the oxonium ion **105**, or by direct hydrolysis, the aldehyde **106** in equilibrium with the hydroxytetrahydrofuran (**107**). Crotonaldehyde **108** and **107** combine with guanosine by

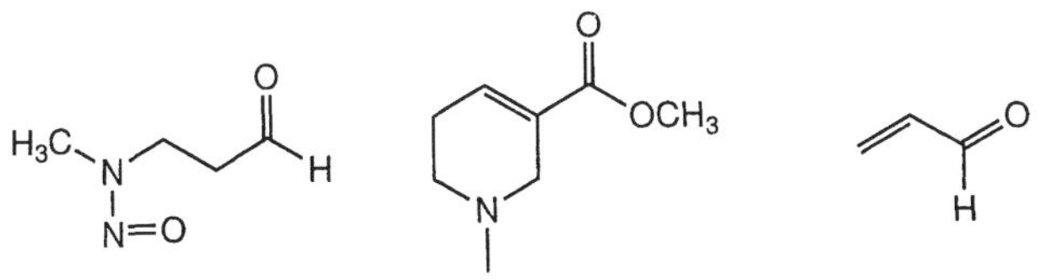

MNPA (**116**) Arecoline (**117**) Acrolein (**118**)

Fig. 4 Structures of arecoline **117**, its *N*-nitroso metabolite MNPA **116**, and acrolein **118**.

reaction with DNA to form a variety of adducts **109–115** (see Scheme 17), isolated as their guanine or guanosine adducts (265,268–270). NPIP **92** undergoes similar bioactivation pathways and leads to the formation of similar intermediates, with one more carbon, and to similar adducts with guanosine.

The 3-(methylnitrosamine)-propionaldehyde (MNPA; **116**), a nitrosamine formed by nitrosation and bioactivation of arecoline (**117**; Fig. 4), a major alkaloid in the areca nut, generates acrolein **118** which forms similar types of adducts with DNA (271).

IV. CONCLUDING REMARKS

It has been 50 years since Boyland reported his observations on the metabolism of naphthalene that eventually led to the characterization of arene oxides and their role in the bioactivation of carcinogenic polycyclic arenes (77,78) and since the Millers reported their first observations on the metabolic bioactivation of aminoazo dyes (5). Today, a major concern in environmental sciences and in the design and development of new therapeutic agents is the metabolism-mediated toxicity of xenobiotics. It is clear that an increasing understanding of the metabolic fates of biologically active compounds will continue to enrich the knowledge base that will be required to provide mechanistic interpretations of metabolism-mediated toxicities and to identify latent functionalities that may mediate toxic effects following metabolic bioactivation. A key to gaining the needed insights is an appreciation of the particular structural features of the molecules of interest that may lead to reactive metabolites. Generally, one can anticipate that the enzymes that catalyze the oxidation, conjugation, and other reactions that xenobiotics undergo lead to structural modifications that result in detoxification. On the other hand, these same biotransformations can yield products that have chemical reactivities appropriate for forming covalent bonds with biomacromolecules and that may lead to toxic outcomes. For other compounds, such as with the parkinsonian-inducing agent MPTP, there are special structural features resulting in enzyme–substrate properties that can lead to toxic metabolites.

In this chapter we have focused on oxidative bioactivation pathways. A relatively few basic reactions—arene oxide formation; quinone, iminoquinone, and quinone methide formation; *N*-oxidation of amino- and amidoarenes; and α-carbon oxidation of cyclic tertiary amines and nitrosamines—need to be considered. All of the examples, except MPTP, examined in this chapter, involve the participation of electrophilic metabolites that are capable of alkylating biomacromolecules in the mediation of toxic outcomes. This ever-expanding knowledge base should improve the opportunities to design needed drugs and industrial chemicals devoid of toxicities mediated by reactive metabolites.

REFERENCES

1. R. T. Williams, *Detoxification Mechanisms*, 2d ed., Wiley, New York, 1971.
2. A. Parkinson, Biotransformation of xenobiotics, *Toxicology the Basic Science of Poisons* (C. D. Klaassen, ed.), McGraw-Hill, New York, 1996, p. 113.
3. N. Castagnoli, Jr. and K. P. Castagnoli, Biotransformation of xenobiotics to chemically reactive metabolites, *Drug Toxicokinetics* (P. G. Welling and F. A. de la Iglesia, eds.), Marcel Dekker, New York, 1993, p. 43.
4. D. E. Green, F. C. Chao, K. O. Loeffler, and S. L. Kanter, Quantitation of delta-9-tetrahydrocannabinol and its metabolites in human urine by probability based matching GC/MS, cannabinoids analysis in physiological fluids. *ACS Symp. Ser.* p. 914.
5. E. C. Miller and J. A. Miller, *Cancer Res. 7*: 468 (1947).
6. J. A. Miller and E. C. Miller, Ultimate chemical carcinogens and their reactions with cellular macromolecules, Cold Spring Harbor Conf. Cell Proliferation, p. 605.
7. J. A. Miller and E. C. Miller, *Biological Reactive Intermediates* (D. J. Jollows, J. J. Kocsis, R. Synder, and H. Vainio, eds.), Plenum Press, New York, 1977.
8. *Drug. Metab. Rev., Chem. Res. Toxicol.*, and *Drug Metab. Disp.*
9. *Biological Reactive Intermediates I–V*, Plenum Press, New York.
10. F. J. Gonzalez and H. V. Gelboin, Role of human cytochromes P-450 in the metabolic activation of chemical carcinogens and toxins, *Drug. Metab. Rev. 26*: 165 (1994).
11. M. Waterman and E. Johnson, *Methods in Enzymology*, 206, Academic Press, New York, 1991.
12. P. Ortiz de Montellano, *Cytochrome P-450: Structure, Mechanism, and Biochemistry*, Plenum Press, New York, 1986.
13. J. Schenkman and D. Kupfer, *Hepatic Cytochrome P-450 Monooxygenase System*, Pergamon, New York, 1982.
14. F. P. Guengerich, *Mammalian Cytochromes P-450*, Vol. 2, CRC Press, Boca Raton, FL, 1987.
15. T. Poulos, Cytochrome P450, *Curr. Opin. Struct. Biol. 5*: 767 (1995).
16. P. Hall, Cytochromes P-450 and the regulation of steroid synthesis, *Steroids 48*: 131 (1986).
17. M. Lawton, R. Glasser, R. Tynes, E. Hodgson, and R. Philpot, The flavin-containing monooxygenase enzyme expressed in rabbit liver and lung are products of related but distinctly different genes, *J. Biol. Chem. 265*: 5885 (1990).
18. R. Glasser and R. Philpot, Primary structures of cytochrome P-450 isozyme 5 from rabbit and rat and regulation of species-dependent expression and induction in lung and liver: Identification of cytochrome P-450 gene subfamily IVB, *Mol. Pharmacol. 35*: 617 (1989).
19. J. Hjelle, G. Hazelton, C. Klaassen, and J. Hjelle, Glucuronidation and sulfation in rabbit kidney, *J. Pharmacol. Exp. Ther. 236*: 150 (1986).
20. L. Tremaine, Quantitative determination of organ contribution to excretory metabolism, *J. Pharmacol. Methods 13*: 9 (1985).
21. J. Dutcher and M. Boyd, Species and strain differences in target organ alkylation and toxicity of 4-ipomeanol, *Biochem. Pharmacol. 46*: 3367 (1979).
22. W. Peters and P. Kremers, Cytochromes P-450 in the intestinal mucosa of man, *Biochem. Pharmacol. 38*: 1535 (1989).
23. D. Adams, A. Jones, C. Plopper, C. Serabjit-Singh, and R. Philpot, Distribution of cytochrome P-450 monooxygenase enzymes in the nasal mucosa of hamster and rat, *Am. J. Anat. 190*: 291 (1991).
24. C. Eriksson and E. Brittebo, Metabolic activation of the herbicide dichlobenil in the olfactory mucosa of mice and rats, *Chem. Biol. Interact. 79*: 165 (1991).
25. W. Khan, S. Park, H. Gelboin, D. Bickers, and H. Mukhtar, Monoclonal antibodies directed

characterization of epidermal and hepatic cytochrome P-450 isozymes induced by skin application of therapeutic crude coal tar, *J. Invest. Dermatol. 93*: 40 (1989).

26. T. Hansson, N. Tindberg, M. Ingelman-Sundberg, and C. Köhler, Regional distribution of ethanol-inducible cytochrome P450 IIE1 in the rat central nervous system, *Neuroscience 34*: 451 (1990).
27. B. Walther, J. Ghersi-Egea, A. Minn, and G. Siest, Brain mitochondrial cytochrome P-450s: Spectral and catalytic properties, *Arch. Biochem. Biophys. Commun. 254*: 592 (1987).
28. M. Warner, C. Köhler, T. Hansson, and J.-Ä. Gustafsson, Regional distribution of cytochrome P-450 in the rat brain: Spectral quantitation and contribution of P-450b,e and P-450c,d, *J. Neurochem. 50*: 1057 (1988).
29. A. Dhawan, D. Parmer, M. Das, and P. Seth, Cytochrome P-450 dependent monooxygenases in neuronal and glial cells: Inducibility and specificity, *Biochem. Biophys. Res. Commun. 170*: 441 (1991).
30. H. Anandetheerthavarada, S. Shankar, and V. Ravindranath, Rat brain cytochromes P-450: Catalytic, immunochemical properties and inducibility of multiple forms, *Brain Res. 536*: 339 (1990).
31. A. Bergh and H. Strobel, Reconstitution of the brain mixed function oxidase system: Purification of NAPH–cytochrome P450 reductase and partial purification of cytochrome P450 from whole rat brain, *J. Neurochem. 59*: 575 (1992).
32. T. Massey, Isolation and use of lung cells in toxicology, *In Vitro Toxicology: Model Systems and Methods* (N. J. Caldwell, ed.), Telford Press, NJ, 1989, p. 35.
33. O. Gotoh and Y. Jujii-Kuriyama, Evolution, structure, and gene regulations of cytochrome P-450, *Frontiers in Biotransformation* (K. Ruckpaul and H. Rein, eds.), Akademie-Verlag, Berlin, 1989, p. 195.
34. D. R. Nelson and H. W. Strobel, Evolution of cytochrome P-450 proteins, *Mol. Biol. Evol. 572* (1987).
35. D. R. Nelson, T. Kamataki, D. J. Waxman, F. P. Guengerich, R. W. Estabrook, R. Feyereisen, F. J. Gonzalez, M. J. Coon, I. C. Gunsalus, O. Gotoh, K. Okuda, and D. W. Nebert, The P450 superfamily: Update on new sequence, gene mapping, accession numbers, early trivial names of enzymes, and nomenclature, *DNA Cell. Biol. 12*: 1 (1993).
36. A. P. Alvares, G. Schilling, W. Levin, and R. Kuntzman, Studies on the induction of CO-binding pigments in liver microsomes by phenobarbital and 3-methylcholanthrene, *Biochem. Biophys. Res. Commun. 29*: 521 (1967).
37. A. H. Conney, E. C. Miller, and J. A. Miller, Substrate-induced synthesis and other properties of benzopyrene hydroxylase in rat liver, *J. Biol. Chem. 228*: 753 (1957).
38. N. E. Sladek and G. J. Mannering, Evidence for a new P-450 hemoprotein in hepatic microsomes from methylcholanthrene-treated rats, *Biochem. Biophys. Res. Commun. 24*: 667 (1966).
39. A. Y. H. Lu, K. W. Junk, and M. J. Coon, Resolution of the cytochrome P-450-containing ω-hydroxylation system of liver microsomes into three components, *J. Biol. Chem. 244*: 3714 (1969).
40. A. Y. H. Lu and W. Levin, Partial purification of cytochrome P-450 and P-448 from rat liver microsomes, *Biochem. Biophys. Res. Commun. 46*: 1334 (1972).
41. T. Omura, R. Sato, D. Y. Cooper, O. Rosenthal, and R. W. Estabrook, Function of cytochrome P-450 of microsomes, *Fed. Proc. 24*: 1181 (1965).
42. F. P. Guengerich, Characterization of human microsomal cytochrome P-450 enzymes, *Annu. Rev. Pharmacol. Toxicol. 29*: 241 (1989).
43. J. A. Peterson, I. Sevrioukova, G. Truan, and S. E. Graham-Lorence, P450M-3; a tale of two domains—or is it three? *Steroids 62*: 117 (1997).
44. D. W. Nebert, D. R. Nelson, M. J. Conn, R. W. Estrabrook, R. Feyerseisen, Y. Fujii-Kuriyama, F. J. Gonzalez, F. P. Guengerich, I. C. Gunsalus, E. F. Johnson, J. C. Loper, R.

Sato, M. R. Waterman, and D. J. Waxman, The P450 superfamily: Update on new sequences, gene mapping, and recommended nomenclature, *DNA Cell. Biol. 10*: 1 (1991).

45. R. J. Smith, J. R. Idle, A. A. Mahgoub, T. P. Sloan, and R. Lancaster, Genetically determined defects of oxidation at carbon centers of drugs, *Lancet 1*: 943 (1978).
46. P. K. Srivastava, C.-H. Yun, P. H. Beaune, C. Ged, and F. P. Guengerich, Separation of human liver microsomal tolbutamide hydroxylase and (*S*)-mephenytoin 4′-hydroxylase cytochrome P-450 enzymes, *Mol. Pharmacol. 40*: 69 (1991).
47. M. Kagimoto, M. Heim, K. Kagimoto, T. Zeugin, and U. A. Meyer, Multiple mutations of the human cytochrome P450IID6 gene (*CYP2D6*) in poor metabolizers of debrisoquine, *J. Biol. Chem. 265*: 17209 (1990).
48. M. Eichelbaum and A. S. Gross, The genetic polymorphism of debrisoquine/sparteine metabolism—clinical aspects, *Pharmacol. Ther. 46*: 377 (1990).
49. R. Kato, Y. Yamazoe, and T. Yasumori, Pharmacogenetics and polymorphism of human P-450 which activates and detoxicates xenobiotics and carcinogens, Proceedings of the 21st International Symposium of the Princess Takamatsu Cancer Research Fund, Tokyo, 1990, p. 45.
50. D. W. Nebert and D. R. Nelson, P450 gene nomenclature based on evolution, *Cytochrome P450. Methods in Enzymology* (M. R. Waterman and E. F. Johnson, eds.), Academic Press, New York, 1991, p. 1.
51. T. D. Porter and M. J. Coon, Cytochrome P-450: Multiplicity of isoforms, substrates, and catalytic and regulatory mechanisms, *J. Biol. Chem. 266*: 13469 (1991).
52. P. R. Ortiz de Montellano, Cytochrome P-450 catalysis: Radical intermediates and dehydrogenation reactions, *Trends Pharmacol. Sci. 10*: 354 (1989).
53. R. E. White, The involvement of free radicals in the mechanisms of monooxygenases, *Pharmacol. Ther. 49*: 21 (1991).
54. P. R. Ortiz de Montellano and R. A. Stearns, Radical intermediates in the cytochrome P-450 catalyzed oxidation of aliphatic hydrocarbons, *Drug Metab. Rev. 20*: 183 (1989).
55. T. J. McMurry and J. T. Groves, Metalloporphyrin models for cytochrome P-450, *Cytochrome P-450* (P. R. Ortiz de Montellano, ed.), Plenum Press, New York, 1986, p. 1.
56. P. W. White, Mechanistic studies and selective catalysis with cytochrome P-450 model systems, *Bioorg. Chem. 18*: 440 (1990).
57. D. Mansuy and P. Battioni, Catalytically active metalloporphyrin models for cytochrome P-450, *Basis and Mechanisms of Regulation of Cytochrome P-450. Frontiers in Biotransformation* (K. Ruckpaul and H. Rein, eds), Akademie-Verlag, Berlin, 1989, p. 66.
58. T. L. Poulos, Modeling of mammalian P450s on the basis of $P450_{cam}$ X-ray structure, *Methods in Enzymology* (M. R. Waterman and E. F. Johnson, eds.), Academic Press, New York, 1991, p. 11.
59. H. Yeom, S. Sligar, H. Li, T. Poulos, and A. Fulco, The role of Thr268 in oxygen activation of cytochrome P450BM-3, *Biochemistry 34*: 14733 (1995).
60. H. Li and T. Poulos, Conformational dynamics in cytochrome P450-substrate interactions, *Biochemie 78*: 695 (1996).
61. S. Govindaraj and T. Poulos, The domain architecture of cytochrome P450BM-3, *J. Biol. Chem. 272*: 7915 (1997).
62. C. W. Abell, R. Stewart, P. Andrews, and S. W. Kwan, Molecular and functional properties of the monoamine oxidases, *Heterocycles 39*: 933 (1994).
63. P. Dostert, M. S. Benedetti, and M. Jalfre, *Monoamine Oxidase—Basic and Clinical Frontiers* (K. Kamijo, E. Usdin, and T. Nagatsu, eds.), Excerpta Medica, Amsterdam, 1982, p. 197.
64. T. P. Singer, R. R. Ramsay, P. K. Sonsalla, W. J. Nicklas, and R. E. Heikkila, Biochemical mechanism underlying MPTP-induced and idiopathic parkinsonism, *Advances in Neurology, Parkinson's Disease from Basic Research to Treatment*, Raven Press, New York, 1993, p. 300.
65. S. Ottoboni, T. J. Carlson, W. F. Trager, K. Castagnoli, and N. Castagnoli, Jr., Studies on

the cytochrome P-450 catalyzed ring α-carbon oxidation of the nigrostriatal toxin 1-methyl-4-phenyl-1,2,3,6-tetrahydropyridine (MPTP), *Chem. Res. Toxicol. 3*: 423 (1990).

66. R. B. Silverman, Radical ideas about monoamine oxidase, *Acc. Chem. Res. 28*: 335 (1995).
67. R. B. Silverman, Electron transfer chemistry of monoamine oxidase, *Advances in Electron Transfer Chemistry* (P. S. Mariano, ed.), JAI Press, Greenwich, 1992, p. 177.
68. N. E. Geacintov, M. Cosman, B. E. Hingerty, S. Amin, S. Broyde, and D. J. Patel, NMR solution structures of stereoisomeric covalent polycyclic aromatic carcinogen–DNA adducts: Principles, patterns, and diversity, *Chem. Res. Toxicol. 10*: 111 (1997).
69. N. Nemoto and H. V. Gelboin, Enzymatic conjugation of benzo[*a*]pyrene oxides, phenols and dihydrodiols with UDP-glucuronic acid, *Biochem. Pharmacol. 25*: 1221 (1976).
70. J. G. Dutton and I. D. E. Storey, Glucuronide-joining enzymes, *Methods in Enzymology* (S. P. Colowick and N. O. Kaplan, eds.), Academic Press, New York, 1962, p. 159.
71. G. J. Dutton, *Glucuronidation of Drugs and Other Compounds*, CRC Press, Boca Raton, FL, 1980.
72. J. D. Gregory and F. Lipmann, The transfer of sulfate among phenolic compounds with 3′,5′-diphosphoadenosine as coenzyme, *J. Biol. Chem. 29*: 1081 (1957).
73. G. M. Cohen, S. H. Haws, B. P. Moore, and J. W. Bridges, Benzo[*a*]pyrene-3-yl hydrogen sulfate, a major ethyl acetate-extractable metabolite of benzo[*a*]pyrene in human, hamster and rat lung cultures, *Biochem. Pharmacol. 25*: 2561 (1976).
74. N. Nemoto, S. Takayama, and H. V. Gelboin, Enzymatic conversion of benzo[*a*]pyrene phenols, dihydrodiols and quinones to sulfate conjugates, *Biochem. Pharmacol. 26*: 1825 (1977).
75. D. M. Jerina, J. W. Daly, B. Witkop, P. Zaltzman-Nirenberg, and S. Udenfriend, 1,2-Naphthalene oxide as an intermediate in the microsomal hydroxylation of naphthalene, *Biochemistry 9*: 147 (1970).
76. P. J. van Bladeren, K. P. Vyas, J. M. Sayer, D. E. Ryan, P. E. Thomas, W. Levin, and D. M. Jerina, Stereoselectivity of cytochrome P-450c in the formation of naphthalene and anthracene 1,2-oxides, *J. Biol. Chem. 259*: 8966 (1984).
77. E. Boyland and P. Sims, Metabolism of polycyclic compounds. XII. An acid-labile precursor of 1-naphthylmercapturic acid and naphthol: An *N*-acetyl-S-(1,2-dihydrohydroxynaphthyl)-L-cysteine, *Biochem. J. 68*: 440 (1958).
78. J. Booth, E. Boyland, and P. Sims, An enzyme from rat liver catalysing conjugation with glutathione, *Biochem. J. 79*: 516 (1960).
79. D. M. Jerina, J. W. Daly, B. Witkop, P. Zaltzman-Niremberg, and S. Udenfriend, The role of arene oxide–oxygen systems in the metabolism of aromatic substrates. 3. Formation of 1,2-naphthalene oxide from naphthalene by liver microsomes, *J. Am. Chem. Soc. 90*: 6525 (1968).
80. J. Booth and E. Boyland, Metabolism of polycyclic compounds. Formation of 1,2-dihydroxy-1,2-dihydronaphthalenes, *Biochem. J. 44*: 361 (1949).
81. G. Gurroff, J. W. Daly, D. M. Jerina, J. Renson, B. Witkop, and S. Udenfriend, Hydroxylation-induced migrations: The NIH shift, *Science 158*: 1524 (1967).
82. J. W. Daly, D. M. Jerina, and B. Witkop, Arene oxides and the NIH shift: The metabolism, toxicity and carcinogenicity of aromatic compounds, *Experientia 28*: 1129 (1972).
83. W. Levin, A. W. Wood, R. L. Chang, D. Ryan, P. E. Thomas, H. Yagi, D. R. Thakker, K. Vyas, C. Boyd, S.-Y. Chu, A. H. Conney, and D. M. Jerina, Oxidative metabolism of polycyclic aromatic hydrocarbons to ultimate carcinogens, *Drug. Metab. Rev. 13*: 555 (1982).
84. D. R. Thakker, H. Yagi, W. Levin, A. W. Wood, A. H. Conney, and D. M. Jerina, Polycyclic aromatic hydrocarbons: Metabolic activation to ultimate carcinogens, *Bioactivation of Foreign Compounds* (M. Anders, ed.), Academic Press, Orlando, FL, 1985, p. 177.
85. D. R. Boyd, G. S. Gadaginamath, A. Kher, J. F. Malone, H. Yagi, and D. M. Jerina, (+)- and (−)-benzo[*a*]pyrene 7,8-oxide: Synthesis, absolute stereochemistry, and stereochemical correlation with other mammalian metabolites of benzo[*a*]pyrene, *J. Chem. Soc., Perkin Trans. 10*: 2112 (1980).

86. W. Levin, M. K. Buening, A. W. Wood, R. L. Chang, D, Kedzierski, R. Thakker, D. R. Boyd, G. S. Gadaginamath, R. N. Armstrong, H. Yagi, J. M. Karle, T. J. Staga, D. M. Jerina, and A. H. Conney, An enantiomeric interaction in the metabolism and tumorigenicity of (+)- and (−)-benzo[*a*]pyrene 7,8-oxide, *J. Biol. Chem. 255*: 9067 (1980).
87. D. R. Thakker, H. Yagi, W. Levin, A. Y. H. Lu, A. H. Conney, and D. M. Jerina, Stereospecificity of microsomal and purified epoxide hydrase from rat liver. Hydration of arene oxides of polycyclic hydrocarbons, *J. Biol. Chem. 252*: 6328 (1977).
88. D. M. Jerina, H. Yagi, O. Hernandez, P. M. Dansette, A. W. Wood, W. Levin, R. L. Chang, P. G. Wislocki, and A. H. Conney, Synthesis and biologic activity of potential benzo[*a*]pyrene metabolites, *Polynuclear Aromatic Hydrocarbons: Chemistry, Metabolism, and Carcinogenesis* (R. I. Freudenthal and P. W. Jones, ed.), Raven Press, New York, 1976, p. 91.
89. D. R. Thakker, H. Yagi, H. Akagi, M. Koreeda, A. Y. H. Lu, W. Levin, A. W. Wood, A. H. Conney, and D. M. Jerina, Stereoselective metabolism of benzo[*a*]pyrene and benzo[*a*]-pyrene 7,8-dihydrodiol to diol epoxides, *Chem. Biol. Interact.* 1977, pp. 281–300.
90. K. Nakanish, H. Kalsai, H. Cho, R. G. Harvey, A. M. Jeffrey, K. W. Jennette, and I. B. Weinstein, Absolute configuration of ribonucleic acid adduct formed in vivo by metabolism of benzo[*a*]pyrene, *J. Am. Chem. Soc. 99*: 258 (1977).
91. A. O. Borgen, H. Darvey, N. Castagnoli, Jr., T. T. Crocker, R. E. Rasmussen, and I. Y. Wong, Metabolic conversion of benzo[*a*]pyrene by Syrian hamster liver microsomes and binding of metabolites to deoxyribonucleic acid, *J. Med. Chem. 16*: 502 (1973).
92. I. B. Weinstein, A. M. Jeffrey, K. W. Jennette, S. H. Blobstein, R. G. Harvey, C. Harris, H. Autrup, H. Kasai, and K. Nakanishi, Benzo[*a*]pyrene diol epoxides as intermediates in nucleic acid binding in vitro and in vivo, *Science 193*: 592 (1976).
93. A. W. Wood, P. G. Wislocki, R. L. Chang, W. Levin, A. Y. H. Lu, H. Yagi, O. Hernandez, D. M. Jerina, and A. H. Conney, Exceptional mutagenicity of a benzo[*a*]pyrene diol epoxide in cultured mammalian cells, *Nature 261*: 52 (1976).
94. E. Huberman, L. Sachs, S. K. Yang, and H. V. Gelboin, Identification of mutagenic metabolites of benzo[*a*]pyrine in mammalian cells, *Proc. Natl. Acad. Sci. USA 73*: 607 (1976).
95. J. Kapitulnik, P. G. Wislocki, W. Levin, H. Yagi, D. M. Jerina, and A. H. Conney, Tumorigenicity studies with diol-epoxides of hydroxy-9a, 10a-epoxy-7,8,9,10-tetrahydrobenzo[*a*]pyrene is an ultimate carcinogen in newborn mice, *Cancer Res. 38*: 354 (1978).
96. J. Kapitulnik, W. Levin, A. Conney, H. Yagi, and D. Jerina, Benzo[*a*]pyrene 7,8-dihydrodiol is more carcinogenic than benzo[*a*]pyrene in newborn mice, *Nature 266*: 378 (1977).
97. A. Jeffrey, K. Jennette, S. Blobstein, I. Weinstein, F. Beland, R. Harvey, H. Kadai, I. Miura, and K. Nakanishi, Benzo[*a*]pyrene–nucleic acid derivative found in vivo: Structure of a benzo[*a*]pyrenetetrahydrodiol epoxide–guanosine adduct, *J. Am. Chem. Soc. 98*: 5714 (1976).
98. M. Koreeda, P. D. Moore, H. Yagi, H. J. C. Yen, and D. M. Jerina, Alkylation of polyguanylic acid at the 2-amino group and phosphate by the potent mutagen (+−)-7β,8α-dihydroxy-9β,10β-epoxy-7,8,9,10-tetrahydrobenzo[*a*]pyrene, *J. Am. Chem. Soc. 74*: 6720 (1977).
99. L. Flowers-Geary, R. G. Harvey, and T. M. Penning, Cytotoxicity of polycyclic aromatic hydrocarbon *o*-quinones in rat and human hepatoma cells, *Chem. Res. Toxicol. 6*: 252 (1993).
100. A. H. Conney, R. L. Chang, D. M. Jerina, and S.-J. C. Wei, Studies on the metabolism of benzo[*a*]pyrene and dose-dependent differences in the mutagenic profile of its ultimate carcinogenic metabolite, *Drug Metab. Rev. 26*: 125 (1994).
101. D. H. Phillips and P. L. Grover, Polycyclic hydrocarbon activation: Bay regions and beyond, *Drug Metab. Rev. 26*: 443 (1994).
102. E. G. Rogan, P. D. Devanesan, N. V. S. RamaKrishna, S. Higginbotham, N. S. Padmavathi, K. Chapman, E. L. Cavalieri, H. Jeong, R. Jankowiak, and G. J. Small, Identification and quantitation of benzo[*a*]pyrene–DNA adducts formed in mouse skin, *Chem. Res. Toxicol. 6*: 356 (1993).
103. L. Chen, P. D. Devanesan, S. Higginbothan, F. Ariese, R. Jankowiak, G. J. Small, E. G.

Rogan, and E. L. Cavalieri, Expanded analysis of benzo[*a*]pyrene–DNA adducts formed in vitro and in mouse skin: Their significance in tumor initiation, *Chem. Res. Toxicol. 9*: 897 (1996).

104. A. P. Reddy, D. Pruess-Schwartz, C. Ji, P. Gorycki, and L. J. Marnett, 32-P postlabeling analysis of DNA adduction in mouse skin following topical administration of (+)-7,8-dihydroxy-7,8-dihydrobenzo[*a*]pyrene, *Chem. Res. Toxicol. 5*: 26 (1992).
105. B. Mao, L. A. Margulis, B. Li, V. Ibanez, H. Lee, R. G. Harvey, and N. E. Geacintov, Direct synthesis and identification of benzo[*a*]pyrene diol epoxide–deoxyguanosine binding sites in modified oligodeoxynucleotides, *Chem. Res. Toxicol. 5*: 773 (1992).
106. N. V. S. RamaKrishna, F. Gao, N. S. Padmavathi, E. L. Cavalieri, E. G. Rogan, R. L. Cerny, and M. L. Gross, Model adducts of benzo[*a*]pyrene and nucleosides formed from its radical cation and diol epoxide, *Chem. Res. Toxicol. 5*: 293 (1992).
107. P. D. Devanesan, N. V. S. RamaKrishna, R. Todorovic, E. G. Rogan, E. L. Cavalieri, H. Jeong, R. Jankowiak, and G. J. Small, Identification and quantitation of benzo[*a*]pyrene–DNA adducts formed by rat liver microsomes in vitro, *Chem. Res. Toxicol. 5*: 302 (1992).
108. Y.-J. Surh and S. R. Tannenbaum, Sulfotransferase-mediated activation of 7,8,9,10-tetrahydro-7-ol, 7,8-dihydrodiol, and 7,8,9,10-tetraol derivatives of benzo[*a*]pyrene, *Chem. Res. Toxicol. 8*: 693 (1995).
109. P. Chary and R. S. Lloyd, Impact of the stereochemistry of benzo[*a*]pyrene 7,8-dihydrodiol 9,10-epoxide-deoxyadenosine adducts on resistance to digestion by phosphodiesterases I and II and translesion synthesis with HIV-1 reverse transcriptase, *Chem. Res. Toxicol. 9*: 409 (1996).
110. I. Pontén, A. Seidel, A. Gräslund, and B. Jernström, Synthesis and characterization of adducts derived from the *syn*-diastereomer of benzo[*a*]pyrene 7,8-dihydrodiol 9,10-epoxide and the 5′-d(CCTATAGATATCC) oligonucleotide, *Chem. Res. Toxicol. 9*: 188 (1996).
111. K. H. Stansbury, J. W. Flesher, and R. C. Gupta, Mechanism of aryl-, alkyl–DNA adduct formation from benzo[*a*]pyrene in vivo, *Chem. Res. Toxicol. 7*: 254 (1994).
112. C. K. SooHoo, K. Singh, P. L. Skipper, S. R. Tannenbaum, and R. R. Dasari, Characterization of benzo[*a*]pyrene *anti*-diol epoxide adducts to human histones, *Chem. Res. Toxicol. 7*: 134 (1994).
113. C. J. Hanrahan, M. D. Bacolod, R. R. Vyas, T. Liu, N. E. Geacintov, E. L. Loechler, and A. K. Basu, Sequence specific mutagenesis of the major (+)-*anti*-benzo[*a*]pyrene diol epoxide–DNA adduct at a mutational hot spot in vitro and in *Escherichia coli* cells, *Chem. Res. Toxicol. 10*: 369 (1997).
114. K. Straub, T. Meehan, A. Burlingame, and M. Calvin, Identification of the major adducts formed by reaction of benzo[*a*]pyrene diol epoxide with DNA in vitro, *Proc. Natl. Acad. Sci. USA 74*: 5285 (1977).
115. A. Dipple, M. A. Piggott, S. K. Agarwal, H. Yagi, J. M. Sayer, and D. M. Jerina, Optically active benzo[*c*]phenanthrene diol epoxides bind extensively to adenine in DNA, *Nature 327*: 535 (1987).
116. Y. Ittah, D. R. Thakker, W. Levin, M. Croisy-Delcey, D. E. Ryn, P. E. Thomas, A. H. Conney, and D. M. Jerina, Metabolism of benzo[*c*]phenanthrene by rat liver microsomes and by purified monooxygenase system reconstituted with different isozymes of cytochrome P450, *Chem. Biol. Interact. 45*: 15 (1983).
117. A. Laryea, M. Cosman, J.-M. Lin, T. Liu, R. Agarwal, S. Smirnov, S. Amin, R. G. Harvey, A. Dipple, and N. E. Geacintov, Direct synthesis and characterization of site-specific adenosyl adducts derived from the binding of a 3,4-dihydroxy-1,2-epoxybenzo[*c*]phenanthrene stereoisomer to an 11-mer oligodeoxyribonucleotide, *Chem. Res. Toxicol. 8*: 444 (1995).
118. R. Agarwal, K. A. Canella, H. Yagi, D. M. Jerina, and A. Dipple, Benzo[*c*]phenanthrene–DNA adducts in mouse epidermis in relation to the tumorigenicities of four configurationally isomeric 3,4-dihydrodiol 1,2-epoxides, *Chem. Res. Toxicol. 9*: 586 (1996).
119. K. A. Canella, K. Peltonen, H. Yagi, D. M. Jerina, and A. Dipple, Identification of individual

benzo[c]phenanthrene dihydrodiol epoxide–DNA adducts by the ^{32}P-postlabeling assay, *Chem. Res. Toxicol. 5*: 685 (1992).

120. P. G. Wislocki, K. M. Gadek, M. M. Juliana, J. S. MacDonald, M. W. Chou, S. K. Yang, and A. Y. H. Lu, *Tumorigenicity and Mutagenicity of 7,12-Dimethylbenz[a]anthracene and Selected Metabolites Including the 3,4-Dihydrodiols*, Batelle Press, Columbus, OH, 1981, p. 675.
121. N. V. S. RamaKrishna, P. D. Devanesan, E. G. Rogan, E. L. Cavalieri, H. Jeong, R. Jankowiak, and G. J. Small, Mechanism of metabolic activation of the potent carcinogen 7,12-dimethylbenz[a]anthracene, *Chem. Res. Toxicol. 5*: 220 (1992).
122. P. D. Devanesan, N. V. S. RamaKrishna, N. S. Padmavathi, S. Higginbotham, E. G. Rogan, E. L. Cavalieri, G. A. Marsh, R. Jankowiak, and G. J. Small, Identification and quantitation of 7,12-dimethylbenz[a]anthracene–DNA adducts formed in mouse skin, *Chem. Res. Toxicol. 6*: 364 (1993).
123. C. M. Utermoehlen, M. Singh, and R. E. Lehr, Fjord region 3,4-diol 1,2-epoxides and other derivatives in the 1,2,3,4- and 5,6,7,8-benzo rings of the carcinogen benzo[g]chrysene, *J. Org. Chem. 52*: 5574 (1987).
124. J. Szeliga, H. Lee, R. G. Harvey, J. E. Page, H. L. Ross, M. N. Routledge, B. D. Hilton, and A. Dipple, Reaction with DNA and mutagenic specificity of *syn*-benzo[g]chrysene 11,12-dihydrodiol 13,14-epoxide, *Chem. Res. Toxicol. 7*: 420 (1994).
125. J. Szeliga, J. E. Page, B. D. Hilton, A. S. Kiselov, R. G. Harvey, Y. M. Dunayevski, P. Vouros, and A. Dipple, Characterization of DNA adducts formed by *anti*- benzo[g]chrysene 11,12-dihydrodiol 13,14-epoxide, *Chem. Res. Toxicol. 8*: 1014 (1995).
126. E. H. Weyand, P. Bryla, Y. Wu, Z.-M. He, and E. J. Lavoie, Detection of the major DNA adducts of benzo[j]fluoranthene in mouse skin: Non classical dihydrodiol epoxides, *Chem. Res. Toxicol. 6*: 117 (1993).
127. E. Silberhorn, H. P. Glauert, and L. W. Robertson, Carcinogenicity of polyhalogenated biphenyls: PCBs and PBBs, *Crit. Rev. Toxicol. 20*: 439 (1990).
128. S. Jensen, A new chemical hazard, *New Sci. 32*: 612 (1966).
129. I. C. T. Nisbet and A. F. Sarofim, Rates and routes of transport of PSBs in the environment, *Environ. Health Perspect. 4*: 21 (1972).
130. D. A. Dankovic and R. E. Billings, The role of 4-bromophenol and 4-bromocatechol in bromobenzene covalent binding and toxicity in isolated rat hepatocytes, *Toxicol. Appl. Pharmacol. 79*: 323 (1985).
131. B. B. Brodie, W. D. Reid, A. K. Cho, G. Sipes, G. Krishna, and J. R. Gillette, Possible mechanism of liver necrosis caused by aromatic organic compounds, *Proc. Natl. Acad. Sci. USA 68*: 160 (1971).
132. E. McConnel, J. E. Huff, M. Hetjmancik, A. C. Peters, and R. Persing, Toxicology and carcinogenesis studies of two grades of pentachlorophenol in B6C3F1 mice, *Fundam. Appl. Toxicol. 17*: 519 (1991).
133. D. E. Slaughter and R. P. Hanzlik, Identification of epoxide- and quinone-derived bromobenzene adducts to protein sulfur nucleophiles, *Chem. Res. Toxicol. 4*: 349 (1991).
134. T. J. Monks, J. A. Hinson, and J. R. Gillette, Bromobenzene and *p*-bromophenol toxicity and covalent binding in vivo, *Life Sci. 30*: 841 (1983).
135. J. R. Gillette, R. H. Menard, and B. Stripp, *Clin. Pharmacol. Therap. 14*: 680 (1973).
136. J. Zheng and R. P. Hanzlik, Dihydroxylated mercapturic acid metabolites of bromobenzene, *Chem. Res. Toxicol. 5*: 561 (1992).
137. S. Hesse, T. Wolff, and M. Metzger, Involvement of phenolic metabolites in the irreversible protein-binding of ^{14}C-bromobenzene catalyzed by rat liver microsomes, *Arch. Toxicol. Suppl. 4*: 358 (1980).
138. N. Narasimhan, P. E. Weller, J. A. Buben, R. A. Wiley, and R. P. Hanzlik, Microsomal metabolism and covalent binding of [^{3}H/^{14}C]-bromobenzene: Evidence for quinones as reactive metabolites, *Xenobiotica 18*: 491 (1988).

139. S. S. Lau, T. J. Monks, and J. R. Gillette, Identification of 2-bromohydroquinone as a metabolite of bromobenzene and *o*-bromophenol: Implications for bromobenzene-induced nephrotoxicity, *J. Pharmacol. Exp. Ther. 230*: 360 (1984).
140. T. J. Monks, S. S. Lau, R. J. Highet, and J. R. Gillette, Glutathione conjugates of 2-bromohydroquinone are nephrotoxic, *Drug Metab. Dispos. 13*: 553 (1985).
141. S. Waidyanatha, P.-H. Lin, and S. M. Rappaport, Characterization of chlorinated adducts of hemoglobin and albumin following administration of pentachlorophenol to rats, *Chem. Res. Toxicol. 9*: 647 (1996).
142. A. R. Amaro, G. G. Oakley, U. Bauer, H. P. Spielmann, and L. W. Robertson, Metabolic activation of PCBs to quinones: Reactivity towards nitrogen and sulfur nucleophiles and influence of superoxide dismutase, *Chem. Res. Toxicol. 9*: 623 (1996).
143. M. R. McLean, U. Bauer, A. R. Amaro, and L. W. Robertson, Identification of catechol and hydroquinone metabolites of 4-monochlorobiphenyl, *Chem. Res. Toxicol. 9*: 158 (1996).
144. I. M. C. M. Rietjens, C. den Besten, R. P. Hanzlik, and P. J. van Bladeren, Cytochrome P450-catalyzed oxidation of halobenzene derivatives, *Chem. Res. Toxicol. 10*: 629 (1997).
145. M. R. McLean, L. W. Robertson, and R. C. Gupta, Detection of PCB adducts by the ^{32}P-postlabeling technique, *Chem. Res. Toxicol. 9*: 165 (1996).
146. K. Randerath, R. E. Haglund, D. H. Philips, and M. V. Reddy, ^{32}P-postlabeling analysis of DNA adducts formed in the livers of animals treated with safrole, estragole and other naturally-occurring alkenylbenzenes. I. Adult female CD-1 mice, *Carcinogenesis 5*: 1613 (1984).
147. R. F. Crampton, T. J. B. Gray, P. Grasso, and D. V. Parke, Long-term studies on chemically induced liver enlargement in the rat. II. Transient induction of microsomal enzymes leading to liver damage and nodular hyperplasia produced by safrole and ponceau, *Toxicology 7*: 307 (1977).
148. T. R. Fennel, J. A. Miller, and E. C. Miller, Characterization of the biliary and urinary glutathione and *N*-acetylcysteine metabolites of the hepatic carcinogen 1′-hydroxysafrole and its oxo metabolite in rats and mice, *Cancer Res. 44*: 3231 (1984).
149. J. A. Miller and E. C. Miller, The metabolic activation and nucleic acid adducts of naturally-occurring carcinogens: Recent results with ethyl carbamate and the spice flavors safrole and estragole, *Br. J. Cancer 48*: 1 (1983).
150. E. W. Boberg, E. C. Miller, J. A. Miller, A. Poland, and A. Liem, Strong evidence from studies with brachymorphic mice and pentachlorophenol that 1′-sulfoxy-safrole is the major ultimate electrophilic and carcinogenic metabolite of 1′-hydroxysafrole in mouse liver, *Cancer Res. 43*: 5163 (1983).
151. J. Klungsoyr and R. R. Scheline, Metabolism of isosafrole and dihydrosafrole in the rat, *Biomed. Mass Spectrom. 9*: 323 (1982).
152. R. R. Scheline, *CRC Handbook of Mammalian Metabolism of Plant Compounds*, CRC Press, Boca Raton, FL, 1991, p. 72.
153. M. S. Benedetti, A. Malnoe, and A. L. Broillet, Absorption, metabolism and excretion of safrole in the rat and mouse, *Toxicology 7*: 69 (1977).
154. J. L. Bolton, L. G. Valerio, and J. A. Thompson, The enzymatic formation and chemical reactivity of quinone methides correlate with alkylphenol-induced toxicity in rat hepatocytes, *Chem. Res. Toxicol. 5*: 816 (1992).
155. J. L. Bolton, N. M. Acay, and V. Vukomanovic, Evidence that 4-allyl-*o*-quinones spontaneously rearrange to their more electrophilic quinone methides: Potential bioactivation mechanism for the hepatocarcinogen safrole, *Chem. Res. Toxicol. 7*: 443 (1994).
156. I. Gardner, P. Bergin, P. Stening, J. G. Kenna, and J. Codwell, Immunochemical detection of covalently modified protein adducts in livers of rats treated with methyleugenol, *Chem. Res. Toxicol. 9*: 713 (1996).
157. S. L. Iverson, L. Q. Hu, V. Vukomanovic, and J. L. Bolton, The influence of the *p*-alkyl substituent on the isomerization of *o*-quinones to *p*-quinones methides: Potential bioactivation mechanism for catechols, *Chem. Res. Toxicol. 8*: 537 (1995).

158. D. C. Thompson, K. Perera, E. S. Krol, and J. L. Bolton, *o*-Methoxy-4-alkylphenols that form quinone methides of intermediate reactivity are the most toxic in rat liver slices, *Chem. Res. Toxicol. 8*: 323 (1995).
159. D. C. Thompson, K. Perera, and R. London, Quinone methide formation from *para* isomers of methylphenol (cresol), ethylphenol, and isopropylphenol: Relationship to toxicity, *Chem. Res. Toxicol. 8*: 55 (1995).
160. J. G. Liehr, T. A. Avitts, E. Randerath, and K. Randerath, Localization of estrogen-induced DNA adducts and cytochrome P450 activity at the site of renal carcinogenesis in the hamster kidney, *Cancer Res. 47*: 2156 (1987).
161. A. Akanni and Y. J. Abul-Hajj, Estrogen–nucleic acid adducts: Reaction of 3,4-estrone-*o*-quinone radical anion with deoxyribonucleosides, *Chem. Res. Toxicol. 10*: 760 (1997).
162. A. Akanni, K. Tabakovic, and Y. J. Abul-Hajj, Estrogen–nucleic acid adducts: Reaction of 3,4-estrone-*o*-quinone with nucleic acid bases, *Chem. Res. Toxicol. 10*: 477 (1997).
163. A. L. Herbst, H. Ulfelder, and D. Z. Poskanzer, Adenocarcinoma of the vagina: Association of maternal stilbesterol therapy with tumor appearance in young women, *N. Engl. J. Med. 284*: 878 (1984).
164. L. Bergkvist, H. O. Adami, I. Persson, R. Hoover, and C. Schairer, The risk of breast cancer after estrogen and estrogen–progestin replacement, *N. Engl. J. Med. 321*: 293 (1989).
165. J. Vana, G. P. Murphy, B. L. Aronoff, and H. W. Baker, Survey of primary liver tumors and oral contraceptive use, *J. Toxicol. Environ. Health 5*: 255 (1979).
166. M. Numazawa, R. Shirao, N. Soeda, and T. Nambara, Properties of enzyme systems involved in the formation of catechol estrogen glutathione conjugates in rat liver microsomes, *Biochem. Pharmacol. 27*: 1833 (1978).
167. S. F. Sarabia, B. T. Zhu, T. Kurosawa, M. Tohma, and J. G. Liehr, Mechanism of cytochrome P450-catalyzed aromatic hydroxylation of estrogens, *Chem. Res. Toxicol. 10*: 767 (1997).
168. M. Numazawa, Y. Tanaka, Y. Monono, and T. Mambara, Occurrence of the cysteine conjugate of 2-hydroxyestrone in rat bile, with special reference to its formation mechanism, *Chem. Pharm. Bull. 22*: 663 (1974).
169. M. Numazawa and T. Nambara, A new mechanism of in vitro formation of catechol estrogen glutathione conjugates by rat liver microsomes, *J. Steroid Biochem. 8*: 835 (1977).
170. K. Tabakovic and Y. J. Abul-Hajj, Reaction of lysine with estrone 3,4-*o*-quinone, *Chem. Res. Toxicol. 7*: 696 (1994).
171. I. Dwivedy, P. Devanesan, P. Cremonesi, E. G. Rogan, and E. Cavalieri, Synthesis and characterization of estrogen 2,3- and 3,4-quinones. Comparison of DNA adducts formed by the quinones versus horseradish peroxidase-activated catechol estrogens, *Chem. Res. Toxicol. 5*: 828 (1992).
172. R. D. Irons and T. Sawahata, Phenols, catechols, and quinones, *Bioactivation of Foreign Compounds* (M. W. Anders, ed.), Academic Press, Orlando, 1985, p. 259.
173. S. I. Iverson, L. Shen, N. Anlar, and J. L. Bolton, Bioactivation of estrone and its catechol metabolites to quinoid–glutathione conjugates in rat liver microsomes, *Chem. Res. Toxicol. 9*: 492 (1996).
174. F. Marks and E. Hecker, Metabolism and mechanism of action of oestrogens. XII. Structure and mechanism of formation of water-soluble and protein-bound metabolites of oestrone in rat liver microsomes in vitro and in vivo, *Biochim. Biophys. Acta 187*: 250 (1969).
175. L. M. Nutter, B. Zhou, E. E. Sierra, Y.-Y. Wu, M. M. Rummel, P. Gutierrez, and Y. Abul-Hajj, Cellular biochemical determinants modulating the metabolism of estrone 3,4-quinone, *Chem. Res. Toxicol. 7*: 609 (1994).
176. L. M. Nutter, Y.-Y. Wu, E. O. Ngo, E. E. Sierra, P. L. Gutierrez, and Y. J. Abul-Hajj, An *o*-quinone form of estrogen produces free radicals in human breast cancer cells: Correlation with DNA damage, *Chem. Res. Toxicol. 7*: 23 (1994).
177. S. D. Nelson, Arylamines and arylamides: Oxidation mechanism, *Bioactivation of Foreign Compounds* (M. W. Anders, ed.), Academic Press, Orlando, 1985, p. 349.

178. P. E. Hanna and R. B. Banks, Arylhydroxylamines and arylhydroxamic acids: Conjugation reactions., *Bioactivation of Foreign Compounds* (M. W. Anders, ed.), Academic Press, Orlando, 1985, p. 375.
179. S. Shibutani and A. Grollman, Molecular mechanism of mutagenesis by aromatic amines and amides, *Mutat. Res. 376*: 71 (1997).
180. A. Bitsch, J. Fecher, M. Jost, P. Klohn, and H. Neumann, Genotoxic and chronic toxic effects in the carcinogenicity of aromatic amines, *Cancer Res. 143*: 23 (1997).
181. F. F. Kadlubar, J. A. Miller, and E. C. Miller, Hepatic metabolism of *N*-hydroxy-*N*-methyl-4-aminoazobenzene and other *N*-hydroxyarylamines to reactive sulfuric acid esters, *Cancer Res. 36*: 2350 (1976).
182. F. F. Kadlubar, B. Ketterer, T. J. Flammang, and L. Christodoulies, Formation of 3-(glutathion-*S*-yl)-*N*-methyl-4-aminoazobenzene and inhibition of aminazo dye–nucleic acid binding in vitro by reaction of glutathione with metabolically generated *N*-methyl-4-aminoazobenzene *N*-sulfate, *Chem. Biol. Interact. 31*: 265 (1980).
183. J. R. Mitchell, D. J. Jollow, W. Z. Potter, C. C. Davis, J. R. Gillette, and B. B. Brodie, Acetaminophen-induced hepatic necrosis. I. Role of drug metabolism, *J. Pharmacol. Exp. Ther. 187*: 185 (1973).
184. D. J. Jollow, J. R. Mitchell, W. Z. Potter, D. C. Davis, J. R. Gillette, and B. B. Brodie, Acetaminophen-induced hepatic necrosis. II. Role of covalent binding in vivo, *J. Pharmacol. Exp. Ther. 187*: 195 (1973).
185. J. R. Mitchell, D. J. Jollow, W. Z. Potter, J. R. Gillette, and B. B. Brodie, Acetaminophen-induced hepatic necrosis. IV. Protective role of glutathione, *J. Pharmacol. Exp. Ther. 187*: 211 (1973).
186. W. Z. Potter, S. S. Thorgeirsson, D. J. Jollow, and J. R. Mitchell, *Pharmacology 12*: 129 (1974).
187. W. Z. Potter, D. C. Davis, J. R. Mitchell, D. J. Jollow, J. R. Gillette, and B. B. Brodie, Acetaminophen-induced hepatic necrosis. III. Cytochrome P450-mediated covalent binding in vivo, *J. Pharmacol. Exp. Ther. 187*: 203 (1973).
188. N. P. E. Vermeulen, J. G. M. Bessems, and R. van de Straat, Molecular aspects of paracetamol-induced hepatotoxicity and its mechanism-based prevention, *Drug Metab. Rev. 24*: 367 (1992).
189. S. D. Nelson, Mechanisms of the formation and disposition of reactive metabolites that can cause acute liver injury, *Drug Metab. Rev. 27*: 147 (1995).
190. J. A. Hinson, N. R. Pumford, and D. W. Roberts, Mechanisms of acetaminophen toxicity: Immunochemical detection of drug–protein adducts, *Drug Metab. Rev. 27*: 73 (1995).
191. D. J. Jollow, S. S. Thorgeirsson, W. Z. Potter, M. Hashimoto, and J. R. Gillette, Acetaminophen-induced hepatic necrosis, *Pharmacology 12*: 251 (1974).
192. C. J. Patten, P. E. Thomas, R. L. Guy, M. Lee, F. J. Gonzalez, F. P. Guengerich, and C. S. Yang, Cytochrome P450 enzymes involved in acetaminophen activation by rat and human liver microsomes and their kinetics, *Chem. Res. Toxicol. 6*: 511 (1993).
193. P. J. Harrison, F. P. Guengerich, M. S. Rashed, and S. D. Nelson, *Chem. Res. Toxicol. 1*: 47 (1988).
194. D. J. Miner and P. T. Kissinger, Evidence for the involvement of *N*-acetyl-*p*-benzoquinone imine in acetaminophen metabolism, *Biochem. Pharmacol. 28*: 3285 (1979).
195. D. C. Dahlin and S. D. Nelson, Synthesis, decomposition kinetics, and preliminary toxicological studies on pure *N*-acetyl-*p*-benzoquinone imine, a proposed toxic metabolite of acetaminophen, *J. Med. Chem. 25*: 885 (1982).
196. J. B. Bartolone, K. Sparks, S. D. Cohen, and E. A. Khairallah, Immunochemical detection of acetaminophen-bound liver proteins, *Biochem. Pharmacol. 36*: 1193 (1987).
197. N. R. Pumford, J. A. Hinson, R. W. Benson, and D. W. Roberts, Immunoblot analysis of protein containing 3-(cystein-*S*-yl)acetaminophen adducts in serum and subcellular liver fractions from acetaminophen-treated mice, *Toxicol. Appl. Pharmacol. 104*: 521 (1990).

198. N. C. Halmes, J. A. Hinson, B. M. Martin, and N. R. Pumford, Glutamate dehydrogenase covalently binds to a reactive metabolite of acetaminophen, *Chem. Res. Toxicol. 9*: 541 (1996).
199. N. R. Pumford, J. A. Hinson, D. W. Potter, K. L. Rowland, R. W. Benson, and D. W. Roberts, Immunochemical quantitation of 3-(cystein-*S*-yl)acetaminophen adducts in serum and liver proteins of acetaminophen-treated mice, *J. Pharmacol. Exp. Ther. 248*: 190 (1989).
200. J. D. Gibson, N. R. Pumford, V. M. Samokyszyn, and J. A. Hinson, Mechanism of acetaminophen-induced hepatotoxicity: Covalent binding versus oxidative stress, *Chem. Res. Toxicol. 9*: 580 (1996).
201. P. C. Burcham and A. W. Harman, Effect of acetaminophen hepatotoxicity on hepatic mitochondrial and microsomal calcium contents in mice, *Toxicol. Lett. 44*: 91 (1988).
202. J. L. Holtzman, The role of covalent binding to microsomal proteins in the hepatotoxicity of acetaminophen, *Drug Metab. Rev. 27*: 277 (1995).
203. J. A. Hinson, N. R. Pumford, and S. D. Nelson, The role of metabolic activation in drug toxicity, *Drug Metab. Rev. 26*: 395 (1994).
204. L. K. Rogers, B. Moorthy, and C. V. Smith, Acetaminophen binds to mouse hepatic and renal DNA at human therapeutic doses, *Chem. Res. Toxicol. 10*: 470 (1997).
205. C. B. Fredericks, J. B. Mays, D. M. Ziegler, F. P. Guengerich, and F. F. Kadlubar, Cytochrome P450 and flavin containing monooxygenase-catalyzed formation of the carcinogen *N*-hydroxy-2-aminofluorene and its covalent binding to nuclear DNA, *Cancer Res. 42*: 2671 (1982).
206. G. R. Hoffmann and R. P. P. Fuchs, Mechanism of frameshift mutations: Insights from aromatic amines, *Chem. Res. Toxicol. 10*: 347 (1997).
207. H. Neumann, A. Bitsch, and P. Klohn, The dual role of 2-acetylaminofluorene in hepatocarcinogenesis: Specific targets for initiation and promotion, *Mutat. Res. 376*: 169 (1997).
208. L. Verna, J. Whysner, and G. Williams, 2-Acetylaminofluorene mechanistic data and risk assessment: DNA reactivity, enhanced cell proliferation and tumor initiation, *Pharmacol. Ther. 71*: 83 (1996).
209. F. Guengerich, W. Humphreys, C. Yun, G. Hammons, F. Kadlubar, Y. Seto, O. Okasaki, and M. Martin, Mechanism of cytochrome P450 1A2-mediated formation of *N*-hydroxylamines and heterocyclic amines and their reaction with guanyl residues, Proceeding of the International Princess Takamatsu Symposium, 1995, p. 78.
210. C. C. Irving, Metabolic activation of *N*-hydroxy compounds by conjugation, *Xenobiotica 1*: 387 (1971).
211. C. C. Irving, Conjugates of *N*-hydroxy compounds, *Metabolic Conjugation and Metabolic Hydrolysis* (W. H. Fishman, ed.), Academic Press, New York, 1970, p. 53.
212. E. Kriek and J. G. Westra, Metabolic activation of aromatic amines and amides and interactions with nucleic acids, *Chemical Carcinogens and DNA* (P. L. Grover, ed.), CRC Press, Boca Raton, FL, 1979, p. 1.
213. L. W. Boteju and P. E. Hanna, Arylamine–nucleoside adduct formation: Evidence for arylnitrene involvement in the reactions of an *N*-acetoxyarylamine, *Chem. Res. Toxicol. 7*: 684 (1994).
214. J. W. Langston, The etiology of Parkinson's disease with emphasis on the MPTP story, *Neurology 47*: S153 (1996).
215. T. Coleman, S. W. Ellis, I. J. Martin, M. S. Lennard, and G. T. Tucker, 1-Methyl-4-phenyl-1,2,3,6-tetrahydropyridine (MPTP) is *N*-demethylated by cytochromes P450 2D6, 1A2 and 3A4—implications for susceptibility to Parkinson's disease, *J. Pharmacol. Exp. Ther. 277*: 685 (1996).
216. J. Weissman, A. Trevor, K. Chiba, L. A. Peterson, P. Caldera, N. Castagnoli, Jr., and T. Baillie, Metabolism of the nigrostriatal toxin 1-methyl-4-phenyl-1,2,3,6-tetrahydropyridine by liver homogenate fraction, *J. Med. Chem. 28*: 997 (1985).
217. S. Narimatsu, M. Tachibana, Y. Masubuchi, and T. Suzuki, Cytochrome P4502D and -2C

enzymes catalyze the oxidative *N*-demethylation of the parkinsonism-inducing substance 1-methyl-4-phenyl-1,2,3,6-tetrahydropyridine in rat liver microsomes, *Chem. Res. Toxicol. 9*: 93 (1996).
218. S. Modi, D. E. Gilham, M. J. Sutcliffe, L. Y. Lian, W. U. Primrose, C. R. Wolf, and G. C. Roberts, 1-Methyl-4-phenyl-1,2,3,6-tetrahydropyridine as a substrate of cytochrome P4502D6: Allosteric effects of NADPH-cytochrome P450 reductase, *Biochemistry 36*: 4461 (1997).
219. K. Chiba, A. Trevor, and N. Castagnoli, Jr., Metabolism of the neurotoxic tertiary amine MPTP by brain monoamine oxidase, *Biochem. Biophys. Res. Commun. 120*: 574 (1984).
220. K. Chiba, L. A. Peterson, K. Castagnoli, A. J. Trevor, and N. Castagnoli, Jr., Studies on the molecular mechanism of bioactivation of the selective nigrostriatal toxin 1-methyl-4-phenyl-1,2,3,6-tetrahydropyridine, *Drug Metab. Dispos. 13*: 342 (1985).
221. J. A. Javitch, R. J. D'Amato, S. M. Strittmetter, and S. H. Snyder, Parkinsonism-inducing neurotoxin, 1-methyl-4-phenyl-1,2,3,6-tetrahydropyridine: Uptake of the metabolite 1-methyl-4-phenylpyridine by dopamine neurons selective toxicity, *Proc. Natl. Acad. Sci. USA 82*: 2173 (1985).
222. C. D. Dungigan and A. E. Shamoo, Identification of the major transport pathway for the parkinsonism inducing neurotoxin 1-methyl-4-phenylpyridinium, *Neuroscience 75*: 37 (1996).
223. L. M. Sayre, M. P. Singh, P. K. Arora, F. Wang, R. J. McPeak, and C. L. Hoppel, Inhibition of mitochondrial respiration by analogues of the dopaminergic neurotoxin 1-methyl-4-phenylpyridinium: Structural requirements for accumulation-dependent enhanced inhibitory potency on intact mitochondria, *Arch. Biochem. Biophys. 280*: 274 (1990).
224. W. J. Nicklas, I. Vyas, and R. E. Heikkila, Inhibition of NADH-linked oxidation in brain mitochondria by 1-methyl-4-phenylpyridine, a metabolite of the neurotoxin 1-methyl-4-phenyl-1,2,3,6-tetrahydropyridine, *Life Sci. 35*: 2503 (1985).
225. W. J. Vyas, R. E. Heikkila, and W. J. Nicklas, Studies on the neurotoxicity of 1-methyl-4-phenyl-1,2,3,6-tetrahydropyridine: Inhibition of NADH-linked substrate oxidation by its metabolite, 1-methyl-4-phenylpyridinium, *J. Neurochem. 46*: 1501 (1986).
226. M. Santiago, L. Granero, A. Machado, and J. Cano, Complex I inhibitor effect on the nigral and striatal release of dopamine in the presence and absence of nomifensine, *Eur. J. Pharmacol. 280*: 251 (1995).
227. M. P. Singh, F. J. Wang, C. L. Hoppel, and L. M. Sayre, Inhibition of mitochondrial respiration by neutral, monocationic, and dicationic bis-pyridines related to the dopaminergic neurotoxin 1-methyl-4-phenylpyridinium cation (MPP^+), *Arch. Biochem. Biophys. 286*: 138 (1991).
228. R. R. Ramsay, J. I. Salach, J. Dadgar, and T. P. Singer, Inhibition of mitochondrial NADPH dehydrogenase by pyridine derivatives and possible relation to experimental and idiopathic parkinsonism, *Biochem. Biophys. Res. Commun. 135*: 269 (1986).
229. D. Di Monte, S. A. Jewell, G. Ekstrom, M. S. Sandy, and M. T. Smith, 1-Methyl-4-phenyl-1,2,3,6-tetrahydropyridine (MPTP) and 1-methyl-4-phenyl-pyridine (MPP^+) cause rapid ATP depletion in isolated hepatocytes, *Biochem. Biophys. Res. Commun. 137*: 310 (1986).
230. W. Koller, Neuroprotective therapy for Parkinson's disease, *Exp. Neurol. 144*: 24 (1997).
231. S. Ambrosio, A. Espino, B. Cutillas, and R. Bartrons, MPP^+ toxicity in rat striatal slices: Relationship between non selective effects and free radical production, *Neurochem. Res. 21*: 73 (1996).
232. K. Sriram, K. Pai, M. Boyd, and V. Ravindranath, Evidence for generation of oxidative stress in brain by MPTP: In vitro and in vivo studies in mice, *Brain Res. 749*: 44–52 (1997).
233. J. Fallon, R. Matthews, B. Hyman, and M. Beal, MPP^+ produces progressive neuronal degeneration which is mediated by oxidative stress, *Exp. Neurol. 144*: 193 (1997).
234. M. Desole, G. Esposito, L. Fresu, R. Bigheli, S. Sircana, R. Delogu, M. Miele, and E. Miele,

Further investigation of allopurinol effects on MPTP induced oxidative stress in the striatum and brain stem of the rat, *Pharmacol. Biochem. Behav. 54*: 377 (1996).

235. Y. Akaneya, M. Takahashi, and H. Hatanaka, Involvement of free radicals in MPP^+ neurotoxicity against rat dopaminergic neurons in culture, *Neurosci. Lett.* 53 (1993).
236. M. Gotz, G. Kunig, P. Riederer, and M. Youdim, Oxidative stress: Free radical production in neural degeneration, *Pharmacol. Exp. Ther. 63*: 37 (1994).
237. D. Kang, K. Miyako, F. Kuribayashi, E. Hasegawa, A. Mitsumoto, T. Nagano, and K. Takeshige, Changes of energy metabolism induced by 1-methyl-4-phenylpyridinium (MPP^+) and related compounds in rat pheochromocytoma PC12 cells, *Arch. Biochem. Biophys. 337*: 75 (1997).
238. J. Schulz, R. Matthews, and M. Beal, Role of nitric oxide in neurodegenerative diseases, *Curr. Opin. Neurol. 8*: 480 (1995).
239. M. Packer, C. Porteous, and M. Murphy, Superoxide production by mitochondria in the presence of nitric oxide forms peroxynitrite, *Biochem. Mol. Biol. Int. 40*: 527 (1996).
240. I. Paakkari and P. Lindsberg, Nitric oxide in the central nervous system, *Ann. Med. 27*: 369 (1995).
241. M. Gerlach, D. Ehler, D. Blumdegen, K. Lange, B. Mayer, H. Reichmann, and P. Riederer, Regional distribution and characterization of nitric oxide synthase activity in the brain of the common marmoset, *Neuroreport 6*: 1141 (1995).
242. V. L. Dawson and T. M. Dawson, Nitric oxide neurotoxicity, *Chem. Neuroanat. 10*: 179 (1996).
243. V. L. Dawson and T. M. Dawson, Nitric oxide actions in neurochemistry, *Neurochem. Int. 29*: 97 (1996).
244. V. L. Dawson and T. M. Dawson, Nitric oxide in neuronal degeneration, *Proc. Soc. Exp. Biol. Med. 211*: 33 (1996).
245. P. Moore and P. Blandward, 7-Nitroindazole: An inhibitor of nitric oxide synthase, *Neth. Enzymol. 268*: 393 (1996).
246. T. S. Smith, R. H. Swerdlow, W. D. Parker, Jr., and J. P. Bennett, Jr., Reduction of MPP^+ induced hydroxyl radical formation and nigrostriatal MPTP toxicity by inhibiting nitric oxide synthase, *Neuroreport 5*: 2598 (1994).
247. J. B. Schulz, R. T. Matthews, M. M. K. Muqit, S. E. Browne, and M. F. Beal, Inhibition of neuronal nitric oxide synthase by 7-nitroindazole protects against MPTP-induced neurotoxicity in mice, *J. Neurochem. 64*: 936 (1995).
248. S. Przedborski, V. Jackson-Lewis, R. Yokoyama, T. Shibata, V. L. Dawson, and T. M. Dawson, Role of neuronal nitric oxide in 1-methyl-4-phenyl-1,2,3,6-tetrahydropyridine (MPTP)-induced dopaminergic neurotoxicity, *Proc. Natl. Acad. Sci. USA 93*: 4565 (1996).
249. P. Hantraye, E. Brouillet, R. Ferrante, S. Palfi, R. Dolan, R. T. Matthews, and M. F. Beal, Inhibition of neuronal nitric oxide synthase prevents MPTP-induced parkinsonism in baboons, *Nature Med. 2*: 1017 (1996).
250. K. Castagnoli, S. Palmer, A. Anderson, T. Bueters, and N. Castagnoli, Jr., The neuronal nitric oxide synthase inhibitor 7-nitroindazole also inhibits the monoamine oxidase-B-catalyzed oxidation of 1-methyl-4-phenyl-1,2,3,6-tetrahydropyridine, *Chem. Res. Toxicol. 10*: 364 (1997).
251. G. Mackenzie, M. Jackson, P. Jenner, and C. Marsden, Nitric oxide synthase inhibition and MPTP-induced toxicity in the common marmoset, *Synapse 26*: 301 (1997).
252. A. R. Mattocks, *Chemistry and Toxicology of Pyrrolizidine Alkaloids*, Academic Press, Orlando, FL 1986.
253. H.-K. Kim, E. R. Stermitz, R. J. Molyneux, D. W. Wilson, D. Taylor, and R. A. Coulombe, Jr., Structural influences on pyrrolizidine alkaloid-induced cytopathology, *Toxicol. Appl. Pharmacol. 122*: 61 (1993).
254. P. C. Chan, J. Mahler, J. R. Bucher, G. S. Travos, and J. B. Reid, Toxicity and carcinogenicity of riddelline following 13 weeks of treatment to rats and mice, *Toxicon 32*: 891 (1994).

255. V. R. Campesato, U. Graf, M. L. Reguly, and H. H. de Andrade, Recominagenic activity of intergerrimine, a pyrrolizidine alkaloid from *Senecio brasiliaensis* in somatic cells of *Drosophila melanogaster, Environ. Mol. Mutagen. 29*: 91 (1997).
256. R. J. Huxtable, C. C. Yan, S. Wild, S. Maxwell, and R. Cooper, Physicochemical and metabolic basis for the differing neurotoxicity of the pyrrolizidine alkaloids, trichdesmine and monocrotaline, *Neurochem. Res. 21*: 141 (1996).
257. A. F. M. Rizk and A. Kamel, Toxicity, carcinogenicity, pharmacology, and other biological activities of pyrrolizidine alkaloids, *Naturally Occuring Pyrrolizidine Alkaloids* (A. F. M. Rizk, ed.), CRC Press, Boca Raton, FL, 1991, p 211.
258. W. G. Chung, C. L. Miranda, and D. R. Buhler, A cytochrome P450B form is the major bioactivation enzyme for the pyrrolizidine alkaloid senecionine in guinea pig, *Xenobiotica 25*: 929 (1995).
259. W. G. Chung and D. R. Buhler, Major factors for the susceptibility of guinea pig to the pyrrolizidine alkaloid jacobine, *Drug Metab. Dispos. 23*: 1263 (1995).
260. H. J. Segall, D. W. Wilson, M. W. Lamé, S. Moein, and X. K. Qinrwe, Metabolism of pyrrolizidine alkaloids, *Handbook of Natural Toxins* (R. F. Keeler and A. T. Tu, eds.), Marcel Dekker, New York, 1991, p. 3.
261. D. R. Buhler and B. Kedzierski, Biological reactive intermediates of pyrrolizidine alkaloids, *Biological Reactive Intermediates* (J. J. Kocsis, D. J. Jollow, C. M. Witmer, J. O. Nelson, and R. Snyder, eds.) Plenum Publishing, New York, 1986, p. 611.
262. A. R. Mattocks and I. P. Bird, Pyrrolic and *N*-oxide metabolites formed from pyrrolizidine alkaloids by hepatic microsomes in vivo: Relevance to in vivo hepatotoxicity, *Chem. Biol. Interact. 43*: 209 (1983).
263. H. Y. Kim, F. R. Stermitz, and R. A. Coulombe, Jr., Pyrrolizidine alkaloid-induced DNA-protein cross-links, *Carcinogenesis 16*: 2691 (1995).
264. S. E. Murphy, D. A. Spina, M. G. Numes, and D. A. Pullo, Glucuronidation of 4-[(hydroxymethyl)nitrosamino]-1-(3-pyridyl)-1-butanone, a metabolically activated form of 4-(methylnitrosamino)-1-(3-pyridyl)-1-butanone, by phenobarbital-treated rats, *Chem. Res. Toxicol. 8*: 772 (1995).
265. R. Young-Sciame, M. Wang, F.-L. Chung, and S. S. Hecht, Reactions of alpha-acetoxy-*N*-nitrosopyrrolidine and alpha-acetoxy-*N*-nitrosopiperidine with deoxyguanosine: Formation of N^2-tetrahydrofuranyl and N^2-tetrahydropyranyl adducts, *Chem. Res. Toxicol. 8*: 607 (1995).
266. S. S. Hecht and D. Hoffmann, Tobacco-specific nitrosamines: An important group of carcinogens in tobacco and tabacco smoke, *Carcinogenesis 9*: 875 (1988).
267. S. G. Carmella, S. S. Kagan, and S. S. Hecht, Evidence that a hemoglobin adduct of 4-(methylnitrosamino)-1-(3-pyridyl)-1-butanone is a 4-(3-pyridyl)-4-oxobutyl carboxylic acid ester, *Chem. Res. Toxicol. 5*: 76 (1992).
268. S. S. Hecht, R. Young-Sciame, and F.-L. Chung, Reaction of alpha-acetoxy-*N*-nitrosopiperidine with deoxyguanosine: Oxygen-dependent formation of 4-oxo-2-pentenal and a 1, N^2-ethenodeoxyguanosine adduct, *Chem. Res. Toxicol. 5*: 706 (1992).
269. M. Wang, A. Nishikawa, and F.-L. Chung, Differential effects of thiols on DNA modifications via alkylation and Michael addition by alpha-acetoxy-*N*-nitrosopyrrolidine, *Chem. Res. Toxicol. 5*: 528 (1992).
270. M. Wang and S. S. Hecht, A cyclic N^7C-8 guanine adduct of *N*-nitrosopyrrolidine (NPYR): Formation in nucleic acids and excretion in the urine of NPYR-treated rats, *Chem. Res. Toxicol. 10*: 772 (1997).
271. F.-L. Chung, J. Krzeminski, M. Wang, H.-J. C. Chen, and B. Prokopczyk, Formation of the acrolein-derived 1, N^2-propanodeoxyguanosine adducts in DNA upon reaction with 3-(*N*-carbethoxy-*N*-nitrosamino)propionaldehyde, *Chem. Res. Toxicol. 7*: 62 (1994).

3

Morphofunctional Aspects of the Hepatic Architecture: Functional and Subcellular Correlates

Felix A. de la Iglesia, Donald G. Robertson, and Jeffrey R. Haskins
Parke-Davis Pharmaceutical Research Division, Warner-Lambert Company, Ann Arbor, Michigan

I. INTRODUCTION

This chapter will provide the reader with a morphological and functional overview of the liver in relation to the many biochemical and enzymatic reactions that take place in the organ. Different cell types and organelles within cells interact with each other and participate in basic metabolic reactions, protein synthesis, and biotransformation of xenobiotics. Although a significant portion of liver activity involves providing the proteins, lipoproteins, and carriers necessary for normal function, the protective effect of the liver is reflected in its ability to detoxify foreign compounds, metabolize drugs, and provide compensatory mechanisms for diverse reactions that may compromise cell survival. Liver cells constitute elementary reactors with separate compartments that synthesize, hydrolyze, conjugate, or oxidize exogenous and endogenous chemicals; therefore, the cells are of primary interest to scientists interested in hepatic drug metabolism. The delicate architecture of the hepatocyte must be renewed frequently so it can perform effectively in response to adverse effects or injury. Above all, this multitude of anabolic or catabolic activities are under control from the nucleus, which houses all replicative and messaging functions necessary for structural integrity. The use of combined microscopic and functional approaches can provide an integrated view on drug metabolism and biotransformation. This approach offers some advantages over the uncertainties that can emerge from studies that disrupt the normal cell architecture, as occurs with techniques of isolation and separation of cellular components.

II. GENERAL ANATOMICAL CONSIDERATIONS

A. Lobule Versus Acinar View of Hepatic Architecture

The arrangement of liver cells is in the form of intertwined trabecular plates or cords of cells in a disposition optimally designed for contact with the blood (Fig. 1). Two primary views of the basic architecture of the liver have arisen over the years. The earliest concept, described by Kiernan (1) was based on morphology and defined the basic functional unit of the liver as a lobule. The lobule consists of a hexagonal cluster of hepatocytes centered on a hepatic venule, known as the central vein, and each corner of the hexagon, called the portal triad, grouped a hepatic arteriole, a portal venule, and a bile ductule (Fig. 2). The area surrounding the central vein is known as the centrilobular region, and the region surrounding the portal triad is known as the periportal region, with the area in between described simply as the midzone. This classification neatly fits the morphology of the liver, and this lobular terminology is most frequently used in defining the location of pathological lesions in the liver.

The second classification of liver architecture revolves around the concept of the hepatic acinus (2) and is based on function, rather than morphology, exclusively. The basic unit of hepatic architecture in this view is the acinus, according to the direction of the blood flow within the liver (Fig. 3). At the center of the acinus are the portal venules and hepatic arterioles from which blood flows, percolating in three-dimensional space through cords of hepatocytes to distal central veins in which the blood is collected and distributed to the systemic circulation. The acinus is divided into three zones, based on their proximity to the hepatic blood supply: Zone 1 is closest to the portal venules and hepatic arterioles; Zone 3 lies closest to the central vein; whereas Zone 2 is in between.

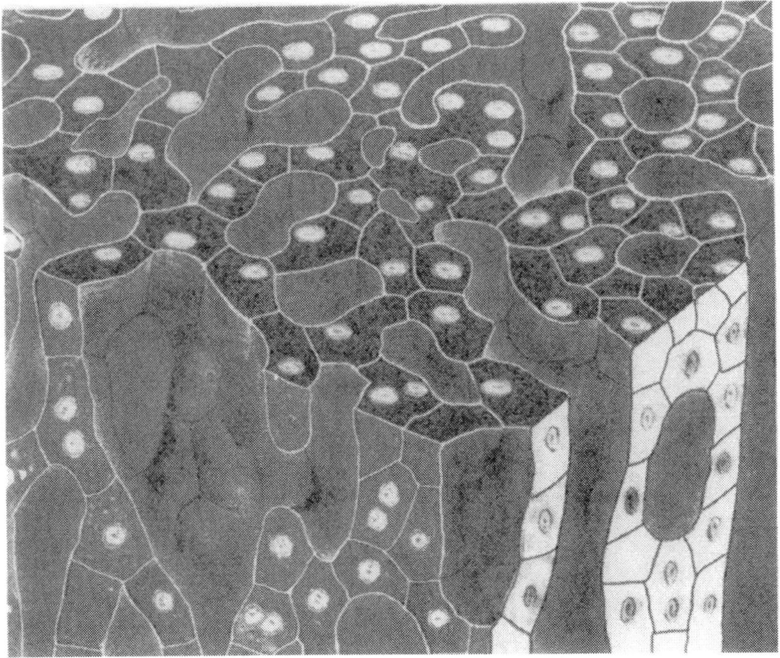

Fig. 1 Anatomical drawing of a portion of liver parenchyma depicting the complete trabecular arrangement of hepatocyte chords.

These subdivisions of the lobule correspond roughly to the periportal, centrilobular, and midzonal regions, respectively.

B. Liver Cell Types

1. Hepatic Parenchymal Cells

Hepatic parenchymal cells, or hepatocytes, are endodermal in origin, arising from outgrowths of the hepatic diverticulum early in neonatal development. In humans, hepatic α-fetoprotein synthesis is evident as early as 25 days after conception, with bile acid synthesis and secretion established by the end of the third month in utero. However, canalicular transport and hepatic excretory function is not fully developed until after birth (3,4), when the system becomes functional (i.e., glucuronidation).

In the adult liver, hepatic parenchymal cells are polyhedral or spherical, are between 28 and 35 μm in diameter, and are approximately 4500–5500 μm^3 in volume. Hepatocytes account for 60% of the cells in the liver and represent 80% or more of the total volume. The human liver contains about 27×10^9 hepatocytes, and the rat liver contains approximately 100 times fewer hepatocytes in absolute numbers. A schematic view of the hepatocyte, which can be compared with an electron microscopic view is shown in Fig. 4.

There are two primary views on the life cycle for the renewal of hepatocytes, which have a lifespan of approximately 200 days in the rat. The cell-streaming view holds that hepatocytes are produced from stem cells in periportal regions and move as continuous sheets of cells toward the region of the central vein, where the cells undergo apoptosis (5,6). An alternative approach is that all hepatocytes are capable of cell division, and

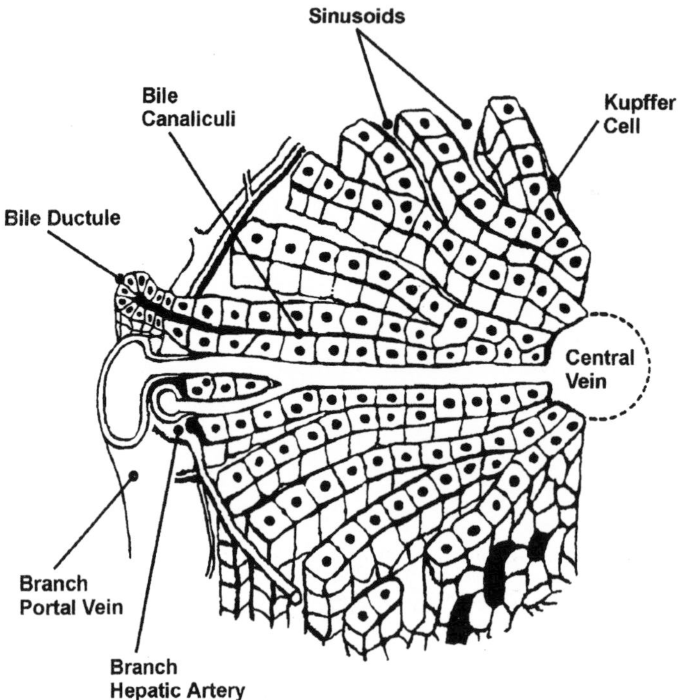

Fig. 2 Diagram of the anatomical structure of the liver lobule: The chords of hepatocytes radiate toward the central vein, which receives sinusoidal blood flowing from the hepatic artery and the portal vein in the periportal area, thus establishing an oxygenation gradient. The flow of bile is toward the portal space, which contains the interlobular collecting bile ductules.

newly formed cells stay in place with populations of cells expanding by clonal expansion in response to various stimuli (7–10).

Hepatocyte cords are bordered by sinusoidal spaces through which portal and hepatic arteriolar blood percolates on its journey to the central vein. The sinusoidal space is lined by a layer of endothelial cells, called sinusoidal cells, which along with the space of Disse, located between the hepatocytes and the sinusoidal cells, serve as a barrier between the sinusoidal space and the hepatocyte. The bile canaliculi provide the route for formed bile to return to the bile ductule, and eventually to the common bile duct. The canaliculi, as their name implies, are not walled vessels, but represent a canal formed by the junctions of adjacent hepatocytes.

A discussion of the many functions of the hepatocyte are beyond the scope of this review, but some can be summarized in general terms. Portal circulation exposes the liver to various dietary constituents and, hence, is a primary site of protein synthesis, carbohydrate and lipid metabolism, and the primary site for phase I and phase II reactions in xenobiotic metabolism.

2. Sinusoidal Endothelial Cells

The sinusoids range in diameter from 10 to 25 μm. They are lined with a continuous sheet of endothelial cells that is punctuated with numerous fenestrae. Little if any basement

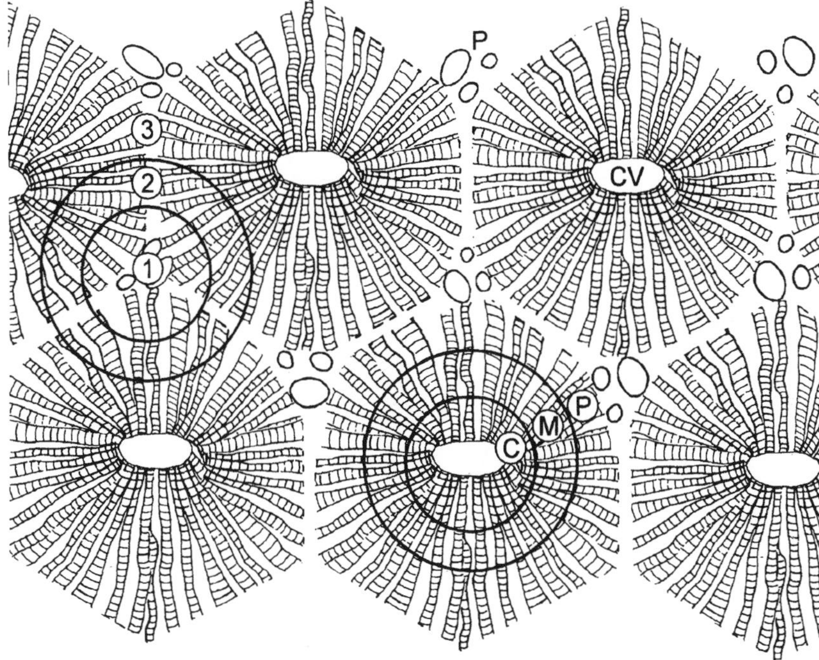

Fig. 3 Diagram of the lobular architecture of the liver: The cords of hepatocytes radiate away from the central vein (CV) to the portal space (P) which contains a branch of the portal vein and hepatic artery and a bile ductule (circles). The concentric circles around the central vein depict the classic lobular architecture, with centilobular (C), midzonal (M), and periportal (P) areas. The corona centered on the portal space represents the vascular zonation according to Rapaport (Zones 1, 2, and 3), with the outermost irrigation toward the central vein.

membrane backs this sheet so the fenestrae provide a direct conduit from the sinusoid to the space of Disse which lies immediately adjacent to the hepatocytes. The fenestrae are approximately 0.1–0.2 μm in diameter and exhibit lobular gradients with increasing pore diameter and pore number from periportal to centrilobular regions. The fenestrae play an important role in the filtration of lipoproteins by the liver, with the pore diameter blocking passage of chylomicrons while allowing passage of smaller chylomicron remnants (11,12). Additionally, the diameter of fenestrae may respond to either endogenous neurohumoral factors or to toxicants (13,14). Indeed, modification of pore diameter by toxicants may play a role in the etiology of fatty liver by altering the lipoprotein-filtering capacity of the fenestrae (15). Alcohol reduces the number and size of fenestrae, together with the development of a subendothelial basal lamina with collagen deposition in the space of Disse (16). The sinusoidal cells themselves exhibit marked endocytotic activity leading to significant lysosomal degradation of taken-up materials, including proteins and lipoproteins (17). The sinusoidal cells are active in the synthesis and release of prostaglandins, endothelin, and various cytokines (18). Sinusoidal endothelial disruption is an important factor in the toxicity of compounds, such as acetaminophen (paracetamol) and carbon tetrachloride (19,20).

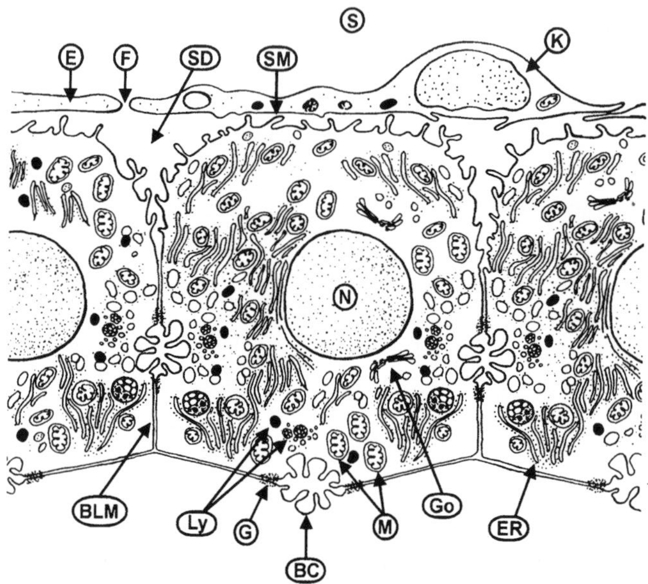

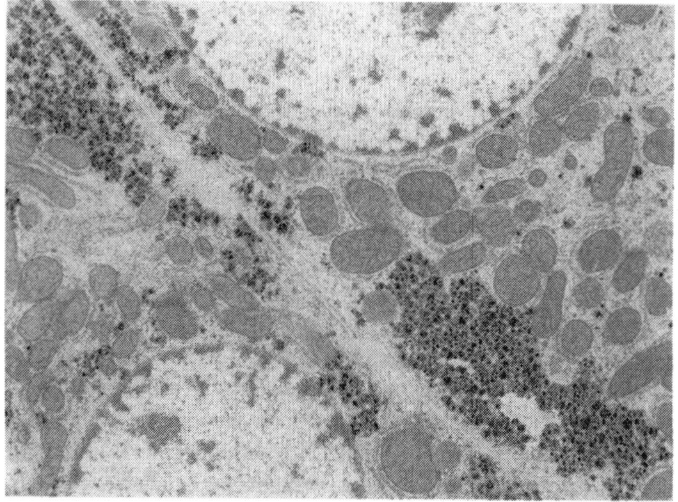

Fig. 4 Idealized schematic of the trabecular cord of hepatocytes: An electron microscopic image of a normal liver cell cytoplasm is shown for interpretative purposes. E, endothelial cell; F, foramen or fenestra; SD, space of Disse between endothelium and hepatocyte; SM, sinusoidal liver; K, Kupffer cell; N, hepatocyte nucleus; BLM, basolateral membrane; Ly, lysosomes; BC, bile canaliculus membrane; G, gap junctions; M, mitochondria; Go, Golgi apparatus; ER, endoplasm reticulum; and speckles represent free ribosome and glycogen particles.

3. Kupffer Cells

Kupffer cells are hepatic resident macrophages attached to the luminal surface of sinusoidal endothelial cells, or they may lie within the endothelial cell layer itself. Kupffer cells represent the largest population of fixed macrophages in the human as well as most other vertebrates (21). The origin of Kupffer cells is uncertain, but they may derive from monocytic precursor cells in the bone marrow (22,23), or they may represent a replicating population of fixed histiocytes (24). Kupffer cells exhibit many of the functions of circulating macrophages and are active in endocytotic removal of particulate, infective, and toxic substance from portal blood (25). Their location in the trabecular structure makes them ideally suited for removal of bacteria and bacterial endotoxins from blood from the intestine (26). The importance of this function is amply demonstrated by the marked proliferation of Kupffer cells in animals experimentally exposed to endotoxin or other inflammatory mediators (27–29). Kupffer cells also have significant secretory functions and are an important source of biologically active mediators, including eicosanoids, cytokines, proteases, tumor necrosis factor (TNF), reactive oxygen species, and nitric oxide (30). The secretory actions of Kupffer cells may play an important role in the regulation of hepatocyte function. The release of cytokines can lead to hepatocyte production of acute-phase reactants (31). Overproduction of these mediators, particularly free radicals and proteases, can lead to hepatocyte dysfunction and necrosis, as observed in severe sepsis (32). Vitamin A may exacerbate carbon tetrachloride toxicity by augmenting the release of active oxygen species from Kupffer cells (33). Hepatocytes may also affect Kupffer cell function, such as the production of prostaglandin E_2 (34,35). The regulatory role of hepatocytes on Kupffer cells can be detrimental. Acetaminophen acts on Kupffer cells by interaction with the hepatocytes, producing a release of active mediators, including superoxide, from the Kupffer cells, leading to hepatocyte disruption (36).

4. *Ito Cells*

The Ito cell is also known as the fat-storing cell, or the stellate cell. Ito cells are located in the space of Disse and were originally thought to be a type of Kupffer cell. In 1952, Ito and Nemoto recognized these stellate-shaped cells as a separate cell population (37) that they believed actively phagocytosed fat. It was later discovered that the cells do not actually take up fat, but store fat synthesized from glycogen (38). Although representing only 5–8% of liver cells (39), Ito cells are regularly dispersed along the hepatocyte chords, and their distribution may be sufficient to permit interactions with the entire hepatic sinusoidal network (40). Extended cytoplasmic processes from Ito cells give the cells their stellate appearance. These processes contact multiple hepatocytes, and a single Ito cell may provide connecting processes to more than one neighboring sinusoid (41). Ito cells express smooth muscle α-actin, suggesting that they are contractile, possibly playing a role in regulation of blood flow through the sinusoid (42). Desmin, an intermediate filament protein, has been proposed as an in vivo marker for Ito cells in rats (43). Ito cells are a primary storage site for vitamin A, storing up to 300 times more retinyl ester per milligram of protein than found in hepatocytes (44). Abundant intracytoplasmic vitamin A droplets are distinguishing morphological features of these cells. The morphology of Ito cells exhibits lobular heterogeneity, with midzonal Ito cells exhibiting the greatest concentration of vitamin A droplets, whereas periportal Ito cells are smaller than in other zones, and centrilobular Ito cells display longer processes and very little vitamin A (38).

In vivo, Ito cells can be activated by hepatotoxic compounds, such as carbon tetrachloride (38), and the cells appear to play a key role in hepatic fibrosis (45–47). Ito cells proliferate in and around areas of acute focal hepatocyte injury, and their normal physiological role is to aid in tissue repair (41). However, chronic liver insult may lead to phenotypic and morphological modulation of Ito cells. The cells change progressively in shape from quiescent compact cells, through a spread transitional stage, with myofilaments and receding vitamin A droplets, eventually resembling a myofibroblast. These phenotypic myoblast or fibroblast spindled cells have pronounced myofilaments and lack vitamin A droplets (40). The fibroblast-like Ito cells are characteristic in experimental animal models of cirrhosis, as well as in human disease (15,48), and are believed to play a key role in alcoholic fibrosis and cirrhosis (49,50).

III. BLOOD CIRCULATION, PATHWAY OF BILE

A. Hepatic Vasculature

The liver receives about 25% of the total cardiac output, representing only 2.5% of the body weight, making liver parenchymal cells the most richly perfused cells in the body (51). Portal blood supplies about two-thirds of the total blood flow to the organ and arterial blood flow accounts for the remainder. The liver does not directly control hepatic portal blood flow; therefore, its intrinsic or extrinsic control occurs through the hepatic artery. There is some disagreement about how homogeneous (portal vs. arterial) blood flow is across the sinusoids. Exquisite control of sinusoidal circulation appears to exist (52). However, blood flow within the normal liver remains remarkably homogeneous, and compounds entering the liver either through the portal vein or hepatic artery are equally distributed (53,54).

B. Microcirculation

Within the liver, the portal vein subdivides into smaller and smaller branches, eventually ending in terminal portal venules that are approximately 20–40 μm in diameter and open into the hepatic sinusoids. Portal blood flow across the sinusoids can be controlled by neurohumoral factors, such as norepinephrine, angiotensin, or histamine, which can activate hepatic venous sphincters at either end of the hepatic sinusoids (51,55). These anatomical structures are more pronounced in dogs. After percolating through the sinusoids, the blood flows into the hepatic venule and out into systemic circulation. Arterial blood is supplied to the liver through the hepatic artery. Hepatic arteries terminate in either the periportal plexus, which distributes around the branches of the portal vein; the peribiliary plexus, which supplies blood to the bile ducts; or into terminal hepatic arterioles. All three branches drain primarily into the hepatic sinusoids.

C. The Biliary System

Bile is a complex, dense viscous fluid, with both organic and inorganic components, including bile acids, phospholipids, cholesterol, glutathione, proteins, metals, and ions (56). Bile serves two primary purposes. First, bile aids in the digestion and absorption of lipids from the intestine. Second, the bile serves as a major route of elimination of many endogenous products as well as various xenobiotics and metabolites. Bile salts are the major constituent of bile, present at a concentration from 2 to 45 mM (57). The amphiphilic nature of bile salts permits the formation of micelles with lipid components, which allows a much greater concentration of the latter in bile than would be anticipated based on aqueous solubility. Bile salts are critical to bile flow because their presence in the fluid creates an osmotic gradient pulling water into the bile, thereby creating bile flow. Bile salts are synthesized in the liver from cholesterol and, subsequently, secreted into the bile.

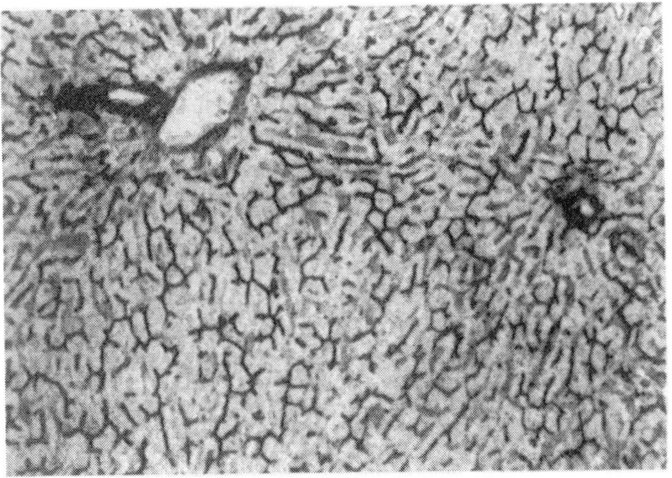

Fig. 5 Histochemical reaction for ATPase in a frozen liver section demonstrating the network of biliary canals in the liver lobule with confluence toward the portal space (top left).

Intestinal reabsorption of bile salts leads back to systemic circulation. Systemic bile salts are rapidly and actively taken up by hepatocytes. In humans, approximately 450 mL of canalicular bile are produced each day (58). Whether synthesized or imported, bile salts exit the hepatocyte at the canalicular membrane. The surface of the hepatocyte membrane forming the bile canaliculi, is covered with microvillar projections. The pericanalicular region of the hepatocyte is largely free of cytoplasmic organelles, but rich in ATP and microfilaments (59). The bile canaliculi can be visualized using a histochemical stain for ATPase (Fig. 5). The tight junctions between hepatocytes were once thought to be impermeable but, subsequently, have been demonstrated to provide a route, known as paracellular diffusion, by which substances may pass from the sinusoids to the bile without going through hepatocytes. Water, neutrally charged organic molecules, and some ions may enter the bile canaliculi by this route (60,61).

The canaliculi themselves form a converging network of anastomotic connections. The ductal system initiates at the canals of Hering, which have a basement membrane and are lined by a combination of ductal cells and hepatocytes (62). These ductules anastomose into larger ducts, eventually leading to the common bile duct. Far from being a passive conduit out of the liver, the biliary epithelial cells modify bile and contribute significantly to bile flow (63,64). Additionally, bile acids may be resorbed from the bile duct and recirculate through the liver by way of the peribiliary plexus (65,66).

IV. LOBULAR GRADIENTS

Considerable heterogeneity exists within the context of the liver lobule or acinus. Metabolic and oxygenation zonations are evident across the lobule in a fashion similar to morphological and phenotypic variations found among cell types within different zones. The metabolic differences can contribute to these gradients, based on the intrinsic requirements for excretion of polar metabolites.

A. Oxygen Gradients

Perhaps the most obvious example of lobular heterogeneity is for oxygen and nutrients along the lateral dimension of the lobule. Cells in the periportal region (or Zone 1) are exposed to blood high in oxygen and nutrient content, whereas cells at the end of the trabecular cords (centrilobular or Zone 3) have to adapt to lower blood oxygen tension and nutrient supply left from the pass over the periportal cells. Periportal oxygen concentrations range from 9 to 13% which diminish to 4–5% by the time blood reaches the central vein (67).

B. Metabolic Gradients

Although the reasons for regional differences in oxygen tension are obvious, less apparent are the zonal differences in cellular enzyme distribution. Significant lobular heterogeneity exists in activities of enzymes involved in carbohydrate, ammonia, lipid, and xenobiotic metabolism (68–71). In general, in the rat, zonation is described by a periportal prevalence of enzymes involved in gluconeogenesis, fatty acid oxidation, urea formation, and albumin synthesis, whereas glycolysis, fatty acid synthesis, glutamine formation, and xenobiotic metabolism are most prevalent in the centrilobular region (72). In humans, a similar zonation appears to exist, although differences in zonal distribution of lipid synthesis, glutamine formation, and ketone formation have been reported (73–75). The reasons for such zonation are unclear, but are likely to involve regional differences in substrate exposure and oxygen tension. Lobular variations in hepatocyte transport of drugs and bile formation are also evident (76). The observed differences are at least partially due to the positioning of cells along the sinusoid, with decreasing oxygen gradients and substrate (drug) concentrations as blood moves from periportal regions toward the central vein. Real differences in transport capacity based on cellular localization have been demonstrated.

Intralobular variations of the cytochrome P450s (CYP) have been well documented, with centrilobular hepatocytes having a greater concentration of the hemoproteins (68). Correspondingly, centrilobular hepatocytes have more smooth endoplasmic reticulum (SER) than do periportal hepatocytes. However, heterogeneity varies according to CYP isozyme. In rats, CYP2E and CYP3A are more prominent in centrilobular areas, whereas CYP1A and CYP2B are equally distributed across the lobule (77,78). CYP distribution across the lobule also changes with age (79). In addition, sensitivity of isozymes to inducing agents also appears to vary across the lobule, in some cases normalizing isozyme activity across the lobule (78,80).

C. Phenotypic and Morphologic Heterogeneity of Hepatocytes

On morphologic examination, either by light or electron microscopy, hepatocytes appear surprisingly homogeneous. Routine stains reveal well-ordered arrays of trabecular cords of hepatocytes radiating from the periportal to the central area. The administration of foreign compounds results in a striking, particular disposition of lesions within the liver lobule. For example, carbon tetrachloride administration produces centrilobular damage as a result of the reductive metabolism by hepatocytes in this area. The early phases of the process of enzyme induction by barbiturates starts in the pericentral hepatocytes, eventually extending to the mid- and periportal zones (81,82). The production of albumin can be a good indicator of protein synthesis by which to follow this induction process. The production of albumin in the rat is about 0.4 mg h^{-1} g^{-1}. Michaelson (83), using immunofluorescent tagging, demonstrated that albumin production is confined to about 1/10 of 1% of all hepatocytes. It is obvious that this albumin secretion must be programmed in hepatocytes that appear to be at random within the liver lobule, whereas other hepatocytes secrete proteins of diverse molecular weights. This functional diversity was demonstrated in the increase of fibrinogen production by parenchymal cells following turpentine injection (84). The increased production of fibrinogen is based on the clonal expansion of fibrinogen-producing hepatocytes after mitotic division. This individualized secretion of serum proteins from hepatocytes is also noticeable during neonatal development. The number of liver cells producing α-fetoprotein decreases with age, whereas albumin, fibrinogen, and other proteins increase (83).

V. SUBCELLULAR ORGANIZATION OF THE HEPATOCYTE: BASIS FOR ORGANELLE PATHOLOGY

According to the level of microscopic visualization, the liver reveals different orders of structural arrangement. For example, the study of liver tissue by routine transmitted microscopy within the visible spectrum provides a survey of well-arranged trabecular parenchymal cells, with a repeating pattern with apparent uniformity. In contrast, the electron microscope reveals a seemingly heterogeneous population of organelles in a diverse network of membrane-bound sacks, vesicles, and microtubules. A conventional view of the hepatocyte by transmission electron microscopy is shown in Fig. 4. The resemblance between hepatocytes from different species is uncanny, although it is recognized that similar organelles do not have similar functionality on a comparative basis. Recent knowledge, gained with embedment-free electron microscopy, suggests unique aspects of the cytoskeleton structure (85). Previously, an approximation of the cytoplasmic infrastructure was described after using high-voltage electron microscopy of unembedded cells (86). Such cytoskeletal network reveals a very important vehicle for organelle interaction and intracytoplasmic motion.

Work at Subramani's laboratory (87) showed that peroxisomes possess bimodal kinetic properties within the hepatocytes. In one mode, peroxisomes display a slow, brownian movement, and in a second or fast mode, the subcellular particles traverse the cytoplasm. These studies were achieved by means of fluorescence-specific analysis combined with real-time kinetic imaging and high-resolution analysis of organelle translations. With use of agents that modulate the stability of microfilaments, it was shown that peroxisomes are closely associated with the microfibrillar network. These important studies revealed unsuspected properties that may lead to further functional investigations contributing to a largely unexplored area of cell biology. These advances were made feasible by the application of converging technologies employing high-resolution fluorescence microscopy and molecular biology.

A. The Plasma Membrane: The Blood–Cytoplasm Interface

The plasma membrane envelopes the hepatic parenchymal cells and constitutes the interface with the surrounding microenvironment (Fig. 6). A specialized structure of the plasma

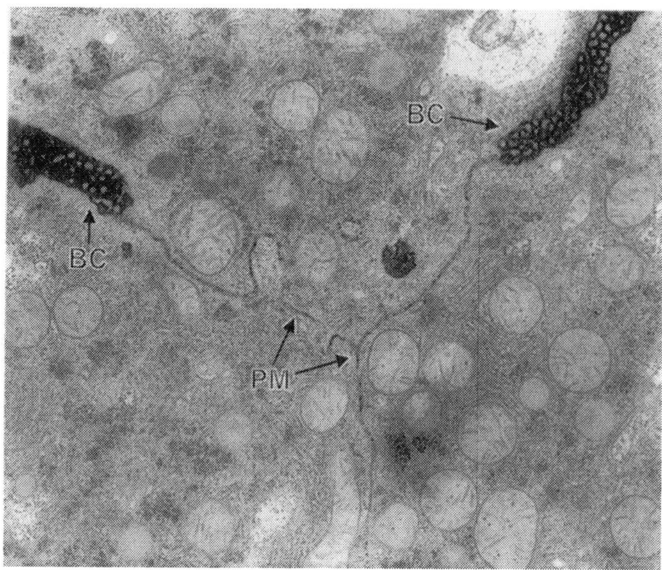

Fig. 6 Electron microscopy cytochemical reaction of ATPase in the plasma membrane (PM) of three contiguous hepatocytes: The ATPase reaction products are evidenced by dark granular deposits of lead phosphate. The light areas surrounded by the reaction product are the microvilli that project into the liver of the bile canaliculus (BC).

membrane can be anticipated based on different functions; specific areas are dedicated to absorption, secretion, and excretion. In a single liver cell, part of the hepatocyte plasma membrane faces the sinusoid, adjoining areas share surfaces with other hepatocytes, and other areas form the biliary canaliculus. Structurally, the bilayer membrane facing each opposing hepatocyte contributes one-half of all specialized structures, such as desmosomes and tight junctions. Each individual hepatocyte has an estimated surface area of 1650–1950 μm^2, whether the idealized stereological model is based on a dodecahedron or a sphere, respectively. The applicability of these models in tissue sections is based on the closest statistical fit and on the correspondence with different measures that contribute to the final shape factor (88). The surface membrane of the hepatocyte contributes most to the sinusoidal interface of 70% in the space of Disse or the space between the hepatocyte and the inner surface of the Kupffer or sinusoid endothelial cells. The remaining membrane is approximately divided between the portion contributing to the bile canalicular tree and to the hepatocyte–hepatocyte interface or basolateral membrane. The membrane exposed to the sinusoid provides specialized transporters to move solutes and nutrients from the plasma into the cell cytoplasm. If a substance is actively transported, it can traverse various membrane interfaces, ranging from the sinusoidal aspect of the endothelial cell, to the hepatocyte facing the space of Disse. Hypothetical avenues of solute transport are schematized in Fig. 7. This absorption process is also supported by active endocytic activity, which occasionally includes receptor internalization. The microvilli that characterize this aspect of the membrane contribute to a significant increase of the extent of surface exposed to the plasma. In contrast, the biliary aspect of the plasma membrane directs excretion products, such as metabolites, bile salts, phospholipids, and cholesterol, across the membrane to a highly concentrated, surface-active environment within the lumen of the bile canaliculus. In turn, the basolateral membrane, because of the tight junctions, provides a

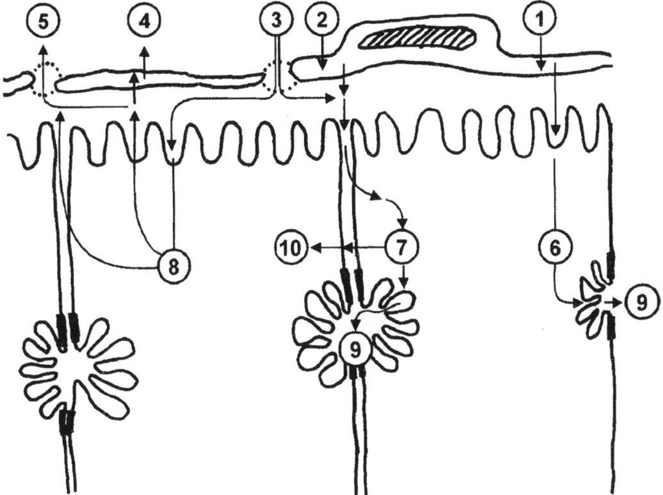

Fig. 7 Schematic drawing to indicate the different pathways (circled numbers) a substance may traverse from the sinusoid into the hepatocyte during absorption (6,7): Compounds may reach liver cells either by passage through the foramen (3) of endothelial and Kupffer cells or through the cell cytoplasm (1,2). Different cytoplasmic interactions may influence the route of excretion (4,5,8). Substances may also be excreted into the bile canaliculus (9) or the basolateral membrane into neighboring cells (10).

dynamic area where contractile microfilaments pulsate the cell membrane, thereby contributing to downstream biliary flow. The biliary membrane also contains digital projections that may enhance the extent of surface area available for excretory functions. In addition, ciliary elements have been found, although their function or purpose is not clearly known. Because of their paucity, they were attributed sensory functions and sometimes a streaming role in biliary flow.

The identification of constitutive elements of the biliary membrane has progressed significantly (89), and specialized canalicular P-glycoproteins, recognized to have a role in multidrug resistance, were located. These P-glycoproteins are ATP-dependent transport proteins that pump a variety of substances with significantly heterogeneous anionic strength. The mdr-2 P-glycoprotein is a phospholipid flippase, with a key role in the elimination of lipids into bile (90). Models of biliary excretion have been evaluated by Oude Elferink (91) in homozygous variants of the *mdr-2* gene in mice. This approach will contribute to the understanding of the physiological basis for bile synthesis and transport and other important functions that modulate drug metabolism and excretion.

B. The Endoplasmic Reticulum: A Membrane-Based Reactor

The ER is the most critical organelle in the hepatocyte for metabolism of drugs. It consists of a labyrinth of tubules and flattened sacs that extend throughout the cytoplasm. The membrane is a continuous sheet that surrounds a common interluminal space called the cisternal space. The ER is the primary location for protein synthesis within the hepatocyte and is responsible for the synthesis of all transmembrane proteins. It is also the primary site of lipid and lipoprotein synthesis and metabolism within the hepatocyte, as well as representing the primary site for xenobiotic metabolism. The ER consists of two distinct membrane populations; the rough ER (RER), which is characterized by a ribosome-studded cytosolic surface, and the smooth ER (SER), which lacks these ribosomes (Fig. 8). Within the hepatocyte, the SER occupies approximately 40% of the total ER membrane; however, this percentage is less in other cell types. The cytosolic surface of the ER is contiguous with the outer nuclear membrane. The ER is the largest single organelle component within the hepatocyte (92) comprising over 24,000 m^2 of surface area in the adult human liver and approximately 160–200 m^2 of surface area in the adult rat liver (93). Quantitative morphometric investigations of ER synthesis suggest that ER membranes increases at a rate of 17 $cm^2\ h^{-1}$ from the postnatal period to maturity, at which time, barring physiological or exogenous stimuli, the membrane content is maintained at a relatively constant level (94). Turnover of the ER membrane is rapid, with rates ranging from 20 to 50 $cm^2\ h^{-1}$ for membrane lipids and from 40 to 140 $cm^2\ h^{-1}$ for membrane proteins (95). Cytochrome P450 inducers can increase the rate of membrane synthesis up to 320 $cm^2\ h^{-1}$, whereas removal of the stimuli for induction can cause an acceleration of membrane elimination, up to 330 $cm^2\ h^{-1}$, until normal physiological membrane content is achieved (96,97). Within the hepatocyte the ER content is closely regulated, and the capacity for the liver to generate membrane appears to decline with age (98). The CYP enzymes are located preferentially in SER, which may explain why, as compared with other cell types, hepatocytes have a higher percentage of cellular membranes in the SER configuration. Smooth ER membranes proliferate rapidly in response to inducing agents. This phenomenon is due to a significant increase in membrane synthesis and not to a decrease in membrane turnover (96). Membrane proliferation in response to induction is a unique

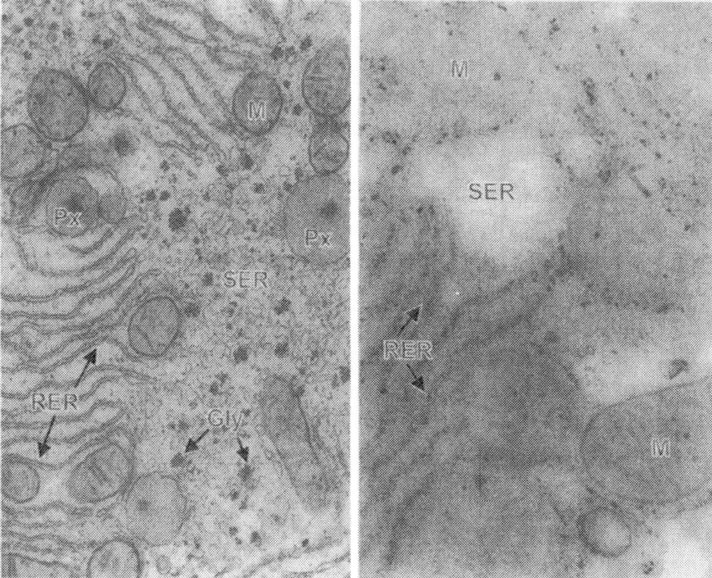

Fig. 8 Appearance of the cytoplasm of a liver cell under conventional electron microscopy with lead and uranium contrast stain at left. At right, electron microscopy cytochemistry for inosine-diphosphatase, revealed by the dark reaction product in the RER from an unstained liver cell. M, mitochondria; RER, rough endoplasmic reticulum; Px, peroxisome; Gly, glycogen aggregate; and SER, smooth endoplasm reticulum.

phenomenon that may be associated with adaptive synthesis of drug-metabolizing enzymes, but may occur with no concurrent increase in enzyme activity, the so-called hyperplastic hypofunctional ER (94,99). Given the role of the ER in metabolism, it seems logical that agents disrupting it can alter metabolism or possibly the response of the ER to an inducing agent. Phospholipids are key components of the ER membrane. Systemic phospholipidosis can cause morphological and functional changes within the membrane and, consequently, alterations in microsomal enzymatic activity (100,101).

C. Mitochondria: Energy Engines (and More)

Other than the nucleus, the mitochondria are the most prominent morphological feature of the normal hepatocyte (Fig. 9). There are approximately 1000 mitochondria per cell; they occupy approximately 20% of the cell volume, or approximately 1000–1400 μm^3 per hepatocyte and they have a half-life of approximately 10 days. The mitochondrial volume fraction within hepatocytes appears to be consistent across several species, including humans (88,94,102,103). However, lobular heterogeneity in the size and distribution of mitochondria exists. In rats, mitochondria are smaller and more numerous in centrilobular regions (104), although the total mitochondrial volume fraction is similar across the lobule (105). The volume of an average rat hepatic mitochondrion is approximately 0.27 μm^3, with an inner membrane surface area of approximately 6.5 μm^2 (106). The mitochondrial envelope consists of an inner and outer membrane that encases the intermembrane space. The inner membrane is characterized by numerous folds, called cristae, that greatly increase the available surface area. In the hepatocyte, approximately two-thirds of the mitochondrial protein is located in the mitochondrial matrix, whereas 20% is located in the inner membrane, with the remainder found in the intermembranous space. In contrast with

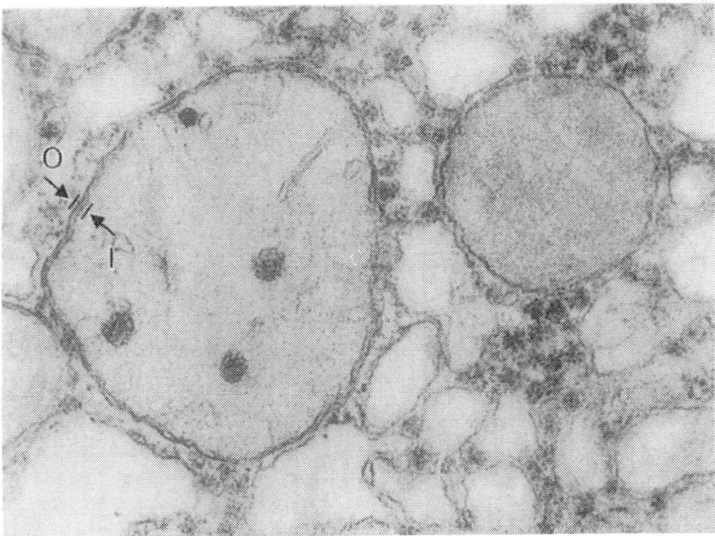

Fig. 9 High magnification of the cytoplasm from a human hepatocyte showing a mitochondrion (left) with dark granules in the matrix and the double membrane (I, inner; and O, outer). The small, homogeneous and round single-membrane body (upper right quadrant) is a typical peroxisome. Note that human peroxisomes do not normally contain a core or nucleoid.

the static electron micrographic depictions of mitochondria, the organelles are actually quite mobile within the hepatocyte and can fuse, split, and rapidly change their shape. Although they are usually thought of as independent organelles, it has been suggested that, in the hepatocyte most, if not all, mitochondria may be associated with the RER, with the ER serving as a framework to cluster similarly sized mitochondria. Furthermore, these clusters of mitochondria appear to exhibit functional heterogeneity (107). When considering the morphology of the organelle, a cautionary note should be made. Some of the information on hepatic mitochondrial morphology obtained by electron microscopic techniques may be misleading, owing to the dramatic changes caused by tissue preparation, fixation, and embedding techniques (108). Hepatic mitochondrial morphological alteration is a frequent sequelae of hepatotoxicity. Chronic alcoholism, drugs that interfere with copper metabolism, ethionine, orotic acid, hypolipidemic agents, and cortisone, all produce enlarged mitochondria (109). Even though hypolipidemic agents produce large mitochondria, the process results in correspondingly fewer mitochondria (110), suggesting that fusion of preexisting mitochondria may lead to the appearance of the enlarged organelle. Additionally, mitochondrial dysfunction has been mechanistically implicated in the hepatic toxicity of the pesticide endosulfan (111), allyl alcohol (112), the bidentate phosphine gold antineoplastic agent SKF 104524 (113), modaline sulfate (unpublished observations; Fig. 10), acetaminophen (114), and the Alzheimer's drug tacrine (115). The mitochondrial impairment induced by acetaminophen is related to the compound's propensity to covalently bind aldehyde dehydrogenase (116).

The primary function of mitochondria is to convert energy from carbohydrates and fats into usable forms, primarily ATP. Insight into mitochondrial function can be gleaned from the application of transport-dependent fluorophores. Rhodamine 123 is an example

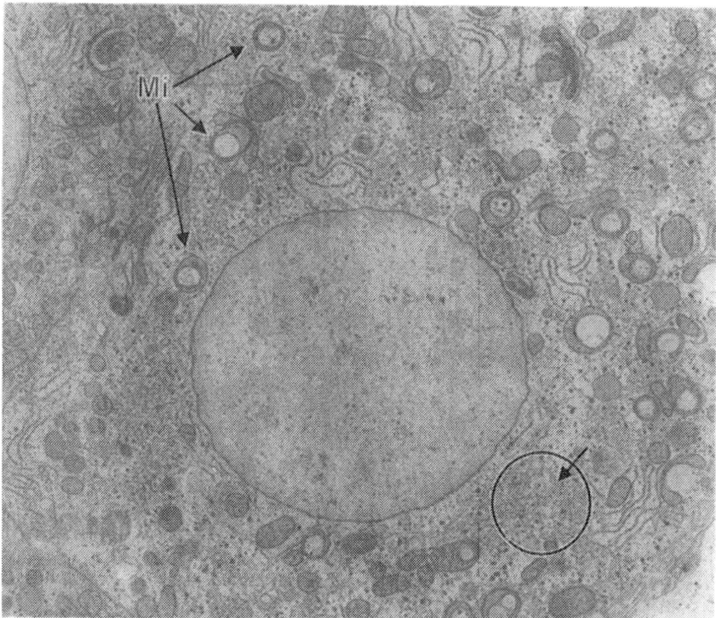

Fig. 10 Electron microscopy of a liver cell from a rat treated with a mitochondrial poison: The altered mitochondria (Mi) are the signet-ring structures engulfing fat droplets or other cytoplasm structures. The thick arrow with a circle indicates areas of smooth endoplasm reticulum proliferation.

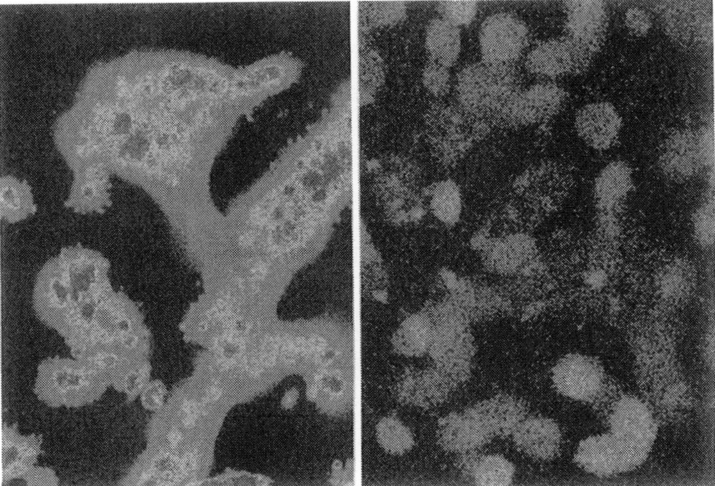

Fig. 11 Confocal fluorescence image of isolated hepatocytes in culture, showing rhodamine 123 uptake into mitochondria (left) and FITC-dextran into lysosomes (right). This technique allows the quantitative evaluation of hepatocyte uptake under real-time conditions.

of such a compound that has been used to estimate mitochondrial functional integrity (117; Fig. 11). Enzymes involved in the β-oxidation of fatty acids and those of the citric acid and urea cycles are found in the mitochondrial matrix, whereas proteins associated with electron transport and oxidative phosphorylation are primarily associated with the inner membrane. The mitochondrion contains its own genome with about five to ten copies per organelle. In humans, mitochondrial DNA (mtDNA) is a circular molecule of 16,569 base pairs, which exclusively code for protein components of the oxidative phosphorylation pathway (118), although most protein components of the mitochondrion are encoded in nuclear DNA and imported into the organelle. Mitochondrial DNA lacks many of the protective histones of nuclear DNA and is constantly subjected to exposure to oxygen radicals generated during oxidative phosphorylation. The effects are coupled with an inefficient DNA repair system found in mitochondria that makes them susceptible to a high mutation rate, estimated to be up to ten times higher than the rate for nuclear DNA (119). Mitochondrial mutations play a role in several human disease states, primarily involving the skeletal muscle or nervous system. No reported human liver disease has been reported as a result of mitochondrial mutations. The antiviral nucleoside analogue fialuridine (FIAU) inhibits mitochondrial DNA polymerase-γ (120), and the compound produced clinically severe hepatic toxicity and liver failure (121). The woodchuck has been proposed as a model for the study of FIAU-induced hepatic injury (122). This raises the intriguing possibility that toxicant-induced mitochondrial DNA changes, or a subset of naturally occurring mitochondrial mutations, may make hepatic mitochondria more susceptible to xenobiotic-induced toxicity, a theory that has recently been proposed as a possible explanation for the hepatotoxicity of the acridine derivative, tacrine (115).

D. Lysosomes

A primary role of the lysosome is disposal or elimination of exogenous or endogenous substances by degradation or solubilization. The organelles contain various esterases and

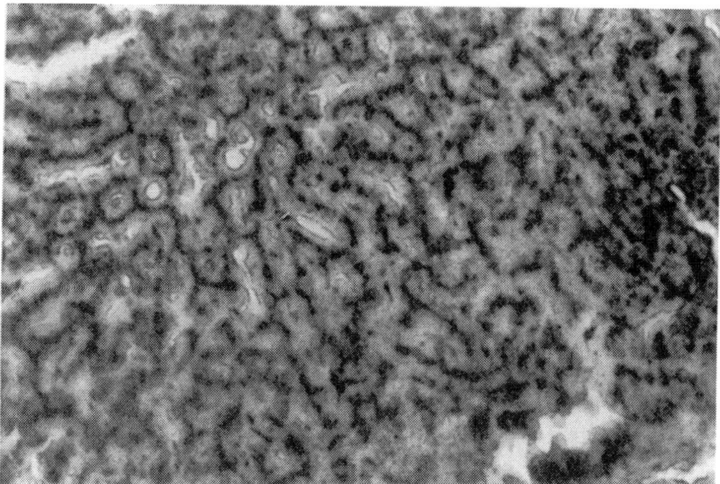

Fig. 12 Histochemical acid phosphatase reaction demonstrating the localization of lysosomes (dark brown granules) in the pericanalicular cytoplasm of hepatocytes.

hydrolases that are active at different pHs (Fig. 12). The average hepatocyte contains approximately 250 lysosomes, which occupy about 1% of the total cell volume and can be traced by the use of fluorophores under confocal microscopy (see Fig. 11). These pleiotrophic organelles can vary substantially in size and shape. Certain physiological and pathological conditions can drastically increase the number of lysosomes within the hepatocytes. Accumulation of lysosomes after exposure to drugs or ethanol may represent an adaptive response to impaired catabolic processes (94). Long-term oral contraceptive administration increases the number of lysosomes in humans (123), and drugs causing phospholipidosis are frequently associated with accumulation of lysosomes resembling myeloid bodies (101). With aging, lysosomes accumulate oxidized fat products resulting in lipofuscin or ceroid deposition (124).

E. Peroxisomes: Evaluation of Fatty Acid Oxidation and Other Enzymes

Peroxisomes constitute a unique subcellular organelle, seen ubiquitously in the cytoplasm of hepatocytes and either in clusters or single units intermeshed in the reticular network (125). In conventional transmission electron micrographs, peroxisomes appear as single, scalloped membrane-bound structures between 0.2 and 0.5 μm, and they are frequently observed containing crystalline nucleiods made up of urate oxidase (Fig. 13). A similar spheroidal shape is derived from the study of pelleted subcellular gradient fraction isolates under the electron microscope (126). These organelles are considered vestigial in terms of evolution because their occurrence is seen in unicellular and multicellular orders, including plants (i.e., glyoxisomes) and higher mammals, with functions genetically modulated according to the level of organization (127). Mature hepatic peroxisomes in mammals contain upward of 40–50 enzymes of intermediary metabolism. The main enzyme component by protein weight is catalase, with preferential β-oxidation of short-chain fatty acids. In higher mammals, peroxisomes may play a role in lipoprotein metabolism (128). Approximately 500–600 peroxisomes are found in the normal hepatocyte, although higher values,

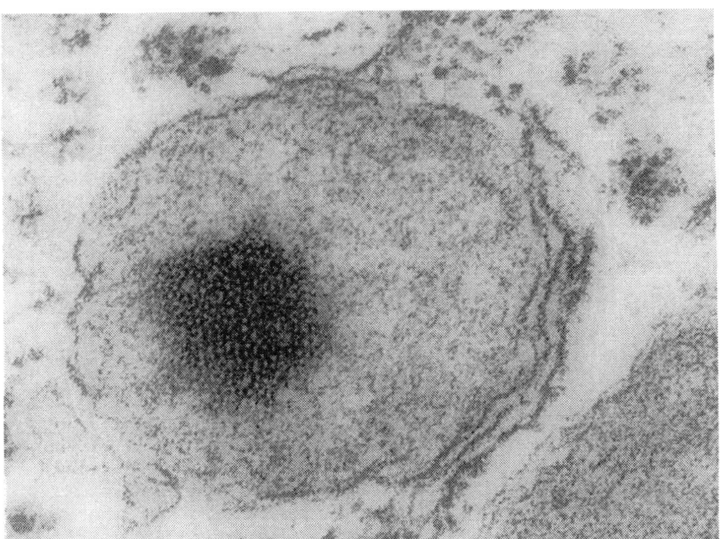

Fig. 13 High-magnification electron microscopy of a typical rodent hepatic peroxisome: The circular single membrane envelopes the proteinaceous matrix and a nucleoid with crystalline arrangement of urate oxidase.

up to 1000, were reported in human liver. The population of peroxisomes is rather stable and seems constant across species with minor differences in lobular disposition. Although there is a normal turnover of peroxisomes, their assembly can proceed very quickly after exposure to certain chemicals, particularly after administration of cholesterol biosynthesis inhibitors (128). Such proliferation is transcriptionally dependent, and several proteins are targeted to the membrane or to the matrix of the peroxisome (129). A certain degree of specificity to the proliferative reaction is appreciable, particularly in rats in which peroxisome enzymes are induced preferentially, whereas urate-oxidase, constitutive of the core, is not. Such preferential enzyme synthesis accounts for changes in ratios of nucleated to nonnucleated peroxisomes. Quantitative morphometry has been used to study the dynamics of peroxisome replication. Proliferation is triggered from a baseline turnover rate of 7.8–18.5 peroxisomes assembled every hour in every one of the 179×10^6 hepatocytes per gram of liver, equivalent to 20–25% of the normal baseline in the cell population. Chemical structure and lipid-regulating activity affect the degree of the response, and once a stimulus is eliminated, the resulting proliferation of peroxisomes quickly recedes at the same rate the excess peroxisomes accumulated. The proliferative response that follows the signal is receptor-mediated (130), and the nuclear receptor involved belongs to the steroid receptor superfamily, recognized as the peroxisome proliferator-activated receptor (PPAR). Different receptor subtypes have been identified, and PPARα is largely responsible for the proliferation after dimerization with the retinoid receptor (RXR). Peroxisome proliferation shows significant species specificity, in parallel with the prevalence of alpha (α), delta (δ), or gamma (γ) subtypes. PPARα is well characterized and mediates a florid response in rodent liver (Fig. 14). PPARα requires dimerization with the RXR receptor for downstream activation and binding, a process that is not yet entirely clear, but a variety of lipid, carbohydrate, and protein metabolic pathways are involved. The nucleotide sequence of the peroxisome proliferator response element (PPRE) appears to have highly

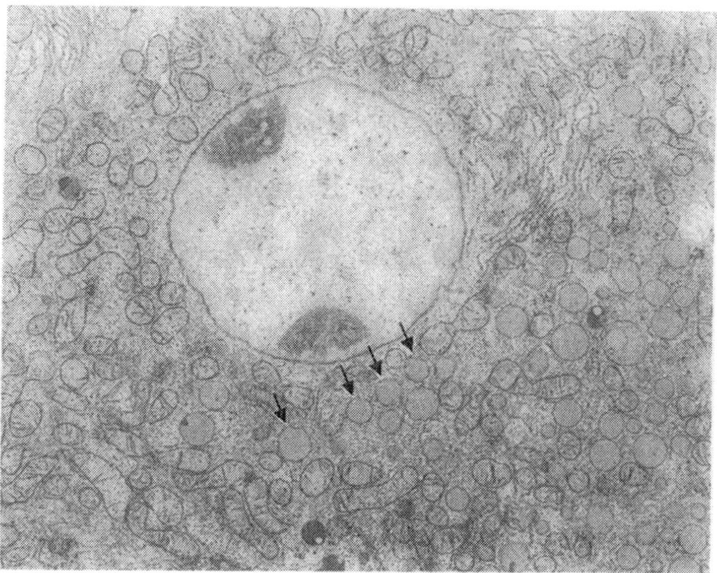

Fig. 14 Liver cell from a rat receiving a fibrate-like hypolipidemic agent: Numerous peroxisomes are observed in the cytoplasm. The arrows point to organelles with homogeneous matrix.

conserved species-specificity and may explain the lack of peroxisome proliferation in humans administered drugs that otherwise cause profound proliferation in laboratory animals. PPARγ has different affinity and localization in organs and tissues, and its activation may play a role in insulin-signaling and carbohydrate metabolism (131). Whereas peroxisome proliferation does not appear to affect humans (94), the lack of one or more peroxisomal functions evokes significant abnormalities in autosomal recessive clinical conditions (129). The most severe forms of peroxisome deficiency constitute the loss of multiple enzymes. Among these diseases is the Zellweger syndrome, neonatal adrenoleukodystrophy, Refsum disease, and hyperpipecolic acidemia. Patients so affected rarely survive to 10 years of age. Other forms of peroxisomal disorders include the lack of one or more enzymes affecting different steps of intermediate metabolism, each with different clinical prognosis. It is not known if peroxisome proliferation induces one or more enzymes; if so, the clinical deficiencies could be eliminated.

VI. CONCLUDING REMARKS

The liver is a central organ for a variety of anabolic and catabolic functions, and as such, it plays a significant role in drug metabolism and toxicity. Central to its polyfacetic functions is the liver architecture, which commands a large role in the physiology of blood clearance. Although hepatocytes appear to represent similarly repeating functional reactors, their diversity is becoming obvious, and it is possible that functions are closely integrated with cell replication. Within liver cells, the different organelles, although having close interrelations, represent functionally discrete compartments. Significant species differences influence their basic metabolic activity. These subcellular compartments reveal unique responses to metabolic or pathological stimuli. Lobular blood flow gradients modulate these responses, and this zonation may relate to nutrient availability or to reactive metabolism. The most important account is the high turnover of structural protein and enzymes contained in these organelles. The relations between hepatocytes and nonparenchymal cells are not yet fully discerned, but reveal a very close metabolic interdependence between different cell populations. This brief functional–anatomical overview may help explain the pharmacodynamic and physiological correlations that significantly influence drug metabolism.

REFERENCES

1. F. Kiernan, The anatomy and physiology of the liver, *Philos. Trans. R. Soc. Lond.* Ser. B: 711–770 (1833).
2. A. M. Rappaport, Z. J. Borowy, W. M. Lougheed, and W. N. Lotto, Subdivision of hexagonal liver lobules into a structural and function unit. Role in hepatic physiology and pathology, *Anat. Rec. 119*: 11–33 (1954).
3. A. M. Dubois, The embryonic liver. *The Liver* (C. L. Rouiller, ed.), Academic Press, New York: 1963, pp. 1–39.
4. F. J. Suchy, J. C. Buscuvalas, and D. A. Novak, Determinants of bile formation during developement: Ontogeny of hepatic bile acid metabolism and transport, *Semin. Liver Dis. 7*: 77–84 (1987).
5. G. Zajicek, R. Oren, and M. Weinreb, Jr., The streaming liver, *Liver 5*: 293–300 (1985).
6. M.-P. Bralet, S. Branchereau, C. Brechot, and N. Ferry, Cell lineage study in the liver using retroviral mediated gene transfer. Evidence against the streaming of hepatocytes in normal liver, *Am. J. Pathol. 144*: 896–905 (1994).
7. N. Arber, G. Zajicek, and I. Ariel, The streaming liver. II. Hepatocyte life history, *Liver 8*: 80–87 (1988).
8. J. W. Grisham, Migration of hepatocytes along hepatic plates and stem cell-fed hepatocyte lineages, *Am. J. Pathol. 144*: 849–854 (1994).
9. S. Sell, Liver stem cells, *Mod. Pathol. 7*: 105–112 (1994).
10. S. Kennedy, S. Rettinger, M. W. Flye, and K. P. Ponder, Experiments in transgenic mice show that hepatocytes are the source for postnatal liver growth and do not stream, *Hepatology 22*: 160–168 (1995).
11. E. Wisse, An electron microscopic study of the fenestrated endothelial lining of rat liver sinusoids, *J. Ultrastruct. Res. 31*: 125–150 (1970).
12. R. Fraser, L. Bowler, and W. Day, The filtration of lipoproteins by the liver, *N.Z. Med. J. 82*: 352 (1975).
13. K. M. Mak and C. S. Lieber, Alterations in endothelial fenestration in liver sinusoids of baboons fed alcohol. A scanning electron microscopic study, *Hepatology 4*: 386–391 (1984).
14. I. M. Arias, The biology of hepatic endothelial fenestrae, *Prog. Liver Dis. 9*: 11–26 (1990).
15. R. Fraser, W. A. Day, and N. S. Fernando, The liver sinusoidal cells. Their role in disorders of the liver, lipoprotein metabolism and atherogenesis, *Pathology 18*: 5–11 (1986).
16. M. H. Witte, P. Borgs, D. L. Way, G. Ramirez, Jr., M. J. Bernas, and C. L. Witte, Alcohol, hepatic sinusoidal microcirculation, and chronic liver disease, *Alcohol 9*: 473–480 (1992).
17. E. Wisse and A. M. De Leeuw, Structural elements determining transport and exchange processes in the liver, *Pharmaceutical, Immunological and Medical Aspects*, 1984, pp. 1–23.
18. H. Reider, K. H. Meyer zum Bushenfelde, and G. Ramadori, Functional spectrum of sinusoidal endothelial liver cells. Filtration, endocytosis, synthetic capacities and intercellular communication, *J. Hepatol. 15*: 237–250 (1992).

19. R. M. Walker, W. J. Racz, and T. F. McElligott, Scanning electron microscopic examination of acetaminophen-induced hepatotoxicity and congestion in mice, *Am. J. Pathol. 113*: 321–330 (1983).
20. C. Rouiller, N. Colombey, and B. Haenni, Modifications of the capillary sinusoids of the liver in acute experimental poisoning, *Rev. Int. Hepatol. 15*: 437–453 (1965).
21. J. M. Singer, L. Adlersberg, E. M. Hoenig, E. Ende, and Y. Tchorsch, Radiolabeled latex particles in the investigation of phagocytosis in vivo: Clearance curves and histological observations. *J. Reticuloendothel. Soc. 6*: 561–589 (1969).
22. R. P. Gale, R. S. Sparkes, and D. W. Golde, Bone marrow origin of hepatic macrophages (Kupffer cells) in humans, *Science 201*: 937–938 (1978).
23. R. van Furth, Monocyte origin of Kupffer cells, *Blood Cells 6*: 87–92 (1980).
24. L. Bouwens, M. Baekeland, and E. Wisse, Importance of local proliferation in the expanding Kupffer cell population of rat liver after zymosan stimulation and partial hepatectomy, *Hepatology 4*: 213–219 (1984).
25. R. S. McCuskey and P. A. McCuskey, Fine structure and function of Kupffer cells, *J. Electron Microsc. Tech. 14*: 237–246 (1990).
26. H. Liehr and M. Grun, Clinical aspects of Kupffer cell failure in liver diseases, *Kupffer Cells and Other Liver Sinusoidal Cells* (E. Wisse and D. L. Knook, eds.), Elsevier/North Holland Biomedical, Amsterdam, 1977, pp. 427–436.
27. L. Bouwens and E. Wisse, Proliferation, kinetics, and fate of monocytes in rat liver during a zymosan-induced inflammation, *J. Leukoc. Biol. 37*: 531–543 (1985).
28. L. Bouwens, D. L. Knook, and E. Wisse, Local proliferation and extrahepatic recruitment of liver macrophages (Kupffer cells) in partial-body irradiated rats, *J. Leukoc. Biol. 39*: 687–975 (1986).
29. A. M. Pilaro and D. L. Laskin, Accumulation of activated mononuclear phagocytes in the liver following lipopolysaccharide treatment of rats. *J. Leukoc. Biol. 40*: 29–41 (1986).
30. P. J. Winwood and M. J. Arthur, Kupffer cells: Their activation and role in animal models of liver injury and human liver disease, *Semin. Liver Dis. 13*: 50–59 (1993).
31. T. R. Billiar and R. D. Curran, Kupffer cell and hepatocyte interactions: A brief overview, *J. Parenter. Enteral. Nutr. 14*: 175s–180s (1990).
32. P. Ghezzi, B. Saccardo, and M. Bianchi, Role of reactive oxygen intermediates in the hepatotoxicity of endotoxin, *Immunopharmacology 12*: 241–244 (1986).
33. A. E. elSisi, D. L. Earnest, and I. G. Sipes, Vitamin A potentiation of carbon tetrachloride hepatotoxicity: Role of liver macrophages and active oxygen species, *Toxicol. Appl. Pharmacol. 119*: 295–301 (1993).
34. T. R. Billiar, T. W. Lysz, R. D. Curran, B. G. Bentz, G. W. Machiedo, and P. L. Simmons, Hepatocyte modulation of Kupffer cell prostaglandin E_2 production in vitro, *J. Leukoc. Biol. 47*: 305–311 (1990).
35. B. G. Harbrecht and T. R. Billiar, The role of nitric oxide in Kupffer cell–hepatocyte interactions [editorial], *Shock 3*: 79–87 (1995).
36. D. L. Laskin, Nonparenchymal cells and hepatotoxicity, *Semin. Liver Dis. 10*: 293–304 (1990).
37. T. Ito and W. Nemoto, Uber die Kupffersche Sternzellen und die Fettspeicherungszellen (fat storing cell) in der Blutkapillarwand der menschlichen Leber, *Okajamas Folia Anat. Jpn. 24*: 243–258 (1952).
38. K. Wake, Perisinusoidal stellate cells (fat-storing cells, interstitial cells, lipocytes), their related structure in and around the liver sinusoids, and vitamin A-storing cells in extrahepatic organs, *Int. Rev. Cytol. 66*: 303–353 (1980).
39. M. P. Giampieri, A. M. Jezequel, and F. Orlandi, The lipocytes in normal human liver. A quantitative study, *Digestion 22*: 165–169 (1981).
40. M. Pinzani, Novel insights into the biology and physiology of the Ito cell, *Pharmacol. Ther. 66*: 387–412 (1995).
41. G. Ramadori, The stellate cell (Ito-cell, fat-storing cell, lipocyte, perisinusoidal cell) of the

liver. New insights into pathophysiology of an intriguing cell, *Virchows Arch. B 61*: 147–158 (1991).

42. T. Ito, T. Itoshima, M. Ukida, et al. Scanning electron microscopy of Ito's fat-storing cells in the rat liver, *Acta Med. Okayama 38*: 1–9 (1984).
43. Y. Yokoi, T. Namihisa, H. Kuroda, et al., Immunocytochemical detection of desmin in fat-storing cells (Ito cells), *Hepatology 4*: 709–714 (1984).
44. H. F. Hendriks, W. A. Verhoofstad, A. Brouwer, A. M. de Leeuw, and D. L. Knook, Perisinusoidal fat-storing cells are the main vitamin A storage sites in rat liver, *Exp. Cell Res. 160*: 138–149 (1985).
45. S. L. Friedman, Cellular sources of collagen and regulation of collagen production in liver, *Semin. Liver Dis. 10*: 20–29 (1990).
46. A. M. Gressner and M. G. Bachem, Cellular sources of noncollagenous matrix proteins: Role of fat-storing cells in fibrogenesis, *Semin. Liver Dis. 10*: 30–46 (1990).
47. J. J. Maher and R. F. McGuire, Extracellular matrix gene expression increases preferentially in rat lipocytes and sinusoidal endothelial cells during hepatic fibrosis in vivo, *J. Clin. Invest. 86*: 1641–1648 (1990).
48. T. W. Lissoos and B. H. Davis, Pathogenesis of hepatic fibrosis and the role of cytokines, *J. Clin. Gastroenterol. 15*: 63–67 (1992).
49. M. Nakano, T. M. Worner, and C. S. Lieber, Perivenular fibrosis in alcoholic liver injury: Ultrastructure and histologic progression, *Gastroenterology 83*: 777–785 (1982).
50. S. W. French, K. Miyamoto, K. Wong, L. Jui, and L. Briere, Role of the Ito cell in liver parenchymal fibrosis in rats fed alcohol and a high fat–low protein diet, *Am. J. Pathol. 132*: 73–85 (1988).
51. W. W. Lautt and C. V. Greenway, Conceptual review of the hepatic vascular bed, *Hepatology 7*: 952–963 (1987).
52. R. S. McCuskey and F. D. Reilly, Hepatic microvasculature: Dynamic structure and its regulation, *Semin. Liver Dis. 13*: 1–12 (1992).
53. C. V. Greenway and G. Oshiro, Intrahepatic distribution of portal and hepatic arterial blood flows in anaesthetized cats and dogs and the effects of portal occlusion, raised venous pressure and histamine, *J. Physiol. Lond. 227*: 473–485 (1972).
54. W. W. Lautt, D. J. Legare, and T. R. Daniels, The comparative effect of administration of substances via the hepatic artery or portal vein on hepatic arterial resistance, liver blood volume and hepatic extraction in cats, *Hepatology 4*: 927–932 (1984).
55. R. S. McCuskey, A dynamic and static study of hepatic arterioles and hepatic sphincters, *Am. J. Anat. 119*: 455–478 (1966).
56. C. D. Klaassen, and J. B. D. Watkins, Mechanisms of bile formation, hepatic uptake, and biliary excretion, *Pharmacol. Rev. 36*: 1–67 (1984).
57. B. Angelin, I. Bjorkhem, K. Einarsson, and S. Ewerth, Hepatic uptake of bile acids in man. Fasting and postprandial concentrations of individual bile acids in portal venous and systemic blood serum, *J. Clin. Invest. 70*: 724–731 (1982).
58. R. C. Strange, Hepatic bile flow, *Physiol. Rev. 64*: 1055–1102 (1984).
59. I. M. Arias, M. Che, Z. Gatmaitan, C. Leveille, T. Nishida, and M. St. Pierre, The biology of the bile canaliculus, *Hepatology 17*: 318–329 (1993).
60. T. J. Layden, E. Elias, and J. L. Boyer, Bile formation in the rat: The role of the paracellular shunt pathway, *J. Clin. Invest. 62*: 1375–1385 (1978).
61. S. E. Bradley and R. Herz, Permselectivity of biliary canalicular membrane in rats: Clearance probe analysis, *Am. J. Physiol. 235*: E570–E576 (1978).
62. R. N. M. MacSween and R. F. Scothorne, Developmental anatomy and normal structure, In: *Pathology of the Liver,* 3rd ed. (R. N. M. MacSween, P. P. Anthony, P. J. Scheuer, A. D. Burt and B. C. Portmann, eds.), Edinburgh: Churchill Livingstone, 1994, pp. 1–49.
63. G. Alphini, R. Lenzi, W.-R. Zhai, et al., Bile secretory function of intrahepatic biliary epithelium in the rat, *Am. J. Physiol. 257*: G124–G133 (1989).

64. M. H. Nathanson and J. L. Boyer, Mechanisms and regulation of bile secretion, *Hepatology 14*: 551–566 (1991).
65. Y. B. Yoon, L. R. Hogey, A. F. Holfmann, et al., Effect of side chain shortening on the physiological properties of bile acids: Hepatic transport and effect on the biliary reaction of 23 urodeoxycholate in rodents, *Gastroenterology 90*: 837–852 (1986).
66. Y. Lamri, S. Erlinger, M. Dumont, A. Roda, and G. Feldmann, Immunoperoxidase localisation of urodeoxycholic acid in rat biliary epithelial cells. Evidence for a cholehepatic circulation, *Liver 12*: 351–354 (1992).
67. M. T. Moslen, Toxic responses of the liver, In: *Toxicology, The Basic Science of Poisons,* 5th ed. (C. D. Klaassen, ed.), New York: McGraw-Hill, 1996, pp. 403–416.
68. K. Jungermann and N. Katz, Functional specialization of different hepatocyte populations, *Physiol. Rev. 69*: 708–764 (1989).
69. J. J. Gumucio, Hepatocyte heterogeneity: The coming of age from the description of a biological curiosity to a partial understanding of its physiological meaning and regulation, *Hepatology 9*: 154–160 (1989).
70. D. Haussinger, W. H. Lamers, and A. F. Moorman, Hepatocyte heterogeneity in the metabolism of amino acids and ammonia, *Enzyme 46*: 72–93 (1992).
71. R. Gebhardt, Metabolic zonation of the liver: Regulation and implications for liver function, *Pharmacol. Ther. 53*: 275–354 (1992).
72. K. Jungermann, Zonal liver cell heterogeneity, *Enzyme 46*: 5–7 (1992).
73. G. E. Hoffmann, H. Andres, L. Weiss, C. Kreisel, and R. Sander, Properties and organ distribution of ATP citrate (pro-3*S*)-lyase, *Biochim. Biophys. Acta 620*: 151–158 (1980).
74. A. F. Moorman, J. L. Vermeulen, R. Charles, and W. H. Lamers, Localization of ammonia-metabolizing enzymes in human liver: Ontogenesis of heterogeneity, *Hepatology 9*: 367–372 (1989).
75. L. Racine Samson, J. Y. Scoazec, et al., The metabolic organization of the adult human liver: A comparative study of normal, fibrotic, and cirrhotic liver tissue, *Hepatology 24*: 104–113 (1996).
76. G. M. Groothuis and D. K. Meijer, Hepatocyte heterogeneity in bile formation and hepatobiliary transport of drugs, *Enzyme 46*: 94–138 (1992).
77. M. Tsutsumi, J. M. Lasker, M. Shimizu, A. S. Rosman, and C. S. Lieber, The intralobular distribution of ethanol-inducible P450IIE1 in rat and human liver, *Hepatology 10*: 437–446 (1989).
78. P. Serasinghe, H. Yamazaki, K. Nishiguchi, S. Serasinghe, and S. Nakanishi, Intralobular localization of different cytochrome P-450 form dependent monooxygenase activities in the liver of normal and inducer-treated rats, *Int. J. Biochem. 24*: 959–965 (1992).
79. D. Ratanasavanh, P. Beaune, F. Morel, J. P. Flinois, F. P. Guengerich, and A. Guillouzo, Intralobular distribution and quantitation of cytochrome P-450 enzymes in human liver as a function of age, *Hepatology 13*: 1142–1151 (1991).
80. R. Buhler, K. O. Lindros, A. Nordling, I. Johansson, and M. Ingelman Sundberg, Zonation of cytochrome P450 isozyme expression and induction in rat liver, *Eur. J. Biochem. 204*: 407–412 (1992).
81. C. Lindamood, Xenobiotic biotransformation, *Hepatotoxicology* (R. G. Meeks, S. D. Harrison, and R. J. Bull, eds.), CRC Press, Boca Raton, 1991, pp. 139–180.
82. J. J. Gumucio, M. May, C. Dvorak, J. Chianale, and V. Massey, The isolation of functionally heterogeneous hepatocytes of the proximal and distal half of the liver acinus in the rat, *Hepatology 6*: 932–944 (1986).
83. J. Michaelson, Cellular selection in the genesis of multicellular organization, *Lab. Invest. 69*: 136–151 (1993).
84. G. Schreiber, A. Tsykin, A. R. Aldred, et al., The acute phase response in the rodent, *Ann. N.Y. Acad. Sci. 557*: 61–85 (1989).
85. S. Penman, Rethinking cell structure, *Proc. Natl. Acad. Sci. USA. 92*: 5251–5257 (1995).

86. K. R. Porter, The cytomatrix: A short history of its study, *J. Cell Biol. 99*: 3s–12s (1984).
87. E. A. Wiemer, T. Wenzel, T. J. Deerinck, M. H. Ellisman, and S. Subramani, Visualization of the peroxisomal compartment in living mammalian cells: Dynamic behavior and association with microtubules, *J. Cell Biol. 136*: 71–80 (1997).
88. E. R. Weibel, W. Staubli, H. R. Gnagi, and F. A. Hess, Correlated morphometric and biochemical studies on the liver cell. I. Morphometric model, stereologic methods and normal morphometric data for rat liver, *J. Cell Biol. 42*: 68–91 (1969).
89. D. Keppler and I. M. Arias, Hepatic canalicular membrane. Introduction: Transport across the hepatocyte canalicular membrane, *FASEB J. 11*: 15–18 (1997).
90. J. J. Smit, A. H. Schinkel, R. P. Oude Elferink, et al., Homozygous disruption of the murine mdr2 P-glycoprotein gene leads to a complete absence of phospholipid from bile and to liver disease, *Cell 75*: 451–462 (1993).
91. R. P. Elferink, G. N. Tytgat, and A. K. Groen, Hepatic canalicular membrane 1: The role of mdr2 P-glycoprotein in hepatobiliary lipid transport, *FASEB J. 11*: 19–28 (1997).
92. G. Feuer, S. D. Cooper, F. A. de la Iglesia, and G. Lumb, Microsomal phospholipids and drug action. Quantitative biochemical and electron microscopic studies, *Int. J. Clin. Pharmacol. Ther. Toxicol. 5*: 389–396 (1972).
93. F. A. de la Iglesia, J. M. Sturgess, and G. Feuer, New approaches for the assessment of hepatotoxicity by means of quantitative functional–morphological interrelationships, *Toxicology of the Liver*, 1982, pp. 47–102.
94. G. Feuer and F. A. de la Iglesia, Subcellular biochemical and pathological correlates in experimental models of hepatotoxicity, *Drug-Induced Hepatotoxicity* (R. Cameron, G. Feuer, and F. A. de la Iglesia, eds.), Springer-Verlag, Heidelberg, 1996, pp. 43–73.
95. G. Dallner and J. L. E. Ericsson, *Molecular Structure and Biological Implication of the Liver Endoplasmic Reticulum*, Grune & Stratton, New York, 1976, pp. 35–50.
96. R. P. Bolender and E. R. Weibel, A morphometric study of removal of phenobarbital-induced membranes from hepatocytes after cessation of treatment, *J. Cell Biol. 51*: 746–776 (1973).
97. W. Staubl, R. Hess, and E. R. Weibel, Correlated morphometric and biochemical studies on the liver cell. II. Effects of phenobarbital on rat hepatocytes, *J. Cell Biol. 42*: 92–112 (1969).
98. D. L. Schmuckler, Age-related changes in hepatic fine structure: A quantitative analysis, *J. Gerontol. 31*: 135–143 (1976).
99. G. Feuer and C. J. DiFonzo, Intrahepatic cholestasis: A review of biochemical–pathological mechanisms, *Drug Metab. Drug Interact. 10*: 1–162 (1992).
100. F. A. de la Iglesia, G. Feuer, A. Takada, and Y. Matsuda, Morphologic studies on secondary phospholipidosis in human liver, *Lab. Invest. 30*: 539–549 (1974).
101. F. A. de la Iglesia, G. Feuer, E. J. McGuire, and A. Takada, Morphological and biochemical changes in the liver of various species in experimental phospholipidosis after diethylaminoethoxyhexestrol treatment, *Toxicol. Appl. Pharmacol. 34*: 28–44 (1975).
102. F. A. Hess, E. R. Weibel, and R. Preisig, Morphometry of dog liver: Normal base-line data, *Virchows Arch. [ZellPathol.] 12*: 303–317 (1973).
103. H. David, *Quantitative Ultrastructural Data of Animal and Human Cells*, Fisher, Stuttgart, 1977.
104. A. Blouin, R. P. Bolender, and E. R. Weibel, Distribution of organelles and membranes between hepatocytes and nonhepatocytes in the rat liver parenchyma. A stereological study, *J. Cell Biol. 72*: 441–455 (1977).
105. D. L. Schmuckler, J. S. Mooney, and A. L. Jones, Stereological analysis of hepatic fine structure in the Fischer 344 rat. Influence of sublobular location and animal age, *J. Cell Biol. 78*: 319–337 (1978).
106. K. Schwerzmann, L. M. Cruz Orive, R. Eggman, A. Sanger, and E. R. Weibel, Molecular architecture of the inner membrane of mitochondria from rat liver: A combined biochemical and stereological study, *J. Cell Biol. 102*: 97–103 (1986).

107. J. Cascarano, P. A. Chambers, E. Schwartz, P. Poorkaj, and R. E. Gondo, Organellar clusters formed by mitochondrial–rough endoplasmic reticulum associations: An ordered arrangement of mitochondria in hepatocytes, *Hepatology 22*: 837–846 (1995).
108. R. C. Candipan and F. S. Sjostrand, An analysis of the contribution of the preparatory techniques to the appearance of condensed and orthodox conformations of the liver mitochondria, *J. Ultrastruct. Res. 89*: 281–294 (1984).
109. R. A. Smith and M. J. Ord, Mitochondrial form and function relationships in vivo: Their potential in toxicology and pathology, *Am. J. Pathol. 83*: 63–134 (1983).
110. E. J. McGuire, J. R. Haskins, J. A. Lucas, and F. A. de la Iglesia, Hypolipidemic-induced quantitative microscopic changes in rat liver mitochondria. *Abstr. VII Int. Cong. Toxicol. 7*: 47–P-7 (1995).
111. R. K. Dubey, M. U. Beg, and J. Singh, Effects of endosulfan and its metabolites on rat liver mitochondrial respiration and enzyme activities in vitro, *Biochem. Pharmacol. 33*: 3405–3410 (1984).
112. J. M. Jacobs, J. V. Rutkowski, B. D. Roebuck, and R. P. Smith, Rat hepatic mitochondria are more sensitive to allyl alcohol than are those of mice, *Toxicol. Lett. 38*: 257–264 (1987).
113. P. F. Smith, G. D. Hoke, D. W. Alberts, et al., Mechanism of toxicity of an experimental bidentate phosphine gold complexed antineoplastic agent in isolated rat hepatocytes, *J. Pharmacol. Exp. Ther. 249*: 944–950 (1989).
114. P. J. Donnelly, R. M. Walker, and W. J. Racz, Inhibition of mitochondrial respiration in vivo is an early event in acetaminophen-induced hepatotoxicity, *Arch. Toxicol. 68*: 110–118 (1994).
115. D. G. Robertson, T. K. Braden, E. R. Urda, N. L. Lalwani, and F. A. de la Iglesia, Elucidation of mitochondrial effects by tetrahydroaminoacridine (tacrine) in rat, dog, monkey and human hepatic parenchymal cells, *Arch. Toxicol. 72*: 362–371 (1998).
116. J. S. Landin, S. D. Cohen, and E. A. Khairallah, Identification of a 54-kDa mitochondrial acetaminophen-binding protein as aldehyde dehydrogenase, *Toxicol. Appl. Pharmacol. 141*: 299–307 (1996).
117. D. K. Monteith, J. C. Theiss, J. R. Haskins, and F. A. de la Iglesia, Functional and subcellular organelle changes in isolated rat and human hepatocytes induced by tetrahydroaminoacridine, *Arch. Toxicol. 72*: 147–156 (1998).
118. D. C. Wallace, Mitochondrial DNA sequence variation in human evolution and disease, *Proc. Natl. Acad. Sci. USA 91*: 8739–8746 (1994).
119. S. P. LeDoux, G. L. Wilson, and V. A. Bohr, Mitochondrial DNA repair and cell injury. *Methods in Toxicology, Mitochondrial Dysfunction*, Vol. 2 (L. H. Lash and D. P. Jones, eds.), Academic Press, New York, 1993, pp. 461–476.
120. W. Lewis, R. R. Meyer, J. F. Simpson, J. M. Colacino, and F. W. Perrino, Mammalian DNA polymerases alpha, beta, gamma, delta, and epsilon incorporate fialuridine (FIAU) monophosphate into DNA and are inhibited competitively by FIAU triphosphate, *Biochemistry 33*: 14620–14624 (1994).
121. W. Lewis and M. C. Dalakas, Mitochondrial toxicity of antiviral drugs, *Nature Med. 1*: 417–422 (1995).
122. W. Lewis, B. Griniuviene, K. O. Tankersley, et al., Depletion of mitochondrial DNA, destruction of mitochondria, and accumulation of lipid droplets result from fialuridine treatment in woodchucks (*Marmota monax*), *Lab. Invest. 76*: 77–87 (1997).
123. K. Stahl, H. Themann, and A. Verhagen, Ultrstructural–morphometric investigations on liver biopsies—the influence of oral contraceptives on the human liver, *Arch. Gynaekol. 223*: 205–211 (1977).
124. K. Miyai, *Structural Organization of the Liver*, CRC Press, Boca Raton, 1991, pp. 1–66.
125. K. Yamamoto and H. Fahimi, Three-dimensional reconstruction of a peroxisomal reticulum in regenerating rat liver: Evidence of interconnections between heterogeneous segments, *J. Cell Biol. 105*: 713–722 (1987).

126. B. A. Afzelius, The occurrence and structure of microbodies, *J. Cell Biol. 26*: 835–841 (1965).
127. C. de Duve, The peroxisome in retrospect, *Ann. N.Y. Acad. Sci. 804*: 1–10 (1996).
128. E. J. McGuire, J. A. Lucas, R. H. Gray, and F. A. de la Iglesia, Peroxisome induction potential and lipid-regulating activity in rats. Quantitative microscopy and chemical structure–activity relationships, *Am. J. Pathol. 39*: 217–229 (1991).
129. S. Subramanl, Protein import into peroxisomes and biogenesis of the organelle, *Annu. Rev. Cell Biol. 9*: 445–478 (1993).
130. T. Lemberger, B. Desvergne, and W. Wahli, Peroxisome proliferator-activated receptors: A nuclear receptor signaling pathway in lipid physiology, *Annu. Rev. Cell Dev. Biol. 12*: 335–363 (1996).
131. A. R. Saltiel and J. M. Olefsky, Thiazolidinediones in the treatment of insulin resistance and type II diabetes, *Diabetes* 45: 1661–1669 (1996).

4

The Cytochrome P450 Oxidative System

Paul R. Ortiz de Montellano
University of California–San Francisco, San Francisco, California

I. INTRODUCTION TO THE CYTOCHROME P450 SYSTEM

Cytochrome P450 enzymes play critical roles in the biogenesis of sterols and other physiological intermediates, the catabolism of endogenous substrates, such as fatty acids and sterols, and the metabolism of most xenobiotics. Cytochrome P450-mediated oxidations are unique in their ability to introduce polar functionalities into systems as unreactive as saturated or aromatic hydrocarbons, and they are particularly critical for the metabolism of lipophilic compounds without functional groups suitable for conjugation reactions. On the negative side, the reactions catalyzed by cytochrome P450 sometimes transform relatively innocuous substrates to chemically reactive toxic or carcinogenic species.

The P450 enzymes involved in the oxidation of drugs, xenobiotics, and endogenous substrates are widely distributed, with particularly high concentrations in the endoplasmic reticulum of the liver, kidney, lung, nasal passages, and gut, and significant concentrations in most other tissues (1). In contrast, the sterol biosynthetic enzymes are primarily found in steroidogenic tissues, such as the adrenals and testes. The P450 enzymes in animals are all membrane-bound, and the solubilization, purification, and reconstitution of the enzymes is technically challenging but is now relatively routine. Because of the insolubility and other physical properties of the membrane-bound proteins, no crystal structure is available for a mammalian cytochrome P450 enzymes. However, many of the bacterial cytochrome P450 enzymes are soluble, and the structures of four bacterial enzymes have been determined: $P450_{cam}$ (CYP101) (2), $P450_{BM-3}$ (CYP102) (3), $P450_{terp}$ (CYP108) (4), and $P450_{eryF}$ (CYP107A1) (5). Most of what we know about the structures of all P450 enzymes is based on these bacterial enzymes.

II. CYTOCHROME P450 GENE FAMILY

The sequences of more than 300 cytochrome P450 enzymes are now known, and additional sequences, particularly those of bacterial, insect, and plant origin, are reported monthly (6). This flood of sequence information has made possible (indeed, has required) the development of a rational nomenclature. The early names assigned to P450 enzymes, based on properties such as electrophoretic mobility, absorption spectrum, or substrate specificity, have given way to a systematic, evolutionary nomenclature system. The current nomenclature is based on the proposal that the extent of sequence and functional identity decreases as a function of evolutionary distance from a common precursor. By using protein sequence identity as the basis for assigning names to the enzymes, the P450s are grouped according to probable structural and functional similarity. Although the cutoff lines are somewhat arbitrary, enzymes with more than 40% sequence identity are assigned to the same family, and those with more than 55% identity to the same subfamily (6). The enzymes are thus identified by a number denoting the family, a letter denoting the subfamily, and a number identifying the specific member of the subfamily (Table 1). For example, cytochrome P450 1A2 is the second member of subfamily A of family 1, and P450 3A4 is the fourth member of subfamily A of family 3. An abbreviated nomenclature is obtained by replacing cytochrome P450 with the letters CYP (i.e., CYP1A2 and CYP3A4). Family numbers below 100 are reserved for eukaryotic enzymes, whereas those above 100 are intended for microbial enzymes. The trivial names of some substrate-specific enzymes

Table 1 Selected Human Cytochrome P450 Isoform Substrates

Enzyme	Substrates	Enzyme	Substrates
CYP1A1	Benzo[*a*]pyrene	CYP2E1	Aniline
	Acetylaminofluorene		Aflatoxin
CYP1A2	2-Acetylaminofluorene		*N*-Nitrosodimethylamine
	Ethoxyresorufin		Acetaminophen
	Phenacetin		Chlorzoxazone
	Acetaminophen		Caffeine
	Caffeine	CYP2F1	Naphthylamine
	Aflatoxin B_1	CYP3A4	Aldrin
CYP2A6	Coumarin		Cortisol
	N-Nitrosodiethylamine		Cyclosporine
CYP2B6	7-Ethoxycoumarin		Diltiazem
CYP2C9	Aminopyrene		Erythromycin
	Benzphetamine		17β-Estradiol
	Hexobarbital		Lidocaine
	Tienilic acid		Nifedipine
	Tolbutamide		Sterigmatocystin
CYP2C19	Mephenytoin		Quinidine
CYP2D6	Bufuralol		Taxol
	Sparteine		Dapsone
	Debrisoquine		Alfentanil
	Desipramine		Warfarin
	Dextromethorphan		Lovastatin
	Propranolol		Ethynyl estradiol
		CYP4A11	Arachidonic acid

Source: Refs. 7 and 8.

continue to be widely used (e.g., $P450_{cam}$), but the isoforms primarily involved in drug metabolism are now generally known by their systematic names.

The cytochrome P450-dependent metabolism of drugs and xenobiotics in humans is mediated primarily by enzymes of the CYP1, CYP2, CYP3, and CYP4 families. In humans, CYP3A4 is the most abundant isoform, representing approximately 30% of the spectroscopically detectable cytochrome P450 in the liver (9). CYP1A2 represents 13%, CYP2A6 4%, the CYP2C enzymes 20%, CYP2D6 2%, and CYP2E1 7% of the total (9). The importance of the isoforms in drug metabolism depends not only on their abundance, but also on the extent to which their substrate specificity coincides with the range of drugs and xenobiotics to which the individual is exposed. As suggested by Table 1, CYP3A4, CYP2D6, and CYP2C enzymes are responsible for the bulk of drug metabolism, although other isoforms can play critical roles with specific substrates. The relative importance of the different enzymes depends, furthermore, on the genetics of the individual and on the history of exposure to environmental factors, such as alcohol or drugs (see Chapter 9). Certain P450 isoforms, notably CYP2C19 and CYP2D6, are polymorphically distributed in the human population (7,8; see Chapter 8). Thus, CYP2D6 levels are low in approximately 7% of the white population, and the ability in this subgroup to metabolize substrates, such as debrisoquin and sparteine, is partially compromised (10). CYP2C19 is in low titer in only 4% of the white, but in 20% of the Asian population, as reflected by the ability of this subgroup to metabolize substrates such as mephenytoin (11). Several criteria are required to unambiguously determine the role of a given P450 enzyme in the in vivo metabolism of a given agent. These include (a) demonstration that the purified enzyme has the required activity, (b) correlation of the activity in question in various liver samples with the activities of substrates considered to be markers for individual P450 enzymes, and (c) inhibition of the activity by isoform-selective or specific inhibitors or antibodies (7,12).

III. SPECTROSCOPIC PROPERTIES OF CYTOCHROME P450 ENZYMES

The cytochrome P450 chromophore provides information on the nature of the iron ligands, the iron oxidation state, and the nature of the heme environment (13,14). The most characteristic P450 spectrum is that of the ferrous–CO complex, which has an absorption maximum at 447–452 nm, indicative of a thiolate-ligated hemoprotein. Denaturation of the enzyme is associated with a shift of the absorption maximum of the ferrous–CO complex to approximately 420 nm, a value similar to that for the ferrous–CO complex of imidazole-ligated proteins, such as myoglobin (14). The thiolate ligand actually gives rise to a split Soret absorption, with maxima at approximately 450 and 370 nm.

The absorption spectrum of ferric cytochrome P450 depends on the ligation state of the iron (13,14). The low-spin state associated with the presence of two strong ligands to the iron has an absorption maximum at approximately 416–419 nm. The high-spin state in which one of the two coordination sites of the iron is either unoccupied, or is occupied by a weak ligand, exhibits an absorption maximum at 390–416 nm. Cytochrome P450 enzymes commonly exist in an equilibrium mixture of the high- and low-spin states. In the P450 enzymes for which crystal structures are available, and presumably also in the membrane-bound mammalian enzymes, the ligand opposite the cysteine thiolate is a water molecule (2–5). The binding of a noncoordinating substrate, as illustrated by the binding of camphor to cytochrome $P450_{cam}$, is accompanied by extrusion of water from the active site, loss of the distal water ligand, and a general decrease in the active site polarity (2,15). The loss of the distal ligand causes a shift from the hexacoordinated to pentacoordinated iron state which, in turn, results in a shift from the low- to the high-spin state. This transition is evidenced by a shift in the absorption maximum from 419 to 390 nm. The spectroscopic shift is usually determined from a difference spectrum in which the absorption of a solution containing everything but the compound of interest is subtracted from the sample also containing the compound (13). The binding of noncoordinating substrates gives rise to what is known as a type I difference spectrum with a maximum at 385–390 nm and a trough at approximately 420 nm. However, if the substrate can coordinate strongly to the iron atom, a type II difference spectrum is observed, with a maximum at approximately 425–435 nm and a trough at 390–405 nm. If the substrate coordinates only weakly to the iron, a variant of the type II spectrum, with a maximum at 420 nm and a trough at 388–390 nm, is obtained. This latter spectrum is known as a type III difference spectrum. Substrate and inhibitor binding to P450 enzymes can thus be monitored spectroscopically, although compounds are known that bind without significantly perturbing the spin state equilibrium and, therefore, do not give rise to a difference spectrum. An example of this is the binding of 2-isopropyl-4-pentenamide to microsomes from phenobarbital-pretreated rats (16).

IV. CATALYTIC CYCLE OF CYTOCHROME P450

The catalytic cycle of cytochrome P450 has been most thoroughly defined for the bacterial enzyme cytochrome $P450_{cam}$ (17), but the same cycle has been confirmed in its essential features for the membrane-bound P450 enzymes (Fig. 1). The spin state change triggered by the binding of a noncoordinating substrate alters the redox potential of the heme and makes it possible for the electron-donor partner to transfer an electron to the iron. The redox potential change for $P450_{cam}$ is from -300 to -170 mV, which makes the hemoprotein reducible by the iron–sulfur protein putidaredoxin ($E_{1/2} = -196$ mV; 18). The catalytic turnover of the protein is thus initiated by the binding of a substrate, a strategy that minimizes uncoupled turnover of the protein. Reduction of the iron is followed by binding of oxygen to give the ferrous–dioxy complex, the last intermediate that has been directly observed under physiological conditions (19,20).

One-electron reduction of the ferrous–dioxy complex produces the activated species that reacts with the substrate. It is thought that the second electron reduces the ferrous–dioxy complex to a ferric–peroxide complex, although the detailed structure of this complex is not known. Nevertheless, the fact that uncoupled turnover of P450 produces H_2O_2 and that catalytic turnover of P450 can be supported by exogenous H_2O_2 (21), support formulation of the initial species obtained on uptake of the second electron as a ferric–peroxide complex of some kind. This complex undergoes heterolytic cleavage of the dioxygen bond, with the uptake of two protons and the loss of a molecule of water to produce what is thought to be an activated ferryl (formally $Fe^{V}{=}O$) species (21). Reaction of the ferryl species with the substrate leads to product formation. In most, but not all, instances the oxygen of the ferryl species is transferred to the substrate to give an oxygenated metabolite. If the substrate is relatively resistant to oxidation, the ferryl species can be directly reduced with the formation of a water molecule. To the extent that superoxide, H_2O_2, and water are produced at the expense of substrate oxidation during the catalytic turnover of cytochrome P450 the reaction is said to be uncoupled.

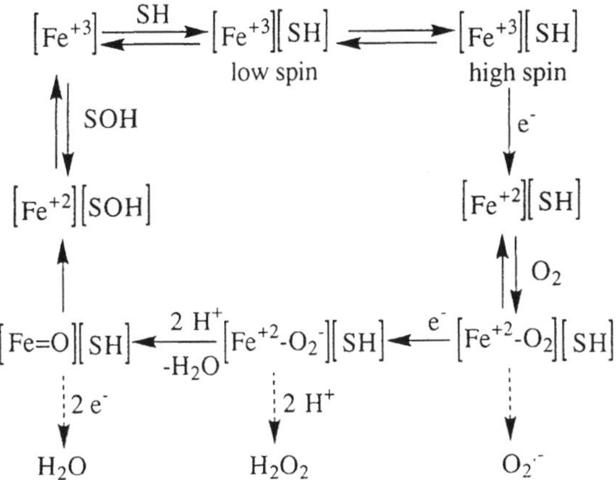

Fig. 1 Catalytic cycle of cytochrome P450 enzymes: The iron in brackets represents the prosthetic heme group of the enzyme and SH represents a substrate. The sites at which the catalytic cycle can be uncoupled to produce $O_2^{\cdot-}$, H_2O_2, or H_2O are indicated.

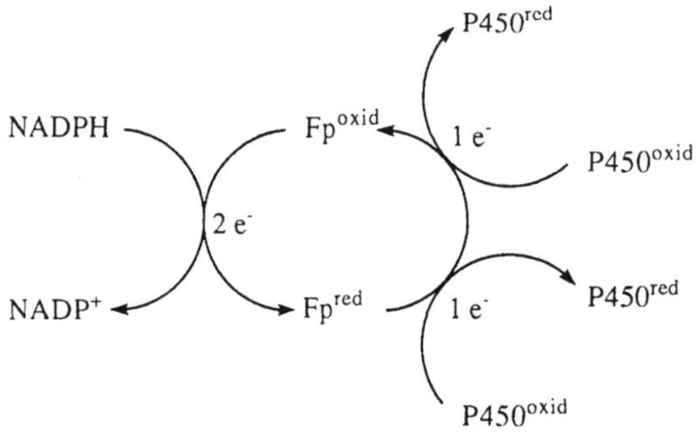

P450^{red}
NADPH Fp^{oxid} 1 e$^-$ P450^{oxid}
2 e$^-$
NADP$^+$ Fp^{red} 1 e$^-$ P450^{red}
P450^{oxid}

Fig. 2 The microsomal system involved in the transfer of electrons from reduced pyridine nucleotides to cytochrome P450: A flavoprotein (Fp) with one FMN and one FAD prosthetic group transfers the electrons from NADPH to cytochrome P450.

V. CYTOCHROME P450 REDUCTASE

The electrons required for catalytic turnover of the xenobiotic-metabolizing isoforms of cytochrome P450 are provided by NADPH–cytochrome P450 reductase (Fig. 2). The reductase, a 78-kDa protein, is anchored to the membrane by an NH_2-terminal hydrophobic peptide (22). Cytochrome P450 reductase binds one FMN and one FAD as prosthetic groups and is reduced by NADPH but not NADH. NADPH transfers its electrons to the FAD group which, in turn, transfers them to the heme via the FMN group. The reductase can be reduced by up to four electrons, but under normal turnover conditions it cycles between the one- and three-electron reduced forms, both of which can transfer electrons to cytochrome P450. Limited tryptic digestion of liver microsomes releases a 72-kDa cytosolic reductase domain that binds NADPH and reduces cytochrome *c* but is no longer able to reduce cytochrome P450 (23). Heterologously expressed P450 reductase without the membrane-binding domain has been crystallized, and its structure should soon be available (24).

The catalytic turnover of cytochrome P450 in some instances can be synergistically increased by cytochrome b_5 (25,26). Cytochrome b_5 can be reduced by either NADH and cytochrome b_5 reductase or NADPH–cytochrome P450 reductase. Cytochrome b_5 is able to deliver the second, but not the first, electron required for the catalytic turnover of cytochrome P450. The synergistic effect of NADH or cytochrome b_5 is due, at least partly, to an increase in the coupling of reduced pyridine nucleotide and oxygen utilization to product formation (i.e., to a decrease in uncoupled reduction of oxygen to give H_2O_2 or water, rather than substrate oxidation).

VI. CYTOCHROME P450 STRUCTURE

Cytochrome P450$_{cam}$ is the most thoroughly characterized of the four cytochrome P450 enzymes for which structures are available (2–5). It is roughly a triangular prism in which 12 helical segments account for approximately 45% of the amino acid residues (2). The heme prosthetic group is bound in P450$_{cam}$ by at least three interactions: (a) coordination of a cysteine thiolate ligand to the iron, (b) the pincer action of two helices on the heme, and (c) hydrogen bonds to the heme propionic acid groups. A cysteine thiolate is the proximal iron ligand in all P450 enzymes. The active site cavity is lined with hydrophobic residues, as might be expected of an enzyme designed for the metabolism of lipophilic substrates. A few residues on the distal (substrate-binding) side of the heme are strongly, but not universally, conserved throughout the P450 family. The most important of these is Thr-252 in $P450_{cam}$. The threonine may help stabilize the ferrous–dioxy complex or to promote heterolysis of the ferric–peroxide intermediate. Sequence alignments indicate that the threonine is conserved in most P450 enzymes, but it is not essential because it is absent in $P450_{eryF}$ and a few other enzymes (5). Nevertheless, the threonine is important because its replacement by non–hydrogen-bonding residues alters catalysis. Specifically, mutation of Thr-252 in $P450_{cam}$ greatly increases the degree of uncoupled turnover (27,28), and mutation of the corresponding residue (Thr-319) in CYP1A2 to an alanine suppresses the ability of the enzyme to oxidize benzphetamine, but not 7-ethoxycoumarin (29). Recent studies with $P450_{eryF}$ indicate that the hydrogen-bonding role of the threonine is played in that enzyme by a hydroxyl group on the substrate (30).

The crystal structure of $P450_{cam}$ with camphor bound in the active site shows that the heme and substrate are buried within the protein (2). Binding of the substrate requires the transient opening of a channel into the active site. Once in the active site, the substrate is positioned approximately 4 Å above the heme iron atom by a hydrogen bond to a tyrosine residue (Tyr-96) and by contacts with a variety of active site residues (2). A weakening of the interactions that position the substrate, either by mutagenesis of the protein or by alteration of the substrate, decreases the regiospecificity of the hydroxylation reaction and increases the extent of uncoupled turnover (27,28). Although only relatively minor changes are observed in the active site of $P450_{cam}$ when the iron is reduced (31), recent nuclear magnetic resonance (NMR) studies indicate that the position of the fatty acid substrate changes by approximately 6 Å when the iron is reduced in $P450_{BM\text{-}3}$ (32). These results indicate that P450 enzymes may or may not undergo major conformational changes as the catalytic cycle is traversed.

Sequence alignments and the available physical, chemical, and biochemical data suggest that the structures of the membrane-bound mammalian P450 enzymes are similar to those of the bacterial enzymes (33). On the basis of sequence alignments with $P450_{cam}$, six domains of the primary sequence of P450 enzymes, known as sequence recognition sequences (SRS), have been proposed to be particularly important in determining substrate specificity (34). Sequence alignments are being used, in conjunction with the templates provided by the bacterial P450 crystal structures, to construct structural models of the mammalian P450 enzymes. For example, models are currently available for mammalian cytochromes P450 1A1 (35), 2B1 (36), and 2D6 (37). These models provide rough three-dimensional structures of the indicated proteins, but caution must be taken not to use them in ways that exceed their validity. Specifically, the models now available are still too unreliable to address detailed mechanistic questions, or to use for the docking of substrates in efforts to determine detailed substrate specificity.

VII. CYTOCHROME P450-CATALYZED REACTIONS

Cytochrome P450-catalyzed reactions produce a diversity of metabolites, many of which are formed by secondary, nonenzymatic decomposition of the products formed by the P450 reaction. Although the range of reactions catalyzed by P450 is growing, most of the

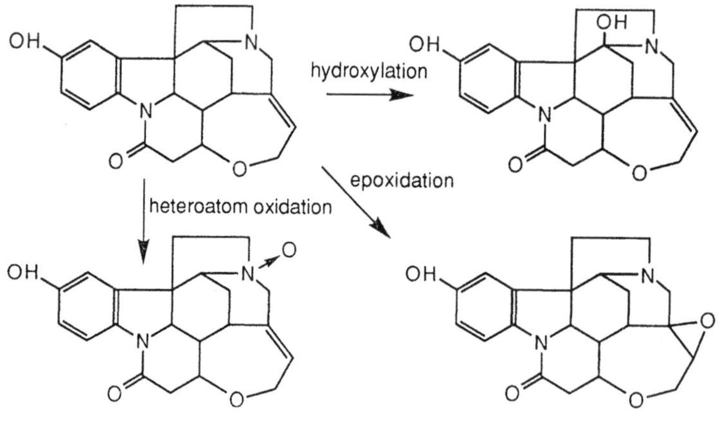

Fig. 3 Hydroxylation, π-bond oxidation, and heteroatom oxidation, the three basic cytochrome P450-catalyzed reactions, intervene in the metabolism of strychnine. (From Ref. 39.)

reactions involve (a) insertion of an oxygen atom into the bond between a hydrogen and a carbon or other heavy atom (hydroxylation), (b) addition of an oxygen atom to a π-bond (epoxidation), or (c) addition of an oxygen atom to the electron pair on a heteroatom (heteroatom oxidation) (38). These reactions are illustrated by the metabolism of strychnine (Fig. 3); (39). Cytochrome P450 enzymes also catalyze reduction reactions, particularly under conditions of low oxygen tension. The catalytic process can be viewed as consisting of two stages: activation of molecular oxygen to the active oxidizing species, followed by reaction of the oxidizing species with the substrate. The catalytic machinery of cytochrome P450 is required for the first stage of the process. In contrast, the enzyme appears to contribute little to the reaction of the activated oxygen with the substrate beyond providing an appropriate environment for the reaction and limiting the sites on the substrate exposed to the activated oxygen. The outcome of the catalytic process, therefore, is determined by the nature and the relative reactivities of the substrate functionalities that are accessible to the activated oxygen species.

VIII. HYDROXYLATION

A. Carbon Hydroxylation

The regio- and stereoselective hydroxylation of unactivated hydrocarbons is one of the most common, but most difficult, reactions catalyzed by cytochrome P450. Isotopic effects provide direct evidence that the reaction outcome depends on the reactivity of the accessible C–H bonds. Thus, hydroxylation of the undeuterated carbon is strongly favored in the oxidation of [1,1-$^{2}H_2$]-1,3-diphenylpropane, a symmetrical molecule in which the hydrogens are replaced by deuteriums on one of the two otherwise identical methylene groups (Fig. 4). A large isotopic effect ($k_H/k_D = 11$) is observed in the intramolecular preference for hydroxylation of the undeuterated methylene, even though only a small kinetic isotopic effect is observed on the net rate of hydroxylated product formation (40). The large intramolecular isotopic effect indicates that the reaction is sensitive to the relative energies required to break competing C–H bonds, and that the reactive species is

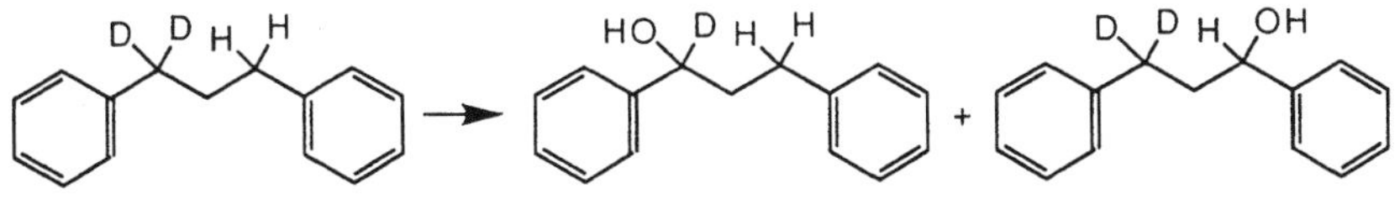

Fig. 4 Competitive hydroxylation of otherwise equivalent deuterated and undeuterated sites in a symmetrical molecule. (From Ref. 40.)

able to choose between the competing sites. Only a small kinetic isotopic effect is observed on the actual rate of product formation, because the rate is determined by steps other than insertion of the oxygen into the C–H bond.

The relation between bond strength and reactivity is confirmed by the finding that the P450-catalyzed oxidation of hydrocarbon C–H bonds decreases in the order tertiary > secondary > primary (41,42). The intrinsic higher reactivity of weaker C–H bonds is often masked by steric effects or by the orientation of the substrate to the activated oxidizing species. If these factors are minimized, however, the intrinsic reactivity of the C–H bonds becomes evident, as illustrated by the hydroxylation of small hydrocarbons for which movement within the P450 active site is less restricted. Thus, the microsomal oxidation of *tert*-butane [$CH(CH_3)_3$] yields 95% *tert*-butanol [$HOC(CH_3)_3$] and only 5% 2-methylpropanol [$CH_3CH(CH_3)CH_2OH$] (41). The sterically hindered, but weaker, tertiary C–H bond is oxidized in preference to the nine relatively unhindered, but primary C–H bonds.

The relation of bond strength to hydroxylation by the enzyme suggests that reactivity is determined by the homolytic C–H bond scission energy, an inference consistent with a nonconcerted ''oxygen-rebound'' mechanism in which hydrogen abstraction by the activated oxygen is followed by recombination of the resulting radical species to give the hydroxylated product (43):

$$[Fe^{V}{=}O] + R_3C{-}H \rightarrow [Fe^{IV}{-}OH] + R_3C^{\cdot} \rightarrow [Fe^{III}] + R_3C{-}OH \quad (1)$$

If a radical intermediate is formed as a transient species, it should be possible to detect it in substrates in which the radical can undergo a rearrangement before recombination to give the product. Indeed, the hydroxylation of *exo*-tetradeuterated norbornane yields, among other products, the *endo*-deuterated alcohol metabolite (44). This inversion of the deuterium stereochemistry requires the formation of an intermediate in which the geometry of the tetrahedral carbon can be inverted. Other examples are known of reactions that proceed with a loss of stereo- or regiochemistry that require an intermediate in the reaction process (38,43). A recent example of such a reaction is provided by the allylic rearrangement that accompanies the P450-catalyzed hydroxylation of taxa-4(5),11(12)-diene during the biosynthesis of taxol (Fig. 5) (45).

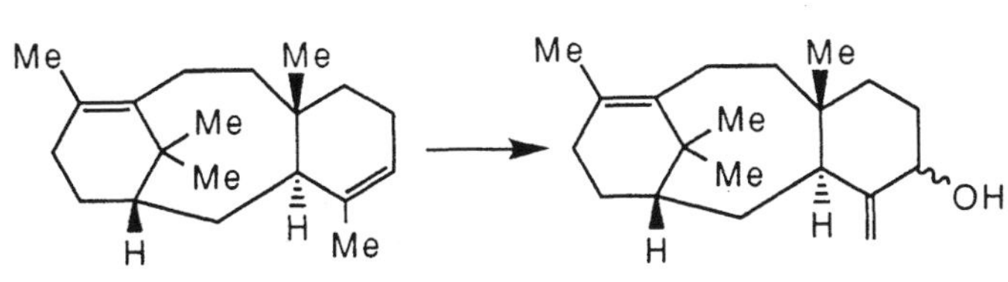

Fig. 5 Rearrangement of the double bond accompanying the cytochrome P450-catalyzed oxidation of an intermediate in the biosynthesis of taxol. (From Ref. 45.)

The cytochrome P450-catalyzed hydroxylation of hydrocarbon chains commonly occurs at the terminal (ω) carbon or at the carbon adjacent to it (ω-1), although the hydroxylation can occur at internal sites in the hydrocarbon chain. ω-Hydroxylation is disfavored relative to ω-1 hydroxylation by the higher strength of the primary C–H bond. Except for the CYP4A family of enzymes, which are specifically designed as fatty acid ω-hydroxylases, most P450 enzymes preferentially catalyze ω-1 hydroxylation. The hydroxylation of C–H bonds that are much stronger than those of a terminal methyl group is rarely observed. Thus, the direct oxidation of a vinylic, acetylenic, or aromatic C–H bond is extremely rare, although the π-bonds themselves are readily oxidized (see next section).

B. Carbon Hydroxylation Followed by Heteroatom Elimination

Hydroxylation adjacent to a heteroatom or a π-bond is highly favored owing to the weaker bond strength of the relevant C–H bonds and, with nitrogen and sulfur, to the availability of alternative mechanisms that lead to the same reaction outcome. Allylic or benzylic hydroxylation produces a stable product but hydroxylation adjacent to a heteroatom yields a product that readily fragments with elimination of the heteroatom (Fig. 6). Oxidation adjacent to an oxygen usually results in *O*-dealkylation, adjacent to a nitrogen in *N*-dealkylation, adjacent to a sulfur in *S*-dealkylation, and adjacent to a halogen in oxidative dehalogenation (Fig. 7).

C. Heteroatom Hydroxylation

The mechanism of cytochrome P450-catalyzed amine hydroxylations is ambiguous because the reaction can proceed by insertion of an oxygen into the N–H bond or oxidation

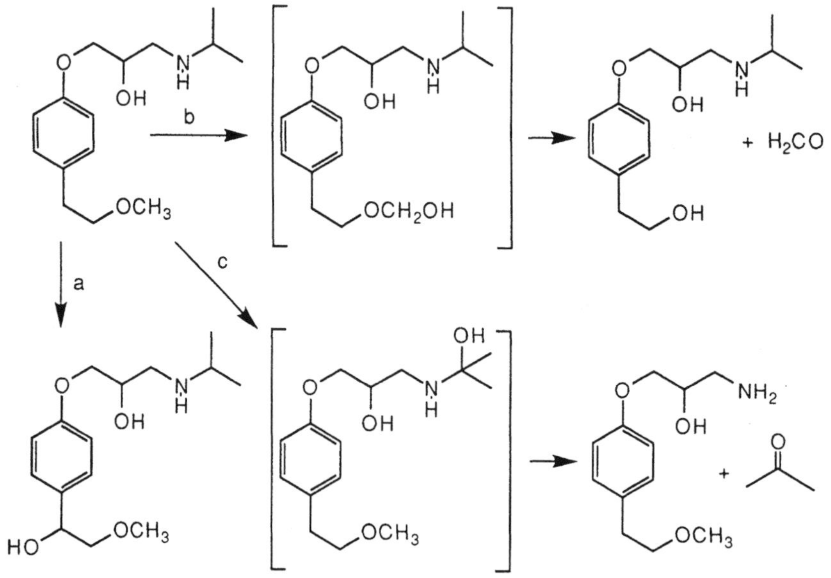

Fig. 6 The hydroxylations of metoprolol illustrate (a) benzylic hydroxylation, (b) hydroxylation adjacent to an oxygen followed by elimination (*O*-dealkylation), and (c) hydroxylation adjacent to a nitrogen followed by elimination (*N*-dealkylation).

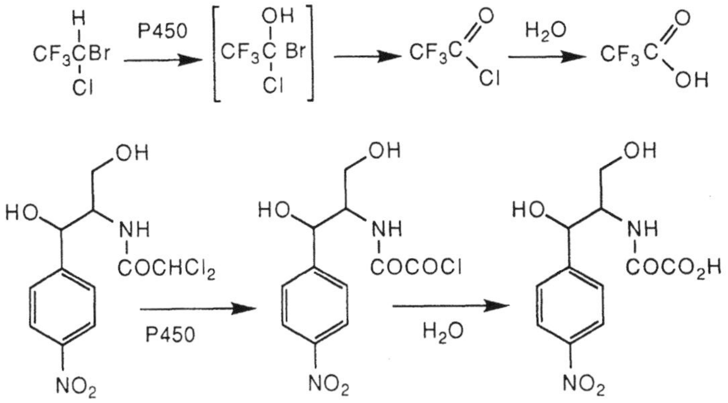

Fig. 7 Hydroxylation adjacent to a halogen followed by elimination (oxidative dehalogenation), as illustrated by the oxidation of halothane and chloramphenicol. The acyl chloride produced from the dihalogenated carbon by the P450 reaction can react with water, as shown, or with other nucleophiles.

of the nitrogen to a nitroxide (see Section X) followed by proton tautomerization to give the hydroxylamine. A similar ambiguity exists in the hydroxylation of sulfhydryl groups. Direct *N*-hydroxylation reactions are most favored in situations in which the nitrogen electron pair is delocalized and, therefore, relatively unavailable. This is true for amides and sulfonamides and the oxidation of these groups to the hydroxylated products, as illustrated by the hydroxylation of *p*-chloroacetanilide (46), probably occurs by a hydroxylation mechanism (Fig. 8).

IX. π-BOND OXIDATION

A. The Oxidation of Aliphatic π-Bonds

The cytochrome P450-catalyzed oxidation of an olefin usually yields the corresponding epoxide (Fig. 9). Retention of the olefin stereochemistry in the reaction suggests that the two carbon–oxygen bonds of the epoxide are formed without the intervention of an intermediate that allows rotation about the carbon–carbon bond (47). The simplest mechanism that satisfies this constraint is that both bonds are formed at the same time, if not necessarily at the same rate. In general, the epoxidation of electron-rich double bonds is favored over that of electron-deficient double bonds. Carbonyl products formed by migration of

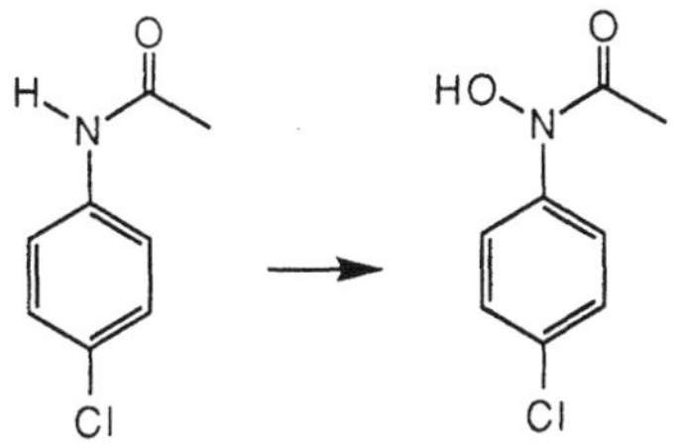

Fig. 8 Nitrogen hydroxylation of *p*-chloroacetanilide.

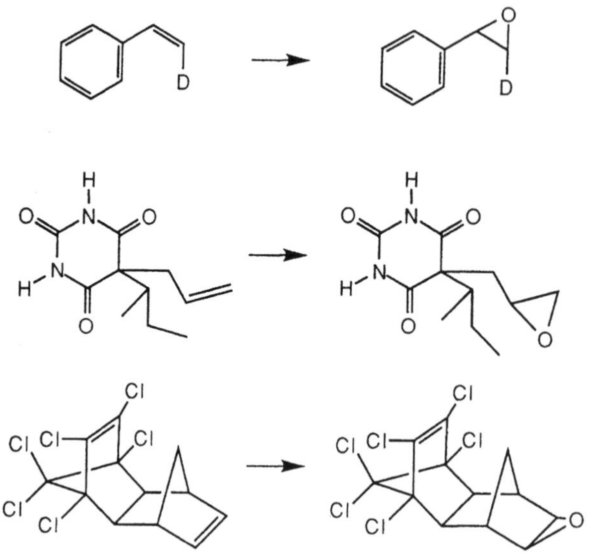

Fig. 9 The epoxidation of 1-deuterated styrene, secobarbital, and aldrin by cytochrome P450. (From Refs. 47, 49, 50.)

a hydrogen or halide from the carbon to which the oxygen is added to the adjacent carbon of the double bond are occasionally obtained as minor metabolites. An example of this is the formation of trichloroacetaldehyde from 1,1,2-trichloroethylene (Fig. 10; 48). The 1,2-migration of a hydrogen or halide implies that positive charge develops at the carbon toward which the migration occurs. Carbonyl products thus appear to result from π-bond oxidations in which the two carbon–oxygen bonds are not simultaneously formed. Although this is a negligible reaction pathway for most olefins, it is an important reaction when the π-bond is part of an aromatic ring (see Section IX.B). Formation of epoxides by concerted oxygen insertion, but of carbonyl products by nonconcerted oxygen transfer, implies either that there are two independent reaction pathways, or a single-reaction manifold, with a branch point leading to concerted (epoxide) or nonconcerted (carbonyl) products.

The oxidation of terminal, unconjugated olefins often results in inactivation of the cytochrome P450 enzyme in addition to formation of the epoxide (47,51). In some instances, inactivation results from reaction of the epoxide with the protein, but in others, inactivation reflects alkylation of a pyrrole nitrogen of the prosthetic heme group by a catalytically activated form of the olefin (Fig. 11). Characterization of the adducts shows that heme alkylation is initiated by oxygen transfer from the enzyme to the heme, but is not due to reaction with the epoxide metabolite. Heme alkylation thus involves an asym-

CCl_3CHO

Fig. 10 The oxidation of trichloroethylene to trichloroacetaldehyde. (From Ref. 48.)

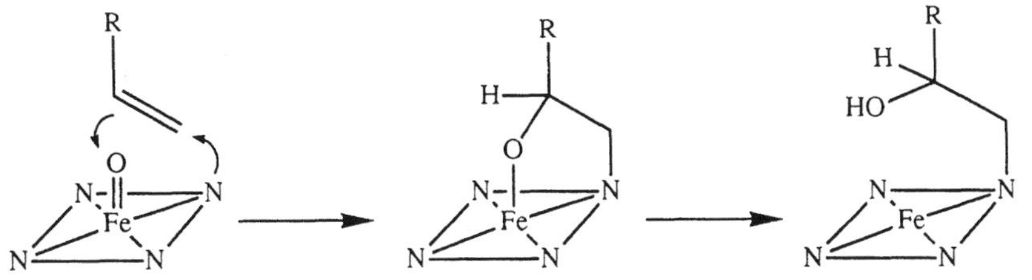

Fig. 11 *N*-Alkylation of the cytochrome P450 prosthetic heme group during terminal olefin oxidation: The P450 heme is represented by the square of nitrogen atoms.

metrical, nonconcerted olefin oxidation pathway that may be part of the same manifold of reactions that produces the epoxide and carbonyl metabolites.

The reaction manifold that leads to the formation of epoxides, carbonyl products, and heme alkylation has not been definitively elucidated. Chemical studies with metalloporphyrin models suggest that the common step in the epoxidation reaction may be the formation of a charge–transfer complex between the ferryl oxygen and the π-bond (52). Decomposition of this complex by one of several pathways leads to the diverse reaction outcomes. The pathways may include concerted epoxide formation (path a) or pathways that proceed by a free radical (path b), cationic (path c), or metallaoxocyclobutane intermediate (path d; Fig. 12). None of these intermediates is supported by unambiguous data except the cationic intermediate that is required for the rearrangement that produces the carbonyl compounds. The various products, however, could also stem from independent reactions, rather than being the result of a single reaction manifold.

B. The Oxidation of Aromatic Rings

The cytochrome P450-catalyzed processing of aromatic systems is a special case of π-bond oxidation. In its simplest form, the oxidation is analogous with the oxidation of an isolated double bond and gives the same product (i.e., the epoxide). However, the epoxides of aromatic systems are unstable and readily rearrange to give phenols. The steps in this reaction are (a) heterolytic scission of one of the strained epoxide carbon–oxygen bonds, (b) migration of the hydrogen (or substituent) on the carbon that retains the epoxide oxygen to the resulting carbocation to give the ketone, and (c) proton tautomerization to give the phenol (Fig. 13). The net result is oxidation of an aromatic ring via the epoxide to the hydroxylated product formally expected from insertion of oxygen into one of the aromatic C–H bonds. This common aromatic "hydroxylation" mechanism is known as the NIH shift (53). The NIH shift of an epoxide can yield two potentially different phenolic products, depending on which of the two epoxide carbon bonds is broken. If the cation formed by breaking one of the bonds is significantly more stable than that obtained by breaking the other, the phenol produced by the lower-energy pathway predominates. Substitution of an electron donating group (e.g., alkoxy) promotes formation of the *ortho*- or *para*-hydroxy metabolite, whereas a strong electron withdrawing substituent (e.g., nitro) favors formation of the *meta*-hydroxy metabolite.

Aromatic π-bond oxidation is subject to mechanistic ambiguities similar to those for the oxidation of simple olefins. Although the epoxides of some aromatic substrates have been isolated and shown to undergo the NIH shift, it is possible that in many instances

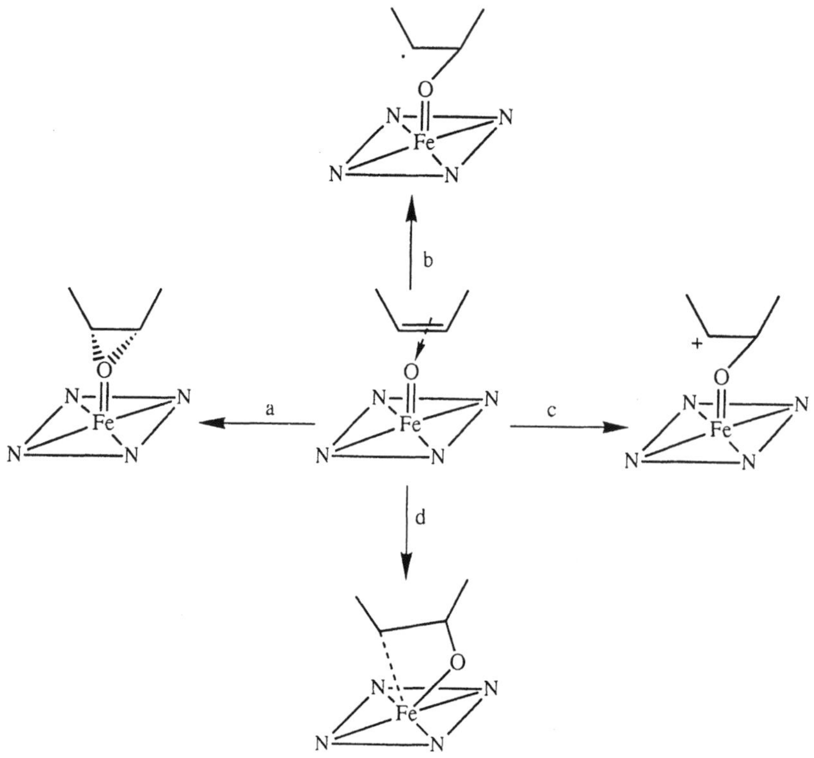

Fig. 12 Possible reaction intermediates in the cytochrome P450-catalyzed oxidation of olefins. The formation of a charge–transfer complex could be followed by (a) concerted epoxide formation, (b) nonconcerted oxygen transfer to give a radical, (c) nonconcerted oxygen transfer to give a cation, or (d) formation of a metallaoxetane intermediate. The pathways are not necessarily independent in that the intermediates can be interconverted.

the epoxide is a not a true intermediate in the reaction trajectory that produces the NIH shift. Asymmetrical transfer of the ferryl oxygen to the aromatic π-bond could yield a cation similar to that expected from cleavage of one of the epoxide carbon–oxygen bonds. This intermediate could close to the epoxide but could also flow directly into the NIH shift (Fig. 14). The reaction manifold can be diverted toward other reaction outcomes. An example of this is the oxidation of pentafluorophenol to 2,3,5,6-tetrafluoroquinone (54). It appears that one-electron abstraction from the polyfluorinated phenol produces a phenoxy radical that combines with the ferryl oxygen at the carbon *para* to the oxygen. The resulting *para*-hydroxylated intermediate eliminates fluoride to give the quinone. (Fig. 15). As always, the reaction with the lowest energy barrier predominates.

X. HETEROATOM OXIDATION

The availability of free electron pairs on nitrogen and sulfur favors their oxidation by the electron-deficient ferryl species of cytochrome P450. Oxygen is not similarly oxidized owing to the higher electronegativity of oxygen. Halogen atoms, similar to oxygen, are

39

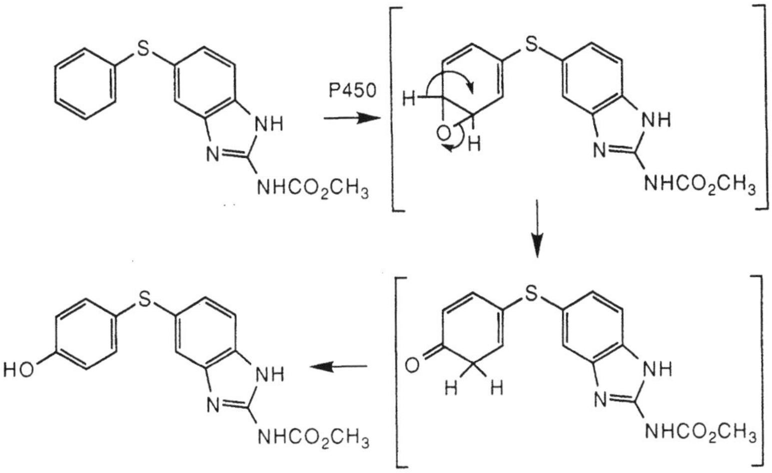

Fig. 13 Aromatic hydroxylation by the NIH shift mechanism as illustrated by the cytochrome P450-catalyzed oxidation of fenbendazole. (From Ref. 55.)

very electronegative and difficult to oxidize, although there is evidence for their oxidation in special circumstances (56).

Several reactions are known to result from the P450-catalyzed oxidation of nitrogen and other heteroatoms. The most straightforward of these reactions is transfer of the electron-deficient ferryl oxygen to a nitrogen to give the *N*-oxide, or to a sulfur to give the sulfoxide. Unless one or more of the substituents is a hydrogen, nitrogen can undergo

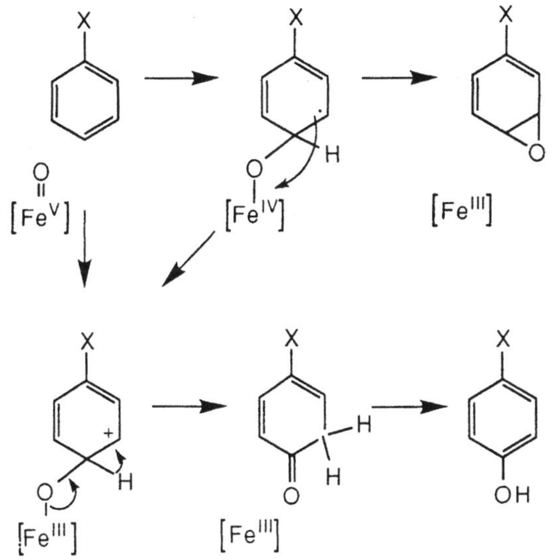

Fig. 14 Hydroxylation of some aromatic rings may occur without the formation of an actual epoxide intermediate. The cytochrome P450 heme iron atom is represented by Fe.

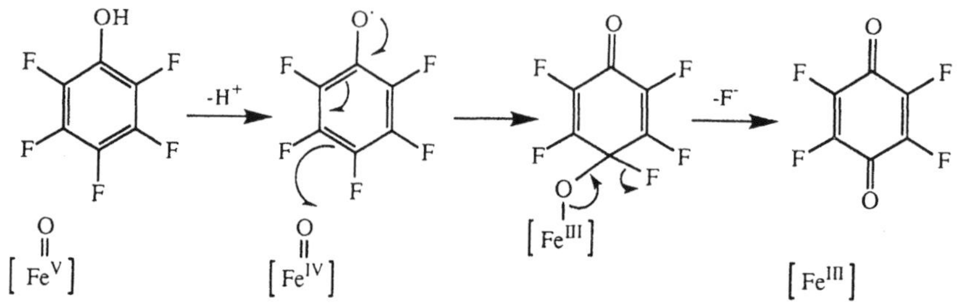

Fig. 15 Proposed mechanism for the oxidation of pentafluorophenol that results in fluoride elimination and formation of tetrafluoro *p*-quinone without the formation of an epoxide intermediate. The cytochrome P450 heme iron atom is represented by Fe. (From Ref. 54.)

only one such oxidation. However, sulfur has two unpaired electrons and can be oxidized twice to give sequentially a sulfoxide ($R_2S{=}O$) and a sulfone (R_2SO_2). The second oxidation is more difficult than the first because the electron pair is less available in a sulfone than in a thioether (57).

A. Nitrogen Oxidation

Tertiary amines, can be oxidized by the endoplastic reticulum to the corresponding *N*-oxides by either cytochrome P450 enzymes or the microsomal flavin monooxygenase (see Chapter 10). In general, but not always (58), cytochrome P450 enzymes catalyze the *N*-dealkylation of alkyl amines rather than the formation of an *N*-oxide. The flavin monooxygenase system forms only the *N*-oxide. It is not possible to attribute the formation of an *N*-oxide to either the P450 or flavoprotein monooxygenase system without evidence that specifically implicates one or the other of these two systems.

The dealkylation of alkyl amines, ethers, thioethers, and other alkyl-substituted heteroatoms is catalyzed by cytochrome P450 and not the flavin monooxygenase. The formal mechanism for this reaction involves introduction of a hydroxyl group adjacent to the heteroatom, followed by intramolecular elimination of the heteroatom. The products of the reaction, therefore, are an aldehyde or ketone and the dealkylated heteroatom-containing fragment of the substrate. In the ethers, the hydroxyl group is introduced by a conventional carbon hydroxylation reaction. The mechanism for introduction of the hydroxyl group adjacent to nitrogen is less well established. One alternative is direct hydroxylation. However, the lower electronegativity of nitrogen than of oxygen makes possible electron transfer from the nitrogen to the ferryl species (Fig. 16). Abstraction of the proton from the adjacent carbon followed by recombination of the iron-bound hydroxyl group with the resulting radical completes the reaction (Fig. 16). Evidence is available for both reaction pathways and the possibility exists that the pathway will be substrate-dependent (38,59).

B. Sulfur Oxidation

The oxidation of thioethers to sulfoxides and the dealkylation of alkyl thioethers is subject to the same considerations as the metabolism of alkyl amines. Sulfoxidation can be catalyzed by both P450 enzymes and flavin monooxygenases. *S*-Dealkylation, almost invari-

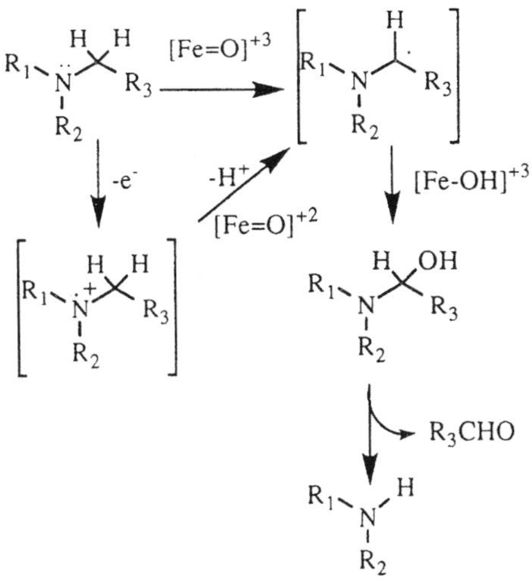

Fig. 16 Mechanistic alternatives for the hydroxylation adjacent to a nitrogen atom.

ably mediated by cytochrome P450, involves carbon hydroxylation, followed by extrusion of the heteroatom.

The P450-catalyzed oxidation of thiocarbonyl groups is unusual in that it leads to elimination of the sulfur to yield a simple carbonyl moiety (Fig. 17). Transfer of the ferryl oxygen to the sulfur of the thiocarbonyl group produces an unstable, unsaturated sulfoxide that decomposes to the observed carbonyl product. The sulfur is eliminated, probably as HSOH, by a hydrolytic mechanism that may be assisted by glutathione (60,61).

XI. UNUSUAL OXIDATIVE REACTIONS

The range of reactions catalyzed by cytochrome P450 continues to expand, although the more novel reactions have been observed to occur only in low yields or in special circumstances. Two of these reactions are described in this section, more as a notice that unusual

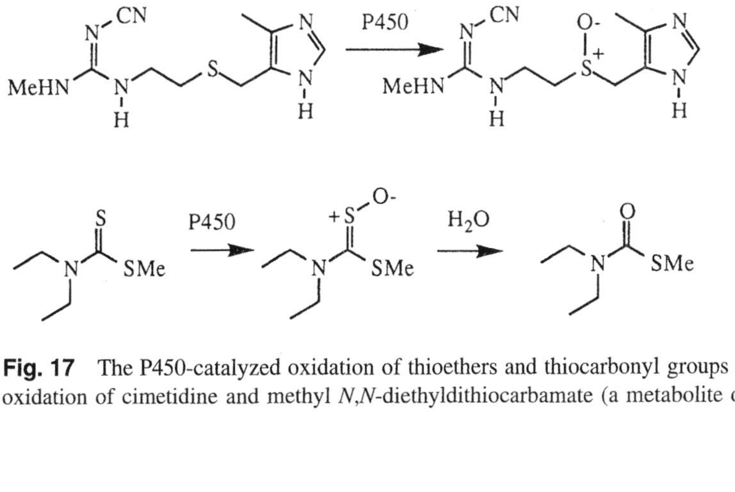

Fig. 17 The P450-catalyzed oxidation of thioethers and thiocarbonyl groups as illustrated by the oxidation of cimetidine and methyl *N,N*-diethyldithiocarbamate (a metabolite of disulfiram).

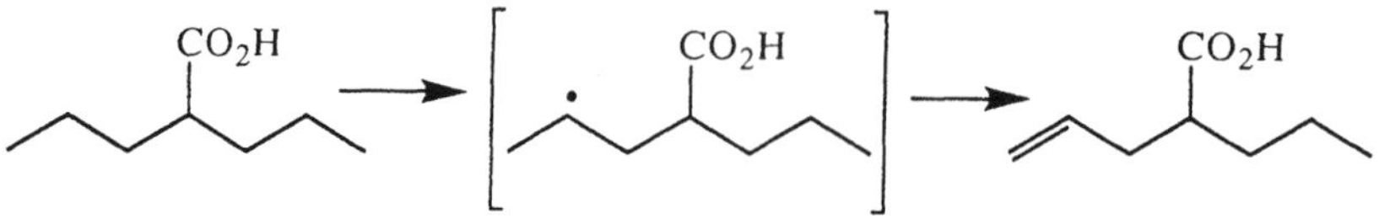

Fig. 18 Cytochrome P450-catalyzed desaturation of valproic acid.

reactions can be expected to occur than because the two reactions can be said, so far, to represent important P450 processes.

One of the P450-catalyzed reactions that has only recently been recognized is the desaturation of hydrocarbon residues. The most thoroughly investigated of these reactions is the desaturation of valproic acid to give the terminal olefin (Fig. 18; 62). Isotopic effect studies suggest that this reaction is initiated by hydrogen abstraction from the ω-1 position. This abstraction is followed, not by recombination to give the 4-hydroxy metabolite (which also occurs), but by a second hydrogen atom abstraction from the terminal carbon. Mammalian cytochrome P450 enzymes similarly catalyze the $\Delta^{6,7}$-desaturation of testosterone (63) and the desaturation of lovastatin (64) and simvastatin (65).

The reaction of aldehydes with cytochrome P450 also proceeds by two pathways. One of them is a relatively conventional oxidation to give the carboxylic acid (66), but the other is a process that results in elimination of the aldehyde group as formic acid (Fig. 19; 67). The carbon–carbon bond cleavage reaction finds strong precedent in the reactions catalyzed by the biosynthetic P450 enzymes lanosterol demethylase, aromatase, and sterol $C^{17,20}$-lyase. Cleavage of a carbon–carbon bond in the reaction catalyzed by each of these three enzymes has been proposed to involve the reaction of an aldehyde or ketone intermediate with the ferrous dioxy intermediate of cytochrome P450 (68). A similar mechanism is proposed for the reactions of simple aldehydes with hepatic P450 enzymes, with the difference that the carbon–carbon bond cleavage process is the major or exclusive reaction in the biosynthetic enzymes, but is a very minor reaction in the oxidation of xenobiotic aldehydes.

XII. REDUCTIVE REACTIONS

In addition to oxidative reactions, cytochrome P450 is known to catalyze reductive reactions (69). These reductive reactions are most important under anaerobic conditions, but can, in some instances, compete with oxidative reactions under aerobic conditions. The principal reductive reaction specifically catalyzed by cytochrome P450 is the dehalogenation of alkyl halides. Cytochrome P450, or at least cytochrome P450 reductase, is also involved in reactions such as the reduction of azo and nitro compounds. The important

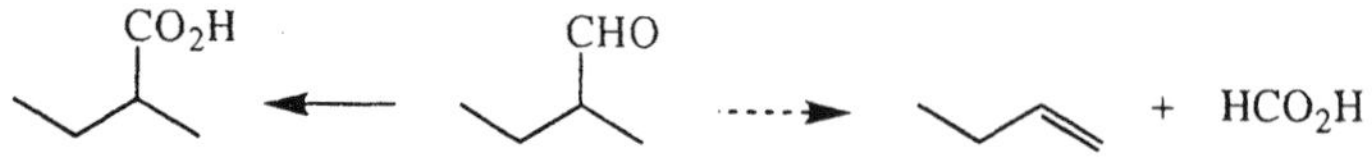

Fig. 19 Aldehyde oxidation by cytochrome P450 yields primarily the corresponding acid but, as a minor pathway, sometimes produces the unsaturated hydrocarbon from which the aldehyde carbon has been eliminated.

catalytic species is the ferrous deoxy intermediate in which the reduced iron is not coordinated to oxygen. Reductive reactions thus compete with the binding and activation of oxygen, a fact that explains the oxygen sensitivity of the pathway. These reactions are discussed in Chapter 11.

REFERENCES

1. H. Vainio and E. Hietanen, Role of extrahepatic metabolism, *Concepts in Drug Metabolism* (P. Jenner and B. Testa, eds.), Marcel Dekker, New York, 1980, pp. 251–284.
2. T. L. Poulos, B. C. Finzel, and A. J. Howard, High-resolution crystal structure of cytochrome $P450_{cam}$, *J. Mol. Biol. 195*: 687 (1987).
3. K. G. Ravichandran, S. S. Boddupalli, C. A. Hasemann, J. A. Peterson, and J. Deisenhofer, Crystal structure of hemoprotein domain of P450BM-3, a prototype for microsomal P450s, *Science 261*: 731 (1993).
4. C. A. Hasemann, K. G. Ravichandran, J. A. Peterson, and J. Deisenhofer, Crystal structure and refinement to cytochrome $P450_{terp}$ at 2.3 Å resolution, *J. Mol. Biol. 236*: 1169 (1994).
5. J. R. Cupp-Vickery, and T. L. Poulos, Structure of cytochrome $P450_{eryF}$ involved in erythromycin biosynthesis, *Struct. Biol. 2*: 144 (1995).
6. D. Nelson, L. Koymans, T. Kamataki, J. Stegeman, R. Feyereisen, D. Waxman, M. Waterman, O. Gotoh, M. Coon, R. Estabrook, L. Gunsalus, and D. Nebert, P450 superfamily: Update on new sequences, gene mapping, accession numbers and nomenclature, *Pharmacogenetics 6*: 1 (1996).
7. F. P. Guengerich, *Cytochrome P450: Structure, Mechanism, and Biochemistry,* 2nd ed. (P. R. Ortiz de Montellano, ed.), Plenum, New York, 1995, pp. 473–535.
8. F. Gonzalez, Cytochrome P450 in humans, *Cytochrome P450* (J. B. Schenkman and H. Greim, eds.), Springer-Verlag, New York, 1993, pp. 239–257.
9. T. Shimada, H. Yamazaki, M. Mimura, Y. Inui, and F. P. Guengerich, Interindividual variations in human liver cytochrome P450 enzymes involved in the oxidation of drugs, carcinogens, and toxic chemicals: Studies with liver microsomes of 30 Japanese and 30 Caucasians, *J. Pharmacol. Exp. Ther. 270*: 414 (1994).
10. K. Nakamura, F. Goto, W. A. Ray, C. B. McAllister, and E. Jacqz, Interethnic differences in genetic polymorphism of debrisoquine and mephenytoin hydroxylation between Japanese and Caucasian populations, *Clin. Pharmacol. Ther. 38*: 402 (1987).
11. W. Kalow, Genetics of drug transformation, *Clin. Biochem. 19*: 76 (1986).
12. M. A. Correia, *Cytochrome P450: Structure, Mechanism, and Biochemistry*, 2nd ed. (P. R. Ortiz de Montellano, ed.), Plenum, New York, 1995, pp. 607–630.
13. J. B. Schenkman, S. G. Sligar, and D. L. Cinti, Substrate interaction with cytochrome P450, *Hepatic Cytochrome P-450 Monooxygenase System* (J. B. Schenkman and D. Kupfer, eds.), Pergamon Press, New York, 1982, pp. 587–615.
14. J. H. Dawson and M. Sono, Cytochrome P450 and chloroperoxidase: Thiolate-ligated heme enzymes. Spectroscopic determination of their active site structures and mechanistic implications of thiolate ligation, *Chem. Rev. 87*: 1255 (1987).
15. T. L. Poulos, B. C. Finzel, and A. J. Howard, Crystal structure of substrate-free *Pseudomonas putida* cytochrome P-450, *Biochemistry 25*: 5324 (1986).
16. G. D. Sweeney, and J. D. Rothwell, Spectroscopic evidence of interaction between 2-allyl-2-isopropylacetamide and cytochrome P-450 of rat liver microsomes, *Biochem. Biophys. Res. Commun. 55*: 798 (1973).
17. E. J. Mueller, P. J. Loida, and S. G. Sligar, Twenty-five years of $P450_{cam}$ research, *Cytochrome P450: Structure, Mechanism, and Biochemistry*, 2nd ed. (P. R. Ortiz de Montellano, ed.), Plenum, New York, 1995, pp. 83–124.

18. S. G. Sligar. Coupling of spin, substrate and redox equilibria in cytochrome P-450, *Biochemistry 15*: 5399 (1976).
19. J. A. Peterson, Y. Ishura, and B. W. Griffin, *Pseudomonas putida* cytochrome P-450: Characterization of an oxygenated form of the hemoprotein, *Arch. Biochem. Biophys. 149*: 197 (1972).
20. C. Bonfils, P. Debey, and P. Maurel, Highly purified microsomal cytochrome P-450: The oxyferro intermediate stabilized at low temperature, *Biochem. Biophys. Res. Commun. 88*: 1301 (1979).
21. P. R. Ortiz de Montellano, Oxygen activation and transfer, *Cytochrome P450: Structure, Mechanism, and Biochemistry* (P. R. Ortiz de Montellano, ed.), Plenum Press, New York, 1986, pp. 217–271.
22. H. W. Strobel, A. V. Hodgson, and S. Shen, NADPH cytochrome P450 reductase and its structural and functional domains, *Cytochrome P450: Structure, Mechanism, and Biochemistry*, 2nd ed. (P. R. Ortiz de Montellano, ed.), Plenum, New York, 1995, pp. 225–244.
23. S. D. Black, and M. J. Coon, Structural features of liver microsomal NADPH–cytochrome P450 reductase: Hydrophobic domain, hydrophilic domain, and connecting region, *J. Biol. Chem. 257*: 5929 (1982).
24. S. Djordevic, D. L. Roberts, M. Wang, T. Shea, M. G. W. Camitta, B. S. S. Masters, and J. J. P. Kim, Crystallization and preliminary x-ray studies of NADPH–cytochrome P450 reductase, *Proc. Natl. Acad. Sci. USA 92*: 3214 (1995).
25. C. Bonfils, C. Balny, and P. Maurel, Direct evidence for electron transfer from ferrous cytochrome b_5 to the oxyferrous intermediate of liver microsomal cytochrome P-450 LM2, *J. Biol. Chem. 256*: 9457 (1981).
26. N. Oshino, Cytochrome b_5 and its physiological significance, *Hepatic Cytochrome P-450 Monooxygenase System* (J. B. Schenkman and D. Kupfer, eds.), Pergamon Press, New York, pp. 407–447 (1982).
27. W. M. Atkins and S. G. Sligar, Molecular recognition in cytochrome P450: Alteration of regioselective alkane hydroxylation via protein engineering, *J. Am. Chem. Soc. 111*: 2715 (1989).
28. W. M. Atkins and S. G. Sligar, The roles of active site hydrogen bonding in cytochrome P-450_{cam} as revealed by site-directed mutagenesis, *J. Biol. Chem. 263*: 18842 (1988).
29. H. Furuya, T. Shimizu, K. Hirano, M. Hatano, Y. Fujii-Kuriyama, R. Raag, and T. L. Poulos, Site-directed mutagenesis of rat liver cytochrome P-450*d*: Catalytic activities toward benzphetamine and 7-ethoxycoumarin, *Biochemistry 28*: 6848 (1989).
30. J. R. Cupp-Vickery, O. Han, C. R. Hutchinson, and T. L. Poulos, Substrate-assisted catalysis in cytochrome P450_{eryF}, *Nature Struct. Biol. 3*: 632 (1996).
31. R. Raag and T. L. Poulos, Crystal structure of the carbon monoxide–substrate–cytochrome P-450cam ternary complex, *Biochemistry 28*: 7586 (1989).
32. S. Modi, M. J. Sutcliffe, W. U. Primrose, L. Lian, and G. C. K. Roberts, The catalytic mechanism of cytochrome P450 BM3 involves a 6 Å movement of the bound substrate on reduction, *Nature Struct. Biol. 3*: 414 (1996).
33. C. Von Wachenfeldt and E. F. Johnson, Structures of eukaryotic cytochrome P450 enzymes, *Cytochrome P450: Structure, Mechanism, and Biochemistry*, 2nd ed. (P. R. Ortiz de Montellano, ed.), Plenum, New York, 1995, pp. 183–223.
34. O. Gotoh, Substrate recognition sites in cytochrome P450 family 2 (CYP2) proteins inferred from comparative analyses of amino acid and coding nucleotide sequences, *J. Biol. Chem. 267*: 83 (1992).
35. J. J. M. Z. Markéta, C. R. Wolf, and M. J. E. Sternberg, A predicted three-dimensional structure of human cytochrome P450: Implications for substrate specificity, *Protein Eng. 4*: 271 (1991).
36. G. Szklarz, Y. He, and J. Halpert, Site-directed mutagenesis as a tool for molecular modeling of cytochrome P450 2B1, *Biochemistry 34*: 14312 (1995).
37. S. Modi, M. J. Paine, M. J. Sutcliffe, L.-Y. Lian, W. U. Primrose, C. R. Wolf, and G. C. K.

Roberts, A model for human cytochrome P450 2D6 based on homology modeling and NMR studies of substrate binding, *Biochemistry 35*: 4540 (1996).

38. P. R. Ortiz de Montellano, Oxygen activation and reactivity, *Cytochrome P450: Structure, Mechanism, and Biochemistry*, 2nd ed. (P. R. Ortiz de Montellano, ed.), Plenum, New York, 1995, pp. 245–303.
39. K. Oguri, Y. Tanimoto, and H. Yoshimura, Metabolite fate of strychnine in rats, *Xenobiotica 19*: 171 (1989).
40. L. M. Hjelmeland, L. Aronow, and J. R. Trudell, Intramolecular determination of primary kinetic isotope effects in hydroxylations catalyzed by cytochrome P-450, *Biochem. Biophys. Res. Commun. 76*: 541 (1977).
41. U. Frommer, V. Ullrich, and H. Staudinger, Hydroxylation of aliphatic compounds by liver microsomes. 1. The distribution of isomeric alcohols, *Hoppe-Seylers Z. Physiol. Chem. 351*: 903 (1970).
42. R. E. White, M.-B. McCarthy, K. D. Egeberg, and S. G. Sligar, Regioselectivity in the cytochromes P-450: Control by protein constraints and by chemical reactivities, *Arch. Biochem. Biophys. 228*: 493 (1984).
43. J. T. Groves and Y.-Z. Han, Models and mechanisms of cytochrome P450 action, *Cytochrome P450: Structure, Mechanism, and Biochemistry*, 2nd ed. (P. R. Ortiz de Montellano, ed.), Plenum, New York, 1995, pp. 3–48.
44. J. T. Groves, G. A. McClusky, R. E. White, and M. J. Coon, Aliphatic hydroxylation by highly purified liver microsomal cytochrome P-450: Evidence for a carbon radical intermediate, *Biochem. Biophys. Res. Commun. 81*: 154 (1978).
45. J. Hefner, S. M. Rubenstein, R. E. B. Ketchum, D. M. Gibson, R. M. Williams, and R. Croteau, Cytochrome P450-catalyzed hydroxylation of taxa-4(5),11(12)-diene to taxa-4(20), 11(12)dien-5α-ol: The first oxygenation step in taxol biosynthesis, *Chem. Biol. 3*: 479 (1996).
46. J. A. Hinson, J. R. Mitchell, and D. J. Jollow, *N*-Hydroxylation of *p*-chloroacetanilide in hamsters, *Biochem. Pharmacol. 25*: 599 (1975).
47. P. R. Ortiz de Montellano, B. L. K. Mangold, C. Wheeler, K. L. Kunze, and N. O. Reich, Stereochemistry of cytochrome P-450 catalyzed epoxidation and prosthetic heme alkylation, *J. Biol. Chem. 258*: 4208 (1983).
48. R. E. Miller and F. P. Guengerich, Oxidation of trichloroethylene by liver microsomal cytochrome P-450: Evidence for chlorine migration in a transition state not involving trichloroethylene oxide, *Biochemistry 21*: 1090 (1982).
49. D. J. Harvey, L. Glazener, D. B. Johnson, C. M. Butler, and M. G. Horning, Comparative metabolism of four allylic barbiturates and hexobarbital by rat and guinea pig, *Drug Metab. Dispos. 5*: 527 (1977).
50. T. Woolf and F. P. Guengerich, Rat liver cytochrome P-450 isozymes as catalysts of aldrin epoxidation in reconstituted monooxygenase systems and microsomes, *Biochem. Pharmacol. 36*: 2581 (1987).
51. P. R. Ortiz de Montellano, and M. A. Correia, Inhibition of cytochrome P450 enzymes, *Cytochrome P450: Structure, Mechanism, and Biochemistry*, 2nd ed. (P. R. Ortiz de Montellano, ed.), Plenum, New York, 1995, pp. 305–364.
52. D. Ostovic, and T. C. Bruice, Mechanism of alkene epoxidation by iron, chromium, and manganese higher valent oxo-metalloporphyrins, *Acct. Chem. Res. 25*: 314 (1992).
53. D. M. Jerina and J. W. Daly, Arene oxides: A new aspect of drug metabolism, *Science 185*: 573 (1974).
54. C. Den Besten, P. J. van Bladeren, E. Duizer, J. Vervoort, and I. M. C. M. Rietjens, Cytochrome P450-mediated oxidation of pentafluorophenol to tetrafluorobenzoquinone as the primary reaction product, *Chem. Res. Toxicol. 6*: 674 (1993).
55. S. A. Barker, L. C. Hsieh, T. R. McDowell, and C. R. Short, Qualitative and quantitative analysis of the anthelminthic febendazole and its metabolites in biological matrices by direct exposure probe mass spectrometry, *Biomed. Environ. Mass Spectrom. 14*: 161 (1987).

56. R. P. Guengerich, Oxidation of halogenated compounds by cytochrome P-450, peroxidases, and model metalloporphyrins, *J. Biol. Chem. 264*: 17198 (1989).
57. J. C. Alvarez and P. R. Ortiz de Montellano, Thianthrene-5-oxide as a probe of the electrophilicity of hemoprotein oxidizing species, *Biochemistry 31*: 8315 (1992).
58. Y. Seto and F. P. Guengerich, Partitioning between *N*-dealkylation and *N*-oxygenation in the oxidation of *N,N*-dialkylarylamines catalyzed by cytochrome P450 2B1, *J. Biol. Chem. 268*: 9986 (1993).
59. S. B. Karki, J. P. Dinnocenzo, J. P. Jones, and K. R. Korzekwa, Mechanism of oxidative amine dealkylation of substituted *N,N*-dimethylanilines by cytochrome P-450: Application of isotope effect profiles, *J. Am. Chem. Soc. 117*: 3657 (1995).
60. B. W. Hart and M. D. Faiman, Bioactivation of *S*-methyl *N,N*-diethylthiolcarbamate to *S*-methyl *N,N*-diethylthiolcarbamate sulfoxide. Implications for the role of cytochrome P450, *Biochem. Pharmacol. 46*: 2285 (1993).
61. A. Madan, T. D. Williams, and M. D. Faiman, Glutathione- and glutathione-*S*-transferase-dependent oxidative desulfuration of the thione xenobiotic diethyldithiocarbamate methyl ester, *Mol. Pharmacol. 46*: 1217 (1994).
62. A. E. Rettie, M. Boberg, A. W. Rettenmeier, and T. A. Baillie, Cytochrome P-450-catalyzed desaturation of valproic acid in vitro. Species differences, induction effects, and mechanistic studies, *J. Biol. Chem. 263*: 13733 (1988).
63. K. R. Korzekwa, W. F. Trager, K. Nagata, A. Parkinson, and J. R. Gillette, Isotope effect studies on the mechanism of the cytochrome P450IIA1-catalyzed formation of Δ^6-testosterone from testosterone, *Drug Metab. Dispos. 18*: 974 (1990).
64. K. P. Vyas, P. H. Kari, S. R. Prakash, and D. E. Duggan, Biotransformation of lovastatin. II. In vitro metabolism by rat and mouse liver microsomes and involvement of cytochrome P-450 in dehydrogenation of lovastatin, *Drug Metab. Dispos. 18*: 218 (1990).
65. S. Vickers and C. A. Duncan, Studies on the metabolic inversion of the 6′ chiral center of simvastatin, *Biochem. Biophys. Res. Commun. 181*: 1508 (1991).
66. K. Watanabe, S. Narimatsu, I. Yamamoto, and H. Yoshimura, Oxygenation mechanism in conversion of aldehydes to carboxylic acid catalyzed by a cytochrome P-450 isozyme, *J. Biol. Chem. 266*: 2709 (1991).
67. E. S. Roberts, A. D. N. Vaz, and M. J. Coon, Catalysis by cytochrome P-450 of an oxidative reaction in xenobiotic aldehyde metabolism: Deformylation with olefin formation, *Proc. Natl. Acad. Sci. USA 88*: 8963 (1991).
68. M. Akhtar, V. C. O. Njar, and J. N. Wright, Mechanistic studies on aromatase and related C–C bond cleaving P-450 enzymes, *J. Steroid Biochem. Mol. Biol. 44*: 375 (1993).
69. A. R. Goeptar, H. Scheerens, and N. P. E. Vermeulen, Oxygen and xenobiotic reductase activities of cytochrome P450, *Crit. Rev. Toxicol. 25*: 25 (1995).

5
Transformation Enzymes: Oxidative; Non-P450

Allan E. Rettie and Michael B. Fisher
University of Washington, Seattle, Washington

I. GENERAL INTRODUCTION TO METABOLISM BY NON-P450 MONOOXYGENASES

Currently, there is considerable interest in the identity of specific human oxidative enzymes responsible for the termination, or unmasking, of a drug's pharmacological effects (1). Such information can facilitate the rational, prospective evaluation of inhibitory drug–drug interactions and improve the safety profile of many substances, particularly those with low therapeutic indices (2). Significant advances in cytochrome P450 research, including the cloning and expression of most of the principal human liver forms of the enzyme, and the identification of P450 isozyme-selective inhibitors and inducers now permit the identification of specific human P450 isozymes involved in a given in vivo biotransformation pathway, with a high degree of confidence (3,4). However, frequently, our understanding of the multiplicity, substrate specificity, and regulation of *non-P450* oxidases, is much less well developed.

This chapter will focus on three mammalian nonheme oxygenases, which can conveniently be categorized by the nature of their active centers as either *flavin-containing* or *molybdenum-containing* enzymes. The flavin-containing monooxygenase (FMO) isoforms are located predominantly in the microsomal cell fraction, whereas the molybdenum-containing enzymes, aldehyde oxidase (AO) and xanthine oxidase (XO), are both located in the soluble fraction. Each of the three enzymes will be discussed separately in terms of their catalytic mechanism, multiplicity, tissue distribution and regulation; their biotransformation reactions; and their relevance to human drug metabolism.

Although other non-P450 enzyme systems, such as the alcohol, aldehyde, and dihydrodiol dehydrogenases, monoamine oxidase, semicarbazide-sensitive amine oxidase, and the enzymes of mitochondrial and peroxisomal β-oxidation, are also clearly capable of carrying out xenobiotic oxidation (5–10), their quantitative contribution to drug metabolism is either too low, or our present knowledge of their basic biochemistry is insufficiently comprehensive to warrant discussion in the format just described for the FMOs and molybdozymes.

II. MICROSOMAL FLAVIN-CONTAINING MONOOXYGENASE

Mammalian FMO is an NADPH-dependent and oxygen-dependent microsomal FAD-containing enzyme system that functions as a sulfur, nitrogen, and phosphorus oxygenase. For many years following its initial purification from hog liver (11) in the early 1970s, only one form of the enzyme (FMO1) was recognized, and extensive studies have been carried out by Ziegler and his associates on the mechanism of action, substrate specificity, toxicological importance, and physiological significance of hog liver FMO1 (12,13, and references therein). In recent years, molecular biology techniques have identified a family of five structurally related isoforms that appear to be expressed in all mammalian species yet examined (14).

A. Catalytic Mechanism

The simplest catalytic cycle consistent with available experimental data involves the sequential binding of NADPH and oxygen to the enzyme to generate an FAD C-4α-hydro-

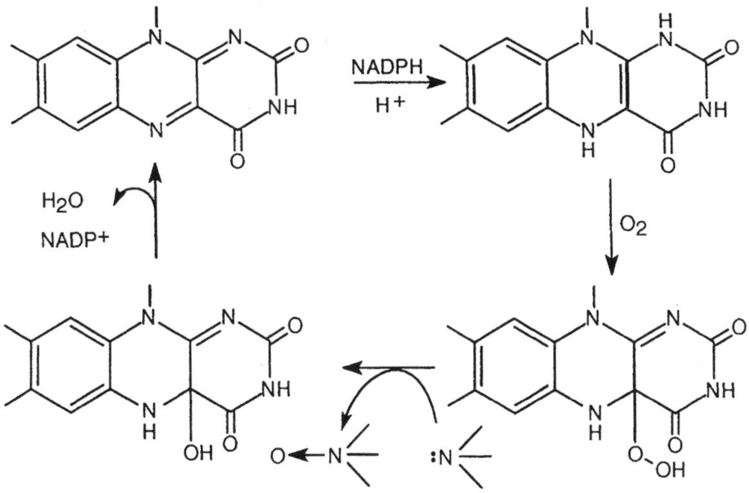

Fig. 1 Catalytic cycle for FMO-dependent oxidation.

peroxide (Fig. 1). Substrate nucleophiles (e.g., trimethylamine) attack the distal oxygen of this hydroperoxide with resultant oxygen transfer to the substrate, and the generation of a hydroxyflavin species. The rate-limiting step in FMO catalysis is considered to be the subsequent decomposition of the hydroxyflavin. Because this step follows transfer of oxygen to the substrate, it provides a rationalization for the observation that the V_{max} for hog liver FMO1-dependent reactions is relatively constant in the absence of uncoupling. This kinetic mechanism has been examined in detail only for hog FMO1 (11), but there is no reason to expect that it will differ significantly among the various isoforms.

B. Multiplicity, Tissue Distribution, and Species Differences for the Mammalian Isoforms

The mammalian FMO gene family is composed of a minimum of five members termed; FMO1, FMO2, FMO3, FMO4, and FMO5 (14). The primary sequences of the five forms are 50–55% identical with each other and each contains 1 mol of noncovalently bound FAD at the enzyme active site. Two conserved glycine-rich regions toward the *N*-terminal region of each isoform are associated with binding of NADPH and FAD.

The most detailed analysis of the tissue distribution of these five isoforms has been carried out in the rabbit. FMO1 mRNA expression is highest in liver and intestine, whereas FMO2 mRNA is highest in lung (15,16). Variable levels of mRNA encoding FMO3, FMO4, and FMO5 are present in rabbit liver and kidney, but low to undetectable levels of these isoforms exist in rabbit lung (15,17). As a general rule, FMO1 is the dominant form of the enzyme in the liver of most experimental animals. However, it is not expressed to any significant extent in adult human liver, where FMO3 appears to be the major isoform (18,19). Because the substrate specificities of the individual FMO isoforms can differ significantly (20,21), it is important to exercise caution when attempting to extrapolate liver microsomal data for FMO catalysis from experimental animals to humans.

C. Enzyme Regulation

Unlike the cytochrome P450s, the FMO system is not induced by exogenously administered xenobiotics, such as the barbiturates or polycyclic hydrocarbons. However, the FMOs do appear to be subject to hormonal and developmental regulation, and it is becoming increasingly evident that dietary constituents can exert considerable effects on the hepatic and extrahepatic levels of certain FMO isoforms.

1. *Dietary Control*

Kaderlik et al. (22) reported that rats that were switched from commercial chow to a chemically defined diet exhibited a pronounced loss of liver FMO activity. This suggested that some components of the ''normal'' diet were capable of FMO induction. More recently, Larsen-Su and Williams (23) have demonstrated that dietary indole-3-carbinol, a plant alkaloid found in cruciferous vegetables, dramatically inhibits the expression of FMO1 in rat liver and intestine. Interestingly, indole-3-carbinol induces hepatic levels of CYP1A1 (and several other P450s) in rats. Therefore, the capacity appears to exist for dietary constituents to alter the hepatic CYP/FMO ratio.

2. *Hormonal Control*

The FMO2 enzyme was induced in late gestation in maternal rabbit lung (24), and subsequent studies showed that this increase was related to elevated levels of progesterone (25). In contrast, FMO2 expression in the kidney appears to be more closely linked to glucocorticoid levels (26). Sex-dependent expression of FMOs is pronounced in mouse liver. Female mice express higher general FMO-dependent liver activities than males, which is due to female-predominant and female-specific hepatic expression of FMO1 and FMO3, respectively (27).

3. *Developmental Control*

Perhaps the most striking example of developmental control concerns the selective expression of FMO1 in fetal human liver and FMO3 in adult human liver. The mRNA for FMO1 is below detectable limits in adult human liver, although it is easily quantified in human fetal liver, whereas the reverse is true for FMO3 mRNA (19). In addition, form-specific antibodies for FMO3 identify this protein in adult human liver microsomes, but not to any significant extent in fetal liver or adult kidney (28). Conclusions drawn from the mRNA and immunochemical studies are paralleled by functional data for differential stereoselective sulfoxide formation catalyzed by adult and fetal human tissue microsomal FMO (29). Further studies are required to determine both the temporal course of hepatic FMO3 up-regulation–FMO1 down-regulation and trigger mechanism(s) for these events.

D. Transformation Reactions

The wide panoply of metabolic transformations of which mammalian FMOs are capable is exemplified by consideration of the substrate specificity of hog liver FMO1. Extensive structure–function studies with this purified form of the enzyme have shown that, in principle, any soft nucleophile that gains access to the active site of FMO will be oxygenated. Several excellent reviews on this topic have appeared (12,13,30,31). Therefore, the present discussion will consider only a few illustrative examples of oxidation at nitrogen and sulfur centers—the most commonly encountered nucleophilic sites in drug molecules.

1. Oxidation at Nitrogen Centers

A very large number of nitrogen-containing functionalities are found in xenobiotics. Given the nucleophilic mechanism of catalysis, it might be expected that FMO substrates would exhibit a minimum basicity requirement. However, predictions of substrate specificity based on pK_a alone do not accommodate all of the experimental observations, and it is clear that both steric effects of the substrates themselves and the relative expression levels of specific FMO and P450 isoforms in a given species will determine the nature of the enzyme system involved in vivo. Nonetheless, xenobiotics that contain an sp^3-hybridized nitrogen and exhibit a pK_a between 5 and 10 are, in the main, candidate substrates for FMO.

A major group of substrates for the FMOs are the *tertiary acyclic and cyclic amines*. Substrates such as trimethylamine (Fig. 1), benzydamine, clozapine, and guanethidine (Fig. 2) are converted to stable *N*-oxides. The levels of *N*-oxide metabolite formed may often be underestimated owing to relatively facile reduction back to the parent amine. Although cytochrome P450s tend to *N*-dealkylate tertiary amines, their formation of *N*-oxides has also been documented (32). FMOs also catalyze the metabolism of secondary and primary amine functionalities (33), but are usually not involved in the oxidation of amides, heteroaromatic amines, benzamidines, or guanidines.

2. Oxidation at Sulfur Centers

Thioethers, such as sulindac sulfide (see Fig. 2) are generally better substrates for FMO than the tertiary amines discussed in the foregoing, owing to the enhanced nucleophilicity of the sulfur atom. Relative nucleophilicity also explains why thioamides are excellent substrates and sulfoxides are usually poorer substrates for FMO. Oxidation of thioethers and thioamides is also carried out readily by cytochrome P450. The relative participation of these two monooxygenase systems in sulfur oxidation will depend on their levels of expression in the metabolizing tissue and the substrate specificity of the FMO isoforms involved.

3. Diagnostic Substrates and Inhibitors

As is evident from the foregoing discussion, both FMO and P450 isoforms can catalyze oxidative reactions involving nitrogen and sulfur centers. Approaches to the differentiation of P450-mediated and FMO-dependent catalysis generally use selective substrates and inhibitors for both enzyme systems. For example, the conversion of *N,N*-dimethylaniline to its *N*-oxide is probably one of the most widely used indicators of FMO catalysis in microsomal preparations and, as long as reactions are carried out at pH 8.5–9, only FMO(s) are likely to be involved. However, at physiological pH, even this most diagnostic of substrates may be turned over, in part, by cytochrome P450s (32,34). The anti-inflammatory drug, benzydamine (see Fig. 2) is an excellent substrate for several human FMO isoforms (35), and the fluorescent *N*-oxide metabolite that is generated has permitted the development of a highly sensitive, simple, endpoint metabolite assay for FMO activity (36).

Chemical inhibitors suffer from the disadvantage that they are rarely, if ever, specific for FMO. Unfortunately, no mechanism-based inhibitors of FMO have been identified. However, 1–aminobenzotriazole, a suicide inhibitor of most P450s at high substrate concentrations, is a useful, indirect indicator of FMO catalysis, if preincubation with this

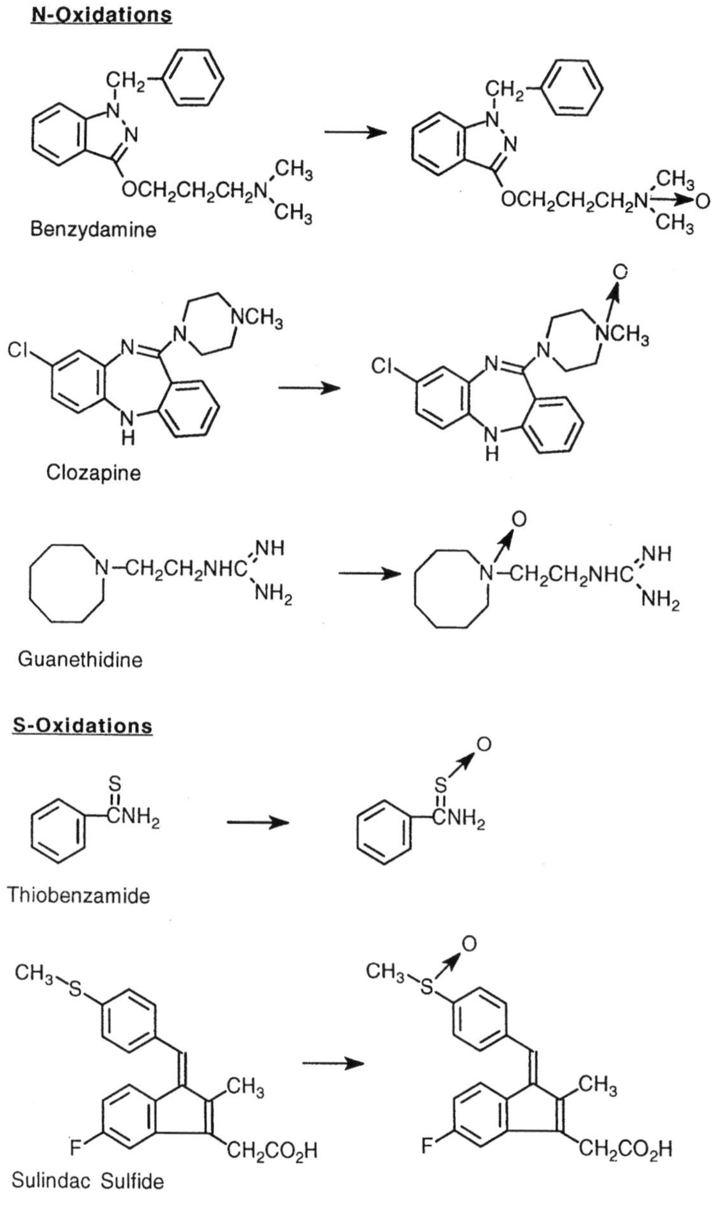

N-Oxidations

Benzydamine

Clozapine

Guanethidine

S-Oxidations

Thiobenzamide

Sulindac Sulfide

Fig. 2 Typical *N*-oxidation and *S*-oxidation reactions catalyzed by FMO.

compound does not decrease the reaction rate that is being monitored. Methimazole is a widely used inhibitor of FMOs, but it also competitively inhibits human CYP2B6, CYP2C9, and CYP3A4 at the relatively low substrate concentrations of 40–100 μM (37). *n*-Octylamine, although an activator of FMO1, either inhibits or has no effect on other FMO isoforms (15,17,38). Therefore, it is prudent to employ a battery of ''selective'' inhibition methods (39) when evaluating the in vitro role of FMO in a given oxidative pathway.

E. Relevance to Human Drug Metabolism

The conversion of tertiary amines to highly polar *N*-oxides is considered the prototypic FMO xenobiotic reaction pathway, and so it is not surprising that FMO has been implicated in the metabolism of a variety of tertiary amine central nervous systems (CNS)-active agents, including nicotine, olanzapine, clozapine, and certain dimethylaminoalkyl phenothiazine derivatives (40–43). As a consequence of these observations, there is considerable current interest in identifying the nature of brain FMO isoforms capable, perhaps, of attenuating the pharmacological activity of such tertiary amines directly at their sites of action within the CNS (44,45).

For nicotine, a deficiency in the *N*-oxide pathway cosegregates with trimethylaminuria (46), a genetic disorder that results from an inability to form trimethylamine *N*-oxide (47). The molecular basis underlying trimethylaminuria may be a P153L mutation in exon 4 of the FMO3 gene (48). However, it appears that the incidence of this polymorphism is very low and, therefore, unlikely to cause significant problems in patient populations.

Sulfoxides are quantitatively significant human metabolites of therapeutic agents such as the H_2-antagonist cimetidine, and the redox-active agent, sulindac sulfide. The latter drug is administered as a sulfoxide prodrug, and relies on metabolic reduction for its pharmacological activity. In vitro microsomal studies indicate that human liver FMO(s) can *S*-oxygenate both cimetidine and sulindac sulfide (49,50), and both of these sulfoxidation reactions proceed stereoselectively in human liver microsomes. Therefore, it may be possible to develop in vivo drug probes for human hepatic FMO, if stereochemical considerations are taken into account, and substrates can be identified that do not undergo significant redox cycling. It can be anticipated that numerous additional examples of human FMO involvement in xenobiotic metabolism will appear with the recent advent of commercial sources of catalytically competent recombinant forms of human FMO3.

III. XANTHINE DEHYDROGENASE–XANTHINE OXIDASE

Xanthine dehydrogenase (XD) and xanthine oxidase (XO) are two forms of a homodimeric, 300-kDa enzyme, for which much of the available information has been derived from detailed studies on the bovine milk form of the enzyme. Extensive studies have been performed in the past two decades on the role that tissue forms of the enzyme play in ischemia–reperfusion injury and in alcohol-related liver injury. Recently, more emphasis has been placed on the molecular characterization of human XD–XO in relation to the other important molybdozyme, aldehyde oxidase (AO), and these developments can be expected to help clarify the role of these enzymes in human drug metabolism and human health.

A. Catalytic Mechanism

Each subunit of XD/XO contains one atom of molybdenum in the form of a molybdopterin cofactor $[Mo^{VI}(=S)(=O)]^{2+}$, one FAD molecule, and two Fe_2–S_2 centers (51). The general reaction catalyzed by these enzymes follows Eq (1);

$$SH + H_2O \rightarrow SOH + 2e^- + 2H^+ \quad (1)$$

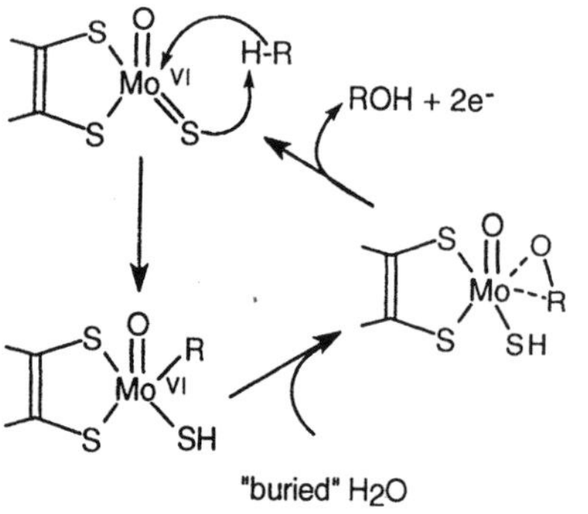

Fig. 3 Reaction mechanism for molybdozyme-dependent oxidation.

where SH is a reduced substrate and SOH is the hydroxylated metabolite. The oxidized molybdopterin provides the first electron acceptor and the source of the oxygen atom, which is derived from water (52). Electron flow in XD/XO continues either to one, or both, iron–sulfur clusters, on to the FAD, and then either to NAD^+ (dehydrogenase activity) or to oxygen (oxidase activity). Recognition that XD/XO can substitute an oxidized substrate for either of the two ultimate electron acceptors NAD^+ or O_2, and act essentially as an oxidoreductase, has considerably clarified the substrate specificities attributed to this enzyme. Compounds that interact with XD/XO can be categorized as either reducing (electron donor at molybdenum center) or oxidizing (electron acceptor from flavin) substrates. Several recent publications address the redox properties of XO (53–57).

For several years, the chemical mechanism of hydroxylation has been considered to entail Mo=S-catalyzed deprotonation of an electrophilic C–H bond in the substrate to provide an Mo^{VI}–SH intermediate (51). Concerted attack of the Mo=O ligand on the substrate carbanion then yields an Mo^{IV}–O–C species, followed by product release and electron transfer. Recently, however, an alternative mechanism has been proposed, based on indirect spectroscopic analysis, which is suggestive of a Mo–C bond and an oxygen insertion mechanism, instead of a traditional oxygen transfer (Fig. 3). Addition of substrate across the Mo=S double bond and incorporation of an active site water (hydroxide) would generate a unique, three-center Mo–C–O bond. Electron transfer and rearrangement could then yield hydroxylated substrate and regenerate the molybdopterin cofactor (58). Interestingly, deuterium and tritium isotopic effect data, coupled with other kinetic experiments, have suggested that hydrogen transfer in the hydroxylation reaction is not rate-limiting in the overall mechanism and, in fact, may be more than an order of magnitude faster than k_{cat} (59). Further work is needed to completely elucidate the general mechanism of XO-catalyzed reactions.

B. Multiplicity, Tissue Distribution, and Species Differences

Xanthine dehydrogenase and XO are two forms of the same enzyme. Much of the early studies were performed with the oxidase form, which was the first to be isolated and was initially thought to be the form found in vivo. In fact, most mammalian forms of this enzyme exist as the dehydrogenase form in vivo (60). In some instances, XO is an artifact of the isolation procedure. If care is not taken to inhibit proteolytic cleavage during purification, the dehydrogenase form of the enzyme is irreversibly converted to the oxidase. A major difference between the dehydrogenase and oxidase forms is the presence of a negative charge in the dehydrogenase form that is thought to participate in the binding of NAD^+ (60,61).

The molybdenum hydroxylases are considered to be cytosolic enzymes, although bovine milk XD/XO is associated with the lipid globules (62). Recently, a human form of XD has been cloned (63,64), and Northern blot analysis revealed human XD mRNA in all tissues examined, with the highest levels in heart, brain, liver, skeletal muscle, pancreas, small intestine, and colon (64). XD/XO mRNA has also been found in human placenta (65), and the gene has been localized to chromosome 2p22 (63,66).

The predicted amino acid sequence of cloned human XD shares 91% identity with rat and mouse XO (64). Comparisons of interspecies sequence alignments with those for *Drosophila* XO, and with the crystal structure of the XO-related fungal aldehyde oxidoreductase (67), have begun to identify important structural features of the enzyme. Molybdopterin- and iron–sulfur cofactor-binding areas have been identified tentatively by comparison with the fungal crystal structure (63,64,67), and a general consensus sequence for molybdopterin binding has been described. Recent studies on mutant rosy strains of *D. melanogaster* have confirmed the location of cofactor-binding sites in XD, the enzyme encoded by the rosy gene (68).

There is as yet little evidence for multiple XD/XO human isoforms and little is known about multiplicity in other species. Although the sequences of two human forms of XD have appeared in the literature, one of these was, in fact, the first human AO to be cloned (69,70).

C. Enzyme Regulation

The XO form of the enzyme transfers electrons ultimately to oxygen, with the concomitant generation of reactive oxygen species that can cause lipid peroxidation. Because the XD to XO conversion occurs under certain pathophysiological conditions, and is associated with a variety of toxicities, notably reperfusion injury, there continues to be much interest in the regulation of the two forms of the enzyme.

Ischemia promotes conversion to the oxidase form, and subsequent tissue reperfusion can lead to increased oxidase activity, significant production of reactive oxygen species, and the potential for increased cellular oxidative stress (71). Regulation, however, is very complex. Aldehydes that are produced during lipid peroxidation, such as 4-hydroxynonenal and malondialdehyde, have sudden and differential regulatory effects on the XD/XO system. By enhancing XD and decreasing XO activity, malondialdehyde acts as a feedback regulator for the generation of reactive oxygen species. Curiously, 4-hydroxynonenal enhances only XO activity (51). Whereas transcriptional activation of XD occurs under ischemic conditions (72), the messenger molecule nitric oxide has been reported to inhibit XO activity under ischemic conditions, possibly attenuating any subsequent oxidative damage on reperfusion (73).

Xanthine oxidase has been implicated in several other toxic responses. Lipopolysaccharide (LPS), an important substance in bacterial infection, enhances XO activity in various mouse tissues by both transcriptional and posttranscriptional mechanisms (74). In fact, XO has been implicated in LPS-induced cytotoxicity, by providing a source of reactive oxygen species. Ethanol facilitates conversion to XO, and this is compounded by the ability of XO to use acetaldehyde, derived from ethanol, as a source of electrons for the reduction of oxygen. This synergistic effect of ethanol may be important in ethanol-induced hepatotoxicity (75,76). Additionally, XO activity is elevated in most cystic fibrosis patients compared with healthy patients (77). The increased generation of reactive oxygen species in such patients was speculated to play a role in the pulmonary complications that arise in these patients.

A sex difference in enzyme activity exists in rats, with mature females possessing about half the XO activity of males (78). However, studies in humans performed either with liver biopsy samples or using caffeine metabolite ratios to assess XO levels (see later discussion), did not reveal a substantial sex difference in human XO activity (79,80). Xanthinuria is a rare genetic disorder in which affected individuals completely lack XO activity (81,82), although some cases may actually result from a combined deficiency in both XO and AO (83).

D. Transformation Reactions

The molybdenum hydroxylases carry out oxidation of electron-deficient sp^2-hybridized carbon atoms found, most commonly, in nitrogen heterocycles, such as purines and pyrimidines. The lowest electron density in these substrates usually occurs at carbons adjacent to the nitrogen atom(s), and so XO-mediated metabolism of reducing substrates, such as purine itself, leads sequentially to the formation of hypoxanthine, xanthine, and ultimately, uric acid (Fig. 4). The initial hydroxyimine products are seldom isolated and normally tautomerize to the α-aminoketone. Most of the substrate specificity studies with mammalian forms of the molybdenum hydroxylases have been conducted with AO and these will be discussed in Section IV. D.

1. *Interaction with Oxidizing Substrates*

The nature of the substrates and reactions involved with oxidizing substrates is less well characterized. So far, only nitrogen-containing functionalities, such as *N*-oxides and nitro groups, have been found to substitute for the ultimate electron acceptors NAD^+ or O_2 (84–86), although the substrate specificity for this newer class of XD/XO-catalyzed reactions needs to be determined. The presence of a reducing substrate is required in investigations of XD-catalyzed reductions to ''charge'' the enzyme with electrons.

2. *Diagnostic Substrates and Inhibitors*

Xanthine Oxidase and AO have overlapping, but not identical substrate and product specificities. Both enzymes hydroxylate purines, but AO appears to be more regiospecific than XO, preferring to catalyze C-8 hydroxylation (51). XO converts 1-methylxanthine, a secondary metabolite of caffeine, to 1-methyluric acid, and the urinary ratio of these two metabolites is an in vivo index of XO activity (87).

Allopurinol, a drug administered to patients suffering from gout, is metabolized by both XO and AO to alloxanthine, which is a potent, and selective, tight-binding inhibitor of XO. Allopurinol- or alloxanthine-dependent inhibition, in vitro or in vivo, can be used to evaluate XO-dependent biotransformation. More recently, the structure–activity relation of phenolic compounds for inhibition of XO has been determined (88,89). Interestingly, the oxidized coumarin nucleus is important for inhibition, with umbelliferone (7-hydroxycoumarin) and esculetin (7,8-dihydroxycoumarin) yielding K_i values of 10^{-6} M.

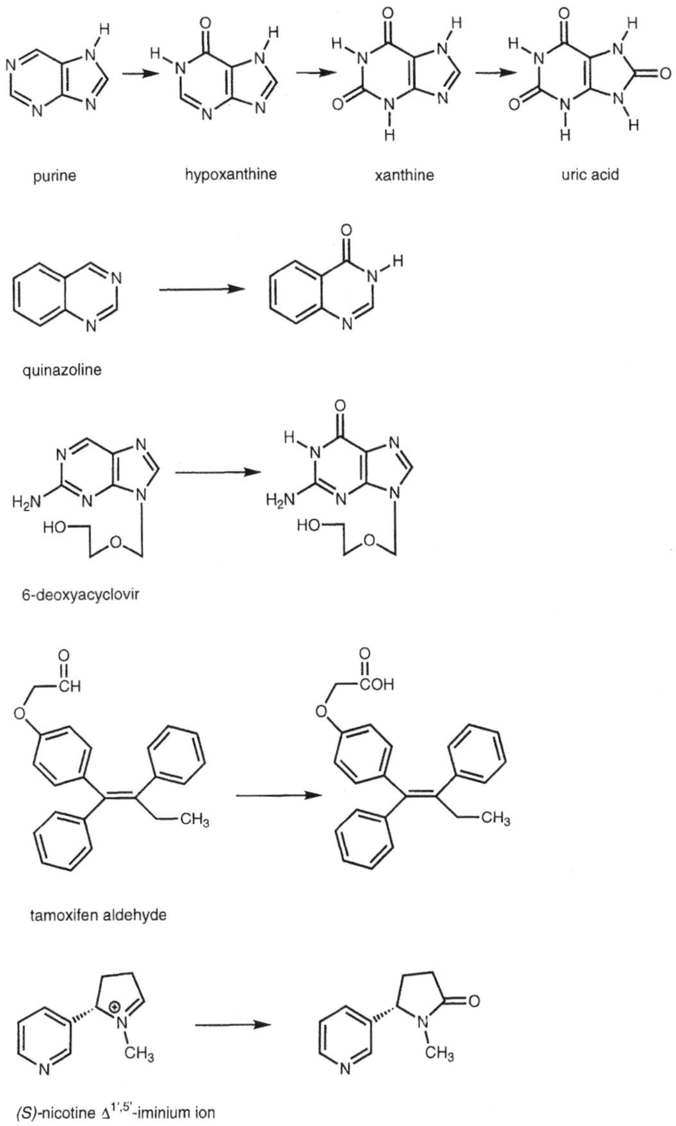

purine hypoxanthine xanthine uric acid

quinazoline

6-deoxyacyclovir

tamoxifen aldehyde

(S)-nicotine $\Delta^{1',5'}$-iminium ion

Fig. 4 Typical oxidative reactions catalyzed by XO and AO.

E. Relevance to Human Drug Metabolism

Relatively high levels of XD have been localized to human liver and small intestine, tissues implicated in the first-pass metabolism of a plethora of agents. XO plays a role in the oxidation of several chemotherapeutic agents and analogs and has been implicated in the bioactivation of mitomycin C, an antineoplastic compound considered to be cytotoxic through DNA alkylation and the generation of reactive oxygen species (90). The enzyme is involved in the first-pass metabolism of 6-mercaptopurine (to 6-thiouric acid), and the coadministration of allopurinol and mercaptopurine has been suggested as a strategy to overcome limited bioavailability (91). XO also appears to be involved in the metabolism of the chemoprotective agent 2,6-dithiopurine (92).

Several purine derivatives are also useful antiviral agents, but are poorly absorbed after oral administration. Attempts to improve bioavailability have centered around the use of more hydrophobic prodrugs, such as 6-deoxyacyclovir (Fig. 4), and 2′-fluoroarabino-dideoxypurine, which can be bioactivated by XO to acyclovir and 2′-fluoroarabino-dideoxyinosine, respectively (93,94).

IV. ALDEHYDE OXIDASE

Despite the parallel in nomenclature with XD/XO, aldehyde oxidase is not an alternative form of aldehyde dehydrogenase. AO is a cytosolic enzyme that shares much similarity with the XD/XO enzymes, but does not participate in a dehydrogenase–oxidase transition. Because many features of the two enzymes are very similar, only significant similarities and differences will be discussed.

A. Catalytic Mechanism

Mechanistic studies have not been carried out as extensively on AO; however, the mechanism(s) described for XO (see Fig. 3) are probably also operative for AO. Indeed, spectroscopic experiments (UV–VIS, EPR) have revealed a remarkable similarity and identity in the cofactor and apoprotein environments of the two molybdozymes (60).

B. Multiplicity, Tissue Distribution, and Species Differences

Aldehyde Oxidase exists solely in its oxidase form, and so the electrons received from a reducing substrate are used by the flavin to reduce dioxygen. Attempts to convert AO back to a dehydrogenase form, such that it will reduce NAD^+ instead of dioxygen, have been unsuccessful. Sequence comparisons between all of the known XD/XOs and AOs have yielded insight into functional differences between the enzymes (60). AO probably lacks the requisite amino acid(s) to either bind NAD^+ or modify the redox potential to allow reduction of NAD^+.

Although the dehydrogenase–oxidase transition does not occur with AO, the same pathophysiological implications that were described for XO exist for AO. Notably, the reactive oxygen species generated during metabolism by AO may react with the acetaldehyde produced during ethanol metabolism to produce longer-lived acetaldehyde radicals and prolong cellular damage (75). This interaction of ethanol and its metabolites with AO and its metabolites may be physiologically important and has been the subject of extensive research (76,95–97).

Similar to XO, AO is expressed in multiple tissues, with high levels reported in bovine liver, lung, and spleen (98). Human AO appears to have a more limited tissue distribution, with significant levels of mRNA detected only in the liver (69). However, a slightly smaller-sized transcript was found in kidney, heart, and other tissues, which may indicate that multiple AO isoforms exist in humans, as they appear to do in plants (99).

A recent comprehensive study compared the substrate specificities of partially purified rabbit, guinea pig, human, and baboon aldehyde oxidases (100). Fifteen quinazoline and 14 phthalazine derivatives were used to probe the active sites of the four species orthologs, and the authors concluded that there were significant differences between the substrate specificities of the animal and human AOs studied, with substrate size being the differentiating factor. The relative volumes of the substrate-binding sites of the four AOs were ranked in order; rabbit < guinea pig < baboon < human. From this study, it would appear that baboon AO would be the best model for human AO, although guinea pig has also been considered to be a useful model for the human isoform (62). Interestingly, AO activity is very low or absent in dog tissues (101).

C. Enzyme Regulation

Similar to XO, androgens and estrogens appear to exert a dual hormonal control of AO activity; androgens causing increased AO activity, whereas estrogens decrease activity (78). A recent report suggests that the sex hormones indirectly affect murine AO enzymatic activity by influencing the secretion pattern of growth hormone (GH) by the hypothalamus–pituitary system (102). Masculinization and feminization experiments further indicated that the AO activity induced in males was kinetically distinguishable from the female form of the enzyme, indicative perhaps of a sex-specific multiplicity of AO in mice.

Human AO activity appears to be rather unstable, for surgically excised and postmortem samples seemed to be devoid of activity (62). Recent data, however, suggests that the instability of human AO could be substrate-dependent. When AO activity in several human liver preparations was examined with multiple substrates, enzyme activity varied by more than 40-fold for *N*-methylnicotinamide, but only by 2.3-fold for 6-methylpurine (103). This may be indicative of the presence of multiple forms of human liver AO that exhibit differences in substrate specificities and stability.

D. Transformation Reactions

The nitrogen heterocyclic substrate specificity of AO, and the molybdenum hydroxylases, in general, has been considered complementary to that of the cytochromes P450 (78). For example, naphthalene is oxidized by cytochrome P450 to phenolic products by unstable epoxides, but it is not a substrate for the molybdenum hydroxylases. As the number of nitrogen atoms in the molecule increases, the metabolite profile and the enzyme specificity changes. Quinazoline (1,3-diazanaphthalene) is oxidized by the molybdenum hydroxylases to quinazolin-4-one (see Fig. 4), whereas only small amounts of P450-derived phenolic products are produced. Pteridine (1,3,5,8-tetraazanaphthalene), which is not a substrate for P450, is oxidized by AO to pteridin-2,4,7-trione. This complementarity and the shift in enzyme–substrate specificity between the two enzyme systems can be rationalized by differences in the catalytic mechanism. Whereas, P450s generally react at positions of high electron density, and usually by hydrogen atom or electron abstraction (104), the molybdenum hydroxylase reaction occurs at positions of decreased electron density.

The substrate specificities of various mammalian AOs have been examined with a range of diverse substrates, including *N*-methylnicotinamide, pyridoxal, 6-methylpurine, methotrexate, quinine, and vanillin (62,78), and in a systematic fashion with a series of

1-substituted phthalazines (105). In general, substrate substituents that increase electronegativity enhance V_{max} values, and substituents that enhance lipophilicity generally increase affinity for AO. Although molecular orbital calculations can often predict the regiochemistry of hydroxylation by the molybdenum hydroxylases in terms of relative electronegativity at the various oxidizable positions, steric factors as well as electronic factors can influence the site of hydroxylation (51).

Together with XO, AO is also capable of oxidizing aldehydes to the corresponding carboxylic acids. However, in vivo this reaction is carried out preferentially by the aldehyde dehydrogenases, rather than the molybdozymes. Nonetheless, AO has been implicated in the metabolism of several important endogenous compounds that were originally considered to be metabolized by aldehyde dehydrogenase. A recent review summarized the role of AO in the metabolism of biogenic amines, such as homovanillyl aldehyde and 5-hydroxy-3-indoleacetaldehyde, the aldehyde metabolites of dopamine and 5-hydroxytryptamine (106). Rabbit liver retinal oxidase, which is the enzyme responsible for the biosynthesis of retinoic acid, has been characterized and is identical with rabbit liver AO (107). Conceivably, tissue AO concentrations can play a role in regulating cellular growth, differentiation, and morphogenesis through the modulation of retinoic acid levels.

1. *Diagnostic Inhibitors*

Menadione and hydralazine are both useful inhibitors of AO-catalyzed reactions. Hydralazine has been used in vivo to implicate AO involvement in the human first-pass metabolism of the cardiac stimulant carbazeran (108). Menadione is perhaps the most widely used in vitro inhibitor of AO, and can be used together with allopurinol to discriminate between AO and XO-catalyzed reactions (109). The latter study demonstrated that the rat is not a good model for human AO-catalyzed reactions, because the 6-oxidation of antiviral deoxyguanine prodrugs was catalyzed exclusively by XO in rat liver, but by AO in human liver.

Methadone has been reported to be an extremely potent inhibitor of rat liver AO, exhibiting K_i values in the range of 10^{-7} M (110). Interestingly, the methadone analogue proadifen HCl (SKF-525A), generally considered to be a broad-spectrum P-450 inhibitor, also inhibits the AO-mediated metabolism of acridine carboxamide (111).

E. Relevance to Human Drug Metabolism

Recent work has implicated AO in several pharmacologically relevant biotransformations. Mammalian liver AO is responsible for metabolism of the nicotine metabolite (*S*)-nicotine $\Delta^{1',5'}$-iminium ion to (*S*)-cotinine (see Fig. 4) in rat (112); the α-*N*-oxidation of the quinoxaline α_2-adrenergic agent, brimonidine, to its 2,3-dioxo derivative, a major metabolite, in humans (113); the metabolism of tamoxifen 4-aldehyde to the corresponding carboxylic acid (see Fig. 4) in rat (114); and the activation of the pyrimidinone prodrug nucleus to uracil in mouse, rabbit, and rat (115,116).

Inhibition by menadione suggests that AO is involved in the human biotransformation of the chemotherapeutic agent O^6 benzylguanine to O^6 benzyl-8-oxoguanine (117). AO involvement in the conversion of the antiviral prodrug famciclovir to penciclovir in humans was also demonstrated recently by menadione and isovanillin inhibition (118).

Finally, as with XO, the potential for AO-catalyzed reduction exists. Mammalian liver AOs appear to be able to catalyze the reduction of carcinogenic hydroxamic acids to amides (119; e.g., *N*-hydroxy-2-acetylaminofluorene), epoxides to olefins (e.g., naphthalene 1,2-oxide and benzo[*a*]pyrene 4,5-oxide; 120), as well as the reduction of the anti-inflammatory sulfoxide prodrug, sulindac, to the active sulfide (121).

ACKNOWLEDGMENTS

FMO research conducted in our laboratory was supported by NIH grant GM43511 (AER). MBF was supported by NIH Training Grant GM07750.

REFERENCES

1. C. C. Peck, R. Temple, and J. M. Collins, Understanding consequences of concurrent therapies, *JAMA 269*: 1550 (1993).
2. D. J. Birkett, P. I. MacKenzie, M. E. Veronese, and J. O. Miners, In vitro approaches can predict human drug metabolism, *Trend Pharmacol. Sci. 14*: 292 (1993).
3. S. A. Wrighton, M. Van den Branden, J. C. Stevens, L. A. Shipley, B. J. Ring, A. E. Rettie, and J. R. Cashman, In vitro methods for assessing human hepatic drug metabolism: Their use in drug development, *Drug Metab. Rev. 25*: 453 (1993).
4. A. D. Rodrigues, Use of in vitro human metabolism studies in drug development. An industrial perspective, *Biochem. Pharmacol. 48*: 2147 (1994).
5. S. J. Yin, Alcohol dehydrogenase: Enzymology and metabolism, *Alcohol Alcohol. Suppl. 2*: 113 (1994).
6. W. Ambroziak and R. Pietruszko, Metabolic role of aldehyde dehydrogenase, *Adv. Exp. Med. Biol. 328*: 5 (1993).
7. T. M. Penning, Dihydrodiol dehydrogenase and its role in polycyclic aromatic hydrocarbon metabolism, *Chem. Biol. Interact. 89*: 1 (1993).
8. G. A. Lyles, Mammalian plasma and tissue-bound semicarbazide-sensitive amine oxidases: Biochemical, pharmacological and toxicological aspects, *Int. J. Biochem. Cell Biol. 28*: 259 (1996).
9. C. M. Dixon, G. R. Park, and M. H. Tarbitt, Characterization of the enzyme responsible for the metabolism of sumatriptan in human liver, *Biochem. Pharmacol. 47*: 1253 (1994).
10. T. F. Woolf, S. M. Bjorge, A. E. Black, A. Holmes, and T. Chang, Metabolism of the acyl-CoA:cholesterol acyltransferase inhibitor 2,2-dimethyl-*N*-(2,4,6-trimethoxyphenyl) dodecanamide in rat and monkey. Omega/beta-oxidation pathway, *Drug Metab. Dispos. 19*: 696 (1991).
11. L. L. Poulsen and D. M. Ziegler, The liver microsomal FAD-containing monooxygenase, *J. Biol. Chem. 254*: 6449 (1979).
12. D. M. Ziegler, Flavin-containing monooxygenases: Catalytic mechanism and substrate specificity, *Drug Metab. Rev. 19*: 1 (1988).
13. D. M. Ziegler, Recent studies on the structure and function of multisubstrate flavin-containing monooxygenases, *Annu. Rev. Pharmacol. Toxicol. 33*: 179 (1993).
14. M. P. Lawton, J. R. Cashman, T. Cresteil, C. T. Dolphin, A. A. Elfarra, R. N. Hines, E. Hodgson, T. Kimura, J. Ozols, I. R. Phillips, R. M. Philpot, L. L. Poulsen, A. E. Rettie, E. A. Shephard, D. E. Williams, and D. M. Ziegler, A nomenclature for the mammalian flavin-containing monooxygenase gene family based on amino acid sequence identities, *Arch. Biochem. Biophys. 308*: 254 (1994).
15. E. Atta-Asafo-Adjei, M. P. Lawton, and R. M. Philpot, Cloning, sequencing, distribution and expression in *Escherischia coli* of flavin-containing monooxygenase 1C1, *J. Biol. Chem. 268*: 9681 (1993).
16. S. E. Shehin-Johnson, D. E. Williams, S. Larsen-Su, D. M. Stresser, and R. N. Hines, Tissue-

specific expression of flavin-containing monooxygenase (FMO) forms 1 and 2 in the rabbit, *J. Pharmacal. Exp. Ther. 272*: 1293 (1995).
17. V. L. Burnett, M. P. Lawton, and R. M. Philpot, Cloning and sequencing of flavin-containing monooxygenases FMO3 and FMO4 from rabbit and characterization of FMO3, *J. Biol. Chem. 269*: 14314 (1994).
18. C. T. Dolphin, T. E. Cullingford, E. A. Shephard, R. L. Smith, and I. R. Phillips, Differential development and tissue-specific regulation of expression of the genes encoding three members of the flavin-containing monooxygenase family of man, *Eur. J. Biochem. 235*: 683 (1996).
19. R. L. Haining, A. P. Hunter, A. J. M. Sadeque, R. M. Philpot, and A. E. Rettie, Baculovirus-mediated expression and purification of human FMO3: Catalytic, immunochemical and structural characterization, *Drug Metab. Dispos. 25*: 790 (1997).
20. A. E. Rettie, M. P. Lawton, A. J. M. Sadeque, G. P. Meier, and R. M. Philpot, Prochiral sulfoxidation as a probe for multiple forms of the microsomal flavin-containing monooxygenase: Studies with rabbit FMO1, FMO2, FMO3, and FMO5 expressed in *Escherichia coli*, *Arch. Biochem. Biophys. 311*: 369 (1994).
21. M. B. Fisher, M. P. Lawton, E. Atta-Asafo-Adjei, R. M. Philpot, and A. E. Rettie, Selectivity of flavin-containing monooxygenase 5 (FMO5) for the (*S*)-sulfoxidation of short-chain aralkyl sulfides, *Drug Metab. Dispos. 23*: 1431 (1996).
22. R. K. Kaderlik, E. Weser, and D. M. Ziegler, Selective loss of liver flavin-containing monooxygenase in rats on chemically-defined diets, *Prog. Pharmacol. Clin. Pharmacol. 3*: 95 (1991).
23. S. Larsen-Su and D. E. Williams, Dietary indole-3-carbinol inhibits FMO activity and the expression of flavin-containing monooxygenase form 1 in rat liver and intestine, *Drug Metab. Dispos. 24*: 927 (1996).
24. D. E. Williams, S. E. Hale, A. S. Muerhoff, and B. S. S. Masters, Rabbit lung flavin-containing monooxygenase: Purification, characterization and induction during pregnancy, *Mol. Pharmacol. 28*: 381 (1984).
25. M. Y. Lee, J. E. Clark, and D. E. Williams, Induction of flavin-containing monooxygenase (FMOB) in rabbit lung and kidney by sex steroids and glucocorticoids, *Arch. Biochem. Biophys. 302*: 332 (1993).
26. M. Y. Lee, S. Smiley, S. Kadkhodyan, R. N. Hines, and D. E. Williams, Developmental regulation of flavin-containing monooxygenase (FMO) isoforms 1 and 2 in pregnant rabbit, *Chem. Biol. Interact. 96*: 75 (1995).
27. J. G. Falls, D. Y. Ryu, Y. Cao, P. E. Levi, and E. Hodgson, Regulation of mouse liver flavin-containing monooxygenases 1 and 3 by sex steroids, *Arch. Biochem. Biophys. 342*: 212 (1997).
28. A. J. M. Sadeque, K. E. Thummel, and A. E. Rettie, Purification of macaque liver flavin-containing monooxygenase: A form of the enzyme related immunochemically to an isozyme expressed selectively in adult human liver, *Biochem. Biophys. Acta 1162*: 127 (1993).
29. A. J. M. Sadeque, A. C. Eddy, G. P. Meier, and A. E. Rettie, Stereoselective sulfoxidation by human flavin-containing monooxygenase, *Drug Metab. Dispos. 20*: 832 (1992).
30. J. R. Cashman, Structural and catalytic properties of the mammalian flavin-containing monooxygenase, *Chem. Res. Toxicol. 8*: 165 (1995).
31. E. Hodgson and P. E. Levi, The role of the flavin-containing monooxygenase (E.C.1.14.13.8) in the metabolism and mode of action of agricultural chemicals, *Xenobiotica 22*: 1175 (1992).
32. Y. Seto and F. P. Guengerich, Partitioning between *N*-dealkylation and *N*-oxygenation in the oxidation of *N,N*-dialkylarylamines catalyzed by cytochrome P450 2B1, *J. Biol. Chem. 268*: 9986 (1993).
33. J. Lin, C. E. Berkman, and J. R. Cashman, *N*-Oxygenation of primary amines and hydoxylamines and retroreduction of hydroxylamines by adult human liver microsomes and adult human flavin-containing monooxygenase 3, *Chem. Res. Toxicol. 9*: 1183 (1996).

34. M. E. McManus, I. Stupans, W. Burgess, J. A. Koenig, P. De La M. Hall, and D. J. Birkett, Flavin-containing monooxygenase activity in human liver microsomes, *Drug Metab. Dispos. 15*: 256 (1987).
35. D. Lang, R. Peter, and A. E. Rettie, unpublished observations.
36. A. Kawaji, K. Ohara, and E. Takabatake, An assay of flavin-containing monooxygenase activity with benzydamine *N*-oxidation, *Anal. Biochem. 214*: 409 (1993).
37. Z. Y. Guo, S. Raeissi, R. B. White, and J. C. Stevens. Orphenadrine and methimazole inhibit cytochrome P450 enzymes in human liver microsomes, *Drug Metab. Dispos. 25*: 390 (1997).
38. K. Itagaki, G. T. Carver and R. M. Philpot, Expression and characterization of a modified flavin-containing monooxygenase 4 from humans, *J. Biol. Chem. 271*: 20102 (1996).
39. A. Grothusen, J. Hardt, L. Brautigam, D. Lang, and R. Bocker, A convenient method to discriminate between cytochrome P450 enzymes and flavin-containing monooxygenases in human liver microsomes, *Arch. Toxicol. 71*: 64 (1996).
40. J. R. Cashman, S. B. Park, Z.-C. Yang, S. A. Wrighton, P. Jacob III, and N. L. Benowitz, Metabolism of nicotine by human liver microsomes: Stereoselective formation of *trans*-nicotine *N′*-oxide, *Chem. Res. Toxicol. 5*: 639–646 (1992).
41. B. J. Ring, J. Catlow, T. J. Lindsay, T. Gillespie, L. K. Roskos, B. J. Cerimelle, S. P. Swanson, M. Hamman, and S. A. Wrighton, Identification of the human cytochromes P450 responsible for the in vitro formation of the major oxidative metabolites of the antipsychotic agent olanzapine, *J. Pharmacol. Exp. Ther. 276*: 658 (1996).
42. M. Tugnait, E. M. Hawes, G. McKay. A. E. Rettie. R. L. Haining, and K. K. Midha, *N*-Oxygenation of clozapine by flavin-containing monooxygenase, *Drug Metab. Dispos. 25*: 524 (1997).
43. J. R. Cashman, Z. Yang, L. Yang, and S. A. Wrighton, Stereo- and regioselective *N*- and *S*-oxidation of tertiary amines and sulfides in the presence of adult human liver microsomes, *Drug Metab. Dispos. 21*: 492 (1993).
44. B. L. Blake, R. M. Philpot, P. E. Levi, and E. Hodgson, Xenobiotic biotransforming enzymes in the central nervous system: An isoform of flavin-containing monooxygenase (FMO4) is expressed in rabbit brain, *Chem. Biol. Interact. 99*: 253 (1996).
45. V. Ravindranath and M. R. Boyd, Xenobiotic metabolism in brain, *Drug Metab. Rev. 27*: 419 (1995).
46. R. Ayesh, M. Al-Waiz, M. J. Crothers, S. Cholerton, J. R. Idle, and R. L. Smith, Deficient nicotine *N*-oxidation in two sisters with trimethylaminuria, *Br. J. Clin. Pharmacol. 25*: 664 (1988).
47. M. Al-Waiz, R. Ayesh, S. C. Mitchell, J. R. Idle, and R. L. Smith, A genetic polymorphism of the *N*-oxidation of trimethylamine in humans, *Clin. Pharmacol. Ther. 42*: 588 (1987).
48. C. T. Dolphin, R. L. Smith, E. A. Shephard, and I. R. Phillips., Molecular genetics of human flavin-containing monooxygenases, Abstracts of the Sixth Stowe Symposium on Drug Metabolism, Grantham, England, 1996, p. S2.
49. J. R. Cashman, S. B. Park, Z. C. Chen, C. B. Washington, D. Y. Gomez, K. M. Giacomini, and C. M. Brett, Chemical, enzymatic, and human enantioselective *S*-oxygenation of cimetidine, *Drug Metab. Dispos. 21*: 587 (1993).
50. M. A. Hamman, S. A. Wrighton, and S. D. Hall, Sulindac metabolism by human liver microsomes, *ISSX Proc.* 6:251 (1994).
51. B. Testa, Oxidations catalyzed by various oxidases and monooxygenases, *The Metabolism of Drugs and Other Xenobiotics: Biochemistry of Redox Reactions* (B. Testa and J. Caldwell, eds.), Academic Press, London, 1995, p. 298.
52. R. Hille and H. Sprecher, On the mechanism of action of xanthine oxidase. Evidence in support of an oxo transfer mechanism in the molybdenum-containing hydroxylases, *J. Biol. Chem. 262*: 10914 (1987).
53. J. H. Kim and R. Hille, Studies of the substrate binding to xanthine oxidase using a spin-labelled analog, *J. Inorg. Biochem. 55*: 295 (1994).

54. J. H. Kim, M. G. Ryan, H. Knaut, and R. Hille, The reductive half-reaction of xanthine oxidase. The involvement of prototropic equilibria in the course of the catalytic sequence, *J. Biol. Chem. 271*: 6771 (1996).
55. G. A. Lorigan, R. D. Britt, J. H. Kim, and R. Hille, Electron spin echo envelope modulation spectroscopy of the molybdenum center of xanthine oxidase, *Biochim. Biophys. Acta 1185*: 284 (1994).
56. J. Hunt and V. Massey, Studies of the reductive half-reaction of milk xanthine dehydrogenase, *J. Biol. Chem. 269*: 18904 (1994).
57. M. S. Mondal and S. Mitra, Kinetics and thermodynamics of the molecular mechanism of the reductive half-reaction of xanthine oxidase, *Biochemistry 33*: 10305 (1994).
58. B. D. Howes, R. C. Bray, R. L. Richards, N. A. Turner, B. Bennett, and D. J. Lowe, Evidence favoring molybdenum–carbon bond formation in xanthine oxidase action: ^{17}O- and ^{13}C-ENDOR and kinetic studies, *Biochemistry 35*: 1432 (1996).
59. S. C. D'Ardenne and D. E. Edmondson, Kinetic isotope effect studies on milk xanthine oxidase and on chicken liver xanthine dehydrogenase, *Biochemistry 29*: 9046 (1990).
60. N. A. Turner, W. A. Doyle, A. M. Ventom, and R. C. Bray, Properties of rabbit liver aldehyde oxidase and the relationship of the enzyme to xanthine oxidase and dehydrogenase, *Eur. J. Biochem. 232*: 646 (1995).
61. R. Hille and T. Nishino, Xanthine oxidase and xanthine dehydrogenase, *FASEB J. 9*: 995 (1995).
62. C. Beedham, Molybdenum hydroxylases: Biological distribution and substrate-inhibitor specificity, *Prog. Med. Chem. 24*: 85 (1987).
63. K. Ichida, Y. Amaya, K. Noda, S. Minoshima, T. Hosoya, S. Osamu, N. Shimizu, and T. Nishino, Cloning of the cDNA encoding human xanthine dehydrogenase (oxidase): Structural analysis of the protein and chromosomal location of the gene, *Gene 133*: 279 (1993).
64. P. Xu, T. P. Huecksteadt, R. Harrison, and J. R. Hoidal, Molecular cloning, tissue expression of human xanthine dehydrogenase, *Biochem. Biophys. Res. Commun, 199*: 998 (1994).
65. A. Many, A. Westerhausen Larson, A. Kanbour Shakir, and J. M. Roberts, Xanthine oxidase/dehydrogenase is present in human placenta, *Placenta 17*: 361 (1996).
66. P. Xu, X. L. Zhu, T. P. Huecksteadt, A. R. Brothman, and J. R. Hoidal, Assignment of human xanthine dehydrogenase gene to chromosome 2p22, *Genomics 23*: 289 (1994).
67. M. J. Romao, M. Archer, I. Moura, J. J. G. Moura, J. LeGall, R. Engh, M. Schneider, P. Hof, and R. Huber, Crystal structure of the xanthine oxidase-related aldehyde oxido-reductase from *D. gigas, Science 270*: 1170 (1995).
68. W. A. Doyle, J. F. Burke, A. Chovnik, F. L. Dutton, J. R. S. Whittle, and R. C. Bray, Properties of xanthine dehydrogenase variants from rosy mutant strains of *Drosophila melanogaster* and their relevance to the enzyme's structure and mechanism, *Eur. J. Biochem. 239*: 782 (1996).
69. R. M. Wright, G. M. Vaitaitis, C. M. Wilson, T. B. Repine, L. S. Terada, and J. E. Repine, cDNA cloning, characterization, and tissue-specific expression of human xanthine dehydrogenase/xanthine oxidase, *Proc. Natl. Acad. Sci. USA 90*: 10690 (1993).
70. A. Glatigny and C. Scazzocchio, Cloning and molecular characterization of *hxA*, the gene coding for the xanthine dehydrogenase (purine hydroxylase I) of *Aspergillus nidulans, J. Biol. Chem. 270*: 3534 (1995).
71. T. Nishino, The conversion of xanthine dehydrogenase to xanthine oxidase and the role of the enzyme in reperfusion injury, *J. Biochem. 116*: 1 (1994).
72. J. J. Lanzillo, F. S. Yu, J. Stevens, and P. M. Hassoun, Determination of xanthine dehydrogenase mRNA by a reverse transcription-coupled competitive quantitative polymerase chain reaction assay: Regulation in rat endothelial cells by hypoxia and hyperoxia, *Arch. Biochem. Biophys. 335*: 377 (1996).
73. C. G. Cote, F. S. Yu, J. J. Zulueta, R. J. Vosatka, and P. M. Hassoun, Regulation of intracellu-

lar xanthine oxidase by endothelial-derived nitric oxide, *Am. J. Physiol. Lung Cell Mol. Physiol. 15*: L869 (1996).

74. M. Kurosaki, M. Li Calzi, E. Scanziani, E. Garattini, and M. Terao, Tissue- and cell-specific expression of mouse xanthine oxidoreductase gene in vivo: Regulation by bacterial lipopolysaccharide, *Biochem. J. 306*: 225 (1995).
75. R. Nordmann, C. Ribiere, and H. Rouach, Implication of free radical mechanisms in ethanol-induced cellular injury, *Free Radic Biol. Med. 12*: 219 (1992).
76. S. Shaw and E. Jayatilleke, The role of cellular oxidases and catalytic iron in the pathogenesis of ethanol-induced liver injury, *Life Sci. 50*: 2045 (1992).
77. B. A. Hamelin, K. Xu, F. Valle, L. Manseau, M. Richer, and M. LeBel, Caffeine metabolism in cystic fibrosis: Enhanced xanthine oxidase activity, *Clin. Pharmacol. Ther. 56*: 521 (1994).
78. C. Beedham, Molybdenum hydroxylases as drug metabolizing enzymes, *Drug Metab. Rev. 16*: 119 (1985).
79. M. V. Relling, J. S. Lin, G. D. Ayers, and W. E. Evans, Racial and gender differences in *N*-acetyltransferase, xanthine oxidase, and CYP1A2 activities. *Clin. Pharmacol. Ther. 52*: 643 (1992).
80. R. Guerciolini, C. Szumlanski, and R. M. Weinshilboum, Human liver xanthine oxidase: Nature and extent of individual variation, *Clin. Pharmacol. Ther. 50*: 663 (1991).
81. A. Nagae, E. Murakami, K. Hiwada, Y. Sato, and M. Kawachi, Asymptomatic hereditary xanthinuria: A case report, *Jpn. J. Med. 29*: 287 (1990).
82. A. K. Daly, S. Cholerton, W. Gregory, and J. R. Idle, Metabolic polymorphisms, *Pharmacol. Ther. 57*: 129 (1993).
83. S. Reiter, H. A. Simmonds, N. Zollner, S. L. Braun, and M. Knedel, Demonstration of a combined deficiency of xanthine oxidase and aldehyde oxidase in xanthinuric patients not forming oxipurinol, *Clin. Chim. Acta 187*: 221 (1990).
84. M. I. Walton, and P. Workman, Enzymology of the reductive bioactivation of SR 4233. A novel benzotriazine di-*N*-oxide hypoxic cell cytotoxin, *Biochem. Pharmacol. 39*: 1735 (1990).
85. S. L. Bauer and P. C. Howard, The kinetics of 1-nitropyrene and 3-nitrofluoranthene metabolism using bovine liver xanthine oxidase, *Cancer Lett. 54*: 37 (1990).
86. P. P. Fu, D. Herreno-Saenz, L. S. Von Tungeln, J. O. Lay, Y. Wu, J. Lai, and F. E. Evans, DNA adducts and carcinogenicity of nitro-polycyclic aromatic hydrocarbons, *Environ. Health Perspect. 102 (Suppl. 6)*: 177 (1994).
87. J. O. Miners and D. J. Birkett, The use of caffeine as a metabolic probe for human drug-metabolising enzymes, *Gen. Pharmacol. 27*: 245 (1996).
88. W. Chang, Y. Chang, F. Lu, and H. Chiang, Inhibitory effects of phenolics on xanthine oxidase, *Anticancer Res. 14*: 501 (1994).
89. W. Chang, and H. Chiang, Structure–activity relationship of coumarins in xanthine oxidase inhibition, *Anticancer Res. 15*: 1969 (1995).
90. C. A. Pristos and D. L. Gustafson, Xanthine dehydrogenase and its role in cancer chemotherapy, *Oncol. Res. 6*: 477 (1994).
91. E. Van Meerten, J. Verweij, and J. H. M. Schellens, Antineoplastic agents. Drug interactions of clinical significance, *Drug Safety 12*: 168 (1995).
92. W. G. Qing, K. L. Powell, G. Stoica, C. L. Szumlanski, R. M. Weinshilboum, and M. C. Macleod, Toxicity and metabolism in mice of 2,6-dithiopurine, a potential chemoprotective agent, *Drug Metab. Dispos. 23*: 854 (1995).
93. D. B. Jones, V. K. Rustgi, D. M. Kornhauser, A. Woods, R. Quinn, J. H. Hoofnagle, and E. A. Jones, The disposition of 6-deoxyacyclovir, a xanthine oxidase-activated prodrug of acyclovir, in the isolated perfused rat liver, *Hepatology 7*: 345 (1987).
94. K. Shanmuganathan, T. Koudriakova, S. Nampalli, J. Du, J. M. Gallo, R. F. Schinazi, and

C. K. Chu, Enhanced brain delivery of an anti-HIV nucleoside 2′-F-ara-ddI by xanthine oxidase mediated biotransformation, *J. Med. Chem. 37*: 821 (1994).

95. H. Rajasinghe, E. Jayatilleke, and S. Shaw, DNA cleavage during ethanol metabolism: Role of superoxide radicals and catalytic iron, *Life Sci. 47*: 807 (1990).
96. S. Shaw, and E. Jayatilleke, The role of aldehyde oxidase in ethanol-induced hepatic lipid peroxidation in the rat, *Biochem. J. 268*: 579 (1990).
97. L. Mira, L. Maia, L. Barreira, and C. F. Manso, Evidence for free radical generation due to NADH oxidation by aldehyde oxidase during ethanol metabolism, *Arch. Biochem. Biophys. 318*: 53 (1995).
98. M. L. Calzi, C. Raviolo, E. Ghibaudi, L. de Giola, M. Salmona, G. Cazzaniga, M. Kurosaki, M. Terao, and E. Garattini, Purification, cDNA cloning, and tissue distribution of bovine liver aldehyde oxidase, *J. Biol. Chem. 270*: 31037 (1995).
99. H. Sekimoto, M. Seo, N. Dohmae, K. Takio, Y. Kamiya, and T. Koshiba, Cloning and molecular characterization of plant aldehyde oxidase, *J. Biol. Chem. 272*: 15280 (1997).
100. C. Beedham, D. J. P. Critchley, and D. J. Rance, Substrate specificity of human liver aldehyde oxidase toward substituted quinazolines and phthalazines: A comparison with hepatic enzyme from guinea pig, rabbit, and baboon, *Arch. Biochem. Biophys. 319*: 481 (1995).
101. C. Beedham, S. E. Bruce, D. J. Critchley, Y. Al-Tayib, and D. J. Rance, Species variation in hepatic aldehyde oxidase activity, *Eur. J. Drug Metab. Pharmacokinet. 12*: 307 (1987).
102. S. Yoshihara and K. Tatsumi, Involvement of growth hormone as a regulating factor in sex differences of mouse hepatic aldehyde oxidase, *Biochem. Pharmacol. 53*: 1099 (1997).
103. A. D. Rodrigues, J. L. Ferrero, M. T. Amann, G. A. Rotert, S. P. Cepa, B. W. Surber, J. M. Machinist, N. R. Tich, J. P. Sullivan, D. S. Garvey, M. Fitzgerald, and S. P. Arneric, The in vitro hepatic metabolism of ABT-418, a cholinergic channel activator, in rats, dogs, cynomolgus monkeys and humans, *Drug Metab. Dispos. 22*: 788 (1994).
104. P. R. Ortiz de Montellano, Oxygen activation and reactivity, *Cytochrome P-450 Structure, Mechanism, and Biochemistry*, 2nd ed. (P. R. Ortiz de Montellano, ed.), Plenum Press, New York, 1995, p. 245.
105. C. Beedham, S. E. Bruce, D. J. Critchley, and D. J. Rance, 1-Substituted phthalazines as probes of the substrate-binding site of mammalian molybdenum hydroxylases, *Biochem. Pharmacol. 39*: 1213 (1990).
106. C. Beedham, C. F. Peet, G. I. Panoutsopoulos, H. Carter, and J. A. Smith, Role of aldehyde oxidase in biogenic amine metabolism, *Prog. Brain. Res. 106*: 345 (1995).
107. S. Tomita, M. Tsujita, and Y. Ichikawa, Retinal oxidase is identical to aldehyde oxidase, *FEBS Lett. 336*: 272 (1993).
108. D. J. P. Critchley, D. J. Rance, and C. Beedham, Biotransformation of carbazeran in guinea pig: Effect of hydralazine pretreatment, *Xenobiotica 24*: 37 (1994).
109. A. W. Harrell, S. M. Wheeler, P. East, S. E. Clarke, and R. J. Chenery Use of rat and human in vitro systems to assess the effectiveness and enzymology of deoxyguanine analogs as prodrugs of an antiviral agent, *Drug Metab. Dispos. 22*: 189 (1994).
110. I. G. C. Robertson and R. S. K. A. Gamange, Methadone: A inhibitor of rat liver aldehyde oxidase, *Biochem. Pharmacol. 47*: 584 (1994).
111. I. G. C. Robertson and T. J. Bland, Inhibition by SKF-525A of the aldehyde oxidase-mediated metabolism of the experimental antitumor agent acridine carboxamide, *Biochem. Pharmacol. 45*: 2159 (1993).
112. C. E. Berkman, S. B. Park, S. A. Wrighton, and J. R. Cashman, In vitro–in vivo correlations of human (*S*)-nicotine metabolism, *Biochem. Pharmacol. 50*: 565 (1995).
113. A. A. Acheampong, D.-S. Chien, S. Lam, S. Vekich, A. Breau, J. Usansky, D. Harcourt, S. A. Munk, H. Nguyen, M. Garst, and D. Tang-Liu, Characterization of brimonidine metabolism with rat, rabbit, dog, monkey, and human liver fractions and rabbit liver aldehyde oxidase, *Xenobiotica 26*: 1035 (1996).

114. P. C. Ruenitz and X. Bai, Acidic metabolites of tamoxifen: Aspects of formation and fate in the female rat, *Drug Metab. Dispos. 23*: 993 (1995).
115. X. Guo, M. Lerner-Tung, H. Chen, C. Chang, J. Zhu, C. Chang, G. Pizzorno, T. Lin, and Y. Cheng, 5-Fluoro-2-pyrimidone, a liver aldehyde oxidase-activated prodrug of 5-fluorouracil, *Biochem. Pharmacol. 49*: 1111 (1995).
116. D. J. T. Porter, J. A. Harrington, M. R. Almond, G. T. Lowen, T. P. Zimmerman, and T. Spector, 5-Ethynyl-2(1*H*)-pyrimidone: Aldehyde oxidase-activation to 5-ethynyluracil, a mechanism-based inactivator of dihydropyrimidine dehydrogenase, *Biochem. Pharmacol. 47*: 1165 (1994).
117. S. K. Roy, K. R. Korzekwa, F. J. Gonzalez, R. C. Moschel, and M. E. Dolan, Human liver oxidative metabolism of O^6-benzylguanine, *Biochem. Pharmacol. 50*: 1385 (1995).
118. S. E. Clarke, A. W. Harrell, and R. J. Chenery, Role of aldehyde oxidase in the in vitro conversion of famciclovir to penciclovir in human liver, *Drug Metab. Dispos. 23*: 251 (1995).
119. S. Kitamura, K. Sugihara, and K. Tatsumi, Reductase activity of aldehyde oxidase toward the carcinogen *N*-hydroxy-2-acetylaminofluorene and the related hydroxamic acids, *Biochem. Mol. Biol. Int. 34*: 1197 (1994).
120. Y. Hirao, S. Kitamura, and K. Tatsumi, Epoxide reductase activity of mammalian liver cytosols and aldehyde reductase, *Carcinogenesis 15*: 739 (1994).
121. S. C. Lee, and A. G. Renwick, Sulphoxide reduction by rat and rabbit tissues in vitro, *Biochem. Pharmacol. 49*: 1557 (1995).

6

Transformation Reactions: Glucuronidation

Brian Burchell

Ninewells Hospital and Medical School, University of Dundee, Dundee, United Kingdom

I. INTRODUCTION

Glucuronidation is a major drug biotransformation reaction. The transfer of glucuronic acid from UDP-glucuronic acid (UDPGA) to an aglycone is catalyzed by a family of UDP-glucuronosyltransferases (UGTs; E.C.2.4.17) (1).

The human UDP-glucuronosyltransferase (UGT) gene family contains coding information for the production of more than 20 known microsomal membrane-bound isoenzymes in humans (1,2). This family of enzymes catalyzes the transfer of glucuronic acid to a multitude of endobiotic and xenbiotic compounds, including drugs, pesticides, and carcinogens. The synthesis of ether, ester, carboxyl, carbamoyl, sulfhuryl, carbonyl, and nitrogenyl glucuronides generally leads to their increase of polarity, water solubility and, hence, suitability for excretion (3 Fig. 1).

The UGT isoforms are present in many species, from fish (4) to humans (5). They are found in liver, lung, skin, intestine, brain, and olfactory epithelium, but the major site of glucuronidation is the liver (3). Bilirubin UGT is highly expressed in human liver, but is absent from human kidney, whereas phenol UGT is highly expressed in both tissues (6). Individual UGTs are subject to differential induction by hormones, which leads to tissue-specific and developmentally regulated expression (3,5). The spectrum of UGT isoforms observed in different tissues can also be differentially altered by exposure to drugs and xenobiotics (5).

Renewed interest in glucuronidation by pharmaceutical firms has focused on development of drugs that avoid glucuronidation as a biotransformation reaction and, thereby, improve the bioavailability. Chemical modification of a well-known drug, such as acetaminophen, by alkylation reduces hepatic glucuronidation of these analogues (M. McPhail and B. Burchell, unpublished work) and does not lead to increased toxicity in rats (7). Thus, an analgesic with longer biological half-life could potentially be produced, although these compounds have not been assessed as new analgesics. This theme is being more widely used to develop new drugs (8).

Additional useful knowledge for this drug development process will be comprehensive quantitative structure–activity relation (QSAR) analysis of drug glucuronidation by animal hepatocytes (9) or individual human UGTs, the latter prospect being considerably improved by cloning and expression of UGTs (10). This work allied with QSAR studies of drug glucuronidation in vivo by high-performance liquid chromatography–nuclear magnetic resonance (HPLC–NMR; 11) will considerably enhance the rate of drug development.

This chapter will focus on drug glucuronidation. The reader should consult more comprehensive reviews for broader literature on drug glucuronidation and UDP-glucuronosyltransferases (1,3,12–14). The literature was reviewed up to Feb. 1997.

A. Nomenclature of UDP-Glucuronosyltransferases

A nomenclature for UDP-glucuronosyltransferases has been proposed based on the divergent evolution of gene sequences. In naming each gene the root symbol *UGT* for humans (*Ugt* for mouse) followed by an Arabic number denoting the family, a letter designating the subfamily, and an Arabic numeral representing the individual gene within the family or subfamily (e.g., human *UGT2B1* and mouse *Ugt2b-1*).

Comparison of amino acid sequences leads to a definition of families and subfamilies. Gene products in the *UGT-1* family are 38–48% identical with the gene products in

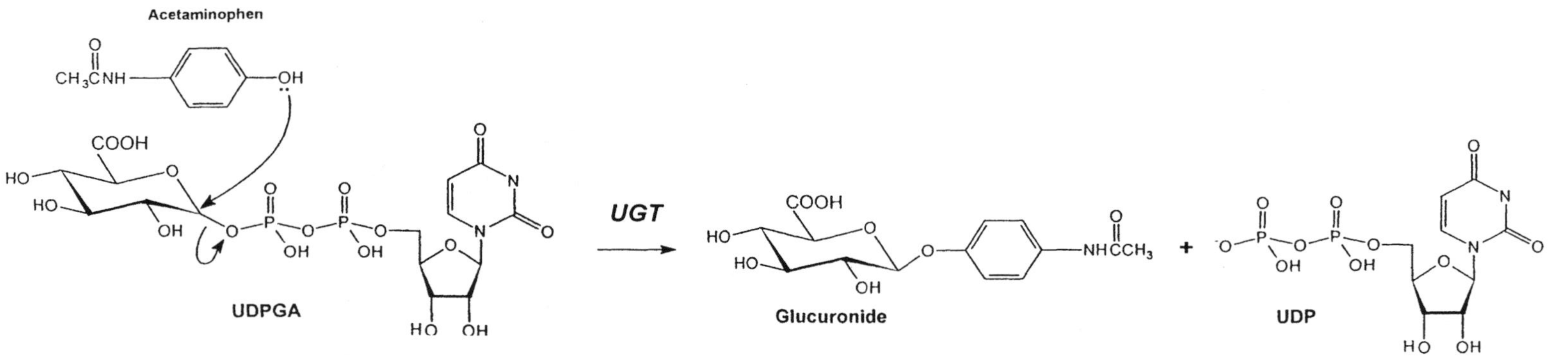

Fig. 1 Glucuronidation of a phenolic compound catalyzed by UDP-glucuronosyltransferase.

the *UGT-2* family. Therefore, a convenient value to choose for family divergence is less than 50% similar. Deduced protein sequences of members of the *UGT-1* family are 62–80% similar; the *UGT-2* family are 57–93% similar. Therefore, within a single family there is greater than 50% similarity; within a subfamily the similarity is increased to more than 60%. The result of the application of this nomenclature has produced the dendogram observed in Fig. 2. The actual numbering of the family members is the result of the chronologic order of discovery of new UGT cDNAs of various species, except in the UGT-1 gene complex for which numbering is based on structural order of the variable exons within the genome (see Fig. 2A).

B. Glucuronidation of Endogenous Compounds

Some key UGTs have evolved to prevent accumulation of potentially toxic endogenous compounds. Bilirubin, the end product of heme catabolism has been the most extensively studied substrate of UGTs. Conjugation of bilirubin with glucuronic acid prevents bilirubin

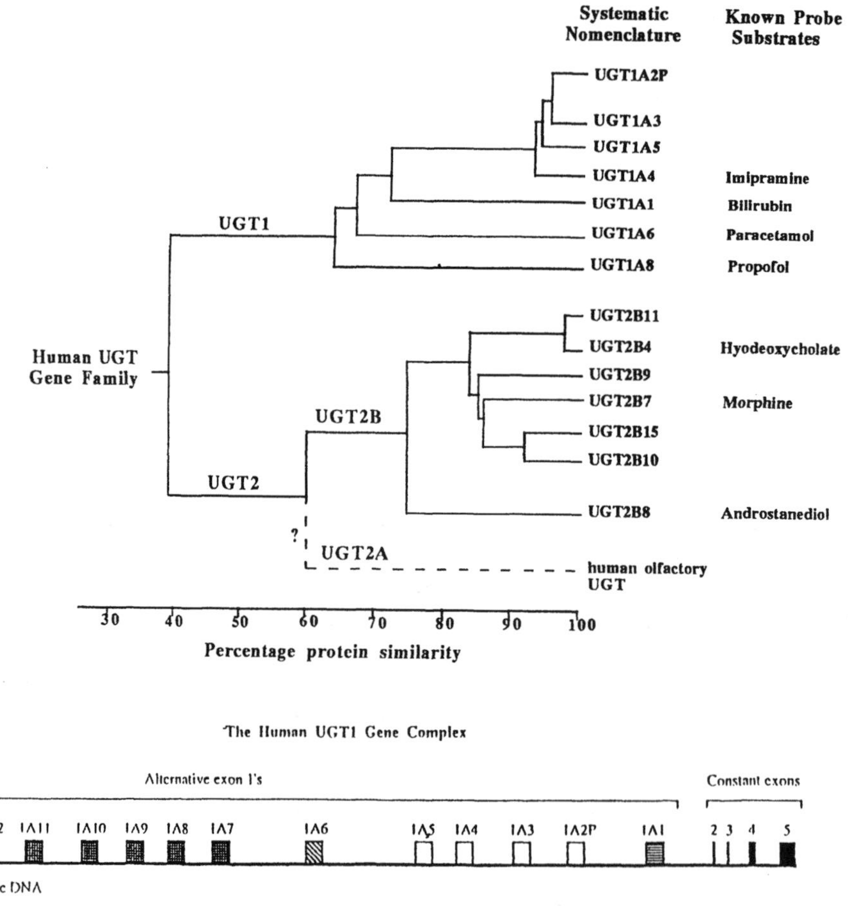

Fig. 2 A dendogram illustrating the human UGT gene family and systematic nomenclature. (A) The human *UGT1* gene complex and systematic nomenclature.

accumulation and toxicity under normal physiological conditions; bilirubin is excreted from the body as biliary bilirubin mono- and diglucuronides (see Ref. 15 for an extensive review).

Other UGTs are concerned with the maintenance of physiological levels of hormones. Thyroxine and triiodothyronine are readily conjugated with glucuronic acid in the liver and excreted in bile (16,17). Thyroxine glucuronide has a markedly reduced plasma-binding capacity when compared with the free hormone (18). Recently, glucuronidation of retinoic acid has been demonstrated to prevent binding of the parent compound to retinoic acid receptor proteins (19).

The biological role of glucuronidation of other endogenous compounds, such as bile acids and steroid hormones, has not been extensively investigated, although a whole subfamily of UGTs exists to serve this primary function (20). It is reasonable to assume that glucuronidation plays mainly a catabolic role, for their glucuronides are found in bile, but determination of their levels in serum, bile, and urine are useful indicators of cholestasis and endocrine disorders. Studies of bile acid glucuronidation have been reviewed (see Ref. 14). Recently, attention has been focused on the estrogen and androgen glucuronidation in humans. Earlier work has been described by Dutton (3). Androsterone glucuronide is the predominant C_{19} steroid glucuronide in plasma (21,22). The marked rise in testosterone during puberty was strongly correlated with increases of androsterone glucuronide and androstane 3α, 17β-diol glucuronide. Serum plasma concentrations of androstanediol and androsterone glucuronides are up to 15-fold higher than their parent substrate in 40-year-old men, but are lowered by 30–50% during aging (23). Measurements of serum androsterone glucuronides are more reliable parameters of global androgen action (23). Both steroid glucuronides are plasma biochemical markers of adrenal hyperandrogenism in hirsuitism in women (21,22) and virilizing congenital adrenal hyperplasion (25,26).

The determination of 17-oxosteroid glucuronides in urine may also be a useful differential diagnosis in adrenal disease. Seven androgen glucuronides can be identified in urine from men. Dehydroepiandrosterone and 11-keto androsterone glucuronides were not observed in aldosteronism. (27)

II. DRUG AND XENOBIOTIC GLUCURONIDATION

The UGTs have evolved to catalyze the glucuronidation of potentially toxic endogenous compounds and routinely encountered environmental chemicals. These compounds obviously do not include recently synthesized (in evolutionary terms) xenobiotics and drugs. In the last century the chemical revolution has outpaced biological evolution; therefore, glucuronidation of modern chemicals is difficult to predict. However, the evolved UGT family have proved to be remarkably effective in xenobiotic metabolism, because the number of adverse effects recorded is relatively small. Nonetheless, there are biologically active glucuronides formed and toxic mechanisms involving glucuronidation are being elucidated.

Drugs from almost all therapeutic classes are glucuronidated. Many of these drugs have narrow therapeutic indices (e.g., morphine and chloramphenicol), and glucuronidation is likely to have important consequences in their clinical use.

Glucuronidation of drug molecules containing a wide range of acceptor groups have been reported, including phenols (e.g., propofol, acetaminophen [paracetamol], naloxone), alcohols (e.g., chloramphenicol, codeine, oxazepam), aliphatic amines (e.g., ciclopiroxolamine, lamotrigine, amitriptyline), acidic carbon atoms (e.g., feprazone, phenylbutazone, sulfinpyrazone), and carboxylic acids (e.g., naproxen, zomepirac, ketoprofen) (1). This indicates the variability of acceptor groups that can be conjugated to glucuronic acid in humans.

Humans and monkeys have the ability to form *N*-linked glucuronides of several tertiary amine drugs (e.g., cyclobenzaprine, cyproheptadine, tripelennamine) (1). This, however, is not true with rats and rabbits, which have an inability to form such quaternary ammonium glucuronides (1), indicating that it is important not to rely exclusively on these rodents for drug metabolism studies.

Another interesting example of species variation has been observed for glucuronidation of tri- and tetraiodothyroacetic acid drugs used in the treatment of pituitary and thyroid disorders (28). These drugs are converted to ester glucuronides by human liver and to ether glucuronides by rat liver (29). The enantioselective and stereoselective glucuronidation of many drugs has been demonstrated in humans; however, the enzymes responsible for this selectivity have yet to be identified (14).

The liver has been established to be the most important site of glucuronidation (3), although the rat gastrointestinal tract makes a major contribution to phenol detoxication in hepatectomized rats (30). Preferential glucuronidation of certain compounds may occur in different organs and tissues. The anesthetic propofol is extensively glucuronidated by kidney and intestine in humans and rats (31,32). These in vivo assessments directly correlated with the organ-specific distribution of UGT1A8 which has been determined to be responsible for the metabolism of propofol (6). The kidney appears to be largely responsible for the glucuronidation and renal clearance of furosemide (33). Similarly, diflusinal phenolic and acyl glucuronides are both formed by human kidney microsomes at about 60–70%, the rate of formation by liver microsomes (34). Therefore, in vitro investigations using liver microsomes may not identify the relevant UGT enzyme catalyzing the drug glucuronidation, because the UGT may reside in extrahepatic tissues.

However, conversely, there are other examples of drug glucuronidation where extensive in vitro investigation could make a useful contribution to our knowledge. Valproate, the antiepileptic agent, is metabolized by a β-oxidation, ω-oxidation and acyl glucuronidation and excreted as a glucuronide in bile (35). The drug exhibits unusual dose-dependent and nonlinear pharmacokinetics in laboratory animals and humans (36). Nonlinear age-dependent disposition has also complicated the investigations (37). A detailed knowledge of the predominant UGT catalyzing the formation of valproate glucuronide would help understand distribution, disposition, and elimination pathways. Our unpublished work demonstrates that valproate is a poor substrate for UGTs, but of the six human UGTs tested it is preferentially metabolized by UGT1A6 (G. Anderson, R. Remmel, B. Burchell, B. Ethell, unpublished work).

The physiological role of drug glucuronidation is the metabolic clearance of drugs from the body in bile and urine. This mechanism terminates the otherwise prolonged and possibly deleterious pharmacological action of many drugs. The pharmacokinetics of drug action are dependent on, among other variables, the rate of glucuronidation. Rates of glucuronide formation vary for different drug substrates owing to the catalytic activity of individual UGTs and factors affecting UGT expression. It will be extremely difficult to forecast accurately UGT's role in the pharmacokinetics of drug metabolism and disposition, until all UGT isoenzymes and factors that control their differential expression in humans are characterized.

A. Biologically Active and Potentially Toxic Glucuronides

Glucuronidation has been described as a safe detoxication process, and glucuronides were never considered to be biologically active intermediates. However, in recent years, the potential toxicity and biological activity of certain glucuronides, such as morphine-6-glucuronide, have been well recognized (38). Compounds range from procarcinogens to cholestatic agents.

More recently, carcinogen glucuronidation has been a focus of investigation. Glucuronidation formation could either serve as a detoxification pathway or a stable transport form of the *N*-glucuronide of 4-aminobiphenyl (39), *N*-acetylbenzidine (49), and the *O*-glucuronide of 4-((hydroxymethyl)nitrosoamino)-1-(3-pyridyl)-1-butanone (NNK); (41). Even if the glucuronide itself has no biological activity, there are stable forms of glucuronides that may be either hydrolyzed by β-glucuronidase present in all tissues after being transported or act themselves and cause toxicity in target tissues (42,43). However, Kim and Wells (44) have suggested that UGTs may be genoprotective for hydroxy-NNK.

B. Drug Acyl Glucuronidation

The mechanism of the reaction catalyzed by UGTs is a S_N2-type reaction, the acceptor group of the substrate involved in a nucleophilic attack of the C-1 of the pyranose acid ring of UDPGA, which results in the formation of a glucuronide, a β-D-glucopyranosiduronic acid conjugate (1). Work with a partially purified rat UGT2B3 provided the first direct evidence for this S_N2-type mechanism in catalysis by UGTs (4,5).

Acyl glucuronides are formed when conjugation with glucuronic acid occurs via a carboxyl group, resulting in ester-type linkage. Ester-type glucuronides are much more unstable than ether-linked glucuronides and can easily undergo nucleophilic substitution. The chemical properties of acyl-linked glucuronides have been extensively reviewed (46).

Glucuronidation is the major pathway for the biotransformation of several acidic drugs, such as nonsteroidal anti-inflammatory drugs (NSAIDS; (47,48)). Under normal conditions, these glucuronides are then excreted in the urine. However, in conditions of renal impairment, excretion may not be as rapid, allowing an accumulation of both the glucuronide and the parent drug, as released by hydrolysis (47,48). The accumulation of acyl glucuronides in the plasma as a result of inefficient excretion may result in considerable acyl migration. Studies have shown that these isomers can bind irreversibly to various proteins in vitro and in vivo.

Smith et al. (49) first documented irreversible drug binding to albumin in vivo, in humans, through an acyl glucuronide. The drug in question was the NSAID zomepirac. In vitro data suggested zomepirac glucuronide was the precursor of adduct formation and that zomepirac itself did not react chemically with albumin (49). Zomepirac was subsequently withdrawn from the pharmaceutical market owing to the high incidence of anaphylaxis following drug administration (49). A number of other drug glucuronides bind irreversibly to proteins in vitro and in vivo, whereas the parent drug alone has been ineffective (48–50). This reactivity is unlikely to be unique for albumin, and formation of adducts with other tissue proteins, leading to potential antigens, might be expected in vitro and in vivo (51). Covalent binding of suprofen to renal tissue in rats and ibuprofen to plasma protein in humans correlated well with formation of the acyl glucuronides (52,53).

One important reaction that acyl glucuronides undergo is acyl migration (Fig. 3), a process whereby the aglycone moves from the 1-hydroxyl group of the glucuronic acid sugar to the 2,3- or 4-hydroxyl groups. This rearrangement of the glucuronide leads to β-glucuronidase-resistant isomers and is completely reversible with one exception; the C-1-glucuronide does not appear to reform from the C-2-isomer (47). The formation of acyl glucuronides is subject to high interspecies variability (62), and the extent of acyl migration may become detectable only when the excretion of conjugates is impaired and their plasma concentrations are raised (69).

The rate of acyl migration differs from compound to compound, and their stability is also highly variable (47). At physiological or slightly alkaline pH, acyl migration and hydrolysis of acyl glucuronides is extensive (47). The analysis and estimation of acyl glucuronides in vivo, therefore, must be approached with caution because in vivo they will be extensively hydrolyzed by esterases and hydrolytic enzymes as well as undergo alkaline hydrolysis (48). The extent of hydrolysis and acyl migration evident in vitro can be kept to a minimum by lowering the pH of the reaction and adding inhibitors of β-glucuronidase such as 1,4-saccharolactone (48).

Identification of positional isomers and measurement of internal migration reaction kinetics of 2-fluorobenzoyl D-glucuronides has been achieved using directly coupled

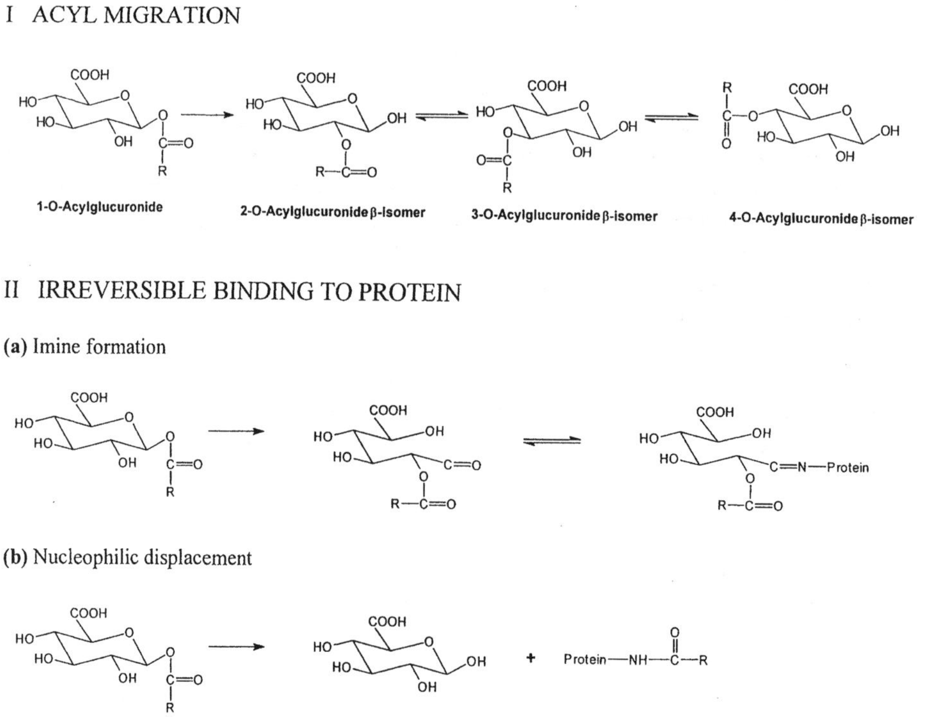

I ACYL MIGRATION

1-O-Acylglucuronide — 2-O-Acylglucuronide β-Isomer — 3-O-Acylglucuronide β-isomer — 4-O-Acylglucuronide β-isomer

II IRREVERSIBLE BINDING TO PROTEIN

(a) Imine formation

(b) Nucleophilic displacement

Protein—NH—C(=O)—R

Fig. 3 Drug–protein binding mediated by glucuronidation and acyl migration.

HPLC–NMR (54,55). This approach provides the opportunity to investigate the molecular physiochemical properties, such as steric hindrance, that influence the acyl migration kinetics of complex drug glucuronide acyl migration reactions of toxicological interest (56).

Various mechanisms have been proposed for this irreversible binding; however, it is not known which of these is principally responsible (47). The first theory involves the nucleophilic displacement of the glucuronic acid moiety, to leave the drug alone irreversible bound to the protein (47). A second mechanism involves opening of the glucuronic acid ring structure, the Amaduri rearrangement (see Fig. 3). Opening the ring structure exposes the aldehyde group on the glucuronic acid, which interacts with a nucleophilic residue, such as lysine, on the protein to form an imine. Although imine formation is reversible, the imine group can be further reduced to a more stable 1-amino-2-keto product. Acyl migration is a prerequisite for the second mechanism. Evidence for both mechanisms exists (47).

Similar events may take place in cholestasis or conjugated hyperbilirubinemia, thereby causing liver damage (15). The endogenous compound bilirubin is glucuronidated in vitro and in vivo to form an acyl glucuronide. This endogenous acyl glucuronide also undergoes acyl migration and irreversible binding to proteins in vitro and in vivo (57).

III. SPECIFICITY OF UDP-GLUCURONOSYLTRANSFERASES

A. Identification of Rat UGTs Catalyzing Glucuronidation of Individual Drugs

The UGT responsible for catalyzing the glucuronidation of a specific drug in rats could be determined by measuring activity in liver from normal Wistar and genetically deficient Gunn and LA rats before and after treatment of the animals with enzyme inducers (see Ref. 14 for a recent review). This approach will considerably reduce the number of possibilities. The assay of the different tissues in the presence or absence of a limited number of specific inhibitors (e.g., novobiocin or triphenylheptanoic acid) should improve specific identification. Finally, the use of recombinant-expressed enzymes in combination with specific inhibitory antibodies should confirm the identity. The specificity of the UGTs is best indicated by endogenous substrates for most enzymes.

The first part of this process is illustrated by Table 1, in trying to determine specific enzymes responsible for glucuronidation of T_3, T_4, and morphine. First, morphine glucuronidation can be examined. The genetic defects in rats considerably narrow down the likely candidate UGTs. The Gunn rat has lost the whole of the UGT-1 family of enzymes. Therefore, glucuronidation of morphine is not catalyzed by a UGT-1 enzyme because morphine glucuronidation is observed in Gunn rat liver. Enzyme induction studies then help improve the identification. Morphine glucuronidation is dramatically induced by phenobarbital, and testosterone glucuronidation is also more specifically induced by phenobarbital, which suggests that a steroid (testosterone) glucuronidating enzyme may be responsible for morphine glucuronidation (see Table 1).

The number of specific or selective inhibitors is very limited, and this is not a generally useful approach (see Ref. 14 for a recent review of inhibitors of glucuronidation). Recently, acyl-CoA has been observed to be a general inhibitor of glucuronidation and

Table 1 Identification of Unknown UGTs Using Xenobiotic Induction and Genetic Defects in Rats

Liver sample[a]	UGT Activity (% Change)						
	Substrate (−) M3G	1-Naphthol	Androsterone	Testosterone	Bilirubin	T_3	T_4
LA(WAG)	—	—	5	—	100	25	87
HA	—	—	100	—	100	100	100
Wistar							
Control	100	100	—	100	100	100	100
3MC	108	230	—	70	86	94	110
PB	608	126	—	205	129	—	—
Clof	55	43	—	100	171	111	160
Gunn							
Control	77	13	—	95	0	111	50
3MC	80	100	—	85	0	—	—
PB	695	13	—	195	0	—	—

[a]3MC, 3-methylcholanthene; PB, phenobarbital; and Clof, cliofibrate indicates pretreatment of animals with these inducing agents.
—, these data not available.
Source: Calculated from Refs. 58–61.

the analogue acyl-3′-dephospho-CoA had no inhibitory effect (62). A preliminary report suggested that the immunosuppressant agent brequinar R may selectively inhibit phenol glucuronidation without affecting morphine glucuronidation (63), although these results need to be substantiated by work with recombinant-expressed enzymes. Recombinant-expressed enzymes often demonstrate the specific ability to catalyze certain glucuronidation reactions, but the K_m of the UGT for the substrate should be determined to be extremely low to indicate specificity.

When considering out logical approach to identify morphine glucuronidation, which seems to be related to testosterone glucuronidation, we observe that three cloned and expressed rat UGTs have been demonstrated to catalyze the glucuronidation of testosterone. Testing of these UGTs has revealed that UGT2B1 catalyzed morphine-3-glucuronidation specifically at a very high rate (49 nmol $min^{-1}mg^{-1}$ protein), although the K_m was 3 mM (64), using a substrate concentration range from 0.5 to 20 mM. Therefore, this enzyme, UGT2B1 would seem to be the main transferase catalyzing morphine glucuronidation in the rat, despite the high K_m.

The UGTs involved in T_3 and T_4 glucuronidation can be identified, using a similar procedure. T_3 glucuronidation is dramatically reduced in LA rats, which have specifically lost the function of the androsterone UGT2B2 gene, which suggested that this enzyme is largely responsible for T_3 glucuronidation (see Table 1). T_4 glucuronidation is reduced to 50% in Gunn rats and inducible specifically by clofibrate, which suggests that a bilirubin UGT probably UGT1A1 is partially responsible for glucuronidation of T_4 in vitro (see Table 1).

Thus, considerable progress has been made in the identification of the functional catalytic ability of individual UGTs, by using this logical approach involving genetics, induction studies, inhibitors, and recombinant-expressed enzymes.

B. Substrate Specificity of Rat UGTs Toward Drugs and Xenobiotics

Investigation of the substrate specificity of the rat UGTs has indicated some overlapping specificity. However, the accumulated information is leading to specific substrate identification of each isoform, but the data shown in Table 2 indicates the limited knowledge of drug glucuronidation by these rat isoforms, where only UGT2B1 seems to have a major role in drug metabolism, although most of the other enzymes are capable of catalyzing the glucuronidation of certain xenobiotics. However, even when there is an observation of a high affinity for a specific substrate, is this individual UGT responsible for the drug glucuronidation in vivo? Therefore, recombinant UGT activities toward various substrates need to be kinetically characterized. Then a panel of specific inhibitory antibodies is required to assess the contribution of each form to a drug's glucuronidation in tissue microsomes. Gene knockout experiments in rats and mice would also provide engineered genetic deficiencies in specific UGTs to study drug glucuronidation in vivo, although significant disruption of endogenous glucuronidation may cause complications to the interpretations.

C. Characterization of the Substrate Specificity of Human UGTs

Studies on the substrate specificity of individual UGTs have been performed using purified transferases or whole-cell homogenates of recombinant cell lines expressing single UGT cDNAs. Whole-cell extracts prepared from these cell lines provide a "constant" source of UGT catalytic material in a membrane environment for such studies. These extracts, together with purified UGT protein prepared from tissue, have led to the determination of the substrate specificity of some individual UGT isoforms. Full-length UGT cDNAs have been either transiently expressed in COS cells, or stably expressed in V79 and HEK 293 fibroblast cells.

Many human UGT isoforms have been identified by gene sequencing and cDNA cloning (20), but not all of these are known to be expressed in vivo. Indeed, one sequence is known to contain stop codons that would prevent functional enzyme expression (65). Consequently, only a few human UGTs have been substantially characterized, and this chapter will further focus on these transferases.

Ten human UGTs have now been characterized and reveal some overlapping substrate specificities; however, certain individual UGTs are specifically responsible for the catalysis of the glucuronidation of individual substrates. The substrate specificity of these isoforms has been extensively reviewed (14). All of the isoforms are capable of catalyzing the glucuronidation of xenobiotics.

Some problems in the use of expressed human drug-metabolizing enzymes in the analysis of drug metabolism and drug–drug interactions have been discussed in a commentary (10). The value of these in vitro systems is their relevance to human drug metabolism in vivo. Recently, attempts have been made to assess the contribution of a specific UGT to the glucuronidation of an endobiotic or xenobiotic in a specific tissue by using inhibitory antibodies. Monospecific polyclonal antibodies raised against the NH_2-terminal portion (14–150 residues) of human liver UGT2B4 protein expressed in *Escherichia coli* were used to immunoinhibit and immunoprecipitate this transferase from human liver and kidney microsomes (66). These experiments demonstrated that UGT2B4 activity was responsible for more than 90% of the hyodeoxycholate 6-*O*-glucuronidation activity in human

Table 2 Drug Glucuronidation by Rat Liver UGTs

UGT isoenzyme[a]						
UGT1A1 (**Bilirubin**)	UGT1A4 (**Bilirubin**)	UGT1A6 (**1-Naphthol**)	UGT2B1 (**Morphine**)	UGT2B2 (**Androsterone**)	UGT2B3 (**Testosterone**)	UGT2B12 (**Borneol**)
Morphine Buprenorphine		Acetaminophen	Chloramphenicol Profens Valproate			Eugenol Hexafluoro-2-propanol[b]

[a]Probe substrates for each isoenzyme are bold in parentheses.
[b]The major metabolite of the new anesthetic agent sevoflurane.
Source: Ref. 88.

liver microsomes, but did not contribute significantly toward the glucuronidation of estriol, 4-hydroxyesterone, 1-naphthol or hyodeoxycholic acid (66).

Similarly, antibodies were raised against the NH_2-terminal half of UGT1A6 expressed in *E. coli* and immunoinhibition analysis of human liver microsomes demonstrated that this isoenzyme represented between 20 and 50% of the total microsomal 1-naphthol glucuronidation (85) UGT activities towards hyodeoxycholic acid, 4-hydroxybiphenyl, 4-*t*-butylphenol, and bilirubin were not inhibited by these specific anti-UGT1A6 antibody.

D. Human UGTs Catalyzing Drug Glucuronidation

Several isoenzymes catalyze the glucuronidation of clinically used drugs (Table 3). UGT1A6 is the most effective catalyst of acetaminophen glucuronidation. UGT1A8 has a remarkably broad specificity toward many compounds (14), including clinically used drugs. When the rate of glucuronidation of drugs by UGT1A8 is compared with the drug glucuronidation by UGT2B7, UGT1A8 is two orders of magnitude more effective than UGT 2B7, based on these rate measurements (67). However, no kinetic analysis has been performed, and comparison of data from two different laboratories using different expression systems is not easily validated, because different amounts of enzyme and substrates may be present in the assays (10). At this stage, determination of the K_m and V/K_m are probably the best parameters to indicate substrate specificity (10), although V/K_m maybe a little misleading, being dependent on the level of heterologously expressed protein.

A human UGT catalyzing the glucuronidation of tertiary amine drugs has recently been identified (68). This transferase, UGT1A4 expressed in HEK293 cells, catalyzes the glucuronidation of naphthylamines, 4-aminobiphenyl, benzidine, imipramine, and other

Table 3 Glucuronidation of Drugs by Human UGTs

UGT isoenzyme[d]					
UGT1A1 (**Bilirubin**)	UGT1A4	UGT1A6 (Planar phenol UGT)	UGT1A10	UGT1A9 (Bulky phenol UGT)	UGT2B7 (Estrogen UGT)
Ethinyl estradiol	**Imipramine**	Acetominophen	Mycophenolic acid[(89)]	**Propofol**	Clofibrate[a]
Buprenorphine[(88)]	Amitryptyline	Bumetanide		Valproate	Fenoprofen
	Chlorpromazine	Ibuprofen		Naproxen	Zomepirac
	Lamotrigine	**1-Naphthol**		Ketoprofen	Diflunisal
	Doxepin	Valproate[b]		Labetalol	Fenoprofen
	Promethazine			Naproxen	Ibuprofen
	Cyproheptidine			Propranolol	Ketoprofen
	Ketotifen			Ethinyl estradiol	Oxazepam
				Dapsone	**Morphine**
				Xanthanone-4 acetic acids	Xanthenone-4 acetic acids
				Mycophenolic acid[c]	

[a]Burchell et al. (unpublished work).
[b]Anderson and Burchell (unpublished work).
[c]Ethell and Burchell (unpublished work).
[d]Probe substrates are indicated in bold type.
Source: See Reference 14 for additional information.

amines (68). Furthermore, UGT27 has been demonstrated as the major catalyst in morphine glucuronidation (69).

IV. POLYMORPHISM OF DRUG GLUCURONIDATION IN HUMANS

Confirmation of the in vivo role of an individual UGT in drug glucuronidation could be obtained from a known genetic polymorphism affecting drug glucuronidation. Unfortunately, no clear-cut examples have yet been identified.

Hereditary hyperbilirubinemias indicate numerous different mutations within the *UGT-1* gene (70). This gene encodes for several UGTs, which are all capable of metabolizing drugs. In vitro analysis of hepatic samples from Crigler-Najjar type I patients with severe hyperbilirubinemia revealed that UGT activities toward propofol, ethinyl estradiol, and phenols are severely reduced (70). A mutation in the common region of the gene will affect the production of several UGTs involved in drug metabolism.

Menthol glucuronidation is considerably reduced in many patients with Crigler-Najjar syndrome, although the accumulated data from family studies suggest that menthol glucuronidation is independently variable within this population (14). It is not surprising that mutations could occur in specific coding exons of the *UGT-1* gene that cause a defect in menthol glucuronidation without affecting bilirubin glucuronidation. Menthol glucuronide is excreted in bile and urine and may provide a relatively harmless test of drug glucuronidation in humans.

A. Gilbert's Patients

Gilbert's disease is a mild familial hyperbilirubinemia. Previously undiagnosed, patients with this disease have been identified by genotyping the Eastern Scottish and Igloolik Inuit of Canada populations in whom 10–13% and 17–19% exhibit the TA insertion genotype, respectively (71,72): 45–51% of these populations are heterozygotes.

This disease provides an opportunity to study variation in drug glucuronidation owing to the prevalence of the familial disorder within the population. There is no obvious indication of impaired drug oxidation, acetylation, or sulfation (73). In early studies, an apparent decrease in menthol glucuronidation was associated with Gilbert's syndrome (74). Decreased clearance of several drugs, such as tolbutamide, rifamycin, josamycin, and acetaminophen has been observed, although decreased clearance was not apparently associated with a decreased rate of glucuronidation measured in overnight urine samples (73,75).

De Morais et al. (76) have reported that acetaminophen glucuronide formation, measured in six Gilbert's patients by clearance from plasma within 2 h, was 31% lower than normal controls. The timing of measurements may be critical in determination of these subtle differences. Overall, the study of drug metabolism in this mild familial disease has produced more controversy than conclusive answers, probably because of the incomplete and heterogeneous nature of the biochemical lesion and a distinct lack of knowledge about the substrate specificity of the UGTs. Further studies require the use of a drug substrate known to be glucuronidated by human bilirubin UGT and family studies.

B. Interindividual Variation of Drug Glucuronidation

Glucuronidation of clomipramine metabolites showed a pronounced 28-fold interindividual difference in a Japanese psychiatric population (77). Histograms and probit plots suggest a normal distribution, although there are obviously slow and fast glucuronidators within this group of patients. It would be interesting to look for variation in the genetics of UGT expression in the slow glucuronidators. Glucuronidation was also clearly affected by gender, use of oral contraceptives and other steroids, and smoking, further indicating the problems in determination of genetic polymorphism (77,78).

Liu et al. (79) reported that the glucuronide excretion of one hypolipidemic drug, clofibrate, in a healthy white population followed a normal distribution, whereas that of fenofibrate appeared to be distributed into the distinct normal groups. However, a follow-up familial study has shown a lack of a genetic polymorphism in the glucuronidation of fenofibrate (80).

Another study of the French population examined variation of dextrorphan glucuronidation (81). Again, a normal distribution was observed. Analysis of the polymorphic variation of diflunisal phenol and its acyl glucuronide showed a unimodal population distribution, especially when excluding females using oral contraceptives (82).

Oxazepam administered as a racemic mixture was preferentially excreted as the (*S*)-glucuronide and *S*/*R* glucuronide ratios were used to assess poor glucuronidation of oxazepam (83). A group of 10% of the whole population was determined to be poor glucuronidators of (*S*)-oxazepam, suggesting a genetic relation to UGT2B7 (83,84). However, oxazepam may not be solely glucuronidated by UGT2B7 in vivo, and this relation requires additional investigation. Nonetheless, this is the most interesting example of a polymorphism of drug glucuronidation to date (see Chapter 7 on Pharmacogenetics by Ann Daly.)

V. FUTURE PERSPECTIVES

A. Identification of Novel Human UGTs

The UGTs that are involved in olfaction and brain function are yet to be characterized. Furthermore, UGTs responsible for the synthesis of macromolecules, such as glycolipids and glycosaminoglycans, are not yet described. It is noteworthy that human isoforms capable of glucuronidating many drugs, such as zidovudine (AZT) and retinoic acids, have yet to be discovered. Therefore, it is obvious that many more UGTs still need to be identified and characterized. Expression of human UGTs in heterologous cell lines will be extremely useful for studies of drug metabolism, drug–drug interactions, and for production of glucuronides (10).

The generation of specific gene knockouts in mice or rats would obviously be interesting to assess the physiological role of UGTs in vivo, but their use in drug metabolism studies may be limited because specificities of apparently homologous forms of UGT observed in rats and humans for structural studies are not identical and, in some instances, such as *N*-glucuronidation there is no similarity at all.

B. Expression of UGTs in Prokaryotes and Insect Cells

Human UGT1A6 has also been expressed in *E. coli* following exchange of the natural signal peptide by bacterial signal peptides of Pel B or Omp T proteins (85). Processing of the precursors into mature proteins was poor, but could be significantly improved by mutagenesis of the first two amino acids of the mature UGT1A6 sequence (85). The difficulties of expressing these proteins in non-mammalian cell systems may be due to the specific posttranslational processing they require, a problem that may be overcome by genetic manipulation of the UGT signal sequences to make them more like those of the host. Further developments of these expression systems should allow synthesis of high levels of UGTs in *E. coli*. Toghrol et al. (86) expressed a mouse UGT in yeast cells that glucuronidated several aglycones at low levels, but no similar studies have since been reported.

Preliminary experiments on the expression of rat phenol UGT using recombinant baculovirus injected in SF9 insect cells have been reported (87). The UGT protein was expressed at a high level and displayed significant activity toward 4-nitrophenol. Sequencing of the NH_2-terminal of the UGT indicated that the signal sequence was not removed and, therefore, the UGT was incorrectly processed by the SF9 cells. In conclusion, expression of human UGTs in prokaryotic or insect cells may provide sufficiently large quantities of protein to permit in-depth structural studies to proceed.

ACKNOWLEDGMENTS

We thank the Wellcome Trust for their support of the research work of this Dundee Laboratory.

REFERENCES

1. D. J. Clarke and B. Burchell, The uridine diphosphate glucuronosyltransferase multigene family: Function and regulation, *Handb. Exp. Pharmcol. 112*:3–43 (1994).
2. C. H. Brierley and B. Burchell, Human UDP-glucuronosyltransferases: Chemical defence, jaundice and gene therapy, *Bioessays 15*:749–754 (1993).
3. G. J. Dutton, *Glucuronidation of Drugs and Other Compounds*, CRC Press, Boca Raton, 1980.
4. D. J. Clarke, S. G. George, and B. Burchell, Multiplicity of UDP-glucuronosyltransferases in fish. Purification and characterisation of a phenol UDP-glucuronosyltransferase from plaice liver, *Biochem. J. 284*:417–423 (1992).
5. B. Burchell and M. W. H. Coughtrie, UDP-glucuronyltransferases. *Genetic Factors Influencing the Metabolism of Foreign Compounds, Pharmacol. Ther. 43*:261–289 (1989).
6. L. Sutherland, T. Ebner, and B. Burchell, The expression of UDP-glucuronosyltransferase of the UGT1 family in human liver and kidney and in response to drugs, *Biochem. Pharmacol. 45*:295–301 (1993).
7. R. Van de Straat, J. de Vries, E. J. Groot, R. Zijl, and N. P. E. Vermeulen, Paracetamol, 3-monoalkyl and 3,5 dialkyl derivatives: Comparison of their in vivo hepatoxicity in mice, *Toxicol Appl. Pharmacol. 89*:183–189 (1987).
8. J. J. Bouska, R. L. Bell, C. L. Goodfellow, A. O. Stewart, R. Hansen, L. Dube, D. W. Brooks, G. W. Carter, and J. Machinist, Utilization of a hepatic microsomal UDP-glucuronosyltransferase assay to improve in vivo duration in monkey and man, *ISSX Proc. 6*:122 (1994).
9. B. C. Cupid, C. R. Beddell, J. C. Lindon, I. D. Wilson, and J. K. Nicholson, Quantitative structure–metabolism relationships for substituted benzoic acids in the rabbit: Prediction of urinary excretion of glycine and glucuronide conjugates, *Xenobiotica 26*:157–176 (1996).
10. R. P. Remmel and B. Burchell, Validation and use of cloned, expressed human drug metabolis-

ing enzymes in heterologous cells for analysis of drug metabolism and drug–drug interactions, *Biochem. Pharmacol. 46*:559–566 (1993).
11. J. K. Nichiolson, E. Holmes, U. G. Sidelmann, J. C. Lindon, and I. D. Wilson, HPLC–NMR spectroscopy: A powerful tool for the investigation of drug metabolism and metabolite reactivity, *Pharm. Sci. 2*:127–130 (1996).
12. H. K. Kroemer and U. Klotz, Glucuronidation of drugs—a reevaluation of the pharmacological significance of the conjugates and modulating factors, *Clin. Pharmacokinet. 23*:292–310 (1992).
13. G. J. Mulder, M. W. H. Coughtrie, and B. Burchell, Glucuronidation, *Conjugation Reactions in Drug Metabolism: An Integrated Approach* (G. J. Mulder, ed.), Taylor Francis, London, 1990, pp. 51–105.
14. B. Burchell, K. McGurk, C. H. Brierley, and D. J. Clarke, UDP-glucuronosyltransferases, *Comprehensive Toxicology* (G. Sipes, et al., eds.), Pergamon Elsevier Science, New York, 1996, pp. 401–435.
15. J. R. Chowdhury, A. W. Wolkoff, N. R. Chowdhury, and I. M. Arias, Hereditary jaundice and disorders of bilirubin metabolism *Metabolic Basis of Inherited Disease* (C. Scriver, et al., eds.), McGraw Hill, New York, 1995, pp. 2161–2208.
16. A. Taurog, I^{131} labeled L-thyroxine. II Nature of the excretion product in bile, *Brookhaven Symp. Biol. 7*:111 (1955).
17. E. V. Flock, J. L. Bollman, C. A. Owens, and P. E. Zollman, Conjugation of thyroid hormones and analogs by the Gunn rat, *Endocrinology 77*:303 (1965).
18. M. T. Hays and L. Hsu, Equilibrium dialysis of plasma binding of thyroxine, tri-iodothyronine and other glucuronide and sulphate conjugates in human and cat plasma, *Endoc. Res. 14*:51–58 (1988).
19. S. F. Sani, A. B. Berau, D. L. Hill, T. W. Shek, and J. A. Olson, Retinoyl β-glucuronide: Lack of binding to receptor proteins of retinoic acid as related to biological activity, *Biochem. Pharmacol. 43*:919–922 (1992).
20. B. Burchell, D. W. Nebert, D. R. Nelson, K. W. Bock, T. Iyanagi, and P. L. M. Jansen, The UDP glucuronosyltransferase gene superfamily: Suggested nomenclature based on evolutionary divergence, *DNA Cell Biol. 10*:487–494 (1991).
21. D. L. Thompson, N. Horton, and R. S. Rittmaster, Androsterone glucuronide is a marker of adrenal hyperandrogenism in hirsute women, *Clin. Endocrinol. 32*:283–292 (1990).
22. K. Salman, R. L. Spielvogel, L. H. Shulman, J. L. Miller, R. E. Vanderlinde, and L. I. Rose, Serum androstanediol glucuronide in women with facial hirsutism, *J. Am. Acad. Dermatol. 26*:411–414 (1992).
23. V. A. Giagulli, R. Giorgino, and A. Vermeulen, Is plasma 5α-androstane 3α, 17β-diol glucuronide a biochemical marker of hirsutism in women? *J. Steroid Biochem. Mol. Biol. 39*:55–61 (1991).
24. A. Belanger, B. Candas, A. Dupont, L. Cusan, P. Diamond, J. L. Gomez, and F. Labrie, Changes in serum concentrations of conjugated and unconjugated steroids in 40- to 80-year-old men, *J. Clin. Endocrinol. Metab. 79*:1086–1090 (1994).
25. J. Montalto, J. W. Funder, A. B. W. Yong, C. B. Whorwood, and J. F. Connelly, Serum levels of 5-androstene-3β-17β-diol sulphate, 5α-androstane-3α,17β-diol sulphate and glucuronide in late onset 21-hydroxylase deficiency, *J. Steroid Biochem. Mol. Biol. 37*:593–598 (1990).
26. S. Pang, M. Macgillivary, M. Wang, S. Jeffries, A. Clark, I. Rosenthal, M. Weigensberg, and L. Riddick, 3a-Androstanediol glucuronide in virilizing congenital adrenal hyperplasia: A useful serum metabolic marker of integrated adrenal androgen secretion, *J. Clin. Endocrinol. Metab. 73*:166–174 (1991).
27. J. Iwata and T. Suga, Direct determination of four sulfates and seven glucuronides of 17-oxosteroids in urine by fluorescence ''high performance'' liquid chromatography, *Clin. Chem. 35*:794–799 (1989).
28. S. I. Sherman and P. W. Ladenson, Organ-specific effects of tiratricol—a thyroid-hormone

analog with hepatic, not pituitary, superagonist effects, *J. Clin. Endocrinol. Metab. 75*:901–905 (1992).

29. M. Moreno, E. Kaptein, F. Goglia, and T. J. Visser, Rapid glucuronidation of tri- and tetraido-thyroacetic acid to ester glucuronides in human liver and to ether glucuronides in rat liver, *Endocrinology 135*:1004–1009 (1994).
30. G. M. Powell, J. J. Miller, A. H. Olavesen, and C. G. Curtis, Liver as major organ of phenol detoxication? *Nature 252*:234–235 (1974).
31. A. A. Raoof, L. J. van Obbergh, J. de Ville de Goyet, and R. K. Verbeeck, Extrahepatic glucuronidation of propofol in man: Possible contribution of gut wall and kidney, *Eur. J. Clin. Pharmacol. 50*:91–96 (1996).
32. A. A. Raoof, P. F. Augustijns, and R. K. Verbeeck, In vivo assessment of intestinal, hepatic, and pulmonary first pass metabolism of propofol in the rat, *Pharm. Res. 13*:891–895 (1996).
33. T. B. Vree, M. Van den Biggelaar-Martea, and C. P. W. G. M. Verwey-Van Wissen, Probenecid inhibits the renal clearance of furosemide and its acyl glucuronide, *Br. J. Clin. Pharmacol. 35*:692–695 (1995).
34. F. M. Brunelle and R. K. Verbeeck, Glucuronidation of diflunisal in liver and kidney microsomes of rat and man, *Xenobiotica 26*:123–131 (1996).
35. C. L. Booth, G. M. Pollack, and K. L. R. Brouwer, Hepatobiliary disposition of valproic acid and valproate glucuronide: Use of a pharmacokinetic model to examine the rate-limiting steps and potential sites of drug interactions, *Hepatology 23*:771–780 (1996).
36. H.-Y. Yu and Y.-Z. Shen, Glucuronidation metabolic kinetics of valproate in guinea pigs: Nonlinear at clinical concentration levels, *Pharm. Res. 13*:1243–1246 (1996).
37. C. Chen, P. W. Slattum, K. L. R. Brouwer, and G. M. Pollack, Influence of age and gender on valproic acid glucuronidation in rats, *Drug Metab. Dispos. 24*:367–369 (1996).
38. D. Paul, K. M. Standifer, C. E. Inturrisi, and G. W. Pasternak, Pharmacological characterization of morphine-6-β-glucuronide, a very potent morphine metabolite, *Pharmacol. Exp. Thera. 251*:477–483 (1989).
39. S. R. Babu, V. M. Lakshmi, G. Pen-Wen Huang, T. V. Zenser, and B. B. Davis, Glucuronide conjugates of 4-aminobiphenyl and its *N*-hydroxy metabolites. pH stability and synthesis by human and dog liver, *Biochem. Pharmacol. 51*:1679–1685 (1996).
40. S. R. Babu, V. M. Lakshmi, F. F. Hsu, T. V. Zenser, and B. B. Davis, Glucuronidation of *N*-hydroxy metabolites of *N*-acetylbenzidine, *Carcinogenesis 16*:3069–3074 (1995).
41. S. E. Murphy, D. A. Spina, M. G. Nunes, and D. A. Pullo, Glucuronidation of 4-(hydroxymethyl)nitrosamino)-1-(3-pyridyl)-1-butanone, a metabolically activated form of 4-(methylnitrosamino)-1-(3-pyridyl)-1-butanone, by phenobarbital-treated rats, *Chem. Res. Toxicol. 8*: 772–779 (1995).
42. K. W. Bock, Roles of UDP-glucuronosyltransferases in chemical carcinogenesis, *Crit. Rev. Biochem. Mol. Biol. 26*:129–150 (1991).
43. U. A. Boelsterli, Specific targets of covalent drug–protein interactions in hepatocytes and their toxicological significance in drug-induced liver-injury, *Drug Metab. Rev. 25*:395–451 (1993).
44. P. M. Kim and P. G. Wells, Genoprotection by UDP-glucuronosyltransferases in peroxidase-dependent, reactive oxygen species-mediated micronucleus initiation by the carcinogens 4-(methylnitrosamino)-1-(3-pyridyl)-1-butanone and benzo[*a*]pyrene, *Cancer Res. 56*:1526–1532 (1996).
45. H. Yin, G. Bennett, and J. P. Jones, Mechanistic studies of uridine diphosphate glucuronosyltransferase, *Chem. Biol. Interact. 90*:47–58 (1994).
46. M. D. Murray and D. C. Brater, Renal toxicity of the nonsteroidal anti-inflammatory drugs, *Annu. Rev. Pharmacol. Toxicol. 32*:435–465 (1993).
47. H. Sphann-Langguth and L. Z. Benet, Acyl glucuronides revisited: Is the glucuronidation process a toxification as well as detoxification mechanism? *Drug Metab. Rev. 24*:5–48 (1992).

48. E. M. Faed, Properties of acyl glucuronides—implications for studies of the pharmacokinetics and metabolism of acidic drugs, *Drug Metab. Rev.* *15*:1213–1249 (1984).
49. P. C. Smith, A. F. McDonagh, and L. Z. Benet, Irreversible binding of zomepirac to plasma protein in vitro and in vivo, *J. Clin. Invest* *77*:934–939 (1986).
50. P. C. Smith and J. H. Liu, Covalent binding of suprofen acyl glucuronide to albumin in vitro, *Xenobiotica* *23*:337–348 (1993).
51. J. C. Ojingwa, H. Spahn-Langguth, and L. Z. Benet, Irreversible binding of tolmetin to macromolecules via its glucuronide: Binding to blood constituents, tissue homogenates and subcellular fractions in vitro, *Xenobiotica* *24*:495–506 (1994).
52. M. Castillo, Y. W. Francis Lam, M. A. Dooley, E. Stahl, and P. C. Smith, Disposition and covalent binding of ibuprofen and its acyl glucuronide in the elderly, *Clin. Pharmacol. Ther.* *57*:636–644 (1995).
53. P. C. Smith and J. H. Liu, Covalent binding of suprofen to renal tissue of rat correlates with excretion of its acyl glucuronide, *Xenobiotica* *25*:531–540 (1995).
54. U. G. Sidelmann, C. Gavaghan, H. A. J. Carless, R. D. Farrant, J. C. Lindon, I. D. Wilson, and J. K. Nichiolson, Identification of the positional isomers of 2-fluorbenzoic acid 1-*O*-acyl glucuronide by directly coupled HPLC–NMR, *Anal. Chem.* *67*:3401–3404 (1995).
55. U. G. Sidelmann, S. H. Hansen, C. Gavaghan, H. A. J. Carless, J. C. Lindon, R. D. Farrant, I. D. Wilson, and J. K. Nichiolson, Measurement of internal acyl migration reaction kinetics using directly coupled HPLC–NMR: Application for the positional isomers of synthetic (2-fluorobenzoyl)-D-glucopyranuronic acid, *Anal. Chem.* *68*:2564–2572 (1996).
56. A. W. Nicholls, K. Akira, J. C. Lindon, R. D. Farrant, I. D. Wilson, J. Harding, D. A. Killick, and J. K. Nicholson, NMR spectroscopic and theoretical chemistry studies on the internal acyl migration reactions of the 1-*O*-acyl-β-D-glucopyranuronate conjugates of 2-, 3-, and 4-trifluormethylbenzoic acids, *Chem. Res. Toxicol.* *9*:1414–1424 (1996).
57. A. Gautam, H. Seligson, E. R. Gordon, D. Seligson, and J. L. Boyer, Irreversible binding of conjugated bilirubin to albumin in cholestatic rats, *J. Clin. Invest.* *73*:873–877 (1984).
58. M. W. H. Coughtrie, B. Burchell, I. M. Shepherd, and J. R. Bend, Defective induction of phenol glucuronidation by 3-methylcholanthrene in Gunn rats is due to the absence of a specific UDP-glucuronosyltransferase isoenzyme, *Mol. Pharmacol.* *31*:585–591 (1987).
59. M. W. H. Coughtrie, B. Ask, A. Rane, B. Burchell, and R. Hume, The enantioselective glucuronidation of morphine in rats and humans, *Biochem. Pharmacol.* *38*:3273–3280 (1989).
60. T. J. Visser, E. Kaptein, H. van Toor, J. A. G. M. van Raaij, K. J. van den Berg, C. Tjong Tjin Joe, J. G. M. van Engelen, and A. Brouwer, Glucuronidation of thyroid hormone in rat liver: Effects of in vivo treatment with microsomal enzyme inducers and in vitro assay conditions, *Endocrinology* *133*:2177–2186 (1993).
61. R. B. Corser, M. W. H. Coughtrie, M. R. Jackson, and B. Burchell, The molecular basis of the inherited deficiency of androsterone UDP-glucuronosyitransferase, *FEBS Lett.* *213*:448–452 (1987).
62. A. Yamashita, T. Nagatsuka, M. Watanabe, H. Kondo, T. Sugiura, and K. Waku, Inhibition of UDP-glucuronosyltransferase activity by fatty acyl-CoA: Kinetic studies and structure–activity relationship, *Biochem. Pharmacol.* *53*:561–570 (1997).
63. S. Diamond and D. D. Christ, Effects of the novel immunosupprestant brequinar on hepatic UDP-glucuronic acid levels and UDP-glucuronosyltransferase activities in the rat, *Drug Metab. Dispos.* *24*:375–376 (1996).
64. M. Pritchard, S. Fournel-Gigleux, G. Siest, P. MacKenzie, and J. Magdalou, A recombinant phenobarbital-inducible rat liver UDP-glucuronosyltransferase (UDP-glucuronosyltransferase 2B1) stably expressed in V79 cells catalyzes the glucuronidation of morphine, phenols and carboxylic acids, *Mol. Pharmacol.* *45*:42–50 (1993).
65. J. K. Ritter, F. Chen, Y. Y. Sheen, H. M. Tran, S. Kimura, M. T. Yeatman, and I. S. Owens, A novel complex locus *UGTI* encodes human bilirubin, phenol and other UDP-glucuronosyltransferase isozymes with identical carboxyl termini, *J. Biol. Chem.* *167*:3257–3261 (1992).

66. T. Pillot, M. Ouzzine, S. Fournel-Gigleux, C. Lafaurie, A. Radominska, B. Burchell, G. Siest, and J. Magdalou, Glucuronidation of hyodeoxycholic acid in human liver—evidence for a selective role for UDP-glucuronosyltransferase-2B4, *J. Biol. Chem. 268*:25636–25642 (1993).
67. B. Burchell, C. H. Brierley, and D. Rance, Specificity of human UDP-glucuronosyltransferases and xenobiotic glucuronidation, *Life Sci. 57*:1819–1831 (1995).
68. M. D. Green, W. P. Bishop, and T. R. Tephly, Expressed human UGT1*4 protein catalyzes the formation of quaternary ammonium-linked glucuronides, *Drug Metab. Dispos. 23*:299 (1995).
69. B. L. Coffmann, G. R. Rios, C. D. King, and T. R. Tephly, Human UGT2B7 catalyses morphine glucuronidation, *Drug Metab. Dispos. 25*:1–4 (1997).
70. D. J. Clarke, N. Moghrabi, G. Monaghan, A. Cassidy, M. Boxer, R. Hume, and B. Burchell, Genetic defects of the UDP-glucuronosyltransferase-1 (UGT1) gene that cause familial non-haemolytic unconjugated hyperbilirubinaemias, *Clin. Chim. Acta 266*:63–74 (1997).
71. G. Monaghan, M. F. Ryan, R. Seddon, R. Hume, and B. Burchell, Genetic variation in bilirubin UDP-glucuronosyltransferase gene promoter and Gilbert's syndrome, *Lancet 347*:578–581 (1996).
72. G. Monaghan, B. Foster, M. Jurima-Romet, R. Hume, and B. Burchell, UGT1*1 genotyping in a Canadian Inuit population, *Pharmacogenetics, 7*:153–156 (1997).
73. D. Ullrich, A. Sieg, R. Blume, K. W. Bock, W. Schrotter, and J. Bircher, Normal pathways for glucuronidation, sulphation and oxidation of paracetamol in Gilbert's syndrome, *Eur. J. Clin. Invest. 17*:237–240 (1987).
74. I. M. Arias, L. M. Gartner, M. Cohen, J. B. Ezzer, and A. J. Levi, Chronic nonhemolytic unconjugated hyperbilirubinemia with glucuronosyltransferase deficiency: Clinical, biochemical, pharmacologic and genetic evidence for heterogeneity, *Am. J. Med. 47*:395–409 (1969).
75. A. F. Macklon, R. L. Savage, and M. D. Rawlins, Gilbert's syndrome and drug metabolism, *Clin. Pharmacokinet. 4*:223–232 (1979).
76. S. M. F. De Morias, J. P. Uetrecht, and P. G. Wells, Decreased glucuronidation and increased bioactivation of acetaminophen in Gilbert's syndrome, *Gastroenterology 102*:577–586 (1992).
77. K. Shimoda, T. Noguchi, Y. Ozeki, S. Morita, M. Shibasaki, T. Someya, and S. Takahashi, Metabolism of clomipramine in a Japanese psychiatric population: Hydroxylation, desmethylation, and glucuronidation, *Neuropsychopharmacology 12*:323–333 (1995).
78. K. W. Bock, D. Schrenk, A. Forster, E. Griese, K. Morike, D. Brockmeier, and M. Eichelbaum, The influence of environmental and genetic factors on CYP2D6, CYP1A2 and UDP-glucuronosyltransferases in man using sparteine, caffeine, and paracetamol as probes, *Pharmacogenetics 4*:209–218 (1994).
79. H. F. Liu, M. Vincentviry, M. M. Galteau, R. Gueguen, J. Magdalou, A. Nicolas, P. Leroy, and G. Siest, Urinary glucuronide excretion of fenofibric and clofibric acid glucuronides in man—is it polymorphic, *Eur. J. Clin. Pharmacol. 41*:153–159 (1991).
80. M. Vincent-Viry, C. Cossy, M. M. Galteau, R. Gueguen, J. Magdalou, A. Nicolas, P. Leroy, and G. Siest, Lack of a genetic polymorphism in the glucuronidation of fenofibric acid, *Pharmacogenetics 5*:50–52 (1995).
81. J. C. Duche, V. Querol-Ferrer, J. Barre, M. Mesangeau, and J. P. Tillement, Dextromethorphan *O*-demethylation and dextrorphan glucuronidation in a French population, *Int. J. Clin. Pharmacol. Ther. Toxicol. 31*:392–398 (1993).
82. R. J. Herman, G. R. Loewen, D. M. Antosh, M. R. Taillon, S. Hussein, and R. K. Verbeeck, Analysis of polymorphic variation in drug metabolism: III. Glucuronidation and sulfation of diflunisal in man, *Clin. Invest. Med. 17*:297–307 (1994).
83. M. Patel, B. K. Tang, D. M. Grant, and W. Kalow, Interindividual variability in the glucuronidation of (*S*) oxazepam contrasted with that of (*R*) oxazepam, *Pharmacogenetics 5*:287–297 (1995).
84. M. Patel, B. K. Tang, and W. Kalow, (*S*)-Oxazepam glucuronidation is inhibited by ketoprofen and other substrates of UGT 2B7, *Pharmacogentics 5*:43–29 (1995).
85. M. Ouzzine, T. Pillot, S. Fournelgigleux, J. Magdalou, B. Burchell, and G. Siest, Expression

and role of the human liver UDP-glucuronosyltransferase UGT1*6 analyzed by specific antibodies raised against a hybrid protein produced in *Escherichia coli, Arch. Biochem. Biophys. 310*:196–204 (1994).

86. F. Tohgrol, T. Kimura, and I. S. Owens, Expression of UDP-glucuronosyltransferase cDNA in *Saccharomyces cerevisiae* as a membrane-bound and as a cytosolic form, *Biochemistry 29*: 2349–2356 (1990).
87. R. N. Armstrong, G. M. Lacourciere, and V. N. Vikharia, Expression of microsomal detoxication enzymes in a recombinant baculovirus system. The case of epoxide hydrolase and UDP-glucuronosyltransferase, *J. Physiol. Pharmacol. 3*:170 (1992).
88. C. D. King, M. D. Green, G. R. Rios, B. L. Coffman, I. S. Owens, W. P. Bishop, and T. R. Tephly, The glucuronidation of exogenous and endogenous compounds by stably expressed rat and human UDP-glucuronosyltransferase 1.1, *Arch. Biochem. Biophys. 332*:92–100 (1996).
89. B. Mojarrobi and P. I. Mackenzie, The human UDP-glucuronosyltransferase, UGT1A10, glucuronidates mycophenolic acid, *Biochem. Biophys. Res. Commun. 238*: 775–778 (1997).

7
Pharmacogenetics

Ann K. Daly
University of Newcastle upon Tyne Medical School,
Newcastle upon Tyne, United Kingdom

I. INTRODUCTION

Pharmacogenetics is a broad term that can be defined as the study of genetically determined variations that are revealed by the effects of drugs and other xenobiotics (1). The subject includes both the areas of drug biotransformation and responses of cells or tissues to drugs, but this chapter deals only with pharmacogenetics of human drug biotransformation. Interindividual variability in drug metabolism can be determined by several different factors but the existence of genetic polymorphisms in the genes encoding metabolizing enzymes and, probably more rarely, in genes encoding transcription factors that regulate the expression of genes encoding metabolizing enzymes, are important factors. Until recently, genetic polymorphisms with functional effects on drug metabolism were detected on the basis of discontinuous variation in phenotype, where phenotype represented either levels of the enzyme or rate of metabolism. This approach led to the detection of a variety of relatively common polymorphisms, but other sources of genetic variation in drug metabolism, such as polygenic effects, for which the variation is controlled by several different genes, or rarer genetic defects that occur at a frequency of fewer than 1:100 may not be detected (1). With the cloning of a variety of genes encoding the enzymes of xenobiotic metabolism, detailed studies on sequence variation have become possible and, occasionally, have resulted in the detection of new pharmacogenetic defects.

Pharmacogenetic polymorphisms in genes encoding xenobiotic-metabolizing enzymes may have a variety of effects, depending on both the type of reaction catalyzed and the type of substrate. With drugs, the consequences of a polymorphism may be toxic plasma concentrations if there is a deficiency in a metabolizing enzyme, or lack of response if activation by a polymorphic enzyme is required for biological activity or if higher than normal levels of a metabolizing enzyme result in too rapid a rate of elimination (2). Whether absence of a metabolizing enzyme results in toxicity will depend on various factors, including the therapeutic margin of safety versus activity and, in particular, the contribution the polymorphic enzyme makes to total metabolism.

Drug-metabolizing enzymes frequently also activate or metabolize other xenobiotics, including procarcinogens and carcinogens. Polymorphisms may, therefore, influence susceptibility to cancer and other diseases associated with chemical exposure (for review see Refs. 3, 4).

II. PHASE I POLYMORPHISMS

Among phase I enzymes, the cytochrome P450 superfamily is the most important group of enzymes in terms of both numbers of drugs metabolized (see Chapter 4) and existence of pharmacogenetic polymorphisms. Consequently, the cytochrome P450 polymorphisms are considered separately from other phase I polymorphisms, which for those relevant to drugs are generally less common or less well understood.

A. Cytochrome P450 Polymorphisms

Polymorphisms have now been detected in several of the genes encoding various cytochrome P450s. Most of these were originally detected by use of population studies involving phenotype determination, often as a means of following up reports of exaggerated clinical responses to drugs. The considerable advances in cytochrome P450 molecular genetics since the early 1980s has enabled the molecular basis of these polymorphisms to be determined and has also resulted in the detection of novel ones.

1. CYP2D6

The cytochrome P450 CYP2D6 is of particular importance because it metabolizes a wide range of commonly prescribed drugs, including antidepressants, antipsychotics, beta-adrenergic blockers, and antiarrythmics (Table 1; for review see Refs. 5, 6). Approximately 5% of Europeans and 1% of Asians lack CYP2D6 activity, and these individuals are known as poor metabolizers. Individuals showing impaired activity and individuals showing particularly high levels of activity (ultrarapid metabolizers) have also been described. Those with activity in the normal range are known as extensive metabolizers.

The molecular basis of variation in CYP2D6 activity is now well understood, and at least 17 different allelic variants of *CYP2D6* have been identified and characterized (Table 2; 7). In the region of 95% of European poor metabolizers can be identified by

Table 1 CYP2D6 Substrates[a]

Drug	Ref.
Psychotrophic agents	
Amiflamine	168
Amitriptyline	169
Brafaromine	170
Chlorpromazine	171
Citalopram	172
Clomipramine	173
Clozapine	174
Desipramine	176
Haloperidol	178
Imipramine	179
Maprotiline	180
Methoxyphenamine	181
Mianserin	182
Minaprine	183
Nortriptyline	184
Paroxetine	185
Perphenazine	186
Thioridazine	188
Tomoxetine	189
Venlafaxine	190
Zuclopenthixol	191
Cardiovascular agents	
Aprinidine	192
Bufuralol	193
Bupranolol	194
Debrisoquin	195
Encainide	196
Flecainide	197
Guanoxan	198
Indoramin	199
Metoprolol	200
Mesiletine	201
N-Propylajamaline	202
Propafenone	203
Propranolol	204
Sparteine	205
Timolol	206
Miscellaneous agents	
Codeine	207
Deprenyl (selegaline)	175
Dexfenfluramine	208
Dextromethorphan	209
Dihydrocodeine	210
Dolasetron	177
Ethylmorphine	211
Hydrocodone	212
Lignocaine	213
Loratadine	214
Methoxyamphetamine	215
Ondansetron	187
Perhexilene	216
Phenformin	217
Tropisetron	187

[a] Compounds listed in this table have been demonstrated to undergo metabolism by CYP2D6, but this may not necessarily be the only or main pathway of oxidative metabolism.

Table 2 *CYP2D6* Alleles

Allele	Nucleotide changes[a]	*Xba*I haplotype (kb)	Trivial name	Effect	Ref.
Alleles associated with normal activity					
*CYP2D6*1*	None	29	Wild-type		217
Alleles associated with increased activity					
*CYP2D6*2XN*	$G_{1749}C$; $C_{2938}T$; $G_{4268}C$	42–175[b]		$R_{296}C$; $S_{486}T$	23
N active genes					
(*N* = 2, 3, 4, 5, or 13)					
Alleles associated with absence of activity					
*CYP2D6*3*	**A_{2637} deletion**	29	*CYP2D6A*	Frameshift	20
*CYP2D6*4*[c]	**$G_{1934}A$**; $C_{188}T$; $G_{4268}C$; and various others ($C_{1085}G$; $C_{1062}A$; $A_{1072}G$; $G_{1749}C$; $T_{3975}C$)	44/29/16+9	*CYP2D6B*	Splicing defect	20, 218, 219
*CYP2D6*5*	***CYP2D6* deleted**	11.5 or 13[d]	*CYP2D6D*	*CYP2D6* deleted	220, 221
*CYP2D6*6*[c]	**T_{1795} deleted;** ($G_{2064}A$)	29	*CYP2D6T*	Frameshift	13–15
*CYP2D6*7*	**$A_{3023}C$**	29	*CYP2D6E*	$H_{324}P$	12
*CYP2D6*8*	$G_{1749}C$; **$G_{1846}T$**; $C_{2938}T$; $G_{4268}C$		*CYP2D6G*	Stop codon	222
*CYP2D6*11*	**$G_{971}C$;** $G_{1749}C$; $C_{2938}T$; $G_{4268}C$	29	*CYP2D6F*	Splicing defect	223
*CYP2D6*12*	**$G_{212}A$**; $G_{1749}C$; $C_{2938}T$; $G_{4268}C$	29		$G_{42}R$; $R_{296}C$; $S_{486}T$	16

*CYP2D6*13*	*CYP2D7P/CYP2D6* hybrid Exon 1 *CYP2D7*, exons 2–9 *CYP2D7*	9 or 11[d]		Frameshift	224, 225
*CYP2D6*14*	$C_{188}T$; $\mathbf{G_{1846}A}$; $C_{2938}T$; $G_{4286}C$	29		$P_{34}S$; $G_{169}R$; $R_{296}C$; $S_{486}T$	226
*CYP2D6*15*	$\mathbf{T_{226}}$ **insertion**	29		Frameshift	227
*CYP2D6*16*	*CYP2D7P/CYP2D6* hybrid Exons 1–7 *CYP2D7P*-related, exons 8–9 *CYP2D6*	11	*CYP2D6D2*	Frameshift	225
*CYP2D6*18*	9-bp insertion in exon 9	29		Insertion	228
*CYP2D6*19*	A_{2627}-T_{2630} deleted	29		Frameshift	229
Alleles associated with impaired activity					
*CYP2D6*2*	$G_{1749}C$; $C_{2938}T$; $G_{4268}C$	29	*CYP2D6L*	$R_{296}C$; $S_{486}T$	21, 23, 230
*CYP2D6*9*	$\mathbf{A_{2701}}$**-**$\mathbf{A_{2703}}$**,** $\mathbf{G_{2702}}$**-**$\mathbf{A_{2704}}$ **or** $\mathbf{A_{2703}}$**-**$\mathbf{G_{2705}}$ **deleted**	29	*CYP2D6C*	K_{281} deleted	231, 232
*CYP2D6*10*[c]	$\mathbf{C_{188}T}$; $G_{1749}C$; $G_{4268}C$; $C_{C1127}T$ and gene conversion to *CYP2D7* in exon 9 also found	29/44	*CYP2D6J* *CYP2D6Ch1* *CYP2D6Ch2*	$P_{34}S$; $S_{486}T$ $P_{34}S$; $S_{486}T$	17–19
*CYP2D6*17*	$C_{1111}T$; $G_{1726}C$; $C_{2938}T$ $G_{4268}C$	29	*CYP2D6Z*	$T_{107}I$; $R_{296}S$; $S_{486}T$	21

[a] Polymorphisms believed to be responsible for the observed effect on activity for each allele are shown in bold.
[b] A range of band sizes are detected by RFLP analysis with *XbaI* depending on precise number of copies of gene.
[c] Several related alleles are seen, all with the same key polymorphism, but other additional polymorphisms may occur. These related alleles can be distinguished on the basis of additional letters (e.g., *CYP2D6*4A*, *CYP2D6*4B*, and so forth).
[d] Two different size estimates for this RFLP have been obtained.
Source: Ref. 7.

screening for the *CYP2D6*3, CYP2D6*4, CYP2D6*5, CYP2D6*6*, and *CYP2D6*7* alleles using polymerase chain reaction (PCR) methods (8–15). The remaining 5% of poor metabolizers are likely to be homozygous or heterozygous for a range of different inactive alleles, with each individual allele relatively rare. Most inactivating mutations in *CYP2D6* are either deletions or point mutations resulting in splicing defects, although two inactivating polymorphisms introducing amino acid substitutions have recently been described (12,16).

Many individuals fall into the category of intermediate metabolizers, which is particularly common among Asians. Intermediate metabolizers may be either heterozygous for one of the inactivating mutations or homozygous for alleles associated with impaired metabolism. The *CYP2D6*10* allele is particularly common among Chinese and Japanese and is associated with reduced CYP2D6 activity owing to an NH_2-terminal proline to serine substitution (17–20). Many Chinese have two tandem copies of *CYP2D6*10* present, with the upstream copy also containing a gene conversion to *CYP2D7P* in part of exon 9 (19). Intermediate metabolizers also occur at high frequencies among African populations, and this is at least partly due to the presence of the *CYP2D6*17* allele that has been detected at a frequency of 0.35 among Zimbabweans and shows reduced activity owing to an amino acid substitution ($Thr_{107}Ile$) (21).

Family studies on DNA samples from ultrarapid metabolizers who were originally identified on the basis of their extremely fast clearance of the antidepressant desmethylimipramine have been reported (22). In one family, several members had 13 copies of *CYP2D6* arranged as tandem repeats, indicating that inherited amplification of the entire gene had occurred (23). Up to 5% of Europeans have one extra copy of the *CYP2D6*2* allele, resulting in faster than average metabolism, and subjects with three or four tandem copies of *CYP2D6*2* have also been detected (24,25).

Poor metabolizers show higher than normal plasma levels of several drugs and may be at increased risk of adverse effects, although this depends very much on the individual drug and the overall contribution of CYP2D6 to its metabolism (5). For perhexilene and phenformin, toxicity in poor metabolizers resulted in withdrawal of these compounds from the market. For those antipsychotics, antidepressants, and antiarrythmics for which most phase I metabolism involves CYP2D6, evidence for elevated plasma levels in poor metabolizers has been well documented (26–28). Codeine is an ineffective analgesic in poor metabolizers owing to the absence of activation to morphine by CYP2D6 (29,30). Low plasma levels and rapid clearance of antidepressants have also been demonstrated in ultrarapid metabolizers (22).

2. CYP2C19

A polymorphism in the metabolism of the anticonvulsant *S*-mephenytoin that was distinct from the *CYP2D6* polymorphism was first identified in the early 1980s (31). Approximately 3% of Europeans are poor metabolizers of mephenytoin, but the deficiency is seen in approximately 20% of Asians (32). The *S*-mephenytoin hydroxylase reaction is catalyzed by CYP2C19, and two mutant alleles associated with the defect have been identified (33–36). The more common mutant allele (*CYP2C19*2*) accounts for approximately 80% of mutant alleles in both Europeans and Japanese. The second allele (*CYP2C19*3*) accounts for the remaining 20% of mutant alleles in Japanese, but appears very rare in Europeans. Both inactivating mutations are single-base–pair substitutions, with an aberrant splice site created in *CYP2C19*2* and a premature stop codon in *CYP2C19*3*.

Drugs known to be CYP2C19 substrates include widely prescribed compounds, such as omeprazole, propranolol, and imipramine (37–39). Omeprezole is a CYP1A2 inducer and high serum omeprazole levels, such as might occur in those deficient in CYP2C19, may result in increased CYP1A2 activity (40). CYP2C19 also appears to be the major enzyme that activates the antimalarial chloroguanide (proguanil) by cyclization; therefore, this compound may be ineffective in deficient individuals (41). Although fewer compounds are known to be metabolized by CYP2C19 than by CYP2D6, that several are widely used and that up to 20% of persons of certain ethnic origins may lack the enzyme means that the *S*-mephenytoin polymorphism is of considerable clinical importance.

3. *CYP2C9*

For some time, a polymorphism in the metabolism of the hypoglycemic agent tolbutamide has been postulated to occur, and there are also reports from the early 1960s onward of some persons showing extremely slow metabolism of phenytoin (42,43). It is only recently that the main cytochrome P450 responsible for hydroxylation of both these drugs has been demonstrated to be CYP2C9 and that some understanding of the molecular basis of polymorphism in the *CYP2C9* gene has been obtained. CYP2C9 metabolizes a range of therapeutically important drugs, including tolbutamide, phenytoin, *S*-warfarin, and a range of nonsteroidal anti-inflammatory drugs, including diclofenac and ibuprofen (37,38).

Comparison of CYP2C9 cDNA clones isolated in several laboratories has demonstrated several single-base–pair substitutions that result in amino acid changes, particularly $Arg_{144}Cys$ (*CYP2C9*2*) and $Ile_{359}Leu$ (*CYP2C9*3*) (44–48). The cysteine-containing form has a markedly lower K_m and V_{max} for *S*-warfarin hydroxylation than the more common arginine-containing enzyme, and similar, but smaller, effects on V_{max} for phenytoin and the nonsteroidal anti-inflammatory agent flurbiprofen have also been described (49–51). The in vitro observations with *S*-warfarin have been confirmed by in vivo studies showing a significant difference in warfarin maintenance dose between different *CYP2C9* genotypes (52). However, a separate study found no difference in kinetic constants in vitro for tolbutamide between the arginine and cysteine-containing variants, suggesting that the effects of this polymorphism may be substrate-dependent (53). In the *CYP2C9*3* allele, the Ile-359 variant shows a fivefold higher activity with phenytoin and tolbutamide, compared with the Leu-359 variant, and the polymorphism also appears to affect warfarin metabolism with substitution of leucine for Ile-359 resulting in alterations in both regiospecificity and stereospecificity (50,53,54).

The frequency of the various *CYP2C9* allelic variants varies between ethnic groups. In whites, the estimated frequency of *CYP2C9*2* is 0.08–0.10 and that for *CYP2C9*3* is 0.06 (53,55). Both variant alleles occur at lower frequencies among African Americans, with *CYP2C9*2* seen a frequency of 0.01 and *CYP2C9*3* at a frequency of 0.005 (53,55). In a Chinese population *CYP2C9*2* was not detected, but the frequency of *CYP2C9*3* was approximately 0.02 (56).

The original study of Scott and Poffenbarger (42) reported a trimodal distribution for tolbutamide elimination in a relatively small population. In a study of tolbutamide clearance in 106 Australians of mainly European origin, no evidence for bi- or trimodality was obtained, and no apparent poor metabolizers were detected (57). However, a single possible poor metabolizer was identified in each of two other studies and the *CYP2C9* gene from both persons has recently been sequenced (53,58,59). One person was homozygous for *CYP2C9*3* and the other showed a *CYP2C9*2*/*CYP2C9*3* genotype. This finding confirms the existence of tolbutamide poor metabolizers. The predicted frequency of this poor metabolizer phenotype in whites is 0.02, if both variant alleles are poor metabolizer associated, or 0.004, if only *CYP2C9*3* gives the poor metabolizer phenotype with tolbutamide, as predicted from in vitro expression studies. Whether poor metabolizers of tolbutamide also show poor metabolism of other CYP2C9 substrates is yet not known.

4. CYP2A6

Cytochrome CYP2A6 catalyzes the 7-hydroxylation of coumarin, a compound that is of importance mainly as a food additive and fragrance, but is also used as an anticancer drug (60). CYP2A6 also has a role in procarcinogen activation and in the metabolism of nicotine (3,61). Some evidence of bimodality in the metabolism of coumarin has been obtained, and two allelic variants termed *CYP2A6*2* and *CYP2A6*3* (also known as *CYP2A6v1* and *CYP2A6v2*) have been identified (62–64). *CYP2A6*2* has an amino acid substitution that results in an inability to 7-hydroxylate coumarin. Whether *CYP2A6*3* encodes a gene product with normal activity is still unclear, but certain exons appear to have undergone gene conversions to the corresponding exons of the related gene *CYP2A7* that encodes an enzyme lacking coumarin 7-hydroxylase activity (64).

5. CYP2E1

The cytochrome P450 CYP2E1 an ethanol-inducible enzyme, metabolizes mainly low molecular weight compounds, such as acetone, ethanol, benzene, and nitrosamines. Because of the nature of its substrate specificity, CYP2E1 is of most interest from the standpoint of toxicology and carcinogenesis (for review see Ref. 65), but it also has a minor role in drug metabolism and is one of several cytochromes P450 demonstrated to convert acetaminophen (paracetamol) to toxic quinones in overdose (66). In a recent study on CYP2E1 knockout mice, these animals were more resistant than normal mice to the hepatotoxic effects of acetaminophen suggesting that CYP2E1 is the main P450 catalyzing quinone production from this compound in the mouse (67). There is evidence of interindividual variation in expression of the enzyme in human livers, and phenotyping studies using the muscle relaxant chlorzoxazone as a probe in white populations have demonstrated two- to threefold variation in levels of activity, but no evidence of bimodality (68–70). It has been suggested that chlorzoxazone may not be a completely specific probe for CYP2E1 because it is also metabolized by CYP1A1 and CYP1A2 (71,72).

Several genetic polymorphisms in *CYP2E1* have been reported, but the majority of these occur in introns and appear to be of no functional significance. However, a polymorphism in the 5′-flanking region within a putative HNF-1 binding site may be of functional significance (73). This polymorphism can be detected with the restriction enzyme *Rsa* I, and the rarer allele occurs at a frequency of 0.27 in Japanese, but only 0.02 in whites (74). In vitro studies suggest that the rarer allele shows an approximately tenfold higher transcriptional activity, but a study on phenotype–genotype relations did not find any evidence for increased activity in vivo in those heterozygous for the polymorphism (69).

6. Other Cytochrome P450 Polymorphisms

The cytochromes P450 CYP1A1, CYP1A2, and CYP3A4 have also been suggested to show polymorphism. For each, some evidence for the existence of a functional polymorphism has been presented, but its existence remains controversial, and an underlying molecular basis has not been identified. Three separate polymorphisms (two in noncoding regions and one in exon 7 giving an isoleucine–valine substitution) have been detected in the *CYP1A1* gene, but although several studies suggest associations between cancer susceptibility and each individual polymorphism, the functional significance of each remains unclear (75–77). CYP1A1 is of little importance in drug metabolism, although it is of considerable importance in the activation of procarcinogens, such as benzo[*a*]pyrene. CYP1A2 is closely related to CYP1A1, but is of greater importance in drug metabolism, with substrates that include theophylline, imipramine, clozapine, phenacetin, and acetaminophen (66,78–81). With caffeine as a probe drug, several population studies investigating polymorphism in the *CYP1A2* gene have been performed (82–85). One of these studies reported a trimodal distribution for CYP1A2 activity (84), but, using different metabolic ratios, the others have obtained a unimodal distribution. There is controversy over which metabolic ratio accurately reflects the contribution made by CYP1A2 to caffeine metabolism (86–88). DNA-sequencing studies on individuals at the extremes of the trimodal distribution have not detected significant differences, casting doubt on the existence of a polymorphism in the *CYP1A2* gene (89). A family study has also failed to show evidence for genetic factors as important determinants of CYP1A2 activity (90). The level of induction of CYP1A2 by aromatic hydrocarbons is less than that for CYP1A1, but it is possible that some of the variation seen in CYP1A2 levels in nonsmokers might reflect polymorphism in induction owing to passive smoking, diet, or other environmental factors.

CYP3A4 is the most abundant cytochrome P450 in most human livers and is also the one with the widest range of drug substrates, which include benzodiazepines, erythromycin, dihydropyridines, and cyclosporine (91). Levels of CYP3A4 activity vary considerably between individuals, but no evidence of a genetic polymorphism in *CYP3A4* has been obtained. However, a closely related gene, *CYP3A5*, shows a polymorphism in its expression, with universal expression in the gut and fetal liver, but detectable expression in only 10–20% of adult livers (92). CYP3A5 shows a similar, but not identical, substrate specificity to CYP3A4 (92,93). The molecular basis of differential CYP3A5 expression is still unclear, but in a recent study, CYP3A5 mRNA appeared to be universally expressed in liver, but a polymorphism introducing an amino acid substitution appeared to be more common in individuals not expressing CYP3A5 protein (94). A third *CYP3A* gene *CYP3A7* is universally expressed in fetal liver, but was also expressed in 7 of 13 adult livers (95). Variable expression of *CYP3A5* and *CYP3A7* may partly account for the degree of variation seen in the metabolism of CYP3A4 substrates.

B. Other Phase I Polymorphisms

Several pharmacogenetic polymorphisms occur in noncytochrome P450-mediated phase I reactions. In general these are either relatively common, but not of great importance in drug metabolism, or are rare and important only in the metabolism of a restricted range of drugs.

1. *Noncytochreme P450-Mediated Mediated Oxidations*

Drug oxidations are carried out by several enzymes other than the cytochromes P450s, including flavin-containing monooxygenases and monoamine oxidases (see Chapter 5). One of the flavin-linked monooxygenase isoforms is subject to a rare polymorphism known as the fish-odor syndrome, which arises from an inability to oxidize trimethylamine (96). Flavin-containing monooxygenases metabolize a range of drugs containing secondary and tertiary amine groups, such as chlorpromazine, morphine, propranolol, and nicotine, but as yet only nicotine-*N*-oxide synthesis has been shown to cosegregate with polymorphic trimethylamine oxidation (97).

2. Esterases

Polymorphisms have been detected in the esterases paraoxonase and cholinesterase (for review see Refs. 98–100). The paraoxonase polymorphism is a common one that is now well understood at the molecular level. However, paraoxonase is not of importance in drug metabolism because it metabolizes mainly organophosphate and carboxylic acid esters. Cholinesterase, however, shows a polymorphism that is important in the hydrolysis of the muscle relaxant succinylcholine and, possibly, substance P and diacetylmorphine (99). There are several different allelic variants known that give rise to a variety of phenotypes, including the atypical enzyme that is found in 2% of the population and shows defective binding of anionic substrates, such as succinylcholine, and the rarer silent variant for which no enzyme is produced. The variety of alleles and their rarity makes population screening for the molecular defects difficult, but affected persons can be identified phenotypically by enzyme assays (99).

3. Epoxide Hydrolases

Four different isoforms of epoxide hydrolase have been demonstrated in humans. Two of these have specific metabolic roles, but the other two, one microsomal and one cytoplasmic, hydrolyze a range of alkene and arene oxides (101). In general, epoxide hydrolases are of minor importance in normal drug metabolism, but are of importance in biotransformation reactions of carcinogens and in the metabolism of toxic intermediates formed from drugs such as s phenytoin and acetaminophen by cytochrome P450-mediated reactions (102). Studies of enzyme activity in lymphocytes gave evidence of interindividual variation in both the microsomal and cytoplasmic enzymes (103,104). Several polymorphisms in the gene encoding the microsomal enzyme (*HYL1*), including three that result in amino acid substitutions, have been described (105,106). Two of the substitutions, $Tyr_{113}His$ and $His_{139}Arg$, appear to affect protein stability, but not enzyme activity in vitro (106). The possibility that these substitutions might be associated with adverse drug reactions to antiepileptic drugs has been investigated, but no evidence for an association has been obtained (105,107).

4. Dehydrogenases

Several polymorphisms in genes encoding aldehyde and alcohol dehydrogenases have been well characterized, but although of importance in determining susceptibility to alcoholism and alcoholic liver disease (for review see Ref. 108), they are not of great importance in the metabolism of commonly prescribed drugs. Another dehydrogenase, dihydropyrimidine dehydrogenase, has a biochemical role in the catabolism of uracil and thymine and is not primarily a drug-metabolizing enzyme (109). However, this enzyme is also responsible for the catabolism of the anticancer drug 5-fluorouracil and interindividual variation in the metabolism of this drug has been correlated with levels of dihydropyrimidine dehydrogenase in peripheral blood mononuclear cells (110). Complete deficiency of dihydropyrimidine dehydrogenase has been linked to various physiological abnormalities (111). An investigation of the molecular basis of the deficiency in one family has shown that the affected person was homozygous for a deletion of 168 bp in the mRNA, apparently caused by exon skipping, although the precise defect in genomic DNA has not yet been identified (112). Both defective copies of the gene in this family were similar, but it is not yet clear whether other occurrences are genotypically similar. It is estimated that up to 3% of the population may be heterozygous for the deficiency and although they do not suffer physiological abnormalities, it appears that these persons are at increased risk of serious toxic effects if given 5-fluorouracil treatment (113,114). Development of genotyping methods for the deficiency, therefore, would be useful because 5-fluorouracil is a commonly used drug in oncology.

III. PHASE II POLYMORPHISMS

In general, the UDP-glucuronsyltranferases and the sulfotransferases are the major phase II-metabolizing enzymes for most commonly prescribed drugs. There is some evidence for the existence of polymorphisms in certain isoforms of these enzyme families, but our understanding of these is still relatively poor. The two most common polymorphisms in genes encoding phase II enzymes occur in *N*-acetyltransferase 2 (NAT2) and glutathione *S*-transferase M1 (GSTM1). NAT2 contributes to the metabolism of a relatively small number of drugs, and few drug substrates metabolized by GSTM1 have been identified. There is also a rare, but well-characterized, polymorphism in the gene encoding thiopurine *S*-methyltransferase, which has an important role in the metabolism of 6-mercaptopurine and related compounds.

A. UDP-Glucuronosyltransferases

Glucuronidation is the most common conjugation reaction in drug metabolism (see Chapter 6). The importance of pharmacogenetic variation in the UDP-glucuronosyltransferases is still unclear, and studies are made difficult by various factors, such as the overlapping substrate specificities of these enzymes, but a few reports of intersubject variation in activity in the general population have appeared. Both Asians and whites were reported to show bimodality in the mean fraction of acetaminofen excreted as the glucuronide, with 5% of subjects showing very low levels of glucuronide excretion (115). In a separate study of codeine glucuronidation, which also involved both white and Chinese, there was a lower overall clearance of the drug in the Chinese group, apparently owing to reduced glucuronidation (116). A population study of the metabolism of fenofibric acid was suggestive of a polymorphism in glucuronidation on the basis of bimodality, but a more detailed family study failed to confirm these findings (117,118).

An inborn error of metabolism, termed Gilbert's syndrome, which is characterized by mild hyperbilirubinemia and affects 2–12% of the population, is also of interest in relation to drug metabolism. Recent reports suggest that there may be two forms of the disease, one common form, with autosomal recessive inheritance that affects approximately 10% of whites, and a more severe form inherited in an autosomal dominant fashion (119,120). The molecular basis of the mild defect appears to be a TA insertion in the TATA box that results in decreased gene expression (119,121). In the more severe disease, patients appear to be heterozygous for polymorphisms in the coding region of the gene (122). In addition to impaired bilirubin metabolism, decreased clearance of several different drugs, including tolbutamide, acetaminofen, and rifampin, has been reported in patients with this syndrome (123,124). The availability of genotyping methods should facilitate studies on the effect of the syndrome on drug metabolism.

B. Sulfotransferases

The sulfotransferases conjugate exogenous and endogenous compounds, including neurotransmitters with sulfate derived from 3′-phosphoadenosine-5′-phosphosulfate (PAPS) and have a role in the metabolism of a range of drugs. Family and population studies have given evidence for polymorphism in at least two of the five known isoforms (125–129), but the molecular basis and pharmacological effects of this variation remains unclear. The complexity of the human sulfotransferase family is demonstrated by data indicating that two separate genes (*STP1* and *STP2*) encode proteins that show 96% homology and that both appear to be phenol sulfotransferases, although it was previously believed that there was a single phenol (or thermostable) sulfotransferase isoform (130). Evidence for the existence of allelic variants of each of the phenol sulfotransferases and for the occurrence of two alternative promoters in *STP1* has also been obtained (131–135). Therefore, it appears that the pharmacogenetics of the sulfotransferase superfamily is particularly interesting, but that further studies are required to understand the precise substrate specificity of each isoform and the significance of the various allelic variants.

C. Acetyltransferases

Acetylation of amino, hydroxyl, and sulfydryl groups is catalyzed in humans by two *N*-acetyltransferases, termed *N*-acetyltransferase 1 (NAT1) and *N*-acetyltransferase 2 (NAT2); (for review see Ref. 1). The existence of a polymorphism in NAT2 has been known for many years, but polymorphism has also been detected recently in NAT1, which was previously often referred to as the monomorphic *N*-acetyltransferase. The polymorphism in NAT2 is the more significant of the two, with substantial numbers of persons completely deficient in this enzyme activity and unable to acetylate a range of drugs, including dapsone, isoniazid, procainide, and sulfamethazine (1,136). The precise percentage of slow acetylators in the population varies with ethnic origin, ranging from 90% in North Africans to less than 10% in many Asian populations, with a frequency of 50% in whites (136). The molecular basis of deficiency in NAT2 activity is well understood, and four variant alleles with low activity, apparently owing to amino acid substitutions, have been identified (Table 3). The frequency of the four low-activity alleles varies between ethnic groups,with *NAT2*7* most common among Japanese, and *NAT2*14* common in individuals of African origin, but not in other ethnic groups.

The *NAT2* polymorphism is of importance from the standpoints of both clinical responses to drugs and disease susceptibility. Although not as significant as CYP2D6 in terms of the range of currently prescribed drugs that are substrates, several NAT2 substrates are clinically important. Several phenotype-associated adverse reactions have been described and include peripheral neuropathy in slow acetylators taking isoniazid, hypersensitivity reactions in response to sulfonamides in slow acetylators, and leukopenia in rapid acetylators taking amonafide, a prodrug used in cancer chemotherapy (137–139).

Although previously believed to be monomorphic, there is now evidence, from phenotyping studies with 4-aminobenzoic acid, for interindividual variation in levels of NAT1 activity (140,141). Several allelic variants showing polymorphism at several positions, but particularly at the 3′-untranslated region have been described (see Table 3; 142,143). Genotyping assays have been developed for detection of these alleles, and an increased susceptibility to colon cancer in individuals with at least one copy of *NAT1*10* has been detected (144). The *NAT*10* allele has a variant polyadenylation signal present, and studies on NAT1 activity in bladder and colon tissue from persons of known genotype suggests that activity is higher in those positive for this allele (145). The relevance of these observations to drug metabolism is still unclear.

Table 3 *N*-Acetyltransferase Alleles

Allele	Nucleotide changes	Effect	Activity
NAT1 alleles			
*NAT1*3*	$C_{1095}A$	Unknown	Unknown
*NAT1*4*	Wild-type sequence		Normal
*NAT1*5*	$G_{350}C$; $G_{351}C$; $G_{497}C$; $G_{498}C$; $G_{499}C$; $A_{884}G$; Δ_{976}; Δ_{1105}	$R_{117}T$; $R_{166}T$; $E_{167}Q$	Unknown
*NAT1*10*	$T_{1088}A$; $C_{1095}A$	Altered poly-adenylation signal (?)	Increased (?)
*NAT1*11*	$C_{-344}T$; $A_{-40}T$; $G_{459}A$; $T_{640}G$; Δ9bp (region of 1066–1088)	$S_{214}A$	Unknown
*NAT1*14*	$G_{560}A$	$R_{187}Q$	Decreased
*NAT1*15*	$C_{559}T$	R_{187}Stop	Decreased
*NAT1*16*	3 bp insertion (position 1091)	Unknown	Unknown
*NAT1*17*	$C_{190}T$	$R_{64}W$	Decreased
NAT2 alleles			
*NAT2*4*	None		Normal
*NAT2*5A*	$T_{341}C$, $C_{481}T$	$I_{114}T$;	Decreased
*NAT2*5B*	$T_{341}C$, $C_{481}T$, $A_{803}G$	$I_{114}T$; $K_{268}R$	Decreased
*NAT2*5C*	$T_{341}C$, $A_{803}G$	$K_{268}R$	Decreased
*NAT2*6A*	$C_{282}T$, $G_{590}A$	$R_{197}Q$	Decreased
*NAT2*6B*	$G_{590}A$	$R_{197}Q$	Decreased
*NAT2*7A*	$G_{857}A$	$G_{286}E$	Decreased
*NAT2*7B*	$C_{282}T$, $G_{857}A$	$G_{286}E$	Decreased
*NAT2*12A*	$A_{803}G$	$K_{268}R$	Normal
*NAT2*12B*	$C_{282}T$, $A_{803}G$	$K_{268}R$	Normal
*NAT2*13*	$C_{282}T$		Normal
*NAT2*14A*	$G_{191}A$	$R_{64}E$	Decreased
*NAT2*14B*	$G_{191}A$, $C_{282}T$	$R_{64}E$	Decreased
*NAT2*17*	$A_{434}C$	$Q_{145}P$	Unknown
*NAT2*18*	$A_{845}C$	$K_{282}T$	Unknown

Source: Refs. 1, 142–145, 233, and 234.

D. Glutathione *S*-Transferases

Glutathione conjugation is an important metabolic pathway for a variety of hydrophobic and electrophilic compounds and is generally detoxifying, with most glutathione conjugates undergoing further metabolism to mercapturic acids before excretion (146). In humans, polymorphisms have been demonstrated and characterized in the class μ-enzyme GSTM1 and in the class θ-enzyme GSTT1. There are also reports of polymorphisms in *GSTM3, GSTP1*, and *GSTA2*. The *GSTM1* polymorphism was originally detected by starch gel electrophoresis, and it results in an absence of GSTM1 activity in 40–50% of persons from a variety of ethnic groups (147). The lack of activity is due to a large deletion in the *GSTM1* gene, and subjects homozygous for the deficiency can be detected by PCR (148). There are also two active allelic variants termed *GSTM1*A* and *GSTM1*B*, but both gene products, which differ by an asparagine or lysine, respectively, at residue 172, appear to be catalytically similar (149). In the *GSTT1* polymorphism, 10–30% of Europeans have a deficiency of this enzyme, again owing to a large gene deletion (150).

The *GSTM1* and *GSTT1* polymorphisms are of more importance in toxicology than in drug metabolism. The main drug substrates for the glutathione *S*-transferases are cytotoxic drugs, with a possible role for GSTM1 in the metabolism of cormustine [1,3-bis(2-chloroethyl)-1-nitrosourea; BCNU] and of nitrogen mustard (151,152). The effect of *GSTM1* genotype on outcome of chemotherapy with these compounds is unclear. GSTT1, however, metabolizes mainly small organic compounds, such as dichloromethane and ethylene oxide, and, therefore, is unlikely to have a significant role in drug metabolism (150). Both polymorphisms may be of importance in determining susceptibility to diseases associated with exposure to xenobiotics (for review see Ref. 4).

Other glutathione *S*-transferase polymorphisms are not well studied, but a polymorphism in the *GSTM3* gene, which lies close to *GSTM1* on chromosome 1, has recently been described. The polymorphism occurs in intron 6, and the variant appears to have a recognition site for the negative transcription factor *YY*1 present (153). The *GSTM3* variant allele appears to be in linkage disequilibrium with the *GSTM1*A* allele. It has also been reported that GSTM3 levels are lower in lung tissue from individuals with a *GSTM1* null genotype, possibly because of the deletion that gives rise to the null allele also affects the regulation of *GSTM3* (154). For *GSTP1*, two polymorphisms in the coding sequence, both resulting in amino acid substitutions in the substrate-binding site and both affecting kinetic constants for the substrate 1-chloro-2,4-dinitrobenzene, have been described (155). Two other polymorphisms in *GSTP1* have also been detected, one a *Bam*HI restriction fragment length polymorphism (RFLP; 156) and the other in an area of pentanucleotide repeats in the promoter region (157), but it is not known whether they are of functional significance. For the *GSTA2* gene, a RFLP has been detected, but again there is no information on functional significance (158).

E. Methyltransferases

Methylation reactions using *S*-adenosylmethionine as a methyl group donor are important in the metabolism of both endogenous molecules, such as neurotransmitters, and of xenobiotics. At least four separate enzymes can carry out either *S*-, *N*-, and *O*-methylation reactions, and some evidence for either heritable variation or polymorphism in each has been obtained (for review see Ref. 159). However, only for thiopurine *S*-methyltransferase is the polymorphism both clearly established and of relevance to drug therapy. Measurement of enzyme activity in erythrocytes has shown that approximately 0.3% of various European populations have undetectable activity and 11.1% intermediate levels (160). The molecular basis of the deficiency has recently been elucidated with the identification of two alleles associated with absence of enzyme activity. The more common of these (*TPMT*3*) results in two amino acid substitutions, which together, result in complete absence of activity and account for approximately 75% of defective alleles (161,162). Another rarer allele, termed *TPMT*2*, also has an amino acid substitution, resulting in loss of activity (163). Thiopurine *S*-methyltransferase is important in the metabolism of the cytotoxic drug 6-mercaptopurine, which is widely used in treatment of childhood acute lymphoblastic leukemia, and it also metabolizes thioguanine and azathioprine, a 6-mercaptopurine precursor that is used as an immunosuppressant. In persons defective in thiopurine *S*-methyltransferase, high concentrations of thioguanine nucleotides will be formed, resulting in toxicities, such as myelosuppression (164,165). Successful 6-mercaptopurine treatment has been reported to have been achieved in a thiopurine *S*-methyltransferase-deficient patient by administration of 6% of the normal dose (166). Azathioprine is frequently used as an immunosuppressant in transplant patients, but there are reports of fatal toxicities in thiopurine *S*-methyltransferase-deficient patients treated with this drug (167). The availability of genotyping assays that can identify most persons with the deficiency, therefore, should result in improved treatment outcomes.

IV. CONCLUDING REMARKS

An understanding of the molecular basis of pharmacogenetic deficiencies has enabled the development of a range of simple genotyping assays usually involving use of the polymerase chain reaction. These assays should assist in more effective prescribing, particularly for compounds that can give rise to serious toxicity in patients with a deficiency in the metabolizing enzyme, such as 6-mercaptopurine and 5-fluorouracil, and also help in recognizing problems with polymorphic drug metabolism during drug development. It is also clear that although the pharmacogenetics of the cytochrome P450s, glutathione *S*-transferases, and certain other genes of importance in drug metabolism has been extensively studied and is now well understood, there are various other genes, such as those encoding the flavin-linked monooxygenases and the sulfotransferases, for which the basis and extent of polymorphism is still poorly understood and on which further studies will be of benefit.

REFERENCES

1. D. A. P. Evans, *Genetic Factors in Drug Therapy. Clinical and Molecular Pharmacogenetics*, Cambridge University Press, Cambridge, 1993.
2. K. Brosen and L. Gram, Clinical significance of the sparteine/debrisoquine oxidation polymorphism, *Eur. J. Clin. Pharmacol 36*:537 (1989).
3. A. K. Daly, S. Cholerton, M. Armstrong, and J. R. Idle, Genotyping for polymorphisms in xenobiotic metabolism as a predictor of disease susceptibility, *Environ. Health Perspect. 102* (Suppl. 9):55 (1994).
4. G. Smith, L. A. Stanley, E. Sim, R. C. Strange, and C. R. Wolf, Metabolic polymorphisms and cancer susceptibility, *Cancer Surv. 25*:27 (1995).
5. S. Cholerton, A. K. Daly, and J. R. Idle, The role of individual human cytochromes P450 in drug metabolism and clinical response, *Trends Pharmacol. Sci. 13*:434 (1992).
6. F. J. Gonzalez and J. R. Idle, Pharmacogenetic phenotyping and genotyping. Present status and future potential, *Clin. Pharmacokinet. 26*:59 (1994).
7. A. K. Daly, J. Brockmoller, F. Broly, M. Eichelbaum, W. E. Evans, F. J. Gonzalez, J.-D. Huang, J. R. Idle, M. Ingelman-Sundberg, T. Ishizaki, E. Jacqz-Aigrain, U. A. Meyer, D. W. Nebert, V. M. Steen, C. R. Wolf, and U. M. Zanger, Nomenclature for human CYP2D6 alleles, *Pharmacogenetics 6*:193 (1996).
8. A. K. Daly, M. Armstrong, S. C. Monkman, M. E. Idle, and J. R. Idle, The genetic and metabolic criteria for the assignment of debrisoquine hydroxylation (cytochrome P450IID6) phenotypes, *Pharmacogenetics 1*:33 (1991).
9. F. Broly, A. Gaedigk, M. Heim, M. Eichelbaum, K. Morike, and U. A. Meyer, Debrisoquine/sparteine hydroxylation genotype and phenotype: Analysis of common mutations and alleles of *CYP2D6* in a European population, *DNA Cell Biol. 10*:545 (1991).

10. W. E. Evans and M. V. Relling, Concordance of P4502D6 (debrisoquine hydroxylase) phenotype and genotype: Inability of dextromethorphan metabolic ratio to discriminate reliably heterozygous and homozygous extensive metabolizers, *Pharmacogenetics 1*:143 (1991).
11. M.-L. Dahl, I. Johansson, M. P. Palmertz, M. Ingelman-Sundberg, and F. Sjoqvist, Analysis of the *CYP2D6* gene in relation to debrisoquin and desipramine hydroxylation in a Swedish population, *Clin. Pharmacol. Ther. 51*:12 (1992).
12. B. Evert, E.-U. Griese, and M. Eichelbaum, A missense mutation in exon 6 of the *CYP2D6* gene leading to a histidine to proline exchange is associated with the poor metabolizer phenotype of sparteine, *Naunyn-Schmiedeberg Arch. Pharmacol. 350*:434 (1994).
13. B. Evert, E.-U. Griese, and M. Eichelbaum, Cloning and sequencing of a new non-functional *CYP2D6* allele: Deletion of T_{1795} in exon 3 generates a premature stop codon, *Pharmacogenetics 4*:271 (1994).
14. A. K. Daly, J. B. S. Leathart, S. J. London, and J. R. Idle, An inactive cytochrome P450 *CYP2D6* allele containing a deletion and a base substitution, *Hum. Genet. 95*:337 (1995).
15. R. Saxena, G. L. Shaw, M. V. Relling, J. N. Frame, D. T. Moir, W. E. Evans, N. Caporaso, and B. Weiffenbach, Identification of a new variant CYP2D6 single base pair deletion in exon 3 and its association with the poor metabolizer phenotype, *Hum. Mol. Genet. 3*:923 (1994).
16. D. Marez, M. Legrand, N. Sabbagh, J.-M. Lo-Guidice, P. Boone, and F. Broly, An additional allelic variant of the *CYP2D6* gene causing impaired metabolism of sparteine, *Hum. Genet. 97*:668 (1996).
17. H. Yokota, S. Tamura, H. Furuya, S. Kimura, M. Watanbe, I. Kanazawa, I. Kondo, and F. J. Gonzalez, Evidence for a new variant allele *CYP2D6J* in a Japanese population associated with lower in vivo rates of sparteine metabolism, *Pharmacogenetics 3*:256 (1993).
18. S.-L. Wang, J.-D. Huang, M.-D. Lai, B.-H. Liu, and M.-L. Lai, Molecular basis of genetic variation in debrisoquin hydroxylation in Chinese subjects: Polymorphism in RFLP and DNA sequence of *CYP2D6, Clin. Pharmacol. Ther. 53*:410 (1993).
19. I. Johansson, M. Oscarson, Q.-Y. Yue, L. Bertilsson, F. Sjoqvist, and M. Ingelman-Sundberg, Genetic analysis of the Chinese cytochrome P4502D locus: Characterization of variant *CYP2D6* genes present in subjects with diminished capacity for debrisoquine hydroxylation, *Mol. Pharmacol. 46*:452 (1994).
20. M. Kagimoto, M. Heim, K. Kagimoto, T. Zeugin, and U. A. Meyer, Multiple mutations of the human cytochrome P450IID6 gene (*CYP2D6*) in poor metabolisers of debrisoquine, *J. Biol. Chem. 265*:17209 (1990).
21. C. Masimirembwa, I. Persson, L. Bertilsson, J. Hasler, and M. Ingelman-Sundberg, A novel mutant variant of the *CYP2D6* gene (*CYP2D6*17*) common in a black African population: Association with diminished debrisoquine hydroxylase activity, *Br. J. Clin. Pharmacol. 42*: 713 (1996).
22. L. Bertilsson, M.-L. Dahl, F. Sjoqvist, A. Aberg-Wistedt, M. Humble, I. Johansson, E. Lundqvist, and M. Ingelman-Sundberg, Molecular basis for rational megaprescribing in ultrarapid hydroxylators of debrisoquine, *Lancet 341*:63 (1993).
23. I. Johansson, E. Lundqvist, L. Bertilsson, M.-L. Dahl, F. Sjoqvist, and M. Ingelman-Sundberg, Inherited amplification of an active gene in the cytochrome P450 *CYP2D* locus as a cause of ultrarapid metabolism of debrisoquine, *Proc. Natl. Acad. Sci. USA 90*:11825 (1993).
24. J. A. G. Agundez, M. C. Ledesma, J. M. Ladero, and J. Benitez, Prevalence of *CYP2D6* gene duplication and its repercussion on the oxidative phenotype in a white population, *Clin. Pharmacol. Ther. 57*:265 (1995).
25. M. Dahl, I. Johansson, L. Bertilsson, M. Ingelman-Sundberg, and F. Sjoqvist, Ultrarapid hydroxylation of debrisoquine in a Swedish population. Analysis of the molecular genetic basis, *J. Pharm. Exp. Ther. 274*:516 (1995).

26. L. Siddoway, K. Thompson, C. McAllister, T. Wang, G. Wilkinson, D. Roden, and R. Woosley, Polymorphism of propafenone metabolism in man, *Circulation 75*:785 (1987).
27. M.-L. Dahl and L. Bertilsson, Genetically variable metabolism of antidepressants and neuroleptic drugs in man, *Pharmacogenetics 3*:61 (1993).
28. M. Jerling, M.-L. Dahl, A. Aberg-Wistedt, B. Liljenberg, N.-E. Landell, L. Bertilsson, and F. Sjoqvist, The *CYP2D6* genotype predicts the oral clearance of the neuroleptic agents perphenazine and zulcopenthixol, *Clin. Pharmacol. Ther. 59*:423 (1996).
29. S. Sindrup, K. Brosen, K. Bjerring, L. Arendt-Nielsen, V. Larsen, H. Angelo, and L. Gram, Codeine increases pain thresholds to copper vapor laser stimuli in extensive but not poor metabolizers of sparteine, *Clin. Pharmacol. Ther. 49*:686 (1991).
30. K. Persson, S. Sjostrom, I. Sigurdardottir, V. Molnar, M. Hammarlundudenaes, and A. Rane, Patient-controlled analgesia (PCA) with codeine for postoperative pain relief in 10 extensive metabolizers and one poor metabolizer of dextromethorphan, *Br. J. Clin. Pharmacol. 39*: 182 (1995).
31. A. Kupfer, B. Dick, and R. Preisig, A new drug hydroxylation polymorphism in man: The incidence of mephenytoin hydroxylation deficient phenotypes in an European population study, *Naunyn-Schmiedebergs Arch. Pharmacol. 321*:33 (1982).
32. G. Wilkinson, F. Guengerich, and R. Branch, Genetic polymorphisms of *S*-mephenytoin hydroxylation, *Pharmacol. Ther. 43*:53 (1989).
33. S. Wrighton, J. Stevens, G. Becker, and M. Van den Branden, Isolation and characterization of human liver cytochrome P450 2C19: Correlation between 2C19 and *S*-mephenytoin 4′-hydroxylation, *Arch. Biochem. Biophys. 306*:240 (1993).
34. J. Goldstein, M. Faletto, M. Romkes-Sparks, T. Sullivan, S. Kitareewan, J. Raucy, J. Lasker, and B. Ghanayem, Evidence that CYP2C19 is the major (*S*)-mephenytoin 4′-hydroxylase in humans, *Biochemistry 33*:1743 (1994).
35. S. M. F. de Morais, G. R. Wilkinson, J. Blaisdell, K. Nakamura, U. A. Meyer, and J. A. Goldstein, The major genetic defect responsible for the polymorphism of *S*-mephenytoin metabolism in humans, *J. Biol. Chem. 269*:15419 (1994).
36. S. M. F. de Morais, G. R. Wilkinson, J. Blaisdell, U. A. Meyer, K. Nakamura, and J. A. Goldstein, Identification of a new genetic defect responsible for the polymorphism of (*S*)-mephenytoin metabolism in Japanese, *Mol. Pharmacol. 46*:594 (1994).
37. A. K. Daly, S. Cholerton, W. Gregory, and J. R. Idle, Metabolic polymorphisms, *Pharmacol. Ther. 57*:129 (1993).
38. J. A. Goldstein and S. M. F. de Morais, Biochemistry and molecular biology of the human *CYP2C* subfamily, *Pharmacogenetics 4*:285 (1994).
39. K. Brosen, S. M. F. de Morais, U. A. Meyer, and J. A. Goldstein, A multifamily study on the relationship between *CYP2C19* genotype and the *S*-mephenytoin oxidation polymorphism, *Pharmacogenetics 5*:312 (1995).
40. K. Rost, H. Brosicke, J. Brockmoller, M. Scheffler, H. Helge, and I. Roots, Increase of cytochrome P-450 1A2 activity by omeprazole: Evidence by the ^{13}C[*N*-3-methyl]-caffeine breath test in poor and extensive metabolizers of *S*-mephenytoin, *Clin. Pharmacol. Ther. 52*:170 (1992).
41. S. A. Ward, N. A. Helsby, E. Skjelbo, K. Brosen, L. F. Gram, and A. M. Breckenbridge, The activation of the biguanide antimalarial proguanil co-segregates with the mephenytoin oxidation phenotype—a panel study, *Br. J. Clin. Pharmacol. 31*:689 (1991).
42. J. Scott and P. Poffenbarger, Pharmacogenetics of tolbutamide metabolism in humans, *Diabetes 28*:41 (1978).
43. H. Kutt, M. Wolk, R. Scherman, and F. McDowell, Insufficient parahydroxylation as a cause of diphenylhydantoin toxicity, *Neurology 14*:542 (1964).
44. D. R. Umbenhauer, M. V. Martin, R. S. Lloyd, and F. P. Guengerich, Cloning and sequence determination of a complementary DNA related to human liver microsomal cytochrome P450 *S*-mephenytoin 4-hydroxylase, *Biochemistry 26*:1094 (1987).

45. M. Romkes, M. B. Faletto, J. A. Blaisdell, J. L. Raucy, and J. A. Goldstein, Cloning and expression of complementary DNAs for multiple members of the human cytochrome P450IIC subfamily, *Biochemistry 30*:3247 (1991).
46. S. Kimura, J. Pastewka, H. V. Gelboin, and F. J. Gonzalez, cDNA and amino acid sequences of two members of the human P450IIC gene subfamily, *Nucleic Acids Res. 15*:10053 (1987).
47. T. Yasumori, S. Kawano, K. Nagata, M. Shimada, Y. Yamazoe, and R. Kato, Nucleotide sequence of a human liver cytochrome P450 related to the rat male-specific form, *J. Biochem. (Tokyo) 102*:493 (1987).
48. R. R. Meehan, J. R. Gosden, D. Rout, N. V. Hastie, T. Friedberg, M. Adesnik, R. Buckland, V. van Heyningen, J. Fletcher, N. P. Spurr, J. Sweeney, and C. R. Wolf, Human cytochrome P450 PB-1: A multigene family involved in mephenytoin and steroid oxidations that maps to chromosome 10, *Am. J. Hum. Genet. 42*:26 (1988).
49. A. E. Rettie, L. C. Wienkers, F. J. Gonzalez, W. F. Trager, and K. R. Korzekwa, Impaired (*S*)-warfarin metabolism catalysed by the R144C allelic variant of CYP2C9, *Pharmacogenetics 4*:39 (1994).
50. M. E. Veronese, C. J. Doecke, P. I. Mackenzie, M. E. McManus, J. O. Miners, D. L. P. Rees, R. Gasser, U. A. Meyer, and D. J. Birkett, Site-directed mutation studies of human liver cytochrome P-450 isoenzymes in the CYP2C subfamily, *Biochem. I. 289*:533 (1993).
51. T. S. Tracy, B. W. Rosenbluth, S. A. Wrighton, F. J. Gonzalez, and K. R. Korzekwa, Role of cytochrome P450 2C9 and an allelic variant in the 4′-hydroxylation of (*R*)- and (*S*)-flurbiprofen, *Biochem. Pharmacol. 49*:1269 (1995).
52. H. Furuya, P. Fernandez-Salguero, W. Gregory, H. Taber, A. Steward, F. J. Gonzalez, and J. R. Idle, Genetic polymorphism of CYP2C9 and its effect on warfarin maintenance dose requirement in patients undergoing anticoagulation therapy, *Pharmacogenetics 5*:389 (1995).
53. T. H. Sullivan-Klose, B. I. Ghanayem, D. A. Bell, Z.-Y. Zhang, L. S. Kaminsky, G. M. Shenfield, J. O. Miners, D. J. Birkett, and J. A. Goldstein, The role of the *CYP2C9*–Leu359 allelic variant in the tolbutamide polymorphism, *Pharmacogenetics 6*:341 (1996).
54. L. S. Kaminsky, S. M. F. de Morais, M. B. Faletto, D. A. Dunbar, and J. A. Goldstein, Correlation of human cytochrome P2502C substrate specificities with primary structure: Warfarin as a probe, *Mol. Pharmacol. 43*:234 (1993).
55. S. J. London, A. K. Daly, J. B. S. Leathart, W. C. Navidi, and J. R. Idle, Lung cancer risk in relation to the *CYP2C9*1/CYP2C9*2* genetic polymorphism among African-Americans and Caucasians in Los Angeles County, California, *Pharmacogenetics 6*:527 (1996).
56. S.-L. Wang, J.-D. Huang, M.-D. Lai, and J.-J. Tsai, Detection of CYP2C9 polymorphism based on the polymerase chain reaction in Chinese, *Pharmacogenetics 5*:37 (1995).
57. M. E. Veronese, J. O. Miners, D. L. P. Rees, and D. J. Birkett, Tolbutamide hydroxylation in humans: Lack of bimodality in 106 healthy subjects, *Pharmacogenetics 3*:86 (1993).
58. J. O. Miners, L. M. H. Wing, and D. J. Birkett, Normal metabolism of debrisoquine and theophylline in a slow tolbutamide metaboliser, *Aust. N. Z. J. Med. 15*:348 (1985).
59. M. A. Page, J. S. Boutagy, and G. M. Shenfield, A screening test for slow metabolisers of tolbutamide, *Br. J. Clin. Pharmacol. 31*:649 (1991).
60. M. E. Marshall, J. L. Mohler, K. Edmonds, B. Williams, K. Butler, M. Ryles, L. Weiss, D. Urban, A. Bueschen, M. Markiewicz, and G. Cloud, An updated review of the clinical development of coumarin (1,2-benzopyrone) and 7-hydroxycoumarin, *J. Cancer Res. Clin. Oncol, 120*(Suppl.):S39 (1994).
61. N. W. McCracken, S. Cholerton, and J. R. Idle, Cotinine formation by cDNA-expressed human cytochromes P450, *Med. Sci. Res. 20*:877 (1992).
62. S. Cholerton, M. E. Idle, A. Vas, F. J. Gonzalez, and J. R. Idle, Comparison of a novel thin-layer chromatographic-fluorescence detection method with a spectrofluorometric method for the determination of 7-hydroxycoumarin in human urine, *J. Chromatogr. 575*:325 (1992).
63. A. Rautio, H. Kraul, A. Kojo, E. Salmela, and O. Pelkonen, Interindividual variation of coumarin 7-hydroxylase in healthy volunteers, *Pharmacogenetics 2*:227 (1992).

64. P. Fernandez-Salguero, S. M. G. Hoffman, S. Cholerton, H. Mohrenweiser, H. Raunio, O. Pelkonen, J. Huang, W. E. Evans, J. R. Idle, and F. J. Gonzalez, A genetic polymorphism in coumarin 7-hydroxylation: Sequence of the human *CYP2A* genes and identification of variant *CYP2A6* alleles, *Am. J. Hum. Genet.* *57*:651 (1995).
65. F. P. Guengerich, D.-H. Kim, and M. Iwasaki, Role of human cytochrome P-450 IIE1 in the oxidation of many low molecular weight cancer suspects, *Chem. Res. Toxicol.* *4*:168 (1991).
66. C. J. Patten, P. E. Thomas, R. L. Guy, M. Lee, F. J. Gonzalez, F. P. Guengerich, and C. S. Yang, Cytochrome P450 enzymes involved in acetaminophen activation by rat and human liver microsomes and their kinetics, *Chem. Res. Toxicol.* *6*:511 (1993).
67. S. T. S. Lee, J. T. M. Buters, T. Pineau, P. Fernandez-Salguero, and F. J. Gonzalez, Role of CYP2E1 in the hepatoxicity of acetaminophen, *J. Biol. Chem.* *271*:12063 (1996).
68. F. P. Guengerich and T. Shimada, Oxidation of toxic and carcinogenic chemicals by human cytochrome P450 enzymes, *Chem. Res. Toxicol.* *4*:391 (1991).
69. R. B. Kim, D. O'Shea, and G. R. Wilkinson, Interindividual variability of chlorzoxazone 6-hydroxylation in men and women and its relationship to *CYP2E1* genetic polymorphisms, *Clin. Pharmacol. Ther.* *57*:645 (1995).
70. E. S. Vesell, T. DeAngelo Seaton, and Y. I. A-Rahim, Studies on interindividual variations of CYP2E1 using chlorzoxazone as an in vivo probe, *Pharmacogenetics* *5*:53 (1995).
71. V. Carriere, T. Goasduff, D. Ratanasavanh, F. Morel, J. C. Gautier, A. Guillouzo, P. Beaune, and F. Berthou, Both cytochromes P450 2E1 and 1A1 are involved in the metabolism of chlorzoxazone, *Chem. Res. Toxicol.* *6*:852 (1993).
72. S. Ono, T. Hatanaka, H. Hotta, M. Tsutsui, T. Satoh, and F. J. Gonzalez, Chlorzoxazone is metabolized by human CYP1A2 as well as by CYP2E1, *Pharmacogenetics* *5*:141 (1995).
73. S. Hayashi, J. Watanabe, and K. Kawagiri, Genetic polymorphisms in the 5′-flanking region change transcriptional regulation of the human cytochrome P450IIE1 gene, *J. Biochem.* *110*: 559 (1991).
74. S. Kato, P. G. Shields, N. E. Caporaso, R. N. Hoover, B. F. Trump, H. Sugimura, A. Weston, and C. C. Harris, Cytochrome P450IIE1 genetic polymorphisms, racial variation, and lung cancer risk, *Cancer Res.* *52*:6712 (1992).
75. K. Kawagiri, K. Nakachi, K. Imai, A. Yoshii, N. Shinoda, and J. Watanabe, Identification of genetically high risk individuals to lung cancer by DNA polymorphisms of the cytochrome P450IA1 gene, *FEBS Lett.* *263*:131 (1990).
76. S. Hayashi, J. Watanabe, K. Nakachi, and K. Kawajiri, Genetic linkage of lung cancer-associated *Msp*I polymorphisms with amino acid replacement in the heme binding region of the human cytochrome P450IA1 gene, J. *Biochem.* *110*:407 (1991).
77. E. Taioli, F. Crofts, J. Trachman, R. Demopoulos, P. Toniolo, and S. J. Garte, A specific African-American *CYP1A1* polymorphism is associated with adenocarcinoma of the lung, *Cancer Res.* *55*:472 (1995).
78. M. A. Butler, M. Iwasaki, F. P. Guengerich, and F. F. Kadlubar, Human cytochrome $P450_{PA}$ (P450IA2), the phenacetin *O*-deethylase, is primarily responsible for the hepatic 3-demethylation of caffeine and *N*-oxidation of carcinogenic arylamines, *Proc. Natl. Acad. Sci. USA* *86*:7696 (1989).
79. U. Fuhr, J. Doehmer, N. Battula, C. Wolfel, C. Kudla, Y. Keita, and A. H. Staib, Biotransformation of caffeine and theophylline in mammalian cell lines genetically engineered for expression of single cytochrome P450 enzymes, *Biochem. Pharmacol.* *43*:225 (1992).
80. A. Lemoine, J. C. Gautier, D. Azoulay, L. Kiffel, C. Belloc, F. P. Guengerich, P. Maurel, P. Beaune, and J. P. Leroux, Major pathway of imipramine metabolism is catalyzed by cytochromes P-450 1A2 and P-450 3A4 in human liver, *Mol. Pharmacol.* *43*:827 (1993).
81. L. Bertilsson, J. A. Carrillo, M. L. Dahl, A. Llerena, C. Alm, U. Bondesson, L. Lindstrom, I. R. Delarubia, S. Ramos, and J. Benitez, Clozapine disposition covaries with CYP1A2 activity determined by a caffeine test, *Br. J. Clin. Pharmacol.* *38*:471 (1994).

82. W. Kalow and B.-K. Tang, Use of caffeine metabolite ratios to explore CYP1A2 and xanthine oxidase activities, *Clin. Pharmacol. Ther. 50*:508 (1991).
83. M. V. Relling, J.-S. Lin, G. D. Ayers, and W. E. Evans, Racial and gender differences in *N*-acetyltransferase, xanthine oxidase and CYP1A2 activities, *Clin. Pharmacol. Ther. 52*: 643 (1992).
84. M. A. Butler, N. P. Lang, J. F. Young, N. E. Caporaso, P. Vineis, R. B. Hayes, C. H. Teitel, J. P. Massengill, M. F. Lawsen, and F. F. Kadlubar, Determination of CYP1A2 and acetyltransferase phenotype in human populations by analysis of caffeine urinary metabolites, *Pharmacogenetics 2*:116 (1992).
85. U. Fuhr and K. Rost, Simple and reliable CYP1A2 phenotyping by the paraxanthine/caffeine ratio in plasma and saliva, *Pharmacogenetics 4*:109 (1994).
86. L. J. Notarianni, S. E. Oliver, P. Dobrocky, P. N. Bennett, and B. W. Silverman, Caffeine as a metabolic probe: A comparison of the metabolic ratios used to assess CYP1A2 activity, *Br. J. Clin. Pharmacol. 39*:65 (1995).
87. A. Rostami-Hodjegan, S. Nurminen, P. R. Jackson, and G. T. Tucker, Caffeine urinary metabolite ratios as markers of enzyme activity: A theoretical assessment, *Pharmacogenetics 6*: 121 (1996).
88. C. P. Denaro, M. Wilson, P. Jacob, and N. L. Benowitz, Validation of urine caffeine metabolite ratios with use of stable isotope-labeled caffeine clearance, *Clin. Pharmacol. Ther. 59*: 284 (1996).
89. M. Nakajima, T. Yokoi, M. Mizutani, S. Shin, F. F. Kadlubar, and T. Kamataki, Phenotyping of CYP1A2 in Japanese population by analysis of caffeine urinary metabolites: Absence of mutation prescribing the phenotype in the *CYP1A2* gene, *Cancer Epidemiol. Biomarkers Prev. 3*:413 (1994).
90. A. Catteau, Y. C. Bechtel, N. Poisson, P. R. Bechtel, and C. Bonaiti-Pellie, A population and family study of CYP1A2 using caffeine urinary metabolites, *Eur. J. Clin. Pharmacol. 47*:423 (1995).
91. P. B. Watkins, Non-invasive tests of CYP3A enzymes, *Pharmacogenetics 4*:171 (1994).
92. T. Aoyama, S. Yamano, D. J. Waxman, D. P. Lapenson, U. A. Meyer, V. Fischer, R. Tyndale, T. Inaba, W. Kalow, H. V. Gelboin, and F. J. Gonzalez, Cytochrome P450 hPCN3, a novel cytochrome P450 IIA gene product that is differentially expressed in adult human liver, *J. Biol. Chem. 264*:10388 (1989).
93. E. M. J. Gillam, Z. Guo, Y.-F. Ueng, H. Yamazaki, I. Cock, P. E. B. Reilly, W. D. Hooper, and F. P. Guengerich, Expression of cytochrome P450 3A5 in *Escherichia coli*: Effects of 5′ modification, purification, reconstitution conditions, and catalytic activities, *Arch. Biochem. Biophys. 317*:374 (1995).
94. Y. Jounaidi, V. Hyrailles, L. Gervot, and P. Maurel, Detection of a CYP3A5 allelic variant: A candidate for the polymorphic expression of the protein? *Biochem. Biophys. Res. Commun. 221*:466 (1996).
95. J. D. Schuetz, D. L. Beach, and P. S. Guzelian, Selective expression of cytochrome P450 CYP3A mRNAs in embryonic and adult human liver, *Pharmacogenetics 4*:11 (1994).
96. I. R. Phillips, C. T. Dolphin, P. Clair, M. R. Hadley, A. J. Hutt, R. R. McCombie, R. L. Smith, and E. A. Shephard, The molecular biology of the flavin-containing monooxygenases of man, *Chem. Biol. Interact. 96*:17 (1995).
97. R. Ayesh and R. L. Smith, Genetic polymorphism of trimethylamine *N*-oxidation, *Pharmacol. Ther. 45*:387 (1990).
98. B. N. La Du, The human-serum paraoxonase arylesterase polymorphism, *Am. J. Hum. Genet. 43*:227 (1988).
99. O. Lockridge, Genetic variants of human serum cholinesterase influence metabolism of the muscle relaxant succinylcholine, *Pharmacol. Ther. 47*:35 (1990).
100. R. Humbert, D. A. Adler, C. M. Disteche, C. Hassett, C. J. Omiecinski, and C. E. Furlong,

The molecular basis of the human serum paraoxonase activity polymorphism, *Nature Genet.* *3*:73 (1993).

101. F. Oesch, C. W. Timms, C. H. Walker, T. M. Guenthner, A. Sparrow, T. Watabe, and C. R. Wolf, Existence of multiple forms of microsomal epoxide hydrolases with radically different substrate specificities, *Carcinogenesis* *5*:7 (1984).
102. T. M. Guenthner, Epoxide hydrolases, *Conjugation Reactions in Drug Metabolism: An Integrated Approach* (G. J. Mulder, ed.), Taylor & Francis, London, 1990, p. 365.
103. D. L. Kroetz, L. V. McFarland, B. M. Kerr, and R. H. Levy, Distribution of microsomal epoxide hydrolase (mEH) activity in healthy subjects, *Clin. Pharmacol. Ther.* *47*:160 (1990).
104. K. M. Norris, T. N. DeAngelo, and E. S. Vesell, Genetic and environmental factors that regulate cytosolic epoxide hydrolase activity in normal human lymphocytes, *J. Clin. Invest.* *84*:1749 (1989).
105. A. Gaedigk, S. P. Spielberg, and D. M. Grant, Characterization of the microsomal epoxide hydrolase gene in patients with anticonvulsant adverse reactions, *Pharmacogenetics* *4*:142 (1994).
106. C. Hassett, L. Aicher, J. S. Sidhu, and C. J. Omiecinski, Human microsomal epoxide hydrolase: Genetic polymorphism and functional expression in vivo of amino acid variants, *Hum. Mol. Genet.* *3*:421 (1994).
107. V. J. Green, M. Pirmohamed, N. R. Kitteringham, A. Gaedigk, D. M. Grant, M. Boxer, B. Burchill, and B. K. Park, Genetic analysis of microsomal epoxide hydrolase in patients with carbamazepine hypersensitivity, *Biochem. Pharmacol.* *50*:1353 (1995).
108. D. P. Agarwal and H. W. Goedde, Pharmacogenetics of alcohol metabolism and alcoholism, *Pharmacogenetics* *2*:48 (1992).
109. G. Milano and M. C. Etienne, Potential importance of dihydropyrimidine dehydrogenase (DPD) deficiency in cancer chemotherapy, *Pharmacogenetics* *4*:301 (1994).
110. R. A. Fleming, G. Milano, A. Thyss, M. C. Etienne, N. Renee, M. Schneider, and F. Demard, Correlation between dihydropyridine dehydrogenase activity in peripheral mononuclear cells and systemic clearance of fluorouracil in cancer patients, *Cancer Res.* *52*:2899 (1992).
111. R. Berger, S. A. Stoker-de Vries, S. K. Wadman, M. Duran, F. A. Beemer, P. K. De Bree, J. J. Weits-Binnerts, T. S. Penders, and J. K. Van der Woude, Dihydropyrimidine dehydrogenase deficiency leading to thymine-uraciluria. An inborn error of pyrimidine metabolism, *Clin. Chim. Acta* *141*:227 (1984).
112. R. Meinsma, P. Fernandez-Salguero, A. B. P. Van Kuilenburg, A. H. Van Gennip, and F. J. Gonzalez, Human polymorphism in drug metabolism: Mutation in the dihydropyrimidine dehydrogenase gene results in exon skipping and thymine uracilurea, *DNA Cell Biol.* *14*:1 (1995).
113. Z. Lu, R. Zhang, and R. B. Diasio, Dihydropyrimidine dehydrogenase activity in human peripheral blood mononuclear cells and liver: Population characteristics, newly discovered deficient patients and clinical implications in 5-fluorouracil chemotherapy, *Cancer Res.* *53*: 5433 (1993).
114. P. Fernandez-Salguero, F. J. Gonzalez, M. Etienne, G. Milano, and S. Kimura, Correlation between catalytic activity and protein-content for the polymorphically expressed dihydropyrimidine dehydrogenase in human lymphocytes, *Biochem. Pharmacol.* *50*:1015 (1995).
115. M. Patel, B. K. Tang, and W. Kalow, Variability of acetaminophen metabolism in Caucasians and Orientals, *Pharmacogenetics* *2*:38 (1992).
116. Q. Yue, J. Svensson, F. Sjoqvist, and J. Sawe, A comparison of the pharmacokinetics of codeine and its metabolites in healthy Chinese and Caucasian extensive hydroxylators of debrisoquine, *Br. J. Clin. Pharmacol.* *31*:643 (1991).
117. H. F. Liu, M. Vincent-Viry, M. Galteau, R. Gueguen, J. Magdalou, A. Nicolas, P. Leroy, and G. Siest, Urinary glucuronide excretion of fenofibric and clofibric acid glucuronides in man. Is it polymorphic? *Eur. J. Clin. Pharmacol.* *41*:153 (1991).

118. M. Vincent-Viry, C. Cossy, M. M. Galteau, R. Gueguen, J. Magdalou, A. Nicolas, P. Leroy, and G. Siest, Lack of a genetic polymorphism in the glucuronidation of fenofibric acid, *Pharmacogenetics 5*:50 (1995).
119. G. Monaghan, M. Ryan, R. Seddon, R. Hume, and B. Burchell, Genetic variation in bilirubin UDP-glucuronosyltransferase gene promoter and Gilbert's syndrome, *Lancet 347*:578 (1996).
120. H. Sato, Y. Adachi, and O. Koiwai, The genetic basis of Gilbert's syndrome, *Lancet 347*: 557 (1996).
121. P. J. Bosma, J. R. Chowdhury, C. Bakker, S. Gantla, A. Deboer, B. A. Oostra, D. Lindhout, G. N. J. Tytgat, P. L. M. Jansen, R. P. J. O. Elferink, and N. R. Chowdhury, The genetic basis of the reduced expression of bilirubin UDP-glucuronosyltransferase 1 in Gilbert's syndrome, *N. Engl. J. Med. 333*:1171 (1995).
122. S. Aono, Y. Adachi, E. Uyama, Y. Yamada, H. Keino, T. Nanno, O. Koiwai, and H. Sato, Analysis of genes for bilirubin UDP-glucuronosyltransferase in Gilbert's syndrome, *Lancet 345*:958 (1995).
123. A. F. Macklon, R. L. Savage, and M. D. Rawlins, Gilbert syndrome and drug metabolism, *Clin. Pharmacokinet. 4*:223 (1979).
124. S. M. de Morais, J. P. Uetrecht, and P. G. Wells, Decreased glucuronidation and increased bioactivation of acetaminophen in Gilbert's syndrome, *Gastroenterology 102*:577 (1992).
125. R. Weinshilboum, Sulphotransferase pharmacogenetics, *Pharmacol. Ther. 45*:93 (1990).
126. R. A. Price, N. J. Cox, R. S. Spielman, J. A. Van Loon, B. L. Maidak, and R. M. Weinshilboum, Inheritance of human platelet thermolabile phenol sulphotransferase (TLPST) activity, *Genet. Epidemiol. 5*:1 (1988).
127. R. A. Price, R. S. Spielman, A. L. Lucena, J. A. Van Loon, B. L. Maidak, and R. M. Weinshilboum, Genetic polymorphism for human platelet thermostable phenol sulphotransferase (TSPST) activity, *Genetics 122*:905 (1989).
128. A. L. Jones, R. C. Roberts, and M. W. H. Coughtrie, The human phenolsulphotransferase polymorphism is determined by the level of expression of the enzyme protein, *Biochem. J. 296*:287 (1993).
129. I. A. Aksoy, V. Sochorova, and R. M. Weinshilboum, Human liver dehydroepiandrosterone sulfotransferase: Nature and extent of individual variation, *Clin. Pharmacol. Ther. 54*:498 (1993).
130. C. Her, R. Raftogianis, and R. M. Weinshilboum, Human phenol sulfotransferase *STP2* gene: Molecular cloning, structural characterization, and chromosomal localization, *Genomics 33*: 409 (1996).
131. T. W. Wilborn, K. A. Comer, T. P. Dooley, I. M. Reardon, R. L. Heinrikson, and C. N. Falany, Sequence analysis and expression of the cDNA for the phenol sulfating form of human liver phenol sulfotransferase, *Mol. Pharmacol. 43*:70 (1993).
132. X. Zhu, M. E. Veronese, L. N. Sansom, and M. E. McManus, Molecular characterization of a human aryl sulfotransferase cDNA, *Biochem. Biophys. Res. Commun. 192*:671 (1993).
133. A. L. Jones, M. Hagen, M. W. H. Coughtrie, R. C. Roberts, and H. Glatt, Human platelet phenolsulfotransferases: cDNA cloning, stable expression in V79 cells and identification of a novel allelic variant of the phenol-sulfating form, *Biochem. Biophys. Res. Commun. 208*: 855 (1995).
134. X. Zhu, M. E. Veronese, P. Iocco, and M. E. McManus, cDNA cloning and expression of a new form of human aryl sulfotransferase, *Int. J. Biochem. Cell Biol. 28*:565 (1996).
135. F. Bernier, P. Soucy, and V. Luu-The, Human phenol sulfotransferase gene contains two alternative promoters: Structure and expression of the gene, *DNA Cell Biol. 15*:367 (1996).
136. D. A. P. Evans, *N*-Acetyltransferase, *Pharmacol. Ther. 42*:157 (1989).
137. H. B. Hughes, J. P. Biehl, A. P. Jones, and L. H. Schmidt, Metabolism of isoniazid in man as related to the occurrence of peripheral neuritis, *Am. Rev. Tuberc. 70*:266 (1954).
138. N. H. Shear, S. P. Spielberg, D. M. Grant, B. K. Tang, and W. Kalow, Differences in metabo-

lism of sulphonamides predisposing to idiosyncratic toxicity, *Ann. Intern. Med. 105*:179 (1986).

139. M. J. Ratain, R. Mick, F. Berezin, L. Janisch, R. L. Schilsky, S. F. Williams, and J. Smiddy, Paradoxical relationship between acetylator phenotype and amonafide toxicity, *Clin. Pharmacol. Ther. 50*:573 (1991).
140. W. W. Weber and K. P. Vatsis, Individual variability in *p*-aminobenzoic acid *N*-acetylation by human *N*-acetyltransferase (NAT1) of peripheral blood, *Pharmacogenetics 3*:209 (1993).
141. A. E. Cribb, R. Isbrucker, T. Levatte, B. Tsui, C. T. Gillespie, and K. W. Renton, Acetylator phenotyping: The urinary caffeine metabolite ratio in slow acetylators correlates with a marker of systemic NAT1 activity, *Pharmacogenetics 4*:166 (1994).
142. K. P. Vatsis and W. W. Weber, Structural heterogeneity of Caucasian *N*-acetyltransferase at the *NAT1* gene locus, *Arch. Biochem. Biophys. 301*:71 (1993).
143. K. P. Vatsis, W. W. Weber, D. A. Bell, J.-M. Dupret, D. A. P. Evans, D. M. Grant, D. W. Hein, H. J. Lin, U. A. Meyer, M. V. Relling, E. Sim, T. Suzuki, and Y. Yamazoe, Nomenclature for *N*-acetyltransferases, *Pharmacogenetics 5*:1 (1995).
144. D. A. Bell, E. A. Stephens, T. Castranio, D. M. Umbach, M. Watson, M. Deakin, J. Elder, C. Hendrickse, H. Duncan, and R. C. Strange, Polyadenylation polymorphism in the acetyltransferase 1 gene (*NAT1*) increases risk of colorectal cancer, *Cancer Res. 55*:3537 (1995).
145. D. A. Bell, A. F. Badawi, N. P. Lang, K. F. Ilett, F. F. Kadlubar, and A. Hirvonen, Polymorphism in the *N*-acetyltransferase 1 (*NAT1*) polyadenylation signal: association of *NAT1*10* allele with higher *N*-acetylation activity in bladder and colon tissue, *Cancer Res. 55*:5226 (1995).
146. B. Mannervik and U. H. Danielson, Glutathione transferases—structure and catalytic activity, *Crit. Rev. Biochem. 23*:281 (1988).
147. P. Board, M. Coggan, P. Johnston, V. Ross, T. Suzuki, and G. Webb, Genetic heterogeneity of the human glutathione transferases: A complex of gene families, *Pharmacol. Ther. 48*: 357 (1990).
148. K. E. Comstock, J. S. Sanderson, G. Claflin, and W. D. Kenner, GST1 gene deletion determined by polymerase chain reaction, *Nucleic Acids Res. 18*:3670 (1990).
149. M. Widersten, W. R. Pearson, A. Engstrom, and B. Mannervik, Heterologous expression of the allelic variant mu-class glutathione transferases mu and psi, *Biochem. J. 276*:519 (1991).
150. S. Pemble, K. R. Schroeder, S. R. Spencer, D. J. Meyer, E. Hallier, H. M. Bolt, B. Ketterer, and J. B. Taylor, Human glutathione *S*-transferase theta (GSTT1): cDNA cloning and the characterization of a genetic polymorphism, *Biochem. J. 300*:271 (1994).
151. M. T. Smith, C. G. Evans, P. Doane-Setzer, V. M. Castro, M. K. Tahir, and B. Mannervik, Denitrosation of 1,3-bis(2-chlorethyl)-1-nitrosourea by class m glutathione transferases and its role in cellular resistance in rat brain tumor cells, *Cancer Res. 49*:2621 (1989).
152. C. G. Evans, W. J. Bodell, D. Ross, P. Doane, and M. T. Smith, Role of glutathione and related enzymes in brain tumor resistance to BCNU and nitrogen mustard, *Proc. Am. Assoc. Cancer Res. 37*:267 (1986).
153. A. Inskip, J. Elexperu-Camiruaga, A. Buxton, P. S. Dias, J. MacIntosh, D. Campbell, P. W. Jones, L. Yengi, J. A. Talbot, R. C. Strange, and A. A. Fryer, Identification of polymorphism at the glutathione *S*-transferase, *GSTM3* locus: Evidence for linkage with *GSTM1*A, Biochem. J. 312*:713 (1995).
154. T. Nakajima, E. Elovaara, S. Anttila, A. Hirvonen, A.-M. Camus, J. D. Hayes, B. Ketterer, and H. Vainio, Expression and polymorphism of glutathione *S*-transferase in human lungs: Risk factors in smoking-related lung cancer, *Carcinogenesis 16*:707 (1995).
155. F. Ali-Osman, N. Akande, and J. Mao, Molecular cloning, characterization and expression of novel functionally different glutathione *S*-transferase-P1 gene variant, *ISSX Proc. 7*:38 (1995).
156. A. Kumar, B. C. Das, and J. K. Sharma, *Bam*HI restriction fragment length polymorphism (RFLP) at the human GST3 gene locus, *Hum. Genet. 94*:107 (1994).

157. S. Harada, T. Nakamura, and S. Misawa, Polymorphism of pentanucleotide repeats in the 5′ flanking region of glutathione *S*-transferase (GST) π gene, *Hum. Genet. 93*:223 (1994).
158. L. Z. Chen and P. G. Board, HgiAI restriction fragment length polymorphism at the human glutathione *S*-transferase 2 locus, *Nucleic Acids Res. 15*:6306 (1987).
159. R. Weinshilboum, Methyltransferase pharmacogenetics, *Pharmacol. Ther. 43*:77 (1989).
160. R. Weinshilboum and S. L. Sladek, Mercaptopurine pharmacogenetics: Monogenic inheritance of erythrocyte thiopurine methyltransferase activity, *Am. J. Hum. Genet. 32*:651 (1980).
161. H.-L. Tai, E. Y. Krynetski, C. R. Yates, T. Loennechen, M. Y. Fessing, N. F. Krynetskaia, and W. E. Evans, Thiopurine *S*-methyltransferase deficiency: Two nucleotide transitions define the most prevalent mutant allele associated with loss of catalytic activity in Caucasians, *Am. J. Hum. Genet. 58*:694 (1996).
162. C. Szumlanski, D. Otterness, C. Her, D. Lee, B. Brandriff, D. Kelsell, N. Spurr, L. Lennard, E. Wieben, and R. Weinshilboum, Thiopurine methyltransferase pharmacogenetics: Human gene cloning and characterization of a common polymorphism, *DNA Cell Biol. 15*:17 (1996).
163. E. Y. Krynetski, J. D. Schuetz, A. J. Galpin, C.-H. Pui, M. V. Relling, and W. E. Evans, A single point mutation leading to loss of catalytic activity in human thiopurine *S*-methyltransferase, *Proc. Natl. Acad. Sci. USA 92*:949 (1995).
164. L. Lennard, J. A. van Loon, and R. M. Weilshilboum, Pharmacogenetics of acute azathioprine toxicity: Relationship to thiopurine methyltransferase genetic polymorphism, *Clin. Pharmacol. Ther. 46*:149 (1989).
165. L. Lennard, J. S. Lilleyman, J. van Loon, and R. M. Weinshilboum, Genetic variation in response to 6-mercaptopurine for childhood acute leukaemia, *Lancet 236*:225 (1990).
166. W. E. Evans, M. Horner, Y. Q. Chu, D. Kalwinsky, and W. M. Roberts, Altered mercaptopurine metabolism. Toxic effects and dosage requirement in a thiopurine methyltransferase-deficient child with acute lymphocytic leukaemia, *J. Pediatr. 119*:985 (1991).
167. E. Schutz, J. Gummert, F. Mohr, and M. Oellerich, Azathioprine-induced myelosuppression in thiopurine methyltransferase deficient heart transplant recipient, *Lancet 341*:436 (1993).
168. G. Alvan, M. Grind, C. Graffner, and F. Sjoqvist, Relationship of *N*-demethylation of amiflamine and its metabolite to debrisoquine hydroxylation polymorphism, *Clin. Pharmacol. Ther. 36*:515 (1984).
169. B. Mellstrom, J. Sawe, L. Bertilsson, and F. Sjoqvist, Amitryptyline metabolism: Association with debrisoquine hydroxylation in nonsmokers, *Clin. Pharmacol. Ther. 36*:369 (1986).
170. N. Feifel, K. Kucher, L. Fuchs, M. Jedrychowski, E. Schmidt, K. H. Antonin, P. R. Bieck, and C. H. Gleiter, Role of cytochrome P4502D6 in the metabolism of brofaromine—a new selective MAO-A inhibitor, *Eur. J. Clin. Pharmacol. 45*:265 (1993).
171. G. Muralidharan, J. K. Cooper, E. M. Hawes, E. D. Korchinski, and K. K. Midha, Quinidine inhibits the 7-hydroxylation of chlorpromazine in extensive metabolizers of debrisoquine, *Eur. J. Clin. Pharmacol. 50*:121 (1996).
172. S. H. Sindrup, K. Brosen, M. G. J. Hansen, T. Aaesjorgensen, K. F. Overo, and L. F. Gram, Pharmacokinetics of citalopram in relation to the sparteine and the mephenytoin oxidation polymorphisms, *Ther. Drug Monit. 15*:11 (1993).
173. A. E. Balant-Gorgia, L. P. Balant, C. Genet, P. Dayer, J. M. Aeschilmann, and G. Garrone, Importance of oxidative polymorphism and levomepromazine treatment on the steady-state concentrations of clomipramine and its major metabolites, *Eur. J. Clin. Pharmacol. 31*:449 (1986).
174. V. Fischer, B. Vogels, G. Maurer, and R. E. Tynes, The antipsychotic clozapine is metabolised by the polymorphic human and microsomal and recombinant cytochrome-P450-2D6, *J. Pharmacol. Exp. Ther. 260*:1355 (1992).

175. J. M. Grace, M. T. Kinter, and T. L. Macdonald, Atypical metabolism of deprenyl and its enantiomer, (*S*)-(+)-*N*, α-dimethyl-*N*-propynylphenethylamine, by cytochrome P450 2D6, *Chem. Res. Toxicol. 7*:286 (1994).
176. L. Bertilsson and A. Aberg-Wistedt, The debrisoquine hydroxylation test predicts steady-state plasma levels of desipramine, *Br. J. Clin. Pharmacol. 15*:388 (1983).
177. P. Sanwald, M. David, and J. Dow, Characterization of the cytochrome-P450 enzymes involved in the in vitro metabolism of dolasetron—comparison with other indole-containing 5-HT3 antagonists, *Drug Metab. Dispos. 24*:602 (1996).
178. R. F. Tyndale, W. Kalow, and T. Inaba, Oxidation of reduced haloperidol to haloperidol—involvement of human P450IID6, *Br. J. Clin. Pharmacol. 31*:655 (1991).
179. K. Brosen, V. Otton, and L. F. Gram, Imipramine demethylation and hydroxylation: Impact of the sparteine oxidation phenotype, *Clin. Pharmacol. Ther. 40*:543 (1986).
180. L. Firkusny and C. H. Gleiter, Maprotiline metabolism appears to cosegregate with the genetically-determined CYP2D6 polymorphic hydroxylation of debrisoquine, *Br. J. Clin. Pharmacol. 37*:383 (1994).
181. S. D. Roy, E. M. Hawes, G. McKay, E. D. Korchinski, and K. K. Midha, Metabolism of methoxyphenamine in extensive and poor metabolisers of debrisoquine, *Clin. Pharmacol. Ther. 38*:128 (1985).
182. M.-L. Dahl, G. Tybring, C. E. Elwin, C. Alm, K. Andreasson, M. Gyllenpalm, and L. Bertilsson, Stereoselective disposition of mianserin is related to debrisoquin hydroxylation polymorphism, *Clin. Pharmacol. Ther. 56*:176 (1994).
183. H. Davi, J. M. Bonnet, and Y. Berger, Disposition of minaprine in animals and in human extensive and limited metabolizers, *Xenobiotica 22*:171 (1992).
184. L. Bertilsson, M. Eichelbaum, B. Mellstrom, J. Sawe, N. V. Schultz, and F. Sjoqvist, Nortryptyline and antipyrine clearance in relation to debrisoquine hydroxylation in man, *Life Sci. 27*:1673 (1980).
185. J. C. Bloomer, F. R. Woods, R. E. Haddock, M. S. Lennard, and G. T. Tucker, The role of cytochrome P4502D6 in the metabolism of paroxetine by human liver microsomes, *Br. J. Clin. Pharmacol. 33*:521 (1992).
186. M. L. Dahl-Puustinen, A. Liden, C. Alm, C. Nordin, and L. Bertilsson, Disposition of perphenazine is related to polymorphic debrisoquine hydroxylation in human beings, *Clin. Pharmacol. Ther. 46*:78 (1989).
187. V. Fischer, A. E. M. Vickers, F. Heitz, S. Mahadevan, J. P. Baldeck, P. Minery, and R. Tynes, The polymorphic cytochrome P-4502D6 is involved in the metabolism of both 5-hydroxytryptamine antagonists, tropisetron and ondansetron, *Drug Metab. Dispos. 22*:269 (1994).
188. C. Von Bahr, G. Morvin, C. Nordin, A. Liden, M. Hammarlund-Udenaes, A. Hedberg, H. Ring, and F. Sjoqvist, Plasma levels of thioridazine and metabolites are influenced by the debrisoquine hydroxylation phenotype, *Clin. Pharmacol. Ther. 49*:234 (1991).
189. M. D. Feher, R. A. Lucas, N. A. Farid, J. R. Idle, R. F. Bergstrom, L. Lemberger, and P. S. Sever, Single dose pharmacokinetics of tomoxetine in poor and extensive metabolisers of debrisoquine, *Br. J. Clin. Pharmacol. 26*:231 (1988).
190. S. V. Otton, S. E. Ball, S. W. Cheung, T. Inaba, R. L. Rudolph, and E. M. Sellers, Venlafaxine oxidation in vitro is catalyzed by CYP2D6, *Br. J. Clin. Pharmacol. 41*:149 (1996).
191. M. L. Dahl, B. Ekqvist, J. Widen, and L. Bertilsson, Disposition of the neuroleptic zuclopenthixol cosegregates with the polymorphic hydroxylation of debrisoquine in humans, *Acta Psychiatr. Scand. 84*:99 (1991).
192. T. Ebner and M. Eichelbaum, The metabolism of aprinidine in relation to the sparteine/debrisoquine polymorphism, *Br. J. Clin. Pharmacol. 35*:426 (1993).
193. A. R. Boobis, S. Murray, C. E. Hampden, and D. S. Davis, Genetic polymorphism in drug oxidation: In vitro studies of human debrisoquine 4-hydroxylase and bufuralol 1′-hydroxylase activities, *Biochem. Pharmacol. 34*:65 (1985).

194. J. Pressacco, R. Muller, and W. Kalow, Interactions of bupranolol with the polymorphic debrisoquine sparteine monooxygenase (CYP2D6), *Eur. J. Clin. Pharmacol. 45*:261 (1996).
195. A. Mahgoub, J. R. Idle, L. G. Dring, R. Lancaster, and R. L. Smith, Polymorphic hydroxylation of debrisoquine in man, *Lancet 2*:584 (1977).
196. R. Woosley, D. M. Roden, G. H. Dai, Y. Tang, D. C. Altenbein, J. Oates, and G. R. Wilkinson, Co-inheritance of the polymorphic metabolism of encainide and debrisoquine, *Clin. Pharmacol. Ther. 39*:282 (1986).
197. G. Mikus, A. S. Gross, J. Beckmann, R. Hertrampf, U. Gundert-Remy, and M. Eichelbaum, The influence of the sparteine debrisoquin phenotype on the disposition of flecainide, *Clin. Pharmacol. Ther. 45*:562 (1989).
198. T. P. Sloan, A. Mahgoub, R. Lancaster, J. R. Idle, and R. L. Smith, Polymorphism of carbon oxidation of drugs and clinical implications, *Br. Med. J. 2*:655 (1978).
199. D. M. Pierce, S. E. Smith, and R. A. Franklin, The pharmacokinetics of indoramin in poor and excessive hydroxylators of debrisoquine, *Eur. J. Clin. Pharmacol. 33*:59 (1987).
200. M. S. Lennard, J. H. Silas, S. Freestone, L. E. Ramsay, G. T. Tucker, and H. F. Woods, Oxidation phenotype: a major determinant of metoprolol metabolism and response, *N. Engl. J. Med. 307*:1558 (1982).
201. F. Broly, N. Vandamme, C. Libersa, and M. Lhermitte, The metabolism of mexiletine in relation to the debrisoquine/sparteine-type polymorphism of drug oxidation, *Br. J. Clin. Pharmacol. 32*:459 (1991).
202. C. Zekorn, G. Achtert, H. J. Hausletter, C. H. Moon, and M. Eichelbaum, Pharmacokinetics of *N*-propylajamaline in relation to polymorphic sparteine oxidation, *Klin. Wochenschr. 63*: 1180 (1985).
203. L. A. Siddoway, K. A. Thompson, C. B. McAllister, T. Tang, G. H. Wilkinson, D. M. Roden, and R. L. Woosley, Polymorphism of propafenone metabolism and disposition in man: Clinical and pharmacokinetic consequences, *Circulation 75*:785 (1987).
204. S. A. Ward, T. Walle, U. K. Walle, G. R. Wilkinson, and R. A. Branch, Propranolol metabolism is determined by both mephenytoin and debrisoquin hydroxylase activities, *Clin. Pharmacol. Ther. 45*:72 (1989).
205. M. Eichelbaum, N. Spannbrucker, B. Steincke, and H. J. Dengler, Defective *N*-oxidation of sparteine in man: A new pharmacogenetic defect, *Eur. J. Clin. Pharmacol. 17*:153 (1979).
206. J. C. McGourty, J. H. Silas, J. J. Fleming, A. McBurney, and J. W. Ward, Pharmacokinetics and β-blocking effects of timolol in poor and extensive metabolisers of debrisoquine, *Clin. Pharmacol. Ther. 38*:409 (1985).
207. O. Mortimer, K. Persson, M. G. Ladora, D. Spalding, U. M. Zanger, U. A. Meyer, and A. Rane, Polymorphic formation of morphine from codeine in poor and extensive metabolisers of dextromethorphan: Relation to the presence of immunoidentified cytochrome P-450IID1, *Clin. Pharmacol. Ther. 47*:27 (1990).
208. A. S. Gross, A. C. Phillips, A. Rieutord, and G. M. Shenfield, The influence of the sparteine/debrisoquine genetic polymorphism on the disposition of dexfenluramine, *Br. J. Clin. Pharmacol. 41*:311 (1996).
209. A. Kupfer, B. Schmid, R. Preisig, and G. Pfaff, Dextromethorphan as a safe probe for debrisoquine hydroxylation polymorphism, *Lancet 2*:517 (1984).
210. M. F. Fromm, U. Hofmann, E. U. Griese, and G. Mikus, Dihydrocodeine—a new opioid substrate for the polymorphic CYP2D6 in humans, *Clin. Pharmacol. Ther. 58*:374 (1995).
211. Z. R. Liu, O. Mortimer, C. A. D. Smith, C. R. Wolf, and A. Rane, Evidence for a role of cytochrome P450 2D6 and 3A4 in ethylmorphine metabolism, *Br. J. Clin. Pharmacol. 39*: 77 (1995).
212. S. V. Otton, M. Schadel, S. W. Cheung, H. L. Kaplan, U. E. Busto, and E. M. Sellers, CYP2D6 phenotype determines the metabolic conversion of hydrocodone to hydromorphone, *Clin. Pharmacol. Ther. 54*:463 (1993).

213. S. W. Ellis, M. S. Ching, P. F. Watson, C. J. Henderson, A. P. Simula, M. S. Lennard, G. T. Tucker, and H. F. Woods, Catalytic activities of human debrisoquine 4-hydroxylase cytochrome P450 (CYP2D6) expressed in yeast, *Biochem. Pharmacol. 44*:617 (1992).
214. N. Yumibe, K. Huie, K. J. Chen, M. Snow, R. P. Clement, and M. N. Cayen, Identification of human liver cytochrome P450 enzymes that metabolize the nonsedating antihistamine loratadine—formation of descarboethoxyloratadine by CYP3A4 and CYP2D6, *Biochem. Pharmacol. 51*:165 (1996).
215. I. Kitchen, J. Tremblay, J. Andre, L. G. Dring, J. R. Idle, R. L. Smith, and R. T. Williams, Interindividual and interspecies variation in the metabolism of the hallucinogen 4-methoxy-amphetamine, *Xenobiotica 9*:397 (1979).
216. R. G. Cooper, D. A. P. Evans, and A. H. Price, Studies on the metabolism of perhexilene in man, *Eur. J. Clin. Pharmacol. 32*:569 (1987).
217. S. Kimura, M. Umeno, R. C. Skoda, U. A. Meyer, and F. J. Gonzalez, The human debrisoquine 4-hydroxylase (*CYP2D*) locus: Sequence and identification of a polymorphic *CYP2D6* gene, a related gene, and a pseudogene, *Am. J. Hum. Genet. 45*:889 (1989).
218. A. C. Gough, J. S. Miles, N. K. Spurr, J. E. Moss, A. Gaedigk, M. Eichelbaum, and C. R. Wolf, Identification of the primary gene defect at the cytochrome P450 *CYP2D* locus, *Nature 347*:773 (1990).
219. N. Hanioka, S. Kimura, U. A. Meyer, and F. J. Gonzalez, The human *CYP2D* locus associated with a common genetic defect in drug oxidation: a G_{1934} to A base change in intron 3 of a mutant *CYP2D6* allele results in an aberrant 3′ splice recognition site, *Am. J. Hum. Genet. 47*:994 (1990).
220. A. Gaedigk, M. Blum, R. Gaedigk, M. Eichelbaum, and U. A. Meyer, Deletion of the entire cytochrome P450 gene as a cause of impaired drug metabolism in poor metabolizers of the debrisoquine/sparteine polymorphism, *Am. J. Hum. Genet. 48*:943 (1991).
221. V. M. Steen, A. Molven, N. K. Aarskog, and A.-K. Gulbrandsen, Homologous unequal cross-over involving a 2.8 kb direct repeat as a mechanism for the generation of allelic variants of the human cytochrome P450 *CYP2D6* gene, *Hum. Mol. Genet. 4*:2251 (1995).
222. F. Broly, D. Marez, J.-M. Lo Guidice, N. Sabbagh, M. Legrand, P. Boone, and U. A. Meyer, A nonsense mutation in the cytochrome P450 *CYP2D6* gene identified in a Caucasian with an enzyme deficiency, *Hum. Genet. 96*:601 (1995).
223. D. Marez, N. Sabbagh, M. Legrand, J. M. Lo-Guidice, P. Boone, and F. Broly, A novel *CYP2D6* allele with an abolished splice recognition site associated with the poor metabolizer phenotype, *Pharmacogenetics 5*:305 (1995).
224. S. Panserat, C. Mura, N. Gerard, M. Vincent-Viry, M. M. Galteau, E. Jacqz-Aigrain, and R. Krishnamoorthy, An unequal cross-over event within the *CYP2D* gene cluster generates a chimeric *CYP2D7/CYP2D6* gene which is associated with the poor metabolizer phenotype, *Br. J. Clin. Pharmacol. 40*:361 (1995).
225. A. K. Daly, K. S. Fairbrother, O. A. Andreassen, J. R. Idle, and V. M. Steen, Characterization and PCR-based detection of two different hybrid *CYP2D7P/CYP2D6* alleles associated with the poor metabolizer phenotype, *Pharmacogenetics 6*:319 (1996).
226. S. Wang, Phenotypes and genotypes of debrisoquine hydroxylation polymorphism in Chinese, dissertation, National Cheng Kung University, Tainan, Taiwan (1992).
227. C. Sachse, J. Brockmoller, S. Bauer, T. Reum, and I. Roots, A rare insertion of T226 in exon 1 of *CYP2D6* causes a frameshift and is associated with the poor metabolizer phenotype: *CYP2D6*17, Pharmacogenetics 6*:269 (1996).
228. T. Yokoi, Y. Kosaka, M. Chiba, K. Chiba, H. Nakamura, T. Ishizaki, M. Kinoshita, K. Sato, F. J. Gonzalez, and T. Kamataki, A new *CYP2D6* allele with a nine base insertion in exon 9 in a Japanese population associated with a poor metabolizer phenotype, *Pharmacogenetics 6*:395 (1996).
229. D. Marez, M. Legrand, N. Sabbagh, J. M. LoGuidice, C. Spire, J. L. Lafitte, U. A. Meyer, and F. Broly, Polymorphism of the cytochrome P450 *CYP2D6* gene in a European population:

characterization of 48 mutations and 53 alleles, their frequencies and evolution. *Pharmacogenetics 7*:193 (1997).

230. S. Panserat, C. Mura, N. Gerard, M. Vincent-Viry, M. M. Galteau, E. Jacqz-Aigrain, and R. Krishnamorthy, DNA haplotype-dependent differences in the amino acid sequence of debrisoquine 4-hydroxylase (CYP2D6): Evidence for two major allozymes in extensive metabolisers, *Hum. Genet. 94*:401 (1994).
231. R. Tyndale, T. Aoyama, F. Broly, T. Matsunaga, T. Inaba, W. Kalow, H. V. Gelboin, U. A. Meyer, and F. J. Gonzalez, Identification of a new *CYP2D6* allele lacking the codon encoding Lys-281: Possible association with the poor metabolizer phenotype, *Pharmacogenetics 1*:26 (1991).
232. F. Broly and U. A. Meyer, Debrisoquine oxidation polymorphism: Phenotypic consequences of a 3-base-pair deletion in exon 5 of the *CYP2D6* gene, *Pharmacogenetics 3*:123 (1993).
233. D. M. Grant, N. C. Hughes, S. A. Janezic, G. H. Goodfellow, H. J. Chen, A. Gaedigk, V. L. Yu, and R. Grewel, Human acetyltransferase polymorphisms, *Mutat. Res. 376*:61 (1997).
234. N. J. Butcher, K. F. Ilett, and R. F. Minchin, Functional polymorphism of the human *N*-acetyltransferase type 1 gene caused by $C^{190}T$ and $G^{560}A$ mutations, *Pharmacogenetics 8*: 67 (1998).

8

Inhibition of Drug Metabolizing Enzymes: Molecular and Biochemical Aspects

F. Peter Guengerich
Vanderbilt University School of Medicine, Nashville, Tennessee

I. INTRODUCTION

The topic of inhibition of the enzymes of drug metabolism is of great interest to chemists, enzymologists, pharmacologists, and clinicians. There are two major practical applications of knowledge of inhibition in the pharmaceutical industry. One is drug–drug interactions (i.e., one drug may inhibit the biotransformation of another when two are taken concurrently; 1). Such interactions can be fatal, and the possibilities are scrutinized by regulatory agencies. The other major interest in enzyme inhibition is based on the selection of enzymes as targets for drug action. For instance, monoamine oxidase and some of the cytochrome P450 (P450) enzymes are targets because the products of their normal reactions can be deleterious under certain conditions. Yeast P450 51 (lanosterol 14α-demethylase) is a fungicide target, and P450 19 is a target in breast cancer therapy because it is involved in estrogen synthesis (2).

II. BASIC MECHANISMS OF ENZYME INHIBITION

The general treatments presented here are rather introductory and the reader is referred to more comprehensive treatments of the subject (3–5). Classifications used here are slanted toward major mechanisms known for the enzymes of drug metabolism. Inhibition has its basis in the enzymology itself, including the field of enzyme kinetics. Overviews of major mechanisms and their principles will be presented, followed by a few prominent examples involving various enzymes.

A. Competitive Inhibition

The classic view of competitive inhibition is that the inhibitor shares structural similarity with the normal substrate (although defining a "normal" substrate for many of the enzymes under consideration is not always easy). The inhibitor may or may not be a substrate itself (i.e., be transformed to a product by the enzyme):

$$E + S \underset{k_{-1}}{\overset{k_1}{\rightleftarrows}} ES + I \underset{k_{-2}}{\overset{k_2}{\rightleftarrows}} ES^* \underset{k_{-3}}{\overset{k_3}{\rightleftarrows}} EP \underset{k_{-4}}{\overset{k_4}{\rightleftarrows}} E + P$$

$$E + I \underset{k_{-1'}}{\overset{k_{1'}}{\rightleftarrows}} EI,$$

where the intermediate ES* may or may not be present. Sometimes k_1, $k_{1'}$ and k_{-4} are termed "k_{on}," and k_{-1}, $k_{-1'}$, and k_4 are "k_{off}" rates. In classic competitive inhibition the steady-state $K_{m,apparent}$ value (for the reaction with the "typical" substrate) increases when the inhibitor is present, and examination of steady-state kinetics with increasing concentrations of inhibitor and concomitant increasing $K_{m,apparent}$ values allow estimation of an inhibition constant K_i because they are related by the expression

$$v = \frac{V_{max}}{K_m(1 + [I]/K_i) + [S]}$$

where $K_{m,apparent} = K_m (1 + [I]/K_i)$. This behavior is readily identified by a common intersection point (ordinate) in a Lineweaver–Burk ($1/v$ vs $1/S$) plot, or various characteristics of other linear transformations of the Michaelis–Menten equation.

The classic approach to characterization of competitive inhibition and the associated parameters is to do enzyme assays with varying concentrations of both the substrate S and inhibitor I, fitting to the foregoing equations or their derivatives. An alternative method is the Dixon plot, in which $1/v$ is plotted versus S (3). Historically, linear transforms were used to do fitting ''by eye'' or linear regression. All of the linear plots have some weighting deficiencies (5), and many convenient and useful nonlinear regression programs are now commercially available. However, screening for inhibition is very common in the pharmaceutical industry today, and new approaches have been introduced to handle the increased load of drug candidates. One statistical experimental approach is usually termed *virtual kinetics* (6).

In principle, K_i is an actual binding constant, as opposed to K_m, which is usually not. When the inhibitor is a substrate itself, the two substrates show competitive inhibition of each other. In ideal cases, if the $K_m \sim K_d$ then $K_m \sim K_i$. However, there are many reasons why such a relation may not be seen with a complex enzyme system.

Competitive inhibition is a relatively commonly encountered phenomenon in drug metabolism work, and there are means of characterizing in vivo situations through pharmacokinetic parameters (7–9). Many of the enzymes under consideration have multiple drug substrates (e.g., P450s 2D6, 3A4) that can compete with each other. Drug–drug interactions of this sort can be expected in individuals who are administered several drugs simultaneously.

B. Noncompetitive Inhibition

In classic noncompetitive inhibition, the inhibitor binds to the enzyme at a site distinct from that of the substrate. The expected result is that a decrease in V_{max} is observed without a change in K_m. For instance, one might expect an electrophilic inhibitor to poison the enzyme in such a manner. Although this example is often presented in introductory biochemistry courses, clear examples of such inhibition are not common and are not often encountered in studies with enzymes of drug metabolism. In vitro, one might expect such results by adding heavy metals to or heating an enzyme. What is often encountered is *mixed inhibition*, in which V_{max} usually decreases and K_m increases. The physical meaning of such changes may vary. For example, such behavior might be observed if one were dealing with two different enzymes in a population (e.g., microsomes) that both catalyzed a reaction and one was inhibited competitively, whereas the other was being inactivated by mechanistic inactivation. Interpretation of such results must be done with a single enzyme system.

C. Uncompetitive Inhibition

As in noncompetitive inhibition, classic uncompetitive inhibition is defined, but is seldom seen. The principle is that the inhibitor binds only to the enzyme–substrate complex. Both V_{max} and K_m are decreased proportionately and the ratio V_{max}/K_m remains constant. For instance, in a Lineweaver–Burk plot parallel lines should be seen in the absence and presence of the inhibitor. The enzyme efficiency (and, by extension, the intrinsic clearance of the drug substrate) would not really change. However, there are few clear examples of this phenomenon in the field.

D. Product Inhibition

Sometimes a product of a reaction of a drug-metabolizing enzyme may inhibit the reaction. For instance, $NADP^+$ is a competitive inhibitor of NADPH–P450 reductase (10). In a cellular system, this example would not exist because there are reduction systems that work well on the oxidized cofactor. However, in other instances the product may not have physical characteristics very different from the substrate and competitively inhibits, sometimes being further transformed. For instance, benzene is oxidized by P450 2E1 to phenol and then on to hydroquinone (11). Thus, benzene and phenol compete with each other.

E. Transition-State Analogs

Transition-state analogs are tight-binding, noncovalently bound inactivators that resemble the transition state for the enzymatic reaction, that is, the transient complex formed in a single step within the catalytic cycle with the maximum free energy. The axiom that the enzyme has the highest affinity for this putative entity (which cannot be directly observed) was developed by Haldane (12) and Pauling (13) and is the basis for the production

$$E + I \underset{k_{off}}{\overset{k_{on}}{\rightleftarrows}} E\text{–}I$$

of catalytic antibodies (14). The k_{on} rate is rapid and k_{off} is slow. Inactivation is rapid and no time dependence is observed under typical assay conditions. Enzyme activity can, at least in principle, be restored by removal of inhibitor using dialysis, gel filtration, centrifugal concentration, or others (15).

F. Slow, Tight-Binding Inhibitors

Such compounds are characterized by relatively slow k_{on} rates and even slower k_{off} rates. The binding can be noncovalent or covalent, but owing to the slowness of binding, the loss of activity may be time-dependent and mistaken for mechanism-based inactivation. Removal of the inhibitor by dialysis, or such, will restore enzyme activity, although the process may be slow.

An example of a characterized slow, tight-binding inhibitor of testosterone 5α-reductase is finasteride (Proscar), which binds to the enzymes at a slow rate, competitively inhibits, and effectively irreversibly inactivates the enzyme (16).

Possible causes for the slow, tight-binding include a conformational change of the enzyme imposed by binding, a change in the protonation state of the enzyme, displacement of a water molecule at the active site, or reversible formation of a covalent bond (15).

G. Mechanism-Based Enzyme Inactivators

Silverman (15) states that a broad definition of this term includes any inactivators that use the enzyme mechanism, but invokes a stricter definition, one that will be adhered to in this chapter. The definition is ''an unreactive compound the structure of which resembles that of either the substrate or the product of the target enzyme, and that undergoes a catalytic transformation by the enzyme to a species that, before release from the active site, inactivates the enzyme'' (15). A key point here is the need to inactivate the enzyme before leaving the active site. The definition, taken as a whole, restricts the grouping from transition-state analogs, affinity labels, and slow, tight-binding inhibitors.

Mechanism-based inactivation is sometimes encountered inadvertently with existing drugs. There are several major intentional uses of these compounds, and this group will be covered in some detail (for more extensive discussion see Refs. 4,15,17–22). Mechanism-based inactivators have been of considerable interest because of their usefulness in the delineation of enzyme mechanisms. They are also of interest in the design of new drugs because, in principle, only the target enzyme will be attenuated. Many of the better diagnostic inhibitors of the drug-metabolizing enzymes are mechanism-based inactivators. For instance, these can be used to gain valuable in vitro information about which of the P450s are involved in a particular reaction (23,24).

The relevant scheme for mechanism-based inactivation is

$$\mathrm{E} + \mathrm{I} \underset{k_1}{\overset{k_1}{\rightleftarrows}} \mathrm{E{-}I} \overset{k_2}{\rightleftarrows} \mathrm{E{-}I^*} \overset{k_4}{\rightarrow} \mathrm{EI'}$$

$$\downarrow k_3$$

$$\mathrm{E} + \mathrm{P}$$

where I is the mechanism-based inactivator, E–I* is an intermediate, derived by transformation of the initial complex, EI′ is inactivated enzyme, and P is a stable product that leaves the enzyme. Sometimes these inhibitors are called ''suicide inactivators'' although the term is not generally accepted by purists. It is the enzyme, not the inhibitor, that is dying.

Several parameters are experimentally determined and used to describe these inhibitors. The ratio k_3/k_4 is the *partition ratio*, which can be thought of as the number of times that the enzyme must cycle, on the average, for one inactivation to occur. However, the ratio can range from several thousand to less than 1, even approaching zero.

The inactivation process shows first-order kinetics, that is, a plot of the logarithm of the remaining enzyme activity versus time gives a straight line (first-order, or single-exponential kinetics; Fig. 1). The half-life, $t_{1/2}$, can be determined at each inhibitor concentration used and used to calculate k, using the relationship $t_{1/2} = 0.693/k_{\mathrm{inact}} + 0.693\ 2\ K_{\mathrm{I}}/k_{\mathrm{inact}}\ \mathrm{I}$ (15). The plot of k versus [I] is hyperbolic, and a linear transformation (e.g., plot of $1/k$ vs. $1/[\mathrm{I}]$) yields k_{inact}, the maximum rate of inactivation, and K_{I}, the concentration of inhibitor required for half-maximal inhibition. In the foregoing scheme, $k_{\mathrm{inact}} = k_2$ if k_2 is rate-limiting in the overall reaction. K_{I} is a complex expression of microscopic rate constants, but is useful in estimating the potential usefulness of an inhibitor.

Several criteria can be used to determine if mechanism-based inactivation is actually occurring. Although not all of these tests are applicable to every situation, the case for mechanism-based inactivation is stronger when several can be demonstrated.

One of the simplest tests is whether or not the typical cofactors are required. For instance, in a P450-dependent reaction, is preincubation with NADPH necessary to see inhibition by the compound under consideration? Usually a mechanism-based inactivator also has a strictly competitive component, and results can be misleading. The generally

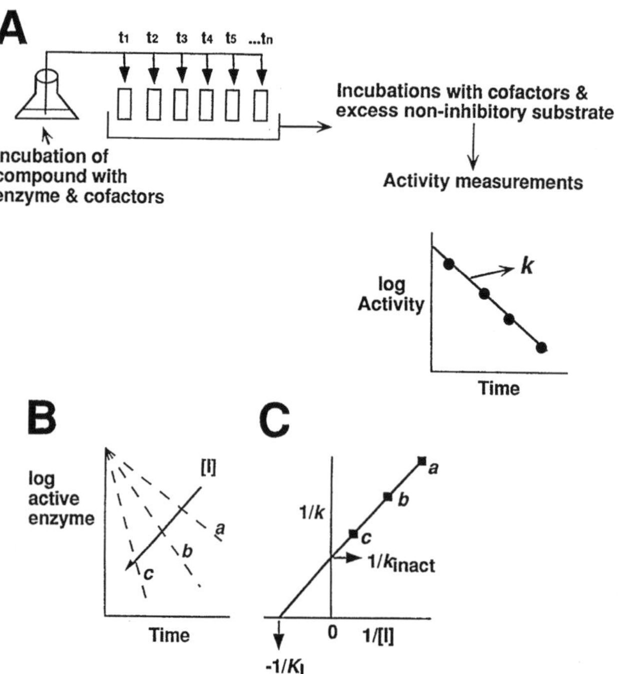

Fig. 1 Determination of k_{inact} for a mechanism-based inactivator: (A) The compound under investigation is incubated with the enzyme (plus all relevant cofactors); at various time points the solution is diluted into an excess of (noninhibitory) substrate to determine the amount of remaining "active enzyme." (B) Plots of $\log_{10}$ [active enzyme] versus time give the plots designated **a, b**, and **c** at various concentrations of the inhibitory substrate I. **a, b**, and **c** have constants k, defined by the slopes ($k = 0.693/t_{1/2}$). (C) The values of k from lines **a, b**, and **c** are related to [I] in the double-reciprocal Kitz–Wilson plot. The intercept on the ordinate in this plot gives k_{inact}, the extrapolated rate of inactivation at infinite inhibitor concentration, and the intercept on the x axis gives K_I, the inhibitor concentration at which a half-maximal rate of inactivation is seen.

accepted way of discerning the two aspects is to preincubate the concentrated enzyme with cofactors and a certain concentration of the inhibitor and then, at a certain time, dilute this enzyme (e.g., 50-fold) into a solution containing a noninhibitory substrate (and cofactors) and assay product formation, either continuously, if possible, or after a set time.

One characteristic of mechanism-based inactivation is the first-order kinetic pattern mentioned earlier. In practice, aliquots are withdrawn at indicated times from an incubation of enzyme, the inhibitor, and any necessary cofactors are diluted into excess substrate. The rate of inactivation k should increase when the experiment is repeated with a higher concentration of inhibitor (see Fig. 1). In a plot of the logarithm of residual enzyme activity versus time, the intercept ($t = 0$) indicates the extent of inhibition that is of a competitive nature.

The enzyme should be protected from inactivation by a normal (noninhibitory) substrate. This criterion may not distinguish a mechanism-based inhibitor from a competitive one, unless the time-dependence of inhibition is examined.

Another criterion is irreversibility. However, occasionally a slow reactivation is seen, usually occurring over a period of days. Exactly what time period of reactivation does or does not constitute mechanism-based inactivation is not specifically defined. In practice, one usually removes excess inhibitor from the enzyme (e.g., dialysis, gel filtration, or centrifugal filtration) and assays the activity of the enzyme to determine if it has been restored.

The inhibitor is usually covalently bound to the enzyme in mechanism-based inactivation, either to the protein or a prosthetic group. The stoichiometry of binding should be unimolecular; that is, only one molecule should be bound per enzyme (subunit). The extent of labeling should be correlated with the degree of inactivation; for example, a ratio of 0.7 labeled inhibitor bound per enzyme subunit should correspond to a 70% loss of activity (corrected for any competitive inhibition). Furthermore, labeling should be specific in the sense that "scavenger" nucleophiles [e.g., glutathione (GSH)] and other proteins (that do have access to the enzyme active site) are not labeled. A general rule of thumb is that a specific radioactivity of 0.5 mCi (^{14}C) per nanomole is needed for labeling studies of this type to obtain sufficient counts for analysis. Other isotopes can be used; tritium is acceptable if the hydrogen is stable in all involved processes. On occasion, it is possible to use other labels, such as fluorescent chromophores.

True mechanism-based inactivators use the same catalytic step in partitioning between product formation and inactivation. Thus, both should show the same cofactor requirements, and any kinetic isotopic effects (e.g., deuterium) should be common to both.

More extensive treatments of the theory and practice of mechanism-based inactivation have appeared elsewhere (4,15,17,18,21,22), as well as examples of applications to various enzymes (15,17–20,25,26).

H. Inhibitors That Generate Reactive Products That Are Covalently Attached to the Enzyme

This group of compounds is often grouped with the mechanism-based inactivators discussed in the foregoing. Many of the same criteria apply, such as the need for cofactors and the irreversible nature of the inactivation. However, labeling may be more extensive and less specific. Also, careful analysis of the kinetics may show critical differences from linearity in plots of log (enzyme activity) versus time (15). For instance, there might be a lag in the inhibition and then an apparent first-order plot, as the concentration of reactive product increases. If fresh enzyme is added to the mixture at this point, then no lag will be observed owing to the buildup of electrophilic product in the medium. However, it should be emphasized that discerning deviations from pseudo–first-order kinetics may not be easy, especially if the assays are subject to considerable error, or if relatively few data points are collected.

Experiments in which mechanism-based inactivation and inhibition of the type discussed here need not be restricted to purified enzymes. They can be done with relatively crude preparations (e.g., microsomes or cytosolic fractions) if appropriate caveats are used. Labeling studies can be done with such preparations to examine the specificity of the process.

III. STIMULATION

One aspect of P450s and some other enzymes that will be mentioned briefly is stimulation, the opposite of inhibition. In the sense used here, stimulation refers to the ability of chemicals to enhance the catalytic activity toward a substrate when added directly to the enzyme, as opposed to induction (in which the amount of the enzyme is increased by treatment of cells or an animal with the chemical). This phenomenon has been recognized for some time, and there are indications that this can occur in vivo, at least in animals (27). The effect seems to occur most often with the P450 3A subfamily enzymes (28,29). Solvents can stimulate P450s (30,31) and epoxide hydrolase (32); hence, care is needed in in vitro assays, particularly with very hydrophobic substrates. Natural products, such as flavonoids, can also stimulate P450s (31,33). The mechanism of such stimulation is unclear, although some of the kinetic results can be interpreted in terms of allosteric mechanisms (23,28).

IV. INHIBITORS OF VARIOUS ENZYMES

This section is not intended to be comprehensive, but to provide some examples of the previously discussed types of enzyme inhibition as they relate to several of the enzymes involved in the biotransformation of drugs and xenobiotics. The reader is referred to other articles for more comprehensive lists of inhibitors.

A. Monoamine Oxidase

There are two forms of monoamine oxidase, termed A and B (34). These have long been recognized to show differences in inhibition by various drugs and are recognized to be the products of different genes. Mechanism-based inactivation is common among many of the inhibitors of these enzymes, and drugs have been developed to treat problems related to the nervous system (15,35).

Monoamine oxidase A oxidizes biogenic amines and is selectively inhibited by the mechanism-based inactivator clorgiline. The B form of the enzyme is involved in the oxidation of noncatecholamines and is inhibited by pargyline and selegiline (deprenyl), also mechanism-based inactivators. These compounds all contain an *N*-propynyl group ($-N-C{\equiv}C-CH_3$), which leads to a covalent adduct (15).

Another inhibitor of monoamine oxidase is the anticonvulsant milacemide, $CH_3(CH_2)_4NHCH_2CONH_2$, which is also a substrate (15). The compound is thought to be oxidized to an aminium radical (one-electron oxidation), with attack on an α-carbon to generate a protein adduct (15).

Another popular group of inhibitors has been ring-strained cycloalkylamines, which rearrange to reactive products following one-electron oxidation (36,37). The antidepressant tranylcypromine was one of the first compounds in this class (Fig. 2). Postulated

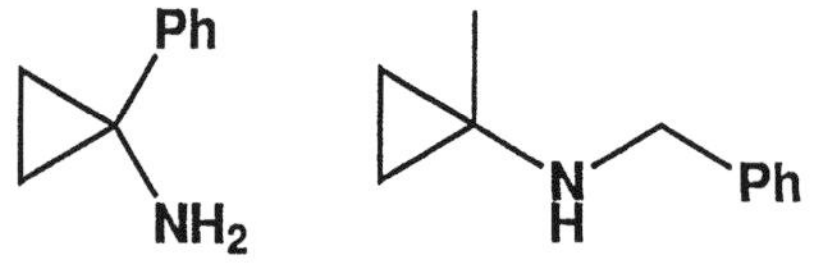

Ph
NH_2
N
H
Ph

Fig. 2 Structures of the monoamine oxidase inhibitors 1-phenylcyclopropylamine and *N*-(1-methylcyclopropyl)benzylamine. (From Ref. 15.)

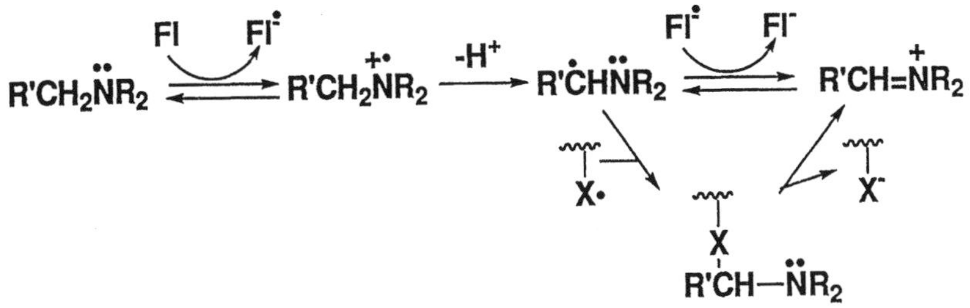

Fig. 3 Proposed mechanism of oxidation of amines by monoamine oxidase and scheme of protein alkylation. The identity of ''X'' on the protein is undefined (possibly Cys) F1 = FAD. (From Ref. 15.)

mechanisms for some of these are shown in Figs. 3 and 4. There are two paths (Figs. 4a,b). One leads to modification of the flavin prosthetic group, which is irreversible. The other pathway leads to modification of a protein cysteine group. This appears to be a reversible process, with the enzyme losing the moiety after 24 h (see Section II. G for discussion of the issue of reversibility). A series of cyclopropyl derivatives of *N*-methyl-4-phenyl-1,2,5,6-tetrahydropyridine have been studied (38,39).

B. Cytochrome P450

Inhibitors of P450s have been studied relative to several aspects. Some aspects are quite basic. First, many different P450 enzymes exist in humans and experimental animals, and

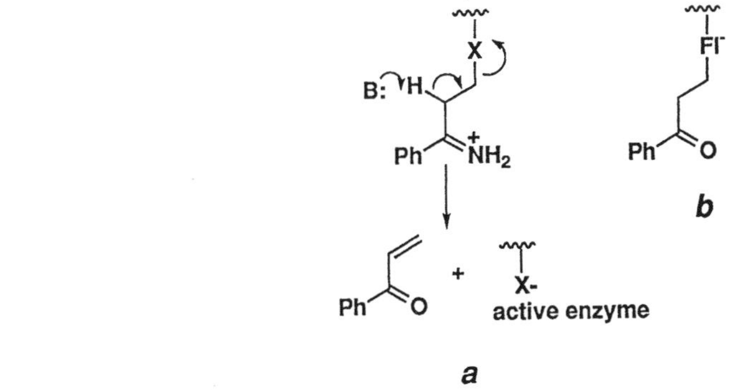

Fig. 4 Proposed scheme for oxidation of 1-phenylcyclopropylamine by monoamine oxidase, with both pathways leading to reversible (a) and irreversible (b) inactivation. The identity of ''X'' on the protein remains undefined (possibly Cys). (From Ref. 15.)

diagnostic inhibitors have been used to ascertain the roles of individual P450s in catalysis of reactions (40). Because of the overlapping catalytic specificity of many of the P450s, inhibitors are usually not totally selective. Nevertheless, the selectivity can be examined and the use of diagnostic inhibition of individual P450s is possible, with appropriate caveats (41,42). A list of generally accepted ''probe'' inhibitors is presented in Table 1 (24,41,42). Moreover, inhibitors have provided considerable insight into function, both in terms of chemical aspects of catalysis (46) and structural details of the proteins that contribute to selectivity (47).

Other aspects of P450 inhibition are more practical (48). For instance, one prominent strategy in therapy of estrogen-dependent tumors is the inhibition of P450 19, the ''aromatase'' that catalyzes the three-step conversion of androgens to estrogens (49,50). Aminoglutethimide is a classic, but not ideal inhibitor, and numerous efforts are in progress to design better drugs (51). Another target in yeast and fungal infections is P450 51, the lanosterol 14α-demethylase (52). Many azoles (e.g., ketoconazole) are used as drugs here (2). Selectivity for the fungal enzyme is observed, but these drugs can also inhibit mammalian P450 51 and other P450s at higher concentrations (53,54). Another example of a P450 inhibitor is piperonyl butoxide, which is used as an insecticide synergist to block the detoxification of chemicals by insect P450s (Fig. 5); (55). As more information becomes available about the P450s present in noxious insects, the design and development of pesticides (and synergists) with more selectivity for the insect enzymes should be possible.

Another practical area in which the development of better inhibitors of P450s may be useful is in cancer prevention. Some inhibitors (e.g., ethynyls) can block rodent and

Table 1 Diagnostic Inhibitors of Human Cytochrome P450 Enzymes

P450	Inhibitor	Apparent mechanism
1A1	7,8-Benzoflavone	Competitive[a]
	Ellipticine	Competitive
1A2	7,8-Benzoflavone	Competitive[a]
	Fluvoxamine	Competitive
	Furafylline	Mechanism-based
2A6	Diethyldithiocarbamate	Mechanism-based
2C9	Sulfaphenazole	Competitive[b]
	Tienilic acid	Mechanism-based
2D6	Quinidine, several others[c]	Competitive[d]
2E1	4-Methylpyrazole	Competitive
	Diethyldithiocarbamate	Mechanism-based[a]
	Many organic solvents	Competitive[a]
3A4, 3A5	Troleandomycin	Conversion to heme ligand
	Erythromycin	Conversion to heme ligand
	Gestodene	Mechanism-based

[a] Known to be oxidized by the enzyme. Also, some of the reported results with these compounds (43) are not supported in subsequent work (41,42).

[b] Apparently not a heme ligand (Mancy, A., Dijols, S., Poli, S., Guengerich, F. P., and Mansuy, D., *Biochemistry 35*: 16205 (1996)).

[c] Ref. 44.

[d] Apparently not oxidized (45).

Source: Ref. 24

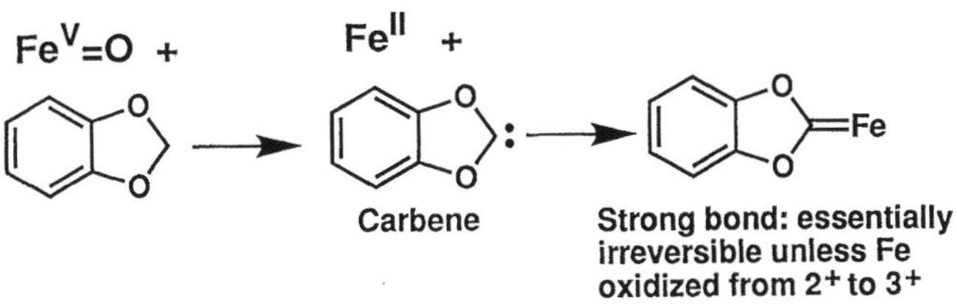

Fig. 5 Oxidation of piperonyl butoxide by P450 to a carbene that yields a stable ferrous ligand. (From Ref. 55.)

human P450s that activate carcinogens (56) and can block carcinogen-induced cancers in rodents (57). The drug oltipraz and the natural compound sulforaphane have been studied in cancer prevention studies because of their abilities to induce conjugation enzymes (e.g., GSH *S*-transferase, quinone reductase; 58,59). More importantly, perhaps, they both block P450s involved in the activation of carcinogens (e.g., aflatoxin B_1; 60).

Several types of inhibition are seen for the P450s. A relatively common mechanism is competitive inhibition. For instance, quinidine and most other P450 2D6 inhibitors seem to act in this way (61); a basic nitrogen in the inhibitor seems to bind in the same site as the substrates (44,62,63). Azoles have already been mentioned and are one example of nitrogen heterocycles that ligand to the heme iron (2,64). Sulfaphenazole is a competitive inhibitor of P450 2C9 and has selectivity (compared with other P450 2C enzymes; 65,66). Many organic solvents inhibit P450 2E1 by competing as substrates (67); thus, care must be taken in experimental designs (especially, in vitro).

Another group of inhibitors are not particularly effective themselves, but are oxidized to metabolic intermediates that bind tightly to the heme and prevent further involvement of the iron atom in catalysis. For instance, some amines are oxidized to *C*-nitroso derivatives (e.g., troleandomycin; 68,69) and piperonyl butoxide yields a carbene (see Fig. 5; 70). This process might be termed mechanism-based inactivation, but perhaps a better description would be transformation by the enzyme(s) to slow, tight-binding inhibitors. The linkages are not covalent in the usual sense, for addition of a strong oxidant [e.g., $Fe(CN)_6^{3-}$] oxidizes such a ferrous complex to ferric and releases the ligand.

Mechanism-based inactivators have been studied extensively, primarily from a basic perspective (23,71). Many vinyl and acetylenic inhibitors have been characterized as yielding heme or protein adducts (Fig. 6). Some of these have practical consequences with drugs (e.g., secobarbital; 71). A dihalomethylene group (72) can also be involved in mechanism-based inactivation, although it is not clear whether chloramphenicol should be classified in this group or among the inhibitors that are converted to a reactive product (73). Inhibition by cyclopropylamines is considered to involve oxidation to an aminium radical that rearranges to a reactive methylene radical (Fig. 7), although adducts remain to be characterized (74–76).

4-Alkyl-1,4-dihydropyridines are readily oxidized by P450s by one-electron oxidation (77). Rearrangement of the putative aminium radical to the pyridine generates an alkyl radical (Fig. 8), which modifies a pyrrole nitrogen of the prosthetic heme (80). If an aromatic group is at the 4-position it is retained, and such compounds are used as drugs (78). This is, strictly speaking, not an example of mechanism-based inactivation, because (a) alkyl radicals can react with spin traps outside of the protein (77), and (b) other P450s that cannot oxidize the 4-alkyl-1,4-dihydropyridines can also be inactivated when an en-

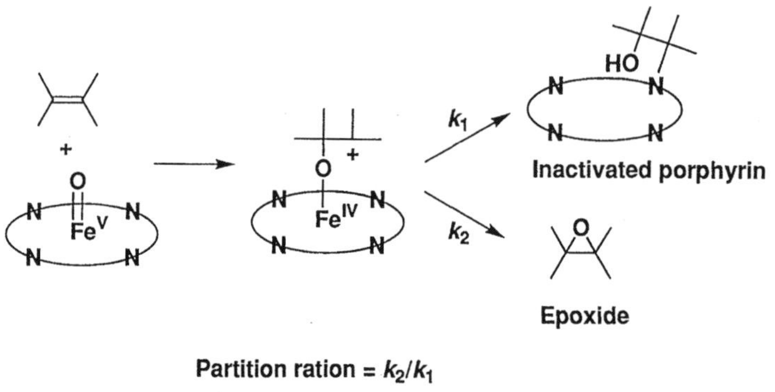

Inactivated porphyrin

Epoxide

Partition ration = k_2/k_1

Fig. 6 Mechanism-based inactivation of P450 by heme destruction during the epoxidation of an olefin. The partition ratio is given by k_2/k_1. Several of the *N*-alkyl porphyrin adducts arising from the heme have been rigorously characterized by NMR and mass spectroscopy. The putative perferryl oxygen is in the formal $Fe^{V}{=}O$ state for convenience. (From Ref. 46.)

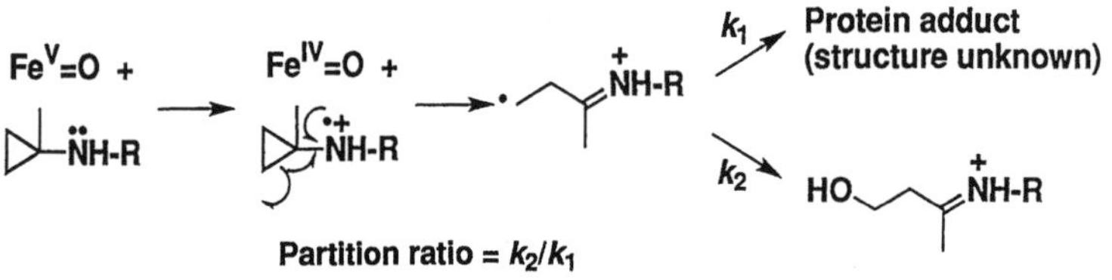

Protein adduct (structure unknown)

Partition ratio = k_2/k_1

Fig. 7 Proposed mechanism of oxidation of a 1-methylcyclopropylamine by P450. The putative perferryl oxygen is shown in the formal $Fe^{V}{=}O$ state for convenience. The partition ratio is k_2/k_1. (From Refs. 74–76.)

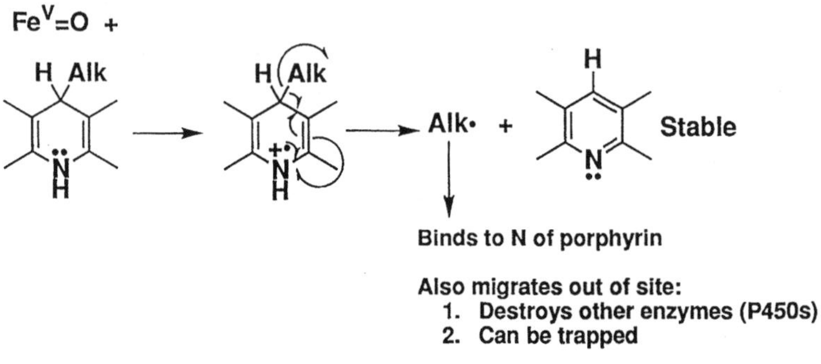

Stable

Binds to N of porphyrin

Also migrates out of site:
1. Destroys other enzymes (P450s)
2. Can be trapped

Fig. 8 Oxidation of a 4-alkyl-1,4-dihydropyridine by P450 to generate a destructive alkyl radical. (From Refs. 46,77–79.)

zyme capable of oxidation (e.g., P450 3A4) is also present. Ortiz de Montellano and his associates have also characterized other compounds (e.g., 1-aminobenztriazole) that are oxidized by P450s to yield covalent heme adducts (46,81,82). Interestingly, Meschter et al. (83) have reported that it is possible to use 1-aminobenztriazole to lower total hepatic P450 levels in rats to less than 30% that of the normal for 13 weeks without significant physiological effects.

C. NADPH–P450 Reductase

This flavoprotein is involved in the transfer of electrons from NADPH to P450s (84). The enzyme also functions in electron transfer to some other hemoproteins (e.g., heme oxygenase; 85).

Inhibition has not been studied extensively. The oxidation product $NADP^+$ is a competitive inhibitor (10). The 2′-phosphate group is important in binding; 2′-AMP is also a competitive inhibitor, and the enzyme is the basis for the use of 2,5′-ADP affinity chromatography in purification (86).

Diphenyliodonium has been reported to be a mechanism-based inactivator of the enzyme (87). The mechanism is postulated to involve one-electron reduction of the iodonium to give an iodide–flavin radical pair that combines to give an N^5-phenylflavin adduct, along with a labeled amino acid (Trp-419).

D. Flavin-Containing Monooxygenase

In contrast with the flavoprotein monoamine oxidase, little information is available about inhibitors of this enzyme. Some are known, but have poor affinities (88). No mechanism-based inactivators have been characterized. The various substrates seem to inhibit each other, at least insofar as they are substrates for the same form of the enzyme (at least five forms can exist in a single animal species; 89,90).

E. Carbonyl Dehydrogenases and Reductases

Disulfiram (Antabuse) is a well-known inhibitor of aldehyde dehydrogenase. This drug has been given to recovering alcoholics to produce unpleasant physiological effects when the individuals consume ethanol. Disulfiram is reduced to diethyldithiocarbamate, which seems to be bound to an enzyme cysteine in disulfide linkage (91). Diethyldithiocarbamate is also an inhibitor of some P450s (especially 2E1 and 2A6) (92). It may have a mechanism-based action, at least as judged by the kinetics seen in limited investigations (11). In vivo, diethyldithiocarbamate is methylated and then oxygenated to yield a more effective inhibitor (93,94).

The sedative chloral (2,2,2-trichloroacetaldehyde) is a potent competitive inhibitor of aldehyde dehydrogenase, with a K_i of 1–10 μM. Apparently a stable thiohemiacetal is formed with Cys-302. Because of the electronegativity of the chlorine atoms, transfer of a hydride ion is effectively blocked.

Similarly, aldose reductase is a target for inhibition in the treatment of certain types of diabetes. The drugs sorbinil, alrestatin, and tolrestat seem to be effective in diabetic rats, but have not been as useful in humans (95).

F. Prostaglandin Synthase–Cyclooxygenase

Inhibitors have been the source of considerable interest owing to the desire to develop drugs for relief of inflammation, analgesics, and such. Aspirin acts mainly by acetylating Ser-530 in the active site of prostaglandin synthase 1 (COX-1) (96). Many nonsteroidal anti-inflammatory drugs (NSAIDs) have been developed by using competitive inhibitors and other approaches.

However, in recent years, the existence of another enzyme, COX-2, has been demonstrated (97–100). In contrast with the "housekeeping" enzyme COX-1, COX-2 is inducible and linked to development of inflammation. It is of interest to develop inhibitors selective for COX-2, in that they not only should be more effective in blocking inflammation, but also devoid of the undesirable gastric effects (e.g., ulcers) of compounds that inhibit COX-1 (101). Two selective inhibitors of the enzymes have been developed that differ only in sulfur oxygenation (Fig. 9; 102). These compounds appear to be acting as slow, tight-binding inhibitors, although extensive characterization of the kinetics apparently has not yet been done. Recently, a single amino acid difference between the two isozymes has been shown to reverse the selectivity of specific inhibitors (103).

G. Esterases and Amidases

The mechanisms of inhibition of acetylcholinesterase have long been of interest because chemical warfare agents (nerve gases) and organophosphate insecticides can interfere with cholinergic transmission and lead to respiratory failure. The basic principles have been known for some time and involve nucleophilic attack by a serine in the active site (104; Fig. 10).

In contrast with most of the examples of drug inhibition, this is a noncompetitive mechanism during which reaction with an active site serine occurs. The reaction is somewhat reversible in that hydrolysis can occur to reactivate the enzyme. However, a rearrangement, usually referred to as "aging," can occur to fix the damage by generation of a nonhydrolyzable linkage (104).

H. Epoxide Hydrolase

The epoxide hydrolases are now recognized to be a subfamily of the α,β-lyase family that have certain features that make epoxides good substrates (105). In recent years the discovery that a covalent ester intermediate is formed has had considerable implication for mechanistic studies (106). Modes of inhibition need to be reexamined in this context.

Several supposedly competitive inhibitors of microsomal epoxide hydrolase have been reported (107). Of these, 3,3,3-trichloropropylene oxide (TCPO) has historically been considered the most diagnostic, although it seems to have disappeared from the commercial market because of unknown regulatory issues. However, if an ester intermediate is formed the rate of hydrolysis of this may be the issue, and this may be better classified as a slow, tight-binding inhibitor. The soluble epoxide hydrolase has inhibitors with lower K_i values, particularly among the chalcone oxides (108).

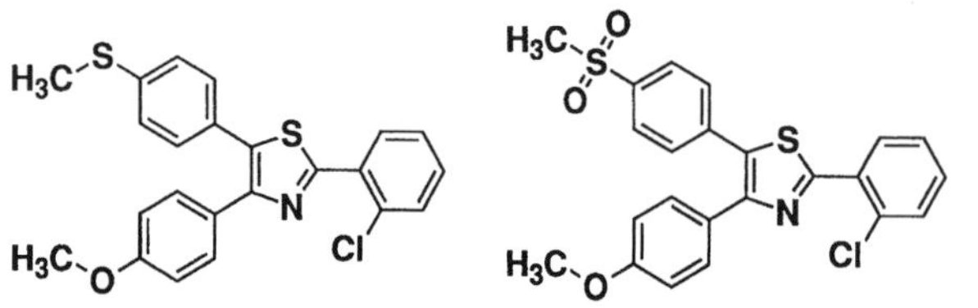

SC-58076 (COX-1) SC-58092 (COX-2)

Fig. 9 Structures of selective inhibitors of COX-1 (SC-58076) and COX-2 (SC-58092)(prostaglandin H synthase isozymes). The ratio of IC_{50} values is 10^3-fold in each case. (From Ref. 102.)

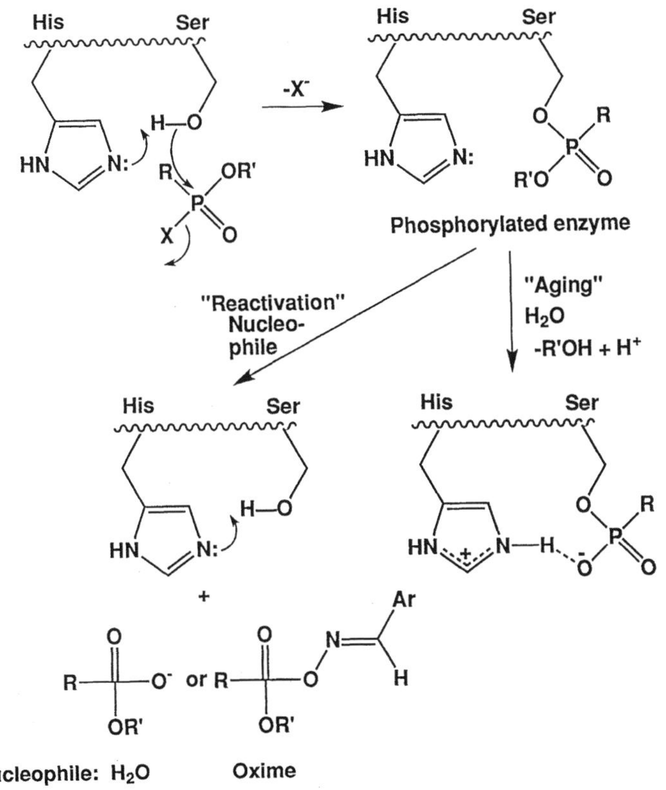

Fig. 10 Inactivation of acetylcholinesterase by organophosphates and enzyme reactivation and "aging." (From Ref. 104.)

I. GSH *S*-Transferases

The GSH *S*-transferase enzymes are very abundant (109) and the crystal structures of many are now known (110). Some hydrophobic compounds are good ligands (hence, one of the original names, "ligandin"; 111), which probably act as competitive inhibitors (109). GSH has a K_d for several of the GSH *S*-transferases of close to 20 μM, facilitating the use of affinity chromatography for purification (112). Replacement of the ⊃CH–CH_2 S^- (⊃CH–CH_2–SH) moiety of GSH with ⊃CH–CO_2^- provides an analog of a key intermediate, which has a K_i of 0.9 μM (113). A somewhat similar approach was used by Mulder, who replaced the entire L-Cys-Gly moiety with D-aminoadipate (K_i 8 μM; 114). Also, a Meisenheimer complex of GSH with 1,3,5-trinitrobenzene (Fig. 11) appears to behave as a transition state analog (115). Product complexes can also be used as inhibitors (116). For instance, *S*-(3′-iodobenzyl)GSH had a K_i of 0.2 μM and was used to solve the structure of GST 3-3, as a source of a heavy atom in diffraction studies. Some GSH conjugates have been used as affinity labels (117). The GSH conjugate 2-(*S*-glutathionyl)-3,5,6-trichloro-1,4-benzoquinone is an effective irreversible inhibitor (K_i < 1 μM, k_{inact} 0.3 min^{-1}) and modifies active site tyrosine groups (118). Because GSH transferases are generally beneficial, the question can be raised of why they should be targets for inhibition. The antischistosomal drug praziquantel is used to inhibit the parasite *Schistosoma japonica*

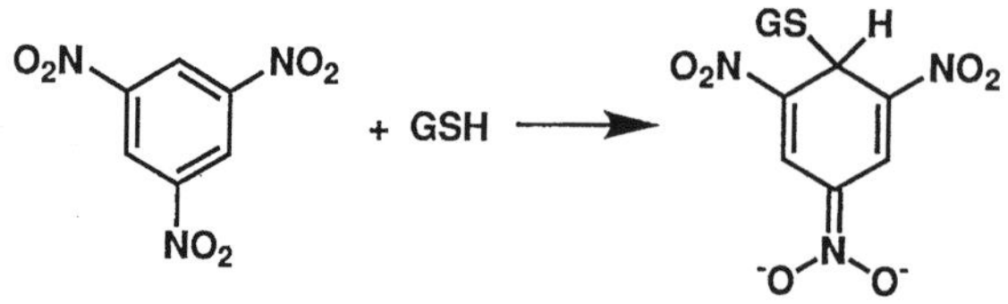

Fig. 11 GSH *S*-transferase-catalyzed conjugation of 1,3,5-trinitrobenzene to a Meisenheimer complex that inhibits the enzyme. (From Ref. 115.)

(119). It appears to compete by binding to a hydrophobic substrate site near the subunit interface of the dimer. Also, GSH *S*-transferase overexpression may contribute to multiple drug resistance in cancer cells and is a potential therapeutic target (119).

J. Sulfotransferases

The phenols 2,6-dichloro-4-nitrophenol and pentachlorophenol have been described as competitive, dead-end inhibitors (120). This is a family of enzymes that is growing in complexity, and the selectivity of these interactions is still a matter of investigation.

K. UDP-Glucuronosyl Transferases

Many drugs have been characterized as competitive inhibitors of the steroid-, bilirubin-, and drug-conjugation activities of UDP-glucuronosyl transferases (see Table 7 of Ref. 121). The roles of these interactions in practical drug metabolism issues is largely unexplored. Both transition-state analog inhibitors (122) and photoaffinity labels (123–125) have been designed and used to characterize these enzymes (121).

V. EXAMPLES OF RELEVANCE OF INHIBITION TO DRUG–DRUG INTERACTIONS

Clinical aspects of drug–drug interaction are covered elsewhere in this book; therefore, only a few classic examples of in vivo problems will be mentioned here. The reader is also referred to other treatments of the subject (126).

As already mentioned, sometimes the enzymes are targets and inhibition is intended. At other times some inhibition may be expected on the basis of preliminary in vitro assays. Nevertheless, most pharmaceutical companies would rather not put a drug with the potential for interaction problems on the market, if another that did not have such potential were available.

What one would like to avoid is the development of a potential inhibition–interaction problem with a drug already on the market (or heavily invested in the developmental process). A few examples will be mentioned.

The H_2 receptor antagonist cimetidine (Tagemet) has been widely prescribed for ulcers. This compound can inhibit P450-catalyzed reactions, although it is a relatively weak inhibitor (127). Nevertheless, a considerable market share was lost to the noninhibitory alternative ranitidine (Zantac) through advertising, on the basis of the prospect of drug–drug interactions.

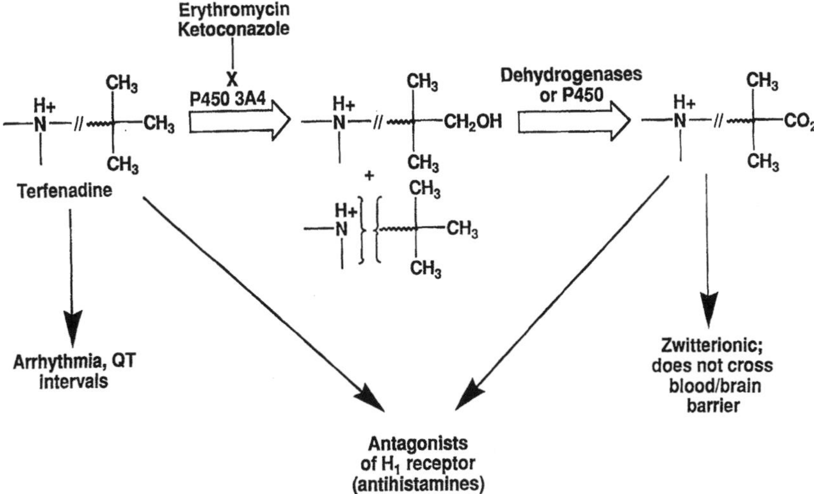

Fig. 12 Events involved in the oxidation of terfenadine by P450 3A4 and consequences of P450 3A4 inhibition by erythromycin or ketoconazole. (From Refs. 128,129.)

Terfenadine (Seldane) was the first nonsedating antihistamine on the market and has been highly successful. Nevertheless, some adverse incidents have been reported, and the basis of some seems to be related to metabolism. Terfenadine is usually extensively oxidized (Fig. 12), and in most persons none of the parent drug is found circulating in plasma. One of the two main oxidation routes yields the inactive *N*-dealkylation products. The product of the other oxidation route, a carboxylic acid, retains its ability to block the histamine receptor. The acid is actually a zwitterion and does not readily cross the blood–brain barrier, so it is nonsedating. If P450 3A4 is inhibited, then terfenadine can accumulate and may cause arrythmias (128,129). Adverse effects have been reported (130), and now warnings advise against the concurrent use of known P450 3A4 inhibitors (e.g., erythromycin and ketoconazole).

Another example of P450 3A4 inhibition involves the progestin gestodene, which has been used with the estrogen 17α-ethynylestradiol in some oral contraceptive formulations. All 17α-acetylenic steroids seem to have some inherent capability of acting as P450 mechanism-based inactivators (131,132), but gestodene appears to be more effective than many others (133). The inhibition of P450 3A4 might explain some of the thrombolytic problems attributed to gestodene (134), because inhibition of 17β-estradiol and 17-ethynyl estradiol oxidation (catalyzed by P450 3A4) could raise estrogen levels, a known factor in thrombolytic problems. However, the levels of gestodene ingested daily do not seem high enough to account for destruction of a substantial fraction of the P450 3A4 pool (133), and the in vivo significance of these phenomena is not yet clear.

VI. CONCLUSIONS

Inhibition of the enzymes usually associated with drug metabolism is a subject of both basic and practical interest. Basic studies involve studies of mechanisms of catalysis and the utilization of selective inhibitors of individual forms of enzymes in multigene families. Practical aspects of inhibition include drug–drug interactions and enzymes as therapeutic targets. Among the more common modes of enzyme inhibition seen are competitive inhibition, product inhibition, slow, tight-binding inhibition, mechanism-based inactivation, and products that become covalently attached. Discrimination among these is necessary for a proper understanding of action. However, for some the classification into a particular mode may not be obvious. A better understanding of inhibition mechanisms and selectivity has led to more efficient screening for drug–drug interactions in the pharmaceutical industry and appreciation of the phenomenon in the regulatory agencies.

ACKNOWLEDGMENTS

Research in the author's laboratory has been supported by USPHS grants R35 CA44353 and P30 ES00267. Thanks are extended to Drs. E. M. J. Gillam and W. W. Johnson for constructive comments and to E. Rochelle for assistance in preparation of the manuscript.

REFERENCES

1. F. P. Guengerich, Role of cytochrome P450 in drug interactions, *Drug Interactions: Scientific and Regulatory Perspectives* (Adv. Pharmacol. Series) (A. P. Li, ed.), Academic Press, San Diego, 1997, Vol. 43, p. 7.
2. H. Vanden Bossche, Inhibitors of P450-dependent steroid biosynthesis: from research to medical treatment, *J. Steroid Biochem. Mol. Biol. 43*: 1003 (1992).
3. S. A. Kuby, *A Study of Enzymes*, Vol. I, *Enzyme Catalysis, Kinetics, and Substrate Binding*, CRC Press, Boca Raton, FL, 1991.
4. R. B. Silverman, *Mechanism-Based Enzyme Inactivation: Chemistry and Enzymology*, CRC Press, Boca Raton, FL, 1988.
5. A. Cornish-Bowden, *Fundamentals of Enzyme Kinetics*, Butterworths, London, 1979.
6. D. D. Bronson, D. M. Daniels, J. T. Dixon, C. C. Redick, and P. D. Haaland, Virtual kinetics: Using statistical experimental design for rapid analysis of enzyme inhibitor mechanisms, *Biochem. Pharmacol. 50*: 823 (1995).
7. A. G. Renwick, Toxicokinetics—pharmacokinetics in toxicology, *Principles and Methods of Toxicology* (A. W. Hayes, ed.), Raven Press, New York, 1994, p. 101.
8. L. L. von Moltke, D. J. Greenblatt, S. X. Duan, J. S. Harmatz, and R. I. Shader, In vitro prediction of the terfenadine–ketoconazole pharmacokinetic interaction, *J. Clin. Pharmacol. 34*: 1222 (1994).
9. D. J. Black, K. L. Kunze, L. C. Wienkers, B. E. Gidal, T. L. Seaton, N. D. Mcdonnell, J. S. Evans, J. E. Bauwens, and W. F. Trager, Warfarin–fluconazole II. A metabolically based drug interaction: In vivo studies, *Drug Metab. Dispos. 24*: 422 (1996).
10. J. L. Vermilion and M. J. Coon, Purified liver microsomal NADPH–cytochrome P-450 reductase: Spectral characterization of oxidation–reduction states, *J. Biol. Chem. 253*: 2694 (1978).
11. F. P. Guengerich, D.-H. Kim, and M. Iwasaki, Role of human cytochrome P-450 IIE1 in the oxidation of many low molecular weight cancer suspects, *Chem. Res. Toxicol. 4*: 168 (1991).
12. J. B. S. Haldane, *Enzymes*. Longmans, Green, London, 1930.
13. L. Pauling, Chemical achievement and hope for the future, *Am. Sci. 36*: 51 (1948).
14. R. A. Lerner and S. J. Benkovic, Principles of antibody catalysis, *Bioessays 9*: 107 (1988).
15. R. B. Silverman, Mechanism-based enzyme inactivators, *Methods Enzymol. 249*: 240 (1995).
16. G. C. Tian, R. A. Mook, M. L. Moss, and S. V. Frye, Mechanism of time-dependent inhibition

of 5alpha-reductases by delta(1)-4-azasteroids: Toward perfection of rates of time-dependent inhibition by using ligand-binding energies, *Biochemistry 34*: 13453 (1995).

17. R. H. Abeles and A. L. Maycock, Suicide enzyme inactivators, *Acc. Chem. Res. 9*: 313 (1976).
18. R. R. Rando, Mechanism-based enzyme inactivators, *Pharmacol. Rev. 36*: 111 (1984).
19. C. T. Walsh, Suicide substrates, mechanism-based enzyme inactivators: Recent developments, *Annu. Rev. Biochem. 53*: 493 (1984).
20. R. H. Abeles and T. A. Alston, Enzyme inhibition by fluoro compounds, *J. Biol. Chem. 265*: 16705 (1990).
21. S. G. Waley, Kinetics of suicide substrates, *Biochem. J. 185*: 771 (1980).
22. S. G. Waley, Kinetics of suicide substrates: Practical procedures for determining parameters, *Biochem. J. 227*: 843 (1985).
23. J. A. Halpert and F. P. Guengerich, Enzyme inhibition and stimulation, *Biotransformation*, Vol. 3, *Comprehensive Toxicology* (F. P. Guengerich, ed.), Elsevier Science, Oxford, 1997, p. 21.
24. M. A. Correia, Rat and human liver cytochromes P450. Substrate and inhibitor specificities and functional markers, *Cytochrome P450: Structure, Mechanism, and Biochemistry* 2nd ed. (P. R. Ortiz de Montellano, ed.), Plenum Press, New York, 1995, p. 607.
25. X. Xiao and G. D. Prestwich, 29-Methylidene-2,3-oxidosqualene: A potent mechanism-based inactivator of oxidosqualene cyclase, *J. Am. Chem. Soc. 113*: 9673 (1991).
26. P. J. Schechter and A. Sjoerdsma, Therapeutic utility of selected enzyme-activated irreversible inhibitors, *Enzymes as Targets for Drug Design*, Academic Press, New York, 1990, p. 201.
27. J. M. Lasker, M-T. Huang, and A. H. Conney, In vivo activation of zoxazolamine metabolism by flavone, *Science 216*: 1419 (1982).
28. G. E. Schwab, J. L. Raucy, and E. F. Johnson, Modulation of rabbit and human hepatic cytochrome P-450-catalyzed steroid hydroxylations by α-naphthoflavone, *Mol. Pharmacol. 33*: 493 (1988).
29. Y.-F. Ueng, T. Shimada, H. Yamazaki, and F. P. Guengerich, Oxidation of aflatoxin B_1 by bacterial recombinant human cytochrome P450 enzymes, *Chem. Res. Toxicol. 8*: 218 (1995).
30. T. Wolff, H. Wanders, and F. P. Guengerich, Organic solvents as modifiers of aldrin epoxidase activity of purified cytochromes P-450 and of microsomes, *Biochem. Pharmacol. 38*: 4217 (1989).
31. F. J. Wiebel, J. C. Leutz, L. Diamond, and H. V. Gelboin, Aryl hydrocarbon (benzo[*a*]pyrene) hydroxylase in microsomes from rat tissues: Differential inhibition and stimulation by benzoflavones and organic solvents, *Arch. Biochem. Biophys. 144*: 78 (1971).
32. J. Seidegård and J. W. DePierre, Benzil, a potent activator of microsomal epoxide hydrolase in vitro, *Eur. J. Biochem. 112*: 643 (1980).
33. J. Kapitulnik, P. J. Poppers, M. K. Buening, J. G. Fortner, and A. H. Conney, Activation of monooxygenases in human liver by 7,8-benzoflavone, *Clin. Pharmacol. Ther. 22*: 475 (1977).
34. A. W. J. Bach, N. C. Lan, D. L. Johnson, C. W. Abell, M. E. Bembenek, S. W. Kwan, P. H. Seeburg, and J. C. Shih, cDNA cloning of human liver monoamine oxidase A and B: Molecular basis of differences in enzymatic properties, *Proc. Natl. Acad. Sci. USA 85*: 4934 (1988).
35. P. L. Dostert, M. S. Benedetti, and K. F. Tipton, Interactions of monoamine oxidase with substrates and inhibitors, *Med. Res. Rev. 9*: 45 (1989).
36. R. B. Silverman, Mechanism of inactivation of monoamine oxidase by *trans*-2-phenylcyclopropylamine and the structure of the enzyme–inactivator adduct, *J. Biol. Chem. 258*: 14766 (1983).
37. R. B. Silverman and C. Z. Ding, Chemical model for a mechanism of inactivation of monoamine oxidase by heterocylic compounds. Electronic effects on acetal hydrolysis, *J. Am. Chem. Soc. 115*: 4571 (1993).
38. K. F. Tipton, J. M. McCrodden, and M. B. H. Youdim, Oxidation and enzyme-activated

irreversible inhibition of rat liver monoamine oxidase-B by 1-methyl-4-phenyl-1,2,3,6-tetrahydropyridine (MPTP), *Biochem. J. 240*: 379 (1986).

39. J. M. Rimoldi, Y. X. Wang, S. K. Nimkar, S. H. Kuttab, A. H. Anderson, H. Burch, and N. Castagnoli, Jr., Probing the mechanism of bioactivation of MPTP type analogs by monoamine oxidase B: Structure–activity studies on substituted 4-phenoxy-, 4-phenyl-, and 4-thiophenoxy-1-cyclopropyl-1,2,3,6-tetrahydropyridines, *Chem. Res. Toxicol. 8*: 703 (1995).
40. P. K. Bridson and C. B. Reese, A novel method for the methylation of heterocyclic amino groups. Conversion of guanosine into its 2-*N*-methyl- and 2-*N*,2-*N*-dimethyl derivatives, *Bioorg. Chem. 8*: 339 (1979).
41. D. J. Newton, R. W. Wang, and A. Y. H. Lu, Cytochrome P450 inhibitors: Evaluation of specificities in the in vitro metabolism of therapeutic agents by human liver microsomes, *Drug Metab. Dispos. 23*: 154 (1994).
42. M. Bourrié, V. Meunier, Y. Berger, and G. Fabre, Cytochrome P450 isoform inhibitors as a tool for the investigation of metabolic reactions catalyzed by human liver microsomes, *J. Pharmacol. Exp. Ther. 227*: 321 (1996).
43. T. K. H. Chang, F. J. Gonzalez, and D. J. Waxman, Evaluation of triacetyloleandomycin, α-naphthoflavone and diethyldithiocarbamate as selective chemical probes for inhibition of human cytochrome P450, *Arch. Biochem. Biophys. 311*: 437 (1994).
44. G. R. Strobl, S. von Kruedener, J. Stöckigt, F. P. Guengerich, and T. Wolff, Development of a pharmacophore for inhibition of human liver cytochrome P-450 2D6: Molecular modeling and inhibition studies, *J. Med. Chem. 36*: 1136 (1993).
45. F. P. Guengerich, D. Müller-Enoch, and I. A. Blair, Oxidation of quinidine by human liver cytochrome P-450, *Mol. Pharmacol. 30*: 287 (1986).
46. P. R. Ortiz de Montellano and N. O. Reich, Inhibition of cytochrome P-450 enzymes, *Cytochrome P-450* (P. R. Ortiz de Montellano, ed.), Plenum Press, New York, 1986, p. 273.
47. J. R. Halpert, Structural basis of selective cytochrome P450 inhibition, *Annu. Rev. Pharmacol. Toxicol. 35*: 29 (1995).
48. H. Vanden Bossche, L. Koymans, and H. Moereels, P450 inhibitors of use in medical treatment: focus on mechanisms of action, *Pharmacol. Ther. 67*: 79 (1995).
49. P. R. Ortiz de Montellano, Oxygen activation and transfer, *Cytochrome P-450: Structure, Mechanism, and Biochemistry* (P. R. Ortiz de Montellano, ed.), Plenum Press, New York, 1986, p. 217.
50. F. P. Guengerich, Human cytochrome P450 enzymes, *Cytochrome P450: Structure, Mechanism and Biochemistry*, 2nd ed. (P. R. Ortiz de Montellano, ed.) Plenum Press, New York, 1995, p. 473.
51. A. M. H. Brodie, Aromatase, its inhibitors and their use in breast cancer treatment, *Pharmacol. Ther. 60*: 501 (1993).
52. Y. Aoyama, Y. Yoshida, T. Nishino, H. Katsuki, U. S. Maitra, V. P. Mohan, and D. B. Sprinson, Isolation and characterization of an altered cytochrome P-450 from a yeast mutant defective in lanosterol 14α-demethylation, *J. Biol. Chem. 262*: 14260 (1987).
53. N. Sonino, The use of ketoconazole as an inhibitor of steroid production, *N. Engl. J. Med. 317*: 812 (1987).
54. S. J. Baldwin, J. C. Bloomer, G. J. Smith, A. D. Ayrton, S. E. Clarke, and R. J. Chenery, Ketoconazole and sulphaphenazole as the respective selective inhibitors of P4503A and 2C9, *Xenobiotica 25*: 261 (1995).
55. J. E. Casida, Mixed function oxidase involvement in the biochemistry of insecticide synergists, *J. Agric. Food Chem. 18*: 753 (1970).
56. G. J. Hammons, W. L. Alworth, N. E. Hopkins, F. P. Guengerich, and F. F. Kadlubar, 2-Ethynylnaphthalene as a mechanism-based inactivator of the cytochrome P-450 catalyzed *N*-oxidation of 2-naphthylamine, *Chem. Res. Toxicol. 2*: 367 (1989).
57. A. Viaje, J. Y. L. Lu, N. E. Hopkins, A. N. Nettikumara, J. DiGiovanni, W. L. Alworth, and T. J. Slaga, Inhibition of the binding of 7,12-dimethylbenz[*a*]anthracene and benzo[*a*]-

pyrene to DNA in mouse skin epidermis by 1-ethynylpyrene, *Carcinogenesis 11*: 1139 (1990).

58. N. E. Davidson, P. A. Egner, and T. W. Kensler, Transcriptional control of glutathione *S*-transferase gene expression by the chemoprotective agent 5-(2-pyrazinyl)-4-methyl-1,2-dithiole-3-thione (oltipraz) in rat liver, *Cancer Res. 50*: 2251 (1990).
59. Y. Zhang, P. Talalay, C. G. Cho, and G. H. Posner, A major inducer of anticarcinogenic protective enzyme from broccoli: Isolation and elucidation of structure, *Proc. Natl. Acad. Sci. USA 89*: 2399 (1992).
60. S. Langouët, B. Coles, F. Morel, L. Becquemont, P. H. Beaune, F. P. Guengerich, B. Ketterer, and A. Guillouzo, Inhibition of CYP1A2 and CYP3A4 by oltipraz results in reduction of aflatoxin B_1 metabolism in human hepatocytes in primary culture, *Cancer Res. 55*: 5574 (1995).
61. L. Koymans, N. P. E. Vermeulen, S. A. B. E. van Acker, J. M. te Koppele, J. J. P. Heykants, K. Lavrijsen, W. Meuldermans, and G. M. Donné-Op den Kelder, A predictive model for substrates of cytochrome P450–debrisoquine (2D6), *Chem. Res. Toxicol. 5*: 211 (1992).
62. T. Wolff, L. M. Distlerath, M. T. Worthington, J. D. Groopman, G. J. Hammons, F. F. Kadlubar, R. A. Prough, M. V. Martin, and F. P. Guengerich, Substrate specificity of human liver cytochrome P-450 debrisoquine 4-hydroxylase probed using immunochemical inhibition and chemical modeling, *Cancer Res. 45*: 2116 (1985).
63. G. Strobl and T. Wolff, Structural models for inhibitors and substrates of cytochrome P-450IID6 based on molecular modeling analysis, Abstracts 7th International Conference: Biochemistry and Biophysics of Cytochrome P-450, Moscow, 28 July–2 August, 1991.
64. M. Murray and C. F. Wilkinson, Interactions of nitrogen heterocycles with cytochrome P-450 and monooxygenase activity, *Chem. Biol. Interact. 50*: 267 (1984).
65. W. R. Brian, P. K. Srivastava, D. R. Umbenhauer, R. S. Lloyd, and F. P. Guengerich, Expression of a human liver cytochrome P-450 protein with tolbutamide hydroxylase activity in *Saccharomyces cerevisiae, Biochemistry 28*: 4993 (1989).
66. J. O. Miners, K. J. Smith, R. A. Robson, M. E. McManus, M. E. Veronese, and D. J. Birkett, Tolbutamide hydroxylation by human liver microsomes: Kinetic characterisation and relationship to other cytochrome P-450 dependent xenobiotic oxidations, *Biochem. Pharmacol. 37*: 1137 (1988).
67. J. S. H. Yoo, R. J. Cheung, C. J. Patten, D. Wade, and C. S. Yang, Nature of *N*-nitrosodimethylamine demethylase and its inhibitors, *Cancer Res. 47*: 3378 (1987).
68. G. Danan, V. Descatoire, and D. Pessayre, Self-induction by erythromycin of its own transformation into a metabolite forming an inactive complex with reduced cytochrome P-450, *J. Pharmacol. Exp. Ther. 218*: 509 (1981).
69. K. H. Jönsson and B. Lindeke, Cytochrome P-455 nm complex formation in the metabolism of phenylalkylamines. XII. Enantioselectivity and temperature dependence in microsomes and reconstituted cytochrome P-450 systems from rat liver, *Chirality 4*: 469 (1992).
70. M. Murray, C. F. Wilkinson, C. Marcus, and C. E. Dubé, Structure–activity relationships in the interactions of alkoxymethylenedioxybenzene derivatives with rat hepatic microsomal mixed-function oxidases in vivo, *Mol. Pharmacol. 24*: 129 (1983).
71. P. R. Ortiz de Montellano and M. A. Correia, Suicidal destruction of cytochrome P-450 during oxidative drug metabolism, *Annu. Rev. Pharmacol. Toxicol. 23*: 481 (1983).
72. J. R. Halpert, C. Balfour, N. E. Miller, and L. S. Kaminsky, Dichloromethyl compounds as mechanism-based inactivators of rat liver cytochromes P-450 in vitro, *Mol. Pharmacol. 30*: 19 (1986).
73. N. E. Miller and J. Halpert, Analogues of chloramphenicol as mechanism-based inactivators of rat liver cytochrome P-450: Modifications of the propanediol side chain, the *p*-nitro group, and the dichloromethyl moiety, *Mol. Pharmacol. 29*: 391 (1986).
74. T. L. Macdonald, K. Zirvi, L. T. Burka, P. Peyman, and F. P. Guengerich, Mechanism of cytochrome P-450 inhibition by cyclopropylamines, *J. Am. Chem. Soc. 104*: 2050 (1982).

75. R. P. Hanzlik and R. H. Tullman, Suicidal inactivation of cytochrome P-450 by cyclopropylamines. Evidence for cation–radical intermediates, *J. Am. Chem. Soc. 104*: 2048 (1982).
76. A. Bondon, T. L. Macdonald, T. M. Harris, and F. P. Guengerich, Oxidation of cycloalkylamines by cytochrome P-450. Mechanism-based inactivation, adduct formation, ring expansion, and nitrone formation, *J. Biol. Chem. 264*: 1988 (1989).
77. O. Augusto, H. S. Beilan, and P. R. Ortiz de Montellano, The catalytic mechanism of cytochrome P-450: Spin-trapping evidence for one-electron substrate oxidation, *J. Biol. Chem. 257*: 11288 (1982).
78. R. H. Böcker and F. P. Guengerich, Oxidation of 4-aryl- and 4-alkyl-substituted 2,6-dimethyl-3,5-bis(alkoxycarbonyl)-1,4-dihydropyridines by human liver microsomes and immunochemical evidence for the involvement of a form of cytochrome P-450, *J. Med. Chem. 29*: 1596 (1986).
79. F. P. Guengerich and R. H. Böcker, Cytochrome P-450-catalyzed dehydrogenation of 1,4-dihydropyridines, *J. Biol. Chem. 263*: 8168 (1988).
80. P. R. Ortiz de Montellano, H. S. Beilan, and K. L. Kunze, *N*-Alkylprotoporphyrin IX formation in 3,5-dicarbethoxy-1,4-dihydrocollidine-treated rats: Transfer of the alkyl group from the substrate to the porphyrin, *J. Biol. Chem. 256*: 6708 (1981).
81. D. Lukton, J. E. Mackie, J. S. Lee, G. S. Marks, and P. R. Ortiz de Montellano, 2,2-Dialkyl-1,2-dihydroquinolines: Cytochrome P-450 catalyzed *N*-alkylporphyrin formation, ferrochelatase inhibition, and induction of 5-aminolevulinic acid synthase activity, *Chem. Res. Toxicol. 1*: 208 (1988).
82. R. A. Stearns and P. R. Ortiz de Montellano, Inactivation of cytochrome P-450 by a catalytically generated cyclobutadiene species, *J. Am. Chem. Soc. 107*: 234 (1985).
83. C. L. Meschter, B. A. Mico, M. Mortillo, D. Feldman, W. A. Garland, J. A. Riley, and L. S. Kaufman, A 13-week toxicologic and pathologic evaluation of prolonged cytochromes P450 inhibition by 1-aminobenzotriazole in male rats, *Fundam. Appl. Toxicol. 22*: 369 (1994).
84. B. S. S. Masters, The role of NADPH–cytochrome *c* (P-450) reductase in detoxication, *Enzymatic Basis of Detoxication*, Vol. I (W. B. Jakoby, ed.), Academic Press, New York, 1980, p. 183.
85. T. Yoshida, S. Takahashi, and G. Kikuchi, Partial purification and reconstitution of the heme oxygenase system from pig spleen microsomes, *J. Biochem. 75*: 1187 (1974).
86. Y. Yasukochi and B. S. S. Masters, Some properties of a detergent-solubilized NADPH–cytochrome *c*(cytochrome P-450) reductase purified by biospecific affinity chromatography, *J. Biol. Chem. 251*: 5337 (1976).
87. D. G. Tew, Inhibition of cytochrome P450 reductase by the diphenyliodonium cation. Kinetic analysis and covalent modifications, *Biochemistry 32*: 10209 (1993).
88. B. Clement, M. Weide, and D. M. Ziegler, Inhibition of purified and membrane-bound flavin-containing monooxygenase 1 by (*N,N*-dimethylamino)stilbene carboxylates, *Chem. Res. Toxicol. 9*: 599 (1996).
89. R. N. Hines, J. R. Cashman, R. M. Philpot, D. E. Williams, and D. M. Ziegler, The mammalian flavin-containing monooxygenases: Molecular characterization and regulation of expression, *Toxicol. Appl. Pharmacol. 125*: 1 (1994).
90. J. R. Cashman, Monoamine oxidase and flavin-containing monooxygenases, *Biotransformation*, Vol. 3, *Comprehensive Toxicology* (F. P. Guengerich, ed.), Elsevier Science, Oxford, 1997, p. 69.
91. D. Petersen and R. Lindahl, Aldehyde dehydrogenase, *Biotransformation*, Vol. 3, *Comprehensive Toxicology* (F. P. Guengerich, ed.), Elsevier Science, Oxford, 1997, p. 97.
92. H. Yamazaki, Y. Inui, C-H. Yun, M. Mimura, F. P. Guengerich, and T. Shimada, Cytochrome P450 2E1 and 2A6 enzymes as major catalysts for metabolic activation of *N*-nitrosodialkylamines and tobacco-related nitrosamines in human liver microsomes, *Carcinogenesis 13*: 1789 (1992).

93. B. W. Hart and M. D. Faiman, Bioactivation of *S*-methyl *N,N*-diethylthiolcarbamate to *S*-methyl *N,N*-diethylthiolcarbamate sulfoxide, *Biochem. Pharmacol. 46*: 2285 (1993).
94. A. Madan, A. Parkinson, and M. D. Faiman, Identification of the human and rat P450 enzymes responsible for the sulfoxidation of *S*-methyl *N,N*-diethylthiolcarbamate (DETC-Me): The terminal step in the bioactivation of disulfiram, *Drug Metab. Dispos. 23*: 1153 (1995).
95. T. G. Flynn, Aldo-ketoreductases: Structure, mechanism, and function, *Biotransformation*, Vol. 3, *Comprehensive Toxicology* (F. P. Guengerich, ed.), Elsevier Science, Oxford, 1997, p. 133.
96. I. Wells and L. J. Marnett, Acetylation of prostaglandin endoperoxide synthase by *N*-acetylimidazole: Comparison to acetylation by aspirin, *Biochemistry 31*: 9520 (1992).
97. W. Xie, J. G. Chipman, D. L. Robertson, R. L. Erikson, and D. L. Simmons, Expression of a mitogen-responsive gene encoding prostaglandin synthase is regulated by mRNA splicing, *Proc. Natl. Acad. Sci. USA 88*: 2692 (1991).
98. D. A. Kujubu, B. S. Fletcher, B. C. Varnum, R. W. Lim, and H. R. Herschman, TIS10, a phorbol ester tumor promoter-inducible mRNA from Swiss 3T3 cells, encodes a novel prostaglandin synthase/cyclooxygenase homologue, *J. Biol. Chem. 266*: 12866 (1991).
99. M. K. O'Banion, V. D. Winn, and D. A. Young, cDNA cloning and functional activity of a glucocorticoid-regulated inflammatory cyclooxygenase, *Proc. Natl. Acad. Sci. USA 89*: 4888 (1992).
100. K. Seibert, Y. Zhang, K. Leahy, S. Hauser, J. Masferrer, W. Perkins, L. Lee, and P. Isakson, Pharmacological and biochemical demonstration of the role of cyclooxygenase 2 in inflammation and pain, *Proc. Natl. Acad. Sci. USA 91*: 12013 (1994).
101. E. A. Meade, W. L. Smith, and D. L. DeWitt, Differential inhibition of prostaglandin endoperoxide synthase (cyclooxygenase) isozymes by aspirin and other non-steroidal anti-inflammatory drugs, *J. Biol. Chem. 268*: 6610 (1993).
102. D. B. Reitz, J. J. Li, M. B. Norton, E. J. Reinhard, J. T. Collins, G. D. Anderson, S. A. Gregory, C. M. Koboldt, W. E. Perkins, K. Weibert, and P. C. Isakson, Selective cyclooxygenase inhibitors: Novel 1,2-diarylcyclopentenes are potent and orally active COX-2 inhibitors, *J. Med. Chem. 37*: 3878 (1994).
103. J. K. Gierse, J. J. Mcdonald, S. D. Hauser, S. H. Rangwala, C. M. Koboldt, and K. Seibert, A single amino acid difference between cyclooxygenase-1 (COX-1) and -2 (COX-2) reverses the selectivity of COX-2 specific inhibitors, *J. Biol. Chem. 271*: 15810 (1996).
104. D. M. Quinn, Esterases of the α/β hydrolase fold family, *Biotransformation*, Vol. 3, *Comprehensive Toxicology* (F. P. Guengerich, ed.), Elsevier Science, Oxford, 1997, p. 243.
105. G. M. Lacourciere and R. N. Armstrong, Microsomal and soluble epoxide hydrolases are members of the same family of C–X bond hydrolase enzymes, *Chem. Res. Toxicol. 7*: 121 (1994).
106. G. M. Lacourciere and R. N. Armstrong, The catalytic mechanism of microsomal epoxide hydrolase involves an ester intermediate, *J. Am. Chem. Soc. 115*: 10466 (1993).
107. F. Oesch, Purification and specificity of a human microsomal epoxide hydratase, *Biochem. J. 139*: 77 (1974).
108. B. D. Hammock, D. F. Grant, D. H. Storms, Epoxide hydrolases, *Biotransformation*, Vol. 3, *Comprehensive Toxicology* (F. P. Guengerich, ed.), Elsevier Science, Oxford, 1997, p. 283.
109. W. B. Jakoby and W. H. Habig, Glutathione transferases, *Enzymatic Basis of Detoxication*, Vol. 2 (W. B. Jakoby, ed.), Academic Press, New York, 1980, p. 63.
110. R. N. Armstrong, Glutathione *S*-transferases: Reaction mechanism, structure, and function, *Chem. Res. Toxicol. 4*: 131 (1991).
111. G. Litwack, B. Ketterer, and I. M. Arias, Ligandin: A hepatic protein which binds steroids, bilirubin, carcinogens, and a number of exogenous anions, *Nature 234*: 466 (1971).
112. P. C. Simons and D. L. Vander Jagt, Purification of glutathione *S*-transferases from human liver by glutathione-affinity chromatography, *Anal. Biochem. 82*: 334 (1977).
113. G. F. Graminski, Y. Kubo, and R. N. Armstrong, Spectroscopic and kinetic evidence for the

thiolate anion of glutathione at the active site of glutathione *S*-transferase, *Biochemistry 28*: 3562 (1989).

114. A. E. P. Adang, J. Brussee, A. van der Gen, and G. J. Mulder, Inhibition of rat liver glutathione *S*-transferase isoenzymes by peptides stabilized against degradation by *g*-glutamyl transpeptidases, *J. Biol. Chem. 266*: 830 (1991).
115. G. F. Graminski, P. Zhang, M. A. Sesay, H. L. Ammon, and R. N. Armstrong, Formation of the 1-(*S*-glutathionyl)-2,4,6-trinitrocyclohexadienate anion at the active site of glutathione *S*-transferase: Evidence for enzymic stabilization of *s*-complex intermediates in nucleophilic aromatic substitution reactions, *Biochemistry 28*: 6252 (1989).
116. R. N. Armstrong, Glutathione transferases, *Biotransformation*, Vol. 3, *Comprehensive Toxicology* (F. P. Guengerich, ed.), Elsevier Science, Oxford, 1997, p. 307.
117. R. M. Katusz, B. Bono, and R. F. Colman, Affinity labeling of Cys[111] of glutathione *S*-transferase, isoenzyme 1-1, by *S*-(4-bromo-2,3-dioxobutyl)glutathione, *Biochemistry 31*: 8984 (1992).
118. J. H. T. M. Ploemen, W. W. Johnson, S. Jespersen, D. Vanderwall, B. van Ommen, J. van der Greef, P. J. van Bladeren, and R. N. Armstrong, Active-site tyrosyl residues are targets in the irreversible inhibition of a class mu glutathione transferase by 2-(*S*-glutathionyl)-3,5,6-trichloro-1,4-benzoquinone, *J. Biol. Chem. 269*: 26890 (1994).
119. M. A. McTigre, D. R. Williams, and J. A. Tainer, Crystal structures of a schistomal drug vaccine target: Glutathione *S*-transferase from *Schistosoma japonica* and its complex with the leading antischistosomal drug praziquantel, *J. Mol. Biol. 246*: 21 (1995).
120. M. W. Duffel, Sulfotransferases, *Biotransformation*, Vol. 3, *Comprehensive Toxicology* (F. P. Guengerich, ed.), Elsevier Science, Oxford, 1997, p. 365.
121. B. Burchell, K. McGurk, C. H. Brierly, and D. J. Clarke, UDP-glucuronosyltransferases, *Biotransformation*, Vol. 3, *Comprehensive Toxicology* (F. P. Guengerich, ed.), Elsevier Science, Oxford, 1997, p. 401.
122. D. Noort, M. W. H. Coughtrie, B. Burchell, G. A. van der Morel, J. H. van Boom, A. van der Gen, and G. J. Mulder, Inhibition of UDP-glucuronosyltransferase activity by possible transition state analogues in rat liver microsomes, *Eur. J. Biochem. 281*: 170 (1990).
123. A. Radominska, P. Paul, S. Treat, H. Towbin, C. Pratt, J. Little, J. Magdalou, R. Lester, and R. Drake, Photoaffinity labeling for evaluation of uridinyl analogs as specific inhibitors of rat liver UDP-glucuronosyltransferase, *Biochim. Biophys. Acta 1205*: 336 (1994).
124. J. Thomassin and T. R. Tephly, Photoaffinity labeling of rat liver microsomal morphine UDP-glucuronosyltransferase by [^{3}H]flunitrazepam, *Mol. Pharmacol. 38*: 294 (1990).
125. R. R. Drake, Y. Igari, R. Lester, A. D. Elbein, and A. Radominska, Application of 5-azido-UDP-glucose and 5-azido-UDP-glucuronic acid photoaffinity probes for the determination of the active site orientation of microsomal UDP-glucosyltransferases and UDP-glucuronosyltransferases, *J. Biol. Chem. 267*: 11360 (1992).
126. A. Li, ed., *Drug Interactions: Scientific and Regulatory Perspectives* (Adv. Pharmacol. Series) Academic Press, San Diego, 1997.
127. R. G. Knodell, D. Browne, G. P. Gwodz, W. R. Brian, and F. P. Guengerich, Differential inhibition of human liver cytochromes P-450 by cimetidine, *Gastroenterology 101*: 1680 (1991).
128. C.-H. Yun, R. A. Okerholm, and F. P. Guengerich, Oxidation of the antihistaminic drug terfenadine in human liver microsomes: Role of cytochrome P450 3A(4) in *N*-dealkylation and *C*-hydroxylation, *Drug Metab. Dispos. 21*: 403 (1993).
129. K. T. Kivistö, P. J. Neuvonen, and U. Klotz, Inhibition of terfenadine metabolism: Pharmacokinetic and pharmacodynamic consequences, *Clin. Pharmacokinet. 27*: 1 (1994).
130. R. L. Woosley, Y. Chen, J. P. Freiman, and R. A. Gillis, Mechanism of the cardiotoxic actions of terfenadine, *JAMA 269*: 1532 (1993).
131. P. R. Ortiz de Montellano, K. L. Kunze, G. S. Yost, and B. A. Mico, Self-catalyzed destruction of cytochrome P-450: Covalent binding of ethynyl sterols to prosthetic heme, *Proc. Natl. Acad. Sci. USA 76*: 746 (1979).

132. F. P. Guengerich, Oxidation of 17α-ethynylestradiol by human liver cytochrome P-450, *Mol. Pharmacol. 33*: 500 (1988).
133. F. P. Guengerich, Mechanism-based inactivation of human liver cytochrome P-450 IIIA4 by gestodene, *Chem. Res. Toxicol. 3*: 363 (1990).
134. C. Jung-Hoffmann and H. Kuhl, Pharmacokinetics and pharmacodynamics of oral contraceptive steroids: Factors influencing steroid metabolism, *Am. J. Obstet. Gynecol. 163*: 2183 (1990).

9

Metabolic Drug Interactions

Wayne D. Hooper

The University of Queensland at Royal Brisbane Hospital, Brisbane, Queensland, Australia

I. INTRODUCTION

A. Background and Clinical Significance

There is a vast body of literature on the subject of drug interactions which, as Gibaldi (1) has pointed out, exceeds that on almost any other topic in the discipline of clinical pharmacology. This probably stems from the recognition, several decades ago, that interactions do occur between coadministered drugs and that such interactions could have important, even lethal, consequences. In an attempt to make available to medical practitioners and pharmacists the great body of information on drug interactions that has been accumulated, reviewers have compiled extensive listings of interactions, sometimes in a quite uncritical manner. However it is now recognized that much of this information has little clinical relevance, because many alleged interactions are based on inadequate data, occur between drugs unlikely to be coadministered, or do not result in significant therapeutic or toxicological consequences. There is, moreover, now a sufficient understanding of the mechanisms underlying drug interactions that many interactions are predictable and, as Stockley (12) and others have pointed out, most are manageable. Indeed, drugs are quite often administered in combinations specifically to take advantage of known interactions to obtain enhanced therapeutic effects. Examples are readily found with antihypertensive drugs, antibiotics, and the combination of levodopa with dopa decarboxylase inhibitors.

Undoubtedly the proliferation of reports of drug interactions was stimulated by the era of burgeoning polypharmacy through the past several decades. Epidemiological studies show that the rate of occurrence of adverse drug interactions increases sharply as more drugs are administered concurrently. In the 1970s surveys based on more than 10,000 hospital patients showed adverse interactions occurred in 4% of patients receiving five or fewer drugs, but in 54% of patients receiving 16–20 drugs (3). Although there is nowadays much lower incidence of uncritical polypharmacy, many patients still receive multiple drugs and, therefore, are exposed to possible interactions. Another study from the 1970s showed that whereas 113 of 2422 patients (4.7%) were taking drug combinations that would interact, only 7 patients (0.3%) showed evidence of interactions (4). The enormous intersubject variability in drug disposition ensures that interactions occurring in some subjects will not necessarily occur in others (5).

B. Concepts and Definitions

The term *drug interactions* is usually understood to refer to drug–drug interactions, although it should be remembered that the term, unqualified, can be taken to include interactions between food constituents or environmental factors and drugs, or even interferences by drugs in clinical laboratory tests. In this chapter the meaning is restricted to drug–drug interactions, but even this requires further qualification because the literal connotation of *interactions* implies reciprocal effects. Many drug interactions, as the term is commonly used, involve an effect of one drug on the action or disposition of another, with no recognizable reciprocal effects.

According to Thomas (6) a drug interaction is "considered to occur when the effects of giving two or more drugs are qualitatively or quantitatively different from the simple sum of the observed effects when the same doses of the drugs are given separately." Implications of this statement are that (a) one drug may either increase or decrease the activity of another in a purely quantitative manner, or (b) the effects of two drugs given concurrently may bear no relation to the properties of the drugs used individually (i.e., there are qualitative effects), and (c) both the sequence of administration and the time interval between taking two drugs may be important determinants of the interaction that occurs.

Most attention is paid to drug interactions that have adverse or undesired consequences, whether this results from potentiation of an unwanted drug effect, diminution of an intended effect, or a qualitative change in the effect of a drug. This emphasis is reflected in the literature in which, over many years, treatises on drug interactions have dealt predominantly or exclusively with the hazards of medication (7) or with adverse drug interactions (2). It is worth stressing that beneficial interactions also occur and, as noted earlier, these may be incorporated into therapeutic practice and consciously utilized.

II. MECHANISMS OF DRUG INTERACTIONS

Numerous mechanisms underlie drug interactions, and these may be systematized in a variety of ways. One of the simplest, but still useful classifications, is in terms of interactions resulting from pharmaceutical, pharmacodynamic, or pharmacokinetic mechanisms, although some interactions do not seem to fit neatly into these categories.

A. Pharmaceutical

Pharmaceutical interactions are chemically based and occur when one drug is physically or chemically incompatible with another. Most often these interactions manifest in the context of parenteral formulations, such as the addition of drugs to intravenous infusion solutions. Stockley (2) has argued that these events should be considered as pharmaceutical incompatibilities, rather than as drug interactions. They are usually predictable and almost always avoidable. More subtly, drugs given orally can sometimes interact within the gastrointestinal tract. Examples include the complexation of tetracyclines or fluoroquinolone antibiotics with metal ions, the adsorption of digoxin or warfarin onto cholestyramine, and the reduction in absorption of tetracyclines caused by administration of antacids. These interactions can usually be avoided by the separate administration of the interacting drugs.

B. Pharmacodynamic

Many of the drug interactions of which use is made in clinical practice involve effects at receptor sites; but, on the other hand, some of the most clinically important adverse reactions also result from pharmacodynamic interactions. Pharmacologists usually view these interactions as synergistic or as antagonistic. Synergistic (additive) effects occur, for example, with coadministered central nervous system (CNS) depressants (including alcohol, which should be considered a drug in this context). Although effects that are purely additive are not embraced by the strict definition of drug interactions, given in the foregoing, these effects should nevertheless be treated as such because of their potential clinical importance. Other synergistic interactions include those between a diverse range of drugs, including tetracyclines, clofibrate, and estrogens, with warfarin, leading to increased anticoagulation (probably involving more than one mechanism), and the increased incidence of cardiac arrhythmias that results from the combination of β-adrenergic antagonists with verapamil. The examples already cited involve interactions at the same receptor sites or, as in some combinations of CNS-depressant drugs, at different sites. Antagonistic interactions occurring (almost by definition) at the same site are best exemplified by the effect of naloxone in reversing the effects of opiates, and the use of physostigmine to counteract arrhythmias resulting from overdosage with tricyclic antidepressants.

Pharmacodynamic interactions can also occur by indirect mechanisms. Drugs that cause gastrointestinal lesions, including ulceration, such as aspirin and the nonsteroidal anti-inflammatory drugs, may provide a focus for bleeding if an anticoagulant is coadministered. An associated problem is the inhibition of platelet aggregation by salicylates and other drugs, leading to enhanced tendency to bleeding when anticoagulants are administered.

An interesting theoretical treatment of pharmacodynamic drug interactions has been provided by Pöch (8).

C. Pharmacokinetic

Pharmacokinetic drug interactions can occur during any of the processes (absorption, distribution, metabolism, and excretion) that contribute to a drug's pharmacokinetic profile; the metabolic interactions, which are the major topic of this chapter, are considered in the next section.

At least four mechanisms can contribute to interactions affecting drug absorption. Chemical interactions, such as chelation and complexation, were noted earlier as pharmaceutical interactions, and may cause reductions in oral bioavailability if inappropriate combinations are administered without sufficient temporal separation. Second, drugs that alter gastrointestinal motility may influence the absorption of coadministered drugs. For example, drugs with anticholinergic effects decrease, and drugs such as metoclopramide increase the rate of gastric emptying. Although such effects may cause alterations in the rate of absorption of other orally administered drugs (not always in a predictable manner; 2,5), the extent of absorption is usually not significantly affected, and the clinical consequences are, therefore, generally unimportant. Third, drugs, such as antacids and H_2-receptor antagonists, alter the pH of gastrointestinal fluids and may consequently affect the solubility and absorption of ionizable drugs. Finally, antibiotics that perturb the bacterial flora in the gastrointestinal tract may affect the absorption of any drugs subject to metabolism by bacterial enzymes.

The major distributional process that may contribute to drug interactions is binding to plasma proteins. Although much has been written on this subject, it has been recognized for some time that even if many drugs displace others from plasma proteins, it is only in certain limited situations that these displacements give rise to significant dispositional consequences (1,2,5,9,10).

The renal excretion of drugs or their metabolites may be affected by a coadministered drug in various ways. Some acidic drugs lower urine pH, whereas some antacids and bases cause an increase in the pH of urine. These changes will have some effect on the excretion of other ionizable compounds that have appreciable renal clearances. Some basic drugs for which renal clearance is appreciably increased by aciduria are amphetamines, tricyclic antidepressants, and antihistamines. Acidic drugs for which renal clearance increases with increasing urine pH include salicylic acid, phenobarbital, and nitrofurantoin.

Drugs that compete for renal transport mechanisms (tubular secretion) show interaction effects that may be significant. Examples include the well-known reduction in renal clearances of penicillins and indomethacin by probenecid, and of methotrexate by salicylates and nonsteroidal anti-inflammatory drugs.

D. Multiple and Indeterminate Mechanisms

Some drugs may interact simultaneously at more than one site, making it difficult to attribute a single mechanism. Perhaps the best example is the ubiquitous aspirin, which may interfere with the absorption, plasma protein binding, or renal tubular secretion of other acidic drugs. In addition, it may potentiate the effects of oral anticoagulants by affecting bleeding time, capillary fragility, platelet adhesiveness or synthesis of clotting factors, and may even cause gastrointestinal ulceration, predisposing to bleeding (5). Despite our extensive understanding of mechanisms of drug–drug interactions, there remain a moderate number of miscellaneous interactions for which mechanisms are indeterminate (6).

III. METABOLIC DRUG INTERACTIONS

Many of the clinically important drug–drug interactions result from perturbations of drug metabolism, involving either induction or inhibition of metabolizing enzymes, principally the cytochrome P450s. The natures of induction and inhibition phenomena are described in the preceding chapters. The present discussion, therefore, is limited to consideration of their roles in mediating drug interactions.

A. Induction Based

The phenomenon of induction of cytochrome P450s captured the interest of pharmacologists even before the recognition of the nature of the P450 enzymes (11). Its contributions as a mediator of metabolic drug interactions have likewise been long recognized.

In general terms, two major scenarios arise with induction-based interactions. The most common situation is that a drug, for which metabolism is induced by another, shows increased metabolic clearance, leading to reduced therapeutic efficacy. The other obvious manifestation is that a drug that is metabolically activated, for example, to yield a toxic metabolite, may show increased toxicity if the relevant pathway is induced.

The implications and consequences from induction-based interactions may differ substantially in relation to temporal factors, especially the sequence and duration of administration of the interacting drugs. For example, the classic inducing agent phenobarbital accelerates the metabolism of many drugs, including warfarin. Addition of phenobarbital to the treatment of a patient already stabilized on a warfarin therapeutic regimen can lead, after several days, to a requirement for increased warfarin dosage—to the extent of a doubling or even a quadrupling of the warfarin dose. Conversely, cessation of phenobarbital administration to a patient stabilized on both drugs may precipitate the hazardous situation of warfarin overdose, again after several days of deinduction. The magnitude of this effect may likewise be substantial, indeed sufficient to expose the patient to risk of fatal hemorrhage if the warfarin dose is not promptly reduced. It is the gradual onset of induction-mediated interactions that can render them particularly insidious and likely to go unnoticed by the unwary prescriber.

Some of the best-recognized and most widely studied drug interactions occur among antiepileptic drugs, which are frequently administered in combinations. Some combinations involve true interactions (reciprocal effects; 5), and some of the consequences, therefore, are quite complex. Detailed discussions are given in a recent treatise on epileptic drugs (12). Close monitoring of plasma concentrations of all drugs following changes in drug regimens, for which facilities are widely available, should enable the consequences of these interactions to be recognized and minimized.

Enormous advances have been made over recent years in characterizing the isoforms of cytochrome P450 enzymes and in establishing which isoforms catalyze particular biotransformation reactions. There is also reasonable definition for which isoforms are induced by particular inducing agents. This increased knowledge has made the prediction of induction-based interactions much more reliable. For example, Pollock (13) has recently reviewed the metabolism of psychotropic drugs and has provided listings of those drugs for which metabolism is known to be catalysed by P4503A4, P4502D6, P4502C19 or P4501A2. Specific inducers of these isozymes are given by Spatzenegger and Jaeger (14), who have also provided extensive listings of the substrates known to be metabolized by particular isozymes. Such compilations are expanding rapidly, and their applicability to the prediction of interactions that may result from induction is clear. Again, however, the clinical consequences of most of these interactions will generally be minor, unless the affected drug has a low therapeutic index.

B. Inhibition Based

Just as drug interactions may result from enzyme induction, they may also arise as a consequence of enzyme inhibition. Many drugs inhibit the metabolism of other drugs in humans. There are similarities and contrasts between interactions caused by inhibition and those caused by induction. Obviously, most inhibitors will cause an increase in the plasma concentrations of affected drugs, which may enhance efficacy if the concentration remains within the "therapeutic range," but which potentially result in drug toxicity. The direct consequences of inhibitory interactions may be more clinically severe than those from induction, which often lead to only diminished efficacy. Moreover, the time delays of at least several days that are typical of induction effects are not seen with inhibitory effects, which may be fully manifested soon after the administration of the inhibitor and, therefore, have the potential to precipitate sudden problems.

Several different mechanisms mediate inhibition-based interactions. Foremost among these is competitive inhibition (substrate competition), but interference with drug transport, alteration of the expression or conformation of the P450 enzyme, and interference with the energy (depletion of glycogen storage) or cofactor (uncoupling of the NADPH–enzyme complex) supply have also been recognized.

The reported inhibition-based interactions are so numerous that it is difficult to select representative examples. Many involve anticoagulants, antiarrhythmics, or anticonvulsants. Although those interactions causing elevation in anticonvulsant concentrations may be debilitating, they are, unlike those involving the other two drug classes, seldom life-threatening. Thus, cimetidine, which inhibits the P450-catalyzed metabolism of drugs in each of these three classes, poses particular hazards when combined with warfarin and other anticoagulants (15), quinidine (16), or lidocaine (17). Another hazardous situation arises with oral hypoglycemic agents, such as tolbutamide, the metabolism of which is inhibited by several drugs, which can under certain circumstances give rise to hypoglycemic crisis (18).

Although it may seem self-evident, it is worth emphasizing that the consequences of enzyme inhibition may be amplified for drugs such as phenytoin for which metabolism is capacity-limited (saturable). Extreme consequences may also be encountered, however, for high-clearance drugs. Inhibition of metabolizing enzymes for such drugs (of which there are numerous examples, including lidocaine and many of the β-adrenergic blockers) may result in a marked increase in the systemic availability from an oral dose, which again can have severe consequences.

Several enzymes other than the cytochromes P450 are also involved in drug metabolism and may be involved in inhibitory interactions. Some of these effects are used to therapeutic advantage, such as the dopa decarboxylase inhibitors (e.g., carbidopa and benzserazide) that reduce the peripheral metabolism of levodopa and increase its efficacy in treating parkinsonism. Other inhibitory effects can result in adverse consequences, one of the best examples of which (though not perhaps in a strict sense a drug–drug interaction) is the hypertensive reaction that occurs in patients who eat foods, such as cheese, with a high tyramine content while taking a monoamine oxidase inhibitor (1,9). Mention should also be made of the more recently recognized strongly inhibitory effects of compounds occurring in grapefruit juice (19).

Although in historical context the ''classic'' inhibitors of cytochrome P450s were proadifen (SK&F 525-A) and piperonyl butoxide, the inhibitors most likely to come to mind nowadays would include cimetidine and quinidine. The capacity of cimetidine to inhibit the oxidative metabolism of many other drugs was discovered early in its use, and even by 1982, was sufficiently studied to warrant a major review (20,21). Just as we noted that considerable understanding of the isoform specificity of P450 inducers is now available, substantial detailed knowledge is also at hand concerning the specificity of inhibitors. Thus, for example, P450 1A is inhibited by ciprofloxacin, enoxacin, norfloxacin, and several related drugs (22); P450 2B by chloramphenicol (23) and secobarbital (24); P450 2C by cimetidine (25) and cannabidiol (26); P450 2D6 by many compounds, including quinidine, fluoxetine, clomipramine, and sertraline (27); and P450 3A4 by nifedipine (28), cimetidine (25), verapamil, erythromycin, ethinyl estradiol (28), and many other drugs. Tabulations of inhibitors may be found in Spatzenegger and Jaeger (14) and Pollock (13), but the lists are growing rapidly. As with the enzyme inducers, this expanding detailed knowledge of the isozyme specificities of inhibitors, together with an appreciation of the isozymes catalyzing the metabolism of certain drugs, makes possible the prediction of many drug interactions.

Some of the interactions resulting from enzyme inhibition may have severe clinical consequences. For example, some of the fluoroquinolone antibiotics and erythromycin inhibit pathways of theophylline metabolism; after a 5-day course of erythromycin in a patient stabilized on theophylline, the plasma theophylline levels may increase about twofold (1). It is widely recognized that theophylline is a difficult drug to manage and that theophylline overdose may result in severe adverse events (29,30). A life-threatening interaction occurred when tamoxifen therapy was commenced in a patient stabilized on warfarin (31). The warfarin dose was decreased from 5 to 1 mg/day to maintain appropriate anticoagulation. Important interactions involving anticancer agents have only recently been recognized (32).

IV. STUDY OF DRUG INTERACTIONS

Much of the published information on drug interactions has been in the form of clinical case reports and anecdotal accounts, which are typically followed up with more systematic studies, both at a laboratory (in vitro) and a clinical (in vivo) level. It is not surprising that knowledge of drug interactions would arise in this manner, rather than through structured investigations, because it would be difficult to anticipate all of the combinations in which a new therapeutic agent might be used. One obvious difficulty that attends this situation is many reported interactions are based on inadequate data, and subsequent systematic studies may fail to confirm the initial report, or may reveal that it has limited therapeutic relevance or arises only in particular circumstances.

The power of in vitro methods to focus on specific mechanisms—for example, by working with purified P450 isoforms—renders such methodology very appealing. It may be true, however, that such methods are more effectively applied to confirming mechanisms of interactions observed clinically, than as screening tests to see whether an interaction occurs between two drugs. As Prescott (5) has stressed, not all interactions observed in vitro or even in animal studies will occur in humans. And, conversely, some interactions that do occur in humans, but that require the prior biotransformation of the administered agent to yield a metabolite that mediates the interaction, may not be observed in vitro if only the parent substance is studied. (An example of this may be the inhibition of theophylline metabolism by fluoroquinolone antibiotics, which is believed to involve the 4-oxo metabolite of the fluoroquinolones; 33).

There are specific situations in which the potential for drug interactions is studied prospectively and systematically, and nowadays at a relatively early stage during new drug development. Perhaps the best developed of these situations arises with antiepileptic drugs. Almost invariably, newly introduced antiepileptic agents will be used in combination with established drugs; hence, the potential for interactions to occur is both predictable and important to assess. These potential interactions are usually evaluated in specifically designed clinical studies (e.g., 34,35), either in healthy volunteers or treated patients (or both), during phases II and III of the clinical drug development program. This represents quite a substantial undertaking, as the new drug could be used in combination with any of perhaps five or six established drugs. Furthermore, it is difficult to design a single interaction study that yields reliable information about both potential interactions for any particular pair of drugs (i.e., effect of new drug on disposition of established drug, and vice versa), and separate studies may be required. Any potentially significant interaction shown in these clinical studies would nowadays be complemented with in vitro studies to establish or confirm the mechanism, and would also be followed up as part of the postmarketing surveillance program.

Whereas the antiepileptic drugs may be the best example to illustrate this process, there are other drug classes in which comparable situations arise. These include anticancer drugs, antibiotics (some combinations of which yield beneficial outcomes), and drugs used for the treatment of human immunodeficiency virus (HIV)-infected patients.

Despite all of the advances in knowledge and methodology in the study of drug interactions, the complexity of the topic dictates that even well-recognized interactions may be poorly understood, despite years of study. We recently added certain details to what is known of the mechanism by which valproate interferes with the metabolism of phenobarbital (36), an interaction that has been known for about 20 years.

V. SUMMARY AND SOURCES OF INFORMATION

In summary, we should stress that the topic of drug metabolism is an important one for clinicians, pharmacists, and researchers. Although the great volume of literature on the subject attests to its importance, a critical reading of this literature is essential. Clinically important drug interactions are known to occur with many drugs, and may arise through many mechanisms. In this chapter the mechanisms have been grouped under the headings pharmaceutical, pharmacodynamic, and pharmacokinetic; pharmacokinetic interactions may manifest at the various stages in drug disposition (absorption, distribution, biotransformation, and excretion). Our emphasis has been on the biotransformation interactions that result from enzyme induction or inhibition.

Of the references cited in this chapter, the reader is particularly referred to Stockley (2) and Thomas (6), the latter of which is unfortunately not widely available outside Australia. An excellent reading list is provided by Prescott (5). Similar to all topics covered in this book, this subject continues to expand rapidly, and up-to-date information will require access to computerized databases and literature-abstracting services, many of which are now freely available through the Internet.

REFERENCES

1. M. Gibaldi, *Biopharmaceutics and Clinical Pharmacokinetics*, Lea & Febiger, Philadelphia, 1991, p. 305.
2. I. H. Stockley, Drug Interactions: *A Source Book of Adverse Interactions, Their Mechanisms, Clinical Importance and Management*, 3rd ed., Blackwell, London, 1994.
3. F. E. May, R. B. Stewart, and L. E. Cluff, Drug interactions and multiple drug administration, *Clin. Pharmacol. Ther. 22*: 323 (1977).
4. W. H. Puckett and J. A. Visconti, An epidemiological study of the clinical significance of drug–drug interaction in a private community hospital, *Am. J. Hosp. Pharm. 28*: 247 (1971).
5. L. F. Prescott, Clinically important drug interactions, *Avery's Drug Treatment: Principles and Practice of Clinical Pharmacology and Therapeutics*, 3rd ed. (T. M. Speight, ed.), ADIS Press, Auckland, 1987, p. 255.
6. J. Thomas, Drug interactions, *Australian Prescriptions Product Guide*, 23rd ed. (J. Thomas, ed.), Australian Pharmaceutical Publishing Co., Melbourne, 1994, p. 62.
7. E. W. Martin, *Hazards of Medication: A Manual on Drug Interactions, Incompatibilities, Contraindications, and Adverse Effects*, J. B. Lippincott, Philadelphia, 1971.
8. G. Pöch, Combined Effects of Drugs and Toxic Agents: *Modern Evaluation in Theory and Practice*, Springer-Verlag, Vienna, 1993.
9. S. M. Pond, Pharmacokinetic drug interactions, *Pharmacokinetic Basis for Drug Treatment* (L. Z. Benet, N. Massoud, and J. G. Gambertoglio, eds.), Raven Press, New York, 1984, p. 195.
10. P. E. Rolan, Plasma protein binding displacement interactions—why are they still regarded as clinically important? *Br. J. Clin. Pharmacol. 37*: 125 (1994).
11. A. B. Okey, Enzyme induction in the cytochrome P-450 system, *Pharmacol. Ther. 45*: 241 (1990).
12. R. H. Levy, R. H. Mattson, and B. S. Meldrum (eds.), *Antiepileptic Drugs*, 4th ed., Raven Press, New York, 1995. See especially Chapters 6, 22, 29, 36, 43, 49, 69, and 79.
13. B. G. Pollock, Recent developments in drug metabolism of relevance to psychiatrists, *Harvard Rev. Psychiatry 2*: 204 (1994).
14. M. Spatzenegger and W. Jaeger, Clinical importance of hepatic cytochrome P450 in drug metabolism, *Drug Metab. Rev. 27*: 397 (1995).
15. M. J. Serlin, R. G. Sibeon, S. Mossman, A. M. Breckenridge, J. R. Williams, J. L. Atwood, and J. M. Willoughby. Cimetidine interactions with oral anticoagulants in man, *Lancet* 2: 317 (1979).
16. B. G. Hardy, I. T. Zador, L. Golden, D. Lalka, and J. J. Schentag, Effect of cimetidine on the pharmacokinetics of quinidine, *Am. J. Cardiol. 52*: 172 (1983).
17. A. B. Knapp, W. Maguire, G. Keren, A. Karmen, B. Levitt, D. S. Miura, and J. C. Somberg, The cimetidine–lidocaine interaction, *Ann. Intern. Med. 98*: 174 (1983).
18. J. M. Hansen and L. K. Christensen, Drug interactions with oral sulphonylurea hypoglycemic drugs, *Drugs 13*: 24 (1977).
19. D. G. Bailey, J. D. Spence, C. Munoz, and J. M. O. Arnold, Interaction of citrus juices with felodipine and nifedipine, *Lancet 337*: 268 (1991).
20. A. Somogyi and R. Gugler, Drug interactions with cimetidine, *Clin. Pharmacokinet. 7*: 23 (1982).
21. A. Somogyi and M. Muirhead, Pharmacokinetic interactions with cimetidine 1987, *Clin. Pharmacokinet. 12*: 321 (1987).

22. U. Fuhr, E. M. Anders, G. Mahr, F. Sorgel, and A. H. Staib, Inhibitory potency of quinolone antibacterial agents against cytochrome P450IA2 activity in vivo and in vitro, *Antimicrob. Agents Chemother. 36*: 942 (1992).
23. P. J. Ciaccio, D. B. Duignan, and J. R. Halpert, Selective inactivation by chloramphenicol of the major phenobarbital-inducible isozyme of dog liver cytochrome P-450, *Drug Metab. Dispos. 15*: 852 (1987).
24. J. M. Lunetta, K. Sugiyama, and M. A. Correia, Secobarbital-mediated inactivation of rat liver cytochrome P-450b: A mechanistic reappraisal, *Mol. Pharmacol. 35*: 10 (1989).
25. T. Chang, M. Levine, and G. D. Bellward, Selective inhibition of rat hepatic microsomal cytochrome P-450. II. Effect of the in vitro administration of cimetidine, *J. Pharmacol. Exp. Ther. 260*: 1450 (1992).
26. S. Narimatsu, K. Watanabe, T. Matsunaga, I. Yamamoto, S. Imaoka, Y. Funae, and H. Yoshimura, Suppression of liver microsomal drug-metabolizing enzyme activities in adult female rats pretreated with cannabidiol, *Biol. Pharm. Bull 16*: 428 (1993).
27. H. K. Crewe, M. S. Lennard, G. T. Tucker, F. R. Woods, and R. E. Haddock, The effect of selective serotonin re-uptake inhibitors on cytochrome P4502D6 (CYP2D6) activity in human liver microsomes, *Br. J. Clin. Pharmacol. 32*: 658 (1991).
28. L. Pichard, I. Fabre, G. Fabre, J. Domergue, B. Saint Aubert, G. Mourad, and P. Maurel, Cyclosporin A drug interactions. Screening for inducers and inhibitors of cytochrome P-450 (cyclosporin A oxidase) in primary cultures of human hepatocytes and in liver microsomes, *Drug Metab. Dispos. 18*: 595 (1990).
29. J. H. G. Jonkman and R. A. Upton, Pharmacokinetic drug interactions with theophylline, *Clin. Pharmacokinet. 9*: 309 (1984).
30. J. H. G. Jonkman, Therapeutic consequences of drug interactions with theophylline pharmacokinetics, *J. Allergy Clin. Immunol. 78*: 736 (1986).
31. P. Tenni, D. L. Lalich, and M. J. Byrne, Life threatening interaction between tamoxifen and warfarin, *Br. Med. J. 298*: 93 (1989).
32. K. T. Kivistö, H. K. Kroemer, and M. Eichelbaum, The role of human cytochrome P450 enzymes in the metabolism of anticancer agents: Implications for drug interactions, *Br. J. Clin. Pharmacol. 40*: 523 (1995).
33. W. J. A. Wijnands, T. B. Vree, and C. L. A. van Herwaarden, The influence of quinolone derivatives on theophylline clearance, *Br. J. Clin. Pharmacol.* 22: 687 (1986).
34. W. D. Hooper, M. C. Kavanagh, G. K. Herkes, and M. J. Eadie. Lack of a pharmacokinetic interaction between phenobarbitone and gabapentin, *Br. J. Clin. Pharmacol. 31*: 171 (1991).
35. W. D. Hooper, M. E. Franklin, P. Glue, C. R. Banfield, E. Radwanski, D. B. McLaughlin, M. E. McIntyre, R. G. Dickinson, and M. J. Eadie, Effect of felbamate on valproic acid disposition in healthy volunteers: Inhibition of β-oxidation, *Epilepsia 37*: 91 (1996).
36. I. Bernus, R. G. Dickinson, W. D. Hooper, and M. J. Eadie, Inhibition of phenobarbitone *N*-glucosidation by valproate, *Br. J. Clin. Pharmacol. 38*: 411 (1994).

10

Induction of Human Drug-Metabolizing Enzymes: Mechanisms and Implications

Martin J. J. Ronis
University of Arkansas for Medical Sciences,
Arkansas Children's Hospital Research Institute, Little Rock, Arkansas

Magnus Ingelman-Sundberg
Institute of Environmental Medicine, Karolinska Institute,
Stockholm, Sweden

I. INTRODUCTION

The rate of metabolism of drugs in clinical use is subject to important interindividual variations because of genetic, pathological, environmental, and physiological factors. As a result, the optimal therapeutic level of any drug will not be achieved in a great proportion of all patients. This might result in adverse effects caused by too-high plasma levels of the drug in question, or no pharmacological effects because of ultrarapid metabolism. It is evident that pathological, environmental, and physiological factors influence the relative extent of gene expression, which thus might be influenced by exposure to a variety of endogenous and exogenous factors, including hormones, bacteria, viruses, dietary factors, alcohol, cigarette smoke, and other drugs. The influence on drug turnover is associated with induction, or occasionally repression of drug-metabolizing enzymes, such as the cytochrome P450(CYP)-dependent microsomal monooxygenases. In this overview we discuss the mechanisms and consequences behind the induction of the drug-metabolizing enzymes.

Induction is defined as an increase in enzyme activity associated with an increase in intracellular enzyme concentration. This increase in enzymic protein is normally (but not always) caused by an increase in transcription of the associated gene. Enzyme induction is dose-dependent, generally with a steep dose–response relation and no clear-cut threshold of no-effect (1).

Induction of drug metabolism contributes to interindividual and intraindividual variation in drug efficacy and potential toxicity associated with drug–drug interactions. However, this is not the only cause. Examination of the variation in the expression of P450 enzymes, which are not genetically polymorphic, reveals that the interindividual differences in different human livers are in the range of 15-to several 100-fold. Comparison of the effect of inducers such as ethanol or specific drugs that the patients have been taking on the expression of specific P450 enzymes in liver specimens reveals a documented inductive effect on, for example, CYP2C, CYP2E1, or CYP3A4 of about two to five fold. It is thus likely that a major factor determining interindividual variation is a mixture of environmental factors in addition to physiological factors, with important interindividual differences in expression.

Clinical effects by a defined inducing agent on the pharmacokinetics of drugs have been frequently described and can influence the successive outcome of treatment. Alterations in drug efficacy will be dependent on the extent of enzyme induction in a particular individual, on the relative importance of the enzyme in multiple pathways of metabolism, and on the therapeutic ratio of substrate and metabolite. Similarly, potential toxicity will be dependent on changes in metabolic pathways associated with an alteration in the balance between drug activation and detoxication (2). The degree to which induction is clinically important varies with the P450 enzyme. Several human P450s, such as CYP1A1, CYP2E1, and CYP3A4, are readily inducible by dietary factors, smoking, drinking, and other drugs (3). In contrast, other human P450 enzymes, such as CYP2D6, appear to undergo little or no induction and interindividual variation in CYP2D6 expression and activity is largely the result of genetic polymorphism (4). However, it is evident that the major part of the human drug-metabolizing P450s are either polymorphic or inducible, explaining the important interindividual variation in P450-catalyzed drug metabolism (Table 1).

There are several good examples of induction-producing clinically significant drug–drug interactions. The antibiotic rifampin induces CYP2C and CYP3A enzymes and UDP-

Table 1 Genetic Polymorphism and Inducibility Among the Most Important Human Drug-Metabolizing Cytochrome P450s

Enzyme	Chromosomal localization	Polymorphism	Inducible	Marker substrate
CYP1A2	15q22	No	Yes	Caffeine
CYP2A6	19q13	Yes	Yes	Coumarin
CYP2C9	10q24	Yes	Yes	Warfarin
CYP2C19	10q24	Yes	Yes	Omeprazole
CYP2D6	22q13	Yes	No	Debrisoquine
CYP2E1	10	(Yes)	Yes	Clorzoxazone
CYP3A4	7q22	No	Yes	(Cortisol)

glucuronyltransferase in humans (5–7). In consequence, patients treated long-term with rifampin have a much decreased bioavailability of cyclosporine, largely owing to enhanced first-pass metabolism of the drug by induced CYP3A4 in the small intestine (8). In addition, an increased dose of oral contraceptives is recommended for women treated over an extended time with rifampin because the drug produces a 42% decrease in bioavailability of both ethinyl estradiol and norethisterone as a consequence of induction of both oxidation and glucuronidation (9). Similarly, there are several good clinical examples of enhanced toxic side effects of drugs as a consequence of induced metabolism. The most well known is the enhanced risk of liver damage produced by consumption of acetaminophen (Tylenol, paracetamol) in heavy drinkers as a consequence of induction of CYP2E1 (10). A similar increased risk of hepatotoxicity and other toxic side effects is observed in drinkers on exposure to the anethetics halothane, enfurane, and isoflurane that are also metabolized by CYP2E1 (11–13).

II. MOLECULAR MECHANISMS UNDERLYING INDUCTION

A. General Aspects

The most commonly described induction response is that which is *substrate-dependent*. A classic example of this is shortened sleep time in rodents associated with multiple exposures to the barbiturate anesthetic phenobarbital. This is associated with the capacity of phenobarbital to induce its own metabolism by transcriptional activation of cytochrome P450 enzymes in gene family 2B (14). This type of induction response is evolutionarily highly conserved and is readily seen in bacteria. For example β-galactosidase is highly induced in *Escherichia coli* when the bacteria is switched to a medium containing lactose as the sole carbon source.

Originally, probably as an adaptative response to diverse energy sources, substrate-mediated induction has evolved in animals as a mechanism for clearance of toxic lipophilic xenobiotics, such as those found in the diet, that might accumulate. This evolution may have been driven by plant–animal chemical warfare in which plants have synthesized more and more complex (and toxic) products of secondary metabolism to discourage herbivores from eating them, and animals have developed more and more complex and inducible detoxification systems to deal with these dietary xenobiotics. This relation is especially clear among insects, in which polyphagous caterpillars have much more complex P450 systems than monophagous species (15) and for which P450s are induced when animals are moved onto new food plants (16). Similarly, cytochrome P450 enzyme content and diversity is much higher among herbivorous and omnivorous vertebrates than in obligate carnivores at the top of food chains, such as birds of prey, big cats, and others (17), and may be induced by a wide variety of dietary phytochemicals (e.g., induction of CYP2B isozymes by diallyl sulfide found in garlic; 18).

A second type of induction response is *receptor-mediated*, by interactions with important endogenous regulatory pathways. Many drug-metabolizing enzymes that are also involved in metabolism of endogenous cellular regulators, such as steroids and eicosanoids, are induced by hormones. In the rat, many P450 enzymes in gene family 2C are sexually dimorphic and under regulation by androgens and growth hormone (19), whereas P450 enzymes in gene family 3A are inducible by glucocorticoids (20,21). There is now clinical evidence to suggest there is a degree of sexual dimporphism in human drug metabolism (22) and that hormones, such as growth hormone, are also capable of altering human cytochrome P450 expression (23). Alterations in endocrine status probably also underlie induction of drug metabolism by pathophysiological conditions, such as starvation, obesity, and diabetes. Commonly, coordinate induction occurs in which several enzymes involved in a particular metabolic pathway or response are induced simultaneously. Such coordinate induction of cytochrome P450 enzymes in gene family 1 (*CYP1A1, CYP1A2, CYP1B1*) and phase II enzymes (*GSTα, UGT1 *06, DT diaphorase*) occurs in response to dioxin (TCDD) and polycyclic aromatic hydrocarbons (PAHs; 24,25). Similar coordinate induction of CYP2B and other P450 enzymes, cytochrome b_5, cytochrome P450 reductase, UDP-glucuronyl transferases, and glutathione *S*-transferases occurs with phenobarbital and other ''phenobarbital-type'' inducers (14,26–28). Coordinate induction has been reported for CYP3A inducers such as rifampin, phenobarbital, clotrimazole, and the P-glycoprotein xenobiotic transporter (29), and ethanol induces CYP2E1, some members of the CYP2C family (30,31), and a form of UDP-glucuronyltransferase (32). The CYP1 family induction response is mediated by the cytosolic aryl hydrocarbon (Ah) receptor (33). This kind of receptor, belonging to the basic helix-loop-helix family of transcription factors (which includes the *Drosophila* proteins ''per'' and ''sim'') are involved in control of circadian rhythm and neurological development (34). It is thought that the Ah receptor has an endogenous endocrine ligand and that it mediates a metabolic pathway critical for normal development, because knockout mice lacking the Ah receptor show disrupted immune system development, liver fibrosis, and early mortality (35).

A third type of induction response is that which is *inhibitor-mediated*. Good examples are induction of CYP1A by methylenedioxyphenyl phytochemicals, such as isosafrole (36); induction of a variety of cytochrome P450 enzymes, especially CYP3As and CYP2Bs by ergosterol biosynthesis-inhibiting imidazole fungicides, such as clotrimazole (26); induction of CYP2E1 by isoniazid (37); and induction of CYP3A1 by macrolide antibiotics (38). In this case interaction with the heme group of cytochrome P450s results in inhibition of endogenous function. Induction of enzyme concentration appears to occur by a mixture of posttranslational protein stabilization and increased transcription in a homeostatic feedback loop response to disruption of endogenous pathways catalyzed by the P450 enzyme in question (26,38,39).

Of the more than 20 gene families of cytochrome P450 enzymes described to date, three—families 1, 2, and 3—appear to be primarily involved in the metabolism of clinically used drugs in humans. Of those, CYP1A2, CYP2C9, CYP2C19, CYP2D6, CYP2E1, and CYP3A4 are the most important. The mechanisms of induction of P450s belonging to these subfamilies have been intensely studied in animal models.

B. The CYP1 Family

Enzymes of the CYP1 family are involved in the metabolism of many planar aromatic drugs, such as tamoxifen, tacrine, antipyrene, and acetaminophen, dietary phytochemicals, and carcinogens, such as the polycylic aromatic hydrocarbons and aromatic amines produced in cooking and found in cigarette smoke (3). Induction of genes in this family (*CYP1A1*, *CYP1A2*, and *CYP1B1*) is under the control of the Ah receptor (34). In the absence of ligand, most of this protein exists as a cytosolic complex, with a heat-shock–chaperone protein Hsp 90 (40,41). Ligand binding dissociates this complex and results in translocation of the Ah receptor to the nucleus, where it forms a heterodimer with a second, nuclear, bHLH protein Arnt (41,42). This heterodimer binds to a common response element present in multiple copies in the 5′-flanking region of the CYP family 1 genes known as the XRE, and functions as a transcriptional enhancer to stimulate gene transcription (33–35,40–42). It is clear that induction mechanisms of CYP1 family members are more complex than the foregoing scheme suggests and are not always coordinated. Induction by TCDD of CYP1A1 and CYP1B1, but not CYP1A2, has been described in rat and human fetal tissues. In humans, CYP1A1 and CYP1B1 are induced in many extrahepatic tissues, whereas expression and induction of CYP1A2 is restricted to the liver (3,45). The induction response may be modulated by various drugs, chemicals, and endogenous factors in a manner that is Ah receptor-dependent, but that apparently does not involve Ah receptor binding. Thus, hepatic CYP1A1 is inducible by carotenoids and benzimidazole antiparasitic drugs, CYP1A2 is inducible by isosafrole, and CYP1A1 is inducible in differentiating keratinocytes in the absence of Ah receptor binding (46–48). Hepatic CYP1A1 induction by TCDD and PAHs is potentiated by glucocorticoids (49) and by the phytoestrogens genistein and daidzein in the human hepatocyte cell line H4IIE (50). In contrast, induction of CYP1A1 by the drug omeprazole in H4-II-E cells and induction of CYP1A1 by TCDD in keratinocytes was almost completely abolished by genistein or daidzein (41). There are also species differences in that the benzimidazoles cannot cause induction of CYP1A1 in Hepa-1 cells derived from mouse liver, whereas dioxin can (43). These contradictory and complex effects suggest several additional elements associated with induction of the CYP1 family. In addition to the XRE, a number of other regulatory elements are present on human CYP1A promoters: a glucocorticoid response element (GRE) has been described and is involved in mediation of glucocorticoid potentiation of CYP1A1 induction by TCDD and PAHs (49). Also, a negative regulatory element has been described on the 5′-flanking region of the human and rat *CYP1A1* genes (51,52) that binds a member of the NF-Y transcription family (52), and an AP-1 site has been described on the human *CYP1A2* gene (53). In addition, other cytosolic proteins designated 4S and 8S selectively bind CYP1A inducers, such as PAHs, although the role they play in the induction response is a subject of controversy (54). There may be more than one Ah receptor subtype (55) and, in addition to Ah receptor–Arnt heterodimerization, these bHLH proteins may also form functional homodimers with altered DNA binding and functionality (44). Ultraviolet light (UV)-cross linking studies have suggested the association of at least one further 115-kDa protein with the Ah–Arnt complex that may be the transcription factor Sp 1 (33,66). Protein phosphorylation also appears to play a role in modulation of CYP1A induction responses. Induction of CYP1A2 by phorbol esters and potentiation of TCDD/PAH induction of CYP1A2 by phorbol esters has been described to be associated with activation of protein kinase C (57). The differentiation-associated induction of CYP1A1 in keratinocytes may also be mediated by effects on protein kinase C associated with cadherin-dependent perturbation of cellular Ca^{2+} flux (48). In addition, tyrosine kinases are involved in phosphorylation of the Ah-receptor–Hsp-90 complex and its subsequent ability to dissociate in the face of ligand binding (41). Their inhibition may partly explain the effects of the phytoestrogens genistein and daidzein on CYP1A1-induction responses (41,50).

C. The CYP2 Family

1. The CYP2B Subfamily

Rodent enzymes of the CYP2B family catalyze the metabolism of many drugs including barbiturates and neuroleptic drugs in laboratory animals. Data on the metabolic capabilities of the human orthologue CYP2B6 is limited and the hepatic expression is low, about 1% of total P450. In contrast with the wealth of mechanistic data on the CYP1 family, induction mechanisms underlying increases in CYP2B expression in mammalian systems in response to barbiturates, pesticides, pollutants, and many structurally diverse drugs have remained largely unknown. Much effort has been expended in the search for a ''barbiturate receptor'' analogous to the Ah receptor, with little or no success. More recently, analysis of the promoter regions of two phenobarbital-inducible P450 enzymes $P450_{BM\text{-}1}$ and $P450_{BM\text{-}3}$ from *Bacillus megaterium* have yielded evidence for the presence of a so-called Barbie box consensus sequence: ATCAAAGCTGGAGG, which has now been found in the 5′-flanking regions of essentially all phenobarbital-inducible genes (58). The potential Barbie box involvement in induction and transcription factor interactions with Barbie box sequences in mammalian CYP2B genes remains the subject of lively debate. Fulco et. al. (59–61) have described a repressor protein, Bm3R1 that binds to the Barbie box of $P450_{BM\text{-}1}$ and $P450_{BM\text{-}3}$ and several other operator sites on these genes. They have demonstrated loss of Bm3R1 binding with phenobarbital treatment associated with the binding of positive regulatry proteins to Bm3R1 independently of protein synthesis, probably involving phenobarbital-stimulated phosphorylation, and hypothesize a similar process for activation of mammalian *CYP2B* genes. In addition they have isolated several phenobarbital-inducible proteins that bind to the Barbie box sequence. There is evidence that mammalian CYP2B induction does involve phosphorylation because elevation of cellular cAMP concentration using forskolin has been reported to inhibit phenobarbital induction of CYP2B1 in hepatocyte cultures, and a similar effect is observed on inhibition of protein phosphatases PP1 and PP2A by okadaic acid (62,63). Although there is evidence for the formation of phenobarbital-inducible DNA-binding complexes in the first 200 bp of the rat *CYP2B1/2* genes, a region that includes a Barbie box sequence (59), other experiments by Omiecinski et al. using transgenic mice incorporating different lengths of the CYP2B2 promoter region linked to various reporter constructs suggest that the most important element required for normal phenobarbital induction in vivo lies between 1.7 and 2.3 kb upsteam from the transcriptional start site (63).

2. CYP2C Subfamily

The CYP2C subfamily is one of the largest and most diverse of gene family 2 and has undergone a burst of recent gene duplication and diversification in mammals. In humans, at least five CYP2C enzymes have been described in liver, CYP2C8–10, CYP2C18, and CYP2C19, and a number of additional extrahepatic forms probably remain to be described

(3). These enzymes are involved in the metabolism of many clinical drugs. CYP2C8 catalyzes the 6α-hydroxylation of the oncological drug paclitaxol (Taxol), and CYP2C9 is responsible for metabolism of tolbutamide, trimethadone, tienilic acid, phenytoin, diclofenac, and hexobarbital (3). CYP2C19 polymorphism is responsible for interindividual variation in metabolism of (*S*)-mephenytoin and nirvanol (64,65). In addition to drugs, these enzymes are involved in the metabolism of retinoids (30) and the rodent othologues CYP2C6, 2C7, 2C11–13 are predominantly associated with the metabolism of sex steroids (3,19). A characteristic of this family of P450 enzymes is a general lack of inducibility by xenobiotics. Some small degree of inducibility of rat CYP2C6/7, human CYP2C8, CYP2C9, and the CYP2C7-related rat CYP protein P450EtOH$_2$ have been described for barbiturates, ethanol, and rifamcin (5,30,31,66). However, the molecular mechanisms for these effects are unknown. In contrast several endogenous endocrine mediators up-regulate CYP2C enzymes, these include retinoids (67), androgens, and growth hormone (GH; 19). In rodents many of these enzymes display sexually dimorphic expression related to sexually dimorphic patterns of GH secretion (19). GH regulation of CYP2C expression appears to be mediated by Jak-Stat kinase phosphorylation cascades, mitiated via phosphorylation of Jak2 kinase catalyzed by the tyrosine kinase activity of the activated GH receptor, that are sensitive to the GH signal pattern (68,69). Rat CYP2C11 appears to be transcriptionally activated by STAT5 via this pathway (68).

3. *CYP2E1*

CYP2E1 is toxicologically one of the most important cytochrome P450 enzymes. It has been implicated in the activation of acetaminophen and organic solvents to hepatotoxic and nephrotoxic intermediates, in the etiology of alcohol-induced liver damage, and in the activation of nitrosamine procarcinogens (3,37). Because of its role in alcohol and acetaminophen hepatotoxicity, the regulation of CYP2E1 has received tremendous attention, and multiple induction mechanisms have been elucidated (37; Fig. 1). Many xenobiotics have the capacity to increase CYP2E1 expression and, in addition, dietary factors and pathophysiological factors are capable of inducing this enzyme (37). Induction occurs at all regulatory levels, from transcription to mRNA stabilization, increases in translational efficiency, and posttranslational protein stabilization (37). Ethanol at low concentrations appears to stabilize the CYP2E1 apoprotein, producing a shift in protein half-life from a biphasic 7-h, 47-h pattern, to a linear 47-h one (70). There is evidence to suggest that the fast phase of CYP2E1 turnover is triggered by protein kinase A-dependent phosphorylation at Ser-129 and subsequent heme loss, followed by degradation within the endoplasmic reticulum by a serine protease or ubiquitin-dependent pathways (71–73). Inhibition of phosphorylation by substrate binding is thought to produce the stabilization and subsequent induction of the enzyme (71–74). Treatment of hepatocytes with hormones, such as glucagon and epinephrine, that stimulate cAMP-dependent phosphorylation cascades, has been demonstrated to significantly shorten CYP2E1 apoprotein half-life. At higher concentrations ethanol produces additional induction by increases in transcription (75). It is thought that this may be an indirect endocrine effect associated with the stress of alcohol intoxication (37). High-lipid–low-carbohydrate diets, obesity, and fasting are also associated with an elevation of CYP2E1 (76–78). The starvation response and CYP2E1 increases observed in hypophysectomized rats are associated with increased CYP2E1 transcription (78,79). In the latter, GH appears to be responsible for down-regulation of CYP2E1 transcription, because GH replacement was capable of restoring CYP2E1 toward control values. The induction of CYP2E1 observed in diabetic rats may be partly associ-

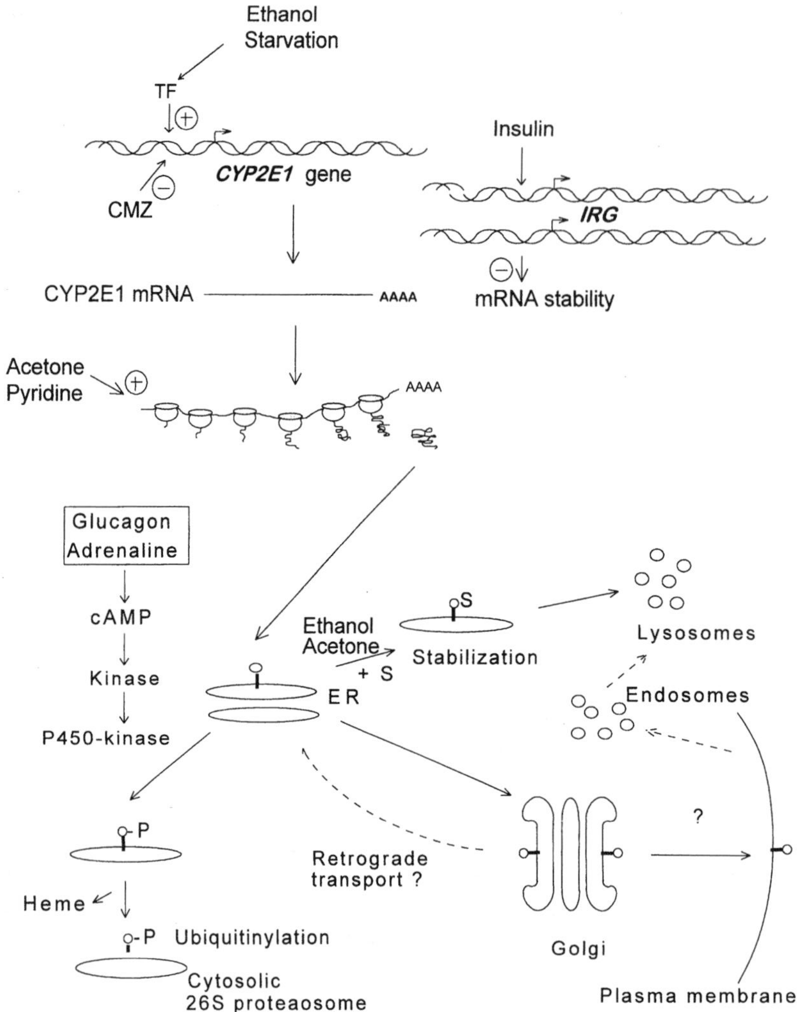

Fig. 1 Multiple mechanisms for regulation of CYP2E1 expression: The level of enzyme expression is influenced at the transcriptional, translational, and posttranslational levels. The mechanisms might act synergistically causing a maximum induction of the rodent enzyme by up to 50-fold. For further explanations, see text. S, substrate; TF, transcriptional factor; P, phosphate; IRG, insulin responsive gene; CMZ, chlomethiazole.

ated with disrupted GH pulses (79); however, this effect has been suggested to be associated with mRNA stabilization, rather than transcriptional effects and is insulin-reversible (79,80). Many small organic molecules other than ethanol appear to induce CYP2E1 by a similar posttranslational stabilization mechanism; however, acetone and pyridine also have a stimulatory effect at the level of translational efficiency, accompanied by shift in polysomal distribution toward a higher density and increased incorporation of radiolabeled amino acids into newly synthesized protein (81,82). Many of these separate induction mechanisms can occur simultaneously. Thus starvation and acetone treatment produce synergistic effects on CYP2E1 expression, whereas administration of pyrazole to streptozocin-induced diabetic rats results in an additive response (83). When taking all induction mechanisms into account, a theoretical maximal induction of 50- to 100-fold is possible (37).

4. *Other CYP Family 2 Members*

Several other CYP2 family members are found in human liver. In particular CYP2A6 and CYP2D6. CYP2A6 has been demonstrated to activate aflatoxin-B_1 to its carcinogenic epoxide metabolite and similar to CYP2E1 is associated with activation of nitrosamine procarcinogens (3). CYP2D6 is responsible for the metabolism of a host of clinical drugs, including debrisoqine, bufuralol, propranolol, and thioridazine (3). Although a major polymorphism has been described in CYP2D6 and the ''poor-metabolizer'' phenotype is of major importance in determination of drug–drug interactions (3,4), neither of these enzymes appear to be highly inducible by xenobiotics, although there have been some reports about phenobarbital-mediated induction of CYP2A6.

D. The CYP3A Subfamily

Arguably the most important P450 enzymes involved in metabolism of clinical drugs, the members of the CYP3 family, of which there are at least four, differentially expressed in human tissues (3), are the most abundant cytochrome P450 forms found in human liver. CYP3A4 alone makes up between 10 and 60% of the total hepatic P450 content and catalyzes the oxidative metabolism of a host of medications and steroids, including antibiotics, cyclosporines, acetaminophen, and contraceptives (84). CYP3A4 is found in all human livers, at high levels in the intestine where it plays a key role in first pass-metabolism (8), and some kidneys (85). CYP3A5 is expressed in the kidney and found in some livers (85), and CYP3A7 is selectively expressed in human fetal liver (86). Many induction studies have focused on CYP3A isoforms, given their high level of human expression and importance in drug metabolism. In animal models CYP3A enzymes have been inducible at the transcriptional level by glucocorticoids such as dexamethasone, antiglucocorticoids such as pregnenolone 16α-carbonitrile, and to some degree by phenobarbital. In addition, CYP3A enzymes are inducible both transcriptionally and by posttranslational stabilization by inhibitors such as the macrolide antibiotics erythromycin and trioleandomycin and the imidazole fungicides clotrimazole and propiconazole (26,38). The mechanisms underlying CYP3A induction have only recently been partially elucidated. Transcriptional regulation has been studied using reporter constructs of chloramphenicol acetyltransferase and secretory placental alkaline phosphatase fused with truncated fragments of the CYP3A promoter regions of the rat, rabbit, and human genes transfected into HepG2 cells and cultured human, rat, and rabbit hepatocytes (87,88). Consensus sequences in the promoter regions of these genes mediate the induction by glucocorticoids, such as dexamethasone (87,88). One of these, a 219-bp region −891 to −1109-bp upstream from the transcriptional start site of CYP3A5 contains no classic GRE, but does contain two GRE ''half-sites'' (TGTTCT) separated by 160 bp (87). Induction of CYP3A5 by dexamethasone appears to involve the cooperative binding of the glucocorticoid receptor to these sequences because the induction response in HepG2 cells required cotransfection of the receptor, was blocked by the glucocorticoid receptor antagonist RU-486, and was blocked by point mutation of either half-site (87). On the other hand, induction responses of CYP3A gene constructs to different classes of inducer—glucocorticoid, antiglucocorticoid, antibiotic—varied depending on the host cell type transfected (88). Thus, the host cellular environment

Table 2 CYP3A4 and P-Glycoprotein (7q22.1)-Inducing Agents

Agent	CYP3A4	P-glycoprotein
Rifampin	Yes	Yes
Phenobarbital	Yes	Yes
Clotrimazole	Yes	Yes
Reserpine	Yes	Yes
Midozolam	No	Yes
Nifedipine	No	Yes

Source: Ref. 29.

plays a large role in determination of the pattern of inducibility. One way in which this has been shown to occur relates to the intracellular inducer concentration. Many CYP3A inducers, such as the antibiotic rifampin, are removed from cells by the P-glycoprotein (Pgp) efflux pump. Thus, in transfected cells, overexpression of Pgp reduces the extent of rifampin induction of CYP3A (89). Many CYP3A inducers also coordinately induce Pgp (Table 2) and thus may modulate their own CYP3A inductive properties by stimulating their rapid removal from the intracellular environment.

III. EVIDENCE FOR INDUCTION OF HUMAN CYTOCHROME P450 ENZYMES

A. The CYP1 Family

Evidence for induction of enzymes in CYP family 1 is well established in humans. The most prevalent factors associated with this induction are smoking and diet. CYP1A1 appears to be largely extrahepatic in humans, with CYP1A2 the major liver enzyme, whereas CYP1B1 is found in many tissues especially the adrenal, glands gonads, and breast (3). All human CYP1 members are inducible in vitro. In explanted embryonic tissue 3-MC and TCDD induced both CYP1A1 and CYP1B1, whereas CYP1A2 was not expressed. Induction of CYP both CYP1A1 and 1A2 and associated methoxy- and ethoxyresorufin *O*-dealkylase activities by classic CYP1A inducers, such as 3-MC, βNF, TCDD, and Arochlor 1254, has been described in hepatocyte-derived cell lines such as HepG2 and H4IIE, in primary human hepatocytes and in precision-cut human liver slices (27,28,50). In addition, induction of CYP1A1 by the drug omeprazole has been described in such systems (50). In vitro induction of CYP1A1 has also been described in human colon and mammary tumor xenografts in immunodeficient mice (90) and in cell lines derived from extrahepatic human tissues, such as ketatinocytes and intestinal cell lines (41,48,91). CYP1A induction-responses in these systems have been demonstrated to be modulated by tyrosine kinase inhibitors, such as genistein, and by cytokines, such as (TNFα) tumor necrosis factor-alpha, interleukin (IL)-1β, and the interferons (IFN) (41,50,92). In vivo, elevated CYP1A1 has been demonstrated in the placenta, lymphocytes, and lung tissues of smokers (2,3,93,94), and a correlation has been noted between expression of pulmonary CYP1A1 and lung cancer (94). Liver CYP1A2 has also been described to be induced in smokers, for an elevation in the CYP1A2 marker activities, caffeine 3-*N*-demethylase and estradiol 2-hydroxylase, have been observed in smokers (2,95). Caffeine 3-*N*-demethylase has also been reported to be elevated following consumption of cruciferous vegetables, such as broccoli that contain the CYP1A inducer 3-methyl-indole (96) and by caffeine consumption itself (97). In addition, the drug omeprazole has been described to induce CYP1A expression 19–167% in duodenal biopsies from volunteers consuming the drug orally (98). Clinically, CYP1A2 has been implicated in the liver damage (elevated serum ALT) reported in 20–50% of patients consuming the anti-Alzheimer's drug tacrine (99). Human CYP1A2 exclusively activates tacrine in vitro, and it has been suggested that CYP1A2 induction may increase the risk of this side effect. In addition, elevated CYP1A2 associated with consumption of charbroiled meats (containing PAH, CYP1A inducers) has been associated with enhanced risk of colon cancer owing to the role of CYP1A2 in the activation of dietary aromatic amine procarcinogens (95).

B. The CYP2 Family

1. CYP2B6

Whereas induction of CYP2B enzymes is a major response to barbiturates and diverse other structurally unrelated drugs, pesticides, and pollutants in rodents and nonhuman primates (58), much less is known concerning the inducibility of the human orthologue CYP2B6. This enzyme is expressed at very low levels, normally less than 1% of total P450 content, and it was detected as expressed in only 4 out of a survey of 60 human livers (3). CYP2B6 was induced slightly by phenobarbital and dexamethesone and in a major fashion by TCPOBOP in human colon and breast tumors xenografted onto immunodeficient mice (90). However, no definitive evidence is currently available for CYP2B6 induction by barbiturates in vivo.

2. CYP2C

Little xenobiotic induction of CYP2C enzymes has been described in rodent models (19). Most of this diverse, sexually dimorphic family are under complex hormonal modulation, especially by androgens, retinoids, and GH (19,30,67). Although the multiple CYP2C enzymes play a major role in metabolism of clinical drugs in humans, there is little evidence of inducibility in humans either. The powerful PB-like inducer TCPOBOP induces CYP2C8 in xenografted human colon tumors (90). However, the only evidence for in vivo CYP2C induction by xenobiotics is that treatment of individuals with barbiturates or rifampin increases the oxidation of the CYP2C9 substrate tolbutamide (5); barbiturate therapy induces the metabolism of methoxyflurane, which is catalyzed by CYP2C among other P450 enzymes (100); and rifampin increases the clearance of the CYP2C19 substrate mephenytoin (101). In addition, examination of patients undergoing GH therapy for adult dwarfism demonstrated that GH could induce clearance of the general P450 substrate antipyrene (23). This suggests that GH may play a role in hormonal regulation of CYP enzymes in humans as well as in rodents.

3. CYP2E1

There is good evidence to show that many of the xenobiotic, dietary, and pathophysiological factors that produce CYP2E1 induction in rodents and other experimental animals also operate in humans (37). Induction of CYP2E1 by ethanol and isopentanol has been described in primary human hepatocytes (102). Ethanol induction has also been described in human fetal hepatocyte cultures (103). In addition, increased concentration of hepatic CYP2E1 has been described in vivo. Ethanol induction of hepatic CYP2E1 has been demonstrated to occur transcriptionally by in situ hybridization analysis of livers from alcoholics (104). Induction of CYP2E1 has been measured in vivo through measurement of the 6-hydroxylation of the CYP2E1 marker substrate chlorzoxazone and is rapidly reversible on cessation of drinking (105); the degree of induction has been described to be a polymorphic response (106). CYP2E1-induction has also been linked to obesity by using chlorzoxazone (107); lymphocyte CYP2E1 has been demonstrated to be elevated in diabetics (108); and prolonged isoniazid treatment increases the clearance of the anesthetic CYP2E1 substrates enflurane and isoflurane (12,13). Exposure to tobacco smoke, which is a complex mixture containing many P450 inducers and substrates, produces induction of CYP2E1 in mouse lung and kidney, but has little effect in liver (109,110), it is not known if this also occurs in humans.

C. The CYP3A Family

Given the high level of CYP3A expression in human liver and intestine and its important role in clearance and first-pass metabolism of many clinical drugs, induction of CYP3A forms (especially CYP3A4) in humans has been well studied. CYP3A4 is expressed at higher levels in women than men, suggesting that there is endogenous hormonal regulation of this enzyme (22). CYP3A4 induction by glucocorticoids, macrolide antibiotics, phenobarbital, the pesticide endosulfan, isopentanol, and ethanol has been demonstrated in primary human hepatocytes and human hepatoma-derived cell lines (66,103,111–113). In addition, the drugs omeprazole and lansoprazole induce CYP3A4 in 50% of primary human hepatocyte cultures, suggesting some degree of induction polymorphism (114). The induction response appears to be mediated, at least partly, by the glucocorticoid receptor (88), although posttranslational stabilization of the apoprotein clearly plays a key role in the induction produced by some macrolide antibiotics and imidazole fungicides (26,38). The induction response has been demonstrated to be suppressed in human hepatocyte cultures by cytokines, such as IL-1β, TNFα and the interferons, in a fashion similar to induction of CYP1A enzymes (92). Induction of CYP3A4 has also been demonstrated in xenografted human colon tumors using the phenobarbital-type inducer TCPOBOB (90). In vivo, induction of CYP3A enzymes has been demonstrated in humans following prolonged treatment with the anticonvulsants phenytoin and carbamazepine and following treatment with the antibiotic rifampin (2,115,117). Carbamazepine treatment increases the clearance of another proposed CYP3A4 substrate clozapine (2), and CYP3A4 induction may play a role in anticonvulsant toxicity (115). Carbamazepine produces a hypersensitivity syndrome in some patients treated over a prolonged period, resulting in skin and liver damage. This may result from CYP3A4-catalyzed formation of an active 10,11-epoxide metabolite from the parent drug (115,117). Several studies have examined the ability of rifampin to induce metabolism of CYP3A4-selective substrates in human volunteers. Rifampin treatment has been reported to increase clearance of dextromethorphan and the benzodiazepine CYP3A substrate midazolam (118). In addition, clearance of orally administered cyclosporine and *S*-verapamil has been demonstrated to be dramatically elevated by rifampin treatment, largely owing to increased first-pass metabolism by CYP3A in the intestine (8,119). For *S*-verapamil, treatment of male volunteers with 600 mg/day rifampin resulted in a 33-fold elevation in clearance and 25-fold decrease in bioavailibility of orally administered *S*-verapamil, but much smaller effects when the drug was administered intravenously. No studies have examined induction of CYP3A by ethanol in humans in vivo. However, increased CYP3A4 expression in ethanol-treated human hepatocytes has led to

Table 3 Known Inducers of Human Drug-Metabolizing P450s

CYP1A2	CYP2A6	CYP2B6	CYP2C9	CYP2C19	CYP2E1	CYP2D6	CYP3A4
Broiled beef (95) Cigarette smoke (95) Cruciferous vegetables (96) Omeprazole (50, 98)	Barbiturates (115)	nk	Rifampicin (5,66,86) Anticonvulsants (66,100)	Rifampin (101)	Ethanol (103,104) Isoniazid (12,13)	nk	Carbamazepine (2,116) Dexzmethasone (66) Phenobarbital (27,66,112) Phenytoin (2,112,116) Rifampin (66,111,112) Ethanol (103) Endosulfan (113)

nk, not known.

speculation that CYP3A4 may play a role in alcohol-associated acetaminophen hepatotoxicity because in addition to CYP2E1, CYP3A4 is also capable of catalyzing the activation of acetaminophen to the quinone intermediate NAPQI (103). Cocaine also induces CYP3A in rabbit hepatocytes, although it is not known if a similar CYP3A induction occurs in human drug abusers (120).

IV. INDUCTION OF PHASE II ENZYMES

In contrast to the considerable amount of information available on induction of cytochrome P450 enzymes in humans and induction mechanisms, far less is known about the potential for induction of phase II drug-metabolizing enzyme systems, such as the UDP-glucuronyltransferases (UDPGTs), the glutathione *S*-transferases (GSTs), and the acetyl- and sulfotransferases. These phase II systems are major routes of detoxification, and there is much interest in induction of these systems for cancer chemoprevention. In experimental animals, UDPGT and GST enzymes have been inducible by phenobarbital-type inducers and Ah receptor ligands (28,29), and an ethanol-inducible UDPGT has been described in the rabbit (32,121). However, the level of induction rarely appears to be greater than two to three fold (122). No evidence for induction of sulfotransferases or acetyltransferases by xenobiotics has been reported, although in the rat, sulfotransferases are sexually dimorphic and are regulated by growth hormone and sex steroids (123). Very few studies have examined induction of phase II enzymes in humans. Bilirubin UDPGT has been reported to be induced by phenobarbital, 1,2-benzanthracene, and rifampin in HepG2 cells, but no other UDPGT activities were affected (27). It has been suggested that both anticonvulsants and rifampin induce human UDPGTs (2). However, the only evidence of UDPGT induction in humans in vivo comes from studies showing increased clearance of oral contraceptives following rifampin therapy in women (6), which can be explained partly by rifampin induction of CYP3A4. GSTs in the alpha class (GST A1 and A2) have been observed to be inducible by TCDD, phenobarbital, 3-MC and the dithiolethiones 1,2-dithiole-3-thione and oltipraz in primary hepatocyte cultures (28,124), and GST alpha levels have been elevated in human blood plasma by consumption of brussels sprouts (125). Oltipraz has recently been demonstrated to elevate GST levels in human peripheral lymphocytes at single doses as low as 125 mg and is now in clinical trials in Qidong, China as a chemopreventive agent for aflatoxin-induced hepatocellular carcinoma (126).

V. SYSTEMS FOR THE STUDY OF HUMAN INDUCTION

The ethical problems associated with human experiments has made the study of induction of drug-metabolizing systems difficult. Only limited in vivo experiments can be done. The clearance of enzyme-specific substrates has been used as probes. One such probe is caffeine. CYP1A2 is the principal enzyme associated with 3-*N*-demethylation and the urinary molar ratio of 3-*N*-demethylated products (17X + 17U) to unmetabolized caffeine (137X) following oral caffeine administration as instant coffee has been used for in vivo CYP1A2 phenotyping (95). With this method Lang et al. (95) have demonstrated a significant two fold elevation in CYP1A2 phenotype in white smokers and an odds ratio of 6.45 for development of colon cancer in individuals with rapid-rapid CYP1A2/*N*-acetyl transferase 2 phenotype and a preference for the consumption of well-done meats. Because charbroiled meats contain high levels of polycyclic aromatic hydrocarbon CYP1A inducers, a preference for well-done meats might in itself be associated with the development of a rapid CYP1A2 phenotype and thus reinforce colon cancer risk. Caffeine has also been suggested to be a potential in vivo probe for CYP2E1 and CYP3A4 in addition to CYP1A2 because CYP2E1 appears to be the principal enzyme catalyzing the metabolism of caffeine, theophylline, and theobromine, whereas CYP3A4 appears to catalyze caffeine 8-hydroxylation selectively (127,128). The centrally acting, skeletal muscle relaxant chlorzoxazone has been used in a similar manner as a specific probe for in vivo phenotyping of CYP2E1 (129). CYP2E1 catalyzes the formation of the 6-hydroxy metabolite, the urinary excretion of which is limited by formation rate (129). The CYP2E1 inhibitor disulfiram and substrate isoniazid substantially reduce chlorzoxazone plasma clearance (130), and clearance is significantly elevated in alcoholics (105). Recently, a polymorphism in human CYP2E1 inducibility has been described using this in vivo probe (106). Many different compounds have been used as in vivo probes for CYP3A4. Urinary levels of 6β-hydroxycortisol, detected by radioimmunoassay and exhalation of $^{14}CO_2$ following administration of [^{14}C] erythromycin have both been developed as commercial CYP3A4 tests (130,131). In addition, clearance of many CYP3A4 substrates have been used, including nifedipine, midazolam, and verapamil (3,118,119). Recently, the over-the-counter antitussive dextromethorphan has been suggested as a dual probe for the phenotyping of both CYP3A4 and CYP2D6 (132,133). CYP3A4 catalyzes the *N*-demethylation and CYP2D6 the *O*-demethylation of this drug and substrate, and metabolites may be readily quantitated in spot urine samples using HPLC with a fluorimetric detector (132,133).

In addition to the use of in vivo probes, in a few instances, metabolism has been studied directly using biopsies of the small bowel taken by endoscopic techniques (98,134). Many groups have examined metabolism in microsomes and liver slices from organ doners; however, relating enzyme levels to clinical histories is often impossible. More recently indirect approaches have been increasingly employed to study human induction. Peripheral lymphocytes express several P450 enzymes, such as CYP1A1, CYP3A, and CYP2E1, that appear to be inducible. Both Western blot and reverse transcription–polymerase chain reaction (RT–PCR) techniques have been applied to examine induction from simple blood samples (134). CYP1A1 is inducible in lymphocytes in vitro by TCDD and in vivo in smokers (135). In addition, CYP2E1 was elevated in peripheral lymphocytes taken from diabetics (108). The relation between P450 induction in lymphocytes and hepatocytes requires further study before these techniques can be applied more generally. However, simultaneous studies of CYP2E1 induction by ethanol in rabbit liver and lymphocytes suggest that the responses were parallel and thus that blood CYP2E1 might be used a phenotypic marker for xenobiotic induction of the liver enzyme in humans (136).

Many studies have been conducted in vitro using primary human hepatocytes, human hepatoma cell lines, or cell lines derived from other human tissues (27,28,41,50,66). Although these have the advantage of being easily manipulable, extrapolation back to intact tissues in vivo must be approached with caution because many regulatory factors may be lost in culture or following cellular transformation. In addition, molecular biological techniques have identified many of the transcription factors that are associated with the promoter regions of many human genes encoding liver drug-metabolizing enzymes (Table 4). Further study of these regulatory proteins will ultimately unravel the general mechanisms underlying induction and allow prediction of inductive properties of new compounds based on chemical structure.

Table 4 Hepatic Transcriptional Factors

Transcription factor	Gene	Ref.
AHR	*CYP1A1/1A2*	33–35,55,56
AP1	*CYP1A2*	53
HNF-1α	*CYP2E1*	137
HNF-1β		138
HNF-3	*CYP2C6*	139
HNF-4	*CYP2C1-3*	140
C/Ebpα/β	*CYP2D5*	141
DBP	*CYP2C6*	142
GABPα/β	*CYP2d9*	143
NF2d9	*CYP2d9*	144

VI. CONCLUSIONS

It is clear that human P450 enzymes (and to some extent phase II drug-metabolizing systems) are inducible and that this contributes to inter- and intrapersonal variation in drug metabolism clearance and toxicity. However, compared with the frequently reported induction in animal models, induction in humans sufficient to produce clinically relevant effects is relatively rare. A few examples of pregnancy, despite taking contraceptive pills, owing to simultaneous treatment with the CYP3A4 inducer rifampin have been described. This is probably because in animals, drugs are tested at relatively high doses. An inducer in a rodent may also be capable of induction in humans, but at the clinical dose, may not be potent enough to produce any discernible effects. In addition, given the overlapping substrate specificity of P450 enzymes, induction of one minor form involved in metabolism of a particular compound may not be sufficient to alter clearance or produce toxicity. The most prevalent inducing conditions in humans are smoking, drinking, and diet. Elevated CYP1A enzymes and CYP2E1 may contribute to increased risks of cancer in smokers, drinkers, and the obese, and induced CYP2E1 probably contributes to alcohol-induced liver damage and acetaminophen hepatotoxicity in drinkers. The most clinically important drug–drug interaction involving induction are probably those involving increased first-pass metabolism of CYP3A substrates such as the cyclosporines following prolonged antibiotic or anticonvulsant treatment.

ACKNOWLEDGMENTS

The research in the authors' laboratories is supported by grants from NIEHS, NIAAA, USDA and the Swedish Medical Research Council.

REFERENCES

1. D. E. Price, A. Mehta, B. K. Park, A. Hay, and M. P. Feely, The effect of low-dose phenobarbitone on three indices of hepatic microsomal enzyme induction, Br. J. Clin. Pharmacol. *22*: 744–747 (1986).

2. B. K. Park, N. R. Kirtteringham, and M. Pirmohamed, Relevance of induction of human drug-metabolizing enzymes: Pharmacological and toxicological Implications, Specificity and Variability of Drug Metabolism (G. Aván, L. P. Balant, P. R. Bechtel, et al., eds.), COST B1, Office for Official Publications of the European Communities EU, Luxembourg, 1995, pp. 169–190
3. F. P. Guengerich, Human cytochrome P450 enzymes, Cytochrome P450: Structure, Mechanisms and Biochemistry (P. R. Ortiz de Montellano, ed.) Plenum Press, New York, 1995 pp. 473–575.
4. F. J. Gonzalez, R. C. Skoda, S. Kimura, M. Umeno, U. M. Zanger, D. W. Nebert, H. V. Gelboin, J. P. Hardwick, and U. A. Meyer, Characterization of the common genetic defect in humans deficient in debrisoquine metabolism, Nature *331*:442–446 (1988).
5. W. Zilly, D. D. Breimer, and E. Richter, Stimulation of drug metabolism by rifampicin in patients with cirrhosis or cholestasis measured by increased hexobartbital and tolbutamide clearance, Eur. J. Clin. Pharmacol. *11*:287–293 (1977).
6. H. M. Bolt, H. Kappus, and U. Bolt, Effect of rifampicin treatment on the metabolism of oestradiol and 17b-ethynyloestradiol by human liver microsomes, Eur. J. Clin. Pharmacol. *8*:301–307 (1975).
7. B. K. Park, and N. R. Kiteringham, Relevance of and means of assessing induction and inhibition of drug metabolism in man, Progress in Drug Metabolism G. G. Gibson, ed.), Taylor & Francis, London, 1989 pp. 1–60.
8. J. C. Kolars, W. M. Awmi, R. M. Merion, and P. B. Watkins, First pass metabolism of cyclosporin by the gut, Lancet 338:1488–1490 (1991).
9. D. J. Back, and M. L. E. Orme, Pharmacology of the Contraceptive Steroids, Raven Press, New York, 1994, pp. 405–427.
10. L. B. Seeff, B. A. Cuccrevini, H. J. Zimmerman, E. Adler, and S. B. Benjamin, Acetaminophen hepatotoxicity in alcoholics: A therapeutic misadventure, Ann. Intern. Med. *104*:399–404 (1986).
11. J. Gut, U. Christen, and J. Huwyler, Mechanisms of halothane toxicity: Novel insights, Pharmacol. Ther. *58*:133–155 (1993).
12. I. S., Gauntlett, D. D. Koblin, M. R. Fahey, K. Greunke, L. D. Waskell, L. and E. I. Eger, Metabolism of enflurane in patients receiving isoniazid, Anesth. Analg. *69*:245–249 (1989).
13. R. I. Mazze, R. E. Woodruff, and M. E. Heerdt, Isoniazid-induced defluorination in humans, Anesthesiology *57*:5–8 (1982).
14. D. J. Waxman, and L. Azaroff, Phenobarbital induction of cytochrome P450 gene expression, Biochem. J. *281*:577–592 (1992).
15. M. J. J. Ronis, and E. Hodgson, Microsomal monooxygenases in insects, Xenobiotica *19*: 1077–1092 (1989).
16. M. A. Schuler, C. F. Hung, J. S. Chen, H. Prapaipong, and M. R. Berenbaum, Adaptation of insects to toxic chemicals in plants, XIth International Symposium on Microsomes and Drug Oxidations, UCLA, 1996, p. 74.
17. C. H. Walker, Species variation in some hepatic microsomal drug metabolizing enzymes, Prog Drug Metab *5*:113–164 (1980).
18. J. Pan, J. Y. Hong, B. L. Ma, S. M. Ning, S. R. Paranawithana, and C. S. Yang, Transcriptional activation of cytochrome P450 2B1/2 genes in rat liver by diallyl sulfide, a compound derived from garlic, Arch. Biochem. Biophys. *302*:337–342 (1993).
19. D. J. Waxman and T. K. H. Chang, Hormonal regulation of liver cytochrome P450 enzymes, Cytochrome P450: Structure, Mechanism and Biochemistry (P.R. Ortiz de Montellano, ed.), Plenum Press, New York, 1995, pp. 391–417.
20. E. G. Schuetz, and P. S. Guzelian, Induction of cytochrome P450 by glucocorticoids in rat liver II: Evidence that glucocorticoids regulate induction of cytochrome P450 by a nonclassical receptor mechanism, J. Biol. Chem. *259*:2007–2012 (1984).
21. S. A. Wrighton E. G. Schuetz, P. B. Watkins, P. Maurel, J., Barwick, B. S. Bailey, H. T.

Hartle, B. Young, and P. S. Guzelian, Demonstration in multiple species of inducible hepatic cytochromes P450 and their mRNAs related to the glucocorticoid-inducible P450 of the rat, Mol. Pharmacol. *28*:312–321 (1985).

22. C. H Gleiter, Gender differences in pharmacokinetics, Eur. J. Drug Metab. Pharmacokinet. *21*:123–128 (1996).
23. N. W. Cheung, C. Liddle, S. Coverdale, J. C. Lou, and S. C. Boyages, Growth hormone treatment increases cytochrome P450-mediated antipyrine clearance in man, J. Clin. Endocrinol. Metab. *81*:1999–2001 (1996).
24. P. M. Fernandez-Salguero, D. M. Hilbert, S. Rudikoff, J. M. Ward, and F. J. Gonzalez, Study of the dioxin-binding Ah receptor by gene targeting, XIth International Symposium on Microsomes and Drug Oxidations, UCLA, 1996, p. 125.
25. A. K. Jaiswal, Regulation of expression and induction of NADP(H):Quinone oxidoreductase (DT diaphorase) gene, XIth International Symposium on Microsomes and Drug Oxidations, UCLA, 1996, p. 141.
26. M. J. J. Ronis, M. Ingelman-Sundberg, and T. M. Badger, Induction, suppression and inhibition of multiple cytochrome P450 isozymes in the male rat and bobwhite quail. (*Colinus virginianus*) by ergosterol biosynthesis inhibiting fungicides, Biochem. Pharmacol. *48*:1953–1965 (1994).
27. H. Doostdar, M. H. Grant, W. T. Melvin, C. R. Wolf, and M. D. Burke, The effects of inducing agents on cytochrome P450 and UDP-glucuronyltransferase activities in human HepG2 hepatoma cells, Biochem. Pharmacol. *46*:629–635 (1993).
28. F. Morel, O. Fardel, D. J. Meyer, S. Langouet, K. S. Gilmore, B. Meunier, C. P. Tu, T. W. Kensler, B. Ketterer, and A. Guillouzo, Preferential increase of glutathione *S*-transferase class alpha transcripts in cultured human hepatocytes by phenobarbital, 3-methylcholanthrene and dithiolethiones, Cancer Res. *53*:231–234 (1993).
29. E. G. Schuetz, W. T. Beck, and J. D. Schuetz. Modulators and substrates of P-glycoprotein and cytochrome P450 3A coordinately up-regulate these proteins in human colon carcinoma cells, Mol. Pharmacol. *49*:311–318 (1996).
30. M. A. Leo, and C. S. Lieber, Retinoic acid metabolism by a system reconstituted with cytochrome P450, Arch. Biochem. Biophys. *234*:302–312 (1984).
31. M. J. J. Ronis, C. K. Lumpkin, A Johansson, M. Ingelman-Sundberg, and T. M. Badger P450-EtOH2: A putative ethanol-inducible member of gene family IIC under different regulation than P450 IIE1, Drug Metabolizing Enzymes: Genetics, Regulation and Toxicology (M. Ingelman-Sundberg, J.-A. Gustafsson, and S. Orrenius, eds.), Karolinska Institutet, Stockholm, 1992, p. 92.
32. R. M. Hutabarat, and G. S. Yost, Purification and characterization of an ethanol-induced UDP-glucuronosyltransferase, Arch. Biochem. Biophys. *273*:16–25 (1989).
33. S. Safe and V. Krishnan, Cellular and molecular biology of aryl hydrocarbon (Ah) receptor-mediated gene expression, Arch. Toxicol *17*:S99–115 (1995).
34. C. Bradfield, Genetic models of the Ah receptor signaling pathway, XIth International Symposium on Microsomes and Drug Oxidations, UCLA, 1996, p. 124.
35. P. Fernandez-Salguero, T. Pineau, D. M. Hilbert, T. McPhail, S. S. Lee, S. Kimura, D. W. Nerbert, S. Rudikoff, J. M. Ward, and F. J. Gonzalez, Immune system impairment and hepatic fibrosis in mice lacking the dioxin binding Ah receptor, Science *268*:722–726 (1995).
36. R. Voorman, and S. D. Aust, Inducers of cytochrome P450*d*: Influence on microsomal catalytic activities and differential regulation by enzyme stabilization, Arch. Biochem. Biophys. *262*:76–84 (1988).
37. M. J. J. Ronis, K. O. Lindros, and M. Ingelman-Sundberg, The CYP2E family, Cytochromes P450: Metabolic and Toxicological Aspects, (C. Ioannides, ed.), CRC Press, Boca Raton, 1996, pp. 211–239.
38. P. B. Watkins, S. A. Wrighton, E. G. Schuetz, P. Maurel, and P. S. Guzelian, Macrolide antibiotics inhibit the degradation of the glucocorticoid-responsive cytochrome P450p in rat

hepatocytes in vivo and in primary monolayer culture, J. Biol. Chem. *261*:6264–6271 (1986).
39. T. A. Kocarek, E. G. Schuetz, and P. S. Guzelian, Regulation of phenobarbital-inducible cytochrome P450 2B1/2 mRNA by lovastatin and oxysterols in primary cultures of adult rat hepatocytes, Toxicol. Appl. Pharmacol. *120*:298–307 (1993).
40. G. H. Perdew, Association of the Ah receptor with the 90 kDa heat shock protein, J. Biol. Chem. *263*:13802–13805 (1988).
41. K. Gradin, M. L. Whitelaw, R. Toftgard, L. Poellenger, A. Berhard. A tyrosine kinase-dependent pathway regulates ligand-dependent activation of the dioxin receptor in human keratinocytes, J. Biol. Chem. *269*:23800–23807 (1994).
42. O. Hankinson, A genetic analysis of processes regulating cytochrome P450 1A1 expression, Adv. Enzyme Regul. *34*:159–171 (1994).
43. H. Kikuchi, H. Kato, M. Mizuno, A. Hossain, S. Ikawa, J. Miyazaki, M. Watanabe, Differences in inducibility of CYP1A1-mRNA by benzimidazole compounds between human and mouse cells: Evidence of a human-specific signal transduction pathway for CYP1A1 induction, Arch. Biochem. Biophys. *334*:235–240 (1996).
44. J. McGuire, K. Gradin, M. Lindebro, I. Pongratz, C. Antonsson, P. Dzeletovic, M. L. Kallio, M. L. Whitelaw, J-A Gustafsson, and L. Poellinger, Mechanism of conditional regulation of bHLH/PAS transcription factors, XIth International Symposium on Microsomes and Drug Oxidations, UCLA, 1996, p. 127.
45. M. R. Juchau, Y. Huang, H. Boutelet-Bochan, and H. L. Yang, Expression of P450 cytochromes in human and rodent embryonic tissues during the period of organogenesis: Implications for teratogenesis, XIth International Symposium on Microsomes and Drug Oxidations, UCLA, 1996, p. 106.
46. J. Foussat, P. Costet, P. Lesca, G. Steiblen, A. Pfouhl-Leszkowicz, C. Hundieker, C. Esser, P. Galtier, and T. Pineau, Use of a deficient mouse line to investigate the role of the aryl hydrocarbon receptor in toxicology and development, XIth International Symposium on Microsomes and Drug Oxidations, UCLA, 1996, p. 95.
47. J. C. Cook, and E. Hodgson, Induction of cytochrome P450 in congenic C57BL/6J mice by isosafrole: Lack of correlation with the Ah locus, Chem. Biol. Interact. *58*:233–240 (1986).
48. B. L. Allen-Hoffmann, M. A. Weitzel, and C. M. Sadek, Adhesion molecule interactions with the Ah receptor in signal transduction, XIth International Symposium on Microsomes and Drug Oxidations, UCLA, 1996, p. 130.
49. J. M. Mathis, W. H. Houser, E. Bresnick, J. A. Cidlowski, R. N. Hines, R. A. Prough, and E. R. Simpson, Glucocorticoid regulation of the rat cytochrome P450c (P4501A1) gene: Receptor binding within intron I, Arch. Biochem. Biopsy. *269*:93–105 (1989)
50. M. Backlund, I. Johansson, and M. Ingelman-Sundberg, Non-ligand dependent activation of the Ah-receptor caused by omeprazole in the rat hepatoma cell line H4-II-E, XIth International Symposium on Microsomes and Drug Oxidations, UCLA, 1996, p. 230.
51. P. D. Boucher, and R. N. Hines, In vitro binding and functional studies comparing human CYP1A1 negative regulatory element with the orthologous sequences from rodent genes, Carcinogenesis *16*:383–392 (1995).
52. P. D. Boucher, M. P. Piechocki, and R. N. Hines, Partial characterization of the human CYP1A1 negatively acting transcription factor and multinational analysis of its cognate DNA recognition sequence, Mol. Cell. Biol. *15*:5144–5151 (1995).
53. L. C. Quattrochi, T. Vu, and R. H. Tukey, The human CYP1A2 gene and induction by 3-methylcholanthrene. A region of DNA which supports Ah receptor binding and promoter specific induction, J. Biol. Chem. *269*:6949–6954 (1994).
54. A. Raha, T. Joyce, S. Gusky, and E. Bresnick, Glycine *N*-methyltransferase is a mediator of cytochrome P450 1A1 gene expression, Arch. Biochem. Biophys. *322*:395–404 (1995).
55. P. A. Harper, J. M. Y. Wong, and A. B. Okey, Variation in structure, function and regulation

of the Ah receptor, XIth International Symposium on Microsomes and Drug Oxidations, UCLA, 1996, p. 131.

56. Y. Fuji-Kuriyama, A. Kobayashi, J. Mimura, M. Ema, M. Morita, Y. Kikuchi, and K. Sogawa, Cooperative interaction between AhR/Arnt and Sp 1 for induction of CYP1A1, XIth International Symposium on Microsomes and Drug Oxidations, UCLA, 1996, p. 128.
57. R. H. Tukey, M. Buck, and Y.-H. Chen, Transcriptional activation of the CYP1A genes are dependent upon protein kinase C, XIth International Symposium on Microsomes and Drug Oxidations UCLA, 1996, p. 132.
58. J. P. Whitlock, and M. S. Denison, Induction of cytochrome P450 enzymes that metabolize xenobiotics, Cytochrome P450: Structure, Mechanism and Biochemistry (P. R. Ortiz de Montellano, ed.), Plenum Press, New York, 1995, pp. 169–190.
59. J. He, and A. J. Fulco, A barbiturate regulated protein binding to a common sequence in the cytochrome P450 genes of rodents and bacteria, J. Biol. Chem. *266*:7864–7869 (1991).
60. G. Shaw, and A. J. Fulco, Inhibition by barbiturates of the binding of Bm3R1 repressor to its operator site on the barbiturate inducible cytochrome P450BM-3 gene of *Bacillus megaterium*, J. Biol. Chem. *268*:2997–3004 (1993).
61. A. J. Fulco, J. S. He, Q. Liang, G-C. Shaw, J. Zhang, and S. Radharkrishnan, Regulation of gene expression by barbiturates and related compounds, XIth International Symposium on Microsomes and Drug Oxidations, UCLA, 1996, p. 62.
62. J. S. Sidhu and C. J. Omiecinski, cAMP-associated inhibition of phenobarbital-inducible cytochrome P450 gene expression in primary hepatocyte cultures, J. Biol. Chem. *270*:12762–12773 (1995).
63. C. J. Omiecinski, R. Ramsden, K. M. Sommer, and J. S. Sidhu, Cytochrome P450 CYP2B1/2 gene induction by phenobarbital: Molecular analysis using transgenic mouse and primary rat hepatocyte models, XIth International Symposium on Microsomes and Drug Oxidations, UCLA, 1996, p. 88.
64. S. A. Wrighton, J. C. Stevens, G. W. Becker, and M. VandenBranden, Isolation and characterization of human liver cytochrome P450 2C19: Correlation between 2C19 and *S*-mephenytoin 4′-hydroxylation, Arch. Biochem. Biophys. *306*:240–245 (1993).
65. J. A. Goldstein, M. B. Faletto, M. Romkessparks, T. Sullivan, S. Kitareewan, J. L. Raucy, J. M. Laser, and B. I. Ghanayem, Evidence that CYP2C19 is the major (*S*) mephenytoin 4′-hydroxylase in humans, Biochemistry, *33*:1743–1752 (1994).
66. E. L. LeCluyse, M. D. Ribadeneira, S. Diamond, J. Forster, S-M. Huang, A. Madan, K. Carroll, and A. Parkinson, Induction of CYP1A, CYP2A, CYP2B, CYP2C and CYP3A by various drugs and prototypical inducers in primary cultures of human heptocytes, ISSX Proc. *10*:196 (1996).
67. S. R. Howell M. A. Shirley, I. S. McIntosh, B. R. Hee, and E. H. Ulm, Effects of five retinoids on rat hepatic cytochrome P450, ISSX Proc. *10*:175 (1996).
68. D. J. Waxman, P. A. Ram, S-H Park, and H. K. Choi, Intermittent plasma growth hormone triggers tyrosine phosphorylation and nuclear translocation of a liver-expressed, Stat 5-related DNA binding protine, J. Biol. Chem. *270*:13262–13270 (1995).
69. P. A. Ram, S-H. Park, H. K. Choi, and D. S. Waxman, Growth hormone activation of Stat 1, Stat 3 and Stat 5 in rat liver, J. Biol. Chem. *271*:5929–5940 (1996).
70. R. E. McGee, Jr., M. J. J. Ronis, R. M. Cowherd, M. Ingelman-Sundberg, and T. M. Badger, Characterization of cytocrhome P450 2el induction in a rat hepatoma FGC-4 cell model by ethanol, Biochem. Pharmacol. *48*:1823–1833 (1994).
71. E. Eliasson, I. Johansson, and M. Ingelman- Sundberg, Substrate-, hormone- and cAMP-regulated cytochrome P450 degradation, Proc. Natl. Acad. Sci. USA *87*:3225–3330 (1990).
72. A. Zhukov, V. Werlinder, and M. Ingelman- Sundberg, Purification and characterization of two membrane bound serine proteases from rat liver microsomes active in the degradation of cytochromes P450, Biochem. Biophys. Res. Commun. 221–225 (1994).
73. E. Eliasson, S. Mkrtchian, and M. Ingelman-Sundberg, Hormone and substrate regulated

intracellular degradation of cytochrome P450 2E1 involving MgATPase-activated rapid proteolysis in the endoplasmic reticulum membrances, J. Biol. Chem. *267*:15765–15770, 1992.

74. D. J. Tierney and D. R. Koop, Degradation of cytochrome P450 2E1—selective loss after labilization of the enzyme, Arch. Biochem. Biophys. *293*:9–16 (1992).

75. M. J. J. Ronis, J. Huang, J. Crouch, C. Mercado, D. Irby, C. Valentine, C. K. Lumpkin, M. Ingelman-Sundberg, and T. M. Badger, Cytochrome P450 2E1 induction during chronic ethanol exposure occurs by a two-step mechanism associated with blood alcohol concentrations in rats, J. Pharmacol. Exp. Ther. *264*:944–950 (1993).

76. J-S. Yoo, S. M. Ning, E. J. Pantuck, and C. S. Yang, Regulation of hepatic microsomal cytochrome P450 IIE1 level by dietary lipids and carbohydrates in rats, J. Nutr. *121*:959–966 (1991).

77. J. L. Raucy, J. M. Lasker, J. C. Kraner, D. E. Salazar, C. S. Lieber, and G. B. Corcoran, Induction of cytochrome P450 IIE1 in the obese, overfed rat, Mol. Pharmacol. *39*:275–280 (1991).

78. I. Johansson, G. Ekstrom, B. Scholte, D. Puzycki, H. Jornvall, and M. Ingelman-Sundberg, Ethanol-, fasting-and acetone-inducible cytochromes P450 in the rat liver: Regulation and characteristics of enzymes belonging to the IIB and IIE subfamilies, Biochemistry *27*:1925–1934 (1988).

79. Y. Yamazoe, N. Murayama, M. Simada, S. Imaoka, Y. Funae, and R. Kato, Suppression of hepatic levels of an ethanol-inducible cytochrome P450 P450DM/j by growth hormone. Relationship between increased level of P450DM/j and depletion of growth hormone in diabetes, Mol. Pharmacol. *36*:716–720 (1989).

80. L. V. Favreau, D. M. Malchoff, J. E. Mole, and J. B. Schenkman, Response to insulin by two forms of rat hepatic microsomal cytochrome P450 that undergo major (RLM6) and minor (RLM5b) elevations in diabetes, J. Biol. Chem. *262*:14319–14327 (1987).

81. S. G. Kim, S. E. Shehin, C. States, and R. F. Novak, Evidence for increased translational efficiency in the induction of P450 IIE1 by solvents: Analysis of P450 IIE1 mRNA polyribosomal distribution, Biochem. Bophys. Res. Commun. *172*:767–774 (1990).

82. J. C. Kraner, J. M. Lasker, J. B. Corcoran, S. D. Ray, and J. L. Raucy, Induction of P450 2E1 in isolated rabbit hepatocytes: Role of increased protein and mRNA synthesis, Biochem. Pharmacol. *45*:1483–1488 (1993).

83. D. Wu, and A. I. Cederbaum, Combined effects of streptozotocin-induced diabetes plus 4-methylpyrazole treatment on rat liver cytochrome P450 2E1, Arch. Biochem. Biophys. *320*:175–180 (1993).

84. A. P. Li, D. L. Kaminski, and A. Rasmussen, Substrates of human heptic cytochrome P450 3A4, Toxicology *140*:1–8 (1995).

85. B. D. Haehner, J. C. Gorski, M. Vandenbranden, S. A. Wrighton, S. K. Janardan, P. B. Watkins, and S. D. Hall, Bimodall distribution of renal cytochrome P450 3A activity in humans, Mol. Pharmacol. *50*:52–59 (1996).

86. M. Komori, K. Nishio, M. Kitada, K. Shiramatsu, K. Muroya, M. Soma, K. Nagashima, and T. Kamataki, Fetus specific expression of a form of cytochrome P450 in human livers, Biochemistry *29*:4430–4430 (1990).

87. J. D. Schuetz, E. G. Schuetz, J. V. Thottassery P. S. Guzelian, S. Strom, and Su, D., Identification of a novel dexamethasone responsive enhancer in the human CYP3A5 gene and its activation in human and rat liver cells, Mol. Pharmacol. *49*:63–72 (1996).

88. J. L. Barwick, L. C. Quattrochi, A. S. Mills, C. Potenza, R. H. Tukey, and P. S. Guzelian, Trans-species gene transfer for analysis of glucocorticoid-inducible transcriptional activation of transiently expressed human CYP3A4 and rabbit CYP3A6 in primary cultures of adult rat and rabbit hepatocytes. Mol. Pharmacol. *50*:100–106 (1996).

89. E. G. Schuetz, A. H. Schinkel, M. V. Relling, and J. D. Schuetz, P-glycoprotein: A major determinant of rifampicin-inducible expression of cytochrome P450 3A in mice and humans, Proc. Natl. Acad. Sci. USA *93*:4001–4005 (1996).

90. G. Smith, D. J. Harrison, N. East, F. Rae, H. Wolf, and C. R. Wolf, Regulation of cytochrome P450 gene expression I human colon and breast tumor xenografts, Br. J. Cancer *68*:57–63 (1993).
91. D. W. Rosenberg, and T. Leff, Regulation of cytochrome P450 in cultured human colonic cells, Arch. Biochem. Biophys. *300*:186–192 (1993).
92. J. Muntane-Relat, J. C. Ourlin, J. Domergue, and P. Maurel, Differential effects of cytokines on the inducible expression of CYP1A1, CYP1A2 and CYP3A4 in human hepatocytes in primary culture, Hepatology *22*:1143–1153 (1995).
93. R. M. Whyatt, S. J. Garte, G. Cosma, D. A. Bell, W. Jedrychowski, J. Wahrendorf, M. C. Randall, T. B. Cooper, R. Ottman, and D. Tang, CYP1a1 messenger mRNA levels in placental tissue as a biomarker of environmental exposure, Cancer Epidemiol. Biomark. Prev. *4*: 147–153 (1995).
94. C. Kiyochara, T. Hirohata, and S. Inutsuka, The relationship between aryl hydrocarbon hydroxylase and polymorphisms of the CYP1A1 gene, Jpn. J. Cancer Res. *87*:18–24 (1996).
95. M. Lang, M. A. Butler, J. Massengill, M. Lawson, R. C. Stotts, M. Hauer-Jensen, and F. F. Kadlubar, Rapid metabolic phenotypes for acetyltransferase and cytochrome P450 1A2 and putative exposure to food-bourne heterocyclic amines increase the risk for colorectal cancer or polyps, Cancer Epidemiol. Biomark. Prev. *3*:675–682 (1994).
96. M. A. Kall, O. Vang, and J. Clausen, Effects of dietary broccoli on human in vivo drug metabolizing enzymes: Evaluation of caffeine, oestrone and chlorzoxazone metabolism, Carcinogenesis *17*:793–799 (1996).
97. E. O. Ayalogu, J. Snelling, D. F. Lewis, S. Talwar, M. N. Clifford, and C. Ioannides, Induction of hepatic CYP1A2 by the oral administration of caffeine to rats: Lack of association with the Ah locus, Biochim. Biophys. Acta *1272*:89–94 (1995).
98. J. Buchthal, K. E. Grund, A. Buchmann, D. Schrenk, P. Beaune, and K. W. Bock, Induction of cytochrome P450 1A by smoking or omeprazole in comparison with UDP-glucuronosyltransferase in biopsies of human duodenal mucosa, Eur. J. Clin. Pharmacol. *47*:431–435 (1995).
99. P. B. Watkins, H. J. Zimmerman, M. J. Knapp, S. I. Gracon, and K. W. Lewes, Hepatotoxic effects of tacrine administration in patients with Alzheimers disease, JAMA *271*:992–998, 1994.
100. R. I. Mazze, J. R. Trudell, and M. J. Cousins, Methoxyflurane metabolism and renal disfunction, Anesthesiology *35*:247–252 (1971).
101. H. H. Zhou, L. B. Anthony, A. J. Wood, and G. R. Wilkinson, Induction of polymorphic 4′-hydroxylation of *S*-mephenytoin by rifampicin. Br. J. Clin. Pharmacol. *30*:471–475 (1990).
102. V. E. Kostrubsky, S. C. Strom, S. G. Wood, S. A. Wrighton, P. R. Sinclair, and J. F. Sinclair, Ethanol and isopentanol increase CYP3A and CYP2E in primary cultures of human hepatocytes, Arch. Biochem. Biophys. *322*:516–520 (1995).
103. S. P. Carpenter, J. M. Lasker, and J. L. Raucey, Expression, induction and catalytic activity of the ethanol-inducible cytochrome P450 (CYP2E1) in human fetal liver and hepatocytes, Mol. Pharmacol. *49*:260–268 (1996).
104. I. T. Takahashi, J. M. Lasker, A. S. Rosman, and C. S. Lieber, Induction of cytochrome P450 2E1 in human liver by ethanol is caused by a corresponding increase in its encoding mRNA, Hepatology *17*:236–245 (1993).
105. D. Lucas, C. Menez, C. Girre, P. Bodenez, E. Hispard, and J. F. Menez, Decrease in cytochrome P450 2E1 as assessed by the rate of chlorzoxazone hydroxylation in alcoholics during the withdrawal phase, Alcohol. Clin. Exp. Res. *19*:362–366 (1995).
106. D. Lucas, C. Menez, C. Girre, F. Berthou, P. Bodeenez, I. Joannet, E. Hispard, L. G. Bardou, and J. F. Menez, Cytochrome P450 2E1 genotype and chlorzoxazone metabolism in healthy and alcoholic Caucasian subjects, Pharmacogenetics *5*:298–304 (1995).
107. D. O'Shea, S. N. Davis, R. B. Kim, and G. R. Wilkinson, Effect of fasting and obesity in

humans on the 6-hydroxylation of chlorzoxazone: A putative probe of CYP2E1 activity, Clin. Pharmacol. Ther. *56*:359–367 (1994).

108. B. J. Song, R. L. Veech, and P. Saenger, Cytochrome P450 IIE1 is elevated in lymphocytes from poorly controlled insulin-dependent diabetics, J. Clin. Endocrinol. Metab. *71*:1036–1040 (1990).
109. P. H. Villard, E. Seree, B. Lararelle, M. C. Therene-Fenoglio, Y. Barra, L. Attolini, B. Bruguerole, A. Durand, and J. Catalin, Effect of cigarette smoke on hepatic and pulmonary cytochromes P450 in mouse: Evidence for CYP2E1 induction in lung, Biochem. Biophys. Res. Commun. *202*:1731–1737 (1994).
110. E. M. Seree, P. H. Villard, J. L. Re, M. DeMeo, B. Larcarelle, L. Attolini, G. Dumenil, J. Catalin, A. Durand, and Y. Barra, High inducibility of mouse renal CYP2e1 gene by tobacco smoke and its possible effect on single strand breaks, Biochem. Biophys. Res. Commun. *219*:429–434 (1996).
111. A. P. Li, A. Rasmussen, L. Xu, and D. L. Kaminski, Rafampicin induction of lidocaine metabolism in cultured human hepatocytes, J. Pharmacol. Exp. Ther. *274*:673–677 (1995).
112. T. A. Kokareck, E. G., Schuetz, S. C. Strom, R. A. Fisher, and P. S. Guzelian, Comparative analysis of cytochrome P450 3A induction in primary cultures of rat, rabbit, and human hepatocytes, Drug Metab. Dispos. *23*:415–421 (1995).
113. M., Dubois, A. Pfohlleszkowicz, I. Dewarziers, and P. Kremers, Selective induction of the CYP3A family by endosulfan and DNA-adduct formation in different hepatic and hepatoma cells, Environ. Toxicol. Pharmacol. *1*:249–256 (1996).
114. R. Curi-Pedrosa, M. Daujat, L. Pichard, J. C., Ourlin, P. Clair, L. Gervot, P. Lesca, J. Domergue, H. Joyeux and G. Fourtanier, Omeprazole and lansoprazole are mixed inducers of CYP1A and CYP3A in human hepatocytes in primary culture, J. Pharmacol. Exp. Ther. *269*:384–392 (1994).
115. H. Raunio, personal communication.
116. N. H. Shear, S. P. Spielberg, M. Cannon, and M. Miller, Anticonvulsant hypersensitivity syndrome: in vitro risk assessment, J. Clin. Invest. *82*:1826–1832 (1988).
117. P. S. Friedmann, I. Sytickland, P. Pirmohamed, and B. K. Park, Investigation of the mechanisms in toxic epidermal necrolysis induced by carbamazepine, Arch. Dermatol. *130*:598–604 (1994).
118. J. T. Backman, K. T. Olkkola, and P. J. Neuvonen, Rifampin drastically reduces plasma concentrations and effects of oral midazolam, Clin. Pharmacol. Ther. *59*:7–13 (1996).
119. M. F. Fromm, D. Busse, H. K. Kroemer, and M. Eichelbaum, Differential induction of prehepatic and hepatic metabolism of verapamil by rifampin, Hepatology *24*:796–801 (1996).
120. P. Pellinen, F. Stenback, A. Kojo, P. Honkakoshi, H. V. Gelboin, and M. Pasanen, Regenerative changes in hepatic morphology and enhanced expression of CYP2B10 and CYP3A during daily administration of cocaine, Hepatology *23*:515–523 (1996).
121. S. S. Narayan, W. L. Hayton, and G. S. Yost, Chronic ethanol consumption causes increased gluronidation of morphine in rabbits, Xenobiotica *21*:515–524 (1991).
122. J. D. Hayes, R. McLeod, D. J. Pulford, and G. E. Neal, Regulation and activity of glutathione *S*-transferases, ISSX Proc. *10*:12 (1996).
123. C. D. Klaasson, L. Liu, and R. T. Dunn II, Regulation of sulfotransferase mRNA expression in male and female rats of various ages, ISSX Proc. *10*:9, (1996).
124. P. A. Egner, T. W. Kensler, T. Prestera, P. Talalay, A. H. Libby, H. H. Joyner, and T. J. Curphey, Regulation of phase 2 enzyme induction by oltipraz and other dithioethiones, Carcinogenesis *15*:177–181 (1994).
125. J. J. Bogaards, H. Verhagen, M. I. Willems, G. van Popel, and P. J. van Bladeren, Consumption of brussels sprouts results in elevated alpha-class glutathione *S*-transferase levels in human blood plasma, Carcinogenesis *15*:1073–1075 (1994).
126. T. W. Kensler, T. Primiano, J. Groopman, Y. Zhang, and P. Talalay, Cancer chemoprotection by inducers of carcinogen detoxication enzymes, ISSX Proc. *10*:18 (1996).

127. L. Gu, F. J. Gonzalez, W. Kalow, and B. K. Tang, Biotransformation of caffeine, paraxanthene, theobromine and theophylline by cDNA expressed human CYP1A2 and CYP2E1, Pharmacogenetics *2*:73–77 (1992).
128. W. Tassaneeyakul, Z. Mohamed, D. J. Birkett, M. E. McManus, M. E. Veronese, R. H. Tukey, L. C. Quattrochi, F. J. Gonzalez, and J. O. Miners, Caffeine as a probe for human cytochromes P450, validation using cDNA expression, immunoinhibition and microsomal kinetic and inhibitor probes, Pharmacogenetics *2*:173–183 (1992).
129. A. W. Dreisbach, N. Ferencz, N. Eddy Hopkins, M. G. Fuentes, A. B. Rege, W. J. George, and J. J. L. Lertora, Urinary excretion of 6-hydroxychlorzoxazone as an index of CYP2E1 activity, Clin. Pharmacol. *58*:498–505 (1995).
130. Y. Hosmans, J. P. Deager, V. Van den Berge, M. Abrassart, and C. Harvengt, Effects of simvastatin and pavastatin on 6b-hydroxycortisol excretion a potential marker of cytochrome P450 3A, Pharmacol. Res. Commun. *28*:243–247 (1993).
131. C. M. Stein, M. T. Kinirons, T. Pincus, G. R. Wilkinson, and A. J. Wood, Comparison of the dapsone recovery ratio and the erythromycin breath test as in vivo probes of CYP3A activity in patients with rheumatoid arthritis receiving cyclosporine, Clin. Pharmacol. Ther. *59*:47–51 (1996).
132. D. R. Jones, J. C. Gorski, M. A. Hamman, and S. Hall, Quantifation of dextromethorphan and metabolites: A dual phenotypic marker for cytochrome P450 3A4/5 and 2D6 activity, J. Chromatogr. *676B*:105–111 (1996).
133. J. Ducharme, S. Abdullah, and I. W. Wainer, Dextromethorphan as an in vivo probe for the simultaneous determination of CYP2D6 and CYP3A4 activity, J. Chromatogr. *678B*:113–128 (1996).
134. P. B. Watkins, K. S. Lown, and R. Fontana, Effects of diet and medications on CYP3A4 and MDR1 in human small bowel, ISSX Proc. *10*:39 (1996).
135. J. P. Vanden Heuvel, G. C. Clark, C. L. Thompson, Z. McCoy, C. R. Miller, G. W. Lucier, and D. A. Bell, CYP1A1 mRNA levels as a human exposure biomarker: Use of quantitative polymerase chain reaction to measure CYP1A1 expression in human peripheral blood lymphocytes, Carcinogenesis *14*:2003–2006 (1993).
136. J. L. Raucy, G. Curley, and S. P. Carpenter, Use of lymphocytes for assessing ethanol-mediated alterations in the expression of hepatic cytochrome P450 2E1, Alcohol Clin. Exp. Res. *19*:1369–1375 (1995).
137. M. Umeno, O. W. MacBride, C. S. Yang, H. V. Gelboin, and F. J. Gonzalez, Human ethanol-inducible P450 2E1: Complete gene sequence, promoter characterization, chromosome mapping and cDNA directed expression, Biochemistry *27*:9006–9012 (1988).

11
Sites of Metabolism: Lung

Garold S. Yost
University of Utah, Salt Lake City, Utah

I. INTRODUCTION

Metabolism of xenobiotic compounds in pulmonary tissues is a complex and only partially characterized issue. Several reviews have been written on the subject of respiratory tract metabolism of xenobiotics (1–5). Complexity arises because of the many different cell types—over 40—in the lung, and the remarkable variety of the various functions of the cells. Functions range from secretory cells with cilia that are designed to move foreign particles up the tracheobronchial tract, to endothelial cells that line the pulmonary vasculature and provide a junction between vascular fluids and respiratory cells. The drug-metabolizing functions of many of the types of cells have not been investigated, and those cells that have been the subject of significant studies still remain only partially characterized. Descriptions of metabolic processes that are derived from whole-lung homogenates are not only overly simplistic, but may frequently lead to erroneous conclusions about the susceptibility of specific cells to toxic insult, or about the capability of certain cells or regions of the tracheobronchial tract to metabolize xenobiotics. Unfortunately, a large majority of the studies concerning xenobiotic metabolism have used whole-lung homogenates. Nevertheless, this chapter will present information that has been obtained on specific cells, regions of the respiratory tract, or whole-lung homogenates, without an attempt to differentiate among cell types or mixtures of cells.

It is doubtful as a general rule that the drug-metabolizing enzymes play an important role in the overall biodistribution of drugs and other xenobiotics (i.e., most of the metabolism of foreign compounds probably occurs in other organs such as the intestine or liver). However, examples of physiological compounds that are subject to accumulation or metabolic clearance by the lung include vasoactive peptides, prostaglandins, endogenous amines, and oligoamines, such as spermine (see review Ref. 5). The accumulation of nitrogen-containing drugs, such as propranolol, fentanyl, and imipramine, by first-pass retention has been observed (see review Ref. 5). One example of the extensive metabolism of a xenobiotic by lung tissues is the metabolism of inhaled butadiene for which both the monoxide and diepoxide were found in higher concentrations in lung tissues than in liver tissues of mice, presumably as a result of metabolism in lung, not by selective accumulation of the metabolites (6).

Xenobiotic metabolism in respiratory tissues can be examined by looking at the chemical classes that are metabolized, or by focusing on the enzymes responsible for the biotransformations. This review will employ the latter approach because the biochemistry of lung enzymes often differs from other organs, in that selective expression of drug metabolism genes is often observed. Frequently, such as for several of the cytochrome P450 genes, one of the flavin-containing monooxygenases genes, and one UDP-glucuronosyltransferase gene, these genes are solely expressed in respiratory tissues.

II. REDOX ENZYMES

A. NADPH Cytochrome P450 Reductase

NADPH cytochrome P450 reductase enzyme is required for the reduction of cytochrome P450 enzymes to permit substrate oxidation. Only one gene product has been identified. Although the reductase enzyme is generally assumed to be tightly coupled to P450 enzymes within the endoplasmic reticulum, the reductase has not been immunochemically detected in certain cells within the lung, type I alveolar cells and vascular endothelial cells, that possess demonstrable P450 protein (7–9). It is doubtful, therefore, that the P450 enzymes in these cells are catalytically active. Similar localizations of the reductase in bronchiolar and bronchial epithelial cells, Clara cells, and alveolar type II cells have been demonstrated in human lung tissues (10).

B. Cytochrome P450

There are a large and growing number of P450 enzymes that are expressed in respiratory tissues, either in addition to expression in other organs, or selectively expressed only in this organ. The selective expression and catalytic participation of certain P450 enzymes in lung cells is a major factor in the selective pneumotoxicity or carcinogenicity of several xenobiotics (11,12), such as naphthalene (13), 4-ipomeanol (14,15), 3-methylindole (16,17), butylated hydroxytoluene (18,19), and 4-(methylnitrosamino)-1-(3-pyridyl)-1-butanone (20). Examples of several of these toxicants and their putative reactive intermediates and P450 enzymes that catalyze the bioactivation processes are shown in Fig. 1.

Cytochrome P450 enzymes that have been identified by immunohistochemical, catalytic, or transcriptional determinations in the respiratory tissues of animals and humans

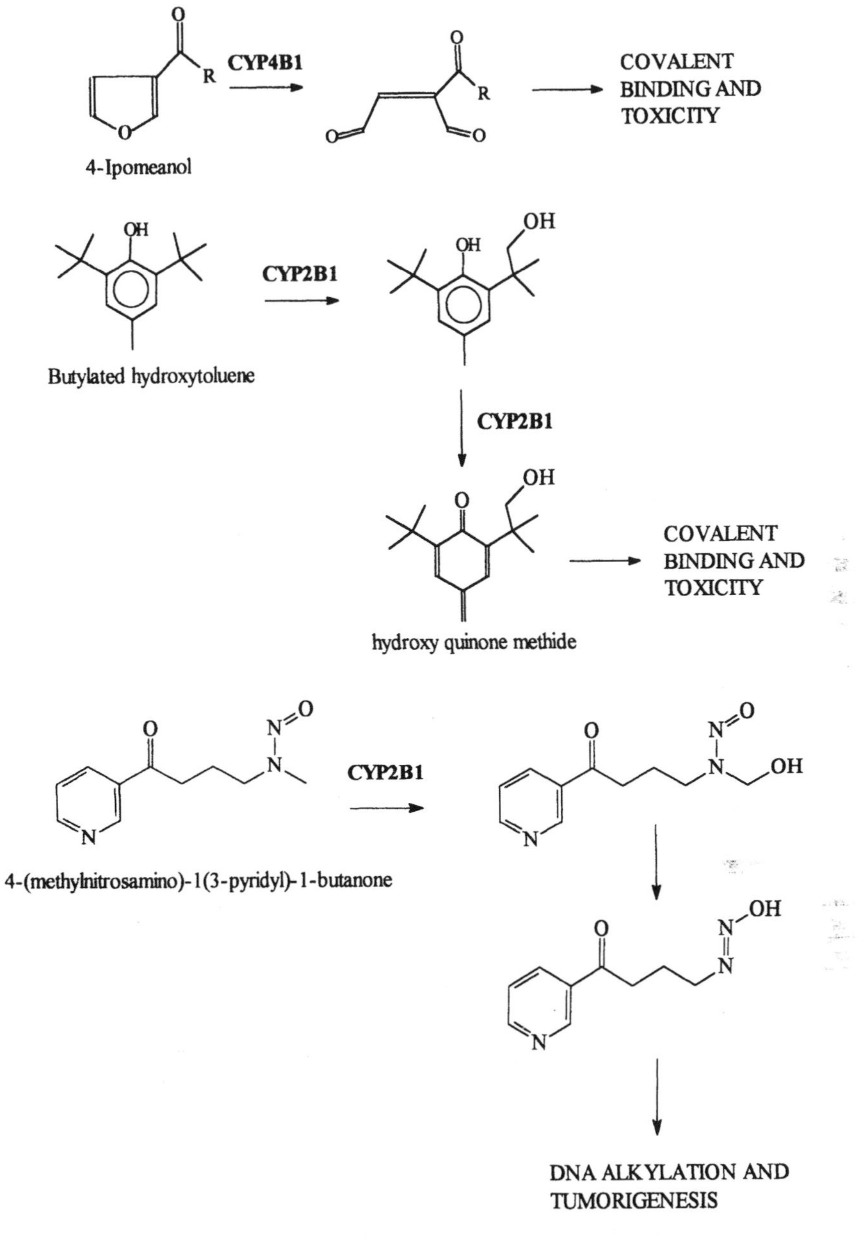

4-Ipomeanol

CYP4B1

COVALENT BINDING AND TOXICITY

Butylated hydroxytoluene

CYP2B1

CYP2B1

hydroxy quinone methide

COVALENT BINDING AND TOXICITY

4-(methylnitrosamino)-1(3-pyridyl)-1-butanone

CYP2B1

DNA ALKYLATION AND TUMORIGENESIS

Fig. 1 Bioactivation of several lung toxicants by cytochrome P450 enzymes from the CYP2B and CYP4B subfamilies. The postulated ultimate reactive intermediates of each toxicant are shown.

include: CYP1A1 (21), CYP1B1 (22), CYP2A (23), CYP2A10 (24), CYP2A11 (24), CYP2B (25), CYP2E1 (26,27), CYP2F1 (28), Cyp2f2 (13), CYP2F3 (29), CYP2G1 (30), CYP3A4 (27), CYP4B1 (25), and CYP4B2 (31). The P450 contents in whole-lung microsomal fractions range from 0.01 nmol/mg microsomal protein for humans (32) to 1.04 nmol/mg protein for goats (33). These values generally are a factor of 10 below the P450 contents for hepatic microsomes from the same species (3). Important contributions of P450 enzymes in the respiratory tract toward the biodistribution of inhaled or ingested drugs or xenobiotics must, therefore, depend on factors such as concentration of specific enzymes in certain cells or regions of the respiratory tract, or on the selective expression of P450 genes in the respiratory tract. It is unlikely that a significant percentage of the total metabolism of most drugs would be produced by P450 enzymes of the respiratory tract, but selective expression of genes could easily lead to the localized production of reactive intermediates that might produce damage to these tissues.

The 2A10, 2A11, and 2G1 enzymes are specifically expressed in nasal tissues, and particularly in the olfactory epithelium, where CYP2A10 and CYP2A11 constitute over 90% of total P450 in this anatomical region of the rabbit. In addition the 2B4 and 4B1 enzymes constitute over 90% of the total P450 content of rabbit lung (25), and the orthologues of these enzymes in other species often appear to be major contributors to the P450 contents and catalytic activities of lung tissues. The enzymes in the 2F subfamily are selectively expressed in lung tissues (28,34), and they have also been identified in human placenta (35).

CYP2B4 from rabbit lung has high substrate selectivity for the *N*-demethylation of benzphetamine (36), and its level in lung tissues is not increased by treatment of rabbits or other animals with phenobarbital, although phenobarbital is an effective inducer of this enzyme in the liver. A highly selective suicide substrate for this enzyme, *N*-α-methylbenzyl-1-aminobenzotriazole has been used to efficiently inhibit this enzyme without significantly affecting other P450 enzymes at low concentrations. This inhibitor has been used in pulmonary microsomal incubations from rabbits (37), guinea pigs (38), and goats (39). The inhibitor was used to demonstrate that CYP2B4 is an efficient catalyst of the formation of epoxyeicosatrienoic acids from arachidonic acid (40). The human *CYP2B7* gene is transcribed in human lung tissues (41).

The prototypical substrate for the CYP4B1 enzyme from rabbit lung is 2-aminofluorene (36), and 4-ipomeanol, a classic pneumotoxicant, is metabolized effectively by this enzyme in rabbit and rat lung tissues. However, the human 4B1 enzyme, expressed in hepatoma cells using a vaccinia virus expression system, did not bioactivate 4-ipomeanol, although the rabbit enzyme expressed in the same system was an efficient catalyst of 4-ipomeanol bioactivation (42). The rabbit cDNA has been expressed in mouse C3H/10T½ cells, and in these cells the enzyme bioactivated 4-ipomeanol to a cytotoxic intermediate (43). CYP4B1 also metabolized 2-aminoanthracene in these cells to produce cytotoxicity.

The enzymes from the 2B and 4B subfamilies are often expressed to a much higher extent in the lung than in the liver, and higher expression often leads to enhanced bioactivation of xenobiotics. Examples include the following:

1. The selective expression of a 2B enzyme in mouse lung that oxidizes butylated hydroxytoluene by subsequent hydroxylation and dehydrogenation steps to produce the ultimate pneumotoxic electrophile (18). This enzyme is not expressed extensively in mouse liver unless the animal is treated with phenobarbital.

2. CYP4B1 is selectively expressed in rat lung, not liver, and this enzyme efficiently catalyzes the bioactivation of 4-ipomeanol (44) to a pneumotoxic electrophile.
3. The pulmonary CYP4B2 enzyme is selectively transcribed in lung tissues of goats (34), and most of the bioactivation of 3-methylindole (3MI) to the electrophilic 3-methyleneindolenine intermediate is catalyzed by this enzyme (29).

Another example of a P450 subfamily that is highly selectively expressed in pulmonary tissues is the 2F subfamily. Three members of this subfamily have been identified from human (28), mouse (13), and goat (29). Rabbits, rats, goats, and humans show high selectivity for transcription in pulmonary tissues, rather than hepatic tissues (28,34). CYP2F1 catalyzes the metabolism of ethoxycoumarin, propoxycoumarin, and pentoxyresorufin (28), and the bioactivation of 3-methylindole (45) and 4-(methylnitrosoamino)-1-(3-pyridyl)-1-butanone (NNK; 46).

Human CYP2F1 has been expressed in hepatoma cells using a vaccinia virus vector and catalyzes the bioactivation of the pneumotoxin 3-methylindole to a covalent-binding intermediate (45). This enzyme possessed higher catalytic activity toward the bioactivation of 3MI than any other human P450 enzyme, including CYP4B1, CYP1A2, CYP3A4, and CYP2E1. The production of 3-methyleneindolenine, the putative electrophilic methylene imine reactive intermediate of 3MI (Fig. 2), was determined with the hepatoma cells that expressed human P450 enzymes. Again, the human CYP2F1 produced the highest

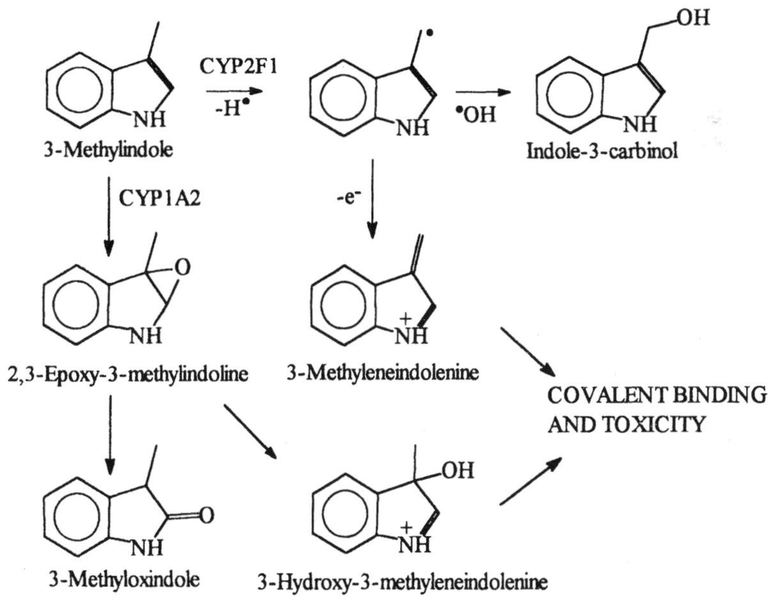

Fig. 2 Bioactivation of the pulmonary toxin, 3-methylindole, by cytochrome P450 enzymes from the CYP2F and CYP1A subfamilies. Three reactive intermediates that may participate in the pneumotoxicity of 3-methylindole are 3-methyleneindolenine, 2,3-epoxy-3-methylindoline, and 3-hydroxy-3-methyleneindolenine. Oxidation of 3-methylindole can proceed through two distinctly different pathways, dehydrogenation or ring oxygenation, that are mediated by different cytochrome P450 enzymes.

amounts of the reactive intermediate when compared with 11 other human P450 enzymes (16). Thus, covalent binding was correlated to the production of the reactive electrophile, and a P450 enzyme that is selectively expressed in lung tissues in humans (28) was the most efficient catalyst of the bioactivation process. Other reactive intermediates have been postulated to be at least partially responsible for the toxicity of 3MI, and it is likely that the epoxide is produced predominantly by other P450 enzymes such as CYP1A2 (see Fig. 2).

The goat 2F subfamily member of P450 genes was cloned from a goat lung library and the cDNA expressed in *Escherichia coli* after suitable modifications at the 5′-end (29). The expressed enzyme demonstrated remarkable catalytic specificity for the dehydrogenation of 3MI, without detectable formation of any oxygenated products. Thus, the organ-selective toxicity of 3MI for lung tissues can be predominantly explained by the selective expression of an enzyme that has efficient catalytic capabilities for the formation of the toxic 3-methyleneindolenine.

The mouse Cyp2f2 enzyme is the primary stereoselective catalyst of the 1*R*,2*S*-naphthalene oxide, the putative reactive epoxide, that produces Clara cell necrosis in mice. This enzyme is highly localized to the Clara cells in distal bronchioles in mouse lung (47), and the production of the reactive epoxide was much higher in the microdissected distal airways from mice than from hamsters or rats, two species with less susceptibilities to naphthalene than mice. These studies also demonstrated that detoxification of the reactive epoxide by glutathione conjugation or epoxide hydrolase activity was not significantly lower in the most susceptible species or in terminal airways where damage was predominantly observed. Thus, this elegant work has demonstrated that the production of the toxic intermediate by a selectively expressed P450 enzyme is the primary mechanism for susceptibility to this toxicant.

Another P450 enzyme that is of primary interest in the metabolism and bioactivation of xenobiotics is CYP2E1, the alcohol-inducible enzyme. Several studies have demonstrated that this enzyme is expressed in lung tissues in animals (48) and in humans (27). Some studies have suggested that this enzyme can be induced by ethanol (49) or pyridine (50) in lung tissues and in nasal tissues. The CYP2E1 enzyme is a primary catalyst of small halogenated hydrocarbons (51), and several of these chemicals cause Clara cell damage in animals. The pulmonary damage is not as highly selective for lung tissue versus liver tissue damage, such as 3MI and naphthalene, but the selective necrosis of Clara cells indicates that expression of 2E1 in these cells could be the primary mechanism of cellular selective toxicity. Indeed, the toxicity of 1,1-dichloroethylene (DCE) to Clara cells has been linked specifically to the expression of 2E1 (52). Metabolism of DCE produces both chloral and the epoxide of DCE, 1,1-dichlorooxirane, but studies with glutathione adducts have demonstrated that the primary ultimate electrophile in the cytotoxic process appears to be the epoxide (53). CYP2E1 is also most likely the catalyst for the oxidation of 1,1-dichloro-2,2-bis(*p*-chlorophenyl) ethane (DDD) to a reactive acyl halide intermediate that causes necrosis in isolated Clara cells and human bronchial epithelial cells (54,55).

C. Prostaglandin H Synthase

Arachidonic acid can be converted by prostaglandin H synthase (PHS) to the hydroperoxide endoperoxide PGG_2, and the subsequent peroxidase activity of PHS that reduces the hydroperoxide to the alcohol can be accomplished by the enzyme complex with concomitant oxidation of xenobiotics to products that are often identical with the P450-catalyzed products. The classic example of such a PHS-mediated cooxidation is the conversion of benzo[*a*]pyrene 7,8-dihydrodiol to the 9,10-epoxide, the ultimate mutagenic and carcinogenic metabolite. This transformation is predominantly catalyzed by PHS, not P450, in type II alveolar cells from rats (56). Human lung explants were used to show that PHS catalyzed the epoxidation of the diol to tetrols (after hydrolysis of the epoxide) at approximately half the rate of P450-mediated turnover (57).

The bioactivation of 3MI by PHS has been proposed to be a contributing factor in the P450-mediated bioactivation of this pneumotoxin (58). However, the contribution of PHS must be a very minor because several in vivo and in vitro experiments have demonstrated that inhibition of P450 enzymes almost completely protects animals or isolated cells from the toxicity of 3MI and inhibits binding of activated 3MI to proteins (16,39,59,60). Although the production of radicals related to 3MI metabolism have been associated with PHS-mediated bioactivation of 3MI, it is unclear that free–radical-mediated tissue damage has any part in 3MI-mediated pneumotoxicity. Conversely, 3MI is a potent antioxidant in lipid peroxidation (61) or other mechanisms of free–radical-mediated toxicities (62).

Arachidonic acid metabolites are formed in isolated human alveolar type II cells, and can be released from these cells by treatment with the calcium ionophore A23187 (63). Thus, the metabolism of xenobiotics by PHS-mediated cooxidation can be an important factor in the distribution or bioactivation of these chemicals.

D. Flavin-Containing Monooxygenases

The flavin monooxygenases in mammals are a family of five genes (64) that predominantly catalyze heteroatom oxidation of xenobiotics (65). Primary amines are good substrates for the enzymes, but several other unusual substrates, such as secondary and tertiary amines, hydrazines, phosphines, iodides, and sulfides, are oxidized by these enzymes (see review, Ref. 65). The expression of these genes can be organ-selective because one gene, *FMO2*, is expressed primarily in the lung tissues of rabbits, hamster, guinea pigs, and mice. This gene is also expressed in rabbit kidney and bladder (66), and in maxilloturbinates and ethmoturbinates (67), but is generally absent from the liver tissues of this species.

The catalytic mechanism of all of the FMO enzymes involves the generation of a ''primed'' enzyme that has an active 4α-hydroperoxyflavin in the catalytic site before substrate binding. The substrate selectivity observed with the lung FMO, FMO2, in comparison with one of the liver FMOs, FMO1, can be explained by the access of substrates to the active site. There are several striking differences between the substrate selectivity of the FMO1 and FMO2 enzymes. Examples include the oxidation of primary amines (FMO2 catalyzes the *N*-oxygenation of these amines, but they are not substrates for FMO1), and the stereoselective *N*-oxidation of nicotine [FMO2 stereospecifically forms *trans*-(*S*)-nicotine*N*-1′-oxide (68), but FMO1 catalyzes an equal mixture of the two isomers (69)].

The participation of the FMO enzymes in the metabolism of xenobiotics has been well documented. In particular, the oxidation of amine- and sulfur-containing chemicals by these enzymes, and in some cases selective metabolism by the lung form of the FMOs, implies that a unique role for the FMO enzymes may exist in the detoxification—and possibly the bioactivation—of xenobiotics, and that lung metabolism may be an important player in this process. In addition, the physiological role of FMO enzymes has only recently been explored, particularly relative to the metabolism of cysteamine, farnesylated proteins, and trimethylamine (see review Ref. 65). It is likely that the pulmonary FMO may contribute significantly to normal homeostatic process as well as the disposition of xenobiotics.

III. HYDROLYSIS AND CONJUGATION ENZYMES

A. Epoxide Hydrolases

The levels of these enzymes are generally significantly lower in whole-lung homogenates than in liver preparations. However, as stated in the introduction, the precise cellular distribution of the enzymes represents a much more accurate description of metabolic events at the cellular level. An example is the tracheobronchial distribution of enzyme activity, with styrene oxide as the substrate, in the beagle dog lung (70). Epoxide hydrolase-mediated hydrolysis of this substrate was highest in the distal airways, and these levels were twice the activity of hepatic preparations. The increase in enzyme activities in the distal airways corresponds to increases in Clara (nonciliated bronchiolar epithelial) cell populations among the epithelial cells of the airways. The activities of epoxide hydrolase are significantly higher in isolated Clara cells than in alveolar type II cells (71), and presumably higher than other epithelial cell types (Table 1). Therefore, it seems likely that detoxification enzymes such as epoxide hydrolase are expressed to a greater extent in cells that generate higher amounts of reactive cytotoxic or genotoxic epoxides. The balance between bioactivation and detoxification in these cells plays a predominant role in overt cytotoxicity or the induction of neoplastic lesions.

B. UDP-Glucuronosyltransferases

The microsomal UDP-glucuronosyltransferases (UGT) provide a primary mechanism of protection against the accumulation of toxic xenobiotics or their oxidative products. There exist a reasonably large number of different genes that belong to a supergene family (72). The *UGT1* gene family is unique among drug-metabolizing enzymes because at least seven different enzymes are encoded by a single large gene the mRNA of which is differentially spliced to produce different UGT enzymes with different enzyme activities. A UGT gene called *UGT2A1* has been cloned from rat nasal tissue; and it is selectively expressed in this tissue, but not in lung, kidney, brain, intestine, or liver tissues (73). It is possible that this enzyme could be responsible for the glucuronidation and inactivation of odorants such as eugenol. It seems feasible that other UGT genes may be selectively expressed in olfactory or pulmonary tissues, but investigators have not yet identified them.

Expression of UGT enzymes in lung microsomal samples is generally lower than in liver microsomes—4-nitrophenol glucuronidation in rat lung microsomes is only 30% of the rat liver microsomal rate of glucuronidation (74). Rat Clara cells glucuronidate 4-methylumbelliferone to a greater extent than do isolated alveolar type II cells (75). Again, the metabolic capacity of the Clara cells is generally higher than other cells in the lung.

A mechanism for the pneumotoxicity and pulmonary carcinogenicity of trichloroethylene has been proposed to be mediated by deficient UGT activities in lung Clara cells (76). The authors showed that isolated Clara cells from mice did not possess sufficient UGT activity to glucuronidate trichloroethanol; isolated hepatocytes formed this metabolite quite efficiently. The lack of glucuronidation of trichloroethanol was proposed to lead

Table 1 Xenobiotic-Metabolizing Enzymes in the Respiratory Tract

Enzyme	Species	Substrate	Localization
Cytochrome P450 Enzymes			
1A1	All	Ethoxyresorufin, polycyclic aromatic hydrocarbons	Olfactory epithelium, Clara cells, type II cells, induced macrophages
1B1	Rat, mouse, human	Polyclic aromatic hydrocarbons	Unknown
2A3	Rat	Unknown	Unkown
2A10/11	Rabbit, rat	Diethylnitrosmine, aflatoxin B_1, NNK	Olfactory mucosa
2B1/4/7	Rabbit, rat, human	Arachidonic acid, BHT	Clara cells
2E1	All	4-Nitrophenol, dimethylnitrosamine, dichloroethylene	Nasal tissues
2F1/2/3	Rabbit, rat, mouse, human, goat	Ethoxycoumarin, naphthalene, 3MI	Clara cells
2GI	Rabbit	Testosterone	Olfactory mucosa
3A	Human	Unknown	Unknown
4B1/2	All	2-Aminofluorene, 4-ipomeanol, valproic acid	Clara cells
Prostaglandin H synthase	All	Benzo[*a*]pyrene-7,8-diol	Endothelial and epithelial cells
Flavin-containing monooxygenases	All	Amines, phosphines, sulfides	Entire respiratory system
Epoxide hydrolases	All	PAH epoxides	Clara cells
UDP-Glucuronosyltransferases	All	1-Naphthol, 4-methylumbelliferone, eugenol	Clara cells
Glutathione transferases	All	1-Cholro-2,4-dinitrobenzene, benzopyrene epoxide	Clara cells and ciliated cells

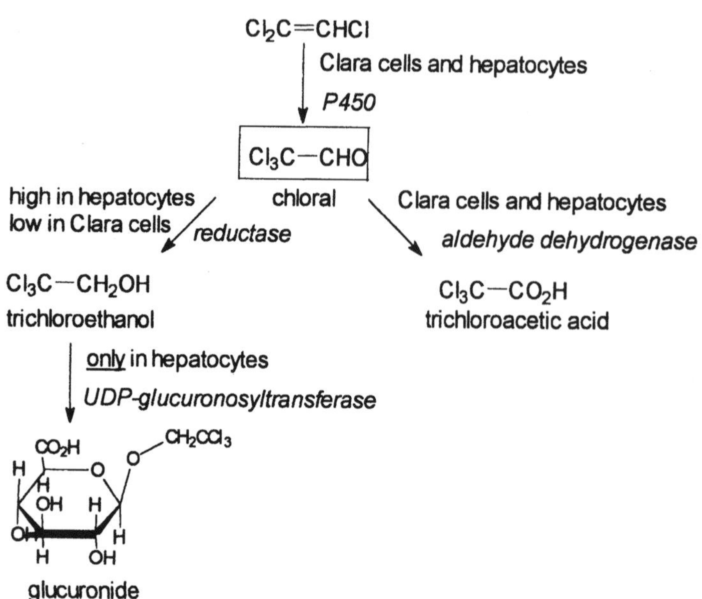

Fig. 3 Mechanism of selective Clara cell damage by trichloroethylene, caused by the lack of detoxification. Although chloral, the putative toxic intermediate from trichloroethylene, can be formed in both mouse lung Clara cells and in hepatocytes, efficient detoxication of this aldehyde by reductase and by UDP-glucuronosyltransferase is lacking in the lung epithelial cells. (From Ref. 76.)

to the buildup of the toxic intermediate, chloral, in Clara cells (Fig. 3). Thus, the lack of detoxication by glucuronidation in susceptible cells was proposed as an operative mechanism for the toxicity of trichloroethylene. It is also possible that bioactivation by glucuronidation may produce toxicities to respiratory tissues, by mechanisms analogous to acyl-linked glucuronide toxicities to hepatic tissues (77) and renal tissues (78).

C. Glutathione Transferases

Both cytosolic and microsomal forms of the glutathione (GSH) transferases can be identified. There are six different enzyme classes of the cytosolic enzymes (79). The expression of GSH transferases in respiratory tissues has been evaluated in animals and in humans to a reasonable extent. The precise elucidation of which GSH transferase exists in which cells of the respiratory system has not been determined, but it appears that at least the α, μ, and π classes have been identified in rat (80) and human (81) lung samples.

The activities of most GSH transferases can be measured by conjugation with 1-chloro-2,4-dinitrobenzene (CDNB), and the distribution of activities with CDNB are slightly higher in proximal airways than in distal airways of the mouse and monkey (82). The π-, μ-, and α-class GSH transferases represent approximately 94, 3, and 3%, respectively, of the total human lung catalytic activity toward CDNB (83). As with many other drug-metabolizing enzymes, Clara cells have considerably higher activities with CDNB than alveolar type II cells (75,84).

Immunochemical staining with antibodies to the transferase enzymes in human lung demonstrated that the α-and π-forms were identified in large and small airway epithelial tissues (81). Several studies have attempted to link polymorphisms of the μ-class GSH transferase enzymes in human lung tissue with susceptibility to lung cancer induced by cigarette smoking (85). A particularly strong association has been shown between the genetic polymorphisms of the *CYP1A1* gene and a deficient genotype of a μ-class GSH transferase, relative to lung cancer induced by cigarette smoking (86). Persons with both genetic alterations possessed the highest risk of lung cancer from smoking.

Mechanisms of pulmonary toxicity ascribed to low GSH levels or decreased GSH transferase activities in lung tissues or cells are rational, but, with the exception of lung cancer, the lack of glutathione-mediated detoxification in respiratory tissues has not proved to be an operative mechanism of pneumotoxicity. Conversely, toxicities occurring by free–radical-induced mechanisms, particularly through the intervention of toxic oxygen species, such as superoxide, hydrogen peroxide, or hydroxyl radical, can be effectively ameliorated by factors that increase pulmonary glutathione (87).

IV. CONCLUSIONS

This chapter has attempted to provide a brief overview of the metabolic enzymes that exist in respiratory tissues, and has made an effort to distinguish this organ system from other anatomical regions that participate to a greater or lesser extent in the metabolism of xenobiotics. The lung should not be viewed as a metabolic organ with lower activities than liver, but rather as an active, dynamic, and often highly selective metabolic tissue with important contributions to xenobiotic disposition and toxicity. The complexity of cellular distribution and function in respiratory tissues provides a unique paradigm for scientists involved in drug metabolism, toxicology, and risk assessment. As we learn more about the mechanisms of selective gene expression in certain lung cells, and the functional consequences of the enzymology of the gene products, predictions of metabolic processes and toxicological consequences will become possible.

REFERENCES

1. A. R. Dahl and J. L. Lewis, Respiratory tract uptake of inhalants and metabolism of xenobiotics, *Annu. Rev. Pharmacol. Toxicol. 32*:383–407 (1993).
2. G. S. Yost, Bioactivation and selectivity of pneumotoxic chemicals, *Tissue Specific Toxicity: Biochemical Mechanisms* (W. Dekant and H.-G. Neumann, eds.), Academic Press, London, 1992, pp. 195–220.
3. A. R. Buckpitt and M. K. Cruikshank, Biochemical function of the respiratory tract: Metabolism of xenobiotics, *Comprehensive Toxicology* (I. G. Sipes, C. A. McQueen, and A. J. Gandolfi, eds), Vol 8, *Toxicology of the Respiratory System* (R. A. Roth, ed.) Elsevier Science, New York, 1997, pp. 159–186.
4. J. A. Bond, Metabolism of xenobiotics by the respiratory tract, *Toxicology of the Lung* (D. E. Gardner, J. D. Crapo, and R. O. McClellan, eds.), Raven Press, New York, 1993, pp. 187–215.
5. H. Foth, Role of the lung in accumulation and metabolism of xenobiotic compounds—implications for chemically induced toxicity, *Crit. Rev. Toxicol. 25*:165–205 (1995).

6. M. W. Himmelstein, B. Asgharian, and J. A. Bond, High concentrations of butadiene epoxides in livers and lungs of mice compared to rats exposed to 1,3-butadiene, *Toxicol. Appl. Pharmacol. 132*:281–288 (1995).
7. L. H. Overby, S. Nishio, A. Weir, G. T. Carver, C. G. Plopper, and R. M. Philpot, Distribution of cytochrome P450 1A1 and NADPH-cytochrome P450 reductase in lungs of rabbits treated with 2,3,7,8-tetrachlorodibenzo-*p*-dioxin: Ultrastructural immunolocalization and in situ hybridization, *Mol. Pharmacol. 41*:1039–1046 (1992).
8. M. J. Lee and D. Dinsdale, The subcellular distribution of NADPH-cytochrome P450 reductase and isoenzymes of cytochrome P450 in the lungs of rats and mice, *Biochem. Pharmacol. 49*: 1387–1394 (1995).
9. C. Serabjit-Singh, E. Wolf, R. Philpot, and C. Plopper, Cytochrome P450: Localization in rabbit lung, *Science 207*:1469–1470 (1980).
10. P. de la M. Hall, I. Stupans, W. Burgess, D. J. Birkett, and M. E. McManus, Immunohistochemical localization of NADPH-cytochrome P450 reductase in human tissues, *Carcinogenesis 10*: 521–530 (1989).
11. G. S. Yost, Selected, nontherapeutic agents, *Comprehensive Toxicology* (I. G. Sipes, C. A. McQueen, and A. J. Gandolfi, eds.), Vol 8, *Toxicology of the Respiratory System* (R. A. Roth, ed.), Elsevier Science, New York, 1997, pp. 591–610.
12. G. S. Yost, Mechanisms of cytochrome P450-mediated formation of pneumotoxic electrophiles, *Biological Reactive Intermediates V* (R. Synder, ed.), Plenum Press, New York, 1996, pp. 221–229.
13. J. K. Ritter, I. S. Owens, M. Negishi, K. Nagata, Y. Y. Sheen, J. R. Gillette, and H. A. Sasame, Mouse pulmonary cytochrome P-450 naphthalene hydroxylase: cDNA cloning, sequence, and expression in *Saccharomyces cerevisiae, Biochemistry 30*:11430–11437 (1991).
14. C. R. Wolf, C. N. Statham, M. G. McMenamin, J. R. Bend, M. R. Boyd, and R. M. Philpot, The relationship between the catalytic activities of rabbit pulmonary cytochrome P-450 isozymes and the lung-specific toxicity of the furan derivative, 4-ipomeanol, *Mol. Pharmacol. 22*:738–744 (1982).
15. R. D. Verschoyle, R. M. Philpot, C. R. Wolf, and D. Dinsdale, CYP4B1 Activates 4-ipomeanol in rat lung, *Toxicol. Appl. Pharmacol. 123*:193–198 (1993).
16. J. Thornton-Manning, M. L. Appleton, F. J. Gonzalez, and G. S. Yost, Metabolism of 3-methylindole by vaccinia-expressed P450 enzymes: Correlation of 3-methyleneindolenine formation and protein-binding, *J. Pharmacol. Exp. Ther. 276*:21–29 (1996).
17. G. L. Skiles and G. S. Yost, Mechanistic studies on the cytochrome P450-catalyzed dehydrogenation of 3-methylindole, *Chem. Res. Toxicol. 9*:291–297 (1996).
18. J. L. Bolton and J. A. Thompson, Oxidation of butylated hydroxytoluene to toxic metabolites, *Drug Metab. Dispos. 19*:467–472 (1991).
19. R. D. Verschoyle, C. R. Wolf, and D. Dinsdale, Cytochrome P450 2B isoenzymes are responsible for the pulmonary bioactivation and toxicity of butylated hydroxytoluene, *O,O,S*-trimethylphosphorothioate and methylcyclopentadienyl manganese tricarbonyl, *J. Pharmacol. Exp. Ther. 266*:958–963 (1993).
20. Z. Guo, T. J. Smith, H. Ishizaki, and C. S. Yang, Metabolism of 4-(methylnitrosamino)-1-(3-pyridyl)-1-butanone (NNK) by cytochrome P450IIB1 in a reconstituted system, *Carcinogenesis 12*:2277–2282 (1991).
21. T. Shimada, C.-H. Yun, H. Yamazaki, J. C. Gautier, P. H. Beaune, and F. P. Guengerich, Characterization of human lung microsomal cytochrome P-4501A1 and its role in the oxidation of chemical carcinogens, *Mol. Pharmacol. 41*:856–864 (1992).
22. K. K. Bhattacharyya, P. Brake, S. Eltom, S. Otto, and C. Jefcoat, Identification of a rat adrenal cytochrome P450 active in polycyclic aromatic hydrocarbon metabolism as rat CYP1B1, *J. Biol. Chem. 270*:11595–11602.
23. T. Su, J. J. Sheng, T. W. Lipinskas, and X. Ding, Expression of CYP2A genes in rodent and human nasal mucosa, *Drug Metab. Dispos. 24*:884–890 (1996).

24. H.-M. Peng, X. Ding, and M. J. Coon, Isolation and heterologous expression of cloned cDNAs for two rabbit nasal microsomal proteins, CYP2A10 and CYP2A11, that are related to nasal microsomal cytochrome P450 form a, *J. Biol. Chem. 268*:17253–17260 (1993).
25. B. Domin, T. R. Devereux, and R. M. Philpot, The cytochrome P450 monooxygenase system of rabbit lung: enzyme components, activities and induction in the nonciliated bronchiolar epithelial (Clara) cell, alveolar type II cell and alveolar macrophage, *Mol. Pharmacol. 30*: 296–303 (1986).
26. P. G. Forkert, CYP2E1 is preferentially expressed in Clara cells of murine lung: Localization by in situ hybridization and immunohistochemical methods, *Am. J. Respir. Cell. Mol. Biol. 12*:589–596 (1995).
27. C. W. Wheeler, S. A. Wrighton, and T. M. Guenthner, Detection of human lung cytochromes P450 that are immunochemically related to cytochrome P450IIE1 and cytochrome P450IIIA, *Biochem. Pharmacol. 44*:183–186 (1992).
28. P. T. Nhamburo, S. Kimura, O. W. McBride, C. A. Kozak, H. V. Gelboin, and F. J. Gonzalez, The human *CYP2F* gene subfamily: Identification of a cDNA encoding a new cytochrome P450, cDNA-directed expression, and chromosome mapping, *Biochemistry 29*:5491–5499 (1990).
29. H. Wang, D. L. Lanza, and G. S. Yost, Cloning and expression of CYP2F3, a cytochrome P450 that bioactivates the selective pneumotoxins 3-methylindole and naphthalene, *Arch. Biochem. Biophys. 349*:329–340 (1998).
30. X. Ding, T. D. Porter, H. M. Peng, and M. J. Coon, cDNA and derived amino acid sequence of rabbit nasal cytochrome P450NMb (P450 IIG1), a unique isozyme possibly involved in olfaction, *Arch. Biochem. Biophys. 285*:120–125 (1991).
31. S. Ramakanth, H. P. Maxwell, and G. S. Yost, Cloning of CYP4B2 from a λZAP II goat lung cDNA library, *Toxicologist 14*:142 (1994).
32. C. W. Wheeler and T. M. Guenthner, Cytochrome P-450-dependent metabolism of xenobiotics in human lung, *J. Biochem. Toxicol. 6*:163–169 (1991).
33. W. Ruangyuttikarn, G. L. Skiles, and G. S. Yost, Identification of a cysteinyl adduct of oxidized 3-methylindole from goat lung and human liver microsomal proteins, *Chem. Res. Toxicol. 5*:713–719 (1992).
34. S. Ramakanth, J. R. Thornton-Manning, H. Wang, H. Maxwell, and G. S. Yost, Correlation between pulmonary cytochrome P450 transcripts and the organ-selective pneumotoxicity of 3-methylindole, *Toxicol. Lett. 71*:77–85 (1994).
35. J. Hakkola, M. Pasanen, J. Hukkanen, O. Pelkonen, J. Mäenpää, R. J. Edwards, A. R. Boobis, and H. Raunio, Expression of xenobiotic-metabolizing cytochrome P450 forms in human full-term placenta, *Biochem. Pharmacol. 51*:403–411 (1996).
36. C. R. Wolf, M. Szutowski, L. Ball, and R. M. Philpot, The rabbit pulmonary monooxygenase system: Characteristics and activities of two forms of pulmonary cytochrome P-450, *Chem. Biol. Interact. 21*:29–43 (1978).
37. J. M. Mathews and J. R. Bend, *N*-Aralkyl derivatives of 1-aminobenzotriazole as potent isozyme-selective mechanism-based inhibitors of rabbit pulmonary cytochrome P450 in vivo, *J. Pharmacol. Exp. Ther. 265*:281–285 (1993).
38. C. J. Sinal and J. R. Bend, Kinetics and selectivity of mechanism-based inhibition of guinea pig hepatic and pulmonary cytochrome P450 by *N*-benzyl-1-aminobenzotriazole and *N*-α-methylbenzyl-1-aminobenzotriazole, *Drug Metab. Dispos. 24*:996–1001 (1996).
39. J. C. Huijzer, J. D. Adams, Jr., J.-Y. Jaw, and G. S. Yost, Inhibition of 3-methylindole bioactivation by the cytochrome P-450 suicide substrates 1-aminobenzotriazole and α-methylbenzylaminobenzotriazole, *Drug Metab. Dispos. 17*:37–42 (1989).
40. L. C. Knickle and J. R. Bend, Bioactivation of arachidonic acid by the cytochrome P450 monooxygenases of guinea pig lung. The ortholog of cytochrome P450 2B4 is solely responsible for formation of epoxyeicosatrienoic acids, *Mol. Pharmacol. 45*:1273–1280 (1994).
41. M. Czerwinski, T. L. McLemore, H. V. Gelboin, and F. J. Gonzalez, Quantification of

CYP2B7, CYP4B1, and CYPOR messenger RNAs in normal human lung and lung tumors, *Cancer Res. 54*:1085–1091 (1994).

42. M. Czerwinski, T. L. McLemore, R. M. Philpot, P. T. Nhamburo, K. Korezekwa, H. V. Gelboin, and F. J. Gonzalez, Metabolic activation of 4-ipomeanol by complementary DNA-expressed human cytochromes P-450: Evidence for species-specific metabolism, *Cancer Res. 51*:4636–4638 (1991).
43. P. B. Smith, H. F. Tiano, S. Nesnow, M. R. Boyd, R. M. Philpot, and R. Langenbach, 4-Ipomeanol and 2-aminoanthracene cytotoxicity in C3H/10T½ cells expressing rabbit cytochrome P450 4B1, *Biochem. Pharmacol. 50*:1567–1575 (1995).
44. R. D. Verschoyle, R. M. Philpot, C. R. Wolf, and D. Dinsdale, CYP4B1 activates 4-ipomeanol in rat lung, *Toxicol. Appl. Pharmacol. 123*:193–198 (1993).
45. J. R. Thornton-Manning, W. Ruangyuttikarn, F. J. Gonzalez, and G. S. Yost, Metabolic activation of the pneumotoxin, 3-methylindole, by vaccinia-expressed cytochrome P450s, *Biochem. Biophys. Res. Commun. 181*:100–107 (1991).
46. T. J. Smith, Z. Guo, F. J. Gonzalez, F. P. Guengerich, G. D. Stoner, and C. S. Yang, Metabolism of 4-(methylnitrosoamino)-1-(3-pyridyl)-1-butanone in human lung and liver microsomes and cytochromes P-450 expressed in hepatoma cells, *Cancer Res. 52*:1757–1763 (1992).
47. A. Buckpitt, A.-M. Chang, A. Weir, L. Van Winkle, X. Duan, R. Philpot, and C. Plopper, Relationship of cytochrome P450 activity to Clara cell cytotoxicity. IV. Metabolism of naphthalene and naphthalene oxide in microdissected airways from mice, rats, and hamsters, *Mol. Pharmacol. 47*:74–81 (1995).
48. M. Shimizu, J. M. Lasker, M. Tsutsumi, and C. S. Lieber, Immunohistochemical localization of ethanol-inducible P450IIE1 in the rat alimentary tract, *Gastroenterology 99*:1044–1053 (1990).
49. X. Ding and M. J. Coon, Induction of cytochrome P-450 isozyme 3a (P-450IIE1) in rabbit olfactory mucosa by ethanol and acetone, *Drug Metab. Dispos. 18*:742–745 (1990).
50. G. P. Carlson and B. J. Day, Induction by pyridine of cytochrome P450IIE1 and xenobiotic metabolism in rat lung and liver, *Pharmacology 44*:117–123 (1992).
51. J. L. Raucy, J. C. Kraner, and J. M. Lasker, Bioactivation of halogenated hydrocarbons by cytochrome P4502E1, *Crit. Rev. Toxicol. 23*:1–20 (1993).
52. R. P. Lee and P. G. Forkert, Pulmonary CYP2E1 bioactivates 1,1-dichloroethylene in male and female mice, *J. Pharmacol. Exp. Ther. 273*:561–567 (1995).
53. T. F. Dowsley, P.-G. Forkert, L. A. Benesch, and J. L. Bolton, Reaction of glutathione with the electrophilic metabolites of 1,1-dichloroethylene, *Chem. Biol. Interact. 95*:227–244 (1995).
54. W. K. Nichols, M. O. Convington, D. C. Seiders, S. Safiullah, and G. S. Yost, Bioactivation of halogenated hydrocarbons by rabbit pulmonary cells, *Pharmacol. Toxicol. 71*:335–339 (1992).
55. W. K. Nichols, C. M. Terry, N. S. Cutler, M. L. Appleton, P. K. Jesthi, and G. S. Yost, Oxidation at C-1 controls the cytotoxicity of 1,1-dichloro-2,2-bis(*p*-chlorophenyl)ethane by rabbit and human lung cells, *Drug Metab. Dispos. 23*:595–599 (1995).
56. K. Sivarajah, K. G. Jones, J. R. Fouts, T. R. Devereux, J. E. Shirley, and T. E. Eling, Prostaglandin synthetase and cytochrome P-450-dependent metabolism of (±)benzo[*a*]pyrene 7,8-dihydrodiol by enriched populations of rat Clara cells and alveolar type II cells, *Cancer Res. 43*:2632–2636 (1983).
57. G. A. Reed, R. C. Grafstrom, R. S. Krauss, H. Artrup, and T. E. Eling, Prostaglandin H synthase-dependent co-oxidation of (±)-7,8-dihydroxy-7,8-dihydrobenzo[*a*]pyrene in hamster trachea and human bronchus explants, *Carcinogenesis 5*:955–960 (1984).
58. K. S. Acton, H. J. Boermans, and T. M. Bray, The role of prostaglandin H synthase in 3-methylindole-induced pneumotoxicity in goat, *Comp. Biochem. Physiol. 101C*:101–108 (1992).
59. J. R. Thornton-Manning, W. K. Nichols, B. W. Manning, G. L. Skiles, and G. S. Yost, Metabolism and bioactivation of 3-methylindole by Clara cells, alveolar macrophages, and subcellular fractions from rabbit lungs, *Toxicol. Appl. Pharmacol. 122*:182–190 (1993).

60. W. K. Nichols, D. N. Larson, and G. S. Yost, Bioactivation of 3-methylindole by isolated rabbit lung cells, *Toxicol. Appl. Pharmacol. 105*:264–270 (1990).
61. J. D. Adams, Jr., M. C. Heins, and G. S. Yost, 3-Methylindole inhibits lipid peroxidation, *Biochem. Biophys. Res. Commun. 149*:73–78 (1987).
62. E. A. Lissi, M. Faure, and N. Montoya, Reactivity of indole derivatives towards oxygenated radicals, *Free Radical Res. Commun. 15*:211–222 (1991).
63. F. J. van Overveld, P. G. Jorens, W. A. De Backer, M. Rampart, L. Bossaert, and P. A. Vermeire, Release of arachidonic acid metabolites from isolated human alveolar type II cells, *Prostaglandins 44*:101–110 (1992).
64. M. P. Lawton, J. R. Cashman, T. Cresteil, C. T. Dolphin, A. A. Elfarra, R. N. Hines, E. Hodgson, T. Kimura, J. Ozols, and I. R. Phillips, A nomenclature for the mammalian flavin-containing monooxygenase gene family based on amino acid sequence identities, *Arch. Biochem. Biophys. 308*:254–257 (1994).
65. J. R. Cashman, Structural and catalytic properties of the mammalian flavin-containing monooxygenase, *Chem. Res. Toxicol. 8*:165–181 (1995).
66. R. Tynes and R. M. Philpot, Tissue- and species-dependent expression of multiple forms of mammalian microsomal flavin containing monooxygenase, *Mol. Pharmacol. 31*:569–574 (1987).
67. S. E. Shehin-Johnson, D. E. Williams, S. Larsen-Su, D. M. Stresser, and R. N. Hines, Tissue-specific expression of flavin-containing monooxygenases (FMO) forms 1 and 2 in the rabbit, *J. Pharmacol. Exp. Ther. 272*:1293–1299 (1995).
68. S. B. Park, P. Jacob III, N. L. Benowitz, and J. R. Cashman, Stereoselective metabolism of (*S*)-(−)-nicotine in humans: Formation of *trans*-(*S*)-nicotine *N*-1′-oxide, *Chem. Res. Toxicol. 6*:880–888 (1993).
69. L. A. Damani, W. F. Pool, P. A. Crooks, R. K. Kaderlik, and D. M. Ziegler, Stereoselectivity in the *N*′-oxidation of nicotine isomers by flavin-containing monooxygenase, *Mol. Pharmacol. 33*:702–705 (1988).
70. J. Bond and V. Russell, Regional distribution of xenobiotic metabolizing enzymes in respiratory airways of dogs, *Drug Metab. Dispos. 16*:116–124 (1988).
71. T. Devereux, J. Diliberto, and J. R. Fouts, Cytochrome P450 monooxygenase, epoxide hydrolase, and flavin monooxygenase activities in Clara cells and alveolar type II cells from the rabbit, *Cell Biol. Toxicol. 1*:57–65 (1985).
72. B. Burchell, D. W. Nebert, D. R. Nelson, K. W. Bock, T. Iyanagi, P. L. M. Jansen, D. Lancet, G. J. Mulder, J. R. Chowdhury, G. Siest, T. R. Tephly, and P. I. Mackenzie, The UDP glucuronosyltransferase gene superfamily: Suggested nomenclature based on evolutionary divergence, *DNA Cell Biol. 10*:487–494 (1991).
73. D. Lazard, K. Zupko, Y. Poria, P. Nef, J. Lazarovits, S. Horn, M. Khen, and D. Lancet, Odorant signal termination by olfactory UDP glucuronosyltransferase, *Nature 349*:790–793.
74. T. Yoshimura, S. Tanaka, and T. Horie, Species difference and tissue distribution of uridine diphosphate-glucuronyltransferase activities toward E6080, 1-naphthol and 4-hydroxybiphenyl, *J. Pharmacobiodyn. 15*:387–393 (1992).
75. K. G. Jones, J. F. Holland, G. L. Foureman, J. R. Bend, and J. R. Fouts, Xenobiotic metabolism in Clara cells and alveolar type II cells isolated from lungs of rats treated with beta-naphthoflavone, *J. Pharmacol. Exp. Ther. 225*:316–319 (1983).
76. J. Odum, J. R. Foster, and T. Green, A mechanism for the development of Clara cell lesions in the mouse lung after exposure to trichloroethylene, *Chem. Biol. Interact. 83*:135–153 (1992).
77. S. J. Hargus, B. M. Martin, J. W. George, and L. R. Pohl, Covalent modification of rat liver dipeptidyl peptidase IV (CD26) by the nonsteroidal anti-inflammatory drug diclofenac, *Chem. Res. Toxicol. 8*:993–996 (1995).
78. K. A. McGurk, R. P. Remmel, V. P. Hosagrahara, D. Tosh, and B. Burchell, Reactivity of mefenamic acid 1-*O*-acyl glucuronide with proteins in vitro and ex vivo, *Drug Metab. Dispos. 24*:842–849 (1996).

79. R. N. Armstrong, Structure, catalytic mechanism, and evolution of the glutathione transferases, *Chem. Res. Toxicol. 10*:2–18 (1997).
80. M. Lee and D. Dinsdale, Immunolocalization of glutathione *S*-transferase isozymes in bronchiolar epithelium of rats and mice, *Am. J. Physiol. 267*:L766–L774 (1994).
81. S. Anttila, A. Hirvonen, H. Vainio, K. Husgafvel-Pursiainen, J. D. Hayes, and B. Ketterer, Immunohistochemical localization of glutathione *S*-transferases in human lung, *Cancer Res. 53*:5643–5648 (1993).
82. X. Duan, A. Buckpitt, and C. G. Plopper, Variation in antioxidant enzyme activities by anatomic subcompartment within the lung: A comparison of the rat and the rhesus monkey, *Toxicol. Appl. Pharmacol. 123*:73–82 (1993).
83. S. S. Singhal, M. Saxena, H. Ahmad, S. Awasthi, A. K. Haque, and Y. C. Awasthi, Glutathione *S*-transferases of human lung: Characterization and evaluation of the protective role of the α-class isozymes against lipid peroxidation, *Arch. Biochem. Biophys. 299*:232–241 (1992).
84. P. G. Forkert, B. A. Geddes, D. W. Birch, and T. E. Massey, Morphologic changes and covalent binding of 1,1-dichloroethylene in Clara and alveolar type II cells isolated from lungs of mice following in vivo administration, *Drug Metab. Dispos. 18*:534–539 (1990).
85. T. Nakajima, E. Elovaara, S. Anttila, A. Hirvonen, A. M. Camus, J. D. Hayes, B. Ketterer, and H. Vainio, Expression and polymorphism of glutathione *S*-transferase in human lungs: Risk factors in smoking-related lung cancer, *Carcinogenesis 16*:707–711 (1995).
86. K. Nakachi, K. Imai, S. Hayashi, and K. Kawajiri, Polymorphisms of the *CYP1A1* and glutathione *S*-transferase genes associated with susceptibility to lung cancer in relation to cigarette dose in a Japanese population, *Cancer Res. 53*:2994–2999 (1993).
87. P. E. Morris and G. R. Bernard, Significance of glutathione in lung disease and implications for therapy, *Am. J. Med. Sci. 307*:119–127 (1994).

12

Applications of Heterologous Expressed and Purified Human Drug-Metabolizing Enzymes: An Industrial Perspective

A. David Rodrigues
Merck Research Laboratories, West Point, Pennsylvania

I. INTRODUCTION

It has become widely accepted that in vitro human metabolism studies can play a useful role in the discovery and development of novel drug entities (NDEs; 1–7). With this in mind, most pharmaceutical companies are now routinely performing in vitro testing of drugs with human tissue. Because the liver is considered the major site of metabolism, most researchers employ primary cultured hepatocytes, precision-cut liver tissue slices, or various hepatic subcellular fractions (e.g., 9000 g supernatant fraction, microsomes, or cytosol). This means that a substantial amount of information can be obtained relatively quickly before administration of a new drug to human subjects, thereby making clinical trials better guided, safer, and more cost-effective. In many instances, however, fresh human tissue is relatively difficult to obtain for ethical, economic, or geographic reasons. Moreover, despite the allure of using animal models to predict drug metabolism in humans, it is now accepted that the drug-metabolizing enzymes (DMEs) in the rodent, dog, and monkey are often distinct from those present in the human liver (7, and references therein). Therefore, with the development of heterologous expression systems (8–10), many investigators have turned to various ''transgenic cell'' lines as additional sources of human DMEs (Fig. 1).

Most of the DME systems in native human liver tissue, such as cytochrome P450 (CYP; EC 1.14.14.1), UDP-glucuronosyltransferase (UDPGT; EC 2.4.17), *N*-acetyltransferase (NAT; EC 2.3.1.5), and NADPH-dependent flavin-containing monooxygenase (FMO; EC 1.14.13.8), are rather complex and comprise at least two different enzymes or ''isoforms'' (11–14). Members of each DME system, or ''enzyme pool,'' can often interact differentially with any number of drugs or xenobiotic agents. Moreover, the expression of these different proteins can be under the control of different genetic or environmental factors, which may contribute to intersubject variability. Therefore, investigators have sought to simplify in vitro metabolism models by opting to heterologously express individual human DMEs in yeast (e.g., *Saccharomyces cerevisiae*), *Escherichia coli* (*E. coli*), COS, V79 (Chinese hamster), B-lymphoblast (AHH-1 TK+/−), human hepatoma (HepG2), *Spodoptera frugiperda* (Sf9), or *Trichoplusia ni* (*T. ni*) cells (7,9,10,15–21).

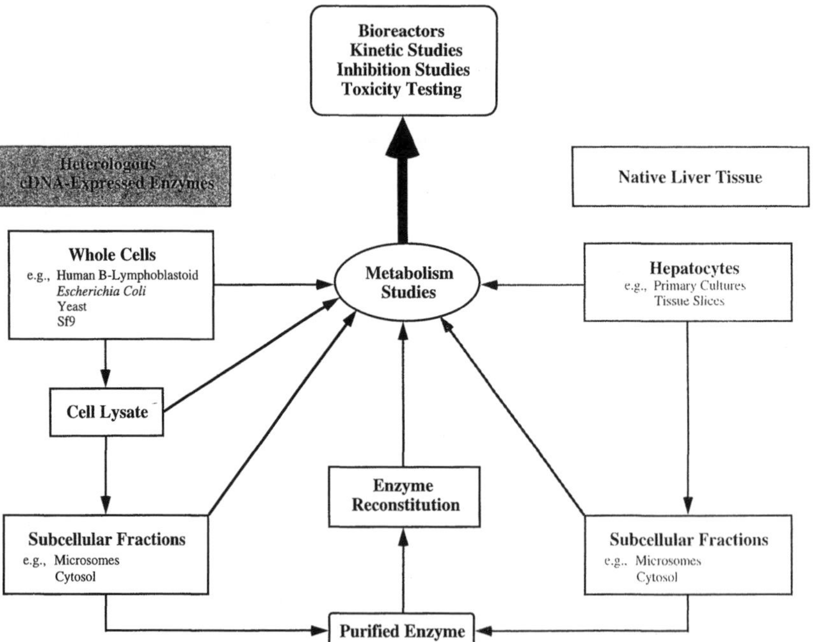

Fig. 1 Heterologous cDNA-expressed systems as an additional source of enzymes for in vitro drug metabolism studies.

A useful feature of heterologous expression systems is that they can often yield enough starting material so that the DME of choice can be purified to electrophoretic homogeneity. As an example, several heterologously expressed CYP proteins (e.g., CYP3A4, CYP3A5, CYP2C9, CYP1A1, and CYP2D6) have been purified and subsequently reconstituted with detergent(s), phospholipid(s), NADPH-CYP reductase (fp_2) and cytochrome b_5 (22–28). Because the level of endogenous CYP is low in many of these cell types, it is possible to simplify the purification procedure and obtain relatively good yields (≥9%) of protein with high (≥5 nmol CYP/mg) specific contents (23–26,28). By comparison, purification of CYP enzymes from native human liver microsomes is difficult and often requires additional steps, especially when attempting to resolve members of the CYP pool that are within the same subfamily, or when trying to purify minor (low-abundance) CYP forms (29–31). Many of these issues will also apply to other enzyme systems (e.g., FMO, UDPGT).

Because of the explosion in technology, researchers can now perform in vitro metabolism studies with commercially available preparations of heterologously expressed human liver DMEs such as CYP, fp_2, cytochrome b_5, FMO, NAT, epoxide hydrolase, and glutathione *S*-transferase (e.g., Gentest Corp, Woburn, MA; Panvera Corp., Madison, WI; Immune Complex Corp., La Jolla, CA; Oxford Biomedical Research Inc., Oxford, MI), and it is only a matter of time before other DMEs (e.g., UDPGT, aldehyde oxidase, and sulfotransferases) also become available. The following describes some of the current and potential applications of heterologous-expressed or purified DMEs. When possible, the discussion will not be limited to the CYP system and will include examples of other, equally important, DME systems.

II. AN "INTEGRATED APPROACH" TO STUDYING IN VITRO DRUG METABOLISM

Although heterologous-expressed DMEs are useful, it is important that they should be used within the framework of an integrated approach (1,7). This involves using the cDNA-expressed enzymes in conjunction with primary cultures of human hepatocytes, precision-cut human liver tissue slices, or native hepatic subcellular fractions (Fig. 2). Such an integrated approach enables one to exploit the strengths of each model system (1,7). It is imperative that an overall metabolic profile is initially obtained with intact cell models, especially when information concerning the metabolism of the drug in vivo is not available. In turn, the data that are obtained with either cDNA-expressed enzymes or subcellular fractions can be viewed in perspective (32–37). For instance, clozapine is metabolized in vitro by cDNA-expressed CYP2D6 (38), whereas in vivo there is no evidence that the enzyme plays a major role in the metabolism of the drug (39). Similarly, oxidation of (*S*)-nicotine to cotinine [via a $\Delta^{1'(5')}$ iminium ion] is mediated by multiple cDNA-expressed CYP proteins (e.g., CYP2A6, CYP2D6, CYP2B6, or CYP2C9), although CYP2A6 is thought to be the major CYP catalyzing the reaction in the presence of native human liver microsomes (40–43). In accordance, phenobarbital, a known CYP2A6 inducer, induces cotinine formation in vivo (44,45). Therefore, in most of the studies conducted in our laboratory, full use has been made of cDNA-expressed and purified enzymes, microsomes, and precision-cut liver tissue slices. In general, we have tended to use heterologous-expressed or purified enzymes as a means of confirming the results obtained with native human liver tissue (1,7,32–37).

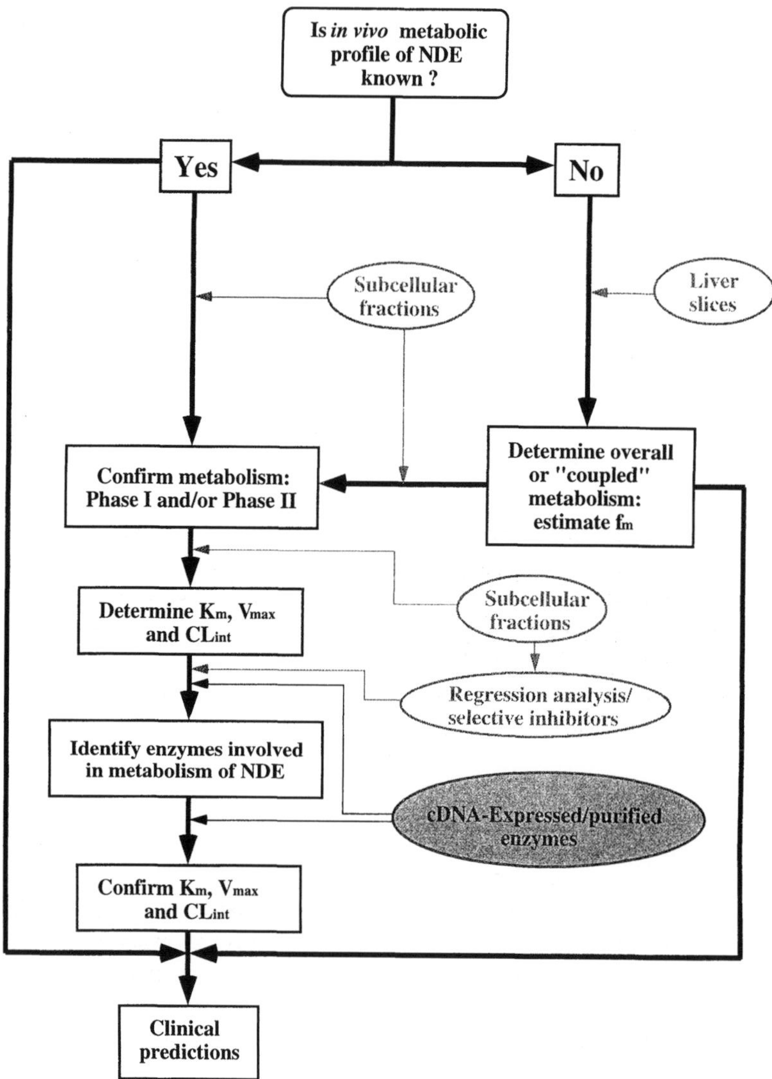

Fig. 2 Integrated scheme for studying the in vitro metabolism of a drug: Where f_{m}, K_{m}, $V_{\max}$, and CL_{int} represent fraction of dose metabolized by each pathway, apparent Michaelis constant, apparent maximal initial reaction velocity, and intrinsic clearance ($V_{\max}/K_{\mathrm{m}}$), respectively. (Adapted from Ref. 7).

Intact cell models, such as precision-cut liver slices, permit "coupled" (phase I and phase II) metabolism of drugs, in additon to minimizing the problems of excessive tissue disruption (1,7). Furthermore, one is able to study the effect of enzyme-inducing agents on the metabolism of a given NDE and also investigate "metabolic shunting" in the presence of inhibitors (1,7). Purified enzymes, either from native human liver tissue or heterologous expression systems, offer different advantages. First, one can study a given biotransformation event in the absence of competing reactions. This is especially important when trying to differentiate between CYP and FMO-catalyzed *N*-oxidations (20,34), or attempting to study oxidation of drugs in the absence of microsomal or cytosolic *N*-oxide, sulfoxide, or epoxide reductases (34,36,46–50). Second, unlike hepatocytes and liver slices, cDNA-expressed enzymes can be stored frozen and are a convenient source of human DMEs. Third, with the use of purified enzymes, concerns about the possible effects of proteolytic enzymes can be minimized. This is important to bear in mind when preparing subcellular fractions from native tissue (1,34,35,51).

A. Studies with ABT-719

Studies with ABT-719, a novel 2-pyridone antimicrobial agent (52), will partly serve as one example to illustrate the integrated approach to studying in vitro drug metabolism. Our department was requested by project chemists and pharmacologists to study the metabolism of ABT-719 and to evaluate the potential of the drug to undergo *N*-acetylation in humans. With this in mind, initial experiments were performed with precision-cut rat and dog liver slices (Fig. 3), which yielded an overall metabolic profile similar to that observed in vivo. The data indicated that [^{14}C]ABT-719 underwent glucuronidation and *N*-acetylation in the rat, whereas no *N*-acetyltransferase (NAT) activity was detected in the dog. It also appeared that the *N*-acetyl metabolite of ABT-719 underwent glucuronidation in the rat (data not shown). These results served to validate the liver slice model and were not surprising, for the dog was known to exhibit relatively low NAT activity (53). In addition, differential NAT activity was confirmed with the appropriate incubations using native liver cytosol and acetyl-CoA (data not shown). In the presence of human liver slices, ABT-719 was metabolized primarily via glucuronidation, and no *N*-acetyl metabolite was detected (see Fig. 3). In fact, we were able to show that ABT-719 underwent glucuronidation when incubated with rat, dog, or human liver microsomes (Fig. 4). Lack of ABT-719 *N*-acetylation in human liver slices could have been because polymorphically expressed NAT2 was involved in the metabolism of ABT-719 and the organ donor in question was a ''poor metabolizer'' (PM) subject (21). In addition, we were made aware of the possibility that NAT1 might be prone to degradation by proteolytic enzymes (D. Grant, personal communication), which might lead to the erroneous conclusion that ABT-719 did not undergo *N*-acetylation in humans. Our fears were substantiated, when a similar result was obtained with the liver of a second organ donor (data not shown). Therefore, lack of appreciable ABT-719 *N*-acetylation was confirmed by incubation of the drug with *E. coli* lysate containing cDNA-expressed human liver cytosolic NAT1 and NAT2 (see Fig. 4). As expected, NAT1 and NAT2 metabolized *p*-aminobenzoic acid (PABA) and sulfamethazine (SMZ), respectively, in the same experiment (21). Therefore, it was concluded that ABT-719 would not be expected to undergo *N*-acetylation in humans.

B. Studies with ABT-418

A bioisostere of (*S*)-nicotine, ABT-418 is a potent and selective neuronal nicotinic acetylcholine receptor ligand, and it has been under development for the treatment of attentional deficit disorder (ADD) and Alzheimer's disease (34,35). In the course of preclinical studies, it was decided to study the metabolism of the drug in multiple species. Overall, the metabolism of ABT-418 was similar to that of nicotine, because the compound underwent CYP- and aldehyde oxidase (AO)-dependent *C*-oxidation to the lactam and FMO-

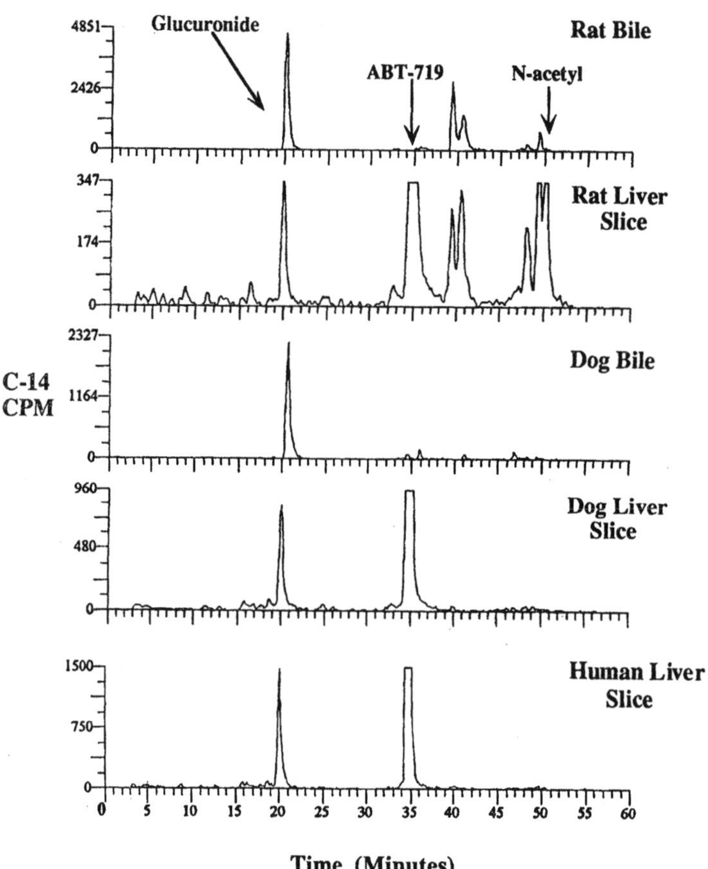

Fig. 3 Metabolism of [^{14}C]ABT-719 in the presence of precision-cut liver tissue slices: ABT-719 (100 μM) was incubated with rat, dog, or human liver slices as previously described (33–35). After 24 h, aliquots of media were removed, extracted, and analyzed by radio-HPLC. Metabolic patterns were also obtained from a bile-cannulated rat and dog, both of which had received a single intravenous dose (1.0 mg/kg) of [^{14}C]ABT-719. The relative retention time of ABT-719 and its glucuronide and *N*-acetyl metabolite is indicated.

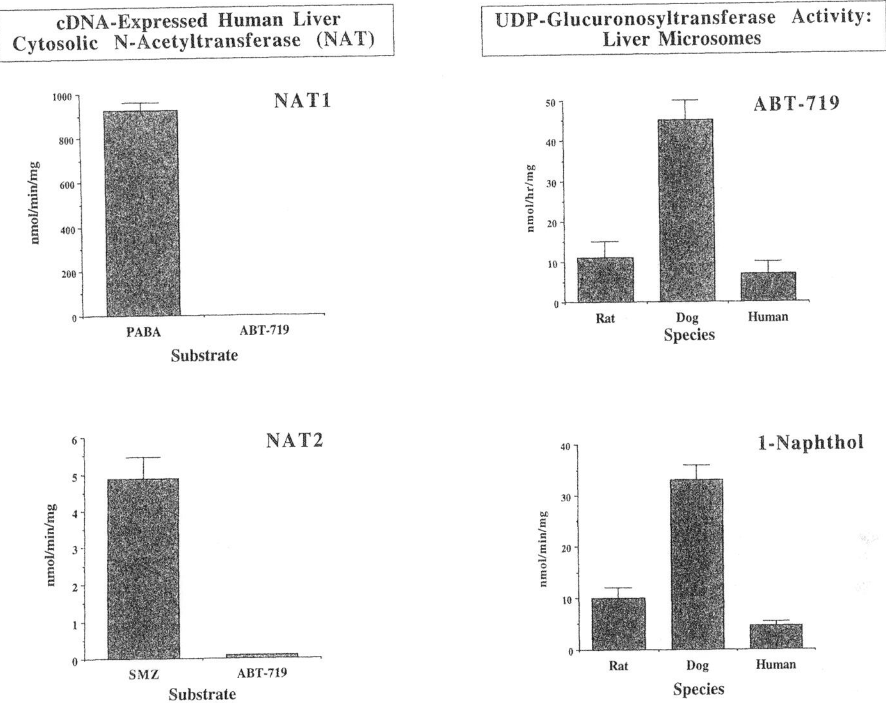

Fig. 4 Metabolism of [^{14}C]ABT-719 in the presence of liver microsomes or cDNA-expressed human liver cytosolic NAT. Glucuronidation of ABT-719 (0.1 mM) was confirmed with liver microsomes (1.0 mg protein/mL), in 0.1 M potassium phosphate buffer (pH 7.4) containing $MgCl_2$ (5 mM), UDP-glucuronic acid (5 mM), and Brij 58 (0.5 mg/mL). Reactions were terminated with an equal volume of methanol and incubates were analyzed by radio-HPLC. For comparison, UDPGT activity with 1-naphthol (50 μM) as substrate is also presented. Lack of ABT-719 *N*-acetylation by human liver NAT enzymes was confirmed using *E. coli* cell lysates containing cDNA-expressed NAT1 or NAT2. Incubations were carried out with [^{14}C]ABT-719 (100 μM), in 0.25 M triethanolamine buffer (pH 7.5) containing EDTA (5 mM), dithiothreitol (5 mM), appropriately diluted cell lysate protein, and an acetyl-CoA (100 μM)-generating system (21). Incubates were analyzed by radio-HPLC. The substrates *p*-aminobenzoic acid (PABA; 100 μM) and sulfamethazine (SMZ; 200 μM) were used as positive controls for NAT1 and NAT2, respectively (21).

dependent *N′*-oxidation (Fig. 5). Owing to concerns about the stability of AO in both dog and human 9000 *g* supernatant fraction (S-9), initial studies were performed with precision-cut liver slices (34,35). The results showed that the rat, dog and chimpanzee were efficient at catalyzing both *C*-oxidation and *N′*-oxidation of ABT-148 (see Fig. 5). The formation of the two metabolites was confirmed in vivo, for both were detected in rat (*N′*-oxide C_{max}, 435 ng/mL; lactam C_{max}, 131 ng/mL) and dog (*N′*-oxide C_{max}, 11.7 μ/mL; lactam C_{max}, 5.3 μg/mL) plasma (T. El-Shourbagy, personal communication). By comparison, only relatively low levels (≤5.8% of total metabolism) of *N′*-oxide were detected in human liver slices after 24 h of incubation.

Although it appeared that *C*-oxidation predominated in man (see Fig. 5), and because the product(s) of *N′*-oxidation might themselves be reduced back to parent drug in the

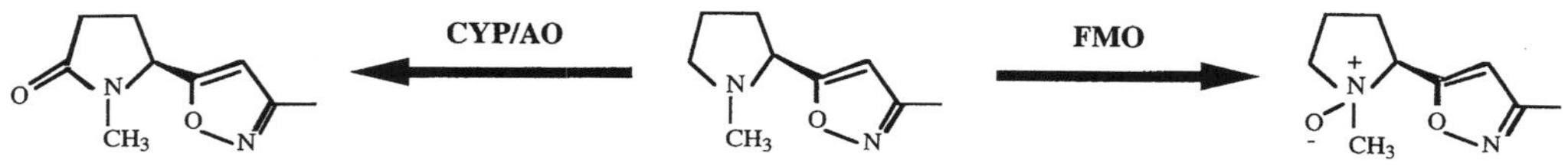

Lactam ← CYP/AO — ABT-418 — FMO → N'-Oxide

Species	Liver Slices (% Total Metabolism)		N'-Oxide Formation (pmol/min/mg)			
	% Lactam	% N'-Oxide	Native Liver Microsomes	FMO1	FMO2	FMO3
Rat	21 ± 3.2	44 ± 3.2	237 ± 22			
Dog	22 ± 2.1	58 ± 0.1	386 ± 67			
Cyno. Monkey	56 ± 11	≤ 1.0	84 ± 20			
Chimpanzee	18 ± 2.0	40 ± 2.0	97 ± 9.8			
Human	53 ± 1.8	≤ 5.8	36 ± 17			650 ± 26
Rabbit			207 ± 6.4		2556 ± 55	
Pig				2467 ± 384		

Fig. 5 In vitro metabolism of ABT-418: [*N*-Methyl^{3}H]ABT-418 (100 μM) was incubated with precision-cut liver slices or native liver microsomes (34,35). The drug was subsequently incubated with purified pig liver microsomal FMO1 (0.40 mg protein/mL), purified rabbit lung microsomal FMO2 (0.23 mg/mL), or *E. coli*-expressed human liver microsomal FMO3 (0.16 mg cell lysate protein/mL), in 50 mM potassium phosphate buffer (pH 8.5) and a NADPH-generating system (34).

presence of cytosolic or microsomal reductase(s) (34,35), it was important to show unambiguously that human liver microsomal FMO was able to catalyze the *N*′-oxidation of ABT-418. Toward this end, the drug was incubated with native human liver microsomes and *E. coli*-expressed FMO3 (34); FMO3 is considered to be the major FMO enzyme in native human liver microsomes (14,20,34). The data indicated that ABT-418 was a substrate for human FMO3, present in native liver microsomes or in *E. coli* cell lysate (see Fig. 5). Moreover, the rate of *N*′-oxidation was relatively low when compared with other sources of FMO. No reduction of ABT-418 *N*-oxides (*cis* or *trans*) was detected in the presence of *E. coli*-expressed FMO3 (34). Therefore, it was concluded that *N*′-oxidation of ABT-418 did occur in humans, but that this was a relatively minor pathway, and that *C*-oxidation to lactam would be the major pathway. This observation has been confirmed in subsequent clinical trials, because no ABT-418 *N*′-oxides(s) have been detected in either the plasma or urine of subjects receiving transdermal (150 mg/day) ABT-418. In the same subjects, it appears that the lactam is the major metabolite detected in plasma (C_{max} = 1.8 μg/mL).

III. METHODOLOGY

A. Perspective

When using cDNA-expressed and purified enzymes, one is essentially dealing with an "artificial" system, in which the protein in question is removed from its native environment. This is especially true for the hepatic microsomal CYP system. In contrast with enzymes, such as FMO and NAT, that are relatively self-contained and do not require additional protein(s) to function enzymatically (20,21,34), CYP enzymes have to be reconstituted in phospholipid with electron donor proteins (e.g., cytochrome b_5, fp_2 or NADH-cytochrome b_5 reductase) and often the molar ratio of the different component proteins in an optimized reconstitution assay bears little resemblance to that present in the native microsomal system (15,22,23,25,26,28,32,33,54). This also extends to studies that make use of microsomes (100,000 *g* membrane fraction) containing heterologous expressed CYP proteins, for which data are often reported without regard for the relative levels of cytochrome b_5 and fp_2 (vs. CYP) present in the membranes. This may become an issue when attempting to establish which CYP forms may be involved in the metabolism of a particular NDE. Therefore, there are three main issues that should be considered when dealing with heterologous-expressed or purified–reconstituted DMEs.

1. *Abundance of Enzymes in Native Tissue*

Irrespective of the DME system being investigated, it is imperative to have some idea of the relative level of each protein in native tissue. By comparison with other DME systems such as FMO and UDPGT (18,19,55), the human liver microsomal CYP pool is relatively well characterized. In this instance, the level of various CYP family and subfamily members has been quantitated immunologically (Table 1) and related to the specific content of total (spectrally detectable) CYP (56,57). Because the specific content of cytochrome b_5 (total CYP/cytochrome b_5 ratio is ~1 : 1) and fp_2 (~100 pmol/mg protein) is also known (58–61), it is now possible to obtain estimates of the ratio of each CYP enzyme to cytochrome b_5 and fp_2 in native human liver microsomes (see Table 1). However, the levels of each CYP form are based on immunological estimates of apoprotein and the amount of holoenzyme in each case is unknown. Moreover, considerable work remains to be done

Table 1 Levels of Various CYP Proteins in Native Human Liver Microsomes

CYP protein	Specific content (pmol/mg protein)[a]	Total CYP[a] (%)	Putative molar ratio[b] fp_2/b_5/CYP
Total[c]	300	—	1:3:3
3A[d]	96	32	1:3:1
2E1	22	7.3	5:14:1
2D6	5.0	1.7	20:60:1
2A6	14	4.7	7:21:1
1A2	42	14	2:7:1
2C9[e]	49	16	2:6:1
2C[f]	11	3.7	9:27:1

[a]Specific content of each CYP form was determined by immunoassay. Data are also presented as percentage of total (spectrally detectable) CYP. *Source*: Refs. 56 and 57.
[b]Molar ratio of NADPH-CYP reductase (fp_2), cytochrome b_5 (b_5) and CYP in native human liver microsomes. Ratio of cytochrome b_5 to total CYP is approximately 1:1 (59–61). Therefore, nominal specific content of cytochrome b_5 is 300 pmol/mg. The mean content of fp_2 in human liver microsomes is 100 pmol/mg (58).
[c]Represents total (spectrally detectable) CYP and does not include cytochrome P420. Typical mean value of total CYP for a bank of human liver microsomes (37,56).
[d]Represents all members of the CYP3A subfamily (e.g., CYP3A4, CYP3A5, and CYP3A7), although CYP3A4 is considered to be the major CYP3A enzyme in adult human liver microsomes.
[e]CYP2C9 represents about 16% of total CYP levels and accounts for about 80% of all CYP2C proteins (56,57).
[f]Various CYP2C proteins (e.g., CYP2C8, CYP2C18, and CYP2C19), not including CYP2C9.

with the human CYP2C subfamily, for which information concerning the relative levels of CYP2C8, CYP2C18, or CYP2C19 (vs. CYP2C9) in native human liver microsomes is not yet available. This lack of information makes data interpretation very difficult when trying to evaluate the selectivity of clinically relevant drugs (e.g., naproxen, tolbutamide, phenytoin, and such) toward different CYP2C proteins (32,62,63). As shown in Table 2, the *O*-demethylation of naproxen can be catalyzed by at least four different *E. coli*-expressed human liver microsomal CYP2C proteins (32).

2. Specific Content of Heterologous-Expressed and Purified Enzyme

It is highly likely that when experiments are performed with purified enzyme that the specific content of that enzyme will be known, because most investigators use this as one of the criteria for purification (25,26,28,63). However, when using whole cells, cell lysate, or subcellular fractions containing recombinant DMEs, this has often not been true, and many investigators have had to express initial rates of reaction on a ''per milligram protein'' basis (40,42,64,65). With the advent of high-level heterologous expression systems, it is now possible to express CYP at levels that can be detected spectroscopically as a ferrous–carbon monooxide complex (15,23,62). This means that one can determine the specific content of a human CYP enzyme heterologously expressed in Sf9, *T. ni*, or B-lymphoblast cell microsomes and express the data on a ''per nanomole CYP'' basis. Rather than relying solely on Western immunoblotting or enzyme-linked immunosorbent assay (ELISA) measurements, which detect apoprotein only, one is now able to obtain a catalytically relevant spectroscopic measurement of holoenzyme. Similarly, it may be prudent to determine the specific content of enzymes, such as FMO (nanomole flavin adenine

Table 2 Metabolism of Various Drugs by Human Liver Cytochrome P450 2C (CYP2C) Subfamily Members

		Rate of product formation (nmol/min/nmol CYP)[a]			
Activity[b]	CYP2C9	CYP2C8	CYP2C18	CYP2C19	Ref.
NAPase[c]	8.0 (3.4)	<0.1 (0.7)	0.5 (0.7)	4.7 (2.1)	32
TAXase[c]	<0.05	3.1	<0.05	<0.05	63
TOLase[c]	1.4	<0.07	0.1	6.0	63
TOLase[d]	6.6	0.5 (0.6)	—	6.4 (2.5)	66
TOLase[d]	~2.0 (~0.7)	—	—	—	28
MEPHase[c]	<0.03	<0.03	<0.03	3.3 (15)	63
MEPHase[d]	<0.1	<0.1	—	1.1 (2.1)	66
MEPHase[e]	0.08	≤0.1	—	6.2	62
MEPHase[d]	<0.01	—	—	—	28

[a]Various CYP2C proteins were reconstituted with NADPH-CYP reductase (fp_2) in phospholipid. In each case, experimental details can be found in the appropriate reference. Data in parentheses indicate activity in the presence of cytochrome b_5.
[b]NAPase, naproxen *O*-demethylase; TOLase, tolbutamide hydroxylase; MEPHase, (*S*)-(+)-mephenytoin 4′-hydroxylase; TAXase, Taxol 6α-hydroxylase.
[c]Purified CYP2C proteins expressed in *E. coli.*
[d]CYP2C proteins purified from native human liver microsomes.
[e]Purified CYP2C proteins expressed in yeast.

dinucleotide or FAD per milligram of protein) and express catalytic activity on a per nanomole FAD basis.

Without doubt, the issue of specific content becomes important when dealing with low-abundance enzymes, such as human liver microsomal CYP2D6 and CYP2B6 (56). At the time of writing, the specific content of CYP2D6, CYP2B6, and CYP3A4, in commercially available preparations of human B-lymphoblast microsomes, was 260 pmol/mg, 160 pmol/mg, and 56 pmol/mg, respectively (data available from Gentest Corp.). For comparison, the specific content of CYP2D6, CYP2B6, and CYP3A4, in native human liver microsomes has been estimated to be 5.0 pmol/mg, 1.0 pmol/mg, and 96 pmol/mg, respectively (56). In this instance, B-lymphoblast microsomes containing cDNA-expressed CYP2D6 would be expected to exhibit high activity with a large number of substrates. Therefore, expressing data on a per milligram protein basis is essentially meaningless, unless all of the enzymes being studied are expressed at identical, or at least similar, levels.

3. Reaction Rate Data: Bridging the Gap Between cDNA-Expressed and Purified Enzymes and Native Tissue

For a given NDE, an ideal scenario would be to determine intrinsic clearance (V_{max}/K_m ratio) in the presence of each heterologous-expressed enzyme and then proceed to multiply this parameter by the abundance of each enzyme in native human liver tissue. However, it is often not possible to carry out formal Michaelis–Menten kinetic analyses with multiple enzymes, because this can be costly and time-consuming when screening large numbers of NDEs. For the CYP system, we have decided on a more simplified approach that involves using a single concentration of test drug: (a) the drug is incubated in the presence of each CYP protein (CYPn), and the rate is expressed on a ''mole product formed per unit time per mole CYPn'' basis, after accounting for background (non-CYP) activity; (b) for each CYP form, the rate is then multiplied by the nominal specific content (mol CYPn/mg) of the enzyme in native human liver microsomes (see Table 1). This yields a normalized rate (NR) for each of the different CYP proteins and is expressed as mole product formed per unit time per milligram protein; (c) the NR values for each CYP form are added to obtain the total normalized rate (TNR); and (d) the NR for each CYP enzyme is then expressed as a percentage of the TNR [Eq. (1)]. Therefore, K_m and V_{max} estimates are obtained only for those heterologous-expressed CYP forms that are primarily involved in the metabolism of a given drug (% of TNR ≥20%). The choice of drug concentration to be used in the analysis is normally based on apparent K_m estimates obtained with native human liver microsomes. In addition, reaction rate is usually expressed as nanomole (nmol) (or picomole; pmol)/unit time per nmol (or pmol) CYP, whereas CYP-specific content is expressed as nmol (or pmol) CYP/mg.

$$\% \text{ of TNR} = \left(\frac{\text{NR}}{\text{TNR}}\right)100 = \frac{(\text{mol/unit time/mol CYPn})(\text{mol CYPn/mg})}{\Sigma[(\text{mol/unit time/mol CYPn})(\text{mol CYPn/mg})]} \quad (1)$$

The data presented in Fig. 6 are illustrative of this approach. In this instance, ''compound A,'' a drug currently in development for the treatment of asthma, was incubated with B-lymphoblast microsomes containing CYP1A2, CYP2A6, CYP2E1, CYP2D6, CYP2C9, CYP3A4, or CYP2C19. Initial data indicated that the drug was metabolized by CYP3A4, CYP2C9, CYP2C19, and CYP2D6 (≥0.41 pmol/h/pmol CYP). In contrast, the rate of hydroxylation in the presence of CYP2E1, CYP2A6, or CYP1A2 was relatively low (≤0.08 pmol/h/pmol CYP). When the data were normalized, relative to the levels of each CYP form in native human liver microsomes, it was estimated that metabolism was primarily mediated by CYP3A4 (% TNR = 67%), although CYP2C9 did play an ancillary role (% TNR = 19%). The estimates were confirmed with native human liver microsomes and CYP-selective inhibitors (see Fig. 6). Lack of appreciable metabolism by CYP1A2 has also been confirmed with a panel of human liver microsomes, because no significant correlation ($r < 0.4$; $p > 0.05$; $n = 11$) between compound A hydroxylation and CYP1A2-selective 7-ethoxyresorufin *O*-deethylase (ERODase) activity is observed (unpublished observations). In accordance, the results of subsequent clinical drug–drug interaction trials have indicated that compound A has no effect on the pharmacokinetics of theophylline (W. Awni, personal communication). This is a clinically relevant finding, because compound A is likely to be coadministered with theophylline in asthma sufferers.

B. Enzyme Reconstitution and Incubation

Because of its complexity, the CYP system can illustrate many of the issues that are faced when using cDNA-expressed and purified enzymes to evaluate NDEs. The reader is advised to search the literature database for more complete descriptions of procedure, technique, and experimental design (8–10,20–37). Moreover, it should be emphasized that other DME systems may also pose unique challenges, such as thermal lability (e.g., FMO and NAT) or latency (e.g., UDPGT) (1,9,34,20,51).

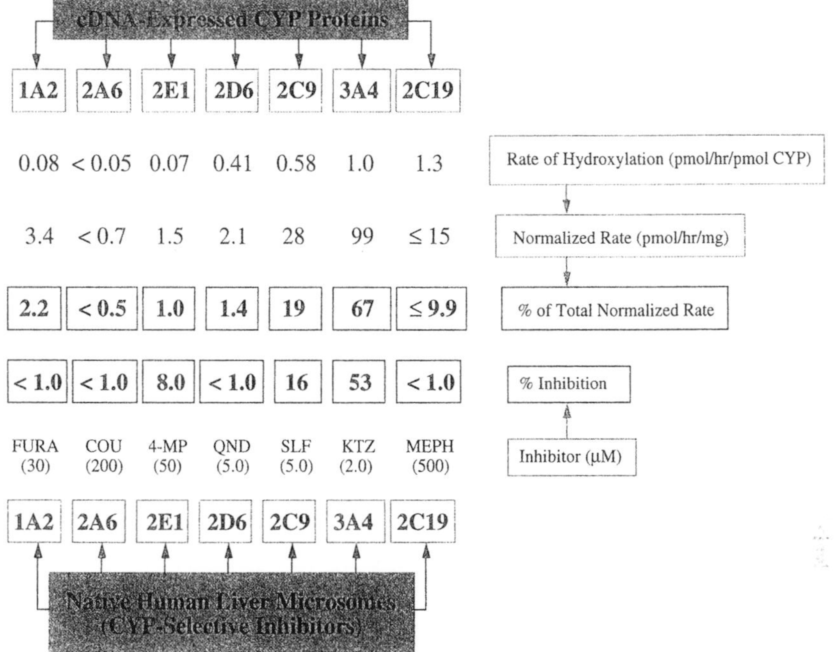

Fig. 6 Hydroxylation of "compound A" in the presence of human B-lymphoblast cell microsomes containing various cDNA-expressed CYP proteins and inhibition of metabolism in native human liver microsomes. Drug (12 μM) was incubated in 0.1 M potassium phosphate buffer (pH 7.4), containing 0.1 mM EDTA, $MgCl_2$ (15 mM), microsomal protein (1.0 mg/mL), and a NADPH-generating system. Incubations with CYP2A6 were carried out in 50 mM Tris-HCl buffer (pH 7.4). For each CYP protein, the rate of hydroxylation is expressed as pmol product/h/pmol CYP. In turn, the data have been normalized relative to the nominal levels of each CYP protein (pmol CYP/mg) in native human liver microsomes (pmol/h/mg microsomal protein). Finally, the normalized rate for each CYP form is expressed as percentage of the sum of the individual normalized rates [see Eq. (1)]. Complementary inhibition studies were carried out with native human liver microsomes and data are expressed as percentage inhibition relative to a (solvent alone) control. Furafylline (FURA), coumarin (COU), 4-methylpyrazole (4-MP), quinidine (QND), sulfaphenazole (SLF), ketoconazole (KTZ), and (*S*)-(+)-mephenytoin (MEPH) were added singularly as CYP1A2-, CYP2A6-, CYP2E1-, CYP2D6-, CYP2C9-, CYP3A-, and CYP2C19-selective inhibitors, respectively.

1. Cellular Membrane Fractions Containing Heterologous-Expressed CYP Proteins

Once transgenic cells are lysed, the various membrane fractions (e.g., 100,000 *g* membrane fraction) are a convenient source of recombinant CYP. However, the phospholipid composition in these membranes may differ considerably from that of native human liver microsomes (60). Differences may also exist for the ratio of CYP to cytochrome b_5 or fp_2, as well as the ratio of CYP to phospholipid. Therefore, it may be necessary to alter standard "native microsome" incubation procedures, to compensate for these differences.

When studying CYP-dependent drug oxidations with native microsomes, a standardized incubation mixture (25–37°C) often comprises of buffer (10–200 mM; phosphate, Tris, or HEPES; pH 7.2–7.8); cofactors, such as NADH, $MgCl_2$, or KCl; substrate; and microsomal protein (0.1–5.0 mg/mL). After a suitable preincubation period (≤5 mins), reactions are usually initiated with NADPH, or a NADPH-generating system (15,28,32,33,36,37,54). However, when using commercially available preparations of B-lymphoblast microsomes (Gentest Corp.), reactions have to be initiated with rapidly thawed (37°C) protein (32,36,37). One is also advised by the manufacturer to carry out incubations, without agitation, in tubes composed of polypropylene. In addition, the choice of buffer (e.g., Tris, phosphate vs. HEPES) can greatly influence the results obtained with microsomes containing CYP2A6 or CYP2C9 (data available from Gentest Corp.). Therefore, the composition of the assay mixture may have to be varied, which can make comparisons between different CYP proteins difficult to interpret. As discussed herein, the problem of interpretation may be alleviated if inhibition studies with CYP-selective inhibitors–native human liver microsomes are carried out in parallel.

Lee et al. (15) have reported that a high level of testosterone 6β-hydroxylase activity (≥23 nmol/min/nmol CYP) could be obtained with "microsomes" prepared from Sf9 and *T. ni.* cells expressing both CYP3A4 and fp_2. The assay conditions employed were similar to those of native human liver microsomes (50 mM potassium phosphate buffer, pH 7.4). By comparison, Buters et al. (24) have reported that they were able to reconstitute activity (e.g., 7-ethoxy-4-trifluoromethylcoumarin *O*-deethylase) by adding exogenous fp_2 and detergent (Emulgen 913) to a Sf9 membrane fraction (100,000 *g*) containing cDNA-expressed CYP1A1. Similarly, Goldstein et al. (62) were able to stimulate (>two fold) both tolbutamide methylhydroxylase (TOLase) and (*S*)-(+)-mephenytoin 4′-hydroxylase (MEPHase) activity by adding exogenous fp_2 to yeast microsomes containing cDNA-expressed CYP2C9. In the same study, the authors reported that the addition of exogenous cytochrome b_5 enhanced (≥75%) MEPHase activity in yeast microsomes containing cDNA-expressed CYP2C9 or CYP2C19. Overall, these data indicate that one may not necessarily have to heterologously coexpress fp_2 and CYP to obtain measurable catalytic activity. However, the applicability of these different approaches to studies involving multiple substrates (e.g., erythromycin, nifedipine, midazolam) or different CYP proteins (e.g., CYP2E1, CYP2A6, CYP2D6), still remains to be explored.

2. *Reconstitution of Purified CYP*

A standard reconstitution experiment typically involves the addition of fp_2 and CYP (± cytochrome b_5) to a suspension of sonicated and buffered phospholipid (≤ 0.2 mL). After incubation for the required length of time (temperature may vary), the mixture is diluted with assay buffer (37°C) containing substrate and cofactors. The reaction is then initiated with NADPH or a NADPH-generating system (22,23,25,26,28,32,33,36,54, 63,66–68). As well as studying the effects of buffer ionic strength, buffer type, and $MgCl_2$ on catalytic activity, investigators have also been able to incubate CYP in the presence of various "artificial" phospholipids or detergents, in addition to varying the ratio of CYP to cytochrome b_5 or fp_2. Although the order of addition of the various components is critical, this type of reconstitution allows for greater flexibility during assay optimization. Therefore, the rate of reaction is limited by CYP or substrate, and not limited by cofactor nor the need for reducing equivalents. However, optimization should be performed judiciously, because this may require relatively large amounts of enzyme. In addition, "optimal" assay conditions for one enzyme or substrate may be suboptimal for a second enzyme or substrate, so that it may be prudent to study the metabolism of a NDE in parallel with well-established model substrate(s). In general, the optimization of a CYP reconstitution is performed with Tris, HEPES, or phosphate buffer (25–100 mM; pH 7.2–7.8), in the absence or presence of $MgCl_2$ (5–15 mM), and largely focuses on three components: (a) phospholipid or detergent; (b) fp_2; and (c) cytochrome b_5.

a. Phospholipid. It has been demonstrated that phosphatidylcholine (PC), phosphatidylethanolamine (PE), and phosphatidylserine (PS) or phosphatidylinositol (PI) account for about 45, 20, and 15% of the total phospholipid present in native human liver microsomes, each condensed with saturated or unsaturated (C_{16}–C_{22}) fatty acids (60,69). Historically, however, dilauroylphosphatidylcholine (DLPC) has been widely used in CYP reconstitution experiments, although many investigators have increasingly opted to use alternative phospholipids, such as dilauroyllecithin or dioleoylphosphatidylcholine (DOPC) (22,23,25,26,28,54,63,67,68). Of the CYP forms so far tested, CYP3A4 is particularly sensitive to the composition of the lipid in the assay mixture, although CYP3A5 may have to be reconstituted under similar conditions (25). In this instance, optimal reconstitution experiments have required the use of a lipid mix (e.g.; DOPC:DLPC:PS; 1:1:1; w/w/w) and the addition of a detergent, such as cholate, Lubrol PX, or CHAPS (22,23,33,54,70). This is an important finding, when one realizes that most drugs and NDEs are metabolized by one or more members of the CYP3A subfamily (1,7,71).

b. NADPH-CYP Reductase (fp_2). One of the most important aspects of reconstitution is the ''titration'' of CYP with fp_2. In this way, the optimal molar ratio of the two proteins is obtained and catalytic activity can be maximized (23,25,26,28,32,33,54, 63,67,68). Even for binary CYP–fp_2 fusion proteins, where the ratio of CYP to fp_2 is fixed (1:1), additional fp_2 may further increase catalytic activity (22,33,70). In fact, many authors have reported ''optimized'' reconstitution assays that require a molar excess (2- to 20-fold) of fp_2 (23,25,26,28,32,33). In most cases, the requirement for excess fp_2 would be consistent with the putative molar ratio of CYP to fp_2 in native human liver microsomes (see Table 1). For instance, we have noted that TOLase activity is optimal when purified CYP2C9 is reconstituted with threefold molar excess of fp_2 (32). On the other hand, reconstitution of optimal CYP3A4-dependent testosterone 6β-hydroxylase activity requires a 9-fold (22) or 40-fold (23) molar excess of fp_2. The exact ratio may vary depending on the type of phospholipid, detergent, and buffer pH or ionic strength used in the experiment. For a given form of CYP, it is possible that some substrates are more sensitive than others to varying CYP/fp_2 ratios.

c. Cytochrome b_5. Although much attention has been focused on the effects of cytochrome b_5 on reconstituted CYP-dependent activity (22,25,26,28,32,63), it should be stated that very few researchers routinely include purified NADH–cytochrome b_5 reductase (fp_1) in their experiments (54). For a given CYP form, it is generally assumed that fp_2 will effectively reduce cytochrome b_5 and thus enhance reduction of CYP. In some instances, an equimolar ratio of CYP to cytochrome b_5 is considered optimal (25,33), although many optimized reconstitutions require a molar excess (three- to sixfold) of cytochrome b_5 (23,28,32,63). However, depending on the form of CYP or experimental conditions, cytochrome b_5 can enhance (22,25,28,32), inhibit (28,32,66), or have no effect (26,32) on CYP-dependent monooxygenase activity. If cytochrome b_5 does stimulate activity, the effect is often dependent on the presence of cofactors (e.g., $MgCl_2$), the type of buffer, buffer pH, or the phospholipid or detergent used in the experiment. As an example, Shet et al. (22) and Gillam et al. (25) have reported that cytochrome b_5 stimulates

CYP3A4 (testosterone 6β-hydroxylase) and CYP3A5 (nifedipine oxidase and testosterone 6β-hydroxylase) activity, when the reconstitution mixture contains detergent (cholate or CHAPS) and $MgCl_2$ (10–15 mM).

Our studies with naproxen (*O*-demethylation), have revealed that cytochrome b_5 stimulates *E. coli*-expressed CYP2C8 (more than sevenfold) activity, but has a minimal effect on similar preparations of CYP2C18 (see Table 2). However, the stimulatory effect (~20%) of cytochrome b_5 on TOLase activity catalyzed by CYP2C8, purified from native human liver microsomes, is rather marginal (see Table 2). In contrast, cytochrome b_5 inhibits (~55%) CYP2C9-dependent naproxen *O*-demethylase (NAPase) activity (see Table 2). A similar effect on TOLase activity has been observed with CYP2C9 purified from native human liver microsomes (28). By comparison, the effect of cytochrome b_5 on reconstituted CYP2C19 activity appears to be rather complex, for it inhibits (~55%) both TOLase and NAPase activity, while stimulating (2- to 4.5-fold) MEPHase activity (see Table 2). Given that CYP2C9, CYP2C8, and CYP2C19 are thought to contain a putative cytochrome b_5-binding site (62), these data are difficult to reconcile and indicate that a "standardized" approach to enzyme reconstitution may be warranted. It will be interesting to determine if these observations apply to other well-known CYP2C substrates, such as ibuprofen, diclofenac, omeprazole, (*S*)-warfarin, hexobarbital, and tienilic acid (71).

IV. APPLICATIONS

A. Enzyme- or Isoform-Selective Biotransformations

The oxidative metabolism of a single drug is often catalyzed by multiple DME systems. Examples include (*S*)-nicotine, ABT-418, quinidine, azapetine, and cyclophosphamide, which are oxidized by cytosolic AO, microsomal CYP, or FMO (34,35,41,43,51,72). Although it is possible to differentially inhibit these various DME systems with selective inhibitors, such as clotrimazole (CYP-selective), thiourea (FMO-selective), and menadione (AO-selective), the involvement of each enzyme can be unambiguously determined with the appropriate preparation of heterologous-expressed and purified protein.

There are numerous examples of problems in drug metabolism that can be potentially solved with the availability of heterologous-expressed and purified DMEs. For instance, studies to determine if trimethylamine is a substrate for FMO3 may finally prove that this enzyme is the locus of the "fish-odor" syndrome (41,55,73). Similarly, the availability of heterologous-expressed UDPGT enzymes (e.g., UGT1.4) may make it possible to identify the form of the enzyme that is involved in the *N*-glucuronidation of various clinically relevant agents, such as amitriptyline, imipramine, clozapine, and losartan (74). Furthermore, the identity of the UDPGT enzyme(s) giving rise to Gilbert's and Crigler-Najjar syndromes may also be confirmed (18). Likewise, in the course of our work with terfenadine, we were able to show that an *E. coli*-expressed, purified, and fully reconstituted CYP3A4–rat fp_2 fusion protein could effectively catalyze the oxidation of the drug to primary (*t*-butylhydroxyterfenadine) and secondary (*t*-butylcarboxyterfenadine) metabolites (Fig. 7). In fact, the conversion of the alcohol to the acid metabolite was confirmed with native human liver microsomes, because the reaction was NADPH-dependent and was inhibited by ketoconazole (≤5.0 μM) (33). This was an important finding, because there appeared to be some confusion in the literature concerning the ability of CYP3A4 to mediate the oxidation of terfenadine to the carboxy metabolite (75,76). It is possible

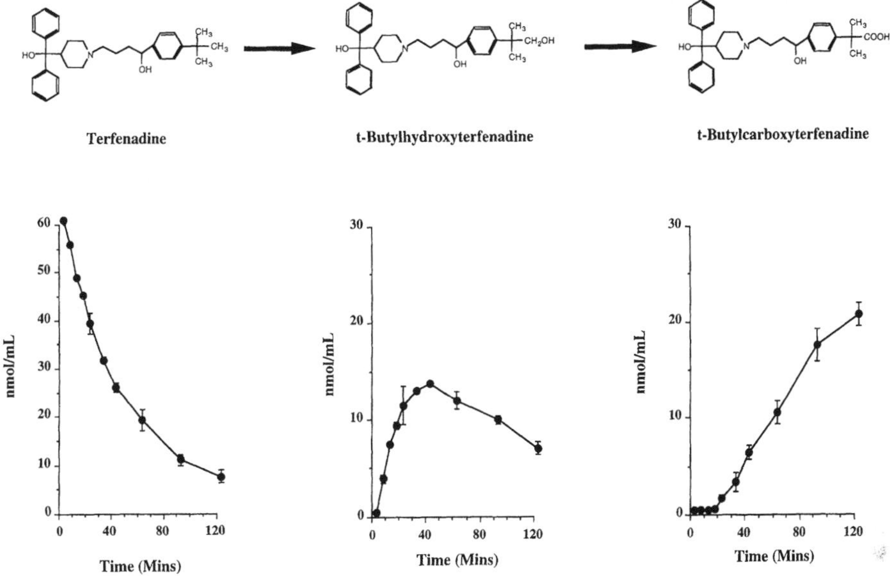

Fig. 7 Metabolism of terfenadine in the presence of an *E. coli*-expressed, purified, and fully reconstituted human liver microsomal CYP3A4–rat fp_2 fusion protein. Terfenadine (60 μM) was incubated with CYP3A4–rat fp_2 fusion protein (1 nmol CYP/mL) in the presence of cytochrome b_5 (1 nmol/mL), rat liver lipid mix (0.63 mg/mL), $MgCl_2$ (10 mM), CHAPS (0.5 mg/mL), and a NADPH-generating system, in 50 mM Tris-HCl buffer (pH 7.5). Aliquots of the incubate were removed and analyzed by radio-HPLC (33).

that the conversion of the alcohol to the acid metabolite involves both microsomal CYP3A4 and cytosolic dehydrogenases (33).

1. Oxidations Involving CYP Enzymes Belonging to Different Gene Families or Subfamilies

Customarily, the CYP forms involved in the metabolism of a given drug are determined using a combination of CYP-selective chemical inhibitors, inhibitory antibodies, correlation analyses, and heterologous-expressed enzymes (see Fig. 2). As examples, the coordinated use of these different approaches has allowed us to confirm that the *O*-demethylation of dextromethorphan (≤20 μM) is primarily mediated by CYP2D6 (37) and that naproxen *O*-demethylase (NAPase) is catalyzed by CYP2C9 and CYP1A2 (32). Overall, these approaches have become widely used, and the number of "reaction phenotyped" compounds continues to grow (5,8,71). Therefore, because this type of technology is so readily available, it has become increasingly necessary for pharmaceutical companies to submit drug applications to regulatory agencies containing some detailed knowledge of the CYP form(s) involved in the metabolism of a NDE (1–7). This is particularly important when a drug is characterized by a relatively narrow therapeutic index and is extensively metabolized (≥50%) by a single CYP enzyme (1,7,71,77,78). Moreover, establishing that a NDE is a substrate for recombinant CYP2D6 or CYP2C19 indicates that clinical trials with phenotyped or genotyped subjects may be warranted. In recent years, we have used a bank of native human liver microsomes to reaction phenotype about 15 drugs within Abbott's pipeline. Most of the drugs have been shown to be metabolized by CYP3A. Two of the drugs (~13%) are also metabolized by CYP2D6 or CYP1A2, whereas CYP2C9 has played an ancillary role in the metabolism of four (~27%) clinical candidates. In every case, we have been able to confirm the data using the appropriate cDNA-expressed or purified CYP proteins (data not shown).

a. Zileuton. Although primarily metabolized (~80%) by direct glucuronidation, zileuton elicits a significant drug interaction with antipyrine and theophylline (36). Therefore, we sought to determine if the drug was metabolized by CYP. Studies with a panel of native human liver microsomes indicated that zileuton underwent CYP1A2-dependent ring hydroxylation, because the activity correlated significantly with the *O*-deethylation of 7-ethoxyresorufin (ERODase) and was inhibited by furafylline (Fig. 8). In contrast, the sulfoxidation of zileuton was significantly correlated with CYP3A-selective erythromycin *N*-demethylase (ERNDase) activity and was inhibited by troleandomycin (TAO). Collectively, these data indicated that the hydroxylation and sulfoxidation of zileuton was catalyzed by CYP1A2 and CYP3A, respectively (36). This was finally confirmed with cDNA-expressed CYP3A4 and CYP1A2 (see Fig. 8).

2. Differentiating Between Different CYP Enzymes of the Same Gene Subfamily

In many instances, a drug will be metabolized by multiple members of a single gene subfamily. This is apparent for CYP, for which it may be difficult to differentiate between the various CYP2C, CYP3A, and CYP1A subfamily members using conventional correlation analyses and chemical inhibitors. Given that CYP-selective immunoinhibitory monoclonal antibodies are not readily available, the most direct way of determining if a reaction is catalyzed by multiple subfamily members is to carry out incubations with the appropriate heterologous-expressed enzymes. However, it is important to emphasize that members of the CYP1A subfamily are expressed in a tissue-specific manner, such that CYP1A2 predominates in native human liver microsomes, whereas CYP1A1 is expressed in extrahepatic tissues (71). Likewise, CYP3A5 appears to be expressed in only about 25% of subjects (67,71). Because CYP3A4 and CYP3A5 can be quantitated by using immunoblotting procedures with polyclonal anti-CYP3A antibodies, it may be possible to screen for native human liver microsomes that contain only CYP3A4 (67). Despite these observations, various workers have used heterologous-expressed, purified CYP proteins to establish the selectivity of a given monooxygenase activity. Zhang et al. (79) have used cDNA-expressed CYP1A1 and CYP1A2 to confirm that the 6-hydroxylation of (*R*)-warfarin is selectively catalyzed by CYP1A2. Similar studies with CYP3A4 and CYP3A5, expressed heterologously or purified from native liver tissue, have revealed that both proteins are equally efficient at catalyzing nifedipine oxidase and testosterone 6β-hydroxylase activity (25,67). Irrespective of the abundance of the two proteins in native liver tissue, CYP3A5 appears to be better than CYP3A4 at catalyzing the *N*-demethylation of ethylmorphine, erythromycin, and benzphetamine (~21 nmol/min/nmol CYP vs. ~6.0 nmol/min/nmol CYP) (25). Likewise, the intrinsic clearance (V_{max}/K_m) of midazolam (1′-hydroxylation pathway) is higher in the presence of CYP3A5 than with similarly reconstituted CYP3A4 (136 μL/min/nmol CYP vs. 36 μL/min/nmol CYP) (67).

a. The CYP2C Subfamily. The increasing availability of heterologous-expressed human CYP2C proteins has greatly facilitated the reaction phenotyping of drugs such as paclitaxol (Taxol) and (*S*)-(+)-mephenytoin. Studies with a panel of native human liver microsomes have indicated that Taxol 6α-hydroxylase (TAXase) is selectively catalyzed by CYP2C8, because this activity correlates well ($r = 0.91$) with the levels of CYP2C8

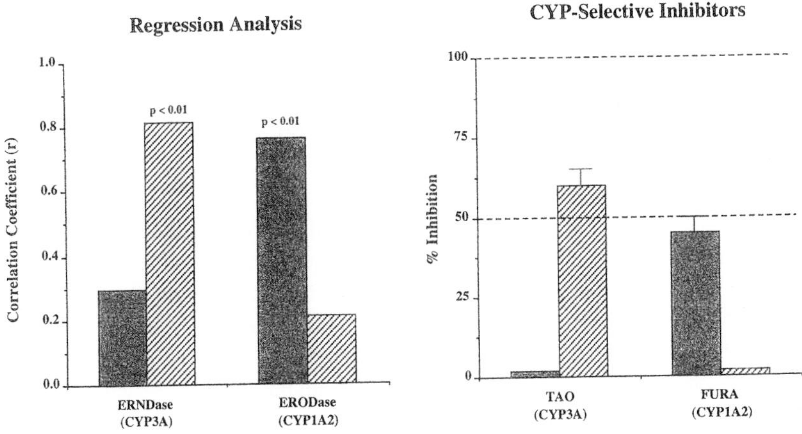

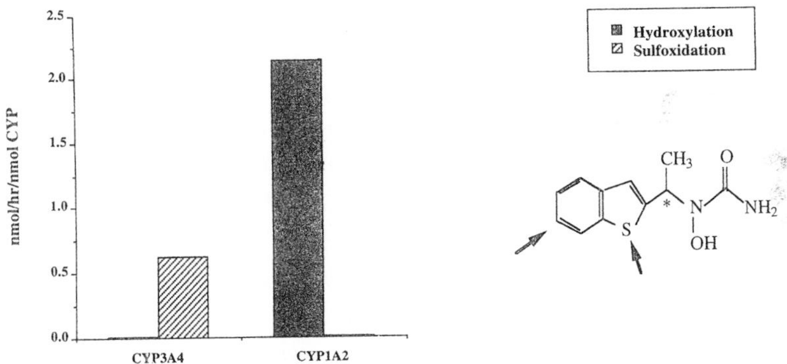

Fig. 8 The hydroxylation and sulfoxidation of zileuton by CYP1A2 and CYP3A, respectively. Zileuton (50 μM) was incubated in the presence of native human liver microsomes (2.0 mg protein/mL) from a panel of 11 different organ donors (36). Unless otherwise indicated, reactions were carried out in 0.1 M potassium phosphate buffer (pH 7.4), containing EDTA (0.1 mM), $MgCl_2$ (15 mM) and a NADPH-generating system. Incubates were subsequently analyzed by radio-HPLC. The rate of hydroxylation and sulfoxidation was then correlated with CYP3A-selective erythromycin *N*-demethylase (ERNDase) and CYP1A2-selective 7-ethoxyresorufin *O*-deethylase (ERODase) activity. The statistical significance of the correlation is indicated. Complementary inhibition studies were carried out with the CYP3A-selective inhibitor troleandomycin (TAO) and the CYP1A2-selective inhibitor furafylline (FURA). Zileuton was subsequently incubated with human B-lymphoblastoid cell microsomes (1.0 mg protein/mL) containing cDNA-expressed CYP1A2 or appropriately reconstituted CYP3A4–rat fp_2 fusion protein (36).

apoprotein (80). By comparison, the correlation with CYP2C9 apoprotein is relatively weak ($r = 0.62$). Subsequent studies with heterologous-expressed CYP2C proteins have revealed that TAXase activity is exclusively catalyzed by CYP2C8 and no activity is detected with similarly incubated CYP2C9, CYP2C19, or CYP2C18, (see Table 2) (80). Likewise, the identity of the polymorphically expressed CYP form catalyzing MEPHase activity in native human liver microsomes has been confirmed (62,63). In this instance, MEPHase activity correlates significantly ($r = 0.72$; $p < 0.01$) with CYP2C19 apoprotein in a panel of native human liver microsomes, and reconstitution experiments have revealed that the reaction is primarily catalyzed by CYP2C19 (see Table 2).

The situation with putative CYP2C9 marker activities [e.g., TOLase, phenytoin *p*-hydroxylase, (*S*)-warfarin 7-hydroxylase, NAPase] is more complicated, despite that all of these activities are highly correlated ($r \geq 0.8$) with each other in different panels of native human liver microsomes (32,81). Hall et al. (81) have reported that TOLase, (*S*)-warfarin 7-hydroxylase, and phenytoin *p*-hydroxylase activity correlates weakly ($r \leq 0.32$) with the levels of CYP2C9 apoprotein. This discrepancy has been attributed to the presence of allelic variants of CYP2C9 or to problems with the procedure used to immunoquantitate the enzyme. In addition, various workers have reported that reactions such as NAPase and TOLase can be catalyzed by multiple heterologous-expressed CYP2C forms other than CYP2C9 (see Table 2). Therefore, definitive phenotyping of reactions such as TOLase and NAPase may require (a) the use of immunoinhibitory monoclonal anti-CYP2C8, -CYP2C9, -CYP2C19, and -CYP2C18 antibodies; (b) correlation experiments with panels of native human liver microsomes, for which the levels of each CYP2C protein have been immunoquantitated with monoclonal or monospecific polyclonal antibodies; and (c) more detailed kinetic studies using cDNA-expressed CYP2C proteins. Toward this end, it has been reported the hydroxylation of tolbutamide in the presence of heterologous expressed CYP2C9 (K_m ~95 μM; V_{max}/K_m = 18 μL/min/nmol CYP) and CYP2C8 (K_m ~500 μM; V_{max}/K_m = 0.7 μL/min/nmol CYP) can be differentiated kinetically (82,83), although this activity conforms to monophasic kinetics in the presence of native human liver microsomes (data not shown). Therefore, when dealing with a NDE that is shown to be metabolized by more than one CYP2C subfamily members, these data indicate that reaction phenotyping should be carried out with caution.

3. *Allelic Variants and Studies Involving Site-Directed Mutagenesis*

Our understanding of the molecular genetics of human DMEs has greatly increased in recent years (84,85). For certain DMEs, the technology has reached a stage where it is now possible to prospectively determine the phenotypic or genotypic status of individuals involved in clinical trials (84–86). This is impressive, when one realizes that many polymorphisms can result from a substitution at one or two amino acid residues. The availability of heterologous-expressed wild-type (wt) and variant (v) forms of these different DMEs has added an extra dimension to NDE-screening procedures, because researchers are now able to carry out confirmatory in vitro studies. Although CYP2C9 is discussed in detail, the reader is advised to consult the literature for other examples (84–86).

a. CYP2C9. There are now reports of at least two clinically relevant allelic variant forms of CYP2C9 (62). In the first, the arginine at position 144 (Arg^{144}) is replaced by cysteine (Cys^{144}). The second allelic variant of CYP2C9 has a leucine (Leu^{359}) in place of an isoleucine (Ile^{359}) residue at position 359, although the gene frequency of this mutation is presently unknown (62,82–84,87). Because these proteins have been heterologously expressed in COS-7, HepG2, yeast, and B-lymphoblastoid cells, it has been possible to directly investigate the effects of these point mutations on monooxygenase activity. As an example, TOLase, phenytoin *p*-hydroxylase, and (*S*)-warfarin 7-hydroxylase activity is significantly reduced (≥50%) in the presence of the allelic variant CYP2C9-Cys^{144}, when compared with the wild-type (CYP2C9-Arg^{144}) form of the enzyme (82,83). Clinically, this has manifest itself as a reduction in the weekly maintenance dose of warfarin in heterozygous (Arg^{144}/Cys^{144}), versus homozygous (Arg^{144}/Arg^{144}), subjects (87). Likewise, a reduction in TOLase (~25%) and phenytoin *p*-hydroxylase (80–91%) activity has also been observed with the recombinant form of CYP2C9-Leu^{359} (62,82). When compared with subjects exhibiting the homozygous wild-type (Ile^{359}/Ile^{359}) genotype, clinical data have indicated that subjects who are homozygous for the Ile^{359} to Leu^{359} point mutation (Leu^{359}/Leu^{359}) exhibit an elevated phenytoin/hydroxyphenytoin plasma ratio (136 vs. 3.7) and decreased urinary metabolite/tolbutamide ratio (163 vs. 324–3033) (88). It appears that metabolism of other CYP2C9 substrates, such as losartan, may also be affected by this mutation, although studies with recombinant wild-type and variant enzyme have not been reported.

b. Site-Directed Mutagenesis. Because no mammalian DMEs such as CYP have been crystallized, putative structural information has been provided by sequence alignment with bacterial enzymes (89,90). However, there are some questions on the precision of this type of approach (91). A second strategy involves the heterologous expression of the products of site-directed mutagenesis, so that information is provided concerning which amino acid residues are required for catalysis, substrate binding, protein folding, or interaction with auxiliary proteins (e.g., cytochrome b_5). Although recent studies have focused on enzymes such as NAT2 (92), dog CYP2B11, and rabbit CYP2B4/5 (93,94), it is probable that researchers will turn their attention to CYP3A4 and CYP2D6 in the near future. The information provided may prove useful, if coupled with computer-assisted active site modeling of clinically relevant enzymes, such as CYP2A6 (95), CYP2C9 (96–98), and CYP2D6 (99,100). As has been reported for murine CYP2a-4 (101), it may also be possible to one day carry out cDNA-directed expression of modified—water-soluble—forms of CYP3A4, CYP1A2, CYP2C9, or CYP2D6. In this way, the detergent solubilization of these proteins can be avoided and suitable crystals obtained as a prelude to X-ray diffraction studies.

B. Enzyme Kinetics

It has become widely accepted that one can attempt to predict the in vivo metabolic (intrinsic) clearance (CL_{int}) of a drug using in vitro estimates of apparent K_m, V_{max}, and V_{max}/K_m (1–7,9,102). In most cases, this involves using microsomes, 9000 *g* supernatant fraction (S-9), precision-cut liver slices, or suspensions of hepatocytes (see Fig. 2). For a given metabolic pathway, these parameter estimates can now be confirmed with heterologous-expressed and purified enzymes (7,32,33,67,79,82,102–105). For instance, the apparent K_m (30 ± 7 μM) characterizing the CYP2E1-dependent hydroxylation of *p*-nitrophenol, in the presence of native human liver microsomes, has been confirmed with COS-7 cell lysate containing cDNA-expressed CYP2E1 (K_m = 21 μM) (104). Similarly, TOLase activity in the presence of native human microsomes and cDNA-expressed CYP2C9 is described by a similar K_m (~0.1 mM) value (82,105).

In the course of our work with naproxen (*O*-demethylase) and terfenadine (*t*-butyl hydroxylase), we have attempted to confirm the estimates of apparent K_m, V_{max}, and V_{max}/K_m obtained with native human liver microsomes (32,33). By correlation analyses and the use of CYP form-selective inhibitors, the oxidation of terfenadine was known to be primarily mediated by CYP3A4 (33,75,76). Similar studies revealed that the *O*-demethylation of naproxen was catalyzed by CYP1A2 and CYP2C9, each enzyme contributing about 50% of the total activity (32). With native human liver microsomes and the corresponding cDNA-expressed and purified CYP protein, the metabolism of both drugs conformed to monophasic Michaelis–Menten kinetics, and we were able to obtain relatively good estimates of apparent K_m in both cases (Table 3). The issue of V_{max} was more difficult, because we had to account for the abundance of each enzyme in native human liver microsomes (see Table 1). For terfenadine *t*-butyl hydroxylation, the V_{max}/K_m ratio obtained with native human liver microsomes was about fourfold higher than that obtained with the reconstituted CYP3A4-rat fp_2 fusion protein (16 vs. 3.7 μL/min/mg). Similar data were obtained with naproxen as substrate, because the V_{max}/K_m ratio in the presence of human liver microsomes was higher than that observed with heterologous-expressed CYP1A2 (13 vs. 3.6 μL/min/mg) or purified and reconstituted CYP2C9 (13 vs. 1.9 μL/min/mg; see Table 3). Therefore, while differences in apparent K_m obtained with native human liver microsomes and heterologous expressed or reconstituted enzyme are relatively small (threefold or less), differences in V_{max}/K_m ratio estimates can be greater (threefold or higher).

The availability of heterologous-expressed enzymes is paramount when dealing with compounds such as theophylline, caffeine, (*S*)-warfarin, and 7-ethoxycoumarin, which exhibit multiphasic kinetics in the presence of native human liver microsomes (79,106–113). This normally manifests itself as nonlinear Eadie-Hofstee plots (initial reaction rate or v versus $v/[S]$), reflecting the involvement of a "low K_m" and "high K_m" enzyme. With some NDEs, metabolism by a relatively low K_m enzyme can be of major clinical importance, especially when that enzyme is expressed polymorphically at relatively low levels (e.g., CYP2D6 vs. CYP3A4), and is readily saturated at therapeutic drug concentrations (1).

As an example, Yamazaki et al, have reported that the *O*-deethylation of 7-ethoxycoumarin (ECODase) conforms to biphasic (low K_m, ~18 μM; high K_m, ~155 μM) kinetics in the presence of native human liver microsomes (106). Studies with *E. coli*-expressed CYP enzymes have revealed CYP1A2 catalyzes low K_m (~21 μM) ECODase activity, whereas the K_m describing ECODase in the presence of CYP2E1 is about 120 μM. Similarly, studies with (*R*)- and (*S*)-warfarin have indicated that CYP2C9 catalyzes low K_m (~4 μM) (*S*)-warfarin 6- and 7-hydroxylase activity (112,113), whereas CYP1A2 mediates relatively high K_m (~0.7 mM) (*R*)-warfarin 6-hydroxylase activity (79,112). Interestingly, the (*R*)-warfarin 6-hydroxylase/(*R*)-warfarin 8-hydroxylase V_{max} ratio, in native human liver microsomes (~5), is similar to that reported for cDNA-expressed CYP1A2 present in human B-lymphoblast microsomes (79,112).

C. Enantio-, Regio-, and Stereoselective Biotransformations

Because enantiomers can often exhibit markedly different pharmacological characteristics, therapeutic indices, or metabolic profiles, most pharmaceutical companies have grown to appreciate the issue of enantio- or stereoselective drug–drug interactions and biotransformation reactions (114–116). Armed with heterologously expressed and purified DMEs, investigators have been afforded the opportunity to unambiguously study the regio- and

Table 3 Estimating Michaelis–Menten Kinetic Parameters in Native Human Liver Microsomes Using Purified or Heterologous cDNA-Expressed CYP Proteins

Substrate	Reaction	Enzyme source	K_m (μM)[a]	V_{max} (nmol/min/nmol CYP)	V_{max} (nmol/min/mg)	V_{max}/K_m[b] (μL/min/mg)
Terfenadine[c]	Hydroxylation	CYP3A4	30	1.1	0.11[d]	3.7
		NHLM	13 ± 3.7	—	0.19 ± 0.05	16 ± 4.2
Naproxen[e]	*O*-Demethylation	CYP1A2	250	24	0.9[d]	3.6
		CYP2C9	430	11	0.8[d]	1.9
		NHLM	160 ± 90	—	4.1 ± 2.8 (2.0)	25 ± 4.9 (13)
Phenacetin[f]	*O*-Deethylation	CYP1A2[g]	46	0.92	0.04[d]	0.90
		NHLM	54 ± 27	—	0.23 ± 0.10	5.0 ± 2.7
Clarithromycin[f]	Hydroxylation	CYP3A4[h]	21	3.6	0.35[d]	17
		NHLM	49 ± 18	—	0.21 ± 0.10	4.2 ± 0.21

[a]Michaelis constant (K_m) and maximal initial reaction velocity (V_{max}) were determined by nonlinear regression analysis (PCNONLIN, version 4.0, Statistical Consultants Inc., Lexington, KY) of the untransformed data (32,33). Microsomal data (mean ± SD) were obtained with liver tissue from at least three different organ donor subjects (mean specific content of total CYP = 0.32 nmol/mg microsomal protein). V_{max} is normalized to total CYP or microsomal protein.

[b]Hepatic intrinsic clearance (V_{max}/K_m).

[c]Terfenadine hydroxylase activity was measured in the presence of native human liver microsomes (NHLM) or a purified and fully reconstituted CYP3A4–rat fp_2 fusion protein. CYP3A(4) is considered to be the major CYP form catalyzing the hydroxylation of terfenadine in human liver microsomes (33).

[d]Predicted V_{max} in human liver microsomes. The V_{max} obtained with purified or cDNA-expressed CYP enzyme has been normalized relative to the abundance of the particular CYP protein in native human liver microsomes (see Table 1).

[e]Naproxen *O*-demethylation was measured in the presence of native human liver microsomes (NHLM), human B-lymphoblastoid cell microsomes containing cDNA-expressed CYP1A2, or purified and fully reconstituted CYP2C9 (32). Data in parentheses represent V_{max} and V_{max}/K_m ratio for CYP1A2 or CYP2C9, assuming that each enzyme equally contributes (~50%) to total naproxen *O*-demethylase activity in native human liver microsomes (32).

[f]Unpublished observations.

[g]Human B-lymphoblastoid cell microsomes containing cDNA-expressed CYP1A2.

[h]Insect cell microsomes containing cDNA-expressed CYP3A4.

stereoselective oxidation of the racemic, pseudoracemic, and individual enantiomeric forms of compounds, such as warfarin, mephenytoin, arachidonic acid, bufuralol and metoprolol (62,63,79,113,116–118). Although there are several examples in the literature, the discussion herein will focus on zileuton and ABT-418.

1. *Zileuton*

Studies with zileuton, a racemic mixture of (*R*)-(+)- and (*S*)-(−)-*N*-(1-benz[*b*]thien-2-ylethyl)-*N*-hydroxyurea (36), indicated that both enantiomers underwent CYP3A-dependent stereoselective sulfoxidation in the presence of native human liver microsomes; (a) the formation of the (*R*)-(+)-enantiomer sulfoxide (S1) was highly correlated ($r = 0.995$; $p < 0.001$; $n = 11$) with the formation of the (*S*)-(−)-enantiomer sulfoxide (S2) in a panel of microsomes; (b) the formation of both S1 and S2 was markedly inhibited (~60%) by TAO, whereas inhibitors of other CYP forms were ineffective (≤ 5% inhibition); (c) S1 and S2 formation was stimulated (~175%) by 7,8-benzoflavone; and (d) the formation of S1 ($r = 0.815$; $p < 0.01$) and S2 ($r = 0.794$; $p < 0.01$) was correlated with CYP3A-selective ERNDase activity (36). However, it appeared that the formation of S1 predominated over S2 (ratio = 2.1) in the presence of native human liver microsomes (0.62 ± 0.44 vs. 0.30 ± 0.21 nmol/h/nmol CYP; mean ± SD; $n = 11$), which suggested that the (*R*)-(+)-enantiomer of zileuton was a better substrate for CYP3A-catalyzed sulfoxidation (Fig. 9). With the availability of a purified CYP3A4–rat fp_2 fusion protein, we were able to confirm that the formation of S1 predominated over S2 (ratio = 1.6). Whether or not S1 and S2 are reduced by microsomal or cytosolic sulfoxide reductase(s) remains to be determined (46,47).

2. *ABT-418*

Depending on the enzyme source, FMO can often catalyze the *N*-oxidation of drugs in a stereoselective manner (20,41,119–121). At the time of our studies with ABT-418, we were aware that the pyrrolidine *N*′-oxidation of (*S*)-nicotine had been investigated with *E. coli* cell lysate containing cDNA-expressed human liver microsomal FMO3 (41,122). Therefore, we sought to study the stereoselective *N*′-oxidation of ABT-418 in a similar manner. Earlier in vitro and in vivo experiments had revealed that the *N*′-oxidation of ABT-418, the (*S*)-enantiomer of 3-methyl-5-(1-methyl-2-pyrrolidinyl)isoxazole, was stereoselective (100% *trans*) in the rat and dog (34,35). By comparison, ABT-418 *N*′-oxidation was not stereoselective when incubated with human (*trans*, ~50%; *cis*, ~50%), cynomolgus monkey (*trans*, 63%; *cis*, 37%), or chimpanzee (*trans*, 26%; *cis*, 74%) liver microsomes (Fig. 10). Interestingly, the *N*′-oxidation of ABT-418 in the presence of human kidney S-9 was stereoselective (100% *trans*), which agreed with the observation that the kidney FMO pool (FMO1 > FMO3) differs from that present in liver microsomes (FMO3 > FMO1) (34,119,120). By using *E. coli*-expressed human FMO3, we were able to finally confirm that the pattern of ABT-418 *N*′-oxidation was not stereoselective and was similar to that of native human liver microsomes (see Fig. 10). In contrast, purified pig liver microsomal FMO1 and rabbit lung microsomal FMO2 catalyzed the *N*′-oxidation of ABT-418 in a stereoselective (100% *trans*) manner (34). This meant that the pattern of *N*′-oxidation of ABT-418 differed from that of (*S*)-nicotine (human FMO3, 100% *trans*; pig FMO1, ~50% *trans*) and more closely resembled that of (*R*)-nicotine (pig FMO1, 100% *trans*) (122,123). At the present time, we believe that the *cis*-*N*′-oxidation of ABT-418 may serve as a useful marker for measuring the levels of FMO3 in monkey and human liver microsomes (34).

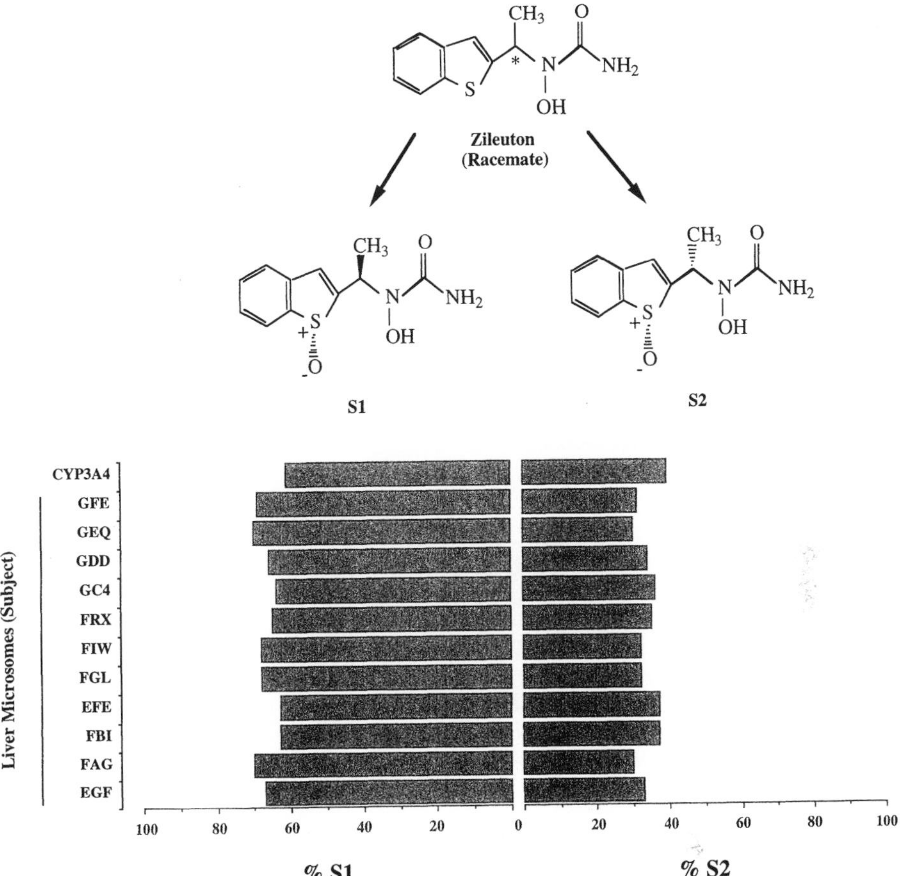

Fig. 9 Stereoselective sulfoxidation of racemic zileuton in the presence of native human liver microsomes and reconstituted CYP3A4–rat fp_2 fusion protein. The sulfoxide diastereomers were resolved by radio-HPLC (36). Data are presented as percentage of total sulfoxides formed.

D. Drug–Drug Interaction (Inhibition) Studies

With DMEs such as CYP, several researchers are routinely using human liver microsomes, primary cultures of hepatocytes, or precision-cut liver slices as a means of predicting clinically relevant drug–drug interactions, although it is now possible to further confirm inhibitory potency with the appropriate heterologous-expressed or purified CYP enzyme (1,7,124–126). In general, characterizing a CYP inhibitor should be carried out with cDNA-expressed enzyme(s) and native human liver microsomes. Afterall, ''true selectivity'' can be confirmed only in the presence of multiple CYP forms. Moreover, when relating native human liver microsomes to cDNA-expressed CYPs, in terms of IC_{50} values (concentration of inhibitor required to decrease activity by 50%), it is important to clearly define the concentration of substrate ([S]) used in the experiment and how this relates to the apparent K_m of substrate (7). This can be avoided by determining the inhibition constant (K_i), which is largely independent of [S] and enzyme concentration.

The availability of cDNA-expressed enzymes has made it possible to retrospectively

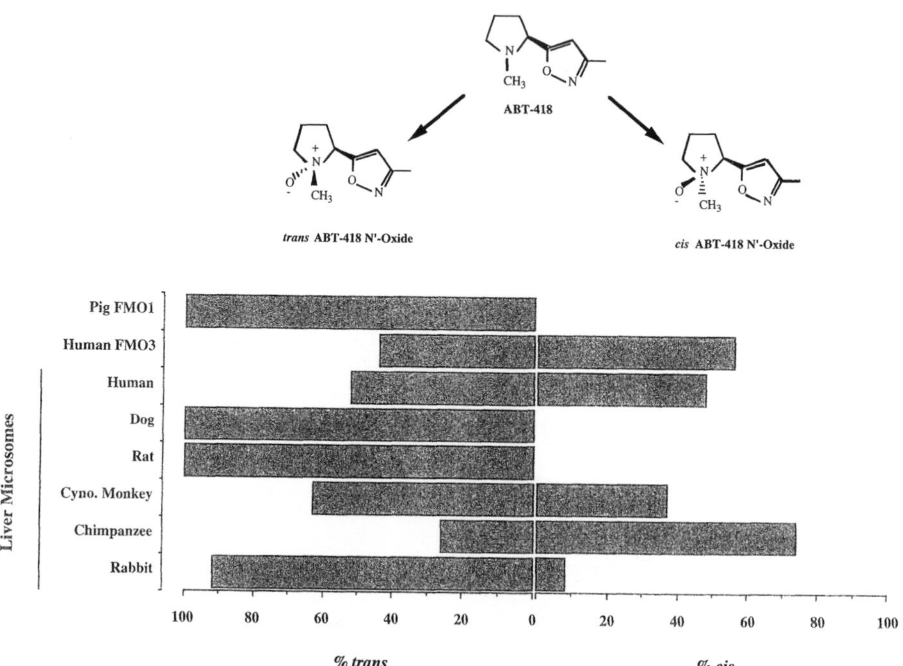

Fig. 10 Stereoselective N'-oxidation of ABT-418 in the presence of native liver microsomes, purified pig liver microsomal FMO1, or *E. coli* cell lysate containing cDNA-expressed human FMO3. [*N*-Methyl^{3}H]ABT-418 (100 μM) was incubated as described in Fig. 5.

confirm or denounce the selectivity of putative ''CYP form-selective'' inhibitors. For instance, 7,8-benzoflavone (ANF) is a more potent inhibitor ($IC_{50} \sim 0.01$ μM) of phenacetin *O*-deethylase activity catalyzed by cDNA-expressed CYP1A2, and native human liver microsomes, than that catalyzed by heterologous-expressed CYP1A1 ($IC_{50} = 0.15$ μM) (126). However, in agreement with the findings obtained with native human liver microsomes (1,127), ANF (≥0.5 μM) appears to also inhibit CYP2C8- and CYP2C9-catalyzed monooxygenase activity (124). Likewise, TAO does not differentiate between members of the CYP3A subfamily, and diethyldithiocarbamate (DDC) is not selective for CYP2E1 (1,124,127).

Data presented in Figure 11 are two examples where we have confirmed that a NDE inhibits a particular form of CYP. In the first example (Fig. 11A), ''compound C'' was a relatively potent inhibitor (IC_{50} ~3 μM) of CYP2D6-catalyzed dextromethorphan *O*-demethylase (DEXase) activity in the presence of native human liver microsomes and human B-lymphoblast microsomes containing cDNA-expressed CYP2D6. In both experiments substrate concentration was equal to apparent K_m. Given that the drug in vitro was a CYP2D6 substrate, it was likely that the mechanism of inhibition was competitive. However, the drug was expected to be a weak inhibitor, because the in vitro IC_{50} value exceeded the therapeutic plasma level (~0.2 μM) of the drug (7). Accordingly, it has been subsequently reported that although the metabolism of compound C cosegregates with the CYP2D6 genotype and phenotype, it fails to inhibit CYP2D6 activity in vivo (S. Wong, unpublished observations). In the second example (see Fig. 11B), the effect of ABT-538 (ritonavir) on CYP2C9-catalyzed monooxygenase activity is presented. ABT-538 is a

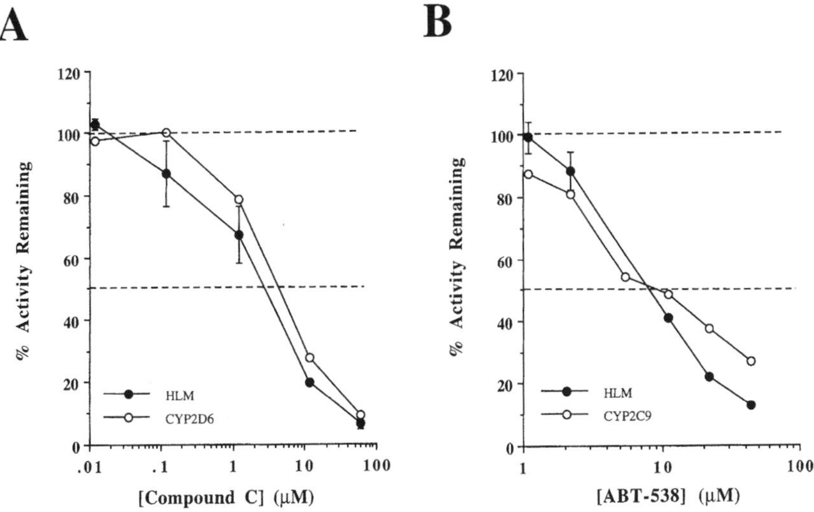

Fig. 11 Inhibition of CYP-selective monooxygenase activity in human liver microsomes (HLM). (A) The effect of "compound C" on dextromethorphan *O*-demethylase (CYP2D6) activity in the presence of native human liver microsomes (filled circles) or human B-lymphoblastoid microsomes containing cDNA-expressed CYP2D6 (open circles). In both cases, the final concentration of dextromethorphan was 10 μM (~K_m) and activity was monitored by radioassay (37). (B) The effect of ABT-538 on tolbutamide methylhydroxylase (TOLase) activity in the presence of native human liver microsomes (filled circles) and naproxen *O*-demethylase (NAPase) activity catalyzed by reconstituted cDNA-expressed CYP2C9 (open circles). Both activities were measured at a substrate concentration that approximated apparent K_m (7,32). Data are presented as percentage activity remaining relative to control (solvent alone) incubations.

HIV-1 protease inhibitor that has been approved for the treatment of AIDS (7). By virtue of its lipophilicity and the presence of a sterically unhindered nitrogen on the 5-substituted thiazole moiety, the drug is a potent inhibitor of CYP3A and CYP2D6 activity in vitro and in vivo (7). However, during the course of our screening experiments, we observed that the drug also inhibited (IC_{50} = 8 μM) TOLase activity in the presence of native human liver microsomes (see Fig. 10). Because the experiments were performed at a final tolbutamide concentration of 0.1 mM (~K_m), it was felt that this effect was reflective of CYP2C9 (vs. CYP2C8 and CYP2C19) inhibition (7,81,82,105). In agreement, the inhibition of *E. coli*-expressed CYP2C9, CYP2C8, and CYP2C19 catalyzed NAPase activity was characterized by an IC_{50} of 8.0 μM, 50 μM, and 30 μM, respectively (7). By comparison with compound C, the therapeutic plasma levels of ABT-538 are relatively high (5–21 μM), so that inhibition of NAPase and TOLase activity may occur in vivo.

E. Toxicity Testing

The concept that xenobiotics can be metabolized to biologically reactive metabolites which, in turn, bind covalently to macromolecules, such as DNA or protein, is now widely accepted. In recent years, the metabolic activation of many procarcinogens and promutagens, as well as the toxicity of agents such as acetaminophen, have been attributed to specific human DMEs (e.g., NAT, CYP3A4, CYP1A2, CYP1A1, and CYP2E1) (128–

130). Because many of these DMEs are readily cloned and stably expressed in various cell lines, it is now possible to coordinately investigate the metabolic activation and appearance of cytotoxicity or mutagenicity as intracellular events (131–134). This is possible because many of the cell types currently in use (e.g., AHH-1 TK +/− and CHO) can be adapted for assaying the induction of micronuclei or chromosomal aberrations. In addition, these cells can be used to assay for the induction of a variety of genotoxicity markers, including mutations at the thymidine kinase (*tk*) and hypoxanthine phosphoribosyltransferase (*hprt*) loci, or for gross cytotoxicity endpoints such as cell survival rates or enzyme leakage (131–135).

Therefore, researchers are able to perform toxicity testing of NDEs with these transgenic cells alone, in parallel with conventional toxicity models such as cultured primary hepatocytes and liver slices, or along side genotoxicity assays employing tester strains of *Salmonella typhimurium* coincubated with native human liver microsomes, S-9, or reconstituted CYPs (136–141). For brevity, several examples are listed in Table 4. Some researchers have managed to coexpress two or more (phase I and phase II) DMEs in a single cell type. This is necessary when the bioactivation of a test compound requires CYP-dependent oxidation and secondary metabolism (e.g., *N*-acetylation) to occur sequentially. Many of these systems are geared toward the bioactivation of drugs. In the future, however, the possibility exists that one may be able to heterologously express a large array of competing (bioactivating and deactivating) DMEs within the same cell, so that these cellular-based screen assays more closely resemble the situation in vivo.

F. Bioreactors

Modern-day drug development has yielded many compounds with relatively complex structures and chemistries, these have included cyclosporine, paclitaxol (Taxol), clarithromycin, various alkaloids and peptidomimetic HIV-1 protease inhibitors, such as saquinavir, indinavir, and ritonavir (80,146,147). In light of the recent advancements in the field of combinatorial chemistry, and their likely effect on drug discovery, it appears the the number of structurally diverse and complex structures making their way through development pipelines will greatly increase in the near future (148,149). In some instances, such as the CYP3A4-dependent conversion of clarithromycin to 14-(*R*)-hydroxyclarithromycin (unpublished observations), it is possible that these entities will be metabolized to products that are themselves pharmacologically active. Toward this end, many pharmaceutical companies are looking to use "bioreactors" as a means of facilitating the "bulk" synthesis of such metabolites. In this way, the power of DMEs to exquisitely carry out regio-, stereo-, and enantioselective biotransformations can be harnessed to generate relatively large amounts of the desired product.

The concept of "cell-based" or "immobilized enzyme-based" bioreactors (IEB) is by no means novel, and different strategies have included (a) covalent attachment to solid supports; (b) adsorption on solid supports; (c) entrapment in polymeric gels; and (d) encapsulation (150–154). The potential advantages of IEB technology include ease of handling, facile separation of products from enzyme incubation mixtures, stabilization of the enzyme toward thermal or oxidative degradation, and the opportunity to recover and reuse the enzyme catalyst (151). For instance, immobilized thermolysin and penicillin amidase have been employed in the biosynthesis of aspartame and 6-aminopenicillanic acid respectively (154, and references therein). Whatever the strategy, it is envisioned that computer-controlled systems will permit on-line monitoring of oxygen tension, pH, temperature, and

Table 4 Examples of In Vitro Toxicity Testing Employing Transgenic Cells Heterologously Expressing One or More Human Drug-Metabolizing Enzymes

Cell type	Cell line	Test compound(s)[a]	Expressed enzyme(s)	Endpoint(s)[b]	Ref.
CHO	Various	PhIP, MeIQ, IQ	CYP1A2, NAT2	Cell survival	132
AHH-1 TK +/−	1A2/Hyg 3A4/Hyg	Aflatoxin B_1	CYP1A2 CYP3A4	*hprt*, DNA binding, and cell survival	133
CHL	Various	IQ, MeIQ	CYP3A7; NAT2	Cell survival	134
HepG2	Mvh2E1-9	Acetaminophen	CYP2E1	LDH leakage and protein adducts	135
AHH-1 TK +/−	Various	Tamoxifen	Epoxide hydrolase, CYP3A4, CYP2E1, CYP1A2, or CYP1A1	Micronuclei test	142
V79	Various	Benzo[*a*]pyrene	CYP1A1	Cellular dye uptake and micronuclei test	143
AHH-1 TK+/−	2D6/Hol	NNK	CYP2D6	Cell survival, *hprt*	144
CHO	AS52	NNK	CYP2A6	*hprt*	145

[a]IQ, 2-amino-3-methylimidazo[4,5-f]quinoline; MeIQ, 2-amino-3,8-dimethylimidazo[4,5-f]quinoline; PhIP, 2-amino-1-methyl-6-phenylimidazo[4,5-b]-pyridine; NNK, 4-(methyl-nitrosamino)-1-(3-pyridyl)-1-butanone

[b]LDH, lactate dehydrogenase; *hprt*, hypoxanthine phosphoribosyltransferase mutation assay.

cofactor concentration within the bioreactor. Moreover, the rate of drug delivery to the system may also be controlled, to circumvent the problem of end-product or substrate inhibition.

Currently available cell-based bioreactors make use of large-scale (≥500 L) fermentation technology, which means that in the future one may be able to incubate the drug of choice in the presence of yeast or *E. coli* cells stably expressing DMEs such as human CYP3A4 and fp_2. Furthermore, it may be possible to entrap or encapsulate these cells in some sort of inert matrix, as opposed to using suspensions of free cells, thereby simplifying product retrieval. Obviously, the intracellular availability of cofactors such as NADPH and UDP-glucuronic acid (UDPGA) would have to be addressed.

On the other hand, it is likely that the development of IEB technology will be more challenging, although various attempts have been made to immobilize native microsomes and solubilized or purified DMEs (155–161). These have included noncovalent immobilization in polymeric gels (e.g., gelatine, alginate; or κ-carrageenan), or covalent coupling onto column or beaded matrices containing CNBr-activated Sepharose 4B or carbonyldiimidazole-activated Sephadex G-150 (155–161). However, many of these methods have relied on the availability of conventionally purified DMEs, such as rabbit CYP2B4 (P450 LM_2) and fp_2 (159,161), while concerns about loss of activity after immobilization have also been reported (155,158,161), and it is not known if these bioreactors are robust enough to generate metabolites (≥1 mg) over extended periods (>24 h). Therefore, advances in IEB technology will require (a) relatively large amounts of purified DMEs; (b) the development of gentle enzyme immobilization procedures; and (c) alternative (inexpensive) sources of reducing equivalents or cofactors. The first of these challenges may be met with currently available high-level heterologous expression systems. For instance, *E. coli* expressed CYP–fp_2 fusion proteins can be readily purified by affinity chromatography on 2′,5′-ADP Sepharose (70,162). Recombinant technology may also permit engineering of DMEs, to achieve strong noncovalent coupling of enzyme to solid supports using relatively mild conditions (163). In the case of CYP, reducing equivalents could be provided by oxygen surrogates, such as cumene hydroperoxide and iodosylbenzene (164), by electrochemical generation of reductants such as cobalt (II) sepulchrate trichloride (162), or by engineered (photoactive) fp_2 (150). Therefore, the possibilities are endless, and the development of this technology could be expanded to include the large-scale disposal of hazardous chemicals, the development of continuous-flow biosensors, or extracorporeal shunt systems for drug detoxification (151,159,162,165).

V. CONCLUSIONS

Advancements in recombinant gene technology have greatly increased our understanding of complex DME systems, such as those present in human liver (e.g., FMO, CYP, and UDPGT). One can now heterologously express many of these DMEs in transgenic cell lines, so that natively minor—low abundance—enzymes can also be expressed at high levels and subsequently purified to electrophoretic homogeneity. Moreover, it is now recognized that strategies employing methods for investigating the metabolism or toxicity of drugs in vitro can play an important role in the discovery and development of NDEs, especially if the different models are used in an integrated fashion (1,7). Given that pharmaceutical companies are turning their attention to "combinatorial methods" and "high-throughput screening" of drugs, the use of heterologous expression systems, and heterolo-

Table 5 Applications of Heterologous-Expressed or Purified Human Drug-Metabolizing Enzymes: Summary

Application	Comment(s)	Refs.
Enzyme/isoform-selective biotransformations	1. Confirm the involvement of a particular DME system in the biotransformation of a NDE (e.g., FMO vs. CYP)	34,41,75
	2. Reaction phenotyping of DME: distinguish between members of *different* gene families or subfamilies	5,8,30,32,37,75–78
	3. Differentiate between members of the *same* DME gene subfamily	25,32,62,63,67,74,82,83
	4. Confirm polymorphic biotransformations and catalysis by allelic (vs. wild-type) forms of DME	82,83,92
Michaelis–Menten kinetics	Confirm apparent K_m, V_{max}, and V_{max}/K_m obtained with native human tissue	79,103,104,106–110,113
Stereo-, regio-, and enantioselective biotransformations	Confirm stereo-, regio-, and enantioselective oxidations in the absence of confounding *N*-oxide, sulfoxide, or epoxide reductases	20,36,113,116–118,122
Inhibition studies	1. Confirm IC_{50} or K_i obtained with native human tissue	124–127
	2. Confirm selectivity of an inhibitor for different DMEs	124–127
Toxicity testing	1. Transgenic cell-based genotoxicity, cytotoxicity and mutagenicity testing of NDEs	131–135,142,145
	2. Genotoxicity testing (e.g., Ames test) using reporter strains of *S. typhimurium* incubated with purified and reconstituted DME(s), or microsomes containing cDNA-expressed DMEs	138,139,140,141
Bioreactors: scaled-up metabolite production	1. Immobilized transgenic cells, containing cDNA-expressed DME(s)	
	2. Immobilized purified DME(s)	
Large-scale antibody production	Purified cDNA-expressed DMEs as antigens for the production of monoclonal or monospecific polyclonal antibodies	
Immunoassays	cDNA-expressed or purified DMEs as standards for the immunoquantitation of DMEs in native tissue	27,56,57
Spectroscopic studies with natively low-abundance DMEs such as CYP2D6	Obtain UV/visible difference spectra (''type I'' or ''type II'') with cDNA-expressed or purified CYP: minimize turbidity, because of high CYP/phospholipid ratio	Unpublished observations

DME, drug-metabolizing enzyme; NDE, new drug entity; IC_{50}, concentration of inhibitor required to decrease activity by 50%; K_i, inhibition constant; K_m, apparent Michaelis constant; V_{max}, apparent maximal reaction velocity; V_{max}/K_m, intrinsic clearance (CLint).

gous expressed DMEs, will continue to increase. Some applications have been described herein, with emphasis placed on using cDNA-expressed DMEs as tools for confirming polymorphic biotransformations, reaction phenotyping, confirming inhibition (e.g., IC_{50}, K_i), or kinetic (e.g., K_m, V_{max}, V_{max}/K_m) data obtained with native human tissue, the development of bioreactors and toxicity and mutagenicity testing (Table 5). Other applications, such as the use of high-level heterologous expression systems to generate protein for the production of commercial quantities of (anti-DME) antibodies, are also likely to be realized. In addition, it may be possible to heterologously express DMEs in various cell lines (e.g., Caco-2), so that NDE absorption, transport (e.g., P-glycoprotein-linked), and metabolism can be coordinately investigated (166,167). Furthermore, numerous DMEs may themselves be considered to be therapeutic targets in the future, which will necessitate relatively large amounts of protein.

Given the speed with which transgenic cell technology has been developed, it is only a matter of time before ''transgenic *whole-animal*'' methods become well established. A number of laboratories are already expressing human DMEs in rodents (168), and in the future it may be possible to extend this technology to encompass nonhuman primates. In addition, exciting possibilities exist for the treatment of various genetic disorders. For instance, the concept that one can ''phenocopy'' subjects with a potent CYP2D6 inhibitor (e.g, quinidine) is now widely accepted. This means that one can convert ''extensive metabolizers'' (EM) into ''poor metabolizers'' (PM) by chemically inhibiting CYP2D6 (169–171). Some day, it may be possible to alter the genotypic and phenotypic status of an individual using ''transgenic'' methodology. In addition, DME-based gene therapy may also become useful in the treatment of genetic deficiencies, such as Gilbert's, Crigler-Najjar, or ''fish odor'' syndrome (18,73). Nevertheless, the eventual implementation of such technology will have to be governed by well-established moral, ethical, and legal guidelines, which are beyond the scope of this discussion.

ACKNOWLEDGMENTS

The author would like to thank Drs. J. Cashman (Seattle Biomedical Research Institute, Seattle, WA), M. Shet and R. Estabrook (University of Texas Southwestern Medical Center, Dallas, TX), J. Lasker (Mount Sinai Medical School, New York, NY), T. Richardson and H. Barnes (Immune Complex Corp., La Jolla, CA) and D. Grant (Hospital For Sick Children, Toronto, Ontario, Canada) for very useful collaborations. The expert technical assistance provided by M. Kukulka, E. Roberts, P. Johnson, and M. Mayer (Abbott Laboratories) is also acknowledged.

REFERENCES

1. A. D. Rodrigues, Use of in vitro human metabolism studies in drug development, *Biochem. Pharmacol. 48*:2147–2156 (1994).
2. S. E. Ball, J. A. Scatina, S. F. Sisenwine, and G. L. Fischer, The application of in vitro models of drug metabolism and toxicity in drug discovery and drug development, *Drug Chem. Toxicol. 18*:1–28 (1995).
3. G. T. Tucker, The rational selection of drug interaction studies: Implications of recent advances in drug metabolism, *Int. J. Clin. Pharmacol. Ther. Toxicol. 30*:550–553 (1992).

4. J. O. Miners, M. E. Veronese, and D. J. Birkett, In vitro approaches for the prediction of human drug metabolism, *Annu. Rep. Med. Chem. 29*:307–316 (1994).
5. S. A. Wrighton, M. Vandenbranden, J. C. Stevens, L. A. Shipley, and B. J. Ring, In vitro methods for assessing human hepatic drug metabolism: Their use in drug development, *Drug Metab. Rev. 25*:453–484 (1993).
6. R. N. Hayes, W. F. Pool, M. W. Sinz, and T. F. Woolf, Recent developments in drug metabolism methodology, *Pharmacokinetics: Regulatory, Industrial, Academic Perspectives* (P. G. Welling and F. L. S. Tse, eds.), Marcel Dekker, New York, 1995, pp. 201–234.
7. A. D. Rodrigues and S. L. Wong, Metabolism-based drug–drug interactions: In vitro–in vivo correlations and the Abbott Laboratories experience, *Advances in Pharmacology 43*:65–101 (1997).
8. T. Kronbach, Hepatic microsomes and heterologous expression systems as in vitro models for human drug metabolism, *Pharmacokinetics: Regulatory, Industrial, Academic Perspectives* (P. G. Welling, and F. L. S. Tse, eds.), Marcel Dekker, New York, 1995, pp. 235–260.
9. R. P. Remmel and B. Burchell, Validation and use of cloned, expressed human drug-metabolizing enzymes in heterologous cells for analysis of drug metabolism and drug–drug interactions, *Biochem. Pharmacol. 46*:559–566 (1993).
10. M. R. Waterman, Heterologous expression of mammalian P450 enzymes, *Adv. Enzymol. 68*: 37–66 (1994).
11. B. Burchell, D. W. Nebert, D. R. Nelson, K. W. Bock, T. Iyanagi, P. L. M. Jansen, D. Lancet, G. J. Mulder, J. R. Chowdury, G. Siest, T. R. Tephly, and P. I. Mackenzie, The UDP-glucuronosyltransferase gene superfamily: Suggested nomenclature based on evolutionary divergence, *DNA Cell Biol. 10*:487–494 (1991).
12. D. R. Nelson, T. Kamataki, D. J. Waxman, F. P. Guengerich, R. W. Estabrook, R. Feyereisen, F. J. Gonzalez, M. J. Coon, I. C. Gunsalus, O. Gotoh, K. Okuda, and D. W. Nebert, The P450 superfamily: Update on new sequences, gene mapping, accession numbers, early trivial names of enzymes, and nomenclature, *DNA Cell Biol. 12*:1–51 (1993).
13. K. P. Vatsis, W. W. Weber, D. A. Bell, J. M. Dupret, D. A. Price Evans, D. M. Grant, D. W. Hein, H. J. Lin, U. A. Meyer, M. V. Relling, E. Sim, T. Suzuki, and Y. Yamazoe, Nomenclature for *N*-acetyltransferases, *Pharmacogenetics 5*:1–17 (1995).
14. M. P. Lawton, J. R. Cashman, T. Cresteil, C. T. Dolphin, A. A. Elfarra, R. Hines, E. Hodgson, T. Kimura, J. Ozols, I. R. Phillips, R. M. Philpot, L. L. Poulsen, A. E. Rettie, E. A. Shepard, D. E. Williams, and D. M. Ziegler, A nomenclature for the mammalian flavin-containing monoxygenase gene family based on amino acid sequence identities, *Arch. Biochem. Biophys. 308*:254–257 (1994).
15. C. A. Lee, S. H. Kadwell, T. A. Kost, and C. J. Serabjit-Singh, CYP3A4 Expressed by insect cells infected with a recombinant baculovirus containing both CYP3A4 and human NADPH–cytochrome P450 reductase is catalytically similar to human liver microsomal CYP3A4, *Arch. Biochem. Biophys. 319*:157–167 (1995).
16. C. L. Crespi, R. Langenbach, and B. W. Penman, Human cell lines, derived from AHH-1 TK+/− human lymphoblasts, genetically engineered for expression of cytochromes P450, *Toxicology 82*:89–104 (1993).
17. J. Doehmer, V79 Chinese hamster cells genetically engineered for cytochrome P450 and their use in mutagenicity and metabolism studies, *Toxicology 82*:105–118 (1993).
18. B. Burchell, C. H. Brierley, and D. J. Clarke, Cloning and expression of human UDP-glucuronosyltransferase genes, *Advances in Drug Metabolism in Man* (G. M. Pacifici and G. N. Fracchia, eds.), European Commission, Brussels, 1995, pp. 181–321.
19. T. Ebner and B. Burchell, Substrate specificities of two stably expressed human liver UDP-glucuronosyltransferases of the *UGT1* gene family, *Drug Metab. Dispos. 21*:50–55 (1993).
20. N. Lomri, Z. Yang, and J. R. Cashman, Regio- and stereoselective oxygenations by adult

human liver flavin-containing monooxygenase 3. Comparison with forms 1 and 2, *Chem. Res. Toxicol. 6*:800–807 (1993).

21. D. W. Hein, M. A. Doll, T. D. Rustan, K. Gray, Y. Feng, R. J. Ferguson, and D. M. Grant, Metabolic activation and deactivation of arylamine carcinogens by recombinant human NAT1 and polymorphic NAT2 acetyltransferases, *Carcinogenesis 14*:1633–1638 (1993).
22. M. S. Shet, K. M. Faulkner, P. L. Holmans, C. W. Fischer, and R. W. Estabrook, The effects of cytochrome b_5, NADPH-P450 reductase, and lipid on the rate of 6β-hydroxylation of testosterone as catalyzed by a human P450 3A4 fusion protein, *Arch. Biochem. Biophys. 318*:314–321 (1995).
23. J. T. M. Buters, K. R. Korzekwa, K. L. Kunze, Y. Omata, J. P. Hardwick, and F. J. Gonzalez, cDNA-directed expression of human cytochrome P450 CYP3A4 using baculovirus, *Drug Metab. Dispos. 22*:688–692 (1994).
24. J. T. M. Buters, M. Shou, J. P. Hardwick, K. R. Korzekwa, and F. J. Gonzalez, cDNA-directed expression of human cytochrome P450 CYP1A1 using baculovirus, *Drug Metab. Dispos. 23*:696–701 (1995).
25. E. M. J. Gillam, Z. Guo, Y.-F. Ueng, H. Yamazaki, I. Cock, P. E. B. Reilly, W. D. Hooper, and F. P. Guengerich, Expression of cytochrome P450 3A5 in *Escherichia coli*: Effects of 5′ modification, purification, spectral characterization, reconstitution conditions, and catalytic activities, *Arch. Biochem. Biophys. 317*:374–384 (1995).
26. E. M. J. Gillam, Z. Guo, M. V. Martin, C. M. Jenkins, and F. P. Guengerich, Expression of cytochrome P450 2D6 in *Escherichia coli*: Purification, and spectral and catalytic characterization, *Arch. Biochem. Biophys. 319*:540–550 (1995).
27. S. Imaoka, T. Yamada, T. Hiroi, K. Hayashi, T. Sakaki, Y. Yabusaki, and Y. Funae, Multiple forms of human P450 expressed in *Saccharomyces cerevisiae*: Systematic characterization and comparison with those of the rat, *Biochem. Pharmacol. 51*:1041–1050 (1996).
28. P. Sandhu, T. Baba, and F. P. Guengerich, Expression of modified cytochrome P450 2C10 (2C9) in *Escherichia coli*, purification, and reconstitution of catalytic activity, *Arch. Biochem. Biophys. 306*:443–450 (1993).
29. J. Combalbert, I. Fabre, G. Fabre, I. Dalet, J. Derancourt, J. P. Cano, and P. Maurel, Metabolism of cyclosporin A: Purification and identification of the rifampicin-inducible human liver cytochrome P450 (cyclosporin A oxidase) as a product of the *P450IIIA* gene subfamily, *Drug Metab. Dispos. 17*:197–207 (1989).
30. F. P. Guengerich, M. V. Martin, P. H. Beaune, P. Kremers, T. Wolff, and D. J. Waxman, Characterization of rat and human liver microsomal cytochrome P-450 forms involved in nifedipine oxidation, a prototype for genetic polymorphism in oxidative drug metabolism, *J. Biol. Chem. 261*:5051–5060 (1986).
31. J. L. Raucy and J. M. Lasker, Isolation of P450 enzymes from human liver, *Methods Enzymol. 206*:577–587 (1991).
32. A. D. Rodrigues, M. J. Kukulka, E. M. Roberts, D. Ouellet, and T. R. Rodgers, [*O*-Methyl ^{14}C]naproxen *O*-demethylase activity in human liver microsomes: Evidence for the involvement of cytochrome P4501A2 and P4502C9/10, *Drug Metab. Dispos. 24*:126–136 (1996).
33. A. D. Rodrigues, D. J. Mulford, R. D. Lee, B. W. Surber, M. J. Kukulka, J. L. Ferrero, S. B. Thomas, M. S. Shet, and R. W. Estabrook, In vitro metabolism of terfenadine by a purified recombinant fusion protein containing cytochrome P4503A4 and NADPH-P450 reductase: Comparison to human liver microsomes and precision-cut liver tissue slices, *Drug Metab. Dispos. 23*:765–775 (1995).
34. A. D. Rodrigues, M. J. Kukulka, J. L. Ferrero, and J. R. Cashman, In vitro hepatic metabolism of ABT-418 in chimpanzee (*Pan troglodytes*): A unique pattern of microsomal flavin-containing monooxygenase-dependent stereoselective *N*′-oxidation, *Drug Metab. Dispos. 23*: 1143–1152 (1995).
35. A. D. Rodrigues, J. L. Ferrero, M. T. Amann, G. A. Rotert, S. P. Cepa, B. W. Surber, J. M. Machinist, N. R. Tich, J. P. Sullivan, D. S. Garvey, M. Fitzgerald, and S. P. Arneric, The

in vitro hepatic metabolism of ABT-418, a cholinergic channel activator, in rats, dogs, cynomolgus monkeys, and humans, *Drug Metab. Dispos.* *22*:788–798 (1994).

36. J. M. Machinist, M. D. Mayer, M. S. Shet, J. L. Ferrero, and A. D. Rodrigues, Identification of the human liver cytochrome P450 enzymes involved in the metabolism of zileuton (ABT-077) and its *N*-dehydroxylated metabolite, Abbott-66193, *Drug Metab. Dispos.* *23*:1163–1174 (1995).
37. A. D. Rodrigues, M. J. Kukulka, B. W. Surber, S. B. Thomas, J. T. Uchic, G. A. Rotert, G. Michel, B. Thome-Kromer, and J. M. Machinist, Measurement of liver microsomal cytochrome P450 (CYP2D6) activity using [*O*-Methyl ^{14}C]dextromethorphan, *Anal. Biochem.* *219*:309–320 (1994).
38. V. Fischer, B. Vogels, G. Maurer, and R. E. Tynes, The antipsychotic clozapine is metabolized by the polymorphic human microsomal and recombinant cytochrome P450 2D6, *J. Pharmacol. Exp. Ther.* *260*:1355–1360 (1992).
39. M.-L. Dahl, A. Llerena, U. Bondesson, L. Lindstrom, and L. Bertilsson, Disposition of clozapine in man: Lack of association with debrisoquine and *S*-mephenytoin hydroxylation polymorphisms, *Br. J. Clin. Pharmacol.* *37*:71–74 (1994).
40. A. M. Flammang, H. V. Gelboin, T. Aoyama, F. J. Gonzalez, and G. D. McCoy, Nicotine metabolism by cDNA-expressed human cytochrome P450s, *Biochem. Arch.* *8*:1–8 (1992).
41. J. R. Cashman, S. B. Park, Z.-C. Yang, S. A. Wrighton, P. Jacob, and N. L. Benowitz, Metabolism of nicotine by human liver microsomes: Stereoselective formation of *trans*-nicotine *N′*-oxide, *Chem. Res. Toxicol.* *5*:639–646 (1992).
42. N. W. McCracken, S. Cholerton, and J. R. Idle, Continine formation by cDNA-expressed human cytochromes P450, *Med. Sci. Res.* *20*:877–878 (1992).
43. C. E. Berkman, S. B. Park, S. A. Wrighton, and J. R. Cashman, In vitro–in vivo correlations of human (*S*)-nicotine metabolism, *Biochem. Pharmacol.* *50*:565-570 (1995).
44. G. A. Kyermaten, M. Morgan, G. Warner, L. F. Martin, and E. S. Vesell, Metabolism of nicotine by hepatocytes, *Biochem. Pharmacol.* *40*:1747–1756 (1990).
45. M. Maurice, S. Emiliani, I. Dalet-Beluche, J. Derancourt, and R. Lange, Isolation and characterization of a cytochrome P450 of the IIA subfamily from human liver microsomes, *Eur. J. Biochem.* *200*:511–517 (1991).
46. P. G. C. Douch and L. L. Buchanan, Some properties of the sulphoxidases and sulphoxide reductases of the cestode *Moniezia expansa*, the nematode *Ascaris suum* and mouse liver, *Xenobiotica* *9*:675–679 (1979).
47. S. Kitamura, K. Tatsumi, and H. Yoshimura, Metabolism in vitro of sulindac. Sulfoxide-reducing enzyme systems in guinea pig liver, *J. Pharm. Dyn.* *3*:290–298 (1980).
48. S. Kitamura and K. Tatsumi, Involvement of liver aldehyde oxidase in the reduction of nicotinamide *N*-oxide, *Biochem. Biophys. Res. Commun.* *120*:602–606 (1984).
49. S. Kitamura and K. Tatsumi, Reduction of tertiary amine *N*-oxides by liver preparations: Function of aldehyde oxidase as a major *N*-oxide reductase. *Biochem. Biophys. Res. Commun.* *121*:749–754 (1984).
50. Y. Hirao, S. Kitamura, and K. Tatsumi, Epoxide reductase activity of mammalian liver cytosols and aldehyde oxidase, *Carcinogenesis* *15*:739–743 (1994).
51. A. D. Rodrigues, Comparison of levels of aldehyde oxidase with cytochrome P450 activities in human liver in vitro, *Biochem. Pharmacol.* *48*:197–200 (1994).
52. G. M. Eliopoulos, C. B. Wennersten, G. Cole, D. Chu, D. Pizzuti, and R. C. Moellering, In vitro activity of A-86719.1, a novel 2-pyridone antimicrobial agent, *Antimicrob. Agents Chemother.* *35*:850–853 (1995).
53. Z. Gregus, J. B. Watkins, T. N. Thompson, M. J. Harvey, K. Rozman, and C, D. Klaassen. Hepatic phase I and phase II biotransformations in quail and trout: Comparison to other species commonly used in toxicity testing, *Toxicol. Appl. Pharmacol.* *67*:430–441 (1983).
54. H. Yamazaki, M. Nakano, Y. Imai, Y.-F. Ueng, F. P. Guengerich, and T. Shimada, Roles of cytochrome b_5 in the oxidation of testosterone and nifedipine by recombinant cytochrome

P450 3A4 and by human liver microsomes, *Arch. Biochem. Biophys. 325*:174–182 (1996).

55. C. T. Dolphin, T. E. Cullingford, E. A Shepard, R. L. Smith, and I. R. Phillips, Differential developmental and tissue-specific regulation of expression of the genes encoding three members of the flavin-containing monooxygenase family of man, FMO1, FMO3 and FMO4, *Eur. J. Biochem. 235*:683–689 (1996).
56. T. Shimada, H. Yamazaki, M. Mimura, Y. Inui, and F. P. Guengerich, Interindividual variations in human liver cytochrome P450 enzymes involved in the oxidation of drugs, carcinogens, and toxic chemicals: Studies with liver microsomes of 30 Japanese and Caucasians, *J. Pharmacol. Exp. Ther. 270*:414–423 (1994).
57. S. Lecoeur, E. Bonierbale, D. Challine, J.-C. Gautier, P. Valadon, P. M. Dansette, R. Catinot, F. Ballet, D. Mansuy, and P. H. Beaune, Specificity of in vitro covalent binding of tienilic acid metabolites to human liver microsomes in relationship to the type of hepatotoxicity: Comparison with two directly hepatotoxic drugs, *Chem. Res. Toxicol. 7*:434–442 (1994).
58. M. E. McManus, P. M. Hall, I. Stupans, J. Brennan, W. Burgess, R. Robson, and D. J. Birkett, Immunohistochemical localization and quantitation of NADPH–cytochrome P450 reductase in human liver, *Mol. Pharmacol. 32*:189–194 (1987).
59. P. Kremers, P. Beaune, T. Cresteil, J. DeGraeve, S. Columelli, J.-P. Leroux, and J. E. Gielen, Cytochrome P450 monooxygenase activities in human and rat liver microsomes, *Eur. J. Biochem. 118*:599–606 (1981).
60. N. A. Petushkova, G. F. Zhirnov, I. P. Kanaeva, O. V. Dobrynina, V. B. Lisitzina, G. I. Bachmanova, and A. I. Archakov, Cytochrome P450-dependent substrate oxidations in autopsy human liver, *In Vitro Toxicol. 8*:229–237 (1995).
61. P. B. Watkins, S. A. Murray, P. E. Thomas, and S. A. Wrighton, Distribution of cytochromes P450, cytochrome b_5 and NADPH–cytochrome P450 reductase in an entire human liver, *Biochem. Pharmacol. 39*:471–476 (1990).
62. J. A. Goldstein, M. B. Faletto, M. Romkes-Sparks, T. Sullivan, S. Kitareewan, J. L. Raucy, J. M. Lasker, and B. I. Ghanayem, Evidence that CYP2C19 is the major (*S*)-mephenytoin 4′-hydroxylase in humans, *Biochemistry 33*:1743–1752 (1994).
63. T. H. Richardson, F. Jung, K. J. Griffen, M. Wester, J. L. Raucy, B. Kemper, L. M. Bornheim, C. Hassett, C. J. Omiecinski, and E. F. Johnson, A universal approach to the expression of human and rabbit cytochrome P450s of the 2C subfamily in *Escherichia coli, Arch. Biochem. Biophys. 323*:87–96 (1995).
64. D. J. Waxman, D. P. Lapenson, T. Aoyama, H, V. Gelboin, F. J. Gonzalez, and K. Korzekwa, Steroid hormone hydroxylase specificities of eleven cDNA-expressed human cytochrome P450s, *Arch. Biochem. Biophys. 290*:160–166 (1991).
65. L. Gu, F. J. Gonzalez, W. Kalow, and B. K. Tang, Biotransformation of caffeine, paraxanthine, theobromine, and theophylline by cDNA-expressed human CYP1A2 and CYP2E1, *Pharmacogenetics 2*:73–77 (1992).
66. J. L. Raucy, M. R. Wester, E. Aramsombatdee, and J. M. Lasker, Purification and characterization of human liver CYP2C19: An *S*-mephenytoin 4′-hydroxylase. Abstract of the Xth International Symposium on Microsomes and Drug Oxidations, Toronto, Canada, July 1994.
67. J. C. Gorski, S. D. Hall, D. R. Jones, M. Vandenbranden, and S. A. Wrighton, Regioselective biotransformation of midazolam by members of the human cytochrome P450 3A (CYP3A) subfamily. *Biochem. Pharmacol. 47*:1643–1653 (1994).
68. J. L. Raucy and J. M. Lasker, Human P450 enzymes associated with the mephenytoin 4-hydroxylation polymorphism, *Human Drug Metabolism: From Molecular Biology to Man* (E. H. Jeffery, ed.), CRC Press, Ann Arbor, 1993, pp. 23–33.
69. K. Ruckpaul and R. Bernhardt, Biochemical aspects of the monooxygenase system in the endoplasmic reticulum of mammalian liver, *Cytochrome P450: Structural and Functional Relationships, Biochemical and Physiochemical Aspects of Mixed Function Oxidases* (K. Ruckpaul and H. Rein, eds.), Akademie-Verlag, Berlin, 1984, pp. 9–57.

70. M. S. Shet, C. W. Fischer, P. L. Holmans, and R. W. Estabrook, Human cytochrome P450 3A4: Enzymatic properties of a purified recombinant fusion protein containing NADPH-P450 reductase, *Proc. Natl. Acad. Sci. USA 90*:11748–11752 (1993).
71. F. P. Guengerich, Cytochromes P450 of human liver. Classification and activity profiles of the major enzymes, *Advances in Drug Metabolism in Man* (G. M. Pacifici and G. N. Fracchia, eds.), European Commission, Brussels, 1995, pp. 181–231.
72. C. Beedham, Molybdenum hydroxylases: Biological distribution and substrate–inhibitor specificity, *Prog. Med. Chem. 24*:85–127 (1987).
73. R. Ayesh, M. Al-Waiz, M. J. Crothers, S. Cholerton, S. C. Mitchell, J. R. Idle, and R. L. Smith, Deficient nicotine *N*-oxidation in two sisters with trimethylaminuria [abstr.], *Br. J. Clin. Pharmacol. 25*:664P–665P (1988).
74. M. D. Green and T. R. Tephly, Glucuronidation of amines and hydroxylated xenobiotics and endobiotics catalyzed by expressed human UGT1.4, *Drug Metab. Dispos. 24*:356–363 (1996).
75. C. H. Yun, R. A. Okerholm, and F. P. Guengerich, Oxidation of the antihistaminic drug terfenadine in human liver microsomes: Role of cytochrome P450 3A(4) in *N*-dealkylation and *C*-hydroxylation, *Drug Metab. Dispos. 21*:403–408 (1993).
76. K. J. King, G. A. Leeson, S. D. Burmaster, R. B. Hook, M. K. Reith, and L. K. Cheng, Metabolism of terfenadine associated with CYP3A(4) activity in human hepatic microsomes, *Drug Metab. Dispos. 23*:631–636 (1995).
77. A. Lemoine, J. C. Gautier, D. Azoulay, L. Kiffel, C. Belloc, F. P. Guengerich, P. Maurel, P. Beaune, and J. P. Leroux, Major pathway of imipramine metabolism is catalyzed by cytochromes P450 1A2 and P450 3A4 in human liver, *Mol. Pharmacol. 43*:827–832 (1993).
78. Y. Masubuchi, S. Hosokawa, T. Horie, T. Suzuki, S. Ohmori, M. Kitada, and S. Narimatsu, Cytochrome P450 isozymes involved in propranolol metabolism in human liver microsomes: The role of CYP2D6 as ring-hydroxylase and CYP1A2 as *N*-desisopropylase, *Drug Metab. Dispos. 22*:909–915 (1994).
79. Z. Zhang, M. J. Fasco, Z. Huang, F. P. Guengerich, and L. S. Kaminsky, Human cytochromes P4501A1 and P4501A2: *R*-Warfarin metabolism as a probe, *Drug Metab. Dispos. 23*:1339–1345 (1995).
80. A. Rahman, K. R. Korsekwa, J. Grogan, F. J. Gonzalez, and J. W. Harris, Selective biotransformation of Taxol to 6α-hydroxytaxol by human cytochrome P450 2C8, *Cancer Res. 54*: 5543–5546 (1994).
81. S. D. Hall, M. A. Hamman, A. E. Rettie, L. C. Wienkers, W. F. Trager, M. Vandenbranden, and S. A Wrighton, Relationships between the levels of cytochrome P4502C9 and its prototypic activities in human liver microsomes, *Drug Metab. Dispos. 22*:975–978 (1994).
82. M. E. Veronese, C. J. Doecke, P. I. MacKenzie, M. E. McManus, J. O. Miners, D. L. P. Rees, R. Gasser, U. A. Mayer, and D. J. Birkett, Site-directed mutation studies of human liver cytochrome P450 isoenzymes in the CYP2C subfamily, *Biochem. J. 289*:533–538 (1993).
83. A. E. Rettie, L. C. Wienkers, F. J. Gonzalez, W. F. Trager, and K. R. Korzekwa, Impaired (*S*)-warfarin metabolism catalysed by the R144C allelic variant of CYP2C9, *Pharmacogenetics 4*:39–42 (1994).
84. M. Ingelman-Sunberg and I. Johansson. The molecular genetics of the human drug metabolizing cytochrome P450s, *Advances in Drug Metabolism in Man* (G. M. Pacifici and G. N. Fracchia, eds.), European Commission, Brussels, 1995, pp. 545–585.
85. F. J. Gonzalez and J. R. Idle, Pharmacogenetic phenotyping and genotyping: Present status and future potential, *Clin. Pharmacokinet. 26*:59–70 (1994).
86. P. Fernandez-Salguero, S. M. G. Hoffman, S. Cholerton, H. Mohrenweiser, H. Raunio, A. Rautio, O. Pelkonen, J.-D. Huang, W. E. Evans, J. R. Idle, and F. J. Gonzalez, A genetic polymorphism in coumarin 7-hydroxylation: Sequence of the human *CYP2A* genes and identification of variant *CYP2A6* alleles, *Am. J. Hum. Genet. 57*:651–660 (1995).
87. H. Furuya, P. Fernandez-Salguero, W. Gregory, H. Taber, A. Steward, F. J. Gonzalez, and

J. R. Idle, Genetic polymorphism of CYP2C9 and its effect on warfarin maintenance dose requirement in patients undergoing anticoagulation therapy, *Pharmacogenetics 5*:389–392 (1995).

88. S. Spielberg, J. McCrea, A. Cribb, T. Rushmore, S. Waldman, T. Bjornsson, M.-W. Lo, and M. Goldberg, A mutation in *CYP2C9* is responsible for decreased metabolism of losartan [abstr.], *Clin. Pharmacol. Ther. 59*:215 (1996).
89. O. Gotoh, Substrate recognition sites in cytochrome P450 family 2 (CYP2) proteins inferred from comparative analyses of amino acid and nucleotide sequences, *J. Biol. Chem. 267*:83–90 (1992).
90. K. R. Korzekwa and J. P. Jones, Predicting the cytochrome P450 mediated metabolism of xenobiotics, *Pharmacogenetics 3*:1–18 (1993).
91. J. A. Peterson, J. Deisenhofer, C. Hasemann, K. G. Ravichandron, S. S. Boddupalli, and S. Graham-Lorence, Structure and function in P450s: Are bacterial P450s good models for eukaryotic P450s? *Cytochrome P450: Biochemistry. Biophysics and Molecular Biology* (M. C. Lechner, ed.), John Libbey and Co., London, 1994, pp. 271–277.
92. J.-M. Dupret and D. M. Grant, Site-directed mutagenesis of recombinant human arylamine *N*-acetyltransferase expressed in *Escherichia coli, J. Biol. Chem. 267*:7381–7385 (1992).
93. J. Liu, Y. A. He, and J. R. Halpert, Role of residue 480 in substrate specificity of cytochrome P450 2B5 and 2B11, *Arch. Biochem. Biophys. 327*:167–173 (1996).
94. G. D. Szklarz, Y. Q. He, K. M. Kedzie, J. R. Halpert, and V. L. Burnett, Elucidation of amino acid residues critical for unique activities of rabbit cytochrome P450 2B5 using hybrid enzymes and reciprocal site-directed mutagenesis with rabbit cytochrome P450 2B4, *Arch. Biochem. Biophys. 327*:308–318 (1996).
95. D. F. V. Lewis and B. G. Lake, Molecular modelling of members of the P4502A subfamily: Application to studies of enzyme specificity, *Xenobiotica 25*:585–598 (1995).
96. J. P. Jones, M. He, W. F. Trager, and A. E. Rettie, Three-dimensional quantitative structure–activity relationship for inhibitors of cytochrome P4502C9, *Drug Metab. Dispos. 24*:1–6 (1996).
97. A. Mancy, P. Broto, S. Dijols, P. M. Dansette, and D. Mansuy, The substrate binding site of human liver cytochrome P450 2C9: An approach using tienilic acid derivatives and molecular modeling, *Biochemistry 34*:10365–10375 (1995).
98. B. C. Jones, G. Hawksworth, V. A. Horne, A. Newlands, J. Morsman, M. S. Tute, and D. A. Smith, Putative active site template model for cytochrome P4502C9 (tolbutamide hydroxylase), *Drug Metab. Dispos. 24*:260–266 (1996).
99. S. Modi, M. J. Paine, M. J. Sutcliffe, L.-Y. Lian, W. U. Primrose, C. R. Wolf, and G. C. K. Roberts, A model for human cytochrome P4502D6 based on homology modeling and NMR studies of substrate binding, *Biochemistry 35*: 4540–4550 (1996).
100. L. Koymans, N. P. E. Vermeulen, S. A. B. E. van Acker, J. M. Koppele, J. J. P. Heykants, K. Lavrijsen, W. Meuldermans, and G. M. Donne-Op den Kelder, A predictive model for substrates of cytochrome P450-debrisoquine (2D6), *Chem. Res. Toxicol. 5*:211–219 (1992).
101. T. Sueyoshi, L. J. Park, R. Moore, R. O. Juvonen, and M. Negishi, Molecular engineering of microsomal P450 2a-4 to a stable, water-soluble enzyme, *Arch. Biochem. Biophys. 322*: 265–271 (1995).
102. J. B. Houston, Utility of in vitro drug metabolism data in predicting in vivo metabolic clearance, *Biochem. Pharmacol. 47*:1469–1479 (1994).
103. G. Appanna, B. K. Tang, R. Muller, and W. Kalow, A sensitive method for determination of cytochrome P4502D6 activity in vitro using bupranolol as substrate, *Drug Metab. Dispos. 24*:303–306 (1996).
104. W. Tassaneeyakul, M. E. Veronese, D. J. Birkett, F. J. Gonzalez, and J. O. Miners, Validation of 4-nitrophenol as an in vitro substrate probe for human liver CYP2E1 using cDNA-expression and microsomal kinetic techniques, *Biochem. Pharmacol. 46*:1975–1981 (1993).

105. C. J. Doecke, M. E. Veronese, S. M. Pond, J. O. Miners, D. J. Birkett, L. N. Sansom, and M. E. McManus, Relationship between phenytoin and tolbutamide hydroxylations in human liver microsomes, *Br. J. Clin. Pharmacol.* *31*:125–130 (1991).
106. H. Yamazaki, K. Inoue, M. Mimura, Y. Oda, F. P. Guengerich, and T. Shimada, 7-Ethoxycoumarin *O*-deethylation catalyzed by cytochromes P450 1A2 and 2E1 in human liver microsomes, *Biochem. Pharmacol.* *51*:313–319 (1996).
107. H. R. Ha, J. Chen, S. Krahenbuhl, and F. Follath, Biotransformation of caffeine by cDNA-expressed human cytochromes P450, *Eur. J. Clin. Pharmacol.* *49*:309–315 (1996).
108. M. E. Campbell, D. M. Grant, T. Inaba, and W. Kalow, Biotransformation of caffeine, paraxanthine, theophylline, and theobromine by polycyclic aromatic hydrocarbon-inducible cytochromes P450 in human liver microsomes, *Drug Metab. Dispos.* *15*:237–249 (1987).
109. H. R. Ha, J. Chen, A. U. Freiburghaus, and F. Follath, Metabolism of theophylline by cDNA-expressed human cytochromes P450, *Br. J. Clin. Pharmacol.* *39*:321–326 (1995).
110. Z.-Y. Zhang and L. S. Kaminsky, Characterization of human cytochromes P450 involved in theophylline 8-hydroxylation, *Biochem. Pharmacol.* *50*:205–211 (1995).
111. J. F. Tjia, J. Colbert, and D. J. Back, Theophylline metabolism in human liver microsomes: Inhibition studies, *J. Pharmacol. Exp. Ther.* *276*:912–917 (1996).
112. A. E. Rettie, A. C. Eddy, L. D. Heimark, M. Gibaldi, and W. F. Trager, Characteristics of warfarin hydroxylation catalyzed by human liver microsomes, *Drug Metab. Dispos.* *17*: 265–270 (1989).
113. A. E. Rettie, K. R. Korzekwa, K. L. Kinze, R. F. Lawrence, A. C. Eddy, T. Aoyama, H. V. Gelboin, F. J. Gonzalez, and W. F. Trager, Hydroxylation of warfarin by human cDNA-expressed cytochrome P450: A role for P4502C9 in the etiology of (*S*)-warfarin–drug interactions, *Chem. Res. Toxicol.* *5*:54–59 (1992).
114. M. Gibaldi, Stereoselective and isozyme-selective drug interactions, *Chirality* *5*:407–413 (1993).
115. D. R. Brocks and F. Jamali, Stereochemical aspects of pharmacotherapy, *Pharmacotherapy* *15*:551–564 (1995).
116. K. Oguri, H. Yamada, and H. Yoshimura, Regiochemistry of cytochrome P450 isozymes, *Annu. Rev. Pharmacol. Toxicol.* *34*:251–279 (1994).
117. B. E. Daikh, J. M. Lasker, J. L. Raucy, and D. R. Koop, Regio- and stereoselective epoxidation of arachidonic acid by human cytochromes P450 2C8 and 2C9, *J. Pharmacol. Exp. Ther.* *271*:1427–1433 (1994).
118. D. S. Mautz, W. L. Nelson, and D. D. Shen, Regioselective and stereoselective oxidation of metoprolol and bufuralol catalyzed by microsomes containing cDNA-expressed human P4502D6, *Drug Metab. Dispos.* *23*:513–517 (1995).
119. A. J. Sadeque, A. C. Eddy, G. P. Meier, and A. E. Rettie, Stereoselective sulfoxidation by human flavin-containing monooxygenase: Evidence for catalytic diversity between hepatic, renal, and fetal forms, *Drug Metab. Dispos.* *20*:832–839 (1992).
120. J. R. Cashman, Structural and catalytic properties of the mammalian flavin-containing monooxygenase, *Chem. Res. Toxicol.* *8*:165–181 (1995).
121. A. E. Rettie, M. P. Lawton, A. J. M. Sadeque, G. P. Meier, and R. M. Philpot, Prochiral sulfoxidation as a probe for multiple forms of the microsomal flavin-containing monooxygenase: Studies with rabbit FMO1, FMO2, FMO3, and FMO5 expressed in *Escherichia coli; Arch. Biochem. Biophys.* *311*:369–377 (1994).
122. S. B. Park, P. Jacob, N. L. Benowitz, and J. R. Cashman, Stereoselective metabolism of (*S*)-(−)-nicotine in humans: Formation of *trans*-(*S*)-(−)-nicotine *N*-1′-oxide, *Chem. Res. Toxicol.* *6*:880–888 (1993).
123. L. A. Damani, W. F. Pool, P. A. Crooks, R. K. Kaderlik, and D. M. Ziegler, Stereoselectivity in the *N*′-oxidation of nicotine isomers by flavin-containing monooxygenase, *Mol. Pharmacol.* *33*:702–705 (1988).
124. T. K. H. Chang, F. J. Gonzalez, and D. J. Waxman, Evaluation of triacetyoleandomycin,

α-naphthoflavone and diethyldithiocarbamate as selective chemical probes for inhibition of human cytochromes P450, *Arch. Biochem. Biophys. 311*:437–442 (1994).

125. M. S. Ching, C. L. Blake, H. Ghabrial, S. W. Ellis, M. S. Lennard, G. T. Tucker, and R. A. Smallwood, Potent inhibition of yeast-expressed CYP2D6 by dihydroquinidine, quinidine, and its metabolites, *Biochem. Pharmacol. 50*:833–837 (1995).
126. W. Tassaneeyakul, D. J. Birkett, M. E. Veronese, M. E. McManus, R. H. Tukey, L. C. Quattrochi, H. V. Gelboin, and J. O. Miners, Specificity of substrate and inhibitor probes for human cytochromes P450 1A1 and 1A2, *J. Pharmacol. Exp. Ther. 265*:401–407 (1993).
127. D. J. Newton, R. W. Wang, and A. Y. H. Lu. Cytochrome P450 inhibitors: Evaluation of specificities in the in vitro metabolism of therapeutic agents by human liver microsomes, *Drug Metab. Dispos. 23*:154–158 (1995).
128. F. P. Guengerich, Bioactivation and detoxication of toxic and carcinogenic chemicals, *Drug Metab. Dispos. 21*: 1–6 (1993).
129. F. P. Guengerich, Metabolic activation of carcinogens, *Pharmacol. Ther. 54*: 17–61 (1992).
130. F. P. Guengerich and T. Shimada, Oxidation of toxic and carcinogenic chemicals by human cytochrome P450 enzymes, *Chem. Res. Toxicol. 4*:391–407 (1991).
131. F. J. Gonzalez, C. L. Crespi, and H. V. Gelboin, cDNA-expressed human cytochrome P450s: A new age of molecular toxicology and human risk assessment, *Mutat. Res. 247*:113–127 (1991).
132. L. H. Thompson, R. W. Wu, and J. S. Felton, Genetically modified Chinese hamster ovary (CHO) cells for studying the genotoxicity of heterocyclic amines from cooked foods, *Toxicol. Lett. 82/83*:883–889 (1995).
133. C. L. Crespi, B. W. Penman, D. T. Steimel, H. V. Gelboin, and F. J. Gonzalez, The development of a human cell line stably expressing human CYP3A4: Role in the metabolic activation of aflatoxin B_1 and comparison to CYP1A2 and CYP2A3, *Carcinogenesis 12*:355–359 (1991).
134. H. Hashimoto, Y. Yanagawa, M. Sawada, S. Itoh, T. Deguchi, and T. Kamataki, Simultaneous expression of human CYP3A7 and *N*-acetyltransferase in Chinese hamster CHL cells results in high cytotoxicity for carcinogenic heterocyclic amines, *Arch. Biochem. Biophys. 320*:323–329 (1995).
135. Y. Dai and A. I. Cederbaum, Cytotoxicity of acetaminophen in human cytochrome P4502E1-transfected HepG2 cells, *J. Pharmacol. Exp. Ther. 273*:1497–1505 (1995).
136. G. M. Hawksworth, Advantages and disadvantages of using human cells for pharmacological and toxicological studies, *Hum. Exp. Toxicol. 13*:568–573 (1994).
137. R. G. Ulrich, J. A. Bacon, C. T. Cramer, G. W. Peng, D. K. Petrella, R. P. Stryd, and E. L. Sun, Cultured hepatocytes as investigational models for hepatic toxicity: Practical applications in drug discovery and development, *Toxicol. Lett. 82/83*:107–115 (1995).
138. H. Yamazaki, Y. Inui, S. A. Wrighton, F. P. Guengerich, and T. Shimada, Procarcinogen activation by cytochrome P450 3A4 and 3A5 expressed in *Escherichia coli* and by human liver microsomes, *Carcinogenesis 16*:2167–2170 (1995).
139. H. Yamazaki, M. Mimura, Y. Oda, Y. Inui, T. Shiraga, K. Iwasaki, F. P. Guengerich, and T. Shimada, Roles of different forms of cytochrome P450 in the activation of the promutagen 6-aminochrysene to genotoxic metabolites in human liver microsomes, *Carcinogenesis 14*: 1271–1278 (1995).
140. T. Aoyama, S. Yamano, P. S. Guzelian, H. V. Gelboin, and F. J. Gonzalez, Five of 12 forms of vaccinia virus-expressed human hepatic cytochrome P450 metabolically activate aflatoxin B_1, *Proc. Natl. Acad. Sci. USA 87*:4790–4793 (1990).
141. T. Shimada, E. M. J. Gillam, P. Sandhu, Z. Guo, R. H. Tukey, and F. P. Guengerich, Activation of procarcinogens by human cytochrome P450 enzymes expressed in *Escherichia coli*. Simplified bacterial systems for genotoxicity assays, *Carcinogenesis 15*:2523–2529 (1994).
142. J. A. Styles, A. Davies, C. K. Lim, F. De Matteis, L. A. Stanley, I. N. H. White, Z.-X. Yuan,

and L. L. Smith, Genotoxicity of tamoxifen, tamoxifen epoxide and toremifene in human lymphoblastoid cells containing human cytochrome P450s, *Carcinogenesis 15*: 5–9 (1994).

143. W. A. Schmalix, H. Maser, F. Kiefer, R. Reen, F. J. Wiebel, F. Gonzalez, A. Seidel, H. Glatt, H. Greim, and J. Doehmer, Stable expression of human cytochrome P450 1A1 cDNA in V79 Chinese hamster cells and metabolic activation of benzo[*a*]pyrene, *Eur. J. Pharmacol. 248*:251–261 (1993).

144. C. L. Crespi, B. W. Penman, H. V. Gelboin, and F. J. Gonzalez, A tobacco smoke-derived nitrosamine, 4-(methylnitrosamino)-1-(3-pyridyl)-1-butanone, is activated by multiple human cytochrome P450s including the polymorphic human cytochrome P4502D6, *Carcinogenesis 12*:1197–1201 (1991).

145. H. F. Tiano, R.-L. Wang, M. Hosokawa, C. Crespi, K. R. Tindall, and R. Langenbach, Human CYP2A6 activation of 4-(methylnitrosamino)-1-(3-pyridyl)-1-butanone (NNK): Mutational specificity in the *gpt* gene of AS52 cells, *Carcinogenesis 15*:2859–2866 (1994).

146. S. C. Piscitelli, L. H. Danziger, and K. A. Rodvold, Clarithromycin and azithromycin: New macrolide antibiotics, *Clin. Pharm. 11*:137–152 (1992).

147. K. T. Chong, Recent advances in HIV-1 protease inhibitors, *Exp. Opin. Invest. Drugs 5*: 115–124 (1996).

148. H. Kubinyi, Strategies and recent technologies in drug discovery, *Pharmazie 50*:647–662 (1995).

149. E. M. Gordon, M. A. Gallop, and D. V. Patel, Strategy and tactics in combinatorial organic synthesis. Applications to drug discovery, *Acc. Chem. Res. 29*:144–154 (1996).

150. R. W. Estabrook, The P450s: Agenda for action, adaptation and application, *Cytochrome P450: Biochemistry, Biophysics and Molecular Biology* (M. C. Lechner, ed.), John Libbey and Co., London, 1994, p. 905–910.

151. D. M. Dulik and C. Fenselau, Use of immobilized enzymes in drug metabolism studies, *FASEB J. 2*:2235–2240 (1988).

152. C. Bucke, Cell immobilization in calcium alginate, *Methods Enzymol. 135*:175–189 (1987).

153. S. Janecek, Strategies for obtaining stable enzymes, *Process Biochem. 28*:435–445 (1993).

154. E. Katchalski-Katzir, Immobilized enzymes—learning from past successes and failures, *Trends Biotechnol. 11*:471–478 (1993).

155. M. Ibrahim, M. Decolin, A.-M. Batt, E. Dellacherie, and G. Siest, Immobilization of pig liver microsomes: Stability of cytochrome P450-dependent monooxygenase activities, *Appl. Biochem. Biotechnol. 12*:199–213 (1986).

156. T. Alebic-Kolbah and I. W. Wainer, Microsomal immobilized-enzyme-reactor for on-line production of glucuronides in a HPLC column, *Chromatographia 37*:608–612 (1993).

157. S. L. Pallante, C. A. Lisek, D. M. Dulik, and C. Fenselau, Glutathione conjugates: Immobilized enzyme synthesis and characterization by fast atom bombardment mass spectrometry, *Drug Metab. Dispos. 14*:313–318 (1986).

158. F. Schubert, D. Kirstein, F. Scheller, and P. Mohr, Immobilization of cytochrome P-450 and its application in enzyme electrodes, *Anal. Lett. 13*:1167–1178 (1980).

159. D. J. King, M. R. Azari, and A. Wiseman, Immobilization of a cytochrome P-450 enzyme from *Saccharomyces cerevisiae, Methods Enzymol. 137*:675–686 (1988).

160. G. Brunner and H. Losgen, Refixation of solubilized and purified microsomal enzymes: Towards an extracorporeal detoxification in liver failure, *Enzyme Eng. 3*:391–396 (1979).

161. G. Brunner, H. Losgen, B. Gawlik, and K. Belsner, Immobilization of solubilized cytochrome P450 and NADPH-cytochrome P450 (*c*) reductase, *Biochemistry, Biophysics and Regulation of Cytochrome P450* (J.-A. Gustafsson, et al., eds.), Elsevier/North Holland, New York, 1980, pp. 573–576.

162. K. M. Faulkner, M. S. Shet, C. W. Fischer, and R. W. Estabrook, Electrocatalytically driven ω-hydroxylation of fatty acids using cytochrome P450 4A1, *Proc. Natl. Acad. Sci. USA 92*: 7705–7709 (1995).

163. G. Stempfer, B. Holl-Neugebauer, E. Kopetzki, and R. Rudolph, A fusion protein designed

for noncovalent immobilization: Stability, enzymatic activity, and use in an enzyme reactor. *Nature Biotechnol. 14*:481–484 (1996).

164. H. Yamazaki, Y.-F. Ueng, T. Shimada, and F. P. Guengerich, Roles of divalent metal ions in oxidations catalyzed by recombinant cytochrome P450 3A4 and replacement of NADPH–cytochrome P450 reductase with other flavoproteins, ferredoxin, and oxygen surrogates, *Biochemistry 34*:8380–8389 (1995).
165. T. Murachi and M. Tabata, Use of immobilized enzyme column reactors in clinical analysis, *Methods Enzymol. 137*:260–271 (1988).
166. E. G. Schuetz, K. N. Furuya, and J. D. Schuetz, Interindividual variation in expression of P-glycoprotein in normal human liver and secondary hepatic neoplasms, *J. Pharmacol. Exp. Ther. 275*:1011–1018 (1995).
167. M. Yamazaki, H. Suzuki, and Y. Sugiyama, Recent advances in carrier-mediated hepatic uptake and biliary excretion of xenobiotics, *Pharm. Sci. 13*:497–513 (1996).
168. A. P. Simula, M. B. Crichton, S. M. Black, S. Pemble, H. F. J. Bligh, J. D. Beggs, and C. R. Wolf, Heterologous expression of drug-metabolizing enzymes in cellular and whole animal models, *Toxicology 82*:3–20 (1993).
169. E. Steiner, E. Dumont, E. Spina, and R. Dahlqvist, Inhibition of desipramine 2-hydroxylation by quinidine and quinine, *Clin. Pharmacol. Ther. 43*:577–581 (1987).
170. M. Schadel, D. Wu, S. V. Otton, W. Kalow, and E. M. Sellers, Pharmacokinetics of dextromethorphan and metabolites in humans: Influence of the CYP2D6 phenotype and quinidine inhibition, *J. Clin. Psychopharmacol. 15*:263–269 (1995).
171. K. Brosen, L. F. Gram, T. Haghfelt, and L. Bertilsson, Extensive metabolizers of debrisoquine become poor metabolizers during quinidine treatment, *Pharmacol. Toxicol. 60*:312–314 (1987).

13

In Vivo and In Vitro Recombinant DNA Technology as a Powerful Tool in Drug Development

Thomas Friedberg, C. J. Henderson, M. P. Pritchard, and C. R. Wolf

Ninewells Hospital and Medical School, University of Dundee, Dundee, United Kingdom

I. INTRODUCTION

Metabolism influences the pharmacological as well as the toxicological actions of drugs (1). To ensure the maximum therapeutic value and safety of medicines in humans, the pharmacotoxicological properties are usually first tested using animals or with in vitro systems that frequently use animal tissues (2). Besides being often considered ethically, and for in vivo studies, economically problematic, animal-based systems are also scientifically inadequate for drug development. This is partially because of large species-specific differences in the enzymes involved in drug metabolism (3,4). Such differences have, for example, resulted in the failure of candidate drugs, such as the antiviral terpenoid carbenoxolone, the antiasthmatic compound FPL 52757, and the cardiotonic drug amirinone in the clinical stages of drug development (5). Therefore, it is essential to develop systems that better reflect the human situation. In addition to providing great insight into the mechanistic aspects of drug metabolism and toxicological reactions, these models should significantly shorten the period between the discovery of a drug and its introduction on the market.

Several models for human drug metabolism have been developed. These can be roughly divided into those that try to mimic the entire cascade of drug-metabolizing events, and those that try to reflect specific steps in these reactions. The latter models (simple drug metabolism systems) predominantly include in vivo and in vitro heterologous expression systems as well as enzymes purified from tissues. Models that try to mimic all steps in human drug metabolism (complex drug metabolism systems) include human hepatocyte and hepatoma cell cultures. A significant disadvantage of the latter models is that a system has not yet been described in which drug metabolism reflects the hepatic situation, because it is altered either owing to culture conditions, or in the case of hepatoma cells, owing to the process of phenotypic transformation (6). It is not the purpose of this chapter to review complex drug metabolism systems extensively, but we will briefly discuss the use of recombinant DNA technology to understand the regulation of xenobiotic-metabolizing enzymes in hepatocytes.

Simple models are useful for characterizing specific steps in the metabolism of a drug and also the enzymes involved (Fig. 1). Furthermore, they provide a powerful means to predict drug–drug interactions, drug toxicity, and polymorphic drug oxidations. These systems also provide recombinant enzymes for the production of antibodies that will allow quantification of drug-metabolizing enzymes in human tissues and antibody inhibition studies in complex systems. Combined with the knowledge about the role of a particular enzyme in drug metabolism, this approach allows one to predict the consequences of polymorphisms in drug metabolism (see Fig. 1). This knowledge is essential for the individual adjustment of clinical drug treatment regimens. Moreover, recombinant DNA technology allows the generation of sufficient material for structural studies on drug-metabolizing enzymes. Microorganisms containing recombinant P450s have the potential to be used as bioreactors (see Fig. 1).

It is important that data obtained with in vitro systems, using purified proteins or heterologous expression, are placed in perspective. The role of an enzyme in a particular metabolic pathway is determined by its kinetic properties and its cellular concentration. For example, if the catalytic activity of an enzyme A toward a specific substrate is tenfold lower, but its concentration tenfold higher, than that of an enzyme B, both proteins will contribute equally to the metabolism of this compound. Therefore, results on the kinetic parameters have to be combined with data on the cellular concentration of an enzyme to

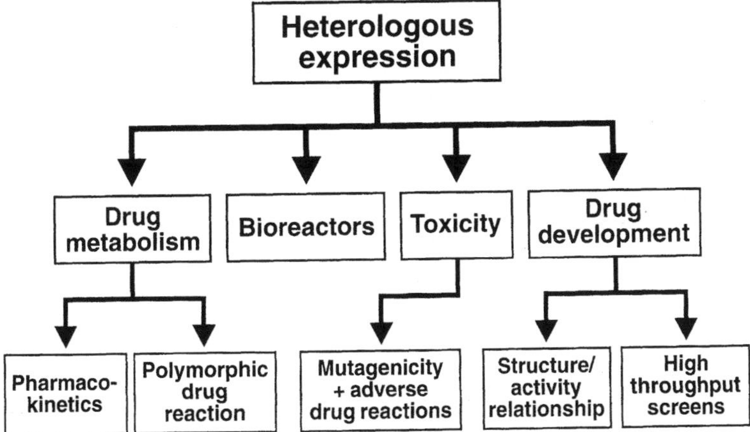

Fig. 1 Flowchart showing the applications of expression systems producing recombinant drug-metabolizing enzymes in drug development and in the biotechnology industry. For details see text.

draw meaningful conclusions. Here again, heterologous expression as a means of obtaining antigens for antibody production is an invaluable tool for the exact quantitation of enzymes in tissues (7–10).

However, heterologous in vitro expression models mimic only certain parts of the processes that determine the metabolism of drugs in vivo. One crucial difference is that drug disposition in vivo is determined by a range of parameters, including biotransformation, absorption, distribution, and excretion. The latter three parameters cannot be mimicked in vitro. In addition, the set of drug-metabolizing enzymes expressed in a given cell type in vitro is very different from the set of enzymes expressed in the same cell type in vivo (11). Therefore, the effects of a heterologously expressed drug-metabolizing enzyme on a metabolic cascade, which is controlled by several enzymes, might be distinct in vivo and in vitro. On the other hand, the expression of an enzyme in a surrounding devoid of other drug-metabolizing enzymes has the advantage that its role can be analyzed in isolation and without interference from other xenobiotic enzymes.

From the foregoing discussion it is evident that there is also a need to develop in vivo models so that the role of specific enzymes in drug metabolism and disposition can be established. This to a degree can be achieved by the use of transgenic animals either for heterologous protein expression or for gene inactivation by homologous recombination (12–14). For the latter approach, the mouse is still the only species for which the necessary technology is available. It is important to take into account that transgenic animal models are limited by the biology of the host; therefore, specific endpoints may not reflect those seen in humans.

Recombinant DNA technology can also be used to study enzyme regulation. In this approach, regulatory regions of a gene are used to drive the expression of an easily detectable indicator gene, such as chloramphenicol acetyltransferase, β-galactosidase, or luciferase. These constructs are then introduced into a suitable cell, and the effects of xenobiotics on the expression of the indicator protein is determined. In this way, patterns of gene regulation as well as the elements important for gene regulation by xenobiotics have been identified (see later discussion). The knowledge gained from these studies will be useful in developing strategies to stabilize the phenotype of complex drug-metabolizing systems as discussed in the following.

It is evident from this general introduction that many of the available methods are complementary and the strength of a study is greatly increased when different approaches are combined. In the following sections an overview of the general technology of the different in vitro heterologous expression systems, together with some examples for their use in drug development and toxicology, will be given. This is followed by a discussion of the merits and limitations of these models.

II. HETEROLOGOUS EXPRESSION IN VITRO

With the advent of recombinant DNA technology, it became feasible to characterize the role of an enzyme in drug metabolism and chemical carcinogenesis by introducing and expressing its genetic information in a cell, thereby circumventing the necessity for purification. This approach is often called ''heterologous expression'' even though the genetic information is sometimes expressed in a syngenic cell. For this purpose, tissue mRNA is isolated and used as a source to synthesize the cDNA encoding the enzyme under investigation. Several methods are available for this approach (15). In earlier studies, the mRNA was used as a source for the generation of a cDNA library. DNA or antibody probes were then employed to isolate the specific cDNA clones. Only a few milligrams of tissue are required to obtain sufficient quantities for the generation of cDNA libraries. However, using reverse transcription–polymerase chain reaction (RT–PCR), it is now possible to isolate a particular cDNA from a single cell (16). The cDNA, recombined with an expression vector containing a selection marker, is introduced into suitable recipient cells. Unlike the parental cells, cells that have taken up the vector survive under selective pressure, and those that express the protein encoded by the cDNA of interest are identified using immunological and enzymatic assays. In the functional analysis of a drug-metabolizing enzyme, the advantages of this approach, as compared with protein purification, are severalfold. Minute amounts of tissue are needed and because genetic information is more easily deciphered than the primary sequence of proteins, the identity of the cDNAs used in different research institutions can be easily verified. Moreover, the amount of the desired protein that can be obtained from recombinant cells is much larger than the quantities that can be isolated with reasonable effort from tissue sources. In addition, the pharmacotoxicological functions of the expressed protein can be evaluated in a cellular environment. Heterologous expression of P450s has been achieved in vitro in mammalian cells, in yeasts, in insect cells, and in bacteria.

A. Mammalian Systems

Generally the expression of foreign mammalian cDNAs in mammalian cells is straightforward, for they will accommodate genetic information derived from another mammalian species. Modifications, therefore, of the cDNA in its noncoding or coding regions are usually not required. In addition, mammalian cells have an endoplasmic reticulum, which is a distinct advantage for the expression of cytochrome P450s (see later). The mechanisms of posttranslational modification, such as membrane targeting, proteolytic processing, protein glycosylation, or protein phosphorylation, are conserved across mammals, however they may be cell type-dependent. The conservation of membrane targeting is particularly important for the expression of P450s (17,18). Cytochrome P450s (CYPs) are, with the exception of the P450 aromatase (19), not glycosylated (17,20,21), but some cytochrome P450s are phosphorylated (10,22). However, the role of phosphorylation in the function of these enzymes is unclear (23).

Heterologous expression of P450s in mammalian cells can be classified as transient or stable. This categorization has direct consequences for their application.

Transient expression is based either on the infection of cells by the vaccinia virus (24) or the transfection of plasmids, containing a SV40 origin of replication, into COS 1 kidney fibroblasts, which express the SV40 T-antigen necessary for the episomal amplification of the plasmid (25,26). In both systems, multiple extrachromosomal copies of the cDNA–cell are present and, consequently, the chances of achieving high expression of the transgene are good. Expression is transient because the cells are either killed by virally induced lysis (vaccinia system) or because the plasmid is not stably maintained in the cells (COS system).

The method for the heterologous expression of cDNAs in COS 1 cells is relatively straightforward. It requires the cloning of the cDNA into a mammalian expression plasmid and the transfection of the resulting construct into COS 1 cells employing either DEAE-dextran, liposomes, or electroporation. Usually the cells are collected after 2–3 days for analysis of the heterologous protein. In principle, several cDNAs located on different plasmids can be cotransfected together to reconstruct a metabolic cascade. A drawback of this approach is that the DNA transfection efficiencies and, consequently, the expression levels of a given protein, may show experimental variations. In addition, this strategy is limited to the use of cells that contain the SV 40 large T-antigen. Technical details for P450 expression in COS 1 cells have been described (25). The first recombinant mammalian model for P450 was the transient expression of CYP17 in COS 1 cells (27). The transfected cells were employed to study the 17α-hydroxylation of progesterone followed by the production of dehydroepiandrosterone by the 17, 20-lyase activity of the recombinant enzyme. The COS cell system is also very suitable for the rapid analysis of the catalytic properties of engineered P450 mutants and has been used for this purpose in some instances (28–30). Further examples of the use of this system for drug metabolism studies are given in Table 1. However, it is evident from this table that expression in COS cells has been only infrequently used in drug metabolism studies, even though it was one of the first P450 expression systems to be established. This is mainly because of the impracticalities associated with the transient nature of expression and to the low expression level of P450s in this model.

Vaccinia virus-based expression has been mainly developed by one group (24,31). The wide use of the vaccinia-based expression technology has been hampered by safety concerns in using an attenuated derivative of a cowpox virus and because the experimental effort for the expression of proteins in this model is quite considerable, for vaccinia virus has a large genome of 187,000 bp. Direct insertion of the cDNA into the virus using standard in vitro methods, therefore, is not possible. To construct recombinant virus, the cDNA must first be cloned into a plasmid, called an insertion vector, in such a way that it is flanked by parts of the vaccinia virus thymidine kinase gene and is under the control of the vaccinia early gene promoter. This plasmid is then used for the recombination of the cDNA by the thymidine kinase gene into the wild-type virus. On infection of mammalian thymidine kinase-deficient cells, cells containing recombinant DNA will survive on bromodeoxyuridine (BUDR), whereas those cells with the wild-type virus will die. The surviving cells can be used to obtain a concentrated virus stock for the infection of a wide

Table 1 Therapeutic Drugs Studied in Recombinant In Vitro Drug Metabolism Models Expressing CYP2D6 or CYP3A4

P450 studied and	Expression model employed[a]						
compounds tested	COS	Vaccinia	EBV	V79	Yeast	Baculovirus	*E. coli*
CYP2D6							
Bufuralol	(177–179)	(180)	(32,34)			(181)	(82,108,182)
Debrisoquine		(180)	(32,108)		(109,183,184)		
Sparteine		(180)			(109)		
Metoprolol			(185)		(109,184)	(181)	
Propafenone					(186)		
Dextromethorphan			(187)		(183,188)		
Imipramine	(189)	(190)					
Desipramine	(189)						
Tropisetron				(191)	(192)		
Ordansetron				(191)			
CYP3A4							
Testosterone		(193)	(9)		(194)	(94)	(195–197)
Nifedipine					(194,198)		(195–197)
Erythromycin					(199)	(94)	(70)
Ergot alkaloids				(200)	(198)		
Lidocaine					(194)		
Epipodo phyllotoxine		(201)					
Flexeril		(202)					
Midazolam		(203)					
Salmeterol						(204)	

[a]Numbers in parentheses refer to reference citations.

variety of mammalian cell lines. This is a distinct advantage of this approach. Table 1 shows that this system has been frequently employed in drug metabolism studies. However in assessing this, one has to take into account that it had been established earlier than bacterial expression systems.

Mammalian models for the stable heterologous expression of proteins are based on Epstein–Barr virus (EBV), retroviral-mediated gene transfer, and on direct gene transfer and chromosomal integration of the transgene.

One to several hundred copies of EBV-based vectors can be maintained episomally, thus increasing the probability of achieving high levels of heterologous protein (26). Because the EBV-based vectors contain selection markers that confer resistance, for example, to l-histidinol or hygromycin B, they can be stably maintained. The cDNA can be cloned directly into the EB-based expression vector, provided the appropriate restriction sites are available. The use of the vector is restricted to mammalian cell lines containing the Epstein–Barr nuclear antigen (EBNA), such as the lymphoblastoid cell line AHH-1, which has been frequently employed for the expression of a variety of P450s (32–34). Some EBV-based vectors allow the insertion of two cDNAs driven by separate promoters. This, together with the possibility of transfecting several vectors containing separate selection markers, allows the coexpression of several proteins (35,36). In the first attempts to use this technology for P450 expression, the P450 levels achieved were relatively low. However, the use of multiple copies of the cDNA in the expression vector improved the expression level of CYP2D6 significantly (32). Some examples for the use of the lymphoblastoid cell line AHH-1 in drug metabolism studies are given in Table 1.

A limited number of P450s have been expressed using retroviruses (37–39). The use of retroviral vectors has the advantage that high-efficiency transfer of genetic information can be achieved into a wide variety of mammalian cells, including primary cells (26). In contrast with the AHH-1 lymphoblastoid expression system, the transfected cDNA is stably integrated into the genome. The retroviral-transfer technology employs a helper cell line containing a crippled retroviral DNA. This DNA encodes all the proteins necessary for the production of viral-packaging proteins, but cannot be packaged itself. These proteins are used to complement a packagable, but for package proteins crippled, retrovirus that had been previously recombined with the cDNA of interest. Improvements in viral-packaging cell lines have included modifications to increase its host range and, important for biosafety considerations, genetic changes leading to a decrease in recombination events resulting in the production of fully functional helper virus. The retrovirus is chromosomally integrated, usually as a single copy, which can limit the capability of the cells to achieve high levels of heterologous protein. However, expression of genes introduced by retroviral expression is usually more efficient than from genes introduced by DNA transfection, apparently because the retroviral DNA is preferentially inserted into transcriptionally active regions of the genome (40). Retroviral integration has the advantage that the cDNA to be expressed can be transferred into a wide variety of cell lines. The system has been used mainly for toxicological studies (38,41). However its widespread use has been limited, mainly because of its complexity and the length of time required to generate a P450 expressing clone, and few drug metabolism studies have been performed using this model (see Table 1).

Finally, a plethora of systems have been developed that use the direct transfer of the P450 cDNA, mainly into V79 Chinese hamster lung fibroblasts, followed by its chromosomal integration (37,42–45). In this approach, the cDNA is recombined with an expression vector, which in its basic form contains an *E. coli* origin of replication for its bacterial propagation, and a mammalian enhancer–promoter region controlling the transcription of a cDNA inserted into a multiple-cloning site. This DNA is then transfected into mammalian cells. To identify cells that have stably integrated the vector, it is necessary to cotransfect the appropriate selection marker, such as the aminoglycoside phosphotransferase or the puromycin *N*-acetyltransferase, which can be located either on the same or on a different vector. Employing this approach, a wide variety of cell lines can be used, the only prerequisite being that a method of gene transfer has been developed for the particular cell line (46). In this method the chromosomal integration event is random and mostly a single copy of the transgene is inserted. Because the genomic localization of the transgene affects its expression, it is often necessary to screen a considerable number of transfectants to identify a cell clone with reasonable heterologous expression levels, and even then the P450 expression levels achieved with these systems is generally low. This is not a major problem in genotoxicity studies that with the appropriate endpoint, can detect even minute amounts of ultimate mutagenic metabolites (see later discussion), but may explain the infrequent use of this model for drug metabolism studies (see Table 1). However, by using the appropriate selection markers, it is now possible to amplify the copy number of the cDNA in the mammalian genome, leading concomitantly to increased expression of recombinant proteins. For example, cotransfection with the dihydrofolate reductase (*dhfr*) gene into a *dhfr* minus CHO cell line will allow selection for cells harboring amplified copies of the transgene, by culturing the transfectants in the presence of increasing concentrations of the *dhfr* inhibitor methotrexate (47). Several hundred copies of the cDNA can be achieved by this method, greatly increasing expression levels. We have used this approach to generate a panel of cell lines that coexpress a range of different P450 isozymes together with P450 reductase. The high expression levels achieved in this system allowed the accurate spectral quantitation of the recombinant P450 (Table 2; Fig. 2), which was

Table 2 Cytochrome P450 Levels and Yield Achieved with Various Heterologous Expression Systems

CYP	Type of expression system	Expression level[a,b]	Yield (pmole/mL)	Ref.
CYP17A	Baculovirus	500×10^5	88	73
CYP17A	*S. cerevisiae*	0.75×10^5	2.5	73
CYP17A	*E. coli*	3.33×10^5	412	73
CYP3A4	EBV	1.7	n.a.	9
CYP3A4	Vaccinia	75	n.a.	76
CYP3A4[c]	CHO-amplified	21	n.a.	Own data
CYP3A4	Baculovirus	460	103	94
CYP3A4	*S. cerevisiae*	100	2–3	205
CYP3A4[c]	*E. coli*	200	215	103
CYP3A4	*E. coli*	300	100	70

[a]CYP17A expression levels were evaluated using total cellular protein. For CYP3A4 the expression levels were determined in membrane fractions, with the exception of CHO cells and the baculovirus-directed expression of P450, which was determined using total cellular protein.

[b]The expression levels of CYP17A are shown as molecules P450 per cell and of CYP3A4 as pmoles P450 per milligram membrane protein, with the exception of the CYP3A4 level in CHO cells, which is given as nmoles P450 per milligram total cellular protein.

[c]CYP3A4 was coexpressed together with human P450 reductase.

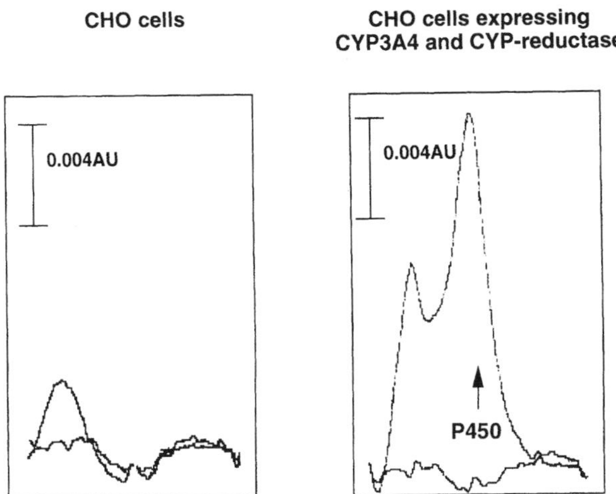

Fig. 2 Spectral analysis of recombinant P450 in mammalian cell lines: CYP3A4 was expressed from an amplifiable expression vector together with P450 reductase in CHO cells. The P450 was analyzed by Fe^{2+} vs. Fe^{2+}-CO difference spectrophotometry in total cell lysate derived from parental CHO cells and from cells expressing the recombinant proteins. The expression level of CYP3A4 was 19 pmol/mg total cellular protein.

difficult for other mammalian cell lines, mainly owing to the extremely low recombinant P450 levels found in these systems.

B. Yeast Systems

The first model to be used for the expression of P450s was yeast (48,49). Similar to mammalian cells, yeasts contains an endoplasmic reticulum, and the mechanisms underlying membrane targeting, as well as membrane topology, are preserved. Therefore, components of the monooxygenase complex, when expressed in *Saccharomyces cerevisiae*, will have the appropriate topology for electron coupling. *S. cerevisiae* also contain posttranslational modification enzymes, such as protein kinases and enzymes for the glycosylation of proteins. However, the posttranslational modifications may be quantitatively and qualitatively distinct in mammals when compared with *S. cerevisiae*. For example, the type of *O*-glycosylation and in some cases also the *N*-glycosylation of proteins in *S. cerevisiae* is not entirely the same as in mammalian cells (50). These differences do not play a role for P450; however, they could be of importance for the expression of UDP-glucuronosyltransferases, which are *N*-glycosylated.

The bakers yeast, *S. cerevisiae*, has been almost exclusively used for the heterologous expression of drug-metabolizing enzymes, even though other yeast strains such as *Pichia pastoris* or *Hansenula polymorphica*, have also been very successfully employed for large-scale production of proteins (51). *S. cerevisiae* has proved suitable for the expression of a wide variety of eukaryotic gene products both for basic research and for industrial and pharmaceutical applications (52) As it is a unicellular organism, many of the manipulations commonly used for bacteria can be readily applied to yeast. Cells can be grown to high cell densities, propagated on simple, defined media, and relatively easily transformed with DNA. In addition a wide variety of simple genetic, molecular, and biochemical approaches have been developed for these organisms.

There are two types of yeast vectors available: episomally replicating vectors and vectors with the potential for genomic integration (53). These have the following features: a selection marker, a bacterial origin of replication for propagation in *E. coli*, and an *S. cerevisiae* promoter region controlling the transcription of an inserted cDNA. The episomal vectors contain either sequences from the 12 μm endogenous yeast plasmid that confer high-copy stable maintenance, or sequences from a yeast ARS element (autonomous replicating sequence; presumed to correspond to a chromosomal origin of replication) that yield lower copy numbers. Integrating plasmids lack a yeast origin of replication, but contain regions of homology with the yeast genome flanking the cDNA-cloning site. These vectors can be propagated in yeast only after their integration into the host genome. Recombination of introduced DNA with the chromosome occurs extremely efficiently and, for all practical purposes, exclusively by homologous recombination. Recombinants are identified by their survival under selective conditions, as dictated by the selection marker, and preferentially also by analyzing for the genomic presence of the transgene. The use of an integrative plasmid is advantageous when trying to generate a yeast strain expressing stable levels of an ancillary protein (e.g., cytochrome P450 reductase) required for the functional heterologous expression of cytochrome P450 (54). Stable integration of the heterologous cDNA also permits the growth of the transformed strains in rich media that enhances the yield of recombinant P450s mainly by the higher biomass achievable under those conditions. This is not possible using expression plasmids carrying auxotrophic markers, because they are not retained in rich medium. *S. cerevisiae*-based expression biotechnology benefits from the availability of numerous selection markers (HIS3, LEU2, LYS2, TRP1, URA3) for gene transfer, making it possible to cotransfect several cDNAs, and from the availability of powerful and regulatable yeast promoters (eg., alcohol dehydrogenase promoter ADH2, metallothionein promoter, galactose-inducible promoters GAL1, GAL7, GAL10) that can yield high levels of heterologous protein expression in *S. cerevisiae*. However, some features of mammalian cDNAs are not optimal for the expression of proteins in *S. cerevisiae*. For example, efficient expression of several proteins, including cytochrome P450s, requires the deletion of most of the 5′ untranslated region of the cDNA (54). For example, a CYP1A1 cDNA under the control of the GAL10-CYC1 promoter containing either 15 bp or 5 bp of 5-untranslated region, yielded 1 and 6 μg of functional CYP1A1 per milligram of microsomal protein, respectively (55). In contrast, the presence of an unmodified 3′-untranslated region seems to be less deleterious for the expression of cytochrome P450s. Drug metabolism catalyzed by yeast-expressed human P450s has been frequently studied (see Table 1).

C. Baculovirus Systems

The eukaryotic model yielding the highest cellular expression level of several mammalian proteins was designed using baculovirus. The baculovirus expression strategy has been described in detail (56–58). Briefly, baculoviruses are a group of over 500 species that infect only insect cells. Baculovirus shares many similarities with vaccinia virus. It contains a large genome, has a lytic life-cycle, and construction of recombinant viruses requires in situ recombination. The virus used mostly for cDNA expression is a nuclear polyhedrosis virus that, in its secreted form, is contained within a capsule of protein called a polyhedrin. This structure protects the virus during its natural life cycle in insects, but is dispensable when the virus is propagated in cell culture. The gene encoding the polyhedrin protein can therefore be used as a site for the insertion of the cDNA to be expressed. Other promoters, however, are also available. The virus used mostly for heterologous expression (*Autographa Californica* nuclear polyhedrosis virus; AcMNPV) has a genome of 130 kb and thus the cDNA cannot be inserted by conventional procedures. It is first cloned into an insertion vector behind the polyhedrin promoter. The vector also contains other polyhedrin sequences that allow its recombination with the wild-type virus. The integration event occurs with a frequency of approximately 0.5% and inactivates the wild-type polyhedrin gene. Cells containing the wild-type virus produce the polyhedrin head and have a different appearance (owing to occlusion) than the recombinant. However the identification of recombinants is not a trivial task and several additional selection markers, such as β-galactosidase, make either the identification less tedious or increase the proportion of cells containing the recombinant virus (57,59). Unlike for *E. coli*- and *S. cerevisiae*-based expression, it is not necessary to modify mammalian cDNAs extensively for their expression in insect cells. However, it is not possible to achieve stable expression of transgenes in the baculovirus system, for infection of the insect cells with the virus leads to a shutdown of housekeeping genes, and ultimately kills the cells.

Insect cells that are used for the baculovirus-based expression are eukaryotic cells and, as such, contain an endoplasmic reticulum. The posttranslational modifications in these cells are more similar to those found in mammalian cells relative to *S. cerevisiae* (57). Some examples where baculovirus-based expression systems have been used for drug metabolism studies are given in Table 1. The practical problems associated with transient expression may have precluded the widespread use of baculovirus expression models for drug metabolism studies. However, the system is suitable for generating sufficient amounts of recombinant P450 for structural and enzymatic studies (see later and Ref. 60).

D. Bacterial Systems

Of all bacterial species, *E. coli* has been most frequently used for the bacterial heterologous expression of drug-metabolizing enzymes. However, glutathione *S*-transferases (GSTs) have been expressed in *Salmonella typhimurium*, generating a valuable tool for mutagenicity studies (61). *Escherichia coli* is a desirable organism for the heterologous expression of proteins (62), for it can be grown to very high cell densities, and fractional abundances of 10%, leading to a yield of more than 1 g of heterologously expressed protein per liter of culture, have often been achieved. In addition, *E. coli* is easily manipulated, and a wide variety of vectors with powerful promoters and multiple cloning sites are available. However, almost as a matter of fact, mammalian cDNAs have to be modified rather extensively before they can be expressed in *E. coli* at high levels (63,64). One reason for this is that *E. coli* and mammalian 5′-untranslated regions are vastly different. The ribosome-binding site (RBS; Shine-Dalgarno sequence) needs to be 9 ± 3 bp in front of the initiation codon of protein synthesis. Usually the RBS is incorporated into the expression vector. However, the 5′-untranslated regions of the mammalian cDNAs to be cloned have to be trimmed appropriately. In addition, the region around the initiation codon of protein biosynthesis should be preferentially free of rigid secondary structures, which can occur in mammalian cDNAs. Because *E. coli* and mammalian cells have a different codon usage, mammalian cDNAs that are rich in rare *E. coli* codons might be expressed only poorly. However, even a cDNA optimized for high translation efficiency does not necessarily guarantee high levels of functional heterologous protein. One reason is that protein biosynthesis in prokaryotes proceeds very rapidly and protein folding in bacterial and mammalian cells may follow different pathways (65). Under these conditions, the heterologous mammalian protein might not adopt the proper conformation. These differences may be even more pronounced for membrane proteins. Even though some basic mechanisms of membrane targeting are conserved in eukaryotes and prokaryotes and NH_2-terminal membrane anchor signal sequences are similar, the kinetics of the processes leading to cotranslational membrane insertion of proteins differ among these organisms (66).

The prerequisite to change the 5′-coding region of a mammalian cDNA for optimal expression in *E. coli* is exemplified by the cDNA modifications that were necessary to achieve high levels of functional cytochrome P450 expression in *E. coli*. Initially, the presence of a hydrophobic cytochrome P450 anchor sequence was deleterious for bacterial expression of mammalian P450s (67,68). In addition, changing codon 2 to an alanine codon (GCT) as well as removing dG- and dC, rich regions, particularly in codons 4 and 5 of the CYP17A, improved the functional expression of this protein (69). A more detailed analysis of the P450 cDNA modifications required for functional expression was performed by Gillam et al. (70).

In this study the CYP3A4 cDNA region encoding the first 35 amino acids was systematically modified by point mutations and deletions. The native CYP3A4 was synthesized in *E. coli* at very low levels. Decreasing the potential of the 5′-coding region to form secondary structures improved the CYP3A4 expression. Similarly, deletion of amino acids 3–24 resulted in an improvement. However, only a combination of both modification strategies was optimal for P450 expression. Similar rules were applied for the expression of CYP2C9 (71). Recently, a more universal approach for the optimization of the P450 cDNA for bacterial expression has been developed (72). In this approach several members of the CYP2C family were expressed at high levels in *E. coli* after replacing their natural NH2-terminal sequence with the first eight amino acid residues of a sequence that had been optimal for the expression of CYP17A (69). However, it is not known if this approach is applicable to P450s outside the CYP2C family.

We have developed a strategy that allows the bacterial expression of a wide range of P450 isoenzymes without modifications in their primary structure. In this approach, we expressed NH_2-terminal fusions of the P450 to bacterial leader sequences such as *ompA* or *pelB*, which usually direct proteins to the periplasmic space. The bacterial signal peptidase cleavage site was optimized to allow efficient intracellular cleavage of the leader sequence from the proprotein to produce the mature, unmodified P450. The main body of the bacterially expressed P450 was most likely localized at the cytosolic side of the inner membrane. We also demonstrated that the resulting P450s were able to couple in intact cells with coexpressed human reductase (see following discussion).

III. CHOOSING THE OPTIMAL IN VITRO HETEROLOGOUS EXPRESSION SYSTEM FOR STRUCTURAL, DRUG METABOLISM, AND CHEMICAL CARCINOGENESIS STUDIES

The choice of the optimal strategy for the expression of cytochrome P450 depends on the type of investigation. In this section, we ask the following basic questions: (a) In which systems did investigators achieve the highest expression of apparently correctly folded cytochrome P450? This model would be most suitable for structural analysis; (b) In which in vitro system can a functional cytochrome P450 monooxygenase system, which mimics the P450 enzyme activity in vivo, be most easily be obtained? This approach would be most valuable for drug metabolism studies; and (c) In which system can the pharmacotoxicological role of cytochrome P450s be best determined in intact cells?

A. P450 Levels Achieved in the Different Models and Structural Analysis on Recombinant P450

To answer the first question, it is appropriate to review the heterologous expression of one particular cytochrome P450 in several expression systems performed by the same group. However, in doing so we have to keep in mind that the models were not necessarily completely optimized for expression. CYP17A is a steroidogenic cytochrome P450 and has been expressed in *S. cerevisiae*, in insect cells, and in *E. coli* by the group of Waterman (73; see Table 2). In these experiments the expression level of CYP17A expressed as molecules P450 per cell were highest in baculovirus. However, the total yield of P450 per milliliter of culture was much higher in *E. coli* than in yeast- and baculovirus-based expression systems. Therefore, baculovirus- as well as *E. coli*-based expression is suitable for obtaining the high yields of P450 needed for structural analysis. However, a distinct disadvantage of bacterial P450 expression systems are the NH_2-terminal modifications of these proteins required for functional expression. It cannot be ruled out that these affect certain features of the P450 tertiary structure even though catalytic activities of the recombinant P450s expressed in *E. coli* and assayed in a reconstituted system are similar to those expressed in mammalian cells (see later discussion). This problem, however, can be overcome by the bacterial expression of unmodified P450s using the leader sequence strategy (see foregoing).

Data are also available on the expression of CYP3A4 in different models. Unlike for CYP17A, the level of CYP3A4 achieved in baculovirus-based systems, *S. cerevisiae*, and bacteria is rather similar. Despite this, baculovirus systems and *E. coli* are superior to yeast in yield per litre, of culture. This drawback of most *S. cerevisiae*-based systems has its root in the type of auxotrophic selection markers that had been employed in the older expression strategies. These did not allow the growth of these organisms in rich medium to high cell densities. Recently developed approaches avoid this problem and yields of more than 500 pmol of cytochrome P450 per milliliter of culture have been reported (74,75). In the mammalian systems high levels of CYP3A4 have been achieved using vaccinia virus-based expression in HepG2 hepatoma cells (see Table 2). These transient levels are similar to those found in human liver microsomes (i.e., 100 pmol/mg microsomal protein; 76,77). By using the amplifiable vector system we have achieved similar levels of stable CYP3A4 expression in CHO cells (see Table 2; note that CYP3A4 expression in CHO cells was determined in total cellular protein, which contains approximately 10–15% microsomal protein derived from the endoplasmic reticulum). In comparison, the stable expression level of CYP3A4 in lymphoblastoid cells (AHH1) using EBV-based vectors is much lower (9). This seems to be a problem that is isozyme-specific, because efficient expression of CYP2D6 (160 pmol/mg microsomal protein) has been achieved using this system (32), the CYP2D6 content being 30-fold higher than that in human liver microsomes (77). Similarly, we have obtained a high expression level of CYP2D6 in CHO cells by using an amplifiable system. To our knowledge, the cytochrome P450 content in mammalian heterologous expression systems employing direct gene transfer and genomic integration without further gene amplification is at the limit of spectral detection, and data are not readily available.

Several approaches have been taken to elucidate the structure–activity relations (SAR) of mammalian P450s. Clearly, SAR would benefit from the availability of three-dimensional structures of mammalian P450s, obtained either by X-ray crystallography or by nuclear magnetic resonance (NMR) spectroscopy. Both techniques require large quantities of highly pure protein. These can be obtained from *E. coli*- or insect cell-expressed P450s that contain a polyhistidine extension (His-tag) at their COOH-terminus. This modification allows the protein to be purified by nickel agarose affinity columns (78,79). The resulting P450 preparations usually contain only trace amounts of contaminating proteins, which can be removed by fast protein liquid chromatography (FPLC) on Mono Q. If necessary, the His-tag can be removed by employing protease recognition sites previously introduced at the NH_2-terminal side of the His-tag.

However, P450s, similar to other integral membrane proteins, are notoriously difficult to crystallize and are too large for NMR studies, which can only resolve the structure of proteins no larger than 25 kDa. Recently, several strategies have been developed to engineer more water-soluble derivatives of mammalian P450s to facilitate crystallization and to avoid the use of detergents during P450 purification, for the presence of these complicates structural analysis. In one approach, a protease recognition site was engineered into the COOH-terminal side of CYP52A3 and the protein was expressed in yeast. After removal of the membrane anchor by protease digestion, the protein was soluble and displayed a typical Fe^{2+} vs Fe^{2+}-CO spectrum (80). A water-soluble CYP2A4 was expressed in *E. coli* by replacing the P450 membrane anchor with an amphiphilic peptide, termed peptitergent, consisting of 24-amino acid residues (81). The resulting P450 was catalytically active and could be purified as an oligomer, with a molecular weight of 450×10^3 in the absence of detergent. In another approach, the membrane anchor of CYP2D6 was replaced by a short hydrophilic peptide that included a His-tag (82). The resulting fusion protein was expressed in *E. coli* and purified by chromatography on nickel agarose. The protein formed water-soluble oligomers in the absence of detergents; however, in the presence of nonethyleneglycol monododecyl ether, a detergent often used in attempts to crystallize membrane proteins, the CYP2D6 was mainly in the monomeric state.

In the absence of a three-dimensional structure for mammalian P450s, molecular modeling of mammalian P450s has been performed. In this approach, the primary structure of the protein is altered by site-directed mutagenesis and the resulting effects on the catalytic properties of the P450 are analyzed. Given these data, and the known crystal structure of some bacterial P450s (83,84), the interaction of the mammalian protein with the substrate is visualized. This for example, resulted in a model for the interaction of rat CYP2B1 with androstenedione and progesterone (85). In a complementary approach, the interaction of codeine with the heme group in CYP2D6 was measured by NMR spectroscopy (79) and combined with the structural information on bacterial P450, a model for the complex of CYP2D6 with codeine was constructed. Recombinant P450s have been also used to study the interaction of substrates with the protoporphyrin ring of the heme moiety (86). Here the reaction products of the P450 heme group with arylhydrazines (or aryldiazenes, the oxidation products of arylhydrazines) are analyzed. Because the extent of migration of the aryl groups toward the different pyrrole ring nitrogens depends on steric effects of the residues in the active site, structural information on substrate–protein interaction can be obtained.

The modeling of P450 SARs, based on site-directed mutagenesis, requires an expression system that permits the rapid evaluation of the resulting functional consequences. Transient expression of the mutant P450s in COS cells (87–89) and expression in *E. coli* (85,90) or *S. cerevisiae* (91,92) has been mainly used for this purpose, because the cloning of the mutated P450 cDNA into the expression vector and the expression of the P450 can be most rapidly achieved using these systems. However, as pointed out later, owing to the absence of P450 reductase, most mammalian P450s expressed in *E. coli* display a very low catalytic activity, making it necessary to reconstitute bacterially expressed mutagenized P450s. To speed up the enzymatic analysis of mutated CYP2B11 proteins, the CYP2B11 has been fused to P450 reductase (P450–P450 reductase fusion proteins, see following section) and has been expressed in *E. coli*. Sonicated bacteria expressing this fusion protein displayed catalytic activities similar to those obtained with reconstituted CYP2B11 preparations (93). When using this strategy, each P450 requires the construction of a new cassette allowing the fusion of the reductase to the P450. Recently, however, we were able coexpress P450 reductase together with P450s. Unsonicated *E. coli* expressing these proteins displayed P450 enzyme activities similar to those achieved in reconstituted systems (M. Pritchard, Dundee, personal communication, see following). We envisage that *E. coli* expressing P450 reductase will be ideally suited for the rapid enzymatic analysis of P450 mutants, thereby greatly improving the speed of molecular modeling.

B. Optimal In Vitro Expression Systems for Drug Metabolism Studies and for Bioreactors

Until recently, bacterial- and baculovirus-based expression did not allow the direct determination of the enzyme activity of the heterologously expressed mammalian P450 in whole cells or in membrane fractions (70,94). An exception appears to be the bacterially expressed CYP17A which couples with a bacterial flavodoxin and its reductase (95). However this activity was very low. For the determination of the enzyme activity, the bacterially expressed mammalian P450s had to be either purified and tested in reconstituted systems or NADPH-P450 reductase was directly added to the bacterial membranes (96). The reconstituted system required for optimal enzymatic function of the purified P450 is complex and requires, for example for CYP3A4, addition of the P450 reductase, cytochrome b_5, lipid, and glutathione (70). To reconstitute the enzyme activity of heterologous P450 from insect cells, it is usually sufficient to add exogenous reductase to the disrupted cells.

Two strategies have been developed to generate a functional mammalian P450-dependent monooxygenase system in intact bacteria. One approach mimicked the *B. megaterium* $P450_{bm3}$, which is a fusion protein between a P450 and a P450 reductase domain (83,84). This approach, which had been originally tried in yeast expression systems (97,98), was later applied to the expression of fusion proteins containing domains derived from different P450s in *E. coli* (99–101). The resulting fusion proteins displayed catalytic activities in bacterial membranes or after their purification that were qualitatively rather similar to those found for the analogous P450s reconstituted with P450 reductase. However, the fusions usually displayed higher P450-dependent monooxygenase activities owing to highly efficient coupling between the two fusion partners. In intact bacteria a CYP17A–P450 reductase fusion protein displayed a 17 α-hydroxylase activity toward steroids. However this activity was approximately tenfold lower than the activity of the fusion protein in membranes isolated from these bacteria. Similarly, a CYP1A1–P450 reductase fusion protein metabolized benzo[*a*]pyrene to several hydroxylated products in intact bacteria (102). However, the rate of metabolism during the 24-h incubation appeared to be rather low.

In an alternative approach, we generated a functional monooxygenase system in *E. coli* by coexpression of P450 and P450 reductase as separate entities, thereby mimicking directly the mammalian P450-dependent monooxygenase system in bacteria (103). For this, the P450s were expressed after modification of their NH_2-terminus, as described earlier (70). To achieve reasonable levels of catalytically active P450 reductase, the NH_2-terminus of the human P450 reductase was fused to a bacterial leader sequence (*pelB*) that usually directs the export of bacterial proteins into the periplasmic space. The P450s and the P450 reductase were expressed under the control of separate IPTG inducible $(tac)_2$ promoters either from two plasmids or from the same plasmid. In the latter both proteins were expressed at a 1:1 molar ratio, whereas in the former, less reductase was expressed. Figure 3 displays the immunological and spectral analysis of the coexpressed proteins. Intact bacteria coexpressing CYP3A4 and P450 reductase from the one plasmid system displayed catalytic center activities (substrate turnover numbers) toward typical CYP3A4 substrates (erythromycin, nifedipine, and testosterone) that were as high as in membranes derived from these strains and similar to the values found in human liver microsomes (see

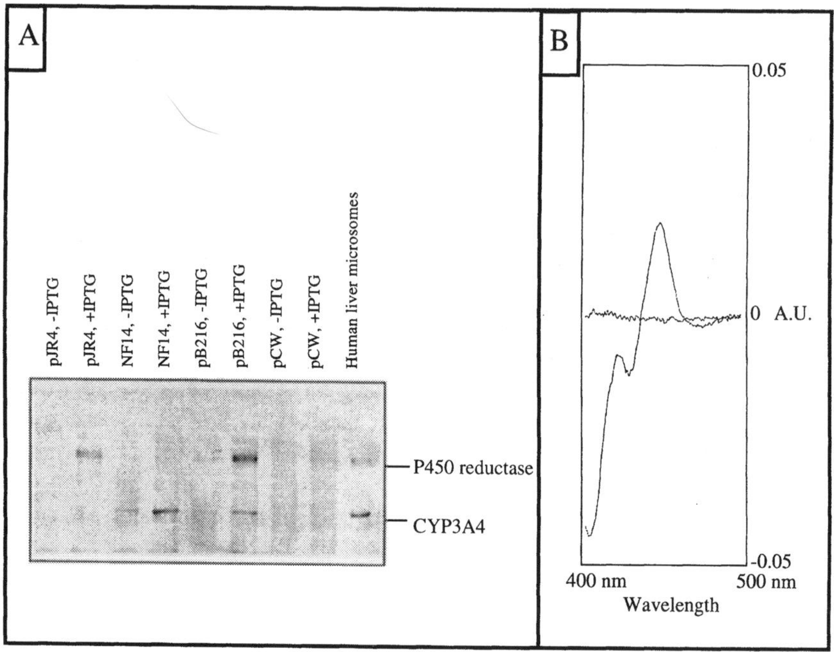

Fig. 3 Generation of a functional human P450 monooxygenase system in *E. coli*: CYP3A4 and P450 reductase were expressed either alone or together using a single expression vector. Expression of the recombinant proteins was verified by (A) immunoblotting or (B) by spectral analysis of bacterial membranes. (A) 10 μg of membranes and of human liver microsomes were analyzed. Expression was induced by IPTG. pCW, empty expression vector; pJR4, plasmid carrying the P450 reductase cDNA; NF14, plasmid carrying the CYP3A4 cDNA; pB216, plasmid carrying CYP3A4 and P450 reductase cDNA. (B) Only the spectral analysis of membranes derived from *E. coli* that coexpress CYP3A4 together with P450 reductase is shown.

Table 3 The Reconstitution of a Functional Cytochrome P450 Monooxygenase System Using Different Heterologous Expression Models

Expression of CYP3A4 in	Additional ancillary proteins added/ coexpressed	Nifedipine oxidase turnover number $(min)^{-1}$	Testosterone 6β-hydroxylation turnover number $(min)^{-1}$	Ref.
CHO-amplified	P450 reductase	15	50	Own data
Vaccinia/HepG2	None	3.9	30	76,94,104
S. cerevisiae	b^5/CYP-reductase[a]	1.8	1.2	206
Baculovirus/ insect cells	b^5/CYP-reductase[b]	n.a.	25	94
E. coli	CYP reductase[a]	15[c]	17[c]	103
E. coli	b^5/CYP-reductase[b]	4.5	10	70

[a]Cellular coexpression of ancillary proteins together with cytochrome P450.
[b]Ancillary proteins added to a cellular extract containing the heterologous P450.
[c]Measured in whole cells.
n.a., data not available.

later discussion and Table 3). Thus, this strategy is ideally suited to generate bacteria containing a highly functional mammalian monooxygenase system for drug metabolism studies. Equally importantly these bacterial strains will be extremely useful as bioreactors for the generation of valuable metabolites because bacterial expression systems can be scaled up to growth in biofermenters with a volume of up to 1000 L. We are now determining the membrane topology of the recombinant P450 and the P450 reductase in the bacterial coexpression system. However, because mammalian P450s that are optimized for bacterial expression are located at the cytosolic site of the inner membrane, and because these P450s couple with the coexpressed reductase, it is likely that the coexpressed proteins are facing the cytosolic side of the membrane.

The enzyme activity of cytochrome P450s expressed in *S. cerevisiae* can be determined directly because of the presence of an endogenous cytochrome P450 reductase that couples with mammalian P450s. However, the coupling appears to be inefficient, and mammalian P450 reductase and cytochrome b_5 should be coexpressed to reconstitute an optimized cellular mammalian monooxygenase system in yeast. For this purpose, it is convenient to express the human P450-reductase and cytochrome b_5 from cDNAs integrated into the genome (54).

It is expected that hepatic cytochrome P450s heterologously expressed in mammalian cells mimic the catalytic properties of the enzyme with the highest fidelity. However, usually, the cell lines used for the expression of P450 are of nonhepatic origin and thus have a different content of cytochrome P450 reductase and cytochrome b_5 compared with liver. That this can affect the enzyme activity was demonstrated for rat CYP2B1 which was expressed in V79 Chinese hamster lung fibroblasts that have a very low cytochrome P450 reductase activity (104). Membranes isolated from these cells showed a testosterone 16β-hydroxylase enzyme activity; however, this was stimulated by adding exogenous rat reductase (105). Only recently, cell lines have been developed in which the P450s were coexpressed together with P450 reductase (106). Similarly, the expression of cytochrome b_5 in human 143 fibroblasts, which are deficient in this flavoprotein, lead specifically to an increase of the heterologous CYP2B1-mediated enzyme activity toward *p*-nitrophenetole, but not toward 7-ethoxycoumarin or 7-pentoxyresorufin (107). These observations demonstrate that the limitations even of mammalian heterologous expression strategies for cytochrome P450s have to be appreciated.

To evaluate whether drug metabolism data obtained with the different in vitro heterologous expression models for cytochromes P450 give comparable results, we will focus our attention on the few substances that have been studied. CYP3A4 has been successfully expressed in mammalian cells, in *S. cerevisiae*, in insect cells, and in *E. coli*. The expression levels achieved in these systems are summarized in Table 2. Data on the catalytic activity toward the substrates nifedipine and testosterone are available for the different CYP3A4 heterologous expression systems (see Table 3). Taking into account that P450 enzyme activities determined for the same sample in different laboratories may show some variation, the catalytic activities of the CYP3A4 expressed in the different systems toward typical CYP3A4 substrates is remarkably similar. Intact CHO cells that coexpressed CYP3A4 and P450 reductase using an amplifiable vector system (S. Ding, Dundee, personal communication) as well as membranes isolated from HepG2 cells, or from insect cells or from *E. coli* expressing recombinant CYP3A4 displayed rather similar P450-dependent enzyme activities toward testosterone and nifedipine. However, membranes derived from the latter two expression systems had to be fortified with optimal concentrations of P450 reductase and cytochrome b_5 to achieve optimal P450 activity. Optimal enzyme activities toward testosterone were observed in the baculovirus expression system at a rather unphysiological ratio of CYP3A4/P450-reductase/b_5 of 1:20:3, whereas in human liver the ratio of total P450/reductase is approximately 4:1 (94). These complications did not arise when using *E. coli* coexpressing CYP3A4 and P450 reductase or membranes derived from them for drug metabolism studies (see Table 3). The high CYP3A4-dependent enzyme activities detected in this system in the absence of mammalian cytochrome b_5, which is essential for several CYP3A4-mediated enzyme activities (97–99), were unexpected and suggest that a bacterial protein may mimic the role of this ancillary protein. CYP3A4 expressed in *S. cerevisiae* displayed a rather low activity toward testosterone, even in the presence of P450-reductase and of cytochrome b_5. The reasons for this are not yet known.

One might then ask: how well does the catalytic activity obtained for heterologously expressed CYP3A4 compare with the catalytic activity of this enzyme in human liver microsomes? Recently, the group of Guengerich (77) analyzed the P450 content and several P450-dependent enzyme activities in 60 individuals. They found that the level of CYP3A proteins in the human liver is about 100 pmol/mg, which is 30% of the total hepatic microsomal cytochrome P450 in humans. In addition one can deduce from the data presented in this reference that the average microsomal nifedipine oxidase enzyme activity is 1.4 nmol/min per nanomole of P450. This activity correlated reasonably well with the CYP3A content in the different microsomal samples. In the human liver, CYP3A-related proteins comprise mainly CYP3A4 and CYP3A5 which display a similar catalytic activity toward nifedipine (76). From the microsomal CYP3A content and from the nifedipine oxidase activity found in human liver microsomes, one can calculate that the catalytic activity of the CYP3A proteins toward nifedipine is roughly 14 nmol/min per nanomole of P450, which is simulated best by recombinant CYP3A4 expressed using vaccinia virus or *E. coli*.

Whereas CYP3A4 catalytic activity in a reconstituted system is critically dependent on the lipid composition and the cytochrome b_5 content, CYP2D6 is less influenced by these parameters. CYP2D6 has been expressed in *S. cerevisiae*, CHO cells (S. Ding,

Table 4 Coexpression of P450s and P450 Reductase in *E. coli* Yields Highly Functional Monooxygenase Systems

Protein expressed[a]	P450 level (pmol/mg)	P450 yield (nmol/L)	Catalytic activity (pmol/min/mg)	Turnover number $(min)^{-1}$
1A2	401	185	660	1.6
2A6	260	140	750	2.9
2C8	1700	440	n.d.	n.d.
2C9	286	498	2400	8.7
2D6	210	370	1200	5.7
3A4	200	215	3500	17.3

[a]The P450 isozymes and the P450 reductase were usually expressed from two separate plasmids, with the exception of CYP3A4 which, together with the reductase, was expressed from one plasmid.
The P450 reductase activity was typically 600–1200 nmol cytochrome *c* reduced per minute per milligram of protein, with the exception of the strain containing CYP1A2, which had a reductase level of 70–120 nmol cytochrome *c* reduced per minute per milligram of protein.
Enzyme activities in human hepatic microsomes (pmol/min/mg): CYP1A2 (7-ethoxyresorufin *O*-dealkylase): 96; CYP2A6 (coumarin-hydroxylase): 30; CYP2C9 (diclofenac-hydroxylase): n.a.; CYP2D6 (bufuralol-hydroxylase): 20; CYP3A4 (testosterone 6β-hydroxylase): 1000–3000.

Dundee, personal communication), and in AHH1 lymphoblastoid cells (32,108), and its bufuralol 1-hydroxylase enzyme activity was determined. In *S. cerevisiae* coexpressing rat P450-reductase and CYP2D6, 1-hydroxylation of bufuralol was catalyzed with a rate of 1.6 nmol/min per nanomole of P450 (M. Bandera, C. R. Wolf, and T. Frieberg, unpublished data). This is similar to the rate found for CYP2D6 expressed in AHH1 cells (4.5 nmol/min per nanomole of P450) and in CHO cells. In addition the catalytic activity of CYP2D6 toward metoprolol has been determined in yeast (109). This substrate is oxidized at two different positions (α-hydroxylation and *O*-demethylation). The ratio of these two reactions was only slightly different from the value found for human liver microsomes. This was attributed to the involvement of other cytochrome P450s in the *O*-demethylation reaction. Therefore, it can be concluded that heterologously expressed CYP2D6 and CYP2D6 in human liver microsomes have similar catalytic properties.

We have demonstrated that not only CYP3A4, but also a variety of other human P450s, can be coexpressed together with P450 reductase in *E. coli* to establish strains that display P450-dependent enzyme activities. We envisage that these strains will be extremely useful as models for drug metabolism studies and have the potential to be valuable as bioreactors. Table 4 lists the catalytic activities of the various isoenzymes in bacterial membranes isolated from the recombinant strains of *E. coli*. Generally, the catalytic activities achieved in these systems are much higher than in human liver microsomes.

C. Optimal In Vitro Heterologous Expression Systems to Study Mutagenesis and Carcinogenesis

If we want to establish the role of P450s in mutagenic reactions, we have to differentiate between models that allow the detection of this event in the P450-expressing target cell (homotopic systems, endogenous metabolic activation systems) and those that measure these events outside the P450-expressing cell (heterotopic systems, exogenous metabolic activation systems). Heterotopic models, such as the hepatocyte-mediated Ames mutagenicity test (110), contributed much to the development of genetic toxicology assays. These approaches, however, have certain limitations. First, both intact cells and cell homogenates generate intermediates external to the target cell in which the genetic endpoint is measured. Thus, highly reactive intermediates may not be stable enough to reach the target cell DNA. Second, the activating cell used in the test may lose certain functions with time. These problems are not encountered with stably expressing homotopic mutagenicity test systems. Truly homotropic models have now been developed for mammalian cells, *S. cerevisiae*, and *S. typhimurium*. The mammalian heterologous expression systems developed for mutagenicity studies are either based on the expression of P450s in V79 Chinese hamster fibroblasts, using direct gene transfer and stable genomic integration of the transgene (104), or on AHH-1 lymphoblastoid cells, employing EBV (111). Both strategies use mutations in the *HPRT* locus as biological endpoints. In principle the same endpoint should be employable for the CHO cell lines expressing drug-metabolizing enzmyes from an amplifiable vector system (see Tables 2 and 3). In addition, mutations can be scored in the thymidine kinase gene in the AHH-1 cells. In *S. cerevisiae*, it is possible to use genotoxin-induced DNA-recombination as a biological endpoint. This system has been used to determine the influence of CYP1A enzymes on the genotoxic effects of aflatoxin B_1 (112). *S. cerevisiae* has been also used to determine the role of CYP2B1 for the mutagenicity of sterigmatocystin (110).

Recently, homotopic bacterial systems have been developed for the heterologous expression of drug-metabolizing enzymes in *S. typhimurium*. The mutagenicity of mitomycin C and of 2-aminofluorene or of halogenated hydrocarbons has been evaluated in *S. typhimurium* expressing a recombinant P450 or a glutathione *S*-transfrase, respectively (61,113). Heterotopic bacterial mutagenicity test systems for several cytochrome P450s have been developed. In one very comprehensive investigation involving several carcinogens (96), membrane fractions derived from *E. coli* expressing human CYP1A1, CYP1A2, and CYP3A4, and fortified with NADPH–P450 reductase were employed as activating system. The mutagenicity was determined in *S. typhimurium*, measuring the induction of SOS-response on exposure to the carcinogen. The patterns of activation of the 14 individual chemicals were consistent with the literature data derived from experiments using human liver microsomes and purified liver P450s. Recently, the group of Guengerich developed a homotopic bacterial mutagenicity system (114). Here, CYP1A2 was expressed in *S. typhimurium* strains containing either high or low endogenous bacterial *N*-acetyltransferase activity. These strains were used to detect the P450-mediated activation of aromatic amines such as 2-amino-3,4-dimethylimidazo(4,5-*f*)quinoline (MeIQ), 2-aminofluorene, or 2-aminoanthracene to mutagens. These compounds were strongly activated only in the *S. typhimurium* strains that coexpressed CYP1A2 and the bacterial *N*-acetyltransferase. However, employing liver S-9 as the activation system, the strain without CYP1A2 expression still provided much greater sensitivity to these mutagens than did the strain that expresses CYP1A2 in the absence of the S-9 mix. This result could be due to a much higher total activity of P450 in the S-9 fraction as compared with the total P450 activity in the bacteria employed in these mutagenicity assays. Alternatively, the coexpression of mammalian P450 reductase may improve the sensitivity of the system for P450-mediated mutagenicity significantly. We have taken this approach and have coexpressed CYP3A4 and P450 reductase in *S. typhimurium*. Membranes isolated from these bacteria displayed CYP3A4 enzyme activities that were qualitatively and quantitatively similar to those found in human liver microsomes. However, we have not as yet employed these *S. typhimurium* strains for drug mutagenicity studies.

One carcinogen that has been studied by several groups using recombinant P450s, is the hepatocarcinogen aflatoxin B_1. P450s catalyze aflatoxin B_1 8,9-epoxidation, which is thought to constitute a toxification event, as well as the 3α-hydroxylation, leading to aflatoxin Q_1, which is a detoxification pathway. Immunoinhibition of cytochrome P450-mediated aflatoxin B_1 metabolism and mutagenicity suggested that CYP3A4 is the predominant enzyme that catalyzes these metabolic reactions and is mainly responsible for the mutagenic effects of aflatoxin B_1 in human liver (115,116).

In addition, other P450s, including CYP1A2 and CYP2A6, were implicated in the mutagenic activation of this compound (116). Complementary to these experiments, the toxification of aflatoxin B_1 by several recombinant human cytochrome P450s has been studied in various systems. In one comprehensive study, 12 recombinant P450s obtained from HepG2 cells were assayed (117). HepG2 lysates served as the metabolic activation system and the mutagenicity was determined in *S. typhimurium* TA 98. Mainly CYP3A4, CYP2A6, and CYP1A2 activated this carcinogen to a mutagen. However, over a wide concentration range, CYP3A4 mediated the mutagenicity of aflatoxin B_1 at least two- to threefold more efficiently than CYP1A2. In contrast, another study, employing AHH1 lymphoblastoid cells as the expression system, reported that CYP1A2 was three- to sixfold more active than CYP3A4 at concentrations of aflatoxin B_1 below 0.1 μM (9). The cytochrome P450-mediated mutagenicity of aflatoxin B_1 was also evaluated using cytochrome P450 expressed in *E. coli* (96,118). Here, *E. coli* membranes containing the recombinant P450 were fortified with cytochrome P450 reductase and cytochrome b_5 and used as the metabolizing system. The mutagenic response was measured in *S. typhimurium*. In agreement with the studies in HepG2 cells, the bacterially expressed CYP3A4 was more active than CYP1A2 in the mutagenic activation of aflatoxin B_1 at all concentrations examined; 1 μM of aflatoxin B_1 being the lowest concentration tested.

Recombinant P450 enzymes have also been used to analyze the metabolism of aflatoxin B_1. In one report aflatoxin B_1 metabolism by CYP3A4 and CYP1A2, expressed in the same AHH1 lymphoblastoid cell line (119) used to establish the relative contribution of these P450s in the mutagenic activation of this carcinogen, was studied (9). As expected from the mutagenicity data, the 8,9-epoxidation of aflatoxin B_1 at low concentrations (16 μM) was almost exclusively catalyzed by CYP1A2 and at higher concentrations (128 μM) by CYP3A4. In addition, it was reported that furafylline, a very potent inhibitor of CYP1A2, inhibited the epoxidation of aflatoxin B_1 at low concentrations by 72% in microsomal samples. At higher concentrations of aflatoxin B_1 this inhibition was less marked. From these data the authors concluded that CYP1A2 is the high-affinity and CYP3A4 the low-affinity cytochrome P450 in aflatoxin B_1 epoxidation. In contrast to the studies employing P450s expressed in AHH1 cells, CYP1A2 and CYP3A4 expressed in *E. coli* catalyzed the epoxidation of aflatoxin B_1 at low concentrations equally well (118). At higher concentrations, the epoxidation was catalyzed mainly by CYP3A4, as was found with CYP3A4 expressed in lymphoblastoid cells. However, it was also shown that CYP3A4 was primarily responsible for the activation of aflatoxin B_1 to the highly mutagenic exoepoxide, whereas CYP1A2 catalyzed the formation of the exo- and the much less mutagenic endoepoxide equally well.

From this synopsis of the CYP1A2- and CYP3A4-mediated mutagenicity and metabolism of aflatoxin B_1, it is evident that in studies employing different heterologous expression systems the relative contribution of these two P450 enzymes may not be consistent. To understand the reasons for these differences, it is imperative to coordinate efforts directed at the validation of the different heterologous expression systems for drug metabolism and mutagenicity studies (120).

IV. IN VIVO SYSTEMS FOR INVESTIGATING THE ROLE OF DRUG-METABOLIZING ENZYMES

Although mammalian, yeast, insect, and bacterial systems represent rapid and simple methods for assessing the role of cytochrome P450s in drug metabolism and chemical carcinogenesis, They are nonetheless in vitro models and may not necessarily reflect the in vivo situation, in which drug disposition, governed by absorption, distribution, and excretion are important. In vivo models give us the opportunity to study the biological effects of gene expression under more physiologically relevant conditions. In addition, they allow the inclusion of pharmacotoxicological endpoints.

One efficient in vivo drug metabolism model that can be rapidly established has been developed for *Drosophila melanogaster* (121). *Drosophila* has been used as an extremely valuable genotoxicity model because several biological endpoints are available for this purpose, and the ease of *Drosophila* genetics makes this organism extremely amenable to genetic manipulation. In our studies the major phenobarbital-inducible cytochrome P450 from rat (CYP2B1) was inserted into a transposable P element vector. P elements are very mobile elements in the *Drosophila* genome and can readily integrate into chromosomes. In addition, the vector contained a phenotypic marker for red eye color. The constructs were injected into fertilized eggs. Transgenes were identified by eye color and by Southern blotting. The larval *Drosophila* promoter (LSP1) directed the expression of CYP2B1 to the fat body of the larvae, as verified by immunoblotting using anti-CYP2B1 antibodies. The catalytic properties of the expressed CYP2B1 toward various alkoxy-resorufin analogues were similar to the values found for the purified enzyme, showing that the P450-reductase coupled well with the mammalian P450. The role of the CYP2B1 in the activation of mutagens was studied using a somatic mutation and recombination test (SMART). In this test, larvae that are heterozygous for different recessive cuticle markers, and express the mammalian P450, are exposed to a potential mutagen. Mutations induced in the larvae become evident in the adult flies as wing spots. One of the mutagens tested in this study was the cytostatic drug cyclophosphamide; CYP2B1 expression led to a more than tenfold increase of genotoxicity when compared with control. In a similar approach, canine CYP1A1 has been expressed in *Drosophila*, and its role in the toxication of 7,12-dimethylbenz[a]anthracene (DMBA) was studied in larvae (122). Premature mortality of the larvae on treatment with DMBA was used as the biological endpoint. The expression of the CYP1A1 in the larvae was under the control of a heat-shock promoter. The toxicity of DMBA was increased in larvae after heat shock owing to the induction of the heterologous CYP1A1. These examples demonstrate that *Drosophila* is a valuable tool to determine the roles of cytochrome P450s in chemical mutagenesis. However, even though *Drosophila* has some phase I and phase II drug-metabolizing enzymes (123), drug metabolism in insects and mammals is rather distinct, and this may affect the fate of the metabolites generated by the recombinant P450.

Important information on the physiological role of an increasing number of mammalian proteins has been elucidated by the use of transgenic animals. Several techniques for gene transfer into animals have been developed (124,125). In one approach, the gene of interest is introduced by microinjection into the germline of mice. Male pronuclei of fertilized mouse oocytes are injected with several copies of the gene of interest and transferred into the oviducts of pseudopregnant foster mice. Here, up to 90% of the injected mice survive microinjection, approximately 20% of these develop to term, and 25% of these carry the injected DNA and, therefore, are transgenic. As the integration of the transgene is random, positional effects of expression and effects on other genes have to be taken into account. In another technique, pluripotent embryonal stem (ES) cells are transfected with the transgene together with a selection marker. Cells that have taken up the transgene are introduced into blastocysts that are implanted into foster mice. Because ES cells and recipient blastocysts are derived from mice strains having different hair color, chimerae can be readily identified by their mosaic hair color pattern. In some of these chimeric mice the transgene incorporates into the germline. By breeding the chimerae, homozygotes for the transgene can be obtained. In another approach, the transgene is introduced into somatic cells in vitro (e.g., by retroviral transfer). Cells with the heterologous gene are then implanted into the adult organism where they produce the protein encoded by the foreign gene (126).

There are as yet few reports on the use of transgenic animals to establish the biological role and the regulation of drug-metabolizing enzymes. We have recently established transgenic mice that specifically express the human glutathione *S*-transferase GSTA1 in the epidermal cells of the mouse skin (113). Expression was achieved by using the epidermal-specific keratin VI promoter, and was confirmed by Western blotting and enzymatically by measuring the alpha-class specific GST isomerase activity toward androsten-3,17-dione as substrate. The skin was chosen as the site of expression of the enzyme for two reasons. First, it is an easily accessible target tissue, both for the presentation of chemicals and for the assessment of toxicological endpoints. Second, for chemically induced skin-tumor formation, the mouse skin is a well-characterized tissue. Skin painting experiments should allow the direct evaluation of the role of this and other transgenes in chemical carcinogenesis. Pilot skin-painting experiments are now being conducted to investigate the protective role of the GST from carcinogen-induced DNA adduct and tumor formation.

Recently, the expression of CYP2B1 in transgenic animals as a very exciting model for tumor gene therapy has been reported (127). Most malignant tumors of the central nervous system (CNS) do not respond to chemotherapy with cyclophosphamide. This agent is metabolized by CYP2B1 to the ultimate DNA-alkylating metabolite which, however, cannot cross the blood–brain barrier. Murine fibroblasts producing a retroviral vector containing a CYP2B1 expression construct were grafted into the glioma-bearing brains of athymic mice. Intrathecal administration of cyclophosphamide into these mice resulted in a partial regression of the tumor mass. This was not true in control mice bearing a *lacZ* transgene.

Several groups have started to characterize the physiological and pharmacotoxicological function of drug-metabolizing enzymes using gene knockouts. CYP1A2 is a highly conserved cytochrome P450 in mammals, which has been reported to be involved in the oxidative metabolism of xenobiotics, including the activation of polycyclic aromatic hydrocarbons (PAHs), which are also capable of inducing the expression of CYP1A2. Two groups have recently established mouse lines in which CYP1A2 has been deleted by homologous recombination in ES cells. Pineau et al. (128) inserted a neomycin-resistance gene into the first coding exon of the CYP1A2 gene, and although heterozygote animals appeared normal, homozygous mutant mice died within an hour of birth, displaying severe respiratory distress. The penetrance of this phenotype was incomplete, with 19/599 animals surviving to adulthood, appearing phenotypically normal and being able to reproduce. Those homozygous mutant mice that died exhibited decreased expression of surfactant apoprotein in type II alveolar cells and a reduced gallbladder size. Liang et al. (129) created their CYP1A2 knockout mouse by replacing part of exon 2, and all of exons 3 to 5, with the hypoxanthine phosphoribosyltransferase gene. In contrast to the previous study, homozygous null mutants were completely viable and appeared phenotypically indistinguishable from their heterozygote or wild-type littermates. Further study of these animals revealed a role for CYP1A2 in zoxazolamine metabolism; paralysis times with this muscle relaxant were increased more than ninefold in the CYP1A2 null mouse. The different phenotypes observed in these two studies is of general interest in the creation of knockout mice. The authors of the second study (129) suggest several possible reasons for the discrepancy, either the genetic background, which was C57BL/6 in their report, and CF-1 or Swiss Black in the first, or the presence of a viral or respiratory pathogen in a genetically susceptible host. Alternatively, the possibility remains that the phenotypic differences lie in the different approaches used to create the targeted allele.

CYP2E1, represented by a single gene in the mouse, is the major P450 responsible for the metabolism of ethanol, and it also carries out the metabolism of a wide range of xenobiotics, including acetaminophen, carbon tetrachloride, pyrazole, and dimethylnitrosamine, as well as endogenous chemicals such as acetone and arachidonic acid. Lee et al. (130) generated a CYP2E1 null mouse by replacing part of intron 1, exon 2, and part of intron 2 with the bacterial neomycin-resistance gene under the control of the phosphoglycerate kinase-1 promoter. Mice homozygous null for CYP2E1 were phenotypically indistinguishable from wild-type mice, but exhibited increased resistance to acetaminophen toxicity, such that at 400 mg/kg acetaminophen, 50% of wild-type mice died compared with none in the null group. Histological examination of liver tissue and estimation of serum bilirubin, creatinine, and alkaline phosphatase confirmed that acetaminophen hepatotoxicity was mediated by CYP2E1. Interestingly, CYP2E1 cannot be the only enzyme involved, because at higher doses of acetaminophen (600 mg/kg) toxicity began to be manifest in the null mice also.

We have also recently generated a mouse line that is nulled for both murine GST pi genes, P1 and P2 (CJ Henderson, manuscript accepted for publication in *Proc. Natl. Acad. Sci. USA*). GST pi expression was originally reported to be greatly elevated in rat preneoplastic hepatic foci (131,132), and in drug-resistant cell lines (133–135). In addition, GST pi is altered in a variety of human tumors, such as ovarian (136,137), colon (138), lung (139,140), bladder (141), gastric (142), and testicular (143), expression being positively correlated with malignant transformation and resistance to chemotherapeutic drugs, and inversely correlated with patient survival. Interestingly, however, there are reports that survival is greater in those patients whose tumors stain positive for GST pi, for example in renal carcinoma (144). In a similar vein, it has also been reported that in human prostatic carcinoma, expression of GST pi is completely absent, whereas in normal prostatic and benign hyperplastic tissue, GST pi expression is unaltered (145). The same authors found that the absence of GST pi expression in malignant prostatic tissue was due to hypermethylation of the promoter. These findings were confirmed in a subsequent study, although GST pi expression was found to correlate with basal cell phenotype, which is absent in prostatic carcinoma, leading to the suggestion that GST pi may be more involved in the process of epithelial differentiation than malignant transformation (146).

Our laboratory reported the cloning and characterization of the murine GST pi genes, of which there are two (147), and we have used this knowledge to develop mouse lines in which either both genes have been deleted, or in which only GST P1 is missing (CJH, manuscript accepted for publication in *Proc. Natl. Acad. Sci. USA*). In the mouse, GST P1 is expressed at much higher levels than GST P2, and displays far greater catalytic activity toward a range of substrates. In addition, the expression of GST P1 is sexually differentiated in the liver, such that male mice have a level approximately an order of magnitude higher than that in females (148).

Initial results with the double knockout (GSTP1/P2) mouse line suggest that GST pi is not essential for survival, since such mice displayed no overt phenotypic changes, and reproduced normally, with litter sizes and sex distribution no different from their wild-type counterparts. Male homozygous null mice had a consistently greater body weight (~10%) than the corresponding wild-types, although the significance of this finding, if any, is unclear. Characterization of these animals is continuing, to discover differences in their ability to metabolize and excrete a number of xenobiotics associated with GST pi. Treatment of wild-type mice with acetaminophen resulted in the observation of a sexual differentiation in terms of the hepatotoxic effects of this compound, as evidenced by histological examination of the liver, and estimation of serum levels of enzymes considered diagnostic for hepatocellular damage, such that male mice were much more sensitive than females. When the GSTP1/P2 null male mice were similarly treated, they exhibited increased resistance to the hepatotoxic effects of the drug, relative to their wild-type male littermates. In humans and mice, acetaminophen is activated to its hepatotoxic metabolite, *N*-acetyl-*p*-benzoquinonimine (NAPQI), by the cytochrome P450 monooxygenase system, principally CYP2E1, although CYP3A4 and CYP1A2 may also play a role (149,150). The sexual dimorphism in renal toxicity observed in mice (151) has been attributed to the sexually differentiated expression of CYP2E1 in this organ (152). However, in the liver, CYP2E1 expression is not sexually differentiated, and this situation is not altered following the deletion of GSTP1/P2. Detoxification and clearance of NAPQI is believed to be achieved by conjugation with glutathione, either spontaneously, or by enzymatic means, with the major GST involved being GST pi, in both rats and humans (153). On this basis, one might expect that a mouse nulled for GST P1 and P2 would exhibit increased sensitivity to the effects of acetaminophen, having a decreased ability to metabolize the hepatotoxic metabolite NAPQI. However, the reverse appears to be true, thus rendering unclear the mechanism of involvement of GST pi in acetaminophen metabolism. However, it is interesting to note that the hepatic level of GST pi may account for the sexual dimorphism in hepatotoxicity seen in the mouse following acetaminophen treatment, because male mice, with much higher levels of GST pi, are more sensitive to the effects of this drug than female mice, and a similar pattern is seen between male wild-type and GST P1/P2 null mice.

Peroxisome proliferators cause a pleiotropic response in the livers of both mice and rats, with an increase in both the number and size of peroxisomes, cellular hypertrophy, and hyperplasia, and the transcriptional induction of enzymes involved in the β-oxidation of fatty acids. It has been suggested that these coordinate events represent an adaptive response in the process of lipid homeostasis, and although the mechanism is still unclear, are indicative of the importance of peroxisome proliferation in the regulation of fatty acid metabolism. A peroxisome proliferator-activated receptor (PPARα), a member of the steroid receptor superfamily was first cloned from the mouse, since then a further five isoforms have been described, with PPAR homologues also being identified in humans, rats, and frogs. The identification of a peroxisome proliferator response element in the promoter region of target genes lends strength to the suggestion that PPARs act as transcription factors, although direct interaction of peroxisome proliferators with PPAR remains to be demonstrated. However, on the basis of structural similarities, it is assumed that a ligand-binding domain exists within PPAR. Lee et al. (154) produced a knockout mouse that was nulled for PPARα by replacing an 83-bp segment of exon 8, within the putative ligand-binding domain, with a neomycin-resistance gene. The resultant homozygous null mice were apparently phenotypically normal; however, when challenged with peroxisome proliferators, these mice failed to display the associated pleiotropic response, leading to the conclusion that PPARα is an essential requirement for the action of these chemicals and the mediation of the effects of peroxisome proliferators.

V. INVESTIGATING THE REGULATION OF DRUG-METABOLIZING ENZYMES USING RECOMBINANT DNA TECHNOLOGY

Enzyme induction was initially discovered by the marked effects that pretreatment with chemicals can have on certain pharmacological responses. For example, pretreatment of rats with phenobarbital shortens the duration of sleeping induced by barbiturates. This effect is due to the induction of cytochrome P450s (155) which enhances the metabolic elimination of barbiturates, and also due to tolerance mediated directly in the CNS. Other examples for which induction of cytochrome P450s by xenobiotics is of clinical significance include the decrease in the plasma level of the anticoagulant warfarin that occurs in patients receiving barbiturates, the increased metabolism of contraceptives during rifampin treatment, and the increased metabolism of cyclosporine in patients receiving phenytoin (156). Induction of cytochromes P450 by xenobiotics can also have profound effects on the metabolism of chemical carcinogens. For example, pretreatment of rats with 3-methylcholanthrene leads to an increased formation of the ultimate carcinogenic bay-region diol-epoxides of several polycyclic aromatic hydrocarbons, whereas pretreatment of the animals with phenobarbital induces the detoxification of these compounds at the K-region (157).

The induction of cytochrome P450 proteins is mainly regulated at the transcriptional level (158); however, for some cytochromes P450, such as CYP2E1, the induction appears to be mainly due to the increased stability of mRNA or protein (159). The transcription of cytochrome P450s is controlled by *cis*-acting elements at the 5′ region of their genes and *trans*-acting factors such as transcriptional repressors or activators that bind to these elements. The identification of regulatory mechanisms for drug-metabolizing enzymes will greatly enhance our understanding of the mechanisms underlying interindividual variations in the level of certain enzymes and may provide a means to generate complex in vitro systems that simulate the in vivo xenobiotic metabolism very closely. In this chapter we will describe some examples illustrating how regulatory paradigms for cytochromes P450 can be characterized in vitro and in vivo.

To identify *cis*-acting elements that control the expression of cytochrome P450, a part of the P450 gene containing these putative elements is ligated to a cDNA encoding a reporter protein that can be easily detected by its enzymatic activity and that is usually not expressed in mammalian cells (e.g., chloramphenicol acetyltransferase [CAT], luciferase, β-galactosidase [β-gal] or green fluorescent protein). The resulting constructs are transfected into mammalian cells and the effects of xenobiotics on the expression of the reporter protein can be rapidly evaluated. The success of this strategy is critically dependent on the availability of a suitable cell type in which the P450 gene under investigation can be induced. Hepatoma cell lines have been used to study the induction of CYP1A1 by compounds such as 2,3,7,8-tetrachlorodibenzo-*p*-dioxin (TCDD) or 3-methylcholanthrene. These compounds bind to the Ah-receptor and mediate its release from the heat-shock protein, hsp 90. The receptor then forms a nuclear complex with the Ah-receptor nuclear receptor translocator (ARNT) protein. By using CYP1A1–CAT gene expression plasmids, *cis*-acting elements (xenobiotic responsive elements; XRE) were identified on the CYP1A1 gene that are essential for the induction of CYP1A1 by xenobiotics by binding the ARNT–AHH receptor heterodimer (158). Several of these XRE elements have been identified on the CYP1A1 gene, but only one or two of these motifs located more than 1-kb upstream from the transcription start site are predominant.

Compared with the paradigms governing the induction of cytochromes P450 by TCDD and planar polycyclic aromatic hydrocarbons, the mechanisms underlying the induction of drug-metabolizing enzymes by barbiturates are much more difficult to study. This difficulty arises, because induction experiments involving reporter constructs are usually performed with immortalized cell lines and these, without exception, are partially refractory to the induction of P450s by barbiturates. For example, whereas P450 forms of the CYP1A family are inducible in dedifferentiated and differentiated Reuber H35 rat hepatoma-derived cell lines, dexamethasone and phenobarbital induce P450s of the CYP2B family only in differentiated cells. However, this induction is markedly weaker than that in the liver (160). Hepatoma cell lines have been used to study the regulation of the rat *CYP2B2* gene by glucocorticoids using a CYP2B2 promoter–CAT reporter construct. In these studies a glucocorticoid responsive element has been identified within a stretch of 25 bp in the 5-flanking region of the *CYP2B2* gene about 1.3-kb upstream from the transcription start site (161). However, studies using transgenic animals show that additional sequences mediating barbiturate induction are located even further upstream on the *CYP2B2* gene. Interestingly, the same reporter construct was used in another study to show that the induction of CYP2B2 could be prevented by the antiglucocorticoid mifepristone (RU486) in vitro, thus indicating that barbiturates act indirectly to cause the accumulation of an endogenous steroid, and that this molecule acting through its receptor may be directly responsible for the induction of CYP2B proteins (162).

The understanding of the factors involved in P450 regulation, may provide a tool for establishing cell lines that retain the properties of their in vivo counterparts. Cultivated hepatocytes or hepatoma cells have been mainly used as complex drug metabolism systems. The disadvantage of hepatocyte- or hepatoma-derived models for drug metabolism studies is the observation that drug metabolism is always altered, either owing to culture conditions (11), or in hepatoma cells owing to the process of phenotypic transformation (163). It is unlikely that these problems can be solved in the near future using a random trial-and-error approach, even though some encouraging attempts have been made to improve the models, by growing cells on specific biomatrices as well as in the presence of specific growth factors (164,165). The heterologous expression of transcription factors, which are required for the appropriate regulation of P450s in cultured hepatocytes may stabilize P450 expression in these cells.

Given that certain P450 forms are difficult to induce in vitro, transgenic animals are a very valuable tool to study the mechanisms leading to the induction of cytochrome P450 enzymes by xenobiotics. To study the paradigms governing the induction of CYP2B2 by barbiturates, the group of Omiecinski generated transgenic mice containing various lengths of the 5′-flanking region of CYP2B2 as well as the entire coding and 3′noncoding region of this gene (166). Interestingly, they found that constructs containing 800 bp of the promoter region, were constitutively expressed in mouse liver and kidney at high levels. However this CYP2B2 construct was not inducible by phenobarbital. Only the inclusion of an additional 19 kb of 5′-flanking region led to a very low basal, but phenobarbital-inducible, hepatic expression of the transgene, analogous to the expression pattern observed for CYP2B2 in rat liver. From their study the authors concluded that sequence elements that had been previously identified 50 to 100-bp 5′ from the transcription start site of CYP2B genes and that had been thought to play an important role in the induction response of CYP2B proteins, are, at least alone, not sufficient to induce this response. On the other hand, it cannot be ruled out that gene-positional effects might have influenced the regulation of the various CYP2B transgenes in these experiments. In a similar approach, the regulatory regions of CYP1A1 were studied in transgenic animals (167). In this study 2.6 kb of the 5′-flanking region of CYP1A1 was recombined with a chloramphenicol acetyltransferase reporter construct; the expression pattern of the reporter was similar to the pattern of CYP1A1 expression in vivo.

Recently, homologous recombination techniques have been used to study the aryl hydrocarbon (Ah) receptor (AHR)-mediated regulation of cytochromes P450 in vivo. Besides being involved in the regulation of certain P450s by xenobiotics, the AHR has been implicated in dioxin-mediated teratogenesis, apoptosis, and immunosuppression. The gene coding for the AHR was inactivated in embryonic stem cells using a positive–negative selection strategy (168). In this strategy one selection marker (neo; positive marker) was interrupting the AHR–gene in the targeting vector and the other selection marker (HSV-thymidine kinase; [TK]; negative marker) was situated several kilobases upstream of the AHR–gene segment. On homologous recombination, the neo selection marker should be retained and the HSV-TK marker lost. Cells with the inactivated copy of the AHR will survive in a medium containing G-418 and gancyclovir, whereas cells in which nonhomologous recombination had taken place should be killed by gancyclovir owing to the expression of the HSV-TK. ES cells containing an inactivated AHR copy were employed to generate $AHR^{+/-}$ heterozygotic animals and these were mated to yield $AHR^{-/-}$ homozygotes. The inactivation of the AHR yielded pleomorphic symptoms. In the first instance, a high proportion of the animals died in the first few days after birth. The precise causes of death could be not established. Dioxin treatment led to the induction of CYP1A1 and *UGT1*6 in* $AHR^{+/+}$ animals but not in $AHR^{-/-}$ mice, demonstrating that a functional AHR is required for the induction in the liver. Interestingly, histological examination of tissues isolated from the mice with the inactivated AHR revealed pronounced liver fibrosis and a highly reduced liver/body weight ratio. The study appeared to demonstrate that AHR also plays an important role in the development and the maintenance of the immune system, as evidenced by a decreased population of the spleen by lymphocytes, in particular in young $AHR^{-/-}$ mice as compared with their normal littermates. However, the mechanisms for this effect are not yet clear.

A second group has inactivated the AHR, this time by replacing exon 2 with a neomycin-resistance gene under the control of an exogenous promoter, the rationale being that this region of the gene encodes the basic/helix-loop-helix domain essential for receptor dimerization and DNA binding (169). Homozygous null mice displayed no overt phenotypical changes from their wild-type counterparts, litters being found in Mendelian proportions indicating no in utero lethality, although the null mice did exhibit a slower growth rate and thus decreased body weight in the first few weeks of life. As with the previous knockout, CYP1A1 was not inducible in these homozygous null mice. Detailed analysis of the AHR null mice revealed that the livers of these animals were significantly smaller (~25%) than their wild-type littermates, even when overall body weights were equivalent after about 1 month of age. Although the livers of the null mice were visibly different in the first 2 weeks of life, being pale in color and of a spongy texture, these features disappeared by 4 weeks. The authors suggested a metabolic defect in hepatocyte function as the underlying cause for these observations, indicating a temporal requirement for AHR expression. After the first few weeks, most of the AHR null mice began to suffer from portal fibrosis and hypercellularity, and approximately 50% of these animals also possessed enlarged spleens consistent with congestive splenomegaly, which the authors speculated may have been a result of decreased liver blood flow. This study could not distinguish whether the phenotypic changes noted in these AHR mice were a direct result of defective liver growth and maturation, or whether the mice were more susceptible to an unidentified environmental agent as a consequence of the gene deletion. No explanation is immediately apparent for the different phenotype observed from the foregoing two reported studies—genetic background, previously cited in other cases where deletion of the same gene by two groups has led to differing phenotypic consequences—is unlikely to be the cause in this case, because both AHR deletions were on a 129 × C57BL/6 background. Equally, the possibility of partial inactivation, a ''leaky'' deletion, or formation of a new gene product at the targeted allele, seems unlikely because both groups showed the absence of functional AHR protein by a lack of an inductive response of CYP1A1 or CYP1A2 to treatment with TCDD, and in the Schmidt et al. study, by an absence of immunoreactive AHR protein on Western blots.

The studies on transgenic animals, as reviewed in Chapters 4 and 5, demonstrate that these technologies are very powerful for elucidating the physiological and pharmacotoxicological role of drug-metabolizing enzymes and their regulatory paradigms. However, equally they demonstrate that this approach also has pitfalls, because the phenotype obtained by inactivation of a certain enzyme or regulatory protein may be different, dependent on the transgenic line generated.

VI. CONCLUSION AND FUTURE DIRECTIONS

From this chapter it is evident that recombinant ''humanized'' models for xenobiotic metabolism will play a pivotal role in drug development. These systems facilitate the prediction of pharmacokinetic and toxic properties of therapeutics already at the preclinical stage, because they allow the evaluation of metabolic pathways catalyzed by human enzymes. In addition they make it feasible to provide models for predicting the metabolism of lead drugs simply from their structure. However, a very important issue for the future applicability of the various heterologous expression systems for xenobiotic-metabolizing enzymes in drug research and development is their appropriate validation (170,171). How well do these systems mimic certain in vivo drug metabolism pathways and drug–drug interactions? Are the data obtained with the different heterologous expression systems really comparable? To answer these and other related questions it is imperative to compare the different systems within a highly coordinated network of teams with the capacity to evaluate these systems with a wide array of xenobiotics using standardized methods. These are some of the issues addressed in an extensive collaboration between us and 13 pharmaceutical companies.

Besides being extremely useful for drug development, organisms that express drug-metabolizing enzymes will have biotechnological applications as bioreactors for the production of valuable metabolites. Bacterial systems would appear to be extremely useful for this purpose, because they express high levels of recombinant protein and can be grown to relatively high cell densities. In addition,scale-up of these systems is straightforward and inexpensive. However, bacterial membranes may be impermeable to some drugs. The use of bacterial strains with altered membrane permeability, such as the *S. typhimurium* strains used in the Ames test, may overcome this problem.

However the application of recombinant bacteria as bioreactors may involve a demanding separation technology, for the metabolites have to be isolated from large volumes of culture medium that constitutes a complex mixture. As an alternative, recombinant P450s can be electrochemically reduced (172,173). This approach uses the transfer of electrons from an electrode via cobalt (II) sepulchrate to P450, thereby obviating the need for the expensive cofactor NADPH. Initially, this system was established for the fatty acid hydroxylase CYP4A1, and the rates of formation of 12-hydroxydodecanoic acid were similar to those obtained using a reconstituted system containing CYP4A1 and P450 reductase. However, until now the stability of the recombinant P450 in this approach is a dilemma because it is inactivated by the H_2O_2 produced during the electrochemical reduction. Addition of catalase and low oxygen tension ameliorated this problem; however, 30% of the P450 became denatured within the first 30 min of the reaction. The highest concentration of substrate used in this electrocatalytically driven P450 reaction was 1 mM, which most likely is appropriate for the small-scale production of valuable metabolites. Scale-up of this system will certainly require further optimizations.

Major progress in understanding the role of drug-metabolizing enzymes will come from transgenic animal research. Here, the inactivation of genes (gene knockout) is a specially elegant and meaningful approach. However, the inactivation of a gene already in the germline may be problematic, for in some cases no viable offspring is obtained. Even though this offers interesting clues to the role of a gene during development, it does not allow the characterization of the function of a protein in the adult organism. In addition, this technique does not permit the evaluation of the tissue-specific role of an enzyme. Recently, an extremely elegant technique has been devised to solve these obstacles (174–176). In this approach, the gene to be inactivated is recombined in ES cells with a targeting construct to yield a product in which the gene is flanked on each side by *loxP* sequences. These are 34-base–pair recognition elements that on expression of Cre recombinase from bacteriophage P1 recombine, thereby excising the intervening gene. Panels of mice expressing Cre under the control of a variety of useful tissue-specific promoters have been developed for use in this extremely versatile strategy.

These few examples of possible future directions in drug metabolism research should serve to illustrate that several powerful approaches employing transgenes are currently being developed. The use of these in vitro and in vivo systems for studying human drug metabolism is likely to expand, thereby increasing the efficiency and the quality of drug research and development.

REFERENCES

1. C. D. Klaassen, M. O. Amdur, and J. Doull, *Toxicology*, MacMillan, New York, 1986.
2. D. A. Smith, K. Beaumont, N. J. Cussans, M. J. Humphrey, S. G. Jezequel, D. J. Rance,

D. A. Stopher and D. K. Walker Bioanalytical data in decision making: Discovery and development, *Xenobiotica 22*: 1195–205 (1992).

3. F. J. Gonzalez, Human cytochromes P450: Problems and prospects, *Trends Pharmacol. Sci. 12*:346–352 (1992).
4. C. R. Wolf, Metabolic factors in cancer susceptibility. *Cancer Surv. 9*:437–474 (1990).
5. C. T. Eason, F. W. Bonner, and D. V. Parke, The importance of pharmacokinetic and receptor studies in drug safety evaluation, *Regul. Toxicol. Pharmacol. 11*:288–307 (1990).
6. A. M. A. Pfeifer, K. E. Cole, D. T. Smoot, A. Weston, J. D. Groopman, P. G. Shields, J. M. Vignaud, M. Juillerat, M. M. Lipsky, and B. F. Trump, Simian virus-40 large tumor antigen-immortalized normal human liver epithelial-cells express hepatocyte characteristics and metábolize chemical carcinogens, *Proc. Natl. Acad. Sci. USA 90*: 5123–5127 (1993).
7. F. Oesch, D. J. Waxman, J. J. Morrisey, W. Honscha, W. Kissel, and T. Friedberg, Antibodies targeted against hypervariable and constant regions of cytochrome P450IIBI and BII, *Arch. Biochem. Biophys. 279*:167–173 (1989).
8. T. Friedberg, W. Kissel, M. Arand, and F. Oesch, Production of site specific P450 antibodies using recombinant fusion proteins as antigens, *Methods Enzymol. 206*:193–201 (1991).
9. C. L. Crespi, B. W. Penman, D. T. Steimel, H. V. Gelboin, and F. J. Gonzalez, The development of a human cell line stably expressing human CYP3A4: Role in the metabolic activation of aflatoxin B_1 and comparison to CYP1A2 and CYP2A3, *Carcinogenesis 12*:355–359 (1991).
10. B. Bartlomowicz, D. J. Waxman, D. Utesch, F. Oesch, and T. Friedberg, Phosphorylation of carcinogen metabolizing enzymes: Regulation of the phosphorylation status of the major phenobarbital inducible cytochromes P450 in hepatocytes, *Carcinogenesis 10*:225–228 (1989).
11. V. Rogiers and A. Vercruysse, Rat hepatocyte cultures and co-cultures in biotransformation studies of xenobiotics, *Toxicology 82*:193–208 (1993).
12. P. Fernandez-Salguero and F. Gonzalez, Targeted disruption of specific cytochrome P450 and xenobiotic receptor genes. *Methods Enzymol. 212*:412–430 (1996).
13. F. J. Gonzalez, P. Fernandez Salguero, S. S. Lee, T. Pineau, and J. M. Ward, Xenobiotic receptor knockout mice, *Toxicol. Lett. 82–83*:117–121 (1995).
14. F. J. Gonzalez, Use of transgenic animals in carcinogenesis studies, *Mol. Carcinog. 16*:63–67 (1996).
15. J. Sambrook, E. F. Fritsch, and T. Maniatis, *Molecular Cloning: A Laboratory Manual*, 2nd ed., Cold Spring Harbor Laboratory Press, Cold Spring Harbor, NY, 1989.
16. J. L. Yang, V. M. Maher, and J. J. McCormick, Amplification and direct nucleotide sequencing of cDNA from the lysate of low numbers of diploid human cells, *Gene 83*:347–354 (1989).
17. S. Monier, P. VanLuc, G. Kreibich, D. D. Sabatini, and M. Adesnik, Signals for the incorporation and orientation of cytochrome P450 in the endoplasmic reticulum, *J. Cell. Biol. 107*: 457–470 (1988).
18. C. H. Yun, T. Shimada, and F. P. Guengerich, Contributions of human liver cytochrome P450 enzymes to the *N*-oxidation of 4,4′-methylene-bis(2-chloroaniline), *Carcinogenesis 13*: 217–222 (1992).
19. O. Shimozawa, M. Sakaguchi, H. Ogawa, N. Harad, K. Mihara, and T. Omura, Core glycosylation of cytochrome P-450(arom), *J. Biol. Chem. 268*:21399–21402 (1993).
20. M. Sakaguchi, R. Tomiyoshi, T. Kuroiwa, K. Mihara, and T. Omura, Functions of signal and signal anchor sequences are determined by the balance between the hydrophobic segment and the N-terminal charge, *Proc. Natl. Acad. Sci. USA 89*:16–19 (1992).
21. E. Szczesna-Skorupa, N. Browne, D. Mead, and B. Kemper, Positive charges at the NH_2 terminus convert the membrane anchor signal peptide of cytochrome P-450 to a secretory signal peptide, *Proc. Natl. Acad. Sci. USA 85*:738–742 (1988).

22. W. Pyerin, and H. Taniguchi, Phosphorylation of hepatic phenobarbital-inducible cytochrome P-450, *EMBO J. 8*:3003–3010 (1989).
23. J. E. Freeman and C. R. Wolf, Evidence against a role of serine 129 in determining murine cytochrome P450 CYP2E1 protein levels, *Biochemistry 33*:13963–13966 (1994).
24. F. J. Gonzalez, T. Aoyama, and H. V. Gelboin, Expression of mammalian cytochrome-P450 using vaccinia virus, *Methods Enzymol. 206*:85–92 (1991).
25. B. J. Clark and M. R. Waterman, Heterologous expression of mammalian P450 in COS cells, *Methods Enzymol. 206*:100–108 (1991).
26. R. J. Kaufman, Vectors used for expression in mammalian cells. *Methods Enzymol. 185*: 487–511 (1990).
27. M. X. Zuber, E. R. Simpson, and M. R. Waterman, Expression of human 17α-hydroxylase cytochrome P450 cDNA in nonsteroidogenic (COS 1) cells, *Science 234*:1258–1261 (1986).
28. Y. Higashi and Y. Fujii-Kuriyama, Functional analysis of mutant P450(C21) genes in COS cell expression system. *Methods Enzymol. 206*:166–173 (1991).
29. T. Kronbach, T. M. Larabee, and E. F. Johnson, Hybrid cytochromes P-450 identify a substrate binding domain in P450IIC5 and P450IIC4, *Proc. Natl. Acad. Sci. USA 86*:8262–8265 (1989).
30. Z. Luo, Y. He, and J. R. Halpert, Role of residues 363 and 206 in the conversion of P450 2B1 from a steroid 16α-hydroxylase to a 15α-hydroxylase. *Arch. Biochem. Biophys. 309*: 52–57 (1994).
31. F. J. Gonzalez, Molecular biology of human xenobiotic metabolizing enzyme: Role of vaccinia virus cDNA expression in evaluating catalytic function, *Toxicology 82*:77–88 (1993).
32. B. W. Penman, J. Reece, T. Smith, C. S. Yang, H. V. Gelboin, F. J. Gonzalez, and C. L. Crespi, Characterization of a human cell line expressing high levels of cDNA derived CYP2D6, *Pharmacogenetics 3*:28–39 (1993).
33. C. L. Crespi, D. T. Steimel, T. Aoyama, H. V. Gelboin, and F. J. Gonzalez, Stable expression of human cytochrome P450IA2 cDNA in a human lymphoblastoid cell line: Role of the enzyme in the metabolic activation of aflatoxin B_1. *Mol. Carcinog. 3*:5–8 (1990).
34. C. L. Crespi, B. W. Penman, H. V. Gelboin, and F. J. Gonzalez, A tobacco smoke-derived nitrosamine, 4-(methylnitrosamino)-1-(3-pyridyl)-1-butanone, is activated by multiple human cytochrome P450s including the polymorphic human cytochrome P4502D6, *Carcinogenesis 12*:1197–1201 (1991).
35. R. Langenbach, P. B. Smith, and C. Crespi, Recombinant-DNA approaches for the development of metabolic systems used in invitro toxicology, *Mutat. Res. 277*:251–275 (1992).
36. C. L. Crespi, R. Langenbach, and B. W. Penman, Human cell-lines, derived from AHH-1 tk+/− human lymphoblasts, genetically-engineered for expression of cytochromes-P450, *Toxicology 82*:89–104 (1993).
37. K. Mace, F. J. Gonzalez, J. R. McConnell, R. C. Garner, O. Avanti, C. C. Harris, and A. M. A. Pfeifer, Activation of promutagens in a human bronchial epithelial-cell line stably expressing human cytochrome-P450 1A2, *Mol. Carcinog. 11*:65–73 (1994).
38. H. F. Tiano, M. Hosokawa, P. C. Chulada, et al., Retroviral mediated expression of human cytochrome P450 2A6 in C3H 10T1/2 cells confers transformability by 4-(methyl-nitrosamino)-1-(3-pyridyl) 1-butanone (NNK), *Carcinogenesis 14*:1421–1427 (1993).
39. N. Battula, Transduction of cytochrome P3-P450 by retroviruses, constitutive expression of enzymatically active microsomal hemoprotein in animal cells, *J. Biol. Chem. 264*:2991–2996 (1989).
40. L. H. S. Hwang and E. Gilboa, Expression of genes introduced into cells by retroviral infection is more efficient than that of genes introduced into cells by DNA transfection, *J. Virol. 50*:417–424 (1984).
41. Y. Dai, J. Rashba-Step, and A. J. Cederbaum, Stable expression of human cytochrome P4502E1 in HepG2 cells: Characterization of catalytic activities and production of reactive oxygen intermediates. *Biochemistry 32*:6928–6937 (1993).

42. D. Lacroix, M. Desrochers, A. Castonguay, and A. Anderson, Metabolism of 4-(methylnitrosamino)-1-(3-pyridyl)-1-butanone (NNK) in human kidney epithelial cells transfected with rat CYP2B1 cDNA, *Carcinogenesis 14*:1639–1642 (1993).
43. J. C. States, T. H. Quan, R. N. Hines, R. F. Novak, and M. Rungemorris, Expression of human cytochrome-P450 1A1 in DNA-repair deficient and proficient human fibroblasts stably transformed with an inducible expression vector, *Carcinogenesis 14*:1643–1649 (1983).
44. J. Doehmer, V79 Chinese-hamster cells genetically-engineered for cytochrome-P450 and their use in mutagenicity and metabolism studies, *Toxicology 82*:105–118 (1993).
45. J. Doehmer, A. R. Goeptar, and N. P. E. Vermeulen, Cytochromes P450 and drug-resistance, *Cytotechnology 12*:357–366 (1993).
46. W. A. Keown, C. R. Campell, and R. S. Kucherlapati, Methods for introducing DNA into mammalian cells. *Methods Enzymol. 185*:527–537 (1990).
47. R. J. Kaufman, Selection and coamplification of heterologous genes in mammalian cells, *Methods Enzymol. 185*:537–566 (1990).
48. T. Sakaki, K. Oeda, M. Miyoshi, and H. Ohkawa, Characterization of rat liver cytochrome P450MC synthesized in *Sácharomyces cerevisiae, J. Biochem. 98*:167–165 (1985).
49. K. Oeda, T. Sakaki, and H. Ohkawa, Expression of rat liver cytochrome P450MC cDNA in *Saccharomyces cerevisiae, DNA Cell. Biol. 4*:167–175 (1985).
50. C. Kreutzfeldt, and W. Witt, Structural biochemistry, *Biotechnology Handbooks*, Vol. 4, (F. Michael, and S. G. Oliver, eds.), Plenum Press, New York, pp. 5–58. 1991.
51. M. A. Romanos, C. A. Scorer, and J. J. Clare, Foreign gene expression in yeast: A review, *Yeast 8*:423–488 (1992).
52. I. Campbell and J. H. Duffus, Yeast: A practical approach. *Practical Approach Series*, (D. Rickwood, and D. Hames, eds.), IRL, Washington, DC, 1988.
53. S. D. Emr, Heterologous gene expression in yeast. *Methods Enzymol. 185*:231–279 (1990).
54. P. Urban, G. Truan, A. Bellamine, R. Laine, J. C. Gautier, and D. Pompon, Engineered yeast simulating P450-dependent metabolism: Tricks, myths and reality, *Drug Metabolism and Interactions*, (N. Kingsley, ed.), Freund Publishing, London, pp. 169–200 1994.
55. C. Cullin and D. Pompon, Synthesis of functional cytochromes P450 P1 and chimeric P-450 P3-1 in the yeast *Saccharomyces cerevisiae, Gene 65*:203–217 (1988).
56. I. M. Kidd and V. C. Emery, The use of baculovirus as expression vectors, *Appl. Biochem. Biotechnol. 42*:137–159 (1993).
57. D. O'Reilly, L. K. Miller, and V. A. Luckov, *Baculovirus Expression Vectors: A Laboratory Manual*. (D. O'Reilly, L. K. Miller, and V. A. Luckov, ed.), Freeman New York, 1992.
58. D. H. L. Bishop, Baculovirus based expression systems. *Membrane Protein Expression Systems: A User Guide* (G. W. Gould, ed.), Portland Press, London, pp. 84–124. 1994.
59. T. C. Peakman, R. A. Harris, and D. R. Gewert, Highly efficient generation of recombinant baculovirus by enzymatically mediated site specific in vitro recombination. *Nucleic Acids Res. 20*:495–500 (1992).
60. C. A. Lee, T. A. Kost, and C. Serabjit-Singh, Recombinant baculovirus strategy for coexpression of functional human cytochrome P450 and P450 reductase, *Methods in Enzymol. 272*: 86–95 (1996).
61. T. P. Simula, M. J. Glancey, and C. R. Wolf, Human glutathione *S*-transferase-expressing *Salmonella-typhimurium* tester strains to study the activation detoxification of mutagenic compounds—studies with halogenated compounds, aromatic-amines and aflatoxin-b(1), *Carcinogenesis 14*:1371–1376 (1993).
62. L. Gold, Expression of heterologous proteins in *Escherichia coli, Methods Enzymol. 185*: 11–37 (1990).
63. H. J. Barnes, Maximizing expression of eukaryotic cytochrome P450s in *Escherichia coli, Methods Enzymol. 272*:3–14 (1996).
64. P. Balbas and F. Bolivar, Design and construction of expression plasmid vectors in *Escherichia coli, Methods Enzymol. 185*:14–37 (1990).

65. M. J. Gething and J. Sambrook, Protein folding in the cell. *Nature 355*:33–45 (1992).
66. M. Taddayon, J. R. Gittins, J. M. Pratt, and J. K. Broome-Smith, Expression of membrane proteins in *Escherichia coli, Membrane Protein Expression Systems*, (G. W. Gould, ed.), London, Portland Press, London, 1992, pp. 30–81.
67. J. R. Larson, M. J. Coon, and T. D. Porter, Alcohol inducible cytochrome P450IIE1 lacking the hydrophobic NH_2-terminal segment retains catalytic activity and is membrane bound when expressed in *Escherichia coli, J. Biol. Chem. 266*:7321–7324 (1991).
68. Y. Chun, and J. Y. L. Chiang, The expression of a catalytically active cholesterol 7α-hydroxylase cytochrome P450 in *Escherichia coli, J. Biol. Chem. 266*:19186–19191 (1991).
69. H. J. Barnes, M. F. Arlotto, and M. R. Waterman, Expression and enzymatic activity of recombinant cytochrome P450 17α-hydroxylase in *Escherichia coli, Proc. Natl. Acad. Sci. USA 88*:5597–5601 (1991).
70. E. M. J. Gillam, T. Baba, B. R. Kim, Ohmori, S. and F. P. Guengerich, Expression of modified human cytochrome P450 3A4 in *Escherichia coli* and purification and reconstitution of the enzyme, *Arch. Biochem. Biophys. 305*:123–131 (1993).
71. P. Sandhu, T. Baba, and F. P. Guengerich, Expression of modified cytochrome P450 2C10-(2C9) in *Escherichia coli*, purification, and reconstitution of catalytic activity, *Arch. Biochem. Biophys. 306*:443–450 (1993).
72. T. H. Richardson F. Jung, K. J. Griffin, M. Wester, J. L. Raucy, B. Kemper, L. M. Bornheim, C. Hassett, C. J. Omiecinski, and E. F. Jonhson, A universal approach for the expression of human and rabbit cytochrome P450s of the 2C subfamily in *Escherichia coli, Arch. Biochem. Biophys. 323*:87–96 (1995).
73. H. J. Barnes, C. M. Jenkins, and M. R. Waterman, Baculovirus expression of bovine cytochrome P450c 17 in Sf9 cells and comparison with expression in yeast, mammalian cells, and *E. coli, Arch. Biochem. Biophys. 315*:489–494 (1994).
74. D. Pompon, B. Louerat, A. Bronine, and P. Urban, Yeast expression of animal and plant P450s in optimized redox environments, *Methods Enzymol. 272*:51–64 (1996).
75. D. Pompon, G. Truan, A. Bellamine and P. Urban, Expression of cytochromes P450 in yeast: Practical aspects; *Assessment of the Use of Single Cytochrome P450 Enzymes in Drug Research* (M. R. Waterman, and M. Hildebrand, eds.) Springer, Berlin, pp. 97–110.
76. T. Aoyama, S. Yamano, D. Waxman, D. Lapenson, U. A. Meyer, V. Fischer, R. Tyndale, T. Inaba, W. Kalow, H. V. Gelboin, and F. J. Gonzalez, Cytochrome P-450hPCN3, a novel cytochrome P450 IIIA gene product that is differentially expressed in adult human liver, *J. Biol. Chem. 264*:10388–10395 (1989).
77. T. Shimada, H. Yamazaki, M. Mimura, Y. Inui, and F. P. Guengerich, Interindividual variations in human liver cytochrome P-450 enzymes involved in the oxidation of drugs, carcinogens and toxic chemicals: Studies with liver microsomes of 30 Japanese and 30 Caucasians, *J. Pharmacol. Exp. Ther. 270*:414–423 (1994).
78. B. Amarneh and E. R. Simpson, Expression of a recombinant derivative of human aromatase P450 in insect cells utilizing the baculovirus vector system, *Mol. Cell Endocrinol. 109*:R1–R5 (1995).
79. S. Modi, M. J. Paine, M. J. Sutcliff, L. Y. Lian, W. U. Primrose, C. R. Wolf, and G. C. K. Roberts, A model for human cytochrome P450 2D6 based on homology modelling and NMR studies of substrate binding, *Biochemistry 35*:4540–4550 (1996).
80. U. Scheller, R. Kraft, K. L. Schroder, and W. H. Schunck, Generation of the soluble and functional cytosolic domain of microsomal cytochrome P450 52A3. *J. Biol. Chem. 269*: 12779–12783 (1994).
81. T. Sueyoshi, L. J. Park, R. Moore, R. O. Juvonen, and M. Negishi, Molecular engineering of microsomal P450 2a-4 to a stable, water soluble enzyme, *Arch. Biochem. Biophys. 322*: 265–271 (1995).
82. A. C. Kempf, U. M. Zanger, and U. A. Meyer, Truncated human P450 2D6: Expression in

Escherichia coli, Ni($^{2+}$)-chelate affinity purification, and characterization of solubility and aggregation, *Arch. Biochem. Biophys. 321*:277–288 (1995).

83. C. A. Hasemann, R. G. Kurumbail, S. S. Boddupalli, J. A. Peterson, and J. Deisenhofer, Structure and function of cytochromes P450: A comparative analysis of three crystal structures, *Structure 3*:41–62 (1995).
84. K. G. Ravichandran, S. S. Boddupalli, C. A. Haseman, J. A. Peterson, and J. Deisenhofer, Crystal structure of hemoprotein domain of P450BM-3, a prototype for microsomal P450s, *Science 261*:731–736 (1993).
85. G. D. Szklarz, Y. A. He, and J. R. Halpert, Site-directed mutagenesis as a tool for molecular modeling of cytochrome P450 2B1, *Biochemistry 34*:14312–14322 (1995).
86. R. Mackman, R. A. Tschirret-Guth, G. Smith, G. P. Hayhurst, S. W. Ellis, M. S. Lennard, G. T. Tucker, C. R. Wolf, and P. R. deMontellano, Active site topologies of human CYP2D6 and its aspartate-301 to glutamate, asparagine, and glycine mutants, *Arch. Biochem. Biophys. 331*:134–140 (1996).
87. R. L. P. Lindberg and M. Negishi, Alteration of mouse cytochrome P450 coh substrate specificity by mutation of a single amino-acid residue, *Nature 339*:632–634 (1989).
88. D. Lin, L. H. Zhang, E. Chiao, and W. L. Miller, Modeling and mutagenesis of the active site of human P450c 17, *Mol. Endocrinol. 8*:392–402 (1994).
89. Y. He, Z. Luo, P. A. Klekotka, V. L. Burnett, and J. R. Halpert, Structural determinants of cytochrome P450 2B1 specificity: Evidence for five substrate recognition sites, *Biochemistry 33*:4419–4424 (1994).
90. A. Wada and M. R. Waterman, Identification by site-directed mutagenesis of two lysine residues in cholesterol side chain cleavage cytochrome P450 that are essential for adrenodoxin binding, *J. Biol. Chem. 267*:22877–22882 (1992).
91. L. S. Kaminsky, S. M. de Morais, M. B. Faletto, D. A. Dunbar, and J. A. Goldstein, Correlation of human cytochrome P4502C substrate specificities with primary structure: Warfarin as a probe, *Mol. Pharmacol. 43*:234–239 (1993).
92. S. W. Ellis, G. P. Hayhurst, G. Smith, T. Lightfoot, M. M. S. Wong, A. P. Simula, M. J. Ackland, M. J. Sternberg, M. S. Lennard, G. T. Tucker, and C. R. Wolf, Evidence that aspartic acid 301 is a critical substrate-contact residue in the active site of cytochrome P450 2D6, *J. Biol. Chem. 270*:29055–29058 (1995).
93. G. R. Harlow and J. R. Halpert, Mutagenesis study of Asp-290 in cytochrome P450 2B11 using a fusion protein with rat NADPH-cytochrome P450 reductase. *Arch. Biochem. Biophys. 326*:85–92 (1996).
94. J. T. M. Buters, K. R. Korzekwa, K. L. Kunze, Y. Omata, J. P. Hardwick, and F. J. Gonzalez, cDNA-directed expression of human cytochrome-P450 CYP3A4 using baculovirus, *Drug Metab. Dispos. 22*:688–692 (1994).
95. C. M. Jenkins and M. R. Waterman, Flavodoxin and NADPH-flavodoxin reductase from *Escherichia coli* support bovine cytochrome P450c17 hydroxylase activities, *J. Biol. Chem. 269*:27401–27408 (1994).
96. T. Shimada, E. M. J. Gillam, P. Sandhu, Z. Y. Guo, R. H. Tukey, and F. P. Guengerich, Activation of procarcinogens by human cytochrome P450 enzymes expressed in *Escherichia coli*—simplified bacterial systems for genotoxicity assays, *Carcinogenesis 15*:2523–2529 (1994).
97. H. Murakami, Y. Yabusaki, T. Sakaki, M. Shibata, and H. Ohkawa, A genetically engineered monooxygenase: Construction of the functional fused enzme between rat cytochrome P450c and NADPH-cytochrome P450 reductase, *DNA 6*:189–197 (1987).
98. Y. Yabusaki, Artificial P450/reductase fusion enzymes: What can we learn from their structures? *Biochimie 77*,594–603 (1995).
99. C. W. Fisher, M. S. Shet, D. L. Caudle, W. C. Martin, and R. W. Estabrook, High-level expression in *Escherichia coli* of enzymatically active fusion proteins containing the domains of mammalian cytochromes P450 and NADPH-P450 reductase flavoprotein, *Proc. Natl. Acad. Sci. USA 89*:10817–10821 (1992).

100. C. W. Fisher, M. Shet, and R. W. Estabrook, Construction of plasmids and expression in *Escherichia coli* of enzymatically active fusion proteins containing the heme domain of a P450 linked to NADPH-P450 reductase, *Methods Enzymol. 272*:15–25 (1996).
101. M. S. Shet, C. W. Fisher, P. L. Holmans, and R. W. Estabrook, Human cytochrome P450 3A4: Enzymatic properties of a purified recombinant fusion protein containing NADPH-P450 reductase. *Proc. Natl. Acad. Sci. USA 90*:11748–11752 (1993).
102. Y. J. Chun, T. Shimada, and F. P. Guengerich, Construction of a human cytochrome P450 1A1:rat NADPH-cytochrome P450 reductase fusion protein cDNA and expression in *Escherichia coli*, purification, and catalytic properties of the enzyme in bacterial cells and after purification, *Arch. Biochem. Biophys. 330*:48–58 (1996).
103. J. A. R. Blake, M. Pritchard, S. Ding, G. C. M. Smith, B. Burchell, C. R. Wolf, and T. Friedberg, Coexpression of a human P450 (CYP3A4) and P450 reductase generates a highly functional monooxygenase system in *Escherichia coli, FEBS Lett. 397*:210–214 (1996).
104. J. Doehmer and F. Oesch, V79 Chinese-hamster cells genetically engineered for stable expression of cytochromes-P450. *Methods Enzymol. 206*:117–123 (1991).
105. D. J. Waxman, J. J. Lapenson, S. S. Morrisey, H. V. Park, J. Gelboin, J. Doehmer, and F. Oesch, Androgen hydroxylation catalyzed by a cell line (SD1) that stably expresses cytochrome P450 PB-4 (IIB1), *Biochem. J. 260*:81–85 (1989).
106. W. A. Schmalix, D. Lang, A. Schneider, R. Bocker, H. Greim, and J. Doehmer, Stable expression and coexpression of human cytochrome-P450 oxidoreductase and cytochrome-P450 1A2 in V79 Chinese-hamster cells—sensitivity to quinones and biotransformation of 7-alkoxyresorufins and triazines, *Drug Metab. Dispos. 24*:1314–1319 (1996).
107. T. Aoyama, K. Nagata, Y. Yamazoe, R. Kato, E. Matsunaga, H. V. Gelboin, and F. J. Gonzalez, Cytochrome b_5 potentiation of cytochrome P-450 catalytic activity demonstrated by a vaccinia virus mediated in situ reconstitution system; *Proc. Natl. Acad. Sci. USA 87*: 5425–5429 (1990).
108. H. Yamazaki, Z. Guo, M. Persmark, M. Mimura, K. Inoue, F. P. Guengerich, and T. Shimada, Bufuralol hydroxylation by cytochrome P450 2D6 and 1A2 enzymes in human liver microsomes; *Mol. Pharmacol. 46*:568–577 (1994).
109. S. W. Ellis, M. S. Ching, P. F. Watson, C. J. Henderson, A. P. Simula, M. S. Lennard, G. T. Tucker, and H. F. Woods, Catalytic activities of human debrisoquine 4-hydroxylase cytochrome P450 (CYP2D6) expressed in yeast. *Biochem. Pharmacol. 44*:617–620 (1992).
110. S. M. Black, S. Ellard, J. M. Parry, and C. R. Wolf, Increased sterigmatocystin-induced mutation frequency in *Saccharomyces cerevisiae* expressing cytochrome-P450 CYP2B1, *Biochem. Pharmacol. 43*:374–376 (1992).
111. C. L. Crespi, Expression of cytochrome-P450 cDNAs in human B lymphoblastoid-cells—applications to toxicology and metabolite analysis, *Methods Enzymol. 206*:123–129 (1991).
112. C. Sengstag and F. E. Würgler, DNA recombination induced by aflatoxin B_1 activated by cytochrome P450 1A enzymes, *Mol. Carcinog. 11*:227–235 (1994).
113. T. P. Simula, M. B. Crichton, S. M. Black, S. Pemble, H. F. J. Bligh, J. D. Beggs, and C. R. Wolf, Heterologous expression of drug-metabolizing-enzymes in cellular and whole animal models, *Toxicology 82*:3–20 (1993).
114. P. D. Josephy, L. S. Debruin, H. L. Lord, J. N. Oak, D. H. Evans, Z. Y. Guo, M. S. Dong, and F. P. Guengerich, Bioactivation of aromatic-amines by recombinant human cytochrome P4501A2 expressed in Ames tester strain bacteria—a substitute for activation by mammalian tissue preparations, *Cancer Res. 55*:799–802 (1995).
115. T. Shimada and F. P. Guengerich, Evidence for cytochrome P450-Nf the nifedipine oxidase, being the principal enzyme involved in the activation of aflatoxins in human liver, *Proc. Natl. Acad. Sci. USA 86*:462–465 (1990).

116. L. M. Forrester, G. E. Neal, D. Judah, M. J. Glancey, and C. R. Wolf, Evidence for the involvement of multiple forms of cytochrome P-450 in aflatoxin B1 metabolism in human liver. *Proc. Natl. Acad. Sci. USA 87*:8306–8310 (1990).
117. T. Aoyama, S. Yamano, P. S. Guzelian, H. V. Gelboin, and F. J. Gonzalez, Five of 12 forms of vaccinia virus-expressed human hepatic cytochrome P450 metabolically activate aflatoxin B1. *Proc. Natl. Acad. Sci. USA 87*:4790–4793 (1990).
118. Y. F. Ueng, T. Shimada H. Yamazaki, and F. P. Guengerich, Oxidation of aflatoxin B_1 by bacterial recombinant human cytochrome P450 enzymes, *Chem. Res. Toxicol. 8*:218–225 (1995).
119. E. P. Gallagher, L. C. Wienkers, P. L. Stapleton, K. L. Kunze, and D. L. Eaton, Role of human microsomal and human complementary DNA-expressed cytochrome-P4501A2 and cytochrome-P4503A4 in the bioactivation of aflatoxin-b(1), *Cancer Res. 54*:101–108 (1994).
120. R. P. Remmel and B. Burchell, Validation and use of cloned, expressed human drug-metabolizing enzymes in heterologous cells for analysis of drug metabolism and drug–drug interactions, *Biochem. Pharmacol. 46*:559–566 (1993).
121. T. Jowett, M. F. F. Wajidi, E. Oxtoby, and C. R. Wolf, Mammalian genes expressed in *Drosophila*: A transgenic model for the study of mechanisms of chemical mutagenesis and metabolism, *EMBO J. 10*:1075–1081 (1991).
122. M. Komori, R. Kitamura, H. Fukuta, H. Inoue, H. Baba, K. Yoshikawa, and T. Kamataki, Transgenic *drosophila* carrying mammalian cytochrome-P-4501A1—an application to toxicology testing, *Carcinogenesis 14*:1683–1688 (1993).
123. S. Y. Fuchs, V. S. Spiegelman, R. D. Safaev, and G. A. Belitsky, Xenobiotic-metabolizing enzymes and benzo[*a*]pyrene metabolism in the benzo[*a*]pyrene-sensitive mutant strain of *Drosophila simulans, Mutat. Res. 269*:185–191 (1992).
124. A. L. Boyd and D. Samid, Molecular biology of transgenic animals, *J. Anim. Sci. 71*:1–9 (1993).
125. G. Kollias and F. Grosveld, The study of gene regulation in transgenic mice, *Transgenic Animals*, (G. Kollias and F. Grosveld, eds.), Academic Press, London, 1992, pp. 79–99.
126. D. Valerio, Retrovirus vectors for gene therapy procedures, *Transgenic Animals*, (G. Kollias and F. Grosveld, eds.), Academic Press, London, 1992, pp. 211–246.
127. M. X. Wei, T. Tamiya, M. Chase, E. J. Boviatsis, T. K. H. Chang N. W. Kowall, F. H. Hochberg, D. J. Waxman, X. O. Breakefield, and E. A. Chiocca, Experimental tumor therapy in mice using the cyclophosphamide-activating cytochrome P450 2B1 gene, *Hum. Gene Ther. 5*:969–978 (1994).
128. T. Pineau, P. Fernandez-Salguero S. S. T. Lee, T. McPhail, J. M. Ward, and F. J. Gonzalez, Neonatal lethality associated with respiratory distress in mice lacking cytochrome P450 1A2, *Proc. Natl. Acad. Sci. USA 92*:5134–5138 (1995).
129. H. C. Liang, H. Li, R. A. McKinnon, J. J. Duffy, S. S. Potter, A. Puga and D. W. Nebert, Cypla2(−/−) null mutant mice develop normally but show deficient drug metabolism, *Proc. Natl. Acad. Sci. USA 93*:1671–1676 (1996).
130. S. S. Lee, J. T. Buters, T. Pineau, Fernandez P. Salguero, and F. J. Gonzalez, Role of CYP2E1 in the hepatotoxicity of acetaminophen, *J. Biol. Chem. 271*:12063–12067 (1996).
131. S. Morimura, T. Suzuki, S. Hochi, A. Yuki, K. Nomura, T. Kitagawa, I. Nagatsu, M. Imagawa, and M. Muramatsu, Trans-activation of glutathione transferase P gene during chemical hepatocarcinogenesis of the rat. *Proc. Natl. Acad. Sci. USA 90*:2065–2068 (1993).
132. T. Kora, M. Sugimoto, and K. Ito, Cellular distribution of glutathione *S*-transferase P gene expression during rat hepatocarcinogenesis by diethylnitrosamine. *Int. Hepatol. Commun 5*: 266–273 (1996).
133. K. H. Cowan, G. Batist, A. Tulpule, B. K. Sinha, and C. E. Myers, Similar biochemical changes associated with multidrug resistance in human breast cancer cells and carcinogen-induced resistance to xenobiotics in rats, *Proc. Natl. Acad. Sci. USA 83*:9328–9332 (1986).

134. C. J. Wareing, S. M. Black, J. D. Hayes, and C. R. Wolf, Increased levels of alpha-class and pi-class glutathione *S*-transferases in cell lines resistant to 1-chloro-2,4-dinitrobenzene, *Eur. J. Biochem. 217*:671–676 (1993).
135. S. M. Black and C. R. Wolf, The role of glutathione-dependent enzymes in drug resistance, *Pharmacol. Ther. 51*:139–154 (1991).
136. J. A. Green, L. J. Robertson, and A. H. Clark, Glutathione *S*-transferase expression in benign and malignant ovarian tumours, *Br. J. Cancer 68*:235–239 (1993).
137. S. Hamada, M. Kamada, H. Furumoto, T. Hirao, and T. Aono, Expression of glutathione *S*-transferase-pi in human ovarian cancer as an indicator of resistance to chemotherapy, *Gynecol. Oncol. 52*:313–319 (1994).
138. T. P. Mulder, H. W. Verspaget, C. F. Sier, H. M. Roelofs, S. Ganesh, G. Griffioen, and W. H. Peters, Glutathione *S*-transferase pi in colorectal tumors is predictive for overall survival, *Cancer Res. 55*:2696–2702 (1995).
139. F. Bai, Y. Nakanishi, M. Kawasaki, K. Takayama, J. Yatsunami, X. H. Pei, N. Tsuruta, K. Wakamatsu, and N. Hara, Immunohistochemical expression of glutathione *S*-transferase-pi can predict chemotherapy response in patients with nonsmall cell lung carcinoma, *Cancer 78*:416–421 (1996).
140. T. Hida, M. Kuwabara, Y. Ariyoshi, T. Takahashi, T. Sugiura, K. Hosoda, Y. Niitsu, and R. Ueda, Serum glutathione *S*-transferase-pi level as a tumor marker for non-small cell lung cancer. Potential predictive value in chemotherapeutic response, *Cancer 73*:1377–1382 (1994).
141. S. V. Singh, B. H. Xu, G. T. Tkalcevic, V. Gupta, B. Roberts, and P. Ruiz, Glutathione-linked detoxification pathway in normal and malignant human bladder tissue, *Cancer Lett. 77*:15–24 (1994).
142. D. Schipper, W. Verspaget, T. P. J. Mulder, C. F. M. Sier, S. Ganesh, H. J. M. Roelofs, and W. H. M. Peters, Correlation of glutathione *S*-transferases with overall survival in patients with gastric carcinoma, *Int. J. Oncol. 9*:357–363 (1996).
143. A. Katagiri, Y. Tomita, T. Nishiyama, M. Kimura, and Sato S. Immunohistochemical detection of P-glycoprotein and GSTP1-1 in testis cancer. *Br. J. Cancer 68*:125–129 (1993).
144. D. J. Grignon, M. Abdel Malak, W. C. Mertens, W. A. Sakr, and R. R. Shepherd, Glutathione *S*-transferase expression in renal cell carcinoma: A new marker of differentiation, *Mod. Pathol. 7*:186–189 (1994).
145. W. H. Lee, R. A. Morton, J. I. Epstein, J. D. Brooks, P. A. Campbell, G. S. Bova, W. S. Hsieh, W. B. Isaacs, and W. G Nelson, Cytidine methylation of regulatory sequences near the pi-class glutathione *S*-transferase gene accompanies human prostatic carcinogenesis, *Proc. Natl. Acad. Sci. USA 91*:11733–11737 (1994).
146. M. Cookson, V. E. Reuter, I. Linkov, and W. R. Fair, Glutathione *S*-transferase pi class expression by immunohistochemistry in benign and malignant prostate tissue. *J. Urol. 157*: 673–676 (1997).
147. T. K. Bammler, C. A. Smith, and C. R. Wolf, Isolation and characterization of two mouse Pi-class glutathione *S*-transferase genes, *Biochem. J. 298*:385–390 (1994).
148. T. K. Bammler, H. Driessen, N. Finnstrom, and C. R. Wolf, Amino acid differences at positions 10, 11, and 104 explain the profound catalytic differences between two murine pi-class glutathione *S*-transferases, *Biochemistry 34*:9000–9008 (1995).
149. C. J. Patten, P. E. Thomas, R. L. Guy, M. Lee F. J. Gonzalez, F. P. Guengerich, and C. S. Yang, Cytochrome P450 enzymes involved in acetaminophen activation by rat and human liver microsomes and their kinetics, *Chem. Res. Toxicol. 6*:511–518 (1993).
150. K. E. Thummnel, C. A. Lee, K. L. Kunze, S. D. Nelson, and J. T. Slattery, Oxidation of acetaminophen to *N*-acetyl-*p*-aminobenzoquinone imine by human CYP3A4, *Biochem. Pharmacol. 45*:1563–1569 (1993).
151. J. J. Hu, M. J. Lee, M. Vapiwala, K. Reuhl, P. E. Thomas, and C. S. Yang, Sex-related differences in mouse renal metabolism and toxicity of acetaminophen, *Toxicol. Appl. Pharmacol. 122*:16–26 (1993).

152. C. J. Henderson, A. R. Scott, C. S. Yang, and C. R. Wolf, Testosterone-mediated regulation of mouse renal cytochrome P-450 isoenzymes. *Biochem. J. 266*:675–681 (1990).
153. B. Coles, I. Wilson, P. Wardman, J. A. Hinson, S. D. Nelson, and B. Ketterer, The spontaneous and enzymatic reaction of *N*-acetyl-*p*-benzoquinonimine with glutathione: A stopped-flow kinetic study, *Arch. Biochem. Biophys. 264*:253–260 (1988).
154. S. S. T. Lee, T. Pineau, J. Drago, E. J. Le, J. W. Owens, D. L. Kroetz, P. M. Fernandez-Salguero, H. Westphal, and F. J. Gonzalez, Targeted disruption of the α isoform of the peroxisome proliferator-activated receptor gene in mice results in abolishment of the pleiotropic effects of peroxisome proliferators, *Mol. Cell. Biol. 15*:3012–3022 (1995).
155. H. Remmer and H. J. Merker, Drug induced changes in the liver endoplasmic reticulum: Association with drug metabolizing enzymes, *Science 142*:1657–1658 (1963).
156. A. B. Okey, Enzyme induction in the cytochrome P450 system. *Pharmacogenetics of Drug Metabolism* (W. Kalow, ed.), Pergamon Press, New York, 1992. pp. 549–608.
157. D. R. Thakker, H. Yagi, W. Levin, A. W. Wood, A. H. Conney, and D. M. Jerina, Polycyclic aromatic hydrocarbons: Metabolic activation to ultimate carcinogens *Bioactivation of Foreign Compounds* (M. W. Anders, ed.), Academic Press, San Diego, 1985, pp. 177–242.
158. F. J. Gonzalez, S. Y. Liu, and M. Yano, Regulation of cytochrome-P450 genes—molecular mechanisms, *Pharmacogenetics 3*:51–57 (1993).
159. F. J. Gonzalez, R. Skoda, J. P. Hardwick, B. J. Song, M. Umeno, O. W. McBride, C. Kozak, E. Matsunaga, T. Matsunaga, S. Kinura, S. S. Park, C. S. Yang, D. W. Nebert, H. V. Gelboin, and U. A. Meyer, (1988) Human and rat debrisoquine and ethanol inducible gene families: Structure, regulation and polymorphisms, *Microsomes and Drug Oxidations* (J. O. Miners, D. J. Birkett, R. Drew, and M. McManus, eds.), Taylor and Francis, London, 1988, pp. 209–215.
160. L. Corcos, and M. C. Weiss, Phenobarbital, dexamethasone and benzanthracene induce several cytochrome P450 mRNAs in rat hepatoma cells, *FEBS Lett. 233*:37–40 (1988).
161. A. K. Jaiswal, T. Haaparanta, P. V. Luc, J. Schembri, and M. Adesnik, (1990) Glucocorticoid regulation of a phenobarbital-inducible cytochrome P450 gene: the presence of a functional glucocorticoid response element in the 5′ flanking region of the CYP2B2 gene. *Nucleic Acids Res. 18*:4237–4242 (1990).
162. P. M. Shaw, M. Adesnik, M. C. Weiss, and L. Corcos, The phenobarbital induced transcriptional activation of cytochrome P-450 genes is blocked by the glucocorticoid-progesterone antagonist RU486, *Mol. Pharmacol. 44*:775–783 (1993).
163. M. T. Donato, F. Goethals, M. J. Gomezlechon, D. Deboyser, I. Decoster, M. Roberfroid, and J. V. Castell Toxicity of the antitumoral drug datelliptium in hepatic cells—use of models in vitro for the prediction of toxicity in vivo. *Toxicol In Vitro 6*:295–302 (1992).
164. Z. Abdelrazzak, L. Corcos, J. P. Campion, and A. Guillouzo, Transforming growth-factor-beta-1 down-regulates basal and polycyclic aromatic hydrocarbon-induced cytochromes p-450 1a1 and 1a2 in adult human hepatocytes in primary culture, *Mol Pharmacol 46*:1100–1110 (1994).
165. A. Guillouzo, F. Morel, O. Fardel, and B. Meunier, Use of human hepatocyte cultures for drug metabolism studies, *Toxicology 82*:209–219 (1993).
166. R. Ramsden, K. M. Sommer, and C. J. Omiecinski, Phenobarbital induction and tissue-specific expression of the rat *CYP2B2* gene in transgenic mice, *J. Biol. Chem. 268*:21722–21726 (1993).
167. S. N. Jones, P. G. Jones, H. Ibarguen, C. T. Caskey, and W. J. Craigen, Induction of the CYP1A1 dioxin-responsive enhancer in transgenic mice, *Nucleic Acids Res. 19*:6547–6551 (1991).
168. P. Fernandez-Salguero, T. Pineau, D. M. Hilbert, T. McPhail, S. S. T. Lee, S. Kimura, D. W. Nebert, S. Rudikoff, J. M. Ward, and F. J. Gonzalez, Immune system impairment and hepatic fibrosis in mice lacking the dioxin binding Ah receptor, *Science 268*:722–726 (1995).
169. J. V. Schmidt, G. H. Su, J. K. Reddy, M. C Simon, and C. A. Bradfield, Characterization

of a murine Ahr null allele: involvement of the Ah receptor in hepatic growth and development, *Proc. Natl. Acad. Sci. USA 93*:6731–6736 (1996).

170. M. H. Tarbit, M. K. Bayliss, D. Herriott, S. R. Hood, J. L. Hutson, G. R. Park, and C. J. Serabjit-Singh, Applications of molecular biology and in vitro technology to drug metabolism studies: An industrial perspective, *Biochem. Soc. Trans. 21*:1018–1023 (1993).
171. A. D Rodrigues, Use of in vitro human metabolism studies in drug development, *Biochem. Pharmacol. 48*:2147–2156 (1994).
172. K. M. Faulkner, M. S. Shet, C. W. Fisher, and R. W. Estabrook, Electrocatalytically driven omega-hydroxylation of fatty acids using cytochrome P450 4A1, *Proc. Natl. Acad. Sci. USA 92*:7705–7709 (1995).
173. R. W. Estabrook, K. M. Faulkner, M. S. Shet, and C. W. Fisher, Application of electrochemistry for P450-catalyzed reactions, *Methods Enzymol. 272*:44–51 (1996).
174. J. Rossant and A. Nagy, Genome engineering: the new mouse genetics. *Nature Med. 1*:592–594 (1995).
175. C. A. Chambers, TKO'ed: Lox, stock and barrel, *BioEssays 16*:865–867 (1994).
176. H. Gu, Y. R. Zou, and K. Rajewski, Independent control of immunological switch recombination at individual switch regions evidenced through Cre-*loxP* mediated gene targeting, *Cell 73*:1155–1164 (1993).
177. M. Kagimoto, M. Heim, K. Kagimoto, T. Zeugin, U. A. Meyer, Multiple mutations of the human cytochrome P450IID6 gene (CYP2D6) in poor metabolizers of debrisoquine. Study of the functional significance of individual mutations by expression of chimeric genes. *J. Biol. Chem. 265*:17209–17214 (1990).
178. E. Matsunaga, U. M. Zanger, J. P. Hardwick, H. V. Gelboin, U. A. Meyer, and F. J. Gonzalez, The CYP2D gene subfamily: Analysis of the molecular basis of the debrisoquine 4-hydroxylase deficiency in DA rats, *Biochemistry 28*:7349–7355 (1989).
179. I. Johansson, M. Oscarson, Q. Y. Yue, L. Bertilsson, F. Sjoqvist, and S. M. Ingelman, Genetic analysis of the Chinese cytochrome P4502D locus: Characterization of variant CYP2D6 genes present in subjects with diminished capacity for debrisoquine hydroxylation, *Mol. Pharmacol. 46*:452–459 (1994).
180. R. Tyndale, T. Aoyama, F. Broly, T. Matsunaga, T. Inaba, W. Kalow, H. V. Gelboin, U. A. Meyer, and F. J. Gonzalez, Identification of a new variant CYP2D6 allele lacking the codon encoding Lys-281: Possible association with the poor metabolizer phenotype. *Pharmacogenetics 1*:26–32 (1991).
181. M. J. I. Paine, D. Gilham, G. C. K. Roberts, and C. R. Wolf, Functional high-level expression of cytochrome-P450 CYP2D6 using baculoviral expression systems, *Arch. Biochem. Biophys. 328*:143–150 (1996).
182. E. M. Gillam, Z. Guo, M. V. Martin, C. M. Jenkins, and F. P. Guengerich, Expression of cytochrome P450 2D6 in *Escherichia coli*, purification, and spectral and catalytic characterization, *Arch. Biochem. Biophys. 319*:540–550 (1995).
183. E. Y. Krynetski, V. L. Drutsa, I. E. Kovaleva, and V. N. Luzikov, High yield expression of functionally active human liver CYP2D6 in yeast cells. *Pharmacogenetics 5*:103–109 (1995).
184. K. Rowland, W. W. Yeo, S. W. Ellis, I. G. Chadwick, I. Haq, M. S. Lennard, P. R. Jackson, L. E. Ramsay, and G. T. Tucker, Inhibition of CYP2D6 activity by treatment with propranolol and the role of 4-hydroxy propranolol, *Br. J. Clin. Pharmacol. 38*:9–14 (1994).
185. D. S. Mautz, W. L. Nelson, and D. D Shen, Regioselective and stereoselective oxidation of metoprolol and bufuralol catalyzed by microsomes containing cDNA-expressed human P4502D6, *Drug. Metab. Dispos. 23*:513–517 (1995).
186. S. Botsch, J. C. Gautier, P. Beaune M. Eichelbaum, and H. K. Kroemer, Identification and characterization of the cytochrome-P450 enzymes involved in *n*-dealkylation of propafenone—molecular-base for interaction potential and variable disposition of active metabolites, *Mol. Pharmacol. 43*:120–126 (1993).
187. A. D. Rodrigues, M. J. Kukulka, B. W. Surber, S. B. Thomas, J. T. Uchic, G. A. Rotert, G.

Michel, K. B. Thome, and J. M. Machinist, Measurement of liver microsomal cytochrome P450 (CYP2D6) activity using [*O*-methyl-^{14}C]dextromethorphan, *Anal Biochem 219*:309–320 (1994).

188. M. S. Ching, C. L. Blake, Ghabrial, S. W. Ellis, M. S. Lennard, G. T. Tucker, and R. A. Smallwood, Potent inhibition of yeast-expressed CYP2D6 by dihydroquinidine, quinidine, and its metabolites, *Biochem. Pharmacol. 50*:833–837 (1995).

189. K. Brosen, T. Zeugin, and U. A. Meyer, Role of P450IID6, the target of the sparteine-debrisoquin oxidation polymorphism, in the metabolism of imipramine, *Clin. Pharmacol. Ther. 49*:609–617 (1991).

190. P. Su, R. T. Coutts, G. B. Baker, and M. Daneshtalab, Analysis of imipramine and 3 metabolites produced by isozyme *CYP2D6* expressed in a human cell-line, *Xenobiotica 23*:1289–1298 (1993).

191. V. Fischer, A. E. M. Vickers, F. Heitz, et al., The polymorphic cytochrome P-4502d6 is involved in the metabolism of both hydroxytryptamin antagonists, tropisetron and ondansetron, *Drug Metabol. Disp. 22*:269–274 (1994).

192. L. Firkusny, H. K. Kroemer, and M. Eichelbaum, In-vitro characterization of cytochrome-P450 catalyzed metabolism of the antiemetic tropisetron, *Biochem. Pharmacol. 49*:1777–1784 (1995).

193. D. J. Waxman, D. P. Lapenson, T. Aoyama, H. V. Gelboin, F. J. Gonzalez, and K. Korzekwa, Steroid hormone hydroxylase specificities of eleven cDNA-expressed human cytochrome P450s, *Arch. Biochem. Biophys. 290*:160–166 (1991).

194. M. A. Peyronneau, J. P. Renaud, G. Truan, P. Urban, D. Pompon, and D. Mansuy, Optimization of yeast-expressed human liver cytochrome P450 3A4 catalytic activities by coexpressing NADPH-cytochrome P450 reductase and cytochrome *b*5, *Eur. J. Biochem. 207*:109–116 (1992).

195. H. Yamazaki, Y. F. Ueng, T. Shimada, and F. P. Guengerich, Roles of divalent metal ions in oxidations catalyzed by recombinant cytochrome P450 3A4 and replacement of NADPH–cytochrome P450 reductase with other flavoproteins, ferredoxin, and oxygen surrogates, *Biochemistry 34*:8380–8389 (1995).

196. H. Yamazaki, M. Nakano, Y. Imai, Y. F. Ueng, F. P. Guengerich, and T. Shimada Roles of cytochrome *b*(5) in the oxidation of testosterone and nifedipine by recombinant cytochrome-P450 3a4 and by human liver-microsomes, *Arch. Biochem. Biophys. 325*:174–182 (1996).

197. H. Yamazaki, M. Nakano, E. M. J. Gillam, L. C. Bell, F. P. Guengerich, and T. Shimada, Requirements for cytochrome *b*(5) in the oxidation of 7-ethoxycoumarin, chlorzoxazone, aniline, and *n*-nitrosodimethylamine by recombinant cytochrome-P450 2el and by human liver-microsomes, *Biochem. Pharmacol. 52*:301–309 (1996).

198. M. A. Peyronneau, M. Delaforge, R. Riviere, J. P. Renaud, and D. Mansuy, High affinity of ergopeptides for cytochromes P450 3A. Importance of their peptide moiety for P450 recognition and hydroxylation of bromocriptine, *Eur. J. Biochem. 223*:947–956 (1994).

199. J. P. Renaud, C. Cullin, D. Pompon, P. Beaune, and D. Mansuy, Expression of human liver cytochrome P450 IIIA4 in yeast. A functional model for the hepatic enzyme, *Eur. J. Biochem. 194*:889–896 (1990).

200. R. Rauschenbach, H. Gieschen, M. Husemann, B. Salomon, and M. Hildebrand, Stable expression of human cytochrome-P450 3a4 in v79 cells and its application for metabolic profiling of ergot derivatives. *Eur. J. Pharmacol. Environ. Toxicol. Pharmacol. 293*:183–190 (1995).

201. M. V. Relling, J. Nemec, E. G. Schuetz, J. D. Schuetz, F. J. Gonzalez, and K. R. Korzekwa, *O*-Demethylation of epipodophyllotoxins is catalyzed by human cytochrome P450 3A4, *Mol. Pharmacol. 45*:352–358 (1994)

202. R. W. Wang, L. D. Liu, and H. Y. Cheng, Identification of human liver cytochrome-P450 isoforms involved in the in-vitro metabolism of cyclobenzaprine, *Drug Metab. Dispos. 24*: 786–791 (1996).

203. C. Wandel, R. Böcker, H. Böhrer, A. Browne, E. Rügheimer, and E. Martin, Midazolam is metabolized by at least three different cytochrome P450 enzymes, *Br. J. Anaesth. 73*:658–661 (1994).
204. G. R. Manchee, P. J. Eddershaw, L. E. Ranshaw, D. Herriott, G. R. Park, M. K. Bayliss, and M. H. Tarbit, The aliphatic oxidation of salmeterol to alpha-hydroxysalmeterol in human liver-microsomes is catalyzed by *CYP3A, Drug Metab. Dispos. 24*:555–559 (1996).
205. M. A. Peyronneau, J. P. Renaud, M. Jaouen, P. Urban, C. Cullin, D. Pompon, and D. Mansuy, Expression in yeast of 3 allelic cDNAs coding for human liver P-450 3A4–different stabilities, binding-properties and catalytic activities of the yeast-produced enzymes, *Eur. J. Biochem. 218*:355–361 (1993).
206. J. P. Renaud, M. A. Peyronneau, P. Urban, G. Truan, C. Cullin, D. Pompon, P. Beaune, and D. Mansuy, Recombinant yeast in drug-metabolism, *Toxicology 82*:39–52 (1993).

14

In Vitro Metabolism: Subcellular Fractions

Sean Ekins*, Jukka Mäenpää†, and Steven A. Wrighton
Lilly Research Laboratories, Eli Lilly & Company, Indianapolis, Indiana

* *Current affiliation:* Central Research Laboratories, Pfizer Inc., Groton, Connecticut.
† *Current affiliation:* Leiras Oy, Helsinki, Finland.
Dedicated to Dr. Patrick J. Murphy on the occasion of his retirement.

I. INTRODUCTION

The past 20 years have seen an increase in the use of in vitro systems for determination of xenobiotic metabolism in the pharmaceutical and toxicological sciences. This is mainly due to the development of such systems by numerous groups worldwide as rapid metabolism screens for pharmacologically active compounds and toxicity endpoints. In addition, in vitro studies decrease animal usage. In vitro models used for metabolism studies can be broadly separated into two groups. The first group consists of subcellular fractions, incorporating microsomes (vesicles of endoplasmic reticulum), cytosol, and S-9 (liver homogenate after removal of nuclei and mitochondria). Included in this group are purified or isolated enzymes that metabolize drugs after reconstitution with appropriate cofactors and coenzymes. The second group of in vitro systems includes intact cells, incorporating freshly isolated and cultured hepatocytes, liver slices, the isolated perfused liver, and cell lines. Unlike the subcellular fractions, none of these whole-cell models require cofactors. The advantages and disadvantages of all cellular models for drug metabolism have been extensively reviewed (1–8) and are discussed elsewhere in this text. As the use of subcellular fractions for metabolism studies has been a popular approach for many years, this present review, although not exhaustive, describes the preparation techniques for the subcellular fraction from various tissues evaluated and the applications of subcellular fractions to drug metabolism in vitro.

II. SUBCELLULAR SYSTEMS

Both hepatic and extrahepatic cells contain a multitude of xenobiotic-and endobiotic-metabolizing enzymes that are usually localized to specific organelles within the cell (Tables 1 and 2), although some of these enzymes may also be expressed in an organ-specific manner. Hence, individual subcellular fractions can be thought of as highly concentrated enzyme sources, with the cytosol, microsomes, and mitochondria containing most such enzymes. The major enzymes involved in drug metabolism; cytochromes P450 (P450), flavin-containing monooxygenases (FMO), epoxide hydrolases (HYL; 9) UDP-glucuronosyltransferases (UDPGT) are predominantly localized within the endoplasmic reticulum (ER; see Tables 1 and 2). Consequently, microsomes isolated from multiple organs and species are the most popular subcellular fraction for use in in vitro metabolism studies (Table 3). Normally, the liver is considered the major drug-metabolizing organ, although depending on the specific enzyme and xenobiotic, other organs may also be important for metabolism.

Advantages of subcellular fractions for xenobiotic metabolism include their ease of preparation, flexibility of incubation conditions (cofactor(s), buffer, pH, and temperature for optimum catalysis), and simple long-term storage protocols (discussed later). Disad-

Table 1 Characteristics of the Phase I Enzymes

Enzyme	Localization	Reaction	Cofactor
Cytochromes P450	Microsomal	Oxidation or reduction	NADPH[a]
Flavin-containing monooxy genases	Microsomal	Oxidation	NADPH
Monoamine–diamine oxi-dases	Mitochondrial	Oxidation	
Alcohol–aldehyde dehydro-genases	Cytosolic	Oxidation	NAD
Prostaglandin synthetase–lipoxygenase	Microsomal	Oxidation	NADPH
Polyamine oxidase	Cytosol	Oxidation	FAD
Xanthine oxidase	Mitochondrial	Oxidation	
Aromatases	Mitochondrial	Oxidation	FAD
Alkylhydrazine oxidase		Oxidation	
Azo and nitro reductases	Cytosolic	Reduction	FAD, FMN
Carbonyl reductase	Microsomal and cytosolic	Reduction	NADPH
Dihydropyrimidinedehydro-genase	Cytosolic	Reduction	NADPH
Esterases	Microsomal and cytosolic	Hydrolysis	
Epoxide hydrolases	Microsomal and cytosolic	Hydration	

[a] Abbreviations: NADPH, β-nicotinamide adenine dinucleotide phosphate, reduced form; NAD, β-nicotinamide adenine dinucleotide; FAD, flavin adenine dinucleotide; FMN, flavin mononnucleotide.

vantages of subcellular fractions include lability of some enzymes during preparation (e.g., FMO), loss of cellular heterogeneity (where the organ consists of multiple cell types), and limitation of sequential metabolism that requires multiple cofactors or multiple subcellular components to be complete. However, the ability to localize metabolism to subcellular compartment(s) aids in the selection of suitable models for further study.

In the crudest sense, isolation of multiple subcellular fractions requires disruption of the cell by homogenization (scissor mince followed by use of Waring Blender, Ultra Turrax, or Potter Elvehjem homogenizers) in a suitable buffer, followed by differential centrifugation (using differences in density and size of particles), and resuspension of the pellet in a suitable storage buffer or isolation of a soluble fraction (Fig. 1). It is possible

Table 2 Characteristics of the Phase II Enzymes

Enzyme	Reaction	Subcellular location	Cofactor
UDP[a]-glucuronosyltransferases	Glucuronidation	Microsomal	UDPGA
Sulfotransferases	Sulfation	Cytosolic	PAPS
Glutathione *S*-transferases	Glutathione conjugation	Cytosolic and microsomal	Glutathione
Acyltransferase	Amino acid conjugation	Mitochondrial	Glycine
N-Acetyltransferases	Acetylation	Cytosolic	Acetyl-coenzyme A
Methyltransferase	Methylation	Cytosolic	*S*-Adenosylmethionine
Acyl-CoA synthetases	Amino acid conjugation	Mitochondrial	Coenzyme A, ATP

[a] Abbreviations: ATP, adenosine triphosphate; PAPS 3′-phosphoadenosine 5′-phosphosulfate; UDP, uridine diphosphate; UDPGA, uridine diphosphate-glucuronic acid.

Table 3 Representative Subcellular Fractionation Methods for Differing Tissues and Species

Subcellular fraction	Organ	Species[a]	Ref.
Microsomes	Liver and adrenal	h (fetal)	10
	Liver	r	11
	Liver	rb	12
	Liver	al	13
	Kidney	h	14
	Gastrointestinal	r	15
	Platelet	h	16
	Platelet	h	17
	Erythrocyte	h	18
	Placenta	m	19
	Nasal	ND	20
	Adrenal cortex	Bov	21
	Lung	r	22
	Testicular	r	23
	Thyroid and thymus	Cow, h, s	24
	Mammary gland	ms	25
	Gingival	h	26
	Whole-body	tb	27
	Skin epidermal	r	28
Mitochondria	Liver and kidney	Any	29
	Heart	r, c, gp, rb, p	30
	Heart	Cow	31
	Skeletal muscle	r, p	32
	Brain	Cow	33
	Brain	d	34
	Mammary gland	gp	35
	Platelet	h	17
Nuclei	Liver	r	36
	Brain	r	37
Synaptosomes	Brain	d	34
	Brain	r	38
Smooth endoplasmic reticulum	Liver	r	39
Rough ribosomes	Liver	r	40
Golgi fractions	Liver	r	41
Endosomes	Liver	ND	42
Plasma membranes	Liver	h	43
Multiple fractions	Whole-body	dm	44
	Proximal tubule	rb	45
	Kidney	ms, r	46
	Kidney	np	47
	Heart	gp	48
	Ocular	ms	49
	Pancreas	gp	48
	Polymorphonuclear leukocyte	h	50
	Vascular	Cow	51
	Brain	gp	52
	Colon mucosa	Cow	53

[a] Species abbreviations: al, alligator; c, cat; d, dog; dm, *Drosophila melanogastar*; gp, guinea pig; h, human; ms, mouse; np, northern pike; p, pigeon; r, rat; rb, rabbit; s, sheep; tb, tobacco budworm; ND, not described.

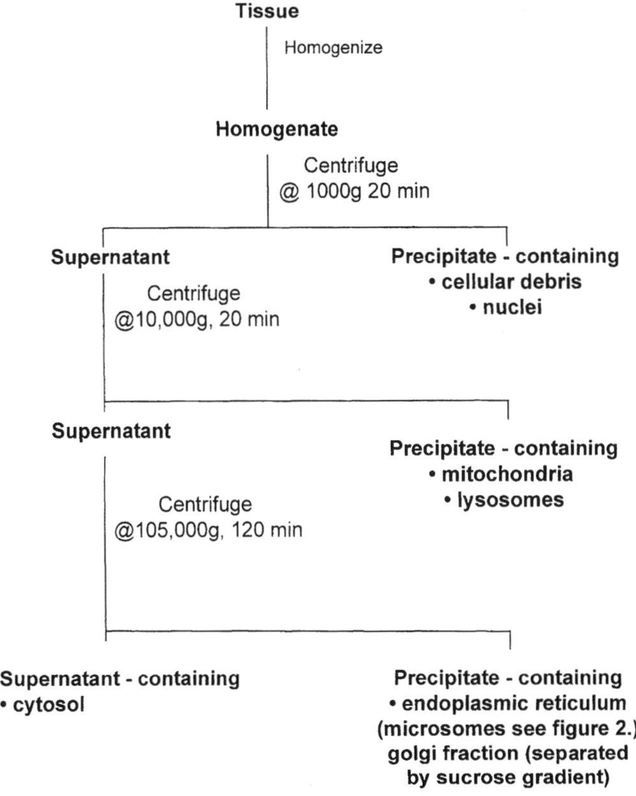

Fig. 1 Generalized subcellular fractionation schematic.

to use variations of this technique with many organs with the minimal of changes to the method (see Table 3). Although extrahepatic organs are suggested as more difficult to fractionate, it has been possible to obtain the desired fraction from numerous tissues (54,55). Data generated with the subcellular fractions from multiple cell types have helped in understanding the relations between metabolism and target organ toxicity. One of the most frequently used fractionation systems for hepatic microsomes is outlined in Fig. 2 (56). Although there have been many variations of this method for the liver and extrahepatic organs, including different buffers, centrifugation time, and speeds, the end products appear to be similar (see Table 3). The principles behind homogenization and centrifugal fractionation have been reviewed by others (57) and will not be discussed here other than to say that there is a choice of methods and buffers available to be used for the isolation of any one particular fraction (see Tables 3 and 4). Differences in metabolism also occur for the routes and rates of metabolism between species and this can complicate overall extrapolations between species. Therefore, multiple substrates have been used to compare metabolic pathways in microsomes with other in vitro and in vivo models (Table 5). Species comparisons using microsomal fractions have been described using multiple substrates (Table 6), and comparisons of xenobiotic metabolism have also been made across numerous subcellular fractions from the same species (Table 7).

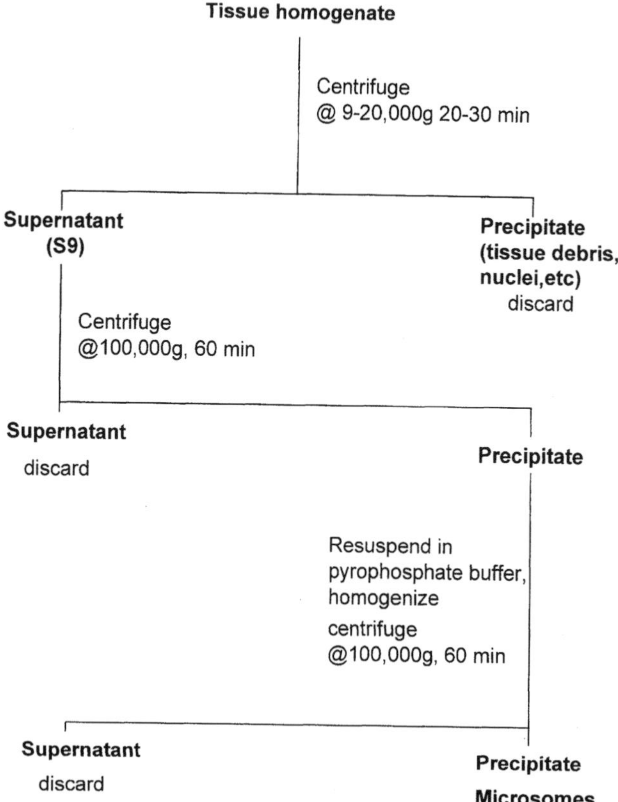

Fig. 2 Schematic of the preparation of hepatic microsomes.

III. REVIEW OF PROTOCOLS FOR PREPARATION

A previous extensive review on subfractionation of the liver cell discussed isolation techniques for each component in detail and linked these with morphometry and biochemical analyses (57). These authors discuss the impurities in each subcellular fraction, the artificial nature of in vitro studies that complicate correlating biochemical data with specific organelles, and then conclude with the disclaimer that in vitro models may not reliably predict the in vivo situation. Additionally, this paper mentions the osmotic requirements of microsomes and considers how ionic changes may imbalance ionic equilibrium of a preparation, although solutions with a low-buffering capacity may be adequate to prevent this imbalance at low temperatures (0–4°C).

There have been many other early reviews of subcellular fractionation that describe different protocols for preparation from many organs. One of the earliest protocols for liver microsomal preparation (48) suggests microsomes should be prepared from livers of fasted animals to eliminate contamination of the fraction with glycogen particles, and it also suggested that a single microsome preparation method may not be universally applicable across all organs. This has been further qualified by the definition of microsomal fractions as consisting of rough and smooth microsomes (with and without ribosomes, respectively; 58). Separation of both types of microsomes from different animals and

Table 4 Selected Characteristic Markers of Subcellular Fractions

Subcellular localization	Marker	Species[a]	Ref.
Microsomal	GDP-mannosyltransferase	mammals	58
	G-6-phosphatase	gp	59
	Neutral α-glucosidase	r	60
	Cytochrome-*c* reductase	r	61
Mitochondrial	Catalase	np	62
	β-Glycerophosphatase		
	p-Nitrophenol-a-manosidase		
	Cytochrome-*c* oxidase		
	Succinate dehydrogenase	gp	59
	Xanthine oxidase		
	Glutamate dehydrogenase	r	61
	Monoamine oxidase		
Cytosol	LDH[b],	np	62
	GST		
	γ-Glutamyl hydrolase	r	60
	Aldehyde oxidase	gp	59
	Alcohol dehydrogenase	r	61
Golgi	Galactosyltransferase	r	41
Ribosomes	Electron microscopy	gp	63
Lysosomal	Acid phosphatase	r	61
	β-Glucuronidase	r	60
	N-Acetyl-β-glucosaminidase	r	60
Peroxisomes	Catalase	r	60
Plasma membrane	5′-Nucleotidase	r	60
	Alkaline phosphatase	r	60
	Alkaline phosphodiesterase	r	60
	γ-Glutamyltransferase	r	60

[a] Species abbreviations: gp, guinea pig; np, northern pike; r, rat.
[b] Abbreviations: LDH, lactate dehydrogenase; GST, glutathione *S*-transferase; ND, not described.

across organs may necessitate modifications in the method for successful isolation. Initially, the investigator must choose between high recovery and high impurity or lower recovery and higher purity of a specific fraction. In addition, excessive homogenization damages rough microsomes and decreases enzyme activity. To try and solve this problem alternative methods were developed. A method using puromycin and potassium chloride has been used for the nondestructive preparation of rat liver rough microsomes, characterization, and separation into ribosomal and membranous components (40).

Rabbit hepatic microsomes have been separated into three fractions following the initial centrifugation: namely, heavy, medium, and light fractions (100). Total and specific enzyme contents of these three fractions were determined using esterase, glucose-6-phosphatase, ATPase, arylsulfatase, aniline hydroxylase, NADH-cytochrome-*c* reductase, NADPH-cytochrome-*c* reductase, cytochrome b_5, and cytochrome P450 (P450, detailed later). P450 and all other enzymes requiring electron transport reactions were more concentrated in the light fraction. The authors located other enzymes in other fractions and concluded that microsomal fractions contained several types of vesicles. Subfractionation of rat liver microsomes using 0.26% sodium deoxycholate in a 1:10 ratio with microsomes, yields smooth ER with no ribosomes (39). Following this procedure, electron mi-

Table 5 Review of Metabolism of Selected Substrates Used with Hepatic Subcellular Fractions and Compared with Other In Vitro or In Vivo Systems

Substrate	Species[a]	In vitro, in vivo models[b]	Summarized findings	Ref.
ABT-418	ch, r, rb, mk, d, h	M, SL	Chimpanzee and rat demonstrate similar metabolism in slices and microsomes, other species are different	64, 65
α-1-Acetylmethadol, butamoxane, *d*-propoxyphene *N,N*-dimethyl-*p*-chloro phenoxyethylamine, ethinimate, 8-methoxybutamoxane	r	S9, HEP	50% of substrates had higher rates of metabolism in hepatocytes	66
Aniline, digitoxin	r	M, H, S9, HEP	Digitoxin metabolism highest in S9; aniline highest in microsomes	67
Benzene	r, ms, h	M, SL	Phase II metabolites important and seen only in slices	68
Biphenyl, thiabendazol, ethylmorphine Benxo[*a*]pyrene	r	M, SL, HEP	Activities maximal in microsomes except biphenyl hydroxylation highest in hepatocytes	69
Caffeine	h	M, SL, HEP	Qualitative metabolic similarity across models	70
Coumarin	ca	M, SL	Metabolism higher in microsomes than slices	71
CQA 206-291	h	M, SL	Microsomes closest to in vivo	72
Deoxyguanine analogues	r, h	M, HEP, CYT	Use of these systems allowed discrimination of xanthine–aldehyde oxidase and P450 activities	73
Diazepam	r	M, HEP	In vivo clearance predicted more closely with hepatocytes	74
Digoxigenin, digoxigenin digitoxosides, digoxin	h	M, HEP	No P450 metabolism of these molecules in either in vitro system	75
Fluarizine	r, ms, d	M, HEP	Highest metabolism in rat, all in vitro profiles similiar to in vivo	76
Metronidazole	r	M, SL, HEP	Microsomes gave highest activity, slices give similiar metabolite ratio to in vivo	77

Table 5 Continued

Substrate	Species[a]	In vitro, in vivo models[b]	Summarized findings	Ref.
N-Containing steroid[c]	r, d, h	M, HEP	Qualitatively similar metabolite profile across species	78
[^{15}N]Nitrosoproline	r	S9, HEP	Metabolized differently from similar nitrosoprolines	79
Palmotoxins B_0 G_0	r, ms, gp, rb, w, du, gt, s, d, l, t	M, SL	Metabolism highest in microsomes across all species	80
Phenytoin, tolubtamide	r	M, HEP	In vivo clearance predicted more closely with hepatocytes	81
Styrene	r	M, S9, HEP	Hepatocytes closest to microsomes, hepatocytes had higher metabolic capacity	82
Terfenadine	h	M, SL, fusion protein	Slices close to in vivo	83
Testosterone	h	M, SL 100,000 g and supernatant fractions	Androstenedione major metabolite in slices and microsomes, but not in soluble fractions	84
Theophylline	h, r	M, SL	Metabolic profiles similar across species	85
Toluene	h, r	M, SL	Quantitative and qualitative differences in metabolism in favor of slices	86
Tris (2-chloroethyl) phosphate	h, r	M, SL	Metabolism greater in slices than microsomes	87
Trospectomycin	r	S9, HEP, in vivo	Rapid accumulation of drug in all systems	88
Verlukast	mk	M, SL, in vivo	Identification of CYP1A1 as responsible for predominant pathway of epoxidation	89

[a] Species abbreviations: ca, calf; ch, chimpanzee; d, dog; du, duck; gp, guinea pig; gt, goat; h, human; l, lizard; mk, monkey; ms, mouse; r, rat; rb, rabbit; s, sheep; t, toad; w, white rock cock.

[b] System abbreviations: CYT, cytosol; H, homogenate; HEP, hepatocytes; M, microsomes; SL, slices; S9, 9000-*g* supernatant.

[c] *N,N*-Diethyl-4-methyl-3-oxo-4-aza-5α-androst-1-ene-17β-carboxamide.

croscopy and enzyme analysis were used to characterize fractions. Membrane-bound enzymes reacted differently to storage and dilution, indicating an enzyme activity could be controlled by external agents acting on the subcellular structure (39). Schenkman and Cinti (11) described another method for rat liver microsome preparation using calcium, which binds the microsomal fraction, negating the need for an ultracentrifuge, and decreas-

Table 6 Review of Hepatic Subcellular Fraction Metabolism of Substrates Across Species

Substrate	Species[a]	Subcellular fractions[b]	Summarized findings	Ref.
Aceclofenac	r, h	H, S9, M, CYT	Esterase activity highest in rat MIC and human CYT.	90
Acetaminophen, benzphetamine, bufuralol, chlorpromazine, coumarin, 3,4-DCNB,[c] 17 α-EE, erythromycin, ethoxycoumarin, ethoxyresorufin, isoniazid, 6-mercaptopurine, midazolam, NDMA, pentoxyresorufin, *S*-mephenytoin, tolbutamide	h, d, mk	M, CYT	Minor differences in metabolism between cynomolgus and rhesus monkey liver samples, but more active than human samples. Metabolism in dog liver samples generally not similiar to human liver samples.	91, 92
Metyrapone	ms, r, gp	MIT, CYT	Carbonyl reduction present in both fractions in all species	93
PhIP	r, ms, h	M, CYT	Species differences in rates of metabolite formation and only humans catalyze the *N* hydroxylation.	94
Zearalone	cow, pig, s, hen, gt	N, MIT, L, M, CYT	Reduction of zearalenone is species-dependent.	95

[a] Species abbreviations: d, dog; gp, guinea pig; gt, goat; h, human; mk, monkey; ms, mouse; r, rat; s, sheep.
[b] System abbreviations: CYT, cytosol; H, homogenate; L, lysosomal; M, microsomes; MIT, mitochondria; N, nuclear; P, submicroscopic particles; S9, 9000-*g* supernatant.
[c] 3,4-DCNB, 3,4-dichloronitrobenzene; 17α-EE, 17α-ethinyl estradiol; NMDA, *N*-nitrosodimethylamine; PhIP, 2-amino-1-methyl-6-phenylimidazo[4,5-*b*]pyridine.

Table 7 Review of Metabolism of Selected Substrates Compared Across Multiple Subcellular Fractions

Substrate	Species[a]	Subcellular fractions[b]	Summarized findings	Ref.
Finasteride	h	M, CYT, MIT	Reduction favored over oxidation	96
Octanoic acid	r	MIT, N, H, P	Metabolism localized to mitochondria	97
Penicillinic acid	ms	MIT, H, N, CYT	Most metabolites cytosolic	98
Glycyrrhetic acid	r	L, MIT, N, M	Oxidation and reduction highest in microsomes	99

[a] Species abbreviations: h, human; ms, mouse; r, rat.
[b] System abbreviations: CYT, cytosol; H, homogenate; L, lysosomal; M, microsomes; MIT, mitochondria; N, nuclear; P, submicroscopic particles.

ing preparation time. Many enzyme activities are stable using this method, although some may decrease. Furthermore, it is also not universally applicable to other organs and species.

The subcellular fractionation of rat liver homogenates into all the major organelles has been described using countercurrent partition or sucrose density centrifugation (60). The countercurrent partition technique enhances the separation of lysosomes and endoplasmic reticulum from the plasma membrane by separation of particles based on their surface properties. Further isolations of various minor subcellular fractions, such as endosomes and Golgi, have been documented. Endosomes are the membrane vacuoles where endogenous ligands are located following transfer from the plasma membrane (42). Endosomes rapidly hydrolyze these ligands, recycle receptors, and amplify cell–cell signals, although the mechanism is not fully characterized. To prepare Golgi fractions, liver homogenates were fractionated to obtain microsomes, then this pellet was gently resuspended and fractionated by sucrose gradient centrifugation. Each fraction was collected and washed in sucrose. Further fractionation of the Golgi illustrated that the light and intermediate fractions possessed highest galactosyltransferase activities. Microsomal enzymes also demonstrated lower activities in Golgi fractions (41). Purity of these fractions was checked by electron microscopy to assess for contaminating organelles. Administration of cycloheximide to rats intravenously depletes the liver of secretory products, allowing Golgi membranes to be isolated in a purer form, free of internal contents (101).

Important considerations required for successful subcellular fractionation from any organ or species include the use of optimal conditions throughout the procedure. Not only do microsomal membranes have a negative surface charge that can be used to aid in fractionation, but also proteins present during subfractionation have an intrinsic pH-buffering capacity. However, despite this intrinsic-buffering capacity it is necessary to use sucrose medium in Tris-HCl (Tris) buffer for homogenization and centrifugation. The generalized protocol for microsomal preparation from common laboratory species may also include fasting the animal for 20 h to reduce glycogen contamination, as described earlier. Other additions to the preparation protocol include using the serine protease–esterase inhibitor, phenylmethylsulfonylfluorine (PMSF) in the homogenization buffer and the chelating agent–antioxidant ethylenediaminetetraacetic acid (EDTA), thiol-protecting agent dithiothreitol, and the antioxidant butylated hydroxytoluene as additives in the microsomal buffer to retard lipid peroxidation and protein degradation. A second high-speed centrifugation step uses pyrophosphate buffer to remove hemoglobin and nucleic acids from the microsomal or subcellular fraction preparation (102).

One of the most-studied enzymes found in liver microsomal fractions is the heme-thiolate protein known as P450. This enzyme was recognized by the seminal studies of an enzyme responsible for the demethylation of azo dyes (103–104) localized within microsomes (105). This was years before identification of P450 in Wistar rat liver microsomes (106), pig liver microsomes (107), and rabbit liver microsomes (12).

The early work by Garfinkel (107) also suggested that additional washing of the microsomes with Ringers solution was necessary to remove hemoglobin interference with the spectral measurements undertaken. The effect of washing rat hepatic microsomes with sucrose solutions, with or without EDTA, has also been evaluated using xenobiotic metabolism as a marker (108). EDTA prevented the decrease in metabolism of some substrates after washing with sucrose alone, stimulating metabolism of other substrates.

Since the realization that P450 was not a single enzyme but a multigene family composed of multiple enzymes with distinct (and in some cases overlapping) substrate specificities (109), it has been possible to identify, quantify, and characterize individual

P450s in a tissue using specific probe substrates (5), inhibitors, or immunoquantification. This approach is useful in the development of a human liver 'microsome bank' (110) that expresses various levels of each of the identified enzymes. The study of effects of genetics and other factors on xenobiotic metabolism is possible in vitro if enough detailed donor histories are kept.

In addition to expression in the endoplasmic reticulum, P450s have been identified in the plasma membrane of rat (111) and human (43) hepatocytes. Recently, the mechanism behind transport of rat P450s and NADPH–cytochrome P450 reductase to the plasma membrane has been suggested as via the Golgi apparatus in an enzyme-selective manner (112). These P450s, located in the Golgi apparatus were catalytically active, and the purity of Golgi fractions used was assessed using marker enzymes.

It may be preferable to use S-9 for initial metabolism screening because the major phase I and phase II pathways are present to point to which system to use in the next phase of xenobiotic evaluation; either microsomes, slices, or hepatocytes, respectively (4). An advantage of subcellular fractions is that unlike intact cell methods (hepatocytes and tissue slices), drug uptake is not an issue with them. The use of subcellular fractions and other in vitro systems obtained from large food animals for metabolism and toxicology studies has been extensively reviewed by Shull et al. (113) and discussed from a regulatory standpoint by Shaw (114). The same advantages and disadvantages in selection of a particular in vitro model are present in these species as in the small laboratory animals. Subcellular fractions from large food animals have been used extensively in metabolism studies with the caveats of required cofactor addition, loss of functional and structural interrelations between organelles, and poor correlation with in vivo studies. The application of whole-cell isolation and slice methodologies have only recently been applied to large food animals in an attempt to achieve in vitro data more comparable with in vivo (71,113).

IV. FRACTIONATION OF HUMAN TISSUE

Besides the use of nontransplanted human organs as a source of tissue for in vitro drug metabolism and toxicology studies, biopsies can provide enough tissue for microsomal preparation. As an example, no significant difference exists in P450 content or catalytic activity of hepatic needle biopsy samples when compared with liver wedge biopsies (as demonstrated in both rat and human; 115). Other groups have suggested liver wedge biopsy enzyme activities may be affected by anesthetics used on the patients (116). In both cases justification for sampling is due to the suspicion of diseased liver or other diagnostic purpose, and these conditions may themselves alter the metabolic characteristics of the whole organ. Incidentally, this may also be true for extrahepatic organs, although in some of these cases tissue biopsy is easier to justify for investigational purposes (e.g., from the intestine).

V. CHARACTERIZATION

The unique localization of particular enzymes to different subcellular fractions allows their use as characteristic markers (see Table 4). In a similar way, certain compounds may be used as selective substrates for individual P450s for characterizing a bank of human liver microsomes. Excellent examples of such characterizations have been reported for the subcellular fractionation of the liver from northern pike (*Esox lucius*; 62) and rat (60). NADPH–cytochrome-*c* reductase, P450, and benzo[*a*]pyrene monooxygenase were concentrated in the microsomal pellet of the pike liver (62). Most of the lactate dehydrogenase and glutathione *S*-transferase (GST) were present in the cytosol. Catalase, β-glycerophosphatase, *p*-nitrophenol-α-mannosidase, and cytochrome oxidase were distributed mainly in the mitochondria of the pike liver. Epoxide hydrolase was distributed between the microsomal and cytosolic fractions (62). The contents of each fraction of the pike hepatocytes were also examined by electron microscopy, which sometimes enables observation of contamination in subcellular fractions, although this cannot be relied on to confirm the absence of any contaminant. For example, the presence of lysosomes and mitochondria were detected in microsomal samples, which may affect quantitative metabolic data by including enzymes from these other fractions. The characterization of control rat liver plasma membrane, lysosomes, mitochondria, endoplasmic reticulum, peroxisomes, and cytosol was also based on well-known marker enzyme activities for individual organelles and compared with phenobarbital treated animals (60; see Table 4).

The use of electron microscopy is probably the ultimate method to characterize subcellular organelle fractions based purely on morphology and contamination with other organelles. Ribosome preparation (see Table 3; Fig. 1) from guinea pig liver microsomes by extraction with deoxycholate has been described (63), and the final fraction analyzed by electron microscopy to demonstrate a relatively pure preparation. In a further example of fractionation characterization by electron microscopy, isolated guinea pig brain nerve endings, myelin fragments, and mitochondria were prepared by sucrose density gradient centrifugation from an initial homogenate, along with nuclei and cell debris, microsomes, and ribosomes (52). Morphological studies may also be carried out using electron micrographs, which are additionally useful in understanding organelle surface area/volume ratios of the whole cell and how these can be related to enzyme activity. The importance of the hepatic ultrastructure has been recognized since the early stereological morphometric studies of rat liver (117,118). These studies indicated that rough endoplasmic reticulum and mitochondria both possessed the largest surface areas out of all the subcellular components in parenchymal cells, and this morphometric data allowed correlation with biochemical studies of phenobarbital induction of microsomal P450 (119). The preceding are examples of the rare occasions when complete characterization of subcellular fractions have been documented, and this could be a criticism of the many groups that inadequately characterize their systems after transfer of one preparation technique across organs and species.

VI. WHOLE-CELL HOMOGENATES

Tissue homogenates are the crudest in vitro system (120), but are useful over subcellular fractionation as they contain most of the enzymes and sometimes cofactors necessary for function. Therefore, liver homogenates may also be used to compare metabolism across many species and to indicate species differences early in drug metabolism research (121). Supernatants of whole-cell homogenates are easily prepared, for example, from extrahepatic organs such as rat lung and kidney, and they have been used to measure carboxylesterase activity using cocaine as a substrate. Carboxylesterase activities were highest in lung and liver in males, rather than females, respectively (122). There are many other examples for which homogenates have been used in drug metabolism. In vivo and in vitro metabolism of the epipodophyllotoxic derivative VP 16-213 was similar using rat liver homogenate and in vivo plasma samples (123). By comparison of metabolism by crude homogenates and microsomal fractions it is possible to determine which models will be necessary for further study. For example, the localization of the enzyme responsible for the major metabolite of the epipodophyllotoxic derivative in rat was in the microsomal fraction (123).

VII. STORAGE

One of the advantages of subcellular fractions is their ability to be frozen and easily stored. Because of the valuable nature of human liver for in vitro metabolism studies, numerous groups have examined the effects of storage of human livers at −80°C as homogenates, cytosol, microsomal pellets, microsomal suspensions, or snap-frozen pieces. Although some had suggested storage at −80°C has little effect on drug metabolism (5) and activities can be well maintained for years with only slight changes (124), it was not until recently that these statements were rigorously tested. In this evaluation, the effects of freezing, thawing, and time in storage (up to 2 years at −80°C) of human liver microsomes was examined. The results indicated no change in CYP1A2, CYP2A6, CYP2C9, CYP2C19, CYP2D6, CYP2E1, CYP3A4/5, or CYP4A9/11 catalytic activities (125). In contrast, an earlier study of a number of human hepatic xenobiotic-metabolizing enzymes stored for various lengths of time as pieces, microsomes, or homogenate showed conjugating enzymes were stable and P450 behavior was unpredictable (126). An even earlier study using human liver microsomes and frozen liver cubes stored at −80°C for 6 months, found little effect on various enzyme activities when compared with fresh tissue, for P450 activities were stable, although glutathione conjugation declined by 20% (110,116). Without a doubt it is acknowledged that to ensure successful storage of human livers as either a subcellular fraction or whole-cell preparation, the starting quality of the tissue should be as high as possible. In the early years of drug metabolism research with human liver, tissue quality was questionable, as demonstrated by enzyme activities (124,126). In recent years, the overall quality and availability of human tissue available for drug metabolism research has improved, probably owing to the increasing network of nonprofit organizations across the United States and Europe that coordinate and distribute donated human tissues under strict ethical committee guidelines. For the liver and kidney specifically, factors such as perfusion with University of Wisconsin buffer have allowed adequate storage time for transportation of viable tissue around the world within 24 h of removal from the donor.

The effects of long-term storage on enzyme activities in hepatic tissues in other species has also been evaluated. Hepatic microsomal, mitochondrial, and cytosolic epoxide hydrolase catalytic activities in the C57B1/6 mouse were stable for approximately 10 h when stored on ice. One period of freeze-thawing did not significantly alter the epoxide hydrolase catalytic activity, whereas six cycles did significantly alter this activity in all subcellular fractions (127). The stability on storage of induced murine liver S-9 fractions has also been studied. P450 decreased over 9 months after being lyophilized and then stored at −80°C or −20°C. However, after storage at −196°C, S-9 fractions demonstrated consistent P450 activities (128).

VIII. METABOLITE IDENTIFICATION

Subcellular systems can be used to generate metabolites of compounds that can then be rapidly identified using modern, highly sensitive analytical techniques as reviewed previously with precision-cut liver slices (2). Examples of these techniques include the following: After brief microsomal incubations, gas chromatography–mass spectrometry (GC–MS) analysis has been used to identify metabolites successfully (129). The identity of metabolites formed by microsomes can also be rapidly determined when this in vitro model is used in conjunction with liquid chromatography–electrospray ionization–mass spectrometry (LC/ESI-MS), a powerful technique for structural identification. A recent example of this technique is the use of LC/ESI-MS to analyze metabolism of paclitaxol (Taxol) after incubation with human microsomes (130). S-9 fractions of liver from rat, man, and monkey have been used to generate metabolites of a new human immunodeficiency virus (HIV)-protease inhibitor for identification by atmospheric pressure chemical ionization liquid chromatography–mass spectrometry (APCI LC/MS) and collisionally activated decomposition liquid chromatography–mass spectrometry–mass spectrometry (CAD–LC/MS/MS; 131). The increasing sensitivity and automation of mass spectrometry and related technology will continue to have a considerable effect on early drug development.

IX. SPECIES COMPARISONS OF XENOBIOTIC METABOLISM

Since the adoption and use of in vitro techniques, such as subcellular fractions, there have been numerous reports of drug metabolism studies with many different species. Therefore, it is not surprising that comprehensive reviews of multiple enzyme–multiple species and sex difference comparisons in metabolism exist. Our group has compared numerous isozyme-specific phase I and II activities in human and rhesus monkey microsomes and cytosol (91; see Table 6). More recently, this work has been updated to include dog and cynomolgus monkey and suggests that the human, in general, is less active in drug metabolism when compared with these species (92; see Table 6). Others have studied azo- and nitro-reductase in numerous species using homogenate or cell fractions from the following species: shark, ray, barracuda, snappers, frogs, toads, turtles, pigeon, mice, rats, and guinea pigs (132). Low azo- and nitro-reductase activity in fish was documented, and addition of cofactor, flavin adenine dinucleotide (FAD) or flavin mononucleotide (FMN) stimulated these activities (132). This group also compared activities across ten strains of mice, illustrating differences in both azo- and nitro-reductase. Pig, goat, sheep, cow, and hen hepatic subcellullar fractions have been compared in their ability to metabolize the estrogenic mycotoxin, zearalenone, a substrate for 3α-hydroxysteroid dehydrogenase (95). Interspecies comparisons of xenobiotic metabolism using a single subcellular fraction are frequently documented. Examples include rat microsomes demonstrating biphasic kinetics for the synthetic opioid analogue sulfentanil, which was not evident in dog and human microsomes, although in general kinetic parameters were similar across all species (133). An example of a single subcellular fraction used across species is the measurement of *N*-acetylation in hamster, guinea-pig, rabbit, dog, human, and murine hepatic cytosol, although cytosolic activity in other organs of hamster and rabbit was found to be much lower (134). There are also examples for which metabolic pathways of selected substrates have been compared across multiple subcellular fractions in a single species (98; see Table 7). In such cases the emphasis of the report is placed on the fraction ultimately responsible for metabolism.

X. XENOBIOTIC ACTIVATION

Interspecies comparisons have also been used to study carcinogen activation by hepatic microsomes, and these have been highly useful in explaining differential susceptibility to such xenobiotic agents. One example is the considerably higher *N*-hydroxylation of the food-borne carcinogen 2-amino-1-methyl-6-phenylimidazo[4,5-*b*]pyridine (PhIP) in human, rather than in rat and mouse in vitro. Whereas the rodent species detoxify PhIP by 4-hydroxylation to an equal extent, humans have little of this activity (94). Mutagenic activation may also be elucidated in both microsomal and cytosolic fractions. Examples of this include, *N*-hydroxy-*n*-2-fluorenylacetamide activation that was mediated by a deacetylase in hepatic microsomes from SD rats, C57BI mice, Hartley guinea pigs, Syrian hamsters, whereas cytosol from these species differed in terms of the enzymes involved in the activation (135). A further example of multiple enzymes involved in mutagenic activation is 2-acetylaminofluorene, which is similarly activated by deacetylation and *N*-hydroxylation in guinea pig liver S-9 fractions (136).

XI. COMPARISONS WITHIN A SINGLE SPECIES

Multiple subcellular fractions have been characterized for metabolic pathways in a single species. A selection of substrates used for comparison of subcellular fractions within a species is presented in Table 7. Subcellular fractions (cytosol, mitochondria, and microsomes) have been prepared from Dunkin-Hartley guinea pig liver homogenates and the following activities were measured: lactate dehydrogenase, succinate dehydrogenase, glucose-6-phosphatase, aldehyde oxidase, and xanthine oxidase, which were predominant in cytosol, mitochondria, microsomes, cytosol, and mitochondria, respectively (59).

Aldehyde dehydrogenase has previously been shown to be predominant in mitochondria and microsomes of rat liver after using numerous marker enzymes to characterize each subcellular fraction (see Table 4; 61). Similarly, others have studied hepatic esterases in various subcellular fractions of SD rat using a range of nicotinate esters and found them to be mainly located in microsomes (38). Acetyl-CoA hydrolase, propionyl-CoA hydrolase, glutamate dehydrogenase, LDH, catalase, urate oxidase, acid phosphatase, NADPH–cytochrome-*c* reductase, and cytochrome-*c* oxidase were measured in rat hepatic nuclear, mitochodrial, peroxisomal, microsomal, and cytosolic fractions (137). Both acetyl-CoA hydrolase and propionyl-CoA hydrolase were present in microsomes and peroxisomes, the marker enzymes suggested contamination of the peroxisomal fraction with mitochondrial and microsomal enzymes. These characterization studies are useful in determining the complement of enzymes present in a particular subcellular fraction that may be involved in drug metabolism in any species and could contribute to interspecies differences.

XII. IN VITRO MODEL COMPARISONS FROM MULTIPLE ORGANS

There are several studies comparing subcellular fractions with other models of in vitro drug metabolism, although there are surprisingly fewer cases for which subcellular fractions from multiple organs are used to study one or more metabolic pathways. Some of the latter comparisons will be briefly described. Tissue homogenates (of rat liver, lung, skin, and blood), mitochondria, nuclei, microsomes, and cytosol from these tissues (138) were used to compare esterase activities. The esterase activities were the highest in liver and plasma. Esterase activity in human liver cytosol and microsomes has been compared with plasma esterase activity (139). Acetyl-CoA thiolase and 3-oxo acid-CoA transferase activities have been assessed in Wistar rat brain, liver, and kidney whole-cell homogenates (140). Thiolase activity was highest in the liver, whereas transferase was higher in kidney than in brain. Styrene metabolism by P450 and epoxide hydrolase has been measured in microsomes from liver, heart, lungs, spleen, and kidneys obtained from rat, mouse, guinea pig, and rabbit (141). Another group assessed epoxide hydrolase in microsomes and GST activity in cytosol of liver, kidney, and lung from mouse, hamster, rat guinea pig, rabbit, dog, pig, baboon, and human (142). Epoxide hydrolase was highest in baboon liver and GST was highest in guinea pig liver when compared with the other tissues and species.

When other in vitro or in vivo models are compared with subcellular fractions in the metabolism of a substrate in a single species, it may give some indication of which fraction contributes toward metabolism to the greatest extent and of which model to use for further studies. The metabolism of various drugs metabolized by different routes in isolated hepatocytes and 9000 *g* supernatant has been compared with in vivo. Metabolism of α-l-acetylmethadol, D-propoxyphene, and *N,N*-dimethylphenoxyethylamine were similar in both isolated hepatocytes and the 9000 *g* supernatant (66; see Table 5). The metabolism of four other agents (*N*,-dimethyl-*p*-chlorophenoxyethylamine, ethinimate, butamoxane, and 8-methoxybutamoxane) was nearly twice as high in the 9000 *g* supernatant as in the hepatocytes. This group suggested isolated hepatocytes gave results closer to those obtained in vivo than did microsomes.

Specific enzymes have been studied in multiple models and organs. For example, glutathione *S*-transferases are mainly present in cytosol (143) or microsomes, although they have also been identified in mitochondria and nuclear fractions of rat (144). The distribution of glutathione is similar in both cytosolic and mitochondrial matrix of fractionated isolated perfused rat liver, with the reduced form predominating at 10 mM (145). Liver and lung interspecies comparisons of P450 and epoxide hydrolase in microsomes, and glutathione *S*-transferase in cytosol indicated considerable differences among species and higher activities in lung (146).

XIII. COMPARISONS WITH OTHER IN VITRO MODELS FOR IN VIVO–IN VITRO CORRELATIONS AND SCALING

Early studies of drug metabolism in humans in vivo had also obtained some in vitro evidence that the liver and, specifically, microsomal enzymes were the site of metabolism (147). In the intervening period, increasingly subcellular fractions have been used to compare the metabolic capability of humans with other species. The extrapolation of in vitro data to the in vivo situation is ultimately desired, although this is a complex process that has been reviewed in depth (148–150). Similarly, the extrapolation of animal pharmacokinetic data to humans is desired and is also a complex procedure (151,152). Several recent studies have examined the relation of the in vitro data derived from various models with that ultimately observed in vivo. The use of rat microsomal data may not be ideal for estimation of clearance, particularly for high-clearance drugs, owing to underestimation of the in vivo value, whereas hepatocyte data is closer to that in vivo (148,149). However, human hepatic microsomes have been used successfully to estimate hepatic clearance (Cl_H) in vivo (153). An additional in vitro model of metabolism is the use of precision-cut liver slices. Comparisons of intrinsic clearance (Cl_{int}) in rat liver slices and hepatocytes have shown a relation whereby high-turnover molecules in hepatocytes demonstrate much lower Cl_{int} in slices, whereas lower-turnover, molecules illustrate less exaggerated differences (154,155) when slice data is expressed in terms of hepatocellularity. There is an increasingly popular proposal that accessibility or diffusion of substrates into slices is a problem with this in vitro model (2,77,155–158). Further studies are required with slices to determine an appropriate scaling factor to account for differences in uptake between drugs and possibly the individual cell types involved. In the meantime microsomes may provide the ideal means to estimate in vivo clearance in humans owing the relative ease of use and availability of human microsomes and in part based on the in vitro–in vivo clearance comparisons presented by Hoener (153).

XIV. DRUG–DRUG INTERACTION STUDIES

Probably all pharmaceutical companies now routinely use human tissue in vitro to investigate metabolism of a drug candidate and drug–drug interactions. One of the reasons for this increased interest in human in vitro studies is the desire by government agencies for such information following the discovery of genetic polymorphisms in drug metabolism. This describes the situation in which populations of poor and extensive metabolizers of drugs exist and is one reason why the examination of interindividual pharmacokinetic variability is necessary along with the study of potential pharmacokinetic interactions of coadministered drugs (6,159). The advantages of conducting such interaction studies in vitro include safety and decreased cost over clinical trials. Drug–drug interaction studies in vitro were initially performed using rat subcellular fractions. The interaction between propranolol and tricyclic antidepressants was one early example (160). More recently, interactions between selective serotonin reuptake inhibitors and individual catalytic probe substrates (161–163) or inhibitors (164) for various P450s have been evaluated in vitro using human microsomal incubations. Similarly in vitro studies by our group on a new antipsychotic agent, olanzapine, have used various techniques to predict that this drug will not interact with coadministered drugs metabolized by any of the major or polymorphic P450s (165). In the future it is possible other in vitro models will be evaluated as suitable for use in interaction studies. As we understand the structure of the active site of individual enzymes (such as P450; 166) to an increasing extent, it seems more likely that computational methods for interaction screening based around inhibitor and enzyme–substrate pharmacophores will supplant the present techniques. In addition, computational methods would allow identification of metabolic pathways that could then be used to predict pharmacokinetic variability.

XV. INCUBATION CONDITIONS AND THEIR EFFECTS ON ENZYME ACTIVITY

A. The Effect of Different Assay Conditions

Numerous different procedures have been used with subcellular fractions to analyze various enzyme activities (102). In this section we will concentrate on describing differences in P450-mediated reactions under different assay conditions. In summary, P450 is reduced by two electrons donated from reduced β-nicotinamide adenine diphosphate (NADPH) and cytochrome P450 reductase which acts as an intermediary molecule in the transfer of electrons from NADPH to P450 (167). It appears that several P450s can receive the second electron from cytochrome b_5 (168). Cytochrome b_5 itself is reduced by b_5 reductase, which is reduced by NADH (169). In vitro incubation conditions require NADPH or an NADPH-generating system using glucose 6-phosphate dehydrogenase (G6PD) to produce NADPH from NADP (102). As an alternative to NADPH–cytochrome P450 reductase plus NADPH, cumene hydroperoxide can be used as a source of active oxygen (170). Recently, many studies have been performed in an effort to improve the catalytic activity of purified or cDNA-expressed CYP3A4 when reconstituted with P450 reductase and lipids (168). Different components, such as cytochrome b_5, phospholipids, glutathione, and divalent cations, enhance the activity of CYP3A4-mediated reactions in reconstituted assay conditions (171–174). Magnesium chloride and calcium chloride stimulate the catalytic activity of recombinant CYP3A4- and CYP3A4-mediated reactions in human liver microsomes (169,175). Recombinant CYP3A4 from a bacterial source had higher activity in phosphate buffer than in Tris buffer (174). However, Shet et al. (176) showed that recombinant fusion protein (CYP3A4 linked to P450 reductase) metabolized substrates in Tris, but not in phosphate buffer. These results indicate that the catalytic activity of the P450 can be greatly influenced by various assay conditions. Thus, caution should be used when examining the metabolism of a compound by these systems.

Increasing ionic strength also enhances P450-mediated activities (170,177,178). Additionally, the catalytic activities of various microsomal P450s may vary significantly with different buffer systems with varying ionic strength; but similar pH (177). Also the ammonium ion present in glucose 6-phosphate dehydrogenase (G6PD) at low buffer concentrations inhibited CYP3A4-mediated midazolam metabolism (178). Ionic components may also have a substantial effect on microsomal membranes, for positively charged phospholipids provide less efficient assay conditions for recombinant CYP3A4 (173).

B. Autoactivation of P450-Mediated Xenobiotic Metabolism

Various P450 substrates give atypical sigmoidal, instead of classic, Michaelis–Menten kinetics, when incubated with human liver microsomes. This sigmoid pattern is indicative of autoactivation of the metabolism of a substrate by the substrate itself (179–183). Autoactivation seems to be characteristic for several human CYP3A4 substrates such as carbamazepine and aflatoxin. However, the recent kinetic analyses of B-lymphoblastoid–expressed CYP2B6- and CYP2E1-mediated 7-ethoxy-4-trifluoromethylcoumarin metabolism also shows autoactivation (184).

It has been suggested that at low concentrations, substrate binds only to the catalytic site of the enzyme. However, as the substrate concentration increases, the substrate may bind to another, allosteric site of the enzyme and may cause conformational changes, which ultimately lead to higher affinity of the substrate with the enzyme. The first suggestion of the involvement of an allosteric mechanism in P450 enzyme kinetics came with the discovery of sex differences in the *N*-demethylation of ethylmorphine by rat liver microsomes (185). In the presence of activators (different molecule to substrate) sigmoidal-type kinetics may also change to hyperbolic (181) as discussed in more detail later. In a sense, the foregoing P450 examples illustrate a classic ''homotropic or heterotropic regulatory enzyme cooperativity'' type behavior. Whether the term *allosteric* should be applied to P450 as absolute proof of the mechanism of interaction of activatable substrates and activators with P450 requires resolution.

C. Stimulation of P450-Mediated Xenobiotic Metabolism

One of the interesting features of CYP3A enzymes is that several chemicals stimulate CYP3A-mediated catalytic activities with human and rodent liver microsomes (186–188). The most potent stimulator of CYP3A catalytic activity found is α-naphthoflavone (αNF), although many other flavones and steroids also stimulate CYP3A4 catalytic activity (180,188,189). However, there is evidence that the catalytic activity of CYP1A1 (190), CYP1A2 (191,192), CYP2A6 (193), and CYP2C9 (194) can also be activated by various compounds in human liver microsomes. Extrahepatic P450 activation occurs with rat intestinal microsomes (195).

The mechanism of stimulation has been examined, but no clear conclusion is yet evident (168). The stimulators may affect the coupling of NADPH–CYTP450 reductase with P450 or enhance affinity of these enzymes for the substrate by an allosteric mechanism. Recently, it has been suggested that the stimulator and the substrate simultaneously bind to different sites in the active center of CYP3A4 (188). The authors suggested that this may influence the active site of CYP3A4, leading to higher turnover in the presence of stimulator. It is also possible that activators stimulate the enzyme activity by using different mechanisms, depending on the substrate and catalytic pathway in question. αNF stimulates benzo[*a*]pyrene and progesterone metabolism by decreasing dramatically the K_m of the substrate or by having no effect on the K_m of the substrate, indicating that there may be two different mechanisms involved (187,189). The stimulation of CYP3A4-mediated reactions can be regioselective. αNF stimulates CYP3A4-mediated aflatoxin B_1 8,9-oxidation and inhibits 3α-hydroxylation (181). Similarly midazolam 1′- and 4-hydroxylation were stimulated regioselectively by αNF and testosterone, respectively (196). This regioselective finding suggests that the αNF functions by binding the allosteric site, rather than increasing the interaction of NADPH–P450 reductase with CYP3A4.

Because there is evidence that stimulation also occurs in rats in vivo and with cultured human hepatocytes in vitro, it may have clinical importance (196,197). Caution, therefore, should be exerted when in vitro–in vivo correlations of stimulation of P450 are made, for assay conditions may affect the degree of stimulation.

D. Protein Binding

Another factor that may affect drug metabolism in vitro is protein binding. Initially, high concentrations of microsomal membrane were shown to inhibit benzo[*a*]pyrene hydroxylation by nonspecific binding of substrate at sites separate from the P450 catalytic site; consequently, decreasing the free substrate concentration (198). More recently, saturable rat liver microsomal binding of ethanol at physiologically relevant concentrations has been found to occur at a number of sites (exceeding those derived from CYP alone), some with high affinity and others with high capacity (199). Microsomal binding of xenobiotics is also demonstrated with coumarin analogues, such as dicoumarol and warfarin, which bind plasma proteins (200,201). Protein binding can also be affected by diet, including lipid intake which is observed on incubation of microsomes from differently treated animals (202). It has been suggested that determination of nonspecific binding in microsomes in vitro is necessary to obtain true kinetic data and thus provide more relevant correlations with in vivo. In fact, this, in turn, might explain the poor in vivo predictions from kinetic constants determined with microsomes as described by others (203; and see Sec XIII).

XVI. OTHER FACTORS AFFECTING ENZYMATIC ACTIVITY

Besides the effects of various assay conditions other factors may affect the enzymatic activity obtained in subcellular fractions, or other in vitro models, for that matter. The distribution of human and rat liver benzo[*a*]pyrene hydroxylase, a P450 activity, was initially identified as the highest in the 12,000 *g* supernatant compared with other subcellular fractions (204). Similarly, others had added soluble supernatant fractions to rat liver microsomes, and this increased microsomal *N*-demethylation (205) owing to increasing the amount of NADPH available for the enzyme.

Pelkonen et al. (204) also compared the levels of hepatic microsomal benzo[*a*]pyrene hydroxylase activity between sexes in human adult, human fetus, rat strains (Sprague-Dawley, Fischer, and Wistar), rabbit, and guinea pig. Enzyme activity was higher in all strains of male rats compared with females. In rabbit and human there was no evident gender difference in metabolism, whereas the female guinea pig possessed higher enzymatic activities than male (204). The variability in metabolism among multiple strains and sexes of animals has also been demonstrated using numerous enzyme activities in subcellular fractions from Swiss and nude mice. The P450-mediated metabolism of ethylmorphine and aminopyrine were highest in both male and female nude mice when compared with Swiss mice, and sex differences for all strains were minimal and restricted to some P450 activities (206).

Another factor that results in differing enzyme activities in vitro is the effect of diet and nutrition, which may also be studied in subcellular fractions obtained from organ tissues removed from treated and untreated animals. Various P450-mediated metabolic pathways (aniline hydroxylase, aminopyrene *N*-demethylase, *p*-nitroanisole *O*-demethylase, and total P450) have been evaluated in guinea pig microsomes from control and after a vitamin C-deficient diet (207). All enzymatic activities were significantly decreased after this vitamin C deficiency.

XVII. EXTRAHEPATIC FRACTIONS

As described earlier, and outlined in Table 3, many organs may be subfractionated using various methods. However, sedimentation of endoplasmic reticulum into the final 100,000-g pellet is variable with different extrahepatic tissues (208). The preparation differences and

P450 activities in many extrahepatic microsomal fractions obtained from multiple species has been thoroughly reviewed elsewhere by Burke and Orrenius (54). Therefore, the following list briefly outlines extrahepatic organs that have been used as a source of subcellular fractions (see Table 3).

When used as a source of subcellular fractions, the kidney is particularly complex to subfractionate because it consists of localized multiple cell types that themselves must first be isolated to provide any meaningful data on cell-specific metabolism. The standard differential centrifugation technique yields microsomal fractions contaminated with other cellular fragments (reviewed in Ref. 54). Enzyme activities are also localized to areas within the kidney, such as the loop of Henle or the proximal tubule, where they are involved in xenobiotic and endobiotic metabolism. P450, GST, and UDPGT activities were lower in microsomes from the whole kidney than from hepatic microsomes from the rat, rabbit, mouse, hamster, and guinea pig (22).

The lung is exposed to high blood flow and also has enzymes localized to various cell types. Besides the high content of connective tissue, which makes homogenization difficult, the preparation of microsomal fractions from rat lungs requires extensive washing to remove hemoglobin. Compared with liver, rabbit lung contains relatively little endoplasmic reticulum (209). Consequently, lung P450 activities are also much lower than liver (210). This difference is similar for GST, *N*-acetyl transferase (NAT), and UDPGT activities between pulmonary and hepatic microsomes of rat, mouse, hamster, guinea pig, and humans (22). Mouse lung contains the following enzymes as major components: carbonyl reductase in the mitochondria, malate dehydrogenase in cytosol, and P450 and NADPH–cytochrome *c* reductase in microsomes (211).

Few metabolism studies use subcellular fractions derived from the heart. However, one example focused on quinone metabolism in rat and guinea pig cardiac subcellular fractions that was localized to the mitochondrial and soluble fraction (212). In contrast, vascular tissue subcellular fractions are more widely used than cardiac tissue. This probably results from the research surrounding nitric oxide generation in this type of tissue. For example, nitric oxide (NO)-generating activities from sodium nitroprusside in bovine coronary artery postnuclear, postmitochondrial, mitochondrial, cytosolic, and microsomal fractions have been measured. This activity was both the highest and most NADPH-dependent in mitochondria and microsomes (51). Similarly, glycerol trinitrate metabolism in the same tissue preparation indicated cytosolic activity was the highest and was inhibited by GST inhibitors (213).

The epidermis is exposed to numerous cosmetics, toxicants, environmental xenobiotics, and is one of the largest organs of the body owing to surface area considerations. Owing to these properties, the skin has enzyme activity toward many chemicals (28). Mitochondria have also been isolated from epidermis after removal of fat from the tissue (32).

The use of nasal tissue from many laboratory animal species has been reviewed relative to metabolic activation by P450 and other drug-metabolizing enzymes (214). Interestingly, mouse nasal microsomes have coumarin 7-hydroxylase activity indicative of CYP2A, and this activity is 37- to 68-fold higher than in murine liver (20).

Localization of enzymes, particularly CYP3A4, within the gastrointestinal tract has only recently been identified as important in the metabolism of orally administered drugs, owing to a major role in first-pass metabolism. Gastrointestinal tract homogenates were originally used for drug metabolism, because stable microsomal preparations were hard to obtain; however, this problem was rectified by the addition of a trypsin inhibitor and glycerol (20%) in isotonic potassium chloride, which act as stabilizers (15). Human colon has also been used to assay 1-naphthol conjugation activity, with supernatants demonstrating more sulfation than glucuronidation (215). S-9 supernatant of intestinal biopsy tissue has been used along with midazolam 1′-hydroxylation to demonstrate an 11-fold variation in CYP3A4 in human intestine (216).

Although not as widely used as other tissues for in vitro metabolism, reproductive tissues are important for investigating metabolites of xenobiotics that are likely to result in teratogenesis (217). Siekevitz (48) used a method similar to that used with liver for the preparation of microsomes from placenta. Placental tissue subcellular fractions have also been prepared and acetylcholinesterase (microsomal) and butyrylcholinesterase (cytosolic) activities measured (19). The standard microsomal preparation protocol (see Fig. 2) is less effective with placental than with liver tissue. The addition of glycerol and heparin in homogenization solutions improves catalytic properties of placental microsomes (218,219). Human placenta expresses several P450s, although many catalytic activities are low compared with liver tissue (219,220). Mouse mammary gland subcellular fractions exhibit epoxide hydrolase and GST activities at much lower levels than in liver subcellular fractions (25). Rat testicular subcellular fractions can be prepared and used to study metabolism, as demonstrated using the Sertoli cell toxicant 1,3-dinitrobenzene, which is reductively metabolized by this S-9 fraction (23).

An additional organ infrequently utilized in drug metabolism is the adrenal cortex; however, mitochondrial P450 11β-hydroxylase activity in adrenal cortex is inhibited by imidazole etomidate (21). Unusually, human fetal adrenal subcellular fractions contain more P450s than fetal liver, for many P450-mediated activities have been identified (208,221). This may result from an endogenous role for these P450 in steroid metabolism during development. Although microsomes from various human fetal tissues are obtainable, their drug-metabolizing enzymatic activities may not be representative of adult (10,221), and also, fetal tissue expresses fetus-specific P450s, such as CYP3A7 (222).

Several studies have used brain tissue from various species for the preparation of subcellular fractions for use in metabolic studies. Rat brain subcellular fractions have been prepared in a method described by Alivisatos et al. (37). Brain subcellular fractions have been obtained and demonstrated higher mitochondrial P450 than microsomal P450 in humans (223). More recently, rat brain subcellular fractions have been isolated as described by Hilliard et al. (224) and used to study anandamide hydrolysis, with the highest activity in microsomes and myelin. Multiple enzymes involved in energy transduction and neurotransmission have also been studied in dog brain (34). Specific brain subcellular fractions have also been isolated and analyzed for the presence of numerous enzymatic activities. A protocol for mitochondrial preparation from homogenized brain tissue in sucrose/EDTA/heparin has also been detailed (33). Sheep brain homogenates have been used to measure distribution of fucosyltransferase, acetylcholinesterase, and cytochrome oxidase (225). Esterase activity in rat brain has been measured in subcellular fractions and in synaptosomes that were isolated by Ficol gradient centrifugation of the crude mitochondrial pellet (38). Isolation of synaptosomes has also been described by Donlon et al. (226).

Various drug-metabolizing enzyme activities have been identified and localized in the bovine eye (49). Subcellular fractions of the ciliary body possess higher activities than the pigmented epithelium. Polyaromatic hydrocarbons also induce enzyme activity in mice using entire eye tissue homogenates (49).

XVIII. FRACTIONS FROM INVERTEBRATES

The preparation of subcellular fractions from invertebrates has been widely employed for studying pesticide metabolism and the role of synergists. In addition invertebrate subcellular fractions have been applied to other fields of study. For example, GST and EH activities were lower in *Drosophila melanogastar* cytosol than in rat liver, but higher when expressed per gram body weight (227). Sulfotransferase, UDPGT, and phosphotransferase have also been studied in this system (228). *Drosophila melanogastar* larvae subcellular fractions can also be used to activate drugs, which can then be used to activate xenobiotics to mutagens in the *Salmonella* test system (44). Similarly, microsomes from a further invertebrate, tobacco budworm (*Heliothis virescens*) larvae, oxidatively metabolize cypermethrin (27).

XIX. USE WITH OTHER IN VITRO TECHNIQUES

Although the previous sections describe in detail how subcellular fractions may be characterized using enzyme or chemical probes and these activities compared across species and other in vitro models, subcellular fractions may also be used with other common in vitro techniques. An example is the use of enzyme inhibitors in numerous subcellular fractions obtained from different species which allows further resolution of a metabolic pathway. The metabolism of guanine analogue prodrugs to BRL 44385 was mediated by xanthine oxidase in rat liver cytosol and aldehyde oxidase in human liver cytosol after using enzyme-specific inhibitors (73). Multiple microsomal enzymes can also be detected using immunoquantification in normal and diseased liver (229) as well as numerous extrahepatic organs (230) using microsomes prepared according to the method of Kremers et al. (231). Immunoquantification has allowed the quantitative comparison of flavin-containing monooxygenase (FMO) in hog, mouse, rat, dog, rabbit, and human microsomal samples, for example (232). Immunohistochemistry is a further alternative technique that allows identification and localization of drug-metabolizing enzymes in individual whole cells, which is particularly useful when the quantity of tissue available is severely limited (233).

XX. SUMMARY AND CONCLUSION

The information discussed in this chapter indicates that subcellular fractions are obtainable from many different cell types, organs, and species and may be stored for a number of years at −80°C and still retain their enzymatic activity reflective of fresh tissue. The unique organ architectures, however, may complicate use of a standard method for preparation of a single subcellular fraction; and this may partially explain the numerous combinations of buffer constituents, homogenization conditions, and centrifugation protocols available both for the same organ as well as across species. Therefore, it is recommended that some form of characterization is performed to ensure the fraction is essentially monoorganellar. Characterization techniques include enzymatic, electron microscopic, or immunochemical techniques. For drug metabolism studies, microsomes are clearly the most widely used subcellular fraction, and the liver is the most completely studied organ for in vitro drug metabolism studies. There is a reasonable relationship using microsomes for in vitro–in vivo correlation of xenobiotic metabolism, although it is not perfect owing to confounding factors present within this system. The use of other fractions, such as cytosol or whole-cell models, may be necessitated if the xenobiotic undergoes metabolism mediated by enzymes found outside of the endoplasmic reticulum fraction. With reference to multiple metabolic pathways in numerous organelles, a single subcellular system will not allow sequential metabolism unless the whole pathway is catalyzed by enzymes found solely in the single preparation. Depending on the enzyme(s) involved, cofactors are required to facilitate metabolism, and this requires prior knowledge of the identity and the amounts of the cofactors necessary for optimal turnover. This represents one of the disadvantages of using this system. When human tissue is available the formation of a fully enzymatically phenotyped bank is possible, which is of enormous value for early definition of metabolic routes for new drugs and ultimately useful to screen out molecules that may present problems to some or all potential patients. If the bank is significantly large, it may also give some idea of the variability in metabolism that may be present in the patient population.

In conclusion, subcellular systems when energized with the required cofactors are a sophisticated representation of only part of the whole cell. They do not take into account the regulatory effects of cell–cell contact–communication nor the heterogeneous cellular nature of many organs. In this respect the use of organ tissue slices, isolated cells in suspension, or culture may be considered advantageous, but this is dependent on the organ and whether the xenobiotic and metabolites of interest can be measured following incubation in these systems. The optimal preparation of extracellular subcellular fractions may be complicated by the nature of the organ, as for instance, the microsomal fractionation method may be required to vary from tissue to tissue. Additional complications in vitro include the multiple incubation condition factors that can have significant effects on enzyme activity in subcellular fractions. These considerations aside, it is clear that subcellular fractions have many useful roles in the pharmaceutical industry, and their niche in the future seems assured.

REFERENCES

1. D. J. Birkett, P. I. Mackenzie, M. E. Veronese, and J. O. Miners, In vitro approaches can predict human drug metabolism, *Trends Pharmacol. Sci. 14*:292 (1993).
2. S. Ekins, Past, present, and future applications of precision-cut liver slices for in vitro xenobiotic metabolism, *Drug Metab. Rev. 28*:591 (1996).
3. P. Olinga, Human liver slices and Isolated hepatocytes in drug disposition and transplantation research, Ph.D. dissertation, The University of Groningen, The Netherlands, 1996.
4. A. D. Rodrigues, Use of in vitro human metabolism studies in drug development. An industrial perspective, *Biochem. Pharmacol. 48*:2147 (1994).
5. S. A. Wrighton, M. VandenBranden, J. C. Stevens, L. A. Shipley, B. J. Ring, A. E. Rettie, and J. R. Cashman, In vitro methods for assessing human hepatic drug metabolism: Their use in drug development, *Drug Metab. Rev. 25*:453 (1993)
6. S. A. Wrighton, B. J. Ring, and M. VandenBranden, The use of in vitro metabolism techniques in the planning and interpretation of drug safety studies, *Toxicol. Pathol. 23*:199 (1995).
7. S. A. Wrighton and P. M. Silber, Screening studies for metabolism and toxicology, In: *Accelerated Drug Discovery and Early Development: Scientific Strategies for Success,* pp 4.1–4.18, Technomic Publishing AG (1997).
8. A. Guillouzo, Acquisition and use of human in vitro liver preparations, *Cell Biol. Toxicol. 11*:141 (1995).

9. J. K. Beetham, D. Grant, M. Arand, J. Garbarino, T. Kiyosue, F. Pinot, W. R. Belknap, K. Shinozaki, and B. D. Hammock, Gene evolution of epoxide hydrolases and recommended nomenclature, *DNA Cell Biol. 14*:61 (1995).
10. G. M. Pacifici, S. S. Park, H. V. Gelboin, and A. Rane, 7-Ethoxycoumarin and 7-ethoxyresorufin *O*-deethylase in human foetal and adult liver: Studies with monoclonal antibodies, *Pharmacol. Toxicol. 63*:26 (1988).
11. J. B. Schenkman and D. L. Cinti, Preparation of microsomes with calcium. *Methods Enzymol. 52*:83 (1978).
12. T. Omura and R. Sato, The carbon monoxide-binding pigment of liver microsomes, *J. Biol. Chem. 239*:2370 (1964).
13. G. W. Winston, M. A. Kirchin, and M. J. J. Ronis, Microsomal activation of benzo[*a*]pyrene by *Alligator mississippiensis*: Mechanisms, mutagenicity and induction, *Biochem. Soc. Trans. 19*:746 (1991).
14. B. D. Haehner, J. C. Gorski, M. VandenBranden, S. A. Wrighton, S. K. Janardan, P. B. Watkins, and S. D. Hall, Bimodal distribution of renal cytochrome P450 3A activity in humans, *Mol. Pharmacol. 50*:52 (1996).
15. S. J. Stohs, R. C. Grafström, M. D. Burke, P. W. Moldeus, and S. G. Orrenius, The isolation of rat intestinal microsomes with stable cytochrome P-450 and their metabolism of benzo(α)-pyrene, *Arch. Biochem. Biophys. 177*:105 (1976).
16. A. G. E. Wilson, H. C. Kung, M. W. Anderson, and T. E. Eling, Covalent binding of intermediates formed during the metabolism of arachidonic acid by human platelet subcellular fractions, *Prostaglandins 18*:409 (1979).
17. H. R. Soliman, D. Dire, P. Boudou, R. Julien, J. Launay, J. Brerault, J. Villette, and J. Fiet, Characterization of estrone sulfatase activity in human thrombocytes, *J. Steroid Biochem. Mol. Biol. 46*:215 (1993).
18. H. Heyn, Y. G. Bankmann, and M. W. Anders, Tissue distribution and stereoselectivity of remacemide-glycine hydrolase, *Drug Metab. Dispos. 22*:973 (1994).
19. C. Simone, L. O. Derewlany, M. Oskamp, D. Johnson, B. Knie, and G. Koren, Acetylcholinesterase and butyrylcholinesterase activity in the human term placenta: Implications for fetal cocaine exposure, *J. Lab. Clin. Invest. 123*:400 (1994).
20. T. Su, J. J. Sheng, T. W. Lipinskas, and X. Ding, Expression of CYP2A genes in rodent and human nasal mucosa, *Drug Metab. Dispos. 24*:884 (1996).
21. H. Vanden Bossche, G. Willemsens, W. Cools, and D. Bellens, Effects of etomidate on steroid biosynthesis in subcellular fractions of bovine adrenals, *Biochem. Pharmacol. 33*: 3861 (1984).
22. C. L. Litterst, E. G. Mimnaugh, R. L. Reagan, and T. E. Gram, Comparison of in vitro drug metabolism by lung, liver, and kidney of several common laboratory species, *Drug Metab. Dispos. 3*:259 (1975).
23. M. K. Ellis and P. M. D. Foster, The metabolism of 1,3-dinitrobenzene by rat testicular subcellular fractions, *Toxicol. Lett. 62* 201 (1992).
24. D. D. Tyler and J. Gonze, The preparation of thyroid and thymus mitochondria, *Methods Enzymol. 10*:101 (1967).
25. M. H. Silva, R. N. Wixtrom, and B. D. Hammock, Epoxide-metabolising enzymes in mammary gland and liver from BALB/c mice and effects of inducers on enzyme activity, *Cancer Res. 48*:1390 (1988).
26. L. X. Zhou, B. Pihlstrom, J. P. Hardwick, S. S. Park, S. A. Wrighton, and J. L. Holtzman, Metabolism of phenytoin by the gingiva of normal humans: The possible role of reactive metabolites of phenytoin in the initiation of gingival hyperplasia, *Clin. Pharmacol. Ther. 60*: 191 (1996).
27. A. R. McCaffery, C. H. Walker, S. E. Clarke, and K. S. Lee, Enzymes and resistance to insecticides in *Heliothis virescens, Biochem. Soc. Trans. 19*:762 (1991).

28. D. R. Bickers, T. Dutta-Choudhury, and H. Mukhtar, Epidermis: A site of drug metabolism in neonatal rat skin. *Mol. Pharmacol. 21*:239 (1982).
29. D. Johnson and H. Lardy, Isolation of liver or kidney mitochondria, *Methods Enzymol. 10*: 94 (1967).
30. D. D. Tyler and J. Gonze, The preparation of heart mitochondria from laboratory animals, *Methods Enzymol. 10*:75 (1967).
31. P. V. Blair, The large-scale preparation and properties of heart mitochondria from slaughterhouse material, *Methods Enzymol. 10*:78 (1967).
32. L. Ernster and K. Nordenbrand, Skeletal muscle mitochondria, *Methods Enzymol. 10*:86 (1967).
33. R. E. Basford, Preparation and properties of brain mitochondria, *Methods Enzymol. 10*:96 (1967).
34. E. Arrigoni, G. Benzi, D. Curti, F. Dagani, S. Gallico, A. Gorini, V. Mandelli, F. Marzatico, A. Moretti, and R. F. Villa, Effect of hypoxia and pharmacological treatment on some enzyme activities in dog brain areas, *Arch. Int. Pharmacodyn. 269*:111 (1984).
35. W. L. Nelson and R. A. Butow, Guinea pig mammary gland mitochondria, *Methods Enzymol. 10*:103 (1967).
36. G. Blobel and V. R. Potter, Nuclei from rat liver: Isolation method that combines purity with high yield, *Science 154*:1662 (1996).
37. S. G. A. Alivisatos, F. Ungar, and S. S. Parmar, Effect of monoamine oxidase inhibitors on the labelling of subcellular fractions of brain and liver by ^{14}C-serotonin, *Biochem. Biophys. Res. Commun. 25*:495 (1966).
38. A. Durrer, B. Walther, A. Racciatti, G. Boss, and B. Testa, Structure-metabolism relationships in the hydrolysis of nicotinate esters by rat liver and brain subcellular fractions, *Pharm. Res. 8*:832 (1991).
39. L. Ernster, P. Siekevitz, and G. E. Palade, Enzyme structure relationships in the endoplasmic reticulum, *J. Cell Biol. 15*:541 (1962).
40. M. R. Adelman, G. Blobel, and D. D. Sabatini, Nondestructive separation of rat liver rough microsomes into ribosomal and membranous components, *Methods Enzymol. 31*:201 (1974).
41. J. J. M. Bergeron, Golgi fractions from livers of control and ethanol-intoxicated rats, *Biochim. Biophys. Acta 555*:493 (1979).
42. W. H. Evans, Hepatic endosomes: Preparation, properties and roles in receptor recycling, *Biochem. Soc. Trans. 14*:170 (1986).
43. J. Loeper, V. Descatoire, M. Maurice, P. Beaune, J. Belghiti, D. Houssin, F. Ballet, G. Feldmann, F. P. Guengerich, and D. Pessayre, Cytochromes P-450 in human hepatocyte plasma membrane: Recognition by several autoantibodies, *Gastroenterology 104*:203 (1993).
44. I. Hällström, A. Sundvall, U. Rannug, R. Grafström, and C. Ramel, The metabolism of drugs and carcinogens in isolated subcellular fractions of *Drosophila melanogaster*. I activation of vinyl chloride, 2-aminanthracene and benzo[*a*]pyrene as measured by mutagenic effects in *Salmonella typhimurium. Chem. Biol. Interact. 34*:129 (1981).
45. J. T. Hjelle, D. R. Peterson, and J. J. Hjelle, Drug metabolism in isolated proximal tubule cells: Aldehyde dehydrogenase, *J. Pharmacol. Exp. Ther. 224*:699 (1983).
46. S. Suzuki and E. Ogawa, Experimental studies on the carbonic anhydrase activity—XI. Effect of adrenocorticosteroids on carbonic anhydrase and Na^+–K^+-activated adenosine triphosphatase from kidney subcellular fractions in normal mice and rats, *Biochem. Pharmacol. 17*: 1855 (1968).
47. L. Balk, S. Maner, A. Bergstrand, and J. W. DePierre, Preparation and characterization of subcellular fractions suitable for studies of drug metabolism from the trunk kidney of the northern pike (*Esox lucius*) and assay of certain enzymes of xenobiotic metabolism in these subfractions, *Biochem. Pharmacol. 33*:2447 (1984).

48. P. Siekevitz, Preparation of microsomes and submicrosomal fractions: Mammalian, *Methods Enzymol.* *5*:61 (1962).
49. H. Shichi and D. W. Nebert, Genetic differences in drug metabolism associated with ocular toxicity, *Environ. Health Perspect.* *44*:107 (1982).
50. G. P. Smith, R. R. MacGregor, and T. J. Peters, Subcellular localisation of leucine aminopeptidase in human polymorphonuclear leukocytes, *Biochim. Biophys. Acta* *728*:222 (1983).
51. E. A. Kowaluk, P. Seth, and H.-L. Fung, Metabolic activation of sodium nitroprusside to nitric oxide in vascular smooth muscle, *J. Pharmacol. Exp Ther.* *262*:916 (1992).
52. E. G. Gray and V. P. Whittaker, The isolation of nerve endings from brain: An electron-microscopic study of cell fragments derived by homogenisation and centrifugation, *J. Anat.* *96*:79 (1962).
53. C. A. Pasternak, The biosynthesis of amino sugars by intestinal mucosa, *Biochem. J.* *78*:25P (1961).
54. M. D. Burke and S. Orrenius, Isolation and comparison of endoplasmic reticulum membranes and their mixed function oxidase activities from mammalian extrahepatic tissues, *Pharmacol. Ther.* *7*:549 (1979).
55. J. R. Bend and C. J. Serabjit-Singh, Xenobiotic metabolism by extrahepatic tissues: Relationship to target organ and cell toxicity, *Drug Metabolism and Drug Toxicity* (J. R. Mitchell and M. G. Horning, eds.), Raven Press, New York, 1984, p. 99.
56. T. A. van der Hoeven and M. J. Coon, Preparation and properties of partially purified cytochrome P-450 and reduced nicotinamide adenine dinucleotide phosphate–cytochrome P-450 reductase from rabbit liver microsomes, *J. Biol. Chem.* *19*:6302 (1974).
57. Y. Moulé and J. Chauveau, The cell components of the liver. Isolation, morphology, biochemistry, *The Liver, Morphology, Biochemistry, Physiology*, (C. Rouille, ed.). Academic Press, New York, 1963, p. 379.
58. G. Dallner, Isolation of microsomal subfractions by use of density gradients, *Methods Enzymol* *52*:71 (1978).
59. D. J. P. Critchley, D. J. Rance, and C. Beedham, Subcellular localisation of guinea pig hepatic molybdenum hydroxylases, *Biochem. Biophys. Res. Commun.* *185*:54 (1992).
60. W. B. Morris, G. D. Smith, and T. J. Peters, Subcellular fractionation of liver organelles from phenobarbital-treated rats by counter-current partition and sucrose gradient centrifugation, *Biochem. Pharmacol.* *35*:2187 (1986).
61. S. O. C. Tottmar, H. Petterson, and K. Kiessling, The subcellular distribution and properties of aldehyde dehydrogenases in rat liver, *Biochem. J.* *135*:577 (1973).
62. L. Balk, J. Meijer, A. Bergstrand, A. Åstrom, R. Morgenstern, J. Seidegard, and J. W. DePierre, Preparation and characterization of subcellular fractions from the liver of the northern pike, *Esox lucius, Biochem. Pharmacol.* *31*:1491 (1982).
63. J. F. Kirsch, P. Siekevitz, and G. E. Palade, Amino acid incorporation in vitro by ribonucleoprotein particles detached from guinea pig liver microsomes. *J. Biol. Chem.* *235*:1419 (1960).
64. A. D. Rodrigues, J. L. Ferrero, M. T. Amann, G. A. Rotert, S. P. Cepa, B. W. Surber, J. M. Machinist, N. R. Tich, J. P. Sullivan, D. S. Garvey, M. Fitzgerald, and S. P. Arneric, The in vitro hepatic metabolism of ABT-418, A cholinergic channel activator, in rats, dogs, cynomolgus monkeys, and humans, *Drug Metab. Dispos.* *22*:788 (1994).
65. A. D. Rodrigues, M. J. Kukulka, J. L. Ferrero, and J. R. Cashman, In vitro hepatic metabolism of ABT-418 in chimpanzee (*Pan troglodytes*). A unique pattern of microsomal flavin-containing monooxygenase-dependent stereoselective *N′*-oxidation, *Drug Metab. Dispos.* *23*: 1143 (1995).
66. R. E. Billings, R. E. McMahon, J. Ashmore, and S. R. Wagle, The metabolism of drugs in isolated rat hepatocytes, *Drug Metab. Dispos.* *5*:518 (1977).
67. M. C. Castle, Digitoxin metabolism by rat liver homogenates, subcellular fractions and iso-

lated hepatocytes: Stimulation by spironolactone and pregnenolone-16α-carbonitrile, *J. Pharmacol. Exp. Ther. 211*:120 (1979).

68. J. I. Brodfuehrer, D. E. Chapman, T. J. Wilke, and G. Powis, Comparative studies of the in vitro metabolism and covalent binding of ^{14}C benzene by liver slices and microsomal fraction of mouse, rat, and human, *Drug Metab. Dispos. 18*:20 (1990).
69. S. Gerayesh-Nejad, R. S. Jones, and D. V. Parke, Comparison of rat hepatic microsomal mixed-function oxidase activities in microsomal preparations, isolated hepatocytes and liver slices, *Biochem. Soc. Trans. 3*:403 (1975).
70. F. Berthou, D. Ratansavanh, C. Riche, D. Picart, T. Voirin, and A. Guillouzo, Comparison of caffeine metabolism by slices, microsomes and hepatocyte cultures from adult human liver, *Xenobiotica 19*:401 (1989).
71. B. G. Lake, M. J. Sauer, F. Esclangon, J. A. Beamand, R. J. Price, and D. G. Walters, Metabolism of coumarin by precision-cut calf liver slices and calf liver microsomes, *Xenobiotica 25*:133 (1995).
72. A. E. M. Vickers, S. Connors, M. Zollinger, W. A. Biggi, A. Larrauri, J. P. W. Vogelaar, and K. Brendel, The biotransformation of the ergot derivative CQA 206–291 in human, dog, and rat liver slices cultures and prediction of in vivo plasma clearance, *Drug Metab. Dispos. 21*:454 (1993).
73. A. W. Harrell, S. M. Wheeler, P. East, S. E. Clarke, and R. J. Chenery, Use of rat and human in vitro systems to assess the effectiveness and enzymology of deoxy-guanine analogues as prodrugs of an antiviral agent, *Drug Metab. Dispos. 22*:124 (1994).
74. K. Zomorodi, D. J. Carlile, and J. B. Houston, Kinetics of diazepam metabolism in rat hepatic microsomes and hepatocytes and their use in predicting in vivo clearance, *Xenobiotica 25*: 907 (1995).
75. B. Lacarelle, R. Rahmani, G. de Sousa, A. Durand, M. Placidi and J. P. Cano, Metabolism of digoxin, digoxigenin digitoxosides and digoxigenin in human hepatocytes and liver microsomes, *Fundam. Clin. Pharmacol. 5*:567 (1991).
76. K. Lavrijsen, J. Van Houdt, D. Van Dyck, J. Hendrickx, M. Bockx, R. Hurkmans, W. Meuldermans, L. Le Jeune, W. Lauwers, and J. Heykants, Comparative metabolism of flunarizine in rats, dogs and man: An in vitro study with subcellular liver fractions and isolated hepatocytes, *Xenobiotica 22*:815 (1992).
77. U. G. Sidelmann, C. Cornett, J. Tjornelund, and S. H. Hansen, A comparative study of precision cut liver slices, hepatocytes, and liver micrsomes from the Wistar rat using metronidazole as a model substance, *Xenobiotica 26*:709 (1996).
78. G. L. Kedderis, L. S. Argenbright, J. S. Walsh, J. L. Smith, R. A. Stearns, B. A. Arison, and G. T. Miwa, Metabolism of a nitrogen-containing steroid by rat hepatocytes and hepatic subcellular fractions, *Drug Metab. Dispos. 17*:606 1989.
79. S. R. Koepke, Y. Tondeur, J. G. Farrelly, M. L. Stewart, C. J. Michejda, and M. B. Kroeger-Koepke, Metabolic studies of ^{15}N-labelled *n*-nitrosoproline in isolated rat hepatocytes and subcellular fractions, *Cancer Lett. 27*:277 (1985).
80. O. Bassir and G. O. Emerole, Species differences in the metabolism of palmotoxins B_0 and G_0 in vitro, *Xenobiotica 5*:649 (1975).
81. E. I. L. Ashworth, D. J. Carlile, R. Chenery, and J. B. Houston, Prediction of in vivo disposition from in vitro systems: Clearance of phenytoin and tolbutamide using rat hepatic microsomal and hepatocyte data, *J. Pharmacol. Exp. Ther. 274*:761 (1995).
82. G. Belvedere, E. Elovaara, and H. Vainio, Activation of styrene to styrene oxide in hepatocytes and subcellular fractions of rat liver, *Toxicol. Lett. 23*:157 (1984).
83. A. D. Rodrigues, D. J. Mulford, R. D. Lee, B. W. Surber, M. J. Kukulka, J. L. Ferrero, S. B. Thomas, M. S. Shet, and R. W. Estabrook, In vitro metabolism of terfenadine by a purified recombinant fusion protein containing cytochrome P4503A4 and NADPH-P450 reductase, *Drug Metab. Dispos. 23*:765 (1995).

84. D. Engelhardt and H. J. Karl, Testosterone metabolism in slices and subcellular fractions of human liver, *Acta Endocrinol. 173 (suppl):132*, (1973).
85. K. L. Salyers, J. Barr, and I. G. Sipes, In vitro metabolism of theophylline by rat and human liver tissue, *Xenobiotica 24*:389 (1994).
86. D. E. Chapman, T. J. Moore, S. R. Michener, and G. Powis, Metabolism and covalent binding of [^{14}C]toluene by human and rat liver microsomal fractions and liver slices, *Drug Metab. Dispos. 18*:929 (1990).
87. D. E. Chapman, S. R. Michener, and G. Powis, Metabolism of the flame retardant plasticizer tris(2-chloroethyl)phosphate by human and rat liver preparations, *Fundam. Appl. Toxicol. 17*:215 (1991).
88. J. W. Cox, G. Dring, L. C. Ginsberg, P. G. Larson, D. A. Constable, and R. G. Ulrich, Distribution and disposition of trospectomycin sulfate in the in vivo rat, perfused liver, and cultured hepatocytes, *Drug Metab. Dispos. 18*:726 (1990).
89. S. J. Grossman, E. G. Herold, J. M. Drey, D. W. Alberts, D. R. Umbenhauer, D. H. Patrick, D. Nicoll-Griffith, N. Chauret, and J. A. Yergey, CYP1A1 specificity of verlukast epoxidation in mice, rats, rhesus monkeys, and humans, *Drug Metab. Dispos. 21*:1029 (1993).
90. R. Bort, X. Ponsoda, E. Carrasco, M. J. Gomez-Lechon, and J. V. Castell, Comparative metabolism of the nonsteroidal antiinflammatory drug, aceclofenac, in the rat, monkey, and human, *Drug Metab. Dispos. 24*:969 (1996).
91. J. C. Stevens, L. A. Shipley, J. R. Cashman, M. VandenBranden, and S. A. Wrighton, Comparison of human and rhesus monkey in vitro phase I and II hepatic drug metabolism activities, *Drug Metab. Dispos. 21*:753 (1993).
92. J. E. Sharer, L. A. Shipley, M. VandenBranden, S. N. Binkley, and S. A. Wrighton, Comparisons of phase I and phase II in vitro hepatic enzyme activities of human, dog, rhesus monkey, and cynomolgus monkey, *Drug Metab. Dispos. 23*:1231 (1995).
93. U. C. T. Oppermann, E. Maser, S. A. Mangoura, and K. J. Netter, Heterogeneity of carbonyl reduction in subcellular fractions and different organs in rodents, *Biochem. Pharmacol. 42*: S189 (1991).
94. D. X. Lin, N. P. Lang, and F. F. Kadlubar, Species differences in the biotransformation of the food-borne carcinogen 2-amino-1-methyl-6-phenylimidazo[4,5-*b*]pyridine by hepatic microsomes and cytosols from humans, rats and mice, *Drug Metab. Dispos. 23*:518 1995.
95. M. Olsen and K. Kiessling, Species differences in zearalenone-reducing activity in subcellular fractions of liver from female domestic animals, *Acta Pharmacol. Toxicol. 52*:287 (1983).
96. S. W. Huskey, D. C. Dean, R. R. Miller, G. H. Rasmusson, and S. L. Chiu, Identification of human cytochrome P450 isozymes responsible for the in vitro oxidative metabolism of finasteride, *Drug Metab. Dispos. 23*:1126 (1995).
97. W. C. Schneider, Intracellular distribution of enzymes III. The oxidation of octanoic acid by rat liver fractions, *J. Biol. Chem. 176*:259 (1948).
98. P. K. Chan and A. W. Hayes, In vitro metabolism of penicillic acid with mouse-liver homogenate fractions, *Food Chem. Toxicol. 20*:61 (1982).
99. T. Akao and K. Kobashi, Metabolism of glycyrrhetic acid by rat liver microsomes: Glycyrrhetinate dehydrogenase, *Biochim. Biophys. Acta 1042*:241 (1990).
100. Y. Imai, A. Ito, and R. Sato, Evidence for biochemically different types of vesicles in the hepatic microsomal fraction, *J. Biochem. 60*:417 (1966).
101. J. A. Taylor, A. R. Limbrick, D. Allan, and J. D. Judah, Isolation of highly purified Golgi membranes from rat liver. Use of cycloheximide in vivo to remove Golgi contents, *Biochim. Biophys. Acta 769*:171 (1984).
102. F. P. Guengerich, Analysis and characterisation of enzymes, *Principles and Methods of toxicology* (A. W. Hayes, ed.), Raven Press, New York, 1994, p. 1259.
103. R. R. Brown, J. A. Miller, and E. C. Miller, The metabolism of methylated aminoazo dyes, *J. Biol. Chem. 209*:211 (1954).
104. A. H. Conney, E. C. Miller, and J. A. Miller, The metabolism of methylated aminoazo dyes

V. Evidence for induction of enzyme synthesis in the rat by 3-methylcholanthrene, *Cancer Res. 16*:450 (1956).

105. A. H. Conney, R. R. Brown, J. A. Miller, and E. C. Miller, The metabolism of methylated aminoazo dyes VI. Intracellular distribution and properties of the demethylase system, *Cancer Res. 17*:628 (1957).
106. M. Klingenberg, Pigments of rat liver microsomes, *Arch. Biochem. Biophys. 75*:376 (1958).
107. D. Garfinkel, Studies on pig liver microsomes. I. Enzymic and pigment composition of different microsomal fractions, *Arch. Biochem. Biophys. 77*:493 (1958).
108. G. Powis and A. R. Boobis, Effect of washing the hepatic microsomal fraction in sucrose solutions and in sucrose solution containing EDTA upon the metabolism of foreign compounds, *Biochem. Pharmacol. 24*:1771, (1975).
109. D. R. Nelson, L. Koymans, T. Kamataki, J. J. Stegeman, R. Feyreisen, D. J. Waxman, M. R. Waterman, O. Gotoh, M. J. Coon, R. W. Estabrook, I. C. Gunsalus, and D. W. Nebert, P450 superfamily: Update on new sequences, gene mapping, accession numbers and nomenclature, *Pharmacogenetics 6*:1 (1996).
110. C. von Bahr, C.-G. Groth, H. Jansson, G. Lundgren, M. Lind, and H. Glaumann, Drug metabolism in human liver in vitro: Establishment of a human liver bank, *Clin. Pharmacol. Ther. 27*:711 (1980).
111. D. Wu and A. I. Cederbaum, Presence of functionally active cytochrome P-450IIE1 in the plasma membrane, *Hepatology 15*:515 (1992).
112. E. P. A. Neve, E. Eliasson, M. A. Pronzato, E. Albano, U. Marinari, and M. Ingelman-Sundberg, Enzyme-specific transport of rat liver cytochrome P450 to the Golgi apparatus, *Arch. Biochem. Biophys. 333*:459 (1996).
113. L. R. Shull, D. G. Kirsch, C. L. Lohse, and J. A. Wisniewski, Application of isolated hepatocytes to studies of drug metabolism in large food animals, *Xenobiotica 17*:345 (1987).
114. I. C. Shaw, The use of in vitro methods for studying drug metabolism to replace animal studies as part of veterinary medicines product license applications, *Toxicol. In Vitro 8*:917 (1994).
115. A. R. Boobis, M. J. Brodie, G. C. Kahn, D. R. Fletcher, J. H. Saunders, and D. S. Davies, Monooxygenase activity of human liver in microsomal fractions of needle biopsy specimens, *Br. J. Clin. Pharmacol. 9*:11 (1980).
116. C. von Bahr, H. Glaumann, B. Mellström, and F. Sjöqvist, In vitro assessment of hepatic drug metabolism in man: a clinical pharmacological perspective, *Drug metabolism and distribution* (J. W. Lamble, ed.), Elsevier Biomedical Press, Amsterdam, 1983, pp. 152.
117. A. V. Loud, A quantitative stereological description of the ultrastructure of normal rat liver parenchymal cells, *J. Cell Biol. 37*:27 (1968).
118. E. R. Weibel, W. Stäubli, H. R. Gnagi, and F. A. Hess, Correlated morphometric and biochemical studies on the liver cell I. Morphometric model, stereologic methods, and normal morphometric data for rat liver, *J. Cell Biol. 42*:68 (1969).
119. W. Stäubli, R. Hess, and E. R. Weibel, Correlated morphometric and biochemical studies on the liver cell, *J. Cell Biol. 42*:92 (1969).
120. V. R. Potter and C. A. Elvehjem, A modified method for the study of tissue oxidations, *J. Biol. Chem. 114*:495 (1996).
121. R. M. Norton, H. L. White, and B. R. Cooper, Metabolism of BW 1370U87 by crude liver homogenates from several species: An in vitro method for preliminary investigation of species differences in metabolism, *Drug Dev. Res. 25*:229 (1992).
122. J. Zhang, R. A. Dean, M. R. Brzezinski, and W. F. Bosron, Gender-specific differences in activity and protein levels of cocaine carboxylesterase in rat tissues, *Life Sci. 59*:1175 (1996).
123. J. M. S. Van Maanen, W. J. V. Oort, and H. M. Pinedo, In vitro and in vivo metabolism of VP16-213 in the rat, *J. Cancer Clin. Oncol. 18*:885 (1982).
124. D. E. Chapman, T. A. Christensen, S. R. Michener, and G. Powis, Xenobiotic metabolism

studies with human liver, *Human Drug Metabolism from Molecular Biology to Man* (E. H. Jeffery, ed.), CRC Press, Boca Raton, 1993, p. 53.
125. R. E. Pearce, C. J. McIntyre, A. Madan, U. Sanzgiri, A. J. Draper, P. L. Bullock, D. C. Cook, L. A. Burton, J. Latham, C. Nevins, and A. Parkinson, Effects of freezing, thawing, and storing human liver microsomes on cytochrome P450 activity, *Arch. Biochem. Biophys. 331*: 145 (1996).
126. G. Powis, I. Jardine, R. Van Dyke, R. Weinshilboum, D. Moore, T. J. Wilke, W. Rhodes, R. Nelson, L. Benson, and C. Szumlanski, Foreign compound metabolism studies with human liver obtained as surgical waste, *Drug Metab. Dispos. 16*:582 (1988).
127. J. Meijer, A. Bergstrand, and J. W. DePierre, Preparation and characterization of subcellular fractions from the liver of C57B1/6 mice, with special emphasis on their suitability for use in studies of epoxide hydrolase activities, *Biochem. Pharmacol. 36*:1139 (1987).
128. C. Bauer, C. Corsi, and M. Paolini, Stability of microsomal monooxygenases in murine liver S9 fractions derived from phenobarbital and B-napthoflavone induced animals under various long-term conditions of storage, *Teratogen. Carcinog. Mutagen. 14*:13 (1994).
129. T. F. Woolf and J. D. Adams, Biotransformation of ketamine, (Z)-6-hydroxyketamine, and (*E*)-6-hydroxyketamine by rat, rabbit, and human liver microsomal preparations, *Xenobiotica 17*:839 (1987).
130. G. K. Poon, J. Wade, J. Bloomer, S. E. Clarke, and J. Maltas, Rapid screening of Taxol metabolites in human microsomes by liquid chromatography/electrospray ionisation–mass spectrometry, *Rapid Commun. Mass Spectrom. 10*:1165 (1996).
131. R. Singh, S. Y. Chang, and L. C. E. Taylor, In vitro metabolism of a potent HIV-protease inhibitor (141W94) using rat, monkey and human liver S9. *Rapid Commun Mass Spectrom. 10*:1019 (1996).
132. R. H. Adamson, R. L. Dixon, F. L. Francis, and D. P. Rall, Comparative biochemistry of drug metabolism by azo and nitroreductase, *Biochemistry 54*:1386 (1965).
133. K. Lavrijsen, J. Van Houdt, D. Van Dyck, J. Hendrickx, W. Lauwers, R. Hurkmans, M. Bockx, C. Janssen, W. Meuldermans, and J. Heykants, Biotransformation of sufentanil in liver microsomes of rats, dogs, and humans, *Drug Metab. Dispos. 18*:704 (1990).
134. T. Glinsukon, T. Benjamin, P. H. Grantham, E. K. Weisburger, and P. P. Roller, Enzymic *N*-acetylation of 2,4-toluenediamine by liver cytosols from various species, *Xenobiotica 5*: 475 (1975).
135. S. Kaneda, T. Seno, and K. Takeishi, Species difference in liver microsomal and cytosolic enzymes involved in mutagenic activation of *n*-hydroy-*n*-2-fluorenylacetamide, *J. Natl. Cancer Inst. 67*:549 (1981).
136. S. Kaneda, T. Seno, and K. Takeishi, Main pathway for mutagenic activation of 2-acetylaminofluorene by guinea pig liver homogenates, *Biochem. Biophys. Res. Commun. 90*:750 (1979).
137. A. Garras, D. K. Asiedu, and R. K. Berge, Subcellular localisation and induction of NADH-sensitive acetyl-CoA hydrolase and propionyl-CoA hydrolase activities in rat liver under lipogenic conditions after treatment with sulphur-substituted fatty acids, *Biochim. Biophys. Acta 1255*:154 (1995).
138. N. W. McCracken, P. G. Blain, and F. M. Williams, Nature and role of xenobiotic metabolising esterases in rat liver, lung, skin and blood, *Biochem. Pharmacol. 45*:31 (1993).
139. N. W. McCracken, P. G. Blain, and F. M. Williams, Human xenobiotic metabolism in esterases in liver and blood, *Biochem. Pharmacol. 46*:1125 (1993).
140. C. Dierks-Ventling and A. L. Cone, Ketone body enzymes in mammalian tissues, *J. Biol. Chem. 246*:5533 (1971).
141. L. Cantoni, M. Salmona, T. Facchinetti, C. Pantarotto, and G. Belvedere, Hepatic and extrahepatic formation and hydration of styrene oxide in vitro in animals of different species and sex, *Toxicol. Lett. 2*:179 (1978).

142. G. M. Pacifici, A. R. Boobis, M. J. Brodie, M. E. McManus, and D. S. Davies, Tissue and species differences in enzymes of epoxide metabolism, *Xenobiotica 11*:73 (1981).
143. R. Morgenstern, J. Meijer, J. W. DePierre, and L. Ernster, Characterization of rat-liver microsomal glutathione *S*-transferase activity, *Eur. J. Biochem. 104*:167 (1980).
144. J. N. M. Commandeur, G. J. Stijntjes, and N. P. E. Vermeulen, Enzymes and transport systems involved in the formation and disposition of glutathione *S*-conjugates, *Pharmacol. Rev. 47*:271 (1995).
145. H. Sies, A. Wahlländer, C. Waydhas, S. Soboll, and D. Haberle, Functions of intracellular glutathione in hepatic hydroperoxide and drug metabolism and the role of extracellular glutathione, *Adv. Enzyme Regul. 18*:303 (1980).
146. J. Lorenz, H. R. Glatt, R. Fleischmann, R. Ferlinz, and F. Oesch, Drug metabolism in man and its relationship to that in three rodent species: monooxygenase, epoxide hydrolase, and glutathione *S*-transferase activities in subcellular fractions of lung and liver, *Biochem. Med. 32*:43, (1984).
147. L. C. Mark, L. Brand, S. Kamvyssi, R. C. Britton, J. M. Perel, M. A. Landrau, and P. G. Dayton, Thiopental metabolism by human liver in vivo and in vitro, *Nature 206*:1117 (1965).
148. J. B. Houston, Utility of in vitro drug metabolism data in predicting in vivo metabolic clearance, *Biochem. Pharmacol. 47*:1469 (1994).
149. J. B. Houston, Relevance of in vitro kinetic parameters to in vivo metabolism of xenobiotics, *Toxicol. In Vitro 8*:507 (1994).
150. T. Iwatsubo, N. Hirota, T. Ooie, H. Suzuki, and Y. Sugiyama, Prediction of in vivo drug disposition from in vitro data based on physiological pharmacokinetics, *Biopharm. Drug Dispos. 17*:273 (1996).
151. J. H. Lin, Species similarities and differences in pharmacokinetics, *Drug Metab. Dispos. 23*: 1008 (1995).
152. J. R. Cashman, B. Y. T. Perotti, C. E. Berkman, and J. Lin, Pharmacokinetics and molecular detoxication, *Environ. Health Perspect. 104* (Suppl 1):23 (1996).
153. B. A. Hoener, Predicting the hepatic clearance of xenobiotics in humans from in vitro data, *Biopharm. Drug Dispos. 15*:295 (1994).
154. P. D. Worboys, A. Bradbury, and J. B. Houston, Kinetics of drug metabolism in rat liver slices. Rates of oxidation of ethoxycoumarin and tolbutamide, examples of high- and low-clearance compounds, *Drug Metab. Dispos. 23*:393 (1995).
155. P. D. Worboys, A. Bradbury, and J. B. Houston, Kinetics of drug metabolism in rat liver slices. II. Comparison of clearance by liver slices and freshly isolated hepatocytes, *Drug Metab. Dispos. 24*:676 (1996).
156. P. Dogterom, Development of a simple incubation system for metabolism studies with precision-cut liver slices, *Drug Metab. Dispos. 21*:699 (1993).
157. S. Ekins, G. I. Murray, M. D. Burke, J. A. Williams, N. C. Marchant, and G. M. Hawksworth, Quantitative differences in phase I and II metabolism between rat precision-cut liver slices and isolated hepatocytes, *Drug Metab. Dispos. 23*:1274 (1995).
158. S. Ekins, J. A. Williams, G. I. Murray, M. D. Burke, N. C. Marchant, J. Engeset, and G. M. Hawksworth, Xenobiotic metabolism in rat, dog, and human precision-cut liver slices, freshly isolated hepatocytes, and vitrified precision-cut liver slices, *Drug Metab. Dispos. 24*: 990 (1996).
159. C. C. Peck, R. Temple, and J. M. Collins, Understanding consequences of concurrent therapies, *JAMA 269*:1550 (1993).
160. D. G. Shand and J. A. Oates, Metabolism of propranolol by rat liver microsomes and its inhibition by phenothiazine and tricyclic antidepressant drugs, *Biochem. Pharmacol. 20*:1720 (1971).
161. K. Brøsen, Drug interactions and the cytochrome P450 system the role of cytochrome P450 1A2, *Clin. Pharmacokinet. 29* (Suppl 1):20 (1995).

162. L. L. Von Moltke, D. J. Greenblatt, J. Schimder, J. S. Harmatz, and R. I. Shader, Metabolism of drugs by cytochrome P450 3A isoforms. Implications for drug interactions in psychopharmacology, *Clin. Pharmacokinet. 29* (Supp 1):33 (1995).
163. B. J. Ring, S. N. Binkley, L. Roskos, and S. A. Wrighton, Effect of fluoxetine, norfluoxetine setraline and desmethyl sertaline on human CYP3A catalysed 1′-hydroxy midazolam formation in vitro, *J. Pharm. Exp. Ther. 275*:1131 (1996).
164. J. C. Stevens and S. A. Wrighton, Interaction of the enantiomers of fluoxetine and norfluoxetine with human liver cytochromes P450, *J. Pharm. Exp. Ther. 266*:964 (1993).
165. B. J. Ring, S. N. Binkley, M. VandenBranden, and S. A. Wrighton, In vitro interaction of the antipsychotic agent olanzapine with human cytochromes P450 CYP2C9, CYP2C19, CYP2D6 and CYP3A, *Br. J. Clin. Pharmacol. 41*:181 (1996).
166. L. Koymans, G. M. Donne-op Den Kelder, J. M. Koppele Te, and N. P. E. Vermeulen, Cytochromes P450: Their active-site structure and mechanism of oxidation, *Drug Metab. Rev. 25*:325 (1993).
167. R. T. Okita and B. S. S. Masters, Biotransformations: The cytochromes P450, *Textbook of Biochemistry, With Clinical Correlations*, (T. M. Devlin, ed.), Wiley–Liss, New York, 1992, p. 981.
168. F. P. Guengerich, Human cytochrome P450 enzymes, Cytochrome P450; Structure, mechanism, and biochemistry (P. R. Ortiz de Montellano, ed.), Plenum Press, New York, 1995, p. 473.
169. H. Yamazaki, M. Nakano, Y. Imai, Y.-U. Ueng, F. P. Guengerich, and T. Shimada, Roles of cytochrome b_5 in the oxidation of testosterone and nifedipine by recombinant cytochrome P450 3A4 and by human liver microsomes, *Arch. Biochem. Biophys. 325*:174 (1996).
170. J. B. Schenkman, A. I. Voznesensky, and I Jansson, Influence of ionic strength on the P450 monooxygenase reaction and role of cytochrome b_5 in the process, *Arch. Biochem. Biophys. 314*:234–241 (1994).
171. S. Imaoka, Y. Imai, T. Shimada, and Y. Funae, Role of phospholipids in reconstituted cytochrome P450 3A form and mechanism of their activation of catalytic activity, *Biochemistry 31*:6063 (1992).
172. M. S. Shet, K. M. Faulkner, P. L. Holmans, C. W. Fisher, and R. W. Estabrook, The effects of cytochrome b_5, NADPH-P450 reductase, and lipid on the rate of 6β-hydroxylation of testosterone as catalyzed by a human P450 3A4 fusion protein, *Arch. Biochem. Biophys. 318*:314 (1995).
173. M. Ingelman-Sundberg, A. L. Hagbjork, Y. F. Ueng, H. Yamazaki, and F. P. Guengerich, High rates of substrate hydroxylation by human cytochrome P450 3A4 in reconstituted membraneous vesicles: Influence of membrane charge, *Biochem. Biophys. Res. Common. 221*: 318 (1996).
174. H. Yamazaki, Y. F. Ueng, T. Shimada, and F. P. Guengerich, Roles of divalent metal ions in oxidations catalyzed by recombinant cytochrome P450 3A4 and replacement of NADPH-cytochrome P450 reductase with other flavoproteins, ferredoxin, and oxygen surrogates, *Biochemistry 34*:8380 (1995).
175. E. M. J. Gillam, Z. Guo, Y. F. Ueng, H. Yamazaki, I. Cock, P. E. B. Reilly, W. D. Hooper, and F. P. Guengerich, Expression of cytochrome P450 3A5 in *Escherichia coli*: Effects of 5′ modification, purification, spectral characterization, reconstitution conditions, and catalytic activities, *Arch. Biochem. Biophys. 317*:374 (1995).
176. M. S. Shet, C. W. Fisher, P. L. Holmans, and R. W. Estabrook, Human cytochrome P450 3A4: Enzymatic properties of a purified recombinant fusin protein containing NADPH-P450 reductase, *Proc. Natl. Acad. Sci. USA 90*:11748 (1993).
177. B. Gemzik, M. R. Halvorson, and A. Parkinson, Pronounced and differential effects of ionic strength and pH on testosterone oxidation by membrane-bound and purified forms of rat liver microsomal cytochrome P-450, *J. Steroid Biochem. 35*:429 (1990).

178. J. Mäenpää, S. D. Hall, and S. A. Wrighton, Human cytochrome P450 3A mediated midazolam metabolism: effect of assay conditions and stimulation by a-naphthoflavone, terfenadine, and testosterone, *Pharmacogenetics 8*:137 (1998).
179. T. Andersson, J. O. Miners, M. E. Veronese, and D. J. Birkett, Diazepam metabolism by human liver microsomes is mediated by both *S*-mephenytoin hydroxylase and CYP3A isoforms, *Br. J. Clin. Pharmacol. 38*:131 (1994).
180. B. M. Kerr, K. E. Thummel, C. J. Wurden, S. M. Klein, D. L. Kroetz, F. J. Gonzalez FJ, and R. H. Levy, Human liver carbamazepine metabolism, *Biochem. Pharmacol. 47*:1969 (1994).
181. Y. F. Ueng, T. Shimada, H. Yamazaki, and F. P. Guengerich, Oxidation of aflatoxin B_1 by bacterial recombinant human cytochrome P450 enzymes, *Chem. Res. Toxicol. 8*:218 (1995).
182. J. Schmider, D. J. Greenblatt, J. S. Harmatz, and R. I. Shader, Enzyme kinetic modelling as a tool to analyse the behaviour of cytochrome P450 catalysed reactions: Application to amitryptyline *N*-demethylation, *Br. J. Clin. Pharmacol. 41*:593 (1996).
183. J. S. Schmider, D. J. Greenblatt, L. L. von Moltke, J. S. Harmatz, S. X. Duan, D. Karsov, and R. I. Shader, Characterisation of six in vitro reactions mediated by human cytochrome P450: Application to the testing of cytochrome P450-directed antibodies, *Pharmacology 52*: 125 (1996).
184. S. Ekins, M. VandenBranden, B. J. Ring, and S. A. Wrighton, Examination of purported probes of human CYP2B6, *Pharmacogenetics 7*:165 (1997).
185. D. S. Davies, P. L. Gigon, and J. R. Gillette, Sex differences in the kinetic constants for the *N*-demethylation of ethylmorphine by rat liver microsomes, *Biochem. Pharmacol. 17*:1865 (1968).
186. F. J. Wiebel, J. C. Leutz, L. Diamond, and H. V. Gelboin, Aryl hydrocarbon (benzo(*a*)pyrene) hydroxylase in microsomes from rat tissues: Differential inhibition and stimulation by benzoflavones and organic solvents, *Arch. Biochem. Biophys. 144*:78 (1971).
187. M. T. Huang, R. L. Chang, J. G. Fortner, and A. H. Conney, Studies on the mechanism of activation of microsomal benzo(*a*)pyrene hydroxylation by flavonoids, *J. Biol. Chem. 256*: 6829 (1981).
188. M. Shou, J. Grogan, J. A. Mancewicz, K. W. Krausz, F. J. Gonzalez, H. V. Gelboin, and K. R. Korzekwa, Activation of CYP3A4: Evidence for the simultaneous binding of two substrates in a cytochrome P450 active site, *Biochemistry 33*:6450 (1994).
189. G. E. Schwab, J. L. Raucy, and E. F. Johnson, Modulation of rabbit and human hepatic cytochrome P-450-catalyzed steroid hydroxylations by a-naphthoflavone, *Mol. Pharmacol. 33*:493 (1988).
190. B. B. Rasmussen, J. Mäenpää, O. Pelkonen, S. Loft, H. E. Poulsen, J. Lykkesfeldt, and K. Brosen, Selective serotonin reuptake inhibitors and theophylline metabolism in human liver microsomes: Potent inhibition by fluvoxamine, *Br. J. Clin. Pharmacol. 39*:151 (1995).
191. D. M. Grant, M. E. Cambell, B. K. Tang, and W. Kalow. Biotransformation of caffeine by microsomes from human liver, *Biochem. Pharmacol. 36*:1251 (1987).
192. W. Tassaeeyakul, Z. Mohamed, D. J. Birkett, M. E. McManus, M. E. Veronese, R. H. Tukey, L. C. Quattrochi, F. J. Gonzalez, and J. O. Miners. Caffeine as a probe for human cytochromes P450: Validation using cDNA-expression, immunoinhibition and microsomal kinetic and inhibitor techniques, *Pharmacogenetics 2*:173 (1992).
193. H. Yamazaki, Y. Inui, C-H. Yun, F. P. Guengerich, and T. Shimada, Cytochrome P450 2E1 and 2A6 enzymes as major catalysts for metabolic activation of *N*-nitrosodialkylamines and tobacco-related nitrosamines in human liver microsomes, *Carcinogenesis 13*:1789 (1992).
194. A. R. Bourrié, R. J. Edwards, D. A. Adams, and D. S. Davies, Dissecting the function of cytochrome P450, *Br. J. Clin. Pharmacol. 42*:81 (1996).
195. S. J. Stohs, R. C. Grafström, M. D. Burke, P. W. Moldeus, and S. G. Orrenius, the isolation of rat intestinal microsomes with stable cytochrome P-450 and their metabolism of benzo(α)-pyrene, *Arch. Biochem. Biophys. 177*:105, (1976).

196. J. Mäenpää, S. D. Hall, B. Ring, S. C. Strom, and S. A. Wrighton, Stimulation in vitro of CYP3A4/5 mediated midazolam metabolism by α-napthoflavone, terfenadine and testosterone, *Pharmacogenetics 8*:137 (1998).
197. J. M. Lasker, M.-T. Huang, and A. H. Conney, In vivo activation of zoxazolamine metabolism by flavone, *Science 216*:1419 (1982).
198. J. Cumps, C. Razzouk, and M. B. Roberfroid, Michaelis–Menten kinetic analysis of the hepatic microsomal benzpyrene hydroxylase from control, phenobarbital- and methyl-3-cholanthrene-treated rats, *Chem. Biol. Interact. 16*:23 (1977).
199. S. Channareddy, S. S. Jose, V. A. Eryomin, E. Rubin, T. F. Taraschi, and N. Janes, Saturable ethanol binding in rat liver microsomes, *J. Biol. Chem. 271*:17625 (1996).
200. A. A. Spector and J. E. Fletcher, Nutritional effects on drug–protein binding, *Symposium on Nutrition and Drug Interrelations* (J. N. Hathcock and J. Coon, eds.), Academic Press, New York, 1978, p. 447.
201. J. H. Lin, D. M. Cocchetto, and D. E. Duggan, Protein binding as a primary determinant of the clinical pharmacokinetic properties of non-steroidal anti-inflammatory drugs. *Clin. Pharmacokinet. 12*:402 (1987).
202. A. E. Wade, W. P. Norred, and J. S. Evans, Lipids in drug detoxication, *Symposium on Nutrition and Drug Interrelations* (J. N. Hathcock and J. Coon, eds.), Academic Press, New York, 1978, p. 475.
203. R. S. Obach, The importance of non-specific binding in in vitro matrices, its impact on enzyme kinetic studies of drug metabolism reactions, and implications for in vitro–in vivo correlations, *Drug Metab. Dispos. 24*:1047, (1996).
204. O. Pelkonen, E. H. Kaltiala, N. T. Karki, K. Jalonen, and K. Pyorala, Properties of benzopyrene hydroxylase from human liver and comparison with the rat, rabbit and guinea-pig enzymes, *Xenobiotica 5*:501 (1975).
205. D. L. Cinti, Explanation of the stimulation of microsomal *N*-demethylation reactions by soluble supernatant fraction, *Res. Commun. Chem. Pathol. Pharmacol. 12*:339 (1975).
206. C. L. Litterst, B. I. Sikic, E. G. Mimnaugh, A. M. Guarino, and T. E. Gram, In vitro drug metabolism in male and female athymic, nude mice. *Life Sci. 22*:1723 (1978).
207. V. G. Zannoni, E. J. Flynn, and M. Lynch, Ascorbic acid and drug metabolism, *Biochem. Pharmacol. 21*:1377 (1972).
208. H. Raunio, M. Pasanen, J. Mäenpää, J. Hakkola, and O. Pelkonen, Expression of extrahepatic cytochrome P450 in humans, *Advances in Drug Metabolism* (G. M. Pacifici, ed.), European Commission Brussels, 1995, p. 233.
209. R. M. Philpot, Modulation of the pulmonary cytochrome P450 system as a factor in pulmonary-selective toxic responses: Fact and fiction. *Am. J. Respir. Cell. Mol. Biol. 9*:347 (1993).
210. T. Matsubara, R. A. Prough, M. D. Burke, and R. W. Estabrook, The preparation of microsomal fractions of rodent respiratory tract and their characterization, *Cancer Res. 34*:2196 (1974).
211. K. Matsuura, A. Hara, H. Sawada, Y. Bunai, and I. Ohya, Localisation of pulmonary carbonyl reductase in guinea pig and mouse: Enzyme histochemical and immunohistochemical studies, *J. Histochem. Cytochem. 38*:217 (1990).
212. M. Floreani and F. Carpendo, Metabolism of simple quinones in guinea pig and rat cardiac tissue, *Gen. Pharmacol. 26*:1757 (1995).
213. D. T-W. Lau, E. K. Chan, and L. Z. Benet, Glutathione *S*-transferase-mediated metabolism of glyceryl trinitrate in subcellular fractions of bovine coronary arteries, *Pharm. Res. 9*:1460 (1992).
214. C. J. Reed, Drug metabolism in the nasal cavity: Relevance to toxicology, *Drug Metab. Rev. 25*:173 (1993).
215. E. M. Gibby and G. M. Cohen, Conjugation of 1-napthol by human colon and tumour tissue using different experimental systems, *Br. J. Cancer 49*:645 (1984).
216. K. S. Lown, J. C. Kolars, K. E. Thummel, J. L. Barnett, K. L. Kunze, S. A. Wrighton, and

P. B. Watkins, Interpatient heterogeneity in expression of CYP3A4 and CYP2A5 in small bowel, *Drug Metab. Dispos. 22*:947 (1994).

217. P. G. Wells and L. M. Winn, Biochemical toxicology of chemical teratogenesis. *Crit. Rev. Biochem. Mol. Biol. 31*:1 (1996).
218. O. Pelkonen and M. Pasanen, Effect of heparin on subcellular distribution of human placental 7-ethoxycoumarin *O*-deethylase activity, *Biochem. Pharmacol. 30*:3254 (1981).
219. M. Pasanen and O. Pelkonen, The expression and environmental regulation of P450 enzymes in human placenta, *Crit. Rev. Toxicol. 24*:211 (1994).
220. J. Hakkola, M. Pasanen, J. Hukkanen, O. Pelkonen, J. Mäenpää, T. Cresteil, and H. Raunio, Expression of xenobiotic-metabolizing cytochrome P450 forms in human full-term placenta, *Biochem. Pharmacol. 51*:403 (1996).
221. O. Pelkonen, Biotransformation of xenobiotics in the fetus, *Pharmacol. Ther. 10*:261 (1980).
222. S. A. Wrighton and J. C. Stevens, The human hepatic cytochrome P450 involved in drug metabolism, *Crit. Rev. Toxicol. 22*:1 (1992).
223. J-F. Ghersi-Egea, R. Perrin, B. leininger-Muller, M-C. Grassoit, C. Jeandel, J. Floquet, G. Cuny, G. Siest, and A. Minn, Subcellular localization of cytochrome P450, and activities of several enzymes responsible for drug metabolism in the human brain, *Biochem. Pharmacol. 45*:647 (1993).
224. C. J. Hillard, D. M. Wilkison, W. S. Edgemond, and W. B. Campbell, Characterization of the kinetics and distribution of *N*-arachidonylethanolamine (anandamide) hydrolysis by rat brain, *Biochim. Biophys. Acta 1257*:249 (1995).
225. P. Broquet, M. Serres-Guillaumond, H. Baubichon-Cortay, M-J. Peschard, and P. Louisot, Subcellular localisation of cerebral fucosyltransferase, *FEBS 174*:43 (1984).
226. M. Donlon, W. Shain, G. S. Tobias, and G. V. Mainetti, Characterisation of an 11,000-dalton B-bungarotoxin: Binding and enzyme activity on rat brain synaptosomal membranes, *Membr. Biochem. 2*:367 (1979).
227. A. J. Baars, M. Jansen, and D. D. Breimer, Xenobiotic metabolizing enzymes in *Drosophila melanogaster, Mutat. Res. 62*:279 (1979).
228. A. J. Baars, J. A. Zijlstra, M. Jansen, E. Vogel, and D. D. Breimer, Biotransformation and spectral interaction of xenobiotics with subcellular fractions from *Drosophila melanogaster, Arch. Toxicol. Suppl. 4*:54 (1980).
229. F. P. Guengerich and C. G. Turvy, Comparison of levels of several human microsomal cytochrome P-450 enzymes and epoxide hydrolase in normal and disease states using immunohistochemical analysis of surgical liver samples, *J. Pharmacol. Exp. Ther. 256*:1189 (1991).
230. I. De Waziers, P. H. Cugnenc, C. S. Yang, J. Leroux, and P. H. Beaune, Cytochrome P450 isoenzymes, epoxide hydrolyase and glutathione transferases in rat and human hepatic and extrahepatic tissues, *J. Pharmacol. Exp. Ther. 253*:387 (1990).
231. P. Kremers, P. Beaune, T. Cresteil, J. De Graeve, S. Columelli, J. Leroux, and J. E. Gielen, Cytochrome P-450 monooxygenase activities in human and rat liver microsomes, *Eur. J. Biochem. 118*:599 (1981).
232. G. A. Dannan and F. P. Guengerich, Immunochemical comparison and quantitation of microsomal flavin-containing monooxygenases in various hog, mouse, rat, rabbit, dog, and human tissues, Mol. Pharmacol. 22:787-(1982).
233. G. I. Murray and M. D. Burke, Immunohistochemistry of drug-metabolizing enzymes, *Biochem. Pharmacol. 50*:895 (1995).

15
In Vitro Metabolism: Hepatocytes

Michael W. Sinz
Parke-Davis Pharmaceutical Research Division, Warner-Lambert Company, Ann Arbor, Michigan

I. INTRODUCTION

Hepatocytes or parenchymal cells are the functional units of the liver and the major site for xenobiotic biotransformations. These cells are capable of providing the endogenous energy resources necessary for the body, as well as performing such tasks as protein synthesis, storage of essential components, and metabolism of both xenobiotics and endogenous substrates. Suspensions or primary cultures of hepatocytes offer a sophisticated environment to study many aspects of toxicology, pharmacology, biochemistry, and drug metabolism. Studies are conducted within a physiological milieu that contains normal concentrations of enzymes and cofactors, as well as the cellular machinery necessary to regulate the synthesis of new proteins (enzyme induction). By analogy with the hepatocyte found in vivo, xenobiotics must cross a biological membrane, interact with cellular organ-

Table 1 Examples of Experiments Employing Hepatocytes to Study Drug Metabolism-Related Issues

Type of experiment	Suspension (S) or primary Culture (C)	Species	Ref.
Metabolite formation	C	Human	20
	C	Duck	12
	C	Pig	21
	S	Rat	22–25
In vitro/in vivo pharmcokinetic studies	C	Rat, dog, rabbit, human	26
	S	Rat	27,28
	C	Human	29
	C	Rat, rabbit, dog, guinea, pig,	30
Species comparison	C	Rat, dog, monkey, human	7
	S	Rat, rabbit, dog, monkey, human	16
	C	Mouse, hamster, rat, guinea pig	8
	S	Rat, rabbit, dog, monkey	6
	C	Rat, rabbit, human	31
Toxicology: formation of toxic metabolites	C	Human	32
	C	Rat	33
	S	Rat, mouse	34
	S	Rat	35
	S	Monkey	36
	C	Rat, human, dog	37
Enzyme induction/inhibition	C	Rat	38,39
	C	Rabbit	11
	C	Trout	13
	C	Pig	15
	C	Dog	40
	C	Human	41
Effects of cofactors (hormones, growth factors, cytokine, ect.)	C	Rat	42
	C	Sheep	14
	S	Rat	43,44
	C	Rat	45
	S	Eel, bullhead, trout	46
Cell membrane transport	S	Rat	47–49
Mechanistic studies	S	Rat	50
Define enzymes involved in metabolism	S	Rat	51
Predicting drug–drug interactions	C	Human	52

elles and receptors, and compete with endogenous substrates for biotransformation. In comparison with many other in vitro techniques, such as isolated enzymes, subcellular fractions, or homogenates, hepatocytes are clearly more akin to the in vivo situation. Hepatocytes have a multitude of applications in the field of drug metabolism from both an industrial perspective, where they are used as a model system to study drugs in discovery and development, as well as use in the pursuit of basic research issues (1–5). Hepatocytes have been isolated from a large number of laboratory animal species, such as rat, mouse, dog, monkey, guinea pig, rabbit, hamster, frog, chicken, quail, fish, sheep, pig, cattle, as well as human liver tissue (1–19). Hepatocytes have been used to study several aspects of drug metabolism, such as metabolic profiling (formation of metabolites for identification purposes), biotransformation pathways (reaction phenotyping), and comparisons of metabolism among species to determine which animal species best represents the human metabolism of a xenobiotic. In suspension or culture, hepatocytes are used to estimate in vivo pharmacokinetic parameters or examine the physiological response of cofactors, regulatory biomolecules, or xenobiotics on hepatocyte function. Hepatocytes are also a useful tool for studying the detoxification or toxification of xenobiotics derived through metabolism. Table 1 lists in greater detail the types of experiments and animal species employed in many of these experiments.

This chapter will attempt to assimilate many of the topics and variables involved in the isolation and use of hepatocytes from both animal and human liver tissues for the beginning investigator. In most cases, a single method will be described in detail, although several references to other methods that are applicable will be noted. In addition, several excellent reference texts concerning hepatocyte isolation and culture conditions are noted (53–56).

II. HEPATOCYTE ISOLATION

Historically, cells from liver tissue were isolated as early as the 1940s (57). Table 2 lists several different isolation techniques employed over the past 50 years. The early methods typically involved some form of mechanical isolation that resulted in mostly unviable cells or cells so severely damaged that they were unable to survive or maintain normal physiological functions. The discovery of proteolytic enzymes, such as collagenase (EC 3.4.24.3 from *Clostridium histolyticum*) resulted in the development of a new generation of hepatocyte isolation procedures that were superior to the mechanical methods. The original enzyme procedure involved slicing the liver tissue, followed by incubation in a buffer containing collagenase and hyaluronidase (58), and was later improved by Fry and Belleman in the 1970s (59–61). In 1969, Berry and Freind established the first protocol involving a liver perfusion technique employing collagenase (53,62). This new perfusion method, commonly referred to as the one-step perfusion method, more efficiently exposed the cells and extracellular matrices to the dissociation buffers and resulted in improved cell yields and viability. Several years later, Seglen refined the perfusion method by introducing the two-step perfusion method (54,55,63). The first step involved perfusing the liver with a medium devoid of Ca^{2+} (with or without chelating agents) to irreversible cleave desmosomal cell-to-cell contacts. The second perfusion medium contained collagenase (along with Ca^{2+} for enzymatic activity) to digest the extracellular matrix (i.e., collagen). Only minor changes have been made to these two original isolation methods over

Table 2 Methods Employed in the Isolation of Hepatocytes

Methodology	Description/advantages
Mechanical/nonenzymatic	Original technique; resulted in mostly damaged cells
Slice/incubate method	Useful for tissue sections that are not amenable to perfusion
One-step perfusion	Improved exposure of digestion buffers containing collagenase to extracellular matrix and cells resulting in a greater yield of viable cells
Two-step perfusion	Improved exposure of digestion buffers containing collagenase to extracellular matrix and cells; initial perfusion (devoid of calcium) to cleave desmosomes and second perfusion (collagenase) to digest collagen extracellular matrix
Biopsy perfusion	Used for small or large biopsy samples of tissue
EDTA perfusion	Perfusion technique not requiring collagenase; may lead to prolonged expression of biochemical functions
Vibration method	Cell dissociation is achieved by vibrating the cells from one another; does not require perfusion of buffers.

the past 25 years. Both methods are currently being employed in most laboratories, although the two-step procedure is somewhat more prevalent.

As hepatocytes began to be isolated from other animal species, especially larger animals and humans, further modifications to the original methods were necessary. In this light, the biopsy perfusion method was developed, for which small or large sections of liver tissue (biopsy sections) could be perfused (54,64). Alternative techniques that have arisen over the years include perfusing the tissue with an EDTA–buffer in place of the enzymatic collagenase buffer (65,66) as well as employing vibration as a means of dissociating cell-to-cell contacts (67). Although unique alternatives to the collagenase perfusion–incubation methods, neither the EDTA or vibrational techniques have been universally employed.

A. Two-Step Procedure for Rat Liver Tissue

The following is a description of the current two-step perfusion method of Seglen employed in our laboratory. Modifications proposed by other investigators are also noted in the standard procedure so that the reader is aware of such modifications and why some investigators may choose to use them. Several buffers or media are necessary at different stages of the isolation and incubation steps, these include (1) first perfusion buffer (wash buffer), (2) second perfusion buffer (collagenase buffer), (3) cell wash or isolation buffer, and (4) the incubation buffer or medium. The initial perfusion buffer is used to remove blood from the liver and deplete the liver of calcium (Ca^{2+}). The lack of Ca^{2+} is responsible for the cleavage of desmosomes or zonulae adherens. Desmosomes are adhering junctions on cell membranes that bind cells tightly together and impart strength to the overall tissue. Irreversible desmosomal cleavage is the first important event in hepatocyte isolation. To irreversibly cleave the desmosomal contacts, the first buffer should be perfused for a minimum of 10 min. Shorter perfusions may result in the desmosomes repairing themselves and result in a poor yield of cells. To remove Ca^{2+} ions from the liver, investigators use a wash buffer devoid of Ca^{2+}, with or without EGTA as a calcium ion chelator. Depending on the quality of your water and reagents, the addition of EGTA may not be necessary. However, the addition of 0.1 mM EGTA can prevent the loss of reduced glutathione from the cells (68). We have found the use of EGTA necessary, particularly with human or monkey tissue, which tend to contain larger amounts of connective tissue and are much more difficult to digest.

The choice of buffering system is also variable among investigators. The two most commonly used buffering systems are HEPES (*N*-2-hydroxyethyl-piperazine-*N*′-2-ethane sulfonic acid) and bicarbonate. The bicarbonate system (e.g., Krebs-Henseleit bicarbonate; KH) is more physiological, but requires continuous bubbling of a 95/5 mixture of oxygen and carbon dioxide throughout the first and second perfusion steps to maintain the appropriate pH. HEPES (10–100 mM), a synthetic buffering system, is more convenient and for some experimental setups may be more appropriate.

1. *Preparation of Buffers*

The composition of the wash buffer routinely employed in our laboratory is as follows: NaCl, 120 mM; KCl, 5.3 mM; $NaHCO_3$, 7.4 mM; Na_2HPO_4, 0.3 mM; glucose, 5.6 mM; HEPES, 10 mM; EGTA, 0.5 mM (final pH 7.4). Typically 400 mL of this buffer is prepared for each isolation experiment. The buffer can also be prepared as a 10× solution and stored at 4°C after sterile filtration.

The composition of the collagenase buffer is similar to that of the wash buffer, except for the addition of calcium chloride and collagenase and the omission of EGTA. The composition of collagenase buffer (final pH 7.4) is as follows; NaCl, 120 mM; KCl, 5.3 mM; $NaHCO_3$, 7.4 mM; Na_2HPO_4, 0.3 mM; glucose, 5.6 mM; HEPES, 10 mM; $CaCl_2$, 2.5 mM; 0.04% collagenase (type IV, Sigma Chemical Co.). Typically 400 mL of this buffer is prepared for each isolation experiment. The buffer can also be prepared as a 10× solution (without addition of $CaCl_2$ and collagenase) and stored at 4°C after sterile filtration. The $CaCl_2$ should be added after the 10× concentrate has been diluted to prevent precipitation of calcium salts, and collagenase should be added just before the start of each experiment.

Both perfusion buffers should be well oxygenated before starting the experiment and during the experiment if possible to prevent cellular hypoxia. The buffers should also be immersed in a water bath set at approximately 40°C. The actual water bath temperature should be predetermined so that the temperature of the perfusion buffer at the inlet cannula to the liver is between 36 and 39°C, the optimal temperature for collagenase activity.

The choices of isolation–wash buffer or medium employed to isolate hepatocytes from the digested tissue are quite varied. We routinely use a complete medium, such as William's E (WE) or Dulbecco's modified Eagle's medium (DMEM) for both the isolation of cells and incubation medium. The use of simple buffers, such as KH or Hank's balanced salt solution (HBSS) are also commonly employed for both the cell isolation and incubation buffer. In general, for incubations of shorter duration (30–90 min) a simple buffer may be adequate for cell viability and biotransformation reactions, but for longer incubations (2–4 h), it is recommended that a more complete medium containing essential amino acids and nutrients be employed to maintain cellular functions, cell viability, and the necessary cofactors of drugs metabolizing enzymes.

2. *Equipment and Supplies*

The equipment and instruments necessary to perform the isolation are categorized in the following:

Water bath: room temperature to 45°C

Pump: peristaltic pump, tubing, tubing end sinkers, bubble trap, two-way switch, lead doughnuts, 20-gauge IV catheter–trocar (catheter and needle assembly, 32 mm), and several 250-mL Erlenmeyer flasks

Surgical supplies: operating board or grate placed over a drain pan, 3-0 surgical silk, surgical scissors (large and small), fine and coarse forceps, sterile saline, cotton-tipped swabs, and cotton gauze

Cell isolation supplies: stainless steel dog comb (or surgical rake), forceps, glass or plastic beakers and Erlenmeyer flasks (125–250 mL), large glass crystallizing dish (125 mL), 50-mL centrifuge tubes, centrifuge, nylon filters (250 and 125 μm), trypan blue solution (0.4%), hemocytometer, and a light microscope

Oxygenated wash and collagenase buffers are poured into 500-mL Erlenmeyer flasks and placed in the water bath with lead doughnuts. Tubing from the peristaltic pump (with attached end sinkers) are put into the flasks and the catheter placed on the opposite end of the tubing. The distance from the water bath to cannula should be minimized to prevent heat loss. A schematic of the equipment setup is shown in Fig. 1.

3. Surgical Procedure: Whole-Liver In Situ Perfusion

Rats between 150 and 200 g are optimal for good cell yields and viability. Larger rats (200–300 g) will result in similar total cell counts, but may have lower cell viability. The animal is anesthetized with sodium pentobarbital (60 mg/kg) by intraperitoneal injection. The unconscious animal is placed on an operating board ventral side up. A large U-shaped incision is made on the ventral surface, exposing the abdominal cavity all the way to the diaphragm. The cut begins at the midsection of the lower abdomen and progresses up toward the right foreleg. The second portion of the U-shape again starts at the same midsection cut and progresses upward toward the left foreleg. The incisions are best made by first cutting through the skin followed by a second U-shaped incision through the muscle layer. Heparin may be injected into the femoral vein to prevent blood clotting during the isolation procedure, but in our hands heparin has not been necessary, for the

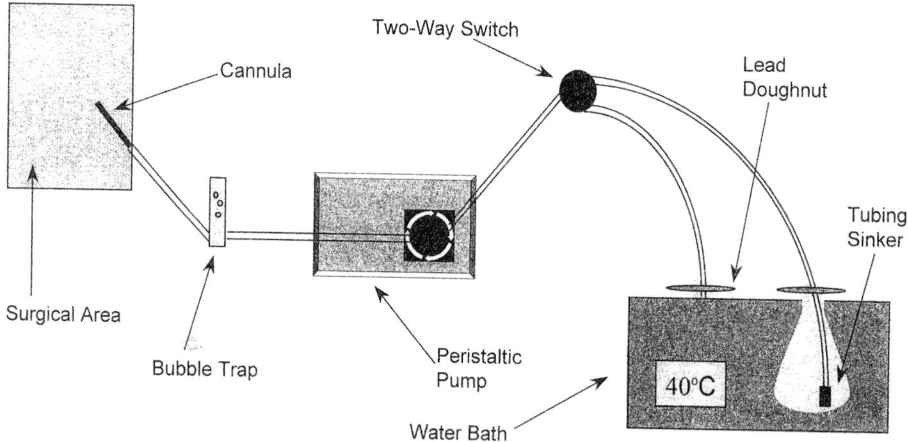

Fig. 1 A schematic diagram of the equipment setup for rat hepatocyte isolations.

wash buffer adequately removes blood from the liver before clotting occurs. The intestines are gently lifted and moved over-and-out of the body cavity to the right side of the animal. This exposes the portal vein, which is then gently teased away from the surrounding fascia with cotton-tipped swabs and forceps. Three ligatures (3-0 silk) are then placed around the portal vein. One ligature is placed near the entrance to the liver, the second approximately 1.0 cm below the first, and the third ligature 1.5 cm below the second. Tie loops are prepared on the first pair of ligatures, but are not tied down. Figure 2 indicates the correct placement of ligatures and cannula.

The portal vein is carefully cannulated with the 20-gauge catheter by gently pulling back on the third ligature to straighten and hold the portal vein in place. The cannula–needle assembly is inserted into the portal vein at a 25-degree angle, and the cannula angle immediately reduced to avoid puncture of the underside of the portal vein as the assembly is pushed forward approximately 2–3 cm. Once inserted, the needle portion of the assembly is carefully removed. After the needle is removed, blood will begin to flow from the back end of the cannula. To prevent an air bubble, the pump tubing is reconnected to the cannula after blood has completely filled the cannula connector. The final position of the cannula end should be just below the first venous branch into a liver lobe. The catheter is immediately secured into place by tightening the first and second ligatures around it and the wash buffer started at a flow rate of 20 mL/min. On starting the wash buffer, the superior vena cava is severed by plunging a pair of fine surgical scissors through the diaphram and severing the vena cava. The wash buffer will exit the vena cava and drain from the animal. During the perfusion, the liver tissue is kept moist and warm at all times by bathing the tissue with warm saline. The wash buffer is perfused for approximately 15 min, at which time collagenase (0.04% w/v) is added to the collagenase buffer. The collagenase buffer perfusion is initiated after dissolution of the collagenase by adjusting the two-way valve and the liver perfused for an additional 10–20 min.

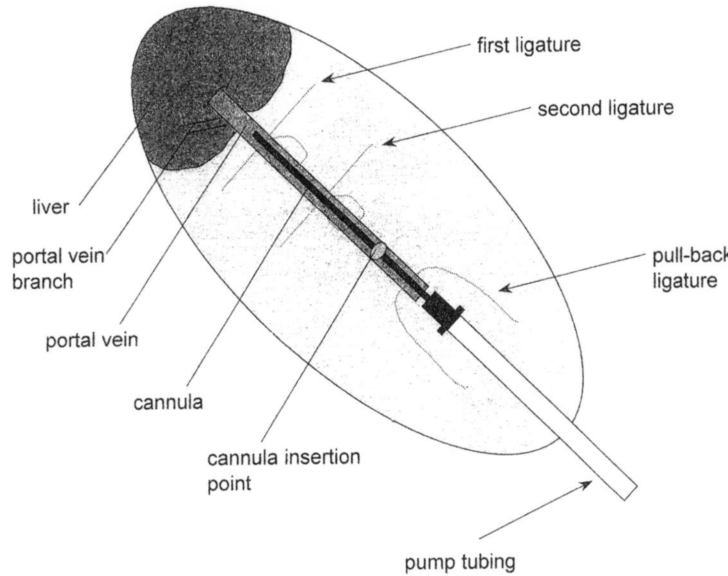

Fig. 2 Diagram of rat portal vein cannulation.

The liver is then carefully removed from the body cavity and placed in a dish containing 50 mL of oxygenated WE. The dog comb is now raked through the tissue to release cells, connective tissue, and any undigested liver tissue. The cells are placed in a shaking water bath for 15 min at 37°C to allow damaged cells to either repair themselves or die (ca. 20–50 rpm). The entire cell suspension is then strained through cheesecloth and centrifuged at 50 *g* for 5 min to sediment liver parenchymal cells from nonparenchymal and damaged cells. Alternatively, the cells can be placed on ice and sedimentation allowed to occur by gravity for 10–20 min. The supernatant is then carefully removed, the cells gently resuspended in medium, and recentrifuged for a total of three centrifugation steps. The final cell pellet is resuspended in 40–50 mL of WE, which is filtered through 250- and 105-μm nylon filters. Smaller filters, 60 μm, may also be used to remove small aggregates of cells. The final cell suspension is used to determine both viability and cell number. Viability of the initial cell suspension is determined by trypan blue exclusion, which determines the intactness of the cell membrane. Only those initial cell suspensions with a percentage viability of more than 85% are deemed acceptable for drug metabolism studies. Typical cell viabilities range from 90 to 100% once the investigator becomes more experienced. The cells are then diluted to approximately 0.5×10^6 to 2×10^6 cells per milliliter in WE (final cell concentration based on the number of live cells present in the suspension) and used for suspension experiments or placed into primary culture. Figure 3 is a phase-contrast photomicrograph of rat hepatocytes cultured on Matrigel and stained with trypan blue.

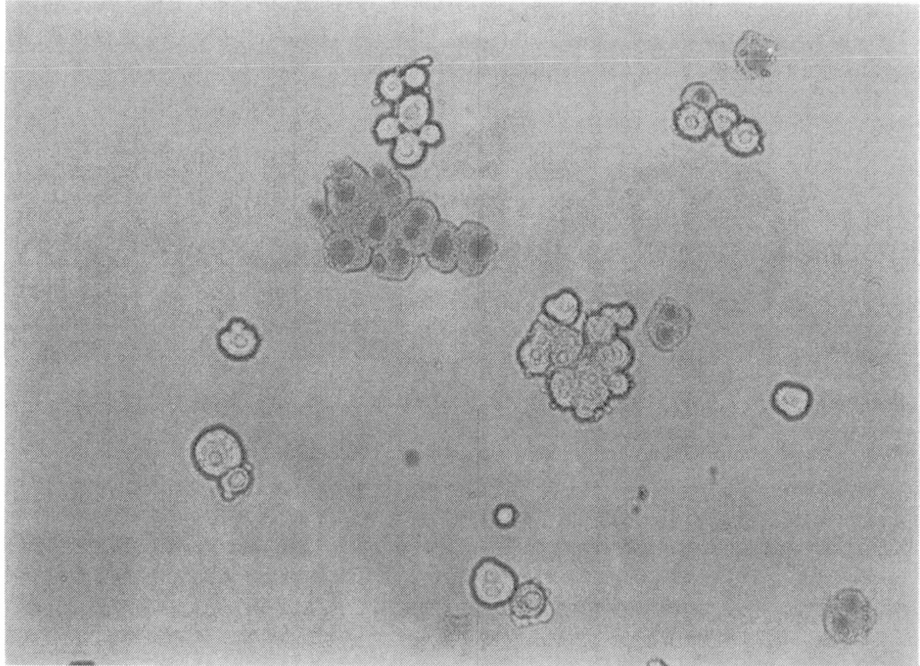

Fig. 3 Phase-contrast photomicrograph (200×) of rat hepatocytes cultured on Matrigel and stained with trypan blue. The figure shows one large grouping of cells that have taken up the dye and are considered damaged. The other cells are clear and have intact cell membranes.

B. Two-Step Procedure for Human Liver Tissue

The following is a description of the two-step perfusion method used in our laboratory for the isolation of human hepatocytes and is based on the biopsy perfusion method with an encapsulated liver lobe of 5–50 g (64,69). The final viability and cell yield from such an experiment are dependent on the medical and social history of the individual donor; the conditions under which the tissue was removed, preserved, and shipped; as well as the conditions under which the cells are isolated (70,107). Freshly excised human livers, perfused with UW solution (Viaspan) and shipped on wet ice in UW solution within 24 h of tissue harvest, generally result in isolations of viable cells. Isolation of cells from livers that are grossly cirrhotic or have a fat content of greater than 30% result in lower cell yields and viability, and cells that do not attach well or survive in primary culture.

1. Preparation of Buffers and Equipment

The perfusion buffers can either be the same as those described for the in situ rat perfusion method or a simpler buffer system containing NaCl, KCl, and HEPES may be substituted. The composition of the simple buffer system is as follows: 5 L of stock buffer are prepared by dissolving 2.5 g of KCl, 41.5 g of NaCl, and 12.0 g of HEPES and adjusting the pH to 7.4 with HCl. The wash buffer is prepared by adding 1.14 g EGTA to 2 L of stock buffer and adjusting the pH to 7.4. The collagenase buffer is prepared by adding 4.0 mL of a 0.37-g/mL solution of calcium chloride to 2.0 L of the stock buffer and adjusting the pH to 7.4 with NaOH. Collagenase (500 mg/L) is added to the collagenase buffer before the start of the experiment. Both buffers should be well oxygenated before the start of the isolation procedure and during the perfusions if logistics permit. The supplies necessary for the biopsy (encapsulated lobe) perfusion are similar to those already described, except that a multichannel perfusion is used and the collagenase buffer is recirculated. The additional 1.0 L of stock buffer can be used to rinse the tissue and flush the tubing.

2. Encapsulated Lobe Perfusion

Both the wash and collagenase buffers are prewarmed in a 45°C waterbath. The encapsulated lobe is placed on the operating board (screen) with a drain pan below. Several perfusion lines (four to seven) are connected to various sizes of mouse and rat oral gavage needles which are placed into exposed vascular openings on a single cut surface of the encapsulated lobe (Fig. 4). The output temperatures of both buffers at the gavage needles should be between 36 and 39°C. Perfusion of the wash buffer is started at 25–30 mL/min per lines with a multichannel peristaltic pump, and continued for 15–20 min, at which time the wash buffer is replaced with the collagenase buffer that is perfused for an additional 15–20 min. The wash buffer is not recirculated, but the collagenase buffer is recirculated to conserve collagenase. It is important that the tissue be warmed to 37–38°C as quickly as possible and that as much of the UW solution be removed from the tissue for maximum effect of buffers and collagenase. Therefore, the tissue is kept moist and warm at all times by bathing with warm saline, except during the perfusion of the collagenase buffer when additional saline will dilute and contaminate the buffer. After 30–40 min of perfusion, the liver is placed in a glass dish and cut into small sections to disperse digested portions of the tissue. Tissue sections are also further disintegrated with a dog comb or tissue rake. The entire tissue may or may not be entirely digested, depending on the size of the tissue and location of the inserted gavage needles, and areas of digested and undigested tissue may be found. The best cell yields are obtained if those sections of tissue which were properly digested are used and undigested areas discarded. Isolated hepato-

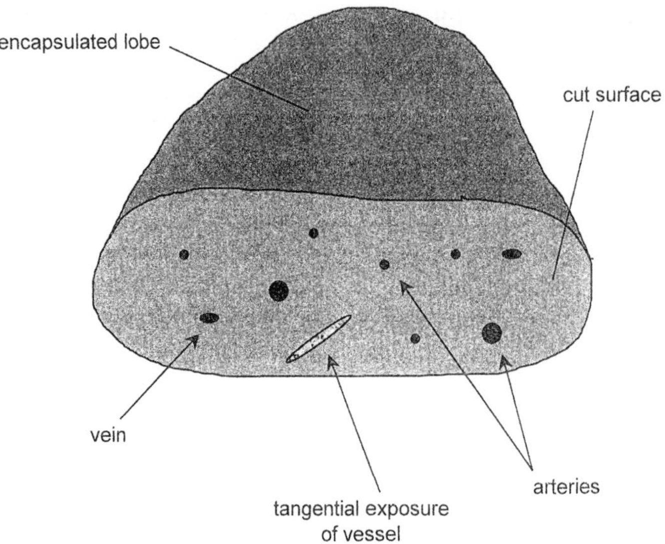

Fig. 4 Illustration of a human encapsulated liver lobe before perfusion. During the perfusion, multiple perfusion lines and cannulas are inserted into the vascular openings.

cytes are washed with freshly oxygenated WE and separated from undigested tissue by filtration through cheesecloth. The hepatocytes are further purified as previously described with rat hepatocytes After the final centrifugation step, the cells are resuspended in 20–200 mL of WE (depending on the approximate cell density) and filtered through 250- and 105-μm nylon filters. They can then be used for suspension experiments or placed into primary culture after cell counting and viability assays.

C. Additional Comments on Hepatocyte Isolation

1. *Isolation of Hepatocytes from Other Animal Species*

The rat in situ or human biopsy perfusion methods are readily adaptable to the isolation of hepatocytes from many different animal species. Whole-liver perfusions are most common with smaller animals, such as mouse, hamster, guinea pig, monkey, or rabbit, and the wedge biopsy or single lobe perfusion methods employed for larger animal species, such as the dog. Whole-liver perfusions can be performed on larger animal species, albeit they are much more expensive in terms of collagenase, and for most experimental needs, the large number of cells isolated from a single liver are not necessary.

Several considerations apply when modifying the previously mentioned methods to other animal species. The cannula size and volume–flow rate of perfusate will vary depending on the size of the tissue. Also, the amount of collagenase and exposure times of wash and collagenase buffers will increase with the size of the tissue. The surgical procedure may vary somewhat between species; for example, the gallbladder in mice, rabbits, guinea pigs, and dogs needs to be carefully isolated and removed without rupture. McQueen has written several excellent reviews on the subject of isolating and culturing hepatocytes from several common laboratory animals (56,71). In addition, there are several detailed experimental protocols noted for the isolation of mouse, rabbit, guinea pig, hamster, dog, monkey, and fish hepatocytes (6,7,9,10,13,16,31,72).

2. Sterile Procedures for Hepatocyte Cultures

Hepatocytes that are to be placed in primary culture require special treatment in both the preparation of buffers and equipment, as well as the actual isolation procedure. Some details are listed in the following that should be considered when isolating cells for primary culture. It is always best to work in a sterile environment as much as possible. All solutions (buffer, medium, and enzyme) should be filtered through a 0.2-μm filter. All surgical apparatus (including gavage needles and surgical thread) and pump tubing should be sterilized before use. After each use, deionized water and a 70% ethanol wash can be used to clean instruments and tubing before sterilization for the next isolation.

If you are working in a completely sterile facility, antibiotics may not be necessary in the perfusion buffers for rat hepatocyte isolations. For human tissue, antibiotics are recommended in the perfusion buffers. In addition, penicillin or streptomycin should be used in the culture medium. If fungal contamination is a concern, gentamicin or amphotericin is also recommended. For maintaining cultures longer than 1 week, it is recommended to check for mycoplasma contamination occasionally; ciprofloxacin works well to eliminate this type of contamination should it occur. Before surgery, it is recommended that the rat abdomen be cleaned with 70% ethanol or providone–iodine (Betadine) to prevent contamination from the outer surface of the animal. In terms of the surgical procedure, cut only the skin first, then change to a new sterile pair of scissors before cutting the muscle layer. This will avoid contaminating the peritoneal cavity. Once the cells have been isolated, all centrifugation steps and any additional cell cleanup steps should be done with sterile tubes and dishes. All of the remaining isolation steps (tissue disintegration, filtration steps, or plating) should be conducted in a laminar flow hood.

3. Percoll Density Gradient for Isolating Hepatocytes

In instances where cell viability is low (70–85%), more elaborate means of cell isolation in addition to the centrifugation (gravity) sedimentation are necessary. This is particularly evident when isolating human hepatocytes, which generally results in the isolation of a large number of unviable cells. Cells with damaged membranes when incubated in suspension or primary culture will release proteolytic enzymes that can be detrimental to intact surrounding cells; therefore, removal of as many damaged cells as possible is beneficial. The use of a Percoll density gradient very effectively separates damaged hepatocytes and other nonparenchymal cells from intact hepatocytes. The procedure involves centrifuging the initial cell suspension (or the cell suspension after one to three low 50 *g*, spins) through a Percoll medium. The resulting cell pellet contains predominately viable hepatocytes, while damaged cells, cellular debris, and nonparenchymal cells remain suspended in the Percoll medium. Hepatocytes isolated by the Percoll method exhibit superior characteristics (longer viability, CYP450 maintenance, membrane transport, and drug metabolism capability) when compared with cells isolated by low-speed centrifugation (73,74). Literature references for the percoll isolation procedure mostly pertain to the isolation of rat hepatocytes, for which a Percoll density gradient of 1.06 g/mL is employed. Direct application of this method has also been demonstrated with monkey hepatocytes. In our laboratory, isolation of human hepatocytes by the referenced procedure was ineffective and resulted in lower cell yields. However, diluting the original Percoll volume by 25–50% with medium and using cell densities of approximately 10×10^6 cells per milliliter or fewer (higher cell density suspensions will not properly sediment) resulted in better yields.

4. *Cryopreservation of Hepatocytes*

There are times when the number of cells isolated from a single liver (larger animals) is so abundant, or the isolated cells are of such significance (human hepatocytes), that long-term storage of cells would be advantageous. The ability to cryopreserve hepatocytes for long time periods and use them when needed would be beneficial, in particular with human hepatocytes for which there is difficulty in obtaining tissue and controlling the frequency at which tissue becomes available.

There are many methods for cryopreserving cells, but all use a cryoprotectant, such as dimethyl sulfoxide (DMSO), and some form of controlled freezing down to −80°C or −196°C (liquid nitrogen). In general, the freezing rate is slow enough to prevent the formation of ice crystals that will damage the cells. When the cells are needed for use, they are rapidly thawed and the cryoprotectant diluted to prevent cell toxicity. Typically, both the freezing and thawing steps are conducted in a nutritional medium with fetal bovine serum (130–132).

The optimal scenario would be the ability to freeze hepatocytes and retrieve them from storage with similar initial viability and cellular functions. Unfortunately, this is only partially true with cryopreserved hepatocytes. The loss of total cell viability ranges from 40 to 60% and varies depending on the species and cryopreservation method employed (133,134). After cryopreservation, cytosolic enzyme function is typically diminished and organelle function will vary from decreased to enhanced activity (133–135). Several of the parameters studied to assess cell performance after cryopreservation include attachment to a substratum, synthesis of new cellular proteins, development of DNA repair systems, functionality of cellular organelles, and presence of drug-metabolizing enzymes (136,137). It is most important to study those cellular parameters that affect the nature of any intended study. The cryopreservation of hepatocytes for long-term storage is an area of intense research. The problems associated with cryopreservation will undoubtedly be resolved in the near future.

III. VIABILITY AND FUNCTIONAL ASSAYS TO DETERMINE CELL COMPETENCE

After the isolation of cells, the functional capabilities and viability of each preparation should be examined. Depending on the type of experiment to be performed, different types of functional assays may be necessary. In any event, the viability of the preparation should be evaluated by one or more different techniques. Trypan blue dye exclusion is by far the most popular method for assessing cell viability. This assay measures the integrity of the cell membrane by excluding trypan blue dye, which will bind to nuclear DNA if the membrane is broken (see Fig. 3). Trypan blue is a fast and simple assay to perform and is generally coupled to the determination of cell number. Hepatocyte preparations that exclude more than 85% of the dye are deemed acceptable for most experimental needs. Lactate dehydrogenase (LDH) enzyme leakage (54,75) or ATP content (85) are also commonly employed tests to determine cell viability. Table 3 lists the range of values for many cellular contents that can be used to determine the viability or functionality of freshly isolated preparations. In general, it is best to use more than one test for cell viability or functionality.

Table 3 Characteristics of Freshly Isolated Rat Hepatocytes[a]

Parameters	Value	Ref.
Cell yield	5.0–5.3 × 10^8 cells/liver	76,77
	26–60 × 10^6 cells/g liver	78–80
CYP450 content	0.17–0.73 nmoles P450/mg protein	25,77,81
	0.23 nmol P450/10^6 cells	82
	9.1–26 pmol P450/mg DNA	75,83
	PB induction: 1.4–1.84 nmol/mg protein	25,82
	CF induction: 0.90 nmol/mg protein	25
	MC induction: 0.82 nmol/mg protein	82
Protein content	1.1–1.7 ng/cell	77,84
DNA content	14.6–19.7 pg/cell	6,77
Glutathione (GSH)	2.3 mmol GSH/mg DNA	83
ATP content	2.44 mmol ATP/g wet weight	85
ATP/ADP	2.42 mM/0.42 mM	86

[a] Collagenase perfusion method: PB, phenobarbital; CF, clofibrate; MC, 3-methylcholanthrene.

A. Quantitation of Cellular Components for Expression of Hepatocyte Activity

In all instances, the data obtained from hepatocyte experiments should be normalized to some factor that allows interpretation of data between different cell preparations. Normalizing the data to the number of cells present, typically *per million cells*, is one of the most common approaches. The cell number can be easily obtained while determining cell viability with trypan blue. Unfortunately, the cell count is also subject to interindividual variability. It is very important to count the cells in a consistent manner and see that all investigators within the laboratory count the cells in a similar manner. Also, greater errors in cell count will occur when fewer than 200 cells per 4 × 4 grid are counted on the hemocytometer or when large aggregates of cells are present. As an alternative, the cells can be also be counted in a Coulter counter if one is available.

Determination of protein content is another common method of normalizing hepatocyte data. The protein content can be either the wet or dry protein weight (87), with the dry weight giving more consistent results and correlation with cell number. As indicated in Table 3, other useful normalization factors include quantitation of cellular DNA (6,11) or total spectral cytochrome P-450 (CYP450; (25,30)).

B. Functional Evaluation of Hepatocyte Enzyme Systems

A measurement of cellular function is often useful to determine depending on your particular experimental needs. There are a variety of biochemical processes that can be measured and related to viability or cytotoxicity, such as glutathione content (19,88), glutathione *S*-transferase activity, peroxisomal β-oxidation (54,89), or fatty acid oxidation (44). In addition, the incorporation of radiolabeled amino acids into protein can be employed to measure protein synthesis (90). For drug metabolism studies, measurement of total cytochrome P-450 content or specific drug-metabolizing enzyme activities can be useful to assess metabolic capability (91). Useful substrates are 7-ethoxycoumarin (54) and *p*-nitrophenol for the measurement of Phase 1- and Phase II-mediated reactions, or testosterone 6β-hydroxylation to measure CYP450 3A activity.

IV. DRUG METABOLISM-RELATED STUDIES WITH ANIMAL AND HUMAN HEPATOCYTES

A. Isolated Hepatocyte Suspensions

Freshly isolated hepatocytes contain nominal levels of drug-metabolizing enzymes, making them suitable for a myriad of drug metabolism studies. These cells when placed in suspension with appropriate nutrients (generally a complete medium) can survive for extended periods, up to 3–4 h, after which significant cell death begins to occur. Hepatocytes, by their nature, survive only when attached to a substratum; therefore, suspension experiments are considered short-term experiments, whereas hepatocytes in culture are considered longer-term experiments. Experiments best suited for suspension type studies include those that require appropriate levels of drug-metabolizing enzymes to form metabolites for identification purposes or to elicit a toxic response. These studies would generally include cross-species comparisons of rate and metabolite formation, in vitro–in vivo estimations of drug clearance or liver extraction, prediction of potential drug–drug interactions, or the formation of toxic metabolites (also see Table 1). Although some of these studies can be performed in culture, it is more prudent to study these phenomena when the cells contain a normal complement and quantity of drug-metabolizing enzymes. Once hepatocytes are placed in culture, the concentrations of various drug-metabolizing enzymes tend to decrease, making hepatocytes in culture a less attractive model system for some studies. More on this subject will be covered in the following section on hepatocytes in primary culture.

Isolated hepatocytes in suspension are a uniquely simple drug-metabolizing system requiring no additional cofactors, except for an adequate buffer system for suspension. In our laboratory we have always chosen to use a complete medium, such as WE or DMEM (wthout pH indicators) that contains many of the nutrients necessary for cell survival. Simple buffers, such as phosphate buffer or HBSS, can be employed for short-term incubations, although we feel cell viability and cofactor-dependent needs are better meet in a complete medium. Incubations are generally performed in either round bottom or Erlenmeyer flasks in a water bath shaker. The volume of suspension in each incubation vessel should not exceed 10–15% of the total vessel volume to permit adequate oxygen transfer into the cell suspension (i.e., 12.5–18.75 mL of suspension in a 125-mL Erlenmeyer flask). The rate of cell shaking is typically about 50 rpm; the rate should be fast enough to allow good oxygenation of the suspension, but not so fast that it damages the hepatocytes. Many laboratories perform suspension incubations under an atmosphere of 95% oxygen–5% carbon dioxide, although additional oxygenation is typically unnecessary, and incubations performed in air are adequate and lead to less evaporation of medium from the incubate.

Experiments performed in hepatocyte suspensions are initiated by the addition of substrate, with sample aliquots removed at appropriate time points. A typical experiment may involve 1.0-mL–sample aliquots removed at 0, 5, 15, 30, 45, 60, 120, and 180 min. Sample preparation and extraction are similar to those employed with other in vitro methods, except for the higher protein and salt content in hepatocyte suspensions that can cause greater problems with extraction, reconstitution, and detection of parent drug or metabolites. As with any in vitro technique, linearity experiments need to be performed relative to time and cell number (or another appropriate normalization factor).

B. Primary Hepatocyte Cultures

When one begins to examine the vast number of research articles describing various techniques and methods of hepatocyte culture, it becomes very apparent that an enormous amount of research has been conducted in this area. This section will attempt to describe some of the important issues necessary for the new investigator wishing to initiate research projects employing primary cultures of hepatocytes. There are several excellent reviews and book chapters devoted to these topics that the reader is directed to for more details on the particular aspects of hepatocyte culture (92–94). Previous sections within this chapter have described some of the special techniques necessary for the isolation of hepatocytes that will be placed in primary culture. In addition to these special methods, one must remember that nothing short of excellent aseptic technique is essential for the successful outcome of any hepatocyte culture experiment.

The application of cultured hepatocytes to the study of drug–metabolism-related issues is somewhat different from those employed with hepatocyte suspension experiments. These differences in application stem from the changes that occur when the cells are placed in culture. When first placed in culture, some of the architectural components of liver structure are restored, such as attachment to a basement membrane, cell-to-cell contacts (both structural and communicational), and regeneration of membrane receptors that may have been removed during cell isolation. A major advantage of employing hepatocytes in primary culture is the ability to study drug metabolism, pharmacological, or toxicological events that take longer than 2–3 h to manifest themselves; for example, the effect of drugs or cytokines on the regulation and expression of drug-metabolizing enzymes (95,96). Whereas, human and animal hepatocytes placed in primary cultures retain the ability to express drug-metabolizing enzymes, they represent a unique model for the study of enzyme induction and its effect on the metabolism of xenobiotics and endogenous cellular functions (97–103).

The most significant change that occurs when hepatocytes are placed in culture is the loss of differentiated function (cellular dedifferentiation; i.e., loss of liver-specific functions). The most dramatic phenotypic change first observed in culture was the rapid decline in cytochrome P450 levels with time. Several investigators have shown that with rat hepatocytes in culture, total cytochrome P450 levels decreased by 50% in the first 24 h and continued to fall by 68 and 85% at 72 and 96 h, respectively (104–106). Each of the cytochrome P450 isoenzymes appeared to have its own half-life, with some decreasing rapidly, whereas others remained reasonably stable. A similar phenomena was observed with the glutathione *S*-transferase isozyme activities (104). Cytochrome P450 levels in primary cultures of rat, monkey, hamster, gerbil, and mouse hepatocytes appear to decline rapidly, whereas hepatocyte cultures of rabbit and human cells decline at a much slower rate (107,108).

Most of the research and modifications to hepatocyte cell culture methods relate to the maintenance of the original differentiated state. The four main factors that have a significant bearing on the maintenance of differentiated hepatocyte function are (1) type of extracellular matrix (ECM) on which the cells attach; (2) three-dimensional configuration in which the cells attach; (3) cell–cell interactions; and (4) medium supplements and modifiers. The currently used ECM materials comprise various extracellular proteins, such as collagens, fibronectin, laminin, and heparin sulfate proteoglycan. The ECM substrata that have proved to be most useful over time are gelled collagen and Matrigel (93,109–111). The ECM with which the cells interact can have profound effects on both cell shape, polarity, and function. For example, Brown et al. found that primary cultures of rat hepatocytes were much more responsive to phenobarbital CYP2B induction when cultured with Matrigel, compared with collagen, type 1 (112). The investigators concluded that the laminin present in Matrigel was in some way responsible for the maintenance of this regulatory phenomenon. The literature on ECMs is considerable, and the only way to determine which ECM provides the most appropriate cellular response for each particular type of experiment and animal species is to examine several different types of ECM. In addition to the type of ECM, the three-dimensional structure or geometry of the ECM surrounding the cells also plays an important role. Earlier and some current methods use cells attached to a substratum of ECM in monolayer culture. A useful technique that results in improved cell morphology, cellular function, and cell–surface interactions, is the sandwich technique with either gelled collagen, Matrigel, or a combination of both ECMs (93,111,113–116). Another technique that improves cell viability and protein expression is the coculture of hepatocytes with another cell line, typically monkey or rat epithelial cells (117,118). The coculture method has been useful in extending CYP450 expression with both rat and human hepatocytes in culture (119,120). These cultures have improved intercellular connections that promote cell survivability and function. The addition of drugs, cofactors, cytokines, growth factors, or hormones to a hepatocyte culture medium have also been shown to sustain a differentiated state for longer periods (93,121,122). Addition of such components as nicotinamide (123), dexamethasone (124), ethanol (125,128), metyrapone (126), DMSO (127), or clofibrate (128) maintain viability, cytochrome P450 levels, as well as Phase II enzyme levels. An additional component often added to hepatocyte cultures, but not often considered, is the vehicle in which substrates are added. Substrates added in a medium or buffer pose little problems, although substrates added in organic solvents may have a significant effect on drug-metabolizing enzymes (125,127–129). In summary, the conditions under which hepatocytes can be cultured is extremely varied and, although several approaches are commonly employed, each investigator is encouraged to examine several different variables when settingup a new model.

V. SUMMARY AND FUTURE DIRECTIONS

The following description and use of isolated or cultured animal or human hepatocytes has attempted to demonstrate the versatility and usefulness of hepatocytes as an in vitro model of drug metabolism. The hepatocyte as a model of drug metabolism, with its complement of enzymes, regulatory elements, and intact cell membrane, is a close approximation to the in vivo functioning liver. Albeit, even the hepatocyte model suffers from many of the issues that confound other in vitro methods. These issues include (1) hepatocytes are a static system and metabolites generated have the ability to accumulate and create toxicity or inhibition, (2) the system examines only hepatic and not extrahepatic metabolism, and (3) the problems that are inherent in primary cultures of hepatocytes (i.e., dedifferentiation and loss of CYP450 activity). Even with these limitations, the hepatocyte serves as an excellent model system for drug metabolism, pharmacology, and toxicology testing. Clearly, hepatocytes are becoming more universal in their application to drug metabolism in such areas as evaluation of potential drug–drug interactions and as a model system to study CYP450 enzyme induction.

Future directions and developments in the area of hepatocytes surround issues such as standardization of conditions, cryopreservation, and improvements in understanding the relations between in vitro and in vivo correlations. A standardization of hepatocyte characterization and culture conditions would permit a more universal cross-referencing of data between different laboratories (138). In addition, improved methods for the cryopreservation of human hepatocytes would allow experiments to be conducted on a more timely basis. And, finally, as our understanding of the use of hepatocytes improves, we should be more successful in our attempts at in vitro–in vivo correlations and interpretations of drug metabolism.

ACKNOWLEDGMENTS

The author would like to acknowledge the kind assistance of Ms Robbie Fox in the preparation of this manuscript and the editorial review comments of Mr. Mark Spahr, Dr. Jasminder Sahi, and Dr. Krishna Iyer. Photographs were graciously supplied by Dr. Jasminder Sahi.

REFERENCES

1. J. R. Fry and J. W. Bridges, The metabolism of xenobiotics in cell suspensions and cell cultures, *Progress in Drug Metabolism*, Vol. 2 (J. W. Bridges and L. F. Chasseaud, eds.), Wiley, New York, 1977, p. 71.
2. R. J. Chenery, The utility of hepatocytes in drug metabolism studies, *Progress in Drug Metabolism*, Vol. 11 (G. G. Gibson, ed.), Taylor & Francis, New York, 1988, p. 217.
3. G. Fabre, J. Combalbert, Y. Berger, and J.-P. Cano, Human hepatocytes as a key in vitro model to improve preclinical drug development, *Eur. J. Drug Metab. Pharmacol. 15*:165 (1990).
4. P. Maurel, The use of adult human hepatocytes in primary culture and other in vitro systems to investigate drug metabolism in man, *Adv. Drug Deliv. Rev. 22*:105 (1996).
5. M. N. Berry, H. J. Halls, and M. B. Grivell, Techniques for pharmacological and toxicological studies with isolated hepatocyte suspension, *Life Sci. 51*:1 (1992).
6. S. J. Gee, C. E. Green, and C. A. Tyson, Comparative metabolism of tolbutamide by isolated hepatocytes from rat, rabbit, dog, squirrel monkey, *Drug Metab. Dispos. 12*:174 (1984).
7. H. G. Oldham, P. Standing, S. J. Norman, T. J. Blake, I. Beattie, P. J. Cox, and R. J. Chenery, Metabolism of temelastine (SK&F93944) in hepatocytes from rat, dog, cynomolgus monkey, and man, *Drug Metab. Dispos. 18*:146 (1990).
8. J. A. Holme, B. Trygg, and E. Soderlund, Species differences in the metabolism of 2-acetylaminofluorene by hepatocytes in primary monolayer, *Cancer Res. 46*:1627 (1986).
9. C. J. Maslansky and G. M. Williams, Primary cultures and the levels of cytochrome P450 in hepatocytes from mouse, rat, hamster, and rabbit liver, *In Vitro 18*:683 (1982).
10. R. P. Remmel and M. W. Sinz, A quaternary ammonium glucuronide is the major metabolite of lamotrigine in guinea pigs: In vitro and in vivo studies, *Drug Metab. Dispos. 19*:630 (1991).
11. J. C. Kraner, J. M. Lasker, G. B. Corcoran, S. D. Ray, and J. L. Raucy, Induction of P4502E1 by acetone in isolated rabbit hepatocytes: Role of increased protein and mRNA synthesis, *Biochem. Pharmacol. 45*:1483 (1993).

12. T. E. Kitos and D. L. J. Tyrrell, Intracellular metabolism of 2′,3′-dideoxynucleosides in duck hepatocyte primary culture, *Biochem. Pharmacol. 49*:1291 (1995).
13. M. Pesonen and T. Anderson, Characterization and induction of xenobiotic metabolizing enzyme activities in a primary culture of rainbow trout hepatocytes, *Xenobiotica 21*:461 (1991).
14. Emmison, L. Agius, and V. A. Zammit, Regulation of fatty acid metabolism and gluconeogenesis by growth hormone and insulin in sheep hepatocyte cultures, *Biochem. J. 274*:21 (1991).
15. L. A. P. Hoogenboom, O. Tomassini, M. B. M. Oorsprong, and H. A. Kuiper, Use of pig hepatocytes to study the inhibition of monoamine oxidase by furazolidone, *Food Chem. Toxicol. 29*:185 (1991).
16. C. E. Green, S. E. LeValley, and C. A. Tyson, Comparison of amphetamine metabolism using isolated hepatocytes from five species including human, *J. Pharmacol. Exp. Ther. 237*: 931 (1986).
17. T. P. Mommsen and K. B. Storey, Hormonal effects on glycogen metabolism in isolated hepatocytes of a freeze-tolerant frog, *Gen. Comp. Endocrinol. 87*:44 (1992).
18. L. R. Shull, D. G. Kirsch, C. L. Lohse, and J. A. Wisniewski, Application of isolated hepatocytes to studies of drug metabolism in large food animals, *Xenobiotica 17*:345 (1987).
19. M. R. Dilloway, J. W. Bridges, P. J. Bunyan, and P. I. Stanley, Species differences in the effects of 1,1-di-(4-chlorophenyl)-2-chloroethylene (DDMU) and 1,1-di-(4-chlorophenyl)-2,2-dichloroethylene (DDE) on glutathione levels in isolated hepatocytes from Japanese quail and rat, *Xenobiotica 12*:489 (1982).
20. J. M. Begue, J. F. Le Bigot, C. Guguen-Guillouzo, J. R. Kiechel, and A. Guillouozo, Cultured human adult hepatocytes: A new model for drug metabolism, *Biochem. Pharmacol. 32*:1643 (1983).
21. J. A. P. Hoogenboom, M. C. J. Berghmans, A. Van Veldhuizen, and H. A. Kuiper, The use of pig hepatocytes to study the biotransformation of β-nortestosterone by pigs, *Drug Metab. Dispos. 18*:999 (1990).
22. D. W. Cragin and T. Shibamoto, Formation of *N*-nitrosothiazolidine metabolites in isolated rat liver hepatocytes, *Toxicol. In Vitro 6*:89 (1992).
23. T. Walle, G. N. Kumar, J. M. McMillan, K. R. Thornburg, and U. K. Walle, Taxol metabolism in rat hepatocytes, *Biochem. Pharmacol. 46*:1661 (1993).
24. R. McCague, I. B. Parr, G. Leclercq, O. T. Leung, and M. Jarman, Metabolism of tamoxifen by isolated rat hepatocytes, *Biochem. Pharmacol. 39*:1459 (1990).
25. M. W. Sinz, A. E. Black, S. M. Bjorge, A. Holmes, B. K. Trivedi, and T. F. Woolf, In vitro and in vivo disposition of 2,2-dimethyl-*N*-(2,4,6-trimethoxyphenyl) dodecanamide (Cl-976): Identification of a novel five-carbon cleavage metabolite in rats, *Drug Metab. Dispos. 25*: 123 (1997).
26. T. Lave, S. Dupin, M. Schmitt, M. Kapps, J. Meyer, B. Morgenroth, R. C. Chou, D. Jaeck, and P. Coassolo, Interspecies scaling of tolcapone, a new inhibitor of catechaol-*O*-methyltransferase (COMT). Use on in vitro data from hepatocytes to predict metabolic clearance in animals and humans, *Xenobiotica 26*:839 (1996).
27. K. A. Hayes, B. Brennan, R. Chenery, and J. B. Houston, In vivo disposition of caffeine predicted from hepatic microsomal and hepatocyte data, *Drug Metab. Dispos. 23*:349 (1995).
28. K. Zomorodi, D. J. Carlile, and J. B. Houston, Kinetics of diazepam metabolism in rat hepatic microsomes and hepatocytes and their use in predicting in vivo hepatic clearance, *Xenobiotica 25*:907 (1995).
29. S. A. Pahernick, J. Schmid, T. Sauter, F. W. Schildberg, and H. G. Koebe, Metabolism of pimobendan in long-term human hepatocyte culture: In vivo–in vivo comparison, *Xenobiotica 25*:811 (1995).
30. R. J. Chenery, A. Ayrton, H. G. Oldham, P. Standring, S. J. Norman, T. Seddon, and R.

Kirby, Diazepam metabolism in cultured hepatocytes from rat, rabbit, dog, guinea pig, and man, *Drug Metab. Dispos. 15*:312 (1987).

31. T. A. Kocarek, E. G. Schuetz, S. C. Strom, R. A. Fisher, and P. S. Guzelian, Comparative analysis of cytochrome P4503A induction in primary cultures of rat, rabbit, and human hepatocytes, *Drug Metab. Dispos. 23*:415 (1995).
32. D. K. Monteith, G. Michalopoulos, and S. C. Strom, Conjugation of chemical carcinogens by primary cultures of human hepatocytes, *Xenobiotica 20*:753 (1990).
33. M. J. Gomez-Lechon, A. Montoya, P. Lopez, T. Donato, A. Larrauri, and J. V. Castell, The potential use of cultured hepatocytes in predicting the hepatotoxicity of xenobiotics, *Xenobiotica 18*:725 (1988).
34. P. Moldeus, Paracetamol metabolism and toxicity in isolated hepatocytes from rat and mouse, *Biochem. Pharmacol. 27*:2859 (1978).
35. D. W. Gottschall, R. A. Wiley, and R. P. Hanzlik, Toxicity of *ortho*-substituted bromobenzenes to isolated hepatocytes: Comparison to in vivo results, *Toxicol. Appl. Pharmacol. 69*: 55 (1983).
36. R. E. Billings and T. R. Tephly, Studies on methanol toxicity and formate metabolism in isolated hepatocytes, *Biochem. Pharmacol. 28*:2985 (1979).
37. C. J. Viau, R. D. Curren, and K. Wallace, Cytotoxicity of tacrine and velnacrine metabolites in cultured rat, dog, and human hepatocytes, *Drug Chem. Toxicol. 16*:227 (1993).
38. T. A. Kocarek and A. B. Reddy, Regulation of cytochrome P450 expression by inhibitors of hydroxymethylglutaryl-coenzyme A reductase in primary cultured rat hepatocytes and in rat liver, *Drug Metab. Dispos. 24*:1197 (1996).
39. L. Liu, E. L. LeCluyse, J. Liu, and C. D. Klaassen, Sulfotransferase gene expression in primary cultures of rat hepatocytes, *Biochem. Pharmacol. 52*:1621 (1996).
40. Y. Nishibe and M. Hirata, Effect of phenobarbital and other model inducers on cytochrome P450 isoenzymes in primary culture of dog hepatocytes, *Xenobiotica 23*:681 (1993).
41. M. T. Doato, J. V. Castell, and M. J. Gomez-Lechon, Effect of model inducers on cytochrome P450 activities of human hepatocytes in primary culture, *Drug Metab. Dispos. 23*:553 (1995).
42. Yukiko Yoshida, N. Kimura, H. Oda, and A. Kakinuma, Insulin suppresses the induction of CYP2B1 and CYP2B2 gene expression by phenobarbital in adult rat cultured hepatocytes, *Biochem. Biophys. Res. Commun. 229*:182 (1996).
43. A. Baquet, A. Lavoinne, and L. Hue, Comparison of the effects of various amino acids on glycogen synthesis, lipogenesis and ketogenesis in isolated rat hepatocytes, *Biochem. J. 273*: 57 (1991).
44. J.-P. Fulgencio, C. Kohl, J. Girard, and J. -P. Peforier, Troglitazone inhibits fatty acid oxidation and esterification and gluconeogenesis in isolated hepatocytes from starved rats, *Diabetes 45*:1556 (1996).
45. K. Z. Ching, K. A. Tenny, J. Chen, and E. T. Morgan, Suppression of constitutive cytochrome P450 gene expression by epidermal growth factor receptor ligands in cultured rat hepatocytes, *Drug Metab. Dispos. 24*:542 (1996).
46. T. W. Moon and T. P. Mommsen, Vasoactive peptides and phenylephrine actions in isolated teleost hepatocytes, *Am. J. Physiol. 259*:E644 (1990).
47. O. Takenaka, T. Horie, H. Suzuki, and Y. Sugiyama, Carrier-mediated active transport of the glucuronide and sulfate of 6-hydroxy-5, 7-dimethyl-2-methylamino-4-(3-pyridylmethyl) benzothiazole (E3040) into rat liver: Quantitative comparison of permeability in isolated hepatocytes, perfused liver, and liver in vivo, *J. Pharmacol. Exp. Ther. 280*:948 (1997).
48. W. E. M. Mol, G. N. Fokkema, B. Weert, and D. K. F. Meijer, Mechanisms for the hepatic uptake of organic cations. Studies with muscle relaxant vecuronium in isolated rat hepatocytes, *J. Pharmacol. Exp. Ther. 244*:268 (1988).
49. M. E. McPhail, R. G. Knowles, M. Salter, J. Dawson, B. Burchell, and C. I. Pegson, Uptake of acetaminophen (paracetamol) by isolated rat liver cells, *Biochem. Pharmacol. 45*:1599 (1993).

50. S. Khan, C. Sood, and P. J. O'Brien, Molecular mechanisms of dibromoalkane cytotoxicity in isolated rat hepatocytes, *Biochem. Pharmacol. 45*:439 (1993).
51. D. DiMonte, T. Shinka, M. S. Sandy, N. Castagnoli, Jr., and M. T. Smith, Quantitative analysis of 1-methyl-4-phenyl-1,2,3,6-tetrahydropyridie metabolism in isolated rat hepatocytes, *Drug Metab. Dispos. 16*:250 (1988).
52. L. Prichard, J. M. Fabre, J. Domergue, G. Fabre, S. -A. Mourad, and P. Maurel, Molecular mechanism of cyclosporine A drug interactions: Inducers and inhibitors of cytochrome P450 screening in primary cultures of human hepatocytes, *Trans. Proc. 23*:978 (1991).
53. M. N. Berry, A. M. Edwards, and G. J. Barritt, Isolated hepatocyte preparation, properties and application, *Laboratory Techniques in Biochemistry and Molecular Biology* (R. H. Burdon and P. H. van Knippenberg, eds.), Elsevier, New York, 1991, p. 1.
54. N. J. DelRaso, In vitro methods for assessing chemical or drug toxicity and metabolism in primary hepatocytes, *Drug Toxicity and Metabolism in Primary Hepatocytes* (R. R. Watson, ed.), CRC Press, Boca Raton, FL, 1992, p. 175.
55. P. O. Seglen, Isolation of hepatocytes by collagenase perfusion, *Methods in Toxicology*, Vol 1A (C. A. Tyson and J. M. Frazier, eds.), Academic Press, New York, 1993, p. 231.
56. C. A. McQueen, Isolation and culture of hepatocytes from different laboratory species, *Methods in Toxicology*, Vol 1A (C. A. Tyson and J. M. Frazier, eds.), Academic Press, New York, 1993, p. 255.
57. W. C. Schneider and V. R. Potter, Assay of animal tissues for respiratory enzymes; succinic dehydrogenase and cytochrome oxidase, *J. Biol. Chem. 217*:149 (1943).
58. R. B. Howard and L. A. Pesch, Respiratory activity of intact, isolated parenchymal cells from rat liver, *J. Biol. Chem. 243*:3105 (1968).
59. P. Belleman, R. Gebhardt, and D. Mecke, An improved method for the isolation of hepatocytes from liver slices, *Anal. Biochem. 81*:408 (1977).
60. J. R. Fry, C. A. Jones, P. Wiekin, P. Belleman, and J. W. Bridges, The enzymic isolation of adult rat hepatocytes in a functional and viable state, *Anal. Biochem. 71*:341 (1976).
61. J. R. Fry, Preparation of mammalian hepatocytes, *Methods in Enzymol. 77*:130 (1981).
62. B. N. Berry and D. S. Friend, High-yield preparation of isolated rat liver parenchymal cells, *J. Cell Biol. 43*:506 (1969).
63. P. O. Seglen, Preparation of isolated rat liver cells, *Methods in Cell Biology* (D. M. Prescott, ed.), Academic Press, New York, 1976, p. 29.
64. K. L. Allen and C. E. Green, Isolation of human hepatocytes by biopsy perfusion methods, *Methods in Toxicology*, Vol 1A (C. A. Tyson and J. M. Frazier, eds.), Academic Press, New York, 1993, p. 262.
65. M. N. Berry, C. Farrington, S. Gay, A. R. Grivell, and P. G. Wallace, Preparation of isolated hepatocytes in good yield without enzymatic digestion, *Isolation, Characterization, and Use of Hepatocytes* (R. A. Harris and N. W. Cornell, eds.), Elsevier Science, New York, 1983, p. 7.
66. M. J. Beredith, Rat hepatocytes prepared without collagenase: Prolonged retention of differentiated characteristics in culture, *Cell Biol. Toxicol. 4*:405 (1988).
67. A. Y. Petrenko and A. N. Sukach, Isolation of intact mitochondria and hepatocytes using vibration, *Anal. Biochem. 194*:326 (1991).
68. J. Vina, R. Hems, and H. A. Krebs, Maintenance of glutathione content in isolated hepatocytes, *Biochem. J. 170*:627 (1978).
69. S. Iqbal, C. R. Elcombe, and E. Elias, Maintenance of mixed-function oxidase and conjugation enzyme activities in hepatocyte cultures prepared from normal and diseased human liver, *J. Hepatol. 12*:336 (1991).
70. A. Guillouzo, P. Gripon, D. Ratanasavanh, B. Clement, and C. Guguen-Guillouzo, Cultured human hepatocytes as a model system for man in pharmacotoxicological research, *Advances in Applied Toxicology* (A. D. Dayan and A. J. Paine, eds.), Taylor & Francis, New York, 1989, p. 165.

71. C. A. McQueen, Hepatocytes in monolayer culture: An in vitro model for toxicity studies, *In Vitro Toxicology Model Systems and Methods* (C. A. McQueen, ed.), Telford Press, New Jersey, 1989, p. 131.
72. R. G. Ulrich, D. G. Aspar, C. T. Cramer, R. F. Kletzien, and L. C. Ginsberg, Isolation and culture of hepatocytes from the cynomolgus monkey (*Macaca fascicularis*), *In Vitro Cell. Dev. Biol. 26*:815 (1990).
73. B. L. Kreamer, J. L. Staecker, N. Sawada, G. L. Sattler, M. T. S. Hsia, and H. C. Pitot, Use of low-speed iso-density Percoll centrifugation method to increase the viability of isolated rat hepatocyte preparations, *In Vitro Cell. Dev. Biol. 22*:201 (1986).
74. C. Dalet, M. Fehlmann, and P. Debey, Use of Percoll density gradient centrifugation for preparing isolated rat hepatocytes having long-term viability, *Anal. Biochem. 122*:119 (1982).
75. E. S. Chao, D. Dunbar, and L. S. Kaminsky, Intracellular lactate dehydrogenase concentration as an index of cytotoxicity in rat hepatocyte primary culture, *Cell Biol. Toxicol. 4*:1 (1988).
76. K. K. Dougherty, S. D. Spilman, C. E. Green, A. R. Stewart, and J. L. Byard, Primary cultures of adult mouse and rat hepatocytes for studying the metabolism of foreign chemicals; *Biochem. Pharmacol. 29*:2117 (1980).
77. M. Dickins and R. E. Peterson, Effects of a hormone-supplemented medium on cytochrome P450 content and mono-oxygenase activities of rat hepatocytes in primary culture, *Biochem. Pharmacol. 29*:1231 (1980).
78. J. A. Reese and J. L. Byard, Isolation and culture of adult hepatocytes from liver biopsies, *In Vitro 17*:935 (1981).
79. J. Bayad, N. Sabolovic, D. Bagrel, J. Magdalou, and G. Siest, Influence of the isolation method on the stability of differentiated phenotype in cultured rat hepatocytes, *J. Pharm. Metab. 25*:85 (1991).
80. M. N. Berry, A. M. Edwards, and G. J. Barritt, Isolated hepatocyte preparation, properties and application, *Laboratory Techniques in Biochemistry and Molecular Biology* (R. H. Burdon and P. H. van Knippenberg, eds.), Elsevier, New York, 1991, p. 53.
81. K. F. Nelson and D. Acosta, Long-term maintenance and induction of cytochrome P-450 in primary cultures of rat hepatocytes, *Biochem. Pharmacol. 31*:2211 (1982).
82. P. O. Seglen, J. Hogberg, and S. Orrenius, Isolation and use of liver cells, *Methods Enzymol. 52*:60 (1978).
83. C. E. Green, J. E. Dabbs, and C. A. Tyson, Functional integrity of isolated rat hepatocytes prepared by whole liver vs biopsy perfusion, *Anal. Biochem. 129*:269 (1983).
84. M. N. Berry, A. M. Edwards, and G. J. Barritt, Isolated hepatocyte preparation, properties and application, *Laboratory Techniques in Biochemistry and Molecular Biology* (R. H. Burdon and P. H. van Knippenberg, eds.), Elsevier, New York, 1991, p. 127.
85. R. A. Page, K. M. Stowell, M. J. Hardman, and K. E. Kitson, The assessment of viability in isolated rat hepatocytes, *Anal. Biochem. 200*:171 (1992).
86. M. N. Berry, A. M. Edwards, and G. J. Barritt, Isolated hepatocyte preparation, properties and application, *Laboratory Techniques in Biochemistry and Molecular Biology* (R. H. Burdon and P. H. van Knippenberg, eds.), Elsevier, New York, 1991, p. 145.
87. M. N. Berry, A. M. Edwards, and G. J. Barritt, Isolated hepatoctyes preparation, properties and application, *Laboratory Techniques in Biochemistry and Molecular Biology* (R. H. Burdon and P. H. van Knippenberg, eds.), Elsevier, New York, 1991, p. 122.
88. P. J. Hissin and F. Hilf, A fluorometric method for determination of oxidized and reduced glutathione in tissues, *Anal. Biochem. 74*:214 (1976).
89. P. B. Lazarow, Assay of peroxisomal beta-oxidation of fatty acids, *Methods Enzymol. 72*: 315 (1981).
90. P. O. Seglen, Incorporation of radioactive amino acids into protein in isolated rat hepatocytes, *Biochim. Biophys. Acta 442*:391 (1976).
91. A. Paine, L. J. Hockin, and R. F. Legg, Relationship between the ability of nicotinamide to

maintain nicotinamide-adenine dinucleotide in rat liver cell culture and its effect on cytochrome P450, *Biochem. J. 193*:461 (1979).

92. A. Guillouzo, Biotransformation of drugs by hepatocytes, *In Vitro Methods in Pharmaceutical Research* (J. V. Castell and M. J. Gomez-Lechon, eds.), Academic Press, New York, 1997, p. 411.
93. E. L. LeCluyse, P. L. Bullock, A. Parkinson, Strategies for restoration and maintenance of normal hepatic structure and function in long-term cultures of rat hepatocytes, *Adv. Drug Deliv. Rev. 22*:133 (1996).
94. A. J. Paine, The maintenance of cytochrome P-450 in rat hepatocyte culture: Some applications of liver cell cultures to the study of drug metabolism, toxicity and the induction of the P450 system, *Chem. Biol. Interace. 74*:1 (1990).
95. M. T. Donato, J. V. Castell, and M. J. Gomez-Lechon, Effect of model inducers on cytochrome P450 activities of human hepatocytes in primary culture, *Drug Metab. Dispos. 23*: 553 (1995).
96. Z. Abdel-Razzak, L. Corcos, A. Fautrel, J. P. Campion, and A. Guillouzo, Transforming growth factor $beta_1$ down-regulates basal and polycyclic aromatic hydrocarbon-induced cytochromes P-450 1A1 and 1A2 in adult hepatocytes in primary culture, *Mol. Pharmacol. 46*: 1100 (1994).
97. F. Morel, P. Beaune, D. Ratansavahnh, J. P. Flinois, F. P. Guengerich, and A. Guillouzo, Effects of various inducers on the expression of cytochromes P-450 IIC8, 9, 10 and IIIA in cultured adult human hepatocytes, *Toxicol. In Vitro 4*:458 (1990).
98. U. Forster, G. Luippold, and L. R. Schwarz, Induction of monooxygenase and UDP-glucuronosyltransferase activities in primary cultures of rat hepatocytes, *Drug Metab. Dispos. 13*:353 (1986).
99. T. Harauchi and M. Hirata, Effect of P-450 on glutathione (GSH) depletion by bromobenzene in primary cultures of dog hepatocytes, *Biol. Pharmcol. Bull. 17*:658 (1994).
100. R. C. Zanger, K. J. Woodcroft, T. A. Kocarek, and R. F. Novak, Xenobiotic-enhanced expression of cytochrome P450 2E1 and 2B1/2 in primary cultured rat hepatocytes, *Drug Metab. Dispos. 23*:681 (1995).
101. O. Sabzevari, M. Hatcher, M. O'Sullivan, P. Kentish, and G. Gibson, Comparative induction of cytochrome P4504A in rat hepatocyte culture by peroxisome proliferators, bifonazole and clofibrate, *Xenobiotica 25*:396 (1995).
102. M. T. Donato, M. J. Gomez-Lechon, and J. V. Castell, Effect of xenobiotics on monooxygenase activities in cultured human hepatocytes, *Biochem. Pharmacol. 39*:1321 (1990).
103. J. S. Sidhu and C. J. Omiecinski, Modulation of xenobiotic-inducible cytochrome P450 gene expression by dexamethasone in primary rat hepatocytes, *Pharmacogeneitcs 5*:24 (1995).
104. T. Croci and G. M. Williams, Activities of several phase I and phase II xenobiotic biotransformation enzymes in cultured hepatocytes from male and female rats, *Biochem. Pharmacol. 34*:3029 (1985).
105. H. M. Wortelboer, C. A. De Kruif, A. A. J. Van lersel, H. E. Falke, J. Noordhoek, and B. J. Blaauboer, The isoenzyme pattern of cytochrome P450 in rat hepatocytes in primary culture, comparing different enzyme activities in microsomal incubations and in intact monolayers, *Biochem. Pharmacol. 40*:2325 (1990).
106. A. R. Steward, G. A. Dannan, P. S. Guzelian, and F. P. Geungerich, Changes in the concentration of seven forms of cytochrome P-450 in primary cultures of adult rat hepatocytes, *Mol. Pharmacol. 27*:125 (1985).
107. A. Guillouzo, F. Morel, O. Fardel, and B. Meunier, Use of human hepatocyte culture for drug metabolism, *Toxicology 82*:209 (1993).
108. L. R. Schwarz and F. J. Wiebel, Cytochrome P450 in primary and permanent liver cell cultures, *Handbook of Experimental Pharmacology*, Vol. 105 (J. B. Schenkman and H. Grein, eds.), Springer-Verlag, New York, 1993, p. 399.
109. H.-G. Koebe, S. Pahernik, P. Elyer, and F. W. Schildberg, Collagen gel immobilization: A

useful cell culture technique for long term metabolic studies on human hepatocytes, *Xenobiotica 24*:95 (1994).

110. H. O. Jaurefui, P. N. McMillan, J. Driscoll, and S. Naik, Attachment and long term survival of adult rat hepatocytes in primary monolayer cultures: Comparison of different substrata and tissue culture media formulations, *In Vitro Cell. Dev. Biol.* 22:13 (1986).
111. P. V. Moghe, F. Berthiaume, R. M. Ezzell, M. Toner, R. G. Tompkins, and M. L. Yarmsuh, Culture matrix configuration and composition in the maintenance of hepatocyte polarity and function, *Biomaterials 17*:373 (1996).
112. S. E. Brown, C. P. Guzelian, E. Schuetz, L. C. Quattrochi, H. K. Kleinman, and P. S. Guzelian, Critical role of extracellular matrix on induction by phenobarbital of cytochrome P450 2B1/2 in primary cultures of adult rat hepatocytes, *Lab. Invest. 73*:818 (1995).
113. J. C. Y. Dunn, M. L. Yarmush, H. G. Koebe, and R. G. Tompkins, Hepatocyte function and extracellular matrix geometry: Long term culture in a sandwich configuration, *FASEB J. 3*: 174 (1989).
114. E. L. LeCluyse, K. L. Audus, and J. H. Hochman, Formation of extensive canalicular networks by rat hepatocytes cultured in collagen-sandwich configuration, *Am. J. Physiol. 266*: C1764 (1994).
115. F. Berthiaume, P. V. Moghe, M. Toner, and M. L. Yarmush, Effect of extracellular matrix topology on cell structure, function, and physiological responsiveness: Hepatocytes cultured in a sandwich configuration, *FASEB J. 10*:1471 (1996).
116. A. Bader, E. Knop, K. H. Boker, O. Crome, N. Fruhauf, A. K. Gonschior, U. Christians, H. Esselman, R. Pichlmayr, and K. F. Sewing, Tacrolimus (FK506) biotransformation in primary rat hepatocytes depends on extracellular matrix geometry, *Naunyn Schmiedebergs Arch. Pharmacol. 353*:461 (1996).
117. J.-M. Begue, C. Guguen-Guillouzo, N. Pasdeloup, and A. Guillouzo, Prolonged maintenance of active cytochrome P-450 in adult rat hepatocytes co-cultured with another liver cell type, *Hepatology 4*:839 (1984).
118. M. T. Donato, M. J. Gomez-Lechon, and J. V. Castell, Rat hepatocytes cultured on a monkey kidney cell line: Expression of biotransformation and hepatic metabolic activities, *Toxicol. In Vitro 5*:435 (1991).
119. C. Niemann, J.-C. Gauthier, L. Richert, M.-A. Ivanov, C. Melcion, and A. Cordier, Rat adult hepatocytes in primary pure and mixed monolayer culture, comparison of the maintenance of mixed function oxidase and conjugation pathways of drug metabolism, *Biochem. Pharmacol. 42*:373 (1991).
120. A. Guillouzo, P. Beaune, M-N. Gascoin, J.-M. Begue, J-P. Campion, F. P. Guengerich, and C. Guguen-Guillouzo, Maintenance of cytochrome P-450 in cultured adult human hepatocytes, *Biochem. Pharmacol. 34*:2991 (1985).
121. V. Rogiers and A. Vercruysse, Rat hepatocyte culture and co-cultures in biotransformation studies of xenobiotics, *Toxicology 82*: 193 (1993).
122. M. Warren and J. R. Fry, Influence of medium composition on 7-alkoxycoumarin O-dealkylase activities of rat hepatocytes in primary maintenance culture, *Xenobiotica 18*:973 (1988).
123. C. Inoue, H. Yamamoto, T. Nakamura, A. Ichihara, and H. Okamoto, Nicotimamide prolongs survival of primary cultured hepatocytes without involving loss of hepatocyte-specific functions, *J. Biol. Chem. 264*:4747 (1989).
124. J. M. McMillan, J. G. Shaddock, D. A. Casciano, M. P. Arlotto, and J. E. A. Leakey, Differential stability of drug-metabolizing enzyme activities in primary rat hepatocytes, cultured in the absence or presence of dexamethasone, *Mutat. Res. 249*:81 (1991).
125. G. Debast, S. Coecke, M. Akrawi, A. Foriers, A. Vercruysse, I. Phillips, E. Shephard, and V. Rogiers, Effect of ethanol on glutathione *S*-transferase expression in co-cultured rat hepatocytes, *Toxicol. In Vitro 9*:467 (1995).
126. B. Goodwin, C. Liddle, M. Murray, M. Tapner, T. Rooney, and G. C. Farrell, Effects of

metyrapone on expression of CYPs 2C11, 3A2, and other 3A genes in rat hepatocytes cultured on Matrigel, *Biochem. Pharmacol. 52*:219 (1996).

127. C. K. Lindsay, R. J. Chenery, and G. M. Hawksworth, Primary culture of rat hepatocytes in the presence of dimethyl sulphoxide, a system to investigate the regulation of cytochrome P450 1A, *Biochem. Pharmacol. 42*:S17 (1991).
128. N. Perrot, C. Chesne, I. De Waziers, J. Conner, P. H. Beaune, and A. Guillouzo, Effects of ethanol and clofibrate on expression of cytochrome P-450 enzymes and epoxide hydrolase in cultures and cocultures of rat hepatocytes, *Eur. J. Biochem. 200*:255 (1991).
129. M. E. Sauers, R. T. Abraham, J. D. Alvin, and M. A. Zemaitis, Effect of drug vehicles on *N*-demethylase activity in isolated hepatocytes, *Biochem. Pharmacol. 29*:2073 (1980).
130. C. Chesne and A. Guillouzo, Cryopreservation of isolated hepatocytes: A critical evaluation of freezing and thawing conditions, *Cryobiology 25*:323 (1988).
131. J. N. Lawrence and D. J. Benford, Development of an optimal method for the cryopreservation of hepatocytes and their subsequent monolayer culture, *Toxicol. In Vitro 5*:39 (1991).
132. L. J. Loretz, A. P. Li, M. W. Flye, and A. G. E. Wilson, Optimization of cryopreservation procedures for rat and human hepatocytes, *Xenobiotica 19*:489 (1989).
133. B. Diener, M. Traiser, M. Arand, J. Leissner, U. Witsch, R. Hohenfellner, F. Fandrich, I. Vogel, D. Utesch, and F. Oesch, Xenobiotic metabolizing enzyme activity in isolated and cryopreserved human liver parenchymal cells, *Toxicol. In Vitro 8*:1161 (1994).
134. D. Utesch, B. Diener, E. Molitor, F. Oesch, and K.-L. Platt, Characterization of cryopreserved rat liver parenchymal cells by metabolism of diagnostic substrates and activities of related enzymes, *Biochem. Pharmacol. 44*:309 (1992).
135. J. A. Coundouris, M. H. Grant, J. Engeset, J. C. Petrie, and G. M. Hawksworth, Cryopreservation of human adult hepatocytes for use in drug metabolism and toxicity studies, *Xenobiotica 23*:1399 (1993).
136. N. J. Swales, C. Loung, and J. Caldwell, Cryopreservation of rat and mouse hepatocytes. I. Comparative viability studies, *Drug Metab. Dispos. 24*:1218 (1996).
137. N. J. Swales, T. Johnson, and J. Caldwell, Cryopreservation of rat and mouse hepatocytes. II. Assessment of metabolic capacity using testosterone metabolism, *Drug Metab. Dispos. 24*:1224 (1996).
138. P. Skett and M. Bayliss, Time for a consistent approach to preparing and culturing hepatocytes? *Xenobiotica 26*:1 (1996).

16

The Isolated Perfused Liver as a Tool in Drug Metabolism Studies

Marián Kukan
Institute of Preventive and Clinical Medicine, Bratislava, Slovakia

I. INTRODUCTION

Many techniques are available for studying drug metabolism, and they have been extensively discussed throughout this handbook. Each of these techniques provides important information for understanding the mechanisms of drug detoxification and bioactivation, yet the use of an isolated perfused organ technique has great advantages. Not only does it permit one to draw definite conclusions about the overall contribution of a given organ to in vivo drug metabolism, it also allows the drug to be delivered to the sites of metabolism by the physiological route. A major advantage of an isolated organ over in vivo experiments is that it can be perfused under precisely defined conditions, such as the composition of the perfusion medium and the perfusion rate and pressure. In addition to the liver (1–5), the following major organs or tissue preparations have been frequently employed for perfusion studies: the heart (6), lung (7,8), kidney (9–11), brain (12,13), intestine (14,15), intestine–liver preparation (16), and pancreas (17).

Because the liver is the major organ involved in metabolism of most drugs, the purpose of this chapter is to provide insight into the methods and techniques of organ isolation and liver perfusion. Detailed information provided here should enable a new-comer to set up the perfusion apparatus, to isolate the liver from the animal body, to plan an appropriate experimental design, and to process as well as interpret the data obtained from drug metabolism studies. In addition, sufficient information will also be provided for studying the effects of the drug on hepatic function at toxicologically relevant concentrations (e.g., concentrations used in toxicokinetic studies).

II. ISOLATED LIVER PREPARATION

Animal species, including monkey (18), pig (19), goat (20), cat, and rabbit (18) have been employed for liver perfusion experiments. The isolated liver of the rat, however, is the most frequently used model in drug metabolism studies. This model has also been successfully applied to explore mechanisms in pharmacological (21), physiological (22–26), toxicological (5), and biochemical (27) investigations.

The perfusion technique through the portal vein appears to be satisfactory because it meets the demands of the liver for oxygen supply. Perfusion techniques through the hepatic artery have also been described (28,29).

A. Apparatus

The perfusion apparatus is relatively easy to set up and establishment of the apparatus will enable one to not only carry out perfusion experiments with rat liver and also kidney but it also can be used to isolate liver cells including those of humans (30) in a cost-effective manner (31). Many new research opportunities are provided by introducing the perfusion method in ones laboratory. A versatile perfusion apparatus used in our laboratory is described in the following. It can be employed for constant pressure perfusion of either rat liver or kidney.

The scheme of the perfusion apparatus is shown in Fig. 1. It consists of a thermostatic control unit, a heater and fan, an oxygenator, a heat exchanger, an upper hydrostatic reservoir with bubble trap, a liver platform, a bottom reservoir, a magnetic stirrer, a peristaltic pump, and a filter. Most of the equipment can be accommodated in a Plexiglass-type perfusion cabinet, and the pump and thermostatic water bath can be located outside the cabinet. The internal measurements of the cabinet depend on the experimental setting. If the kidney is to be perfused by using a constant-pressure system, the height of the cabinet should be taken into account; it should be approximately 180 cm.

1. Thermostatic Control Unit

The thermostatic control unit ensures a constant temperature of the perfusion medium and the liver platform. Two Y-shaped connectors form the outlet and inlet for the circulation of water, which is maintained at 37°C. The outlet of the water bath is connected by appropriate tubings to the heat exchanger and to the liver platform. The inlet drives the water back into the thermostat to be reheated.

2. Heater and Fan

This device helps heat the inside parts of the perfusion apparatus as well as maintain the inside atmosphere of the perfusion cabinet at a constant temperature of 37°C.

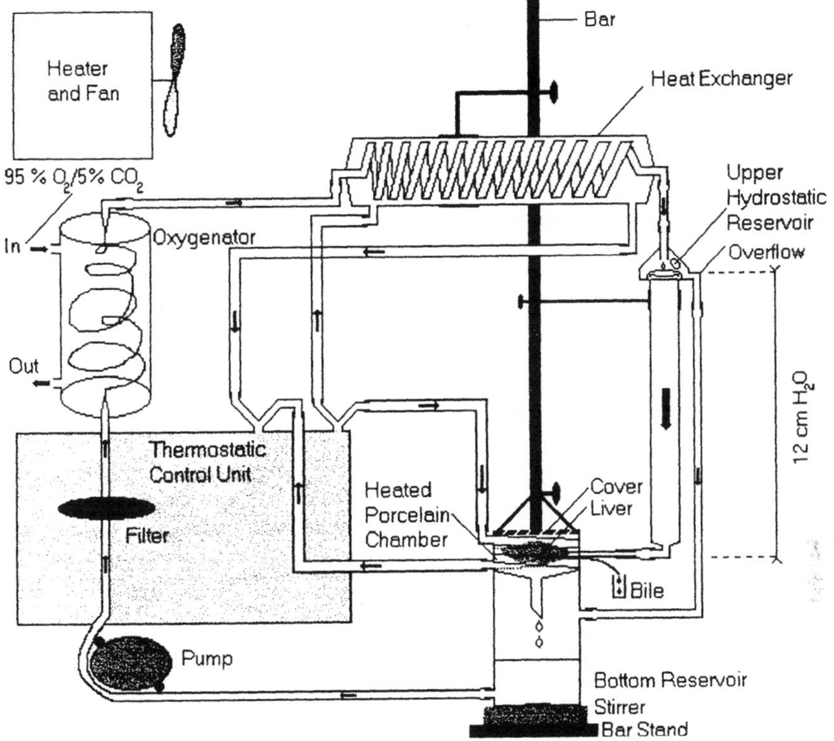

Fig. 1 Schematic representation of the perfusion apparatus used in the perfusion of the isolated rat liver.

3. Oxygenator

There are several possibilities to saturate the perfusion medium with oxygen. A membrane oxygenator containing 3–4 m of silastic tubing with 0.147-cm ID, 0.196-cm OD (Dow-Corning, Midland, Mich.) may be used to supply the perfusate with oxygen. The tubing is coiled in a closed container into which 95% O_2–5% CO_2 flows at a rate of 0.2–0.4 L/min.

A glass multibulb oxygenator can be easily constructed by connecting approximately five *to* six 100-mL–round-bottomed flasks (5). The perfusate flows from the top of the glass multibulb oxygenator to the bottom, there by forming a thin film that allows maximal gaseous exchange. The oxygen input is located at the bottom of the oxygenator, whereas output for the gases is at the top.

Importantly, the gases employed for saturation of the perfusate should be humidified before entering the oxygenator, otherwise the perfusate is dried and concentrated. A Dreschel bottle (containing warm saline) is one type of device, that should be inserted between the gas supply and the oxygenator.

4. Heat Exchanger

As seen in Fig. 1, the heat exchanger consists of a glass spiral and a glass jacket. The perfusate travels through the spiral in which it is heated to the temperature desired. Heat for the perfusate is provided by water pumped from the thermostatic water bath. This device can be made by a glassware technician.

5. Hydrostatic Reservoir

From the physiological point of view, a constant-pressure system of 10–12 cmH_2O is suitable for perfusing the rat liver. Constant pressure is maintained by the hydrostatic reservoir, which is connected to a PE 240 liver cannula by flexible tubing. With 10- to 12-cm pressure and a PE 240 cannula, the flow may range from 5 to 8 mL $min^{-1}g^{-1}$ liver. To control the flow it is advisable to fit the flexible tubing with an appropriate clamping screw so that the flow can be reduced to a desirable, 3–4 mL $min^{-1}g^{-1}$ liver. This is important, especially when performing single-pass experiments.

A side arm of the hydrostatic reservoir is connected to the bottom reservoir; thus, the overflow mechanism allows the perfusate to flow back into the bottom reservoir. A small hole is made on the top of the hydrostatic reservoir to enable direct sampling of the portal inflow perfusate.

The hydrostatic reservoir shown in Fig. 1 has some advantage over a constant head device descried by Miller et al. (2), because after repositioning of the height to 110–150 cm, it can be used for constant-pressure perfusion of the rat kidney. For this, the side arm as well as the bottom outlet of the hydrostatic reservoir can be connected with the bottom reservoir and with the renal artery cannula by appropriate flexible tubings. This piece of equipment can also be made by the glassware technician.

6. Liver Platform

The isolated liver can be housed in a porcelain chamber made from a Büchner suction filter. The porcelain chamber drains into the reservoir. Heat for the chamber is provided by the thermostatic control unit. A hollow spiral of any material may be employed to entwine the porcelain chamber. The liver is covered by a Plexiglass cover.

7. Other items of the Perfusion Apparatus

The glass bottom reservoir has two inputs and one output for the perfusate.

A commercially available magnetic stirrer is used to mix the perfusion medium in the reservoir.

The peristaltic pump should be able to pump the perfusate at a rate of 70–80 mL min^{-1}.

Filtration of the perfusate is necessary to remove broken cells, clots, and any debris. This may be accomplished by inserting a white silk or nylon filter between two Lucite disks.

B. Perfusion Media

A variety of perfusion media have been used to perfuse isolated liver preparations (21–26,32–36). They can be divided into three groups:

1. Basic Medium

As an artificial medium, hemoglobin and albumin-free Krebs-Henseleit bicarbonate buffer (pH 7.4), containing (in mmol/L) NaCl, 113; KCl, 4.7; $CaCl_2 \cdot 2H_2O$, 2.5; KH_2PO_4, 1.2; $MgSO_4 \cdot 7H_2O$, 1.2; $NaHCO_3$, 25; and glucose, 10; or Krebs-Ringer bicarbonate buffer, are commonly used for liver perfusion. The electrolytes may be prepared in stock solutions as follows: 1.3% $NaHCO_3$, 1.15% KCl, 2.11% KH_2PO_4, 3.83% $MgSO_4 \cdot 7H_2O$, and 2.28% $CaCl_2 \cdot 2H_2O$, and stored in a refrigerator for up to 2 weeks. Before perfusion, 100 parts of saline, 21 parts of 1.3% $NaHCO_3$, 4 parts of 1.15% KCl, and 1 part each of 2.11% KH_2PO_4, 3.83% $MgSO_4 \cdot 7H_2O$, and 2.28% $CaCl_2 \cdot 2H_2O$ are mixed. To avoid the formation of milky solution, $CaCl_2 \cdot 2H_2O$ should be added as the last component. This artificial perfusate may be employed for both single-pass and recirculating perfusion experiments. Cell and albumin-free media should be used when comparing the metabolic rate of the isolated microsomal fraction with that of the whole liver, as well as for comparison of isolated liver cells with the whole organ. The absence of oxygen carriers in an artificial medium is compensated by the high oxygenation of the perfusate (approximately 400–500 mmHg), and by high portal flow rates.

Albumin can also be added to this type of artificial perfusate. However, the addition of albumin to Krebs-Henseleit buffer can affect drug clearance (see Sec. II. F).

2. *Blood- or Erythrocyte-Containing Medium*

Whole rat blood can be an advantage for extrapolating drug elimination from the whole organ to an intact animal in terms of clearances. However, to supply the reservoir with only whole blood, at least 80–100 mL of blood will be required for just one perfusion experiment. Some investigators, therefore, have used freshly collected heparinized whole rat blood (35 mL) diluted with 65 mL of Krebs-Ringer buffer containing glucose (10 mmol L^{-1}) and bovine serum albumin (20 g L^{-1}) (23). The use of this type of blood perfusate is not without problems, for it may vary in its content of hormones and substrates. To avoid such variability, prewashed bovine or human erythrocytes can be added to the perfusion medium (26,36).

3. *Fluorocarbon Emulsions*

Emulsions of fluorocarbon (33–35) represent an artificial medium with a high ability to carry oxygen. Several biochemical and functional parameters were investigated to compare fluorocarbon-containing media with those containing erythrocytes. These studies revealed that fluorocarbon emulsions can replace erythrocytes (33,34). Nevertheless, the use of this type of medium may bias the results obtained; for example, lipophilic drugs may be bound to fluorocarbon, leading to an underestimation of the metabolic rate.

C. Surgical Procedure

Schematic representation of the abdominal cavity of the rat and of the procedure for isolating the liver are shown in Figs. 2 and 3a and b, respectively. Hepatectomy can be performed under barbiturate, ether, or halothane anesthesia. Pentobarbital is preferred as an anesthetic agent. It can be administered by an intraperitoneal injection (50–60-mg kg^{-1} body weight) before surgery.

1. *Preparation of the Working Surgical Area*

When the animal is properly anesthetized, the limbs are fixed in extension onto a surgical platform. The fur is removed from the anterior abdomen and the area is cleaned with 75% alcohol. A midline incision of the skin from the pubis to the upper part of the chest is made. A small incision is then made in the peritoneum on the midline. This incision is extended to the pubis and to the xiphoid. To achieve easy access to the abdominal cavity of the animal, it is necessary to cut down (approximately midway down the incision) the

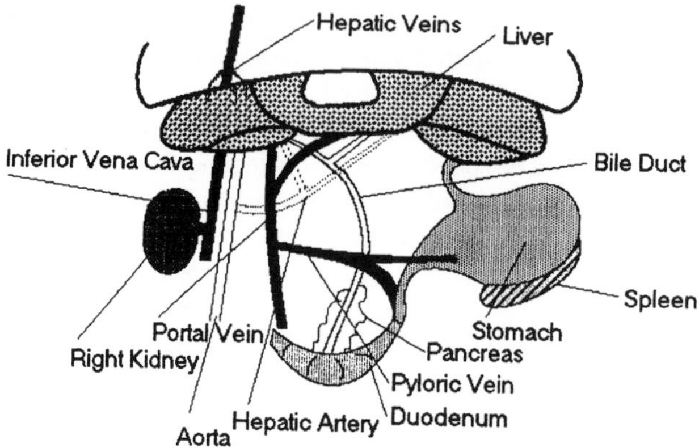

Fig. 2 Schematic representation of the abdominal cavity of the rat.

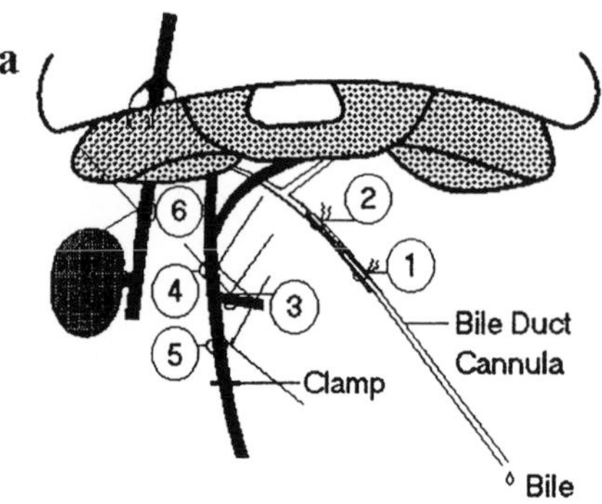

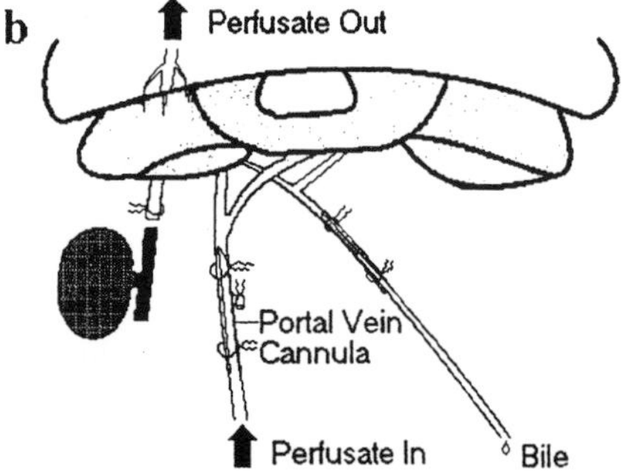

Fig. 3 The position of ligatures in the preparation for (a) cannulation of the liver; (b) in situ perfusion of the liver.

surface of the peritoneum and the skin toward the dorsal area. This procedure is made both on the left- and right-hand side of the animal body. The generated flaps can be fixed by surgical threads to the needles pierced into the platform. Finally, the gastrointestinal tract is retracted to the animal's left and covered with a wet saline gauze. A small piece of wet gauze is also placed below the liver lobes to allow access to the bile duct.

2. Cannulation of the Bile Duct

Because the bile duct is embedded into the pancreatic tissue (see Fig. 2), the first step of the procedure involves piercing the tissue (as close as possible to the duct), approximately halfway between the liver and duodenum. To expose the bile duct, periductular tissue is removed by teasing the tissue around the duct. Thereafter, two ligatures are placed around the bile duct (see Fig. 3a, ligatures 1 and 2), one proximate to the liver, the other one near to duodenum. Before cannulation, it is advisable to place a handle of forceps beneath the duct. The duct is clamped above the second ligature and a small incision is made with pointed scissors between the second ligature and the clamp. A cannula (PE 10–50) is inserted into the bile duct and the ligatures are tied.

3. Heparinization

The rat can be heparinized with 1000 U kg^{-1} of sodium heparin by injecting the solution through the right iliolumbare vein. Before administration of the drug, the vein is cleared from adhering tissue. Some investigators administer heparin into the inferior vena cava; however, for this, the inferior vena cava is ligated.

4. Cannulation of the Portal Vein

The portal vein, the pyloric vein, the inferior vena cava, and the hepatic artery are cleared from perivascular tissues. The position of ligatures is shown in Fig. 3a. One ligature is placed around the pyloric vein (ligature 3) and tied. Two ligatures are placed on the portal vein. The first above the pyloric vein (ligature 4) and the second approximately 1 cm distally (ligature 5). One ligature is placed around the inferior vena cava (ligature 6). Thereafter, one ligature is placed around the hepatic artery. The portal vein is clamped above the second, distal, ligature with an atraumatic Glover bulldog clamp (Roboz Surgical Instrument Co. Washington, D.C.). A small incision is made with pointed scissors between the clamp and the second ligature, and a cannula (PE 240) filled with heparinized Krebs-Henseleit buffer is introduced into the portal vein. To minimize problems with cannulation of the collapsed vein, it is also advisable to place some instrument beneath the vein, as suggested earlier for the bile duct.

5. In Situ Perfusion of the Liver

Once the cannula has been inserted into the portal vein, the liver can be perfused in situ (see Fig. 3b) at a constant pressure of 12 cm, using the upper hydrostatic reservoir (see Fig. 1) fixed to the laboratory stand. (For this purpose it is reasonable to have two or three pieces of hydrostatic reservoirs prepared). About 100 mL of preheated oxygenated Krebs-Henseleit buffer can be infused into the liver to clear it of blood. At iniation of the liver perfusion, the inferior vena cava is cut proximally to the kidney, just below the ligature. Immediately thereafter the inferior vena cava is cut between the heart and liver and the ligatures are completed. The cannula is clamped and, as soon as possible, the liver is dissected free from its peritoneal attachments, transferred into warm saline, where it can be cleared of blood clots and debris.

The inferior vena cava can also be cannulated. In this case, a cannula (PE 240) is inserted into the vein through the right atrium.

D. Perfusion Experiment

Following its isolation, the liver is placed on the platform in the perfusion chamber; 20–30 min are needed for the preparation to become stabilized after surgery. During this time, the portal flow can be adjusted to a desired value. This can be accomplished with a clamping screw positioned on the flexible tubing connecting the upper hydrostatic reservoir with the portal cannula. To clear the liver from remnants of blood constituents, some investigators employ the single-pass perfusion mode during the stabilization period. If the bile does not flow, it is advisable to turn the bile duct cannula very carefully to the right or to the left.

After the stabilization period, depending on the mode of perfusion, the drug is either added into the reservoir or various concentrations of the drug are infused through the portal vein into the liver. Perfusate samples are taken as often as necessary to determine pharmacokinetic parameters of the drug being investigated. The bile can be collected into preweighed Eppendorf tubes and sampled. Hepatic tissue can also be sampled for determination of drug and metabolite concentrations.

Relative to the liver perfusion method, there are no significant advantages or disadvantages between the single-pass mode and a recirculation experiment. The choice between these two experimental techniques will depend on the nature of the problem being studied.

E. Viability and Functional Parameters

Several parameters can be used to assess viability and functionality of the isolated perfused liver preparation. Although functional parameters have not been routinely used in drug metabolism studies, they would be very useful for evaluating the effects of drugs on liver function in toxicokinetic studies.

1. Appearance of the Liver

The liver's appearance is an easily recognizable viability criterion. When the perfusion is inadequate, a blotchy surface on the liver is observed, indicating anoxia. Uniform appearance of the liver, on the other hand, indicates stability of the preparation.

2. Perfusion Flow Rate

The flow rate through the liver is an easily measurable criterion of viability. At a hydrostatic pressure of 12 cm H_2O, a flow rate of approximately 6 mL min^{-1} g^{-1} liver can be expected using a Krebs-Henseleit buffer medium. When increasing the viscosity of the perfusion medium, the flow becomes somewhat lower. For example, perfusion employing 30% rat blood perfusate will typically result in a flow of 4 mL min^{-1} g^{-1} liver (23). A decrease in the perfusion flow rate with time, is an indication of increased resistance in the hepatovasculature. Thus, for example, Kupffer cells, resident macrophages of the liver, may be activated and release vasoactive mediators, such as thromboxanes and platelet-activating factor, which may cause vasoconstriction of the hepatovasculature (37).

3. Oxygen Consumption

Although oxygen consumption is often considered a criterion of viability, it need not reflect the actual state of the liver. For instance, Chazouillères et al. (21) found no changes in oxygen consumption in livers that underwent cold ischemic storage for 18 hs, in spite of deterioration of liver viability and functional parameters. In addition, a recent study has suggested that Kupffer cells can modulate oxygen consumption of the liver (26). Thus, caution should be used when using data on oxygen consumption as a viability index of the perfused liver. For a cellular perfusion medium, Chazouillères et al. (21) reported a hepatic oxygen consumption of 2.2 μmol min^{-1} g^{-1} liver, whereas for Krebs-Henseleit buffer this value reached more than 5.0 μmol min^{-1} g^{-1} liver (38).

4. Bile Flow

Bile flow is a reliable indicator of global liver function. A bile flow of approximately. 1–1.4 μL min^{-1} g^{-1} liver can be expected from a normal liver. This value, however, drops with perfusion time, because of the interruption of enterohepatic recycling of bile salts and the depletion of the endogenous bile acid pool. To maintain bile flow during the whole perfusion period, taurocholate can be infused into the reservoir perfusate at a rate of 15–18 μmol h^{-1}.

5. Redox and Energy Status of the Liver

Determination of lactate and pyruvate concentrations in the perfusate can provide an indication of the respiratory state of the isolated liver. The lactate/pyruvate ratio can also be used to calculate cytosolic NAD^+ redox state. In a normal, viable liver, this ratio is near 10, whereas anoxia generally leads to a severalfold increase.

The ATP concentration in liver tissue of a viable preparation is higher than 2.0 μmol g^{-1} liver and the ATP/ADP ratio is slightly higher than 3 (5). Because the ATP content decreases very rapidly under anoxic conditions, the liver needs to be immediately frozen by a freeze-clamping procedure using aluminium-based tongs cooled down in liquid nitrogen. The freeze-clamped tissue should be stored at a very low temperature (approximately $-70°$ C) before preparation of the sample for ATP, ADP, and AMP analysis.

6. Hepatocellular Membrane Integrity

Release of enzymes into the perfusate, such as lactate dehydrogenase and aminotransferases (aspartate transaminase and alanine transaminase), can be used to evaluate the integrity of the liver cell membrane. However, one should be aware that the use of erythrocytes in the perfusion medium may lead to increased levels of transaminases owing to hemolysis of erythrocytes. To overcome this problem, sorbitol dehydrogenase, which is not present in erythrocytes, can be used as marker of plasma membrane injury. Potassium ion levels in the perfusate can also be used as further index of membrane perturbation. In an appropriate preparation, K^+ release into the perfusion medium should be lower than 0.3 μmol h^{-1} (23).

Acid phosphatase, *N*-acetyl β-glucosaminidase, and β-glucuronidase represent lysosomal enzymes, which may also be released into the perfusion medium by lysosomal and cell membrane dysfunctions.

7. Functional Parameters of the Liver

A brief review of the methods used in investigating functional liver parameters is given next.

a. Proteosynthetic Function. One of the important functions of the liver is to synthesize plasma proteins. The rate of plasma protein synthesis by the isolated liver can be determined by the addition of radiolabeled amino acids, such as [^{3}H]leucine or [^{14}C]lysine into the perfusate at the beginning of the perfusion period, and subsequent analysis of the protein-derived radioactivity in perfusate samples (2,4).

b. Sugar Degradation. The disappearance of glucose from the perfusate can yield useful information on the metabolism of sugar in the preparation.

c. Hepatobiliary Excretory Functions. A decrease in bile flow can be prevented by constant infusion of taurocholate (TC) into the reservoir perfusate. If the bile production drops below these conditions, it is likely that some functional parameter of the liver, such as uptake, transcellular transport, or TC excretion through the canalicular membrane was injured during the perfusion period. Thus, measurements of TC concentration in the portal inflow, the hepatic outflow, and in bile can be used to assess the rat liver's clearing capability for its major bile salt. For this purpose, a tracer amount of [^{14}C]TC and cold TC are infused into the reservoir (21,26). From this measurement the intrinsic clearance of TC can be calculated (see Sect. II. G).

Another valuable marker of hepatic excretory function is phenolphthalein glucuronide, which is a highly sensitive measure of hepatobiliary dysfunction (39). Phenolphthalein glucuronide can be employed also in livers perfused for toxicological studies (39,40).

Biliary excretion of bromosulfophthalein and indocyanine green are additional indices of liver function that can be useful in studying toxic effects of drugs on the excretory function of the liver.

d. Functional Integrity of Tight Junctions. To assess the functional integrity of tight junctions between hepatocytes, specific markers such as sucrose (41), inulin (42), or horseradish peroxidase (43) can be added or infused into the reservoir or into the portal inflow, and concentrations of these markers can be determined in bile. Thus, for example, horseradish peroxidase activity peaks approximately 2 min after its infusion into the portal vein. This peak reflects the paracellular pathway of horseradish peroxidase excretion into bile (43), whereas transcellular transport is marked by a peak of enzyme activity in bile observed at about 20 min after infusion.

After removal of sucrose from the perfusate, the decrease of sucrose concentration in bile exhibits a rapid initial decrease ($t_{1/2}$ = 3.3 min), followed by a much slower decrease ($t_{1/2}$ = 29 min) phase. The rapid initial decrease reflects the paracellular pathway of sucrose movement into bile, whereas the slower decrease represents its transcellular transport (41).

e. Function and Integrity of Biliary Epithelial Cells. Glucose is believed to be reabsorbed from bile by biliary epithelial cells (44,45). If the function of these cells is impaired, biliary glucose levels would be higher than those in normal livers. The different distribution of γ-glutamyltransferase between hepatocytes and biliary epithelial cells (46) allows assessment of the biliary epithelial cell membrane integrity.

8. *Histological Evaluation of the Liver*

Liver fragments can be fixed in formalin and stained with hematoxylin–eosin (H & E) and evaluated by light microscopy. Infusion of trypan blue into the portal vein at the end of the experiment and fixing the tissue in formalin, followed by staining with eosin, is also a suitable way to determine zone-specific damage to the liver (22).

F. Application

The usefulness of the isolated perfused rat liver preparation for characterizing drug metabolism and determining elimination kinetics has been documented in several studies. For practical reasons, only a few examples will be given.

1. Models of Hepatic Elimination

The rationale of using the isolated perfused liver for assessing drug elimination in terms of models has been extensively discussed (47–55). Two well-established models of hepatic elimination (i.e., the well-stirred model and the parallel-tube model) are commonly used for predicting drug elimination kinetics by varying the rate of flow through the liver or the free-drug fraction in the perfusate. Several studies have shown a preference for either the well-stirred model (36,49,50) or the parallel-tube model (51–53). When input concentration was varied, data supported the parallel-tube model (19,54–56). The use of logarithmic average drug concentration (the parallel-tube model) for liver substrate concentrations in the determination of kinetic parameters is also appropriate for hepatic removal of substrate by competing pathways (57). Thus, determination of kinetic parameters from logarithmic average concentrations (the parallel–tube model) and from the rate of metabolism seems to be more appropriate than determinations based on the concentration of the drug leaving or entering the organ.

2. In Vitro and In Vivo Comparison of Drug Clearance

When a drug is completely absorbed after oral administration and its clearance is governed an IX by hepatic metabolism, then the kinetic parameters derived from the isolated liver preparations would predict oral drug clearance in vivo (58). However, other tissues can also be involved in drug metabolism. Therefore, correlations between in vitro and in vivo experiments will depend on the metabolic properties of a particular drug. To determine kinetic parameters for formation of individual metabolites, single-pass perfusion experiments are preferred. Various concentrations of a drug can be infused into the portal vein and the concentrations of its metabolites in the hepatic effluent can be measured (59).

In a study of the Alzheimer's disease drug tacrine, we demonstrated that kinetic parameters obtained from hepatocyte studies were well suited for predicting tarcrine's hepatic extraction ratio in the isolated rat liver as well as its oral clearance in the intact rat (60).

By using intrinsic clearance values calculated from metabolism of several drugs, Rane et al. (61) demonstrated that clearance of drugs in the isolated, perfused rat liver can be predicted from studies involving microsomal preparations.

3. Nonlinear Drug Elimination

Ethimizol studies provide a good example of the use of the isolated perfused liver for exploring nonlinear drug elimination (62–64). In rats, increased intravenous doses of ethimizol led to a disproportionate increase in the area under the drug concentration versus time curve (AUC) for parent drug and primary metabolites (62). To explain the mechanism for this nonlinearity, isolated rat liver preparations employing recirculating perfusate were used (63). Analysis of data from both the in vivo and the in vitro study showed that ethimizol did not obey Michealis–Menten kinetics and the nonlinearity observed appeared to be a result of competitive product inhibition. To test the hypothesis that metabolism of the parent drug was inhibited by its metabolites, various concentrations of primary metabolites of ethimizol were infused into the single-pass rat liver preparation, and the hepatic extraction ratio was determined. Indeed, the results, given in Table 1, clearly showed that elimination of the parent drug was inhibited by its metabolites (64).

4. Proteins and Drug Clearance

The effects of proteins on drug clearance can be assessed by changing the composition of the perfusion medium. Thus, Øie and Fiori (32) demonstrated that addition of albumin

Table 1 Effect of the Primary Metabolites of Ethimizol[a] M_I, and M_V on Its Hepatic Extraction Ratio at Steady-State (ER_{ss})

Study no.	ER_{ss}	M_I(μM) C_{In}	Percent of inhibition[b]	Study no.	ER_{ss}	M_V(μM) C_{In}	Percent of inhibition
1	0.496[c]	0	—	1	0.438	0	—
	0.451	30	13		0.383	10	12
	0.305	100	41		0.317	30	27
	0.487	10	6		0.298	50	32
	0.541	0	—		0.433	0	—
2	0.340	0	—	2	0.385	0	—
	0.289	20	16		0.218	100	42
	0.297	10	13		0.271	30	28
	0.304	5	11		0.357	10	6
	0.345	0	—		0.373	0	—
3	0.336	0	—	3	0.370	0	—
	0.146	100	56		0.207	30	42
	0.204	30	38		0.197	50	45
	0.267	10	19		0.274	10	23
	0.326	0	—		0.342	0	—
4	0.307	0	—	4	0.283	0	—
	0.295	30	4		0.223	50	26
	0.232	100	24		0.259	30	14
	0.305	0	—		0.278	10	8
					0.319	0	—

[a] A constant ethimizol input concentration of 5 μM, with and without various concentrations of its primary metabolites (M_I or M_V), was delivered at fixed perfusate flow (18 mL min^{-1} per liver) to each single-pass rat liver.

[b] The decrease of ethimizol ER_{ss} was compared with its mean value when input perfusate to the liver contained no metabolites

[c] Calculated from the four determinations of ethimizol at steady state.

Source: Ref. 64.

(1.4 gL^{-1}) to the Krebs-Henseleit buffer markedly increased the unbound clearance of prazosin, a drug bound to albumin. The clearance of antipyrine, a drug not bound to albumin was also increased by albumin. On the other hand, α1-acid glycoprotein decreases the unbound clearance of prazosin, without affecting antipyrine clearance (32). Studies by Weisiger et al. (65,66) showed that albumin can facilitate the uptake of fatty acids and bromosulfophthalein by liver cells by the albumin receptor. On the other hand, elimination of diazepam, a drug highly bound to albumin, was greatly dependent on its free fraction in the perfusate (51). These results strongly suggest that the relation between protein binding and elimination is a complex process.

5. *Enzyme Induction Studies*

It is often desirable to establish which enzymes play a role in metabolic detoxification or bioactivation of a given drug. This can be achieved by pretreating the animal with some well-established inducer [e.g., phenobarbital (PB) or 3-methylcholanthrene (3-MC)] and studying the elimination either in the single-pass or in an experiment with the recirculating perfusate (67,68). One example of the effects of both PB and 3-MC pretreatment on the concentration of ethimizol and its metabolites in the reservoir perfusate is shown in Fig.

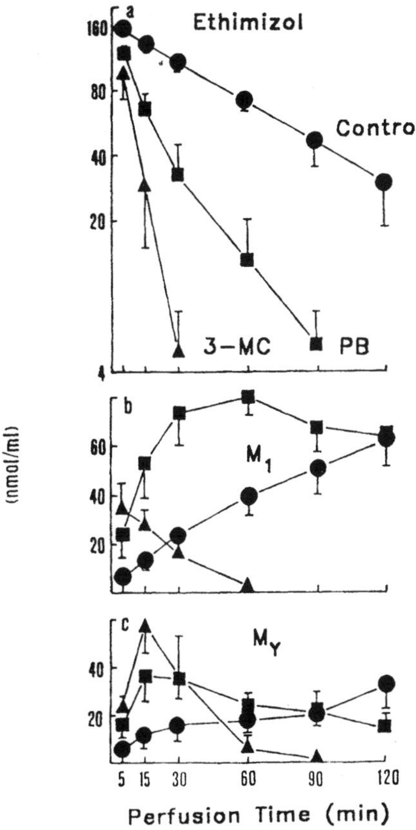

Fig. 4 Recirculating rat liver perfusion with [^{14}C]ethimizol. The effect of phenobarbital (PB) and 3-methylcholanthrene (3-MC) pretreatment on concentration in the reservoir perfusate of ethimizol and ethimizol metabolites M_1 (4-carbamoyl-1-ethyl-5-methylcarbamoyl-imidazole) and M_V [4,5-di(-methylcarbamoyl)-imidazole] is shown. Top, ethimizol; middle, metabolite M_1; bottom metabolite M_V. Livers from control (untreated; ●), PB-treated (■), and 3-MC-treated (▲) rats were used. Each point represents the mean of 3, 6, and 4 values ± SEM for the control, PB, and 3-MC groups, respectively. [^{14}C] ethimizol plus cold ethimizol was introduced into the reservoir (size, 110 mL; initial concentration, 165 μM) simulating intravenous bolus. (From Ref. 67.)

4 (67). As seen, both PB and 3-MC increased ethimizol elimination, but the effect of 3-MC was considerably greater. 3-MC was also more effective than PB in enhancing conversion of ethimizol primary metabolites, suggesting that 3-MC-inducible enzymes play a decisive role in the metabolism of both the parent drug and its primary metabolites.

6. *Biosynthesis of Drug Metabolites by the Liver*

The great advantage of having established the liver perfusion technique in a laboratory of drug metabolism is the possibility to biosynthesize drug metabolites. Large concentrations of a drug can be used to obtain milligram quantities of metabolites (59,69). For such purposes it is advisable to employ cell-free perfusate, which facilitates analysis of drug metabolites. Although the chemical structure of each metabolite should be verified by an independent mode of synthesis, biosynthesis of metabolites by the liver and their subsequent analysis may speed up other metabolic studies.

G. Data Processing

Some basic equations used for processing data obtained in drug metabolism and clearance studies follow: Additional information can be found elsewhere (47). Sophisticated methods, involving computer-assisted analysis of data using differential equations, have been described by Pang (70).

The rate of elimination across the liver under steady-state conditions is given by Eq. (1):

$$v = Q\,(C_{\mathrm{In}} - C_{\mathrm{Out}}) = \text{rate of metabolism} + \text{rate of biliary excretion} \tag{1}$$

where Q is the rate of flow through the liver and C_{In} and C_{Out} denote concentration of the drug entering and leaving the liver (47).

Hepatic extraction ratio at steady-state (ER) can be calculated from Eq. (2):

$$\mathrm{ER} = \frac{C_{\mathrm{In}} - C_{\mathrm{Out}}}{C_{\mathrm{In}}} \tag{2}$$

Clearance (CL) is the product of Q and ER:

$$\mathrm{CL} = Q\mathrm{ER} \tag{3}$$

Intrinsic clearance ($\mathrm{CL_I}$) can be defined as the volume of liver water cleared of substrate in unit time (61):

$$\mathrm{CL_I} = \frac{V_{\mathrm{max}}}{K_{\mathrm{M}}} \tag{4}$$

The following relation was derived for $\mathrm{CL_I}$, ER, and CL, assuming the parallel-tube model of elimination (47):

$$\mathrm{CL_I} = -\,Q\ln(1 - \mathrm{ER}) \tag{5}$$

$$\mathrm{CL} = Q(1 - e^{-\mathrm{CL_I}/Q}) \tag{6}$$

CL can also be calculated by dividing the dose introduced into the reservoir and the area under the drug concentration in the perfusate versus time curve (AUC):

$$\mathrm{CL} = \frac{\mathrm{dose}}{\mathrm{AUC}} \tag{7}$$

Equations (2) through (7) are valid under conditions of linear pharmacokinetics (i.e., when the concentration of the substrate is well below K_{M}; 47).

Assuming the parallel-tube model of elimination, the rate of metabolism of a substrate can be calculated as follows:

$$V = \frac{V_{\mathrm{max}}\mathrm{C}}{K_{\mathrm{M}} + C} \tag{8}$$

where C is the logarithmic average concentration of substrate in the liver:

$$C = (C_{\text{In}} - C_{\text{Out}}) \ln \left(\frac{C_{\text{In}}}{C_{\text{Out}}} \right) \tag{9}$$

III. CONCLUSION

Each of the in vitro techniques outlined in the chapters of this book can provide important information on the mechanisms of drug detoxification as well as bioactivation. The isolated perfused organ technique, however, is a model that offers advantages over other in vitro systems because it provides information on the contribution of a given organ to in vivo drug metabolism. Because of the complex nature of the perfusion model, in vitro techniques can be readily applied in addressing some particular issues. Compared with in vivo experiments, the major advantage of the isolated perfused organ is that the experiments can be performed under precisely defined conditions. This chapter has outlined the usefulness of the isolated perfused rat liver, as a model for drug metabolism studies.

REFERENCES

1. T. G. Brodie, The perfusion of surviving organs, *J. Physiol. 29*:266 (1903).
2. L. L. Miller, C. G. Bly, M. N. Berry, and H. A. Krebs, The dominant role of the liver in plasma protein synthesis: A direct study of the isolated perfused rat liver with the aid of lysine-^{14}C, *J. Exp. Med. 94*:431 (1951).
3. H. Sies, The use of perfusion liver and other organs for the study of microsomal electron-transport and cytochrome P-450 systems, *Methods Enzymol. 52*:48 (1978).
4. R. S. Jones, Liver perfusion techniques in toxicology, *Biochemical Toxicology* (K. Snell and B. Mullock eds.), IRL Press, Oxford, 1987, p. 23.
5. H. M. Mehendale, Application of isolated organ perfusion techniques in toxicology, *Principles and Methods of Toxicology* (A. W. Hayes, ed.), Raven Press, New York, 1994, p. 1157.
6. M. B. Enser, F. Kunz, J. Borensztajn, L. H. Opie, and D. S. Robinson, Metabolism of triglyceride fatty acids by the perfused rat heart, *Biochem. J. 104*:306 (1967).
7. R. W. Niemeier and E. Bingham, An isolated perfused lung preparation for metabolic studies, *Life Sci. 11*:807 (1972).
8. F. A. Baciewicz, M. Arredondo, and B. Chaudhuri, Pharmacokinetics and toxicity of isolated perfusion of lung with doxorubicin, *J. Surg. Res. 50*:124 (1991).
9. A. Nizet, The isolated perfused kidney: Possibilities, limitations and results, *Kidney Int. 7*:1 (1975).
10. S. L. Linas, D. Wittenburg, and J. E. Repine, Role of neutrophil-derived oxidants and elastase in lipopolysaccharide-mediated renal injury, *Kidney Int. 39*:618 (1991).
11. I. Bekersky, Use of the isolated perfused kidney as a tool in drug disposition studies, *Drug Metab. Rev. 14*:931 (1983).
12. R. K. Andjus, K. Suhara, and H. A. Sloviter, An isolated perfused brain preparation, its spontaneous and stimulated activity, *J. Appl. Physiol. 22*:1033 (1967).
13. A. M. Thompson, R. C. Robertson, and T. A. Bauer, A rat head-perfusion technique developed for the study of brain uptake of materials, *J. Appl. Physiol. 24*:407 (1968).
14. H. Kavin, N. W. Levin, and M. M. Stanley, Isolated perfused rat small bowel technique: Studies of viability, glucose absorption, *J. Appl. Physiol. 22*:604 (1967).

15. K. S. Pang, V. Yuen, S. Fayz, J. M. Tekopple, and G. J. Mulder, Absorption and metabolism of acetaminophen by the in situ perfused rat small intestine preparation, *Drug Metab. Dispos. 14*:102 (1986).
16. H. Hirayama, X. Xu, and K. S. Pang, Viability of the vascularly perfused rat intestine and intestine–liver preparations, *Am. J. Physiol. 257*:G249 (1989).
17. K. Iwatsuki, K. Ikeda, and S. Chiba, Effects of nitroprusside on pancreatic juice secretion in the blood perfused canine pancreas, *Eur. J. Pharmacol. 79*:53 (1982).
18. W. H. H. Andrews, R. Hecker, and B. G. Maegraith, The action of adrenaline, noradrenaline, acetylcholine and histamine on the perfused liver of the monkey, cat and rabbit, *J. Physiol. (Lond.) 132*:509 (1956).
19. S. Keiding, S. Johansen, K. Winkler, K. Tonnesen, and N. Tygstrup, Michaelis–Menten kinetics of galactose elimination by isolated perfused pig liver, *Am. J. Physiol. 230*:1302 (1976).
20. R. D. McCarthy, J. C. Shaw, and S. Lakshmanan, Metabolism of volatile fatty acids by the perfused goat liver, *Proc. Soc. Exp. Biol. Med. 99*:560 (1958).
21. O. Chazouillères, F. Ballet, Y. Chrétien, P. Marteau, C. Rey, D. Maillard, and R. Poupon, Protective effect of vasodilators on liver function after long hypothermic preservation: A study in the isolated perfused rat liver, *Hepatology 9*:824 (1989).
22. B. U. Bradford, M. Marotto, J. J. Lemasters, and R. G. Thurman, New simple models to evaluate zone-specific damage due to hypoxia in the perfused rat liver: Time course and effect of nutritional state, *J. Pharmacol. Exp. Ther. 236*:263 (1986).
23. F. Ballet, Y. Chrétien, C. Rey, and R. Poupon, Differential response of normal and cirrhotic liver to vasoactive agents. A study in the isolated perfused liver, *J. Pharmacol. Exp. Ther. 244*:283 (1988).
24. C. A. Brass and T. G. Roberts, Hepatic free radical production after cold storage: Kupffer cell-dependent and -independent mechanism in rats, *Gastroenterology 108*:1167 (1995).
25. J. J. Lemasters and R. G. Thurman, Hypoxia and reperfusion injury to liver, *Progress in Liver Diseases* (J. L. Boyer and R. K. Ockner, eds.), W. B. Saunders, Philadelphia, 1993, p. 85.
26. H. Imamura, F. Sutto, A. Brault, and P. -M. Huet, Role of Kupffer cells in cold ischemia/reperfusion injury of rat liver, *Gastroenterology 109*:189 (1995).
27. C. Tiribelli, G. C. Lunazzi, and G. L. Sottocasa, Biochemical and molecular aspects of the hepatic uptake of organic anions, *Biochim. Biophys. Acta 1031*:261 (1990).
28. S. Sabin and M. Rowland, Development of an optimal method for the dual perfusion of the isolated rat liver, *J. Pharmacol. Toxicol. Methods 39*:35 (1998).
29. G. Powis, Perfusion of rat liver with blood: Transmitter owerflows and gluconeogenesis, *Proc. R. Soc. Lond. B 174*:503 (1970).
30. K. Dorko, P. D. Freeswick, F. Bartoli, L. Cicalese, B. A. Bardsley, A. Tzakis, and A. K. Nussler, A new technique for isolating and culturing human hepatocytes from whole or split livers not used for transplantation. *Cell Transplant. 3*:387 (1994).
31. P. O. Seglen, Preparation of isolated rat liver cells, *Methods Cell Biol. 13*:29 (1976).
32. S. Øie and F. Fiori, Effects of albumin and alpha-1 acid glycoprotein on elimination of prazosin and antipyrine in the isolated perfused rat liver, *J. Pharmacol. Exp. Ther. 234*:636 (1985).
33. L. Triner, M. Verosky, D. Habif, and G. G. Nahas, Perfusion of isolated liver with fluorocarbon emulsion, *Fed. Proc. 29*:1778 (1970).
34. M. N. Goodman, R. Parilla, and C. J. Toews, Influence of fluorocarbon emulsions on hepatic metabolism in perfused rat liver, *Am. J. Physiol. 225*:1384 (1973).
35. N. Ballatori and A. T. Truong, Cholestasis, altered junctional permeability, and inverse changes in sinusoidal and biliary glutathione release by vasopressin and epinephrine, *Mol. Pharmacol. 38*:64 (1990).
36. K. S. Pang and M. Rowland, Hepatic clearance of drugs. II. Experimental evidence for acceptance of the ''well-stirred'' model over the ''parallel-tube'' model using lidocaine in perfused rat liver in situ preparation, *J. Pharmacokinet. Biopharm. 5*:655 (1977).

37. F. Ballet, Hepatic microcirculation: Potential for therapeutic intervention, *Pharmacol. Ther. 47*:281 (1990).
38. C. M. Pastor and T. R. Billiar, Nitric oxide causes hyporeactivity to phenylephrine in isolated perfused livers from endotoxin-treated rats, *Am. J. Physiol. 268*:G177 (1995).
39. H. M. Mehendale, Pesticide-induced modification of hepatobiliary function; hexachlorobenzene, DDT and toxaphene, *Food Cosmet. Toxicol. 16*:19 (1978).
40. H. M. Mehendale, S. A. Swensson, C. Baldi, and S. Orrenius, Accumulation of Ca^{2+} induced by cytotoxic levels of menadione in the isolated perfused rat liver, *Eur. J. Biochem. 149*:201 (1985).
41. H. Jaeschke, H. Krell, and E. Pfaff, Quantitative estimation of transcellular and paracellular pathways of biliary sucrose in isolated perfused rat liver, *Biochem. J. 241*:635 (1987).
42. J. A. Handler, D. C. Kossor, and R. S. Goldstein, Assessment of hepatobiliary function in vivo and ex vivo in the rat, *J. Pharmacol. Toxicol. Methods 31*:11 (1994).
43. J. Llopis, G. E. N. Kass, S. K. Duddy, G. F. Farell, A. Gahm, and S. Orrenius, Mobilization of the hormone-sensitive calcium pool increases hepatocyte tight junctional permeability in the perfused rat liver, *FEBS Lett. 280*:84 (1991).
44. M. Sellinger and J. L. Boyer, Physiology of bile secretion and cholestasis, *Progress in Liver Diseases* (H. Popper and F. Schaffner, eds.), W. B. Saunders, Philadelphia, 1990, p. 237.
45. M. Lira, C. D. Schteingart, J. H. Steinbach, K. Lambert, J. A. McRoberts, and A. F. Hofmann, Sugar absorption by the biliary ductular epithelium of the rat: Evidence for two transport systems, *Gastroenterology 102*:563 (1992).
46. M. Parola, K. H. Cheeseman, M. E. Biocca, M. U. Dianzani, and T. F. Slater, Biochemical studies on bile duct epithelial cells isolated from rat liver, *J. Hepatol. 10*:341 (1990).
47. K. S. Pang and M. Rowland, Hepatic clearance of drugs. I. Theoretical consideration of a ''well-stirred'' model and a ''parallel'' tube model. Influence of hepatic blood flow, plasma and blood cell binding, and the hepatocellular enzymatic activity on hepatic drug clearance, *J. Pharmacokinet. Biopharm. 5*:625 (1977).
48. B. A. Saville, M. R., Gray, and Y. K. Tam, Models of hepatic drug elimination, *Drug Metab. Rev. 24*:49 (1992).
49. K. S. Pang and M. Rowland, Hepatic clearance of drugs. III. Additional experimental evidence for acceptance of the ''well-stirred'' model over the ''parallel-tube'' model using metabolite (MEGX) data generated from lidocaine under varying hepatic blood flows in the rat liver perfused in situ preparation. *J. Pharmacokinet. Biopharm. 5*:681 (1977).
50. D. B. Jones, D. J. Morgan, G. W. Mihaly, L. K. Webster, and R. A. Smallwood, Discrimination between venous equilibrium and sinusoidal models of hepatic drug elimination in the isolated perfused rat liver by perturbation of propranolol protein binding. *J. Pharmacol. Exp. Ther. 229*:522 (1984).
51. M. Rowland, D. Leitch, G. Fleming, and B. Smith, Protein binding and hepatic clearance: Discrimination between models of hepatic clearance with diazepam, a drug of high intrinsic clearance, in the isolated perfused rat liver preparation, *J. Pharmacokinet. Biopharm. 12*:129 (1984).
52. S. Keiding and E. Steiness, Flow dependence of propranolol elimination in the perfused rat liver. *J. Pharmacol. Exp. Ther. 230*:474 (1984).
53. A. J. Byrne, D. J. Morgan, P. M. Harrison, and A. J. McLean, Variation in hepatic extraction ratio with unbound drug fraction: Discriminations between models of hepatic elimination, *J. Pharm. Sci. 74*:205 (1985).
54. L. Bass, S. Keiding, K. Winkler, and N. Tygstrup, Enzymatic elimination of substrates flowing through the intact liver, *J. Theor. Biol. 61*:393 (1976).
55. S. Keiding and E. Chiarantini, Effect of sinusoidal perfusion on galactose elimination kinetics in the perfused rat liver, *J. Pharmacol. Exp. Ther. 205*:465 (1978).
56. M. Kukan, T. F. Woolf, M. Meluš, and Š. Bezek, Characterization of nonlinear elimination of the xanthine-related drug ethimizol in 3-methylcholanthrene-induced rat liver by the ''parallel-tube'' model, *Drug Metab. Dispos. 21*:547 (1993).

57. K. S. Pang, X. Xu, M. E. Morris, and V. Yuen, Kinetic modeling of conjugation reactions in liver, *Fed. Proc. 46*:2439 (1987).
58. G. R. Wilkinson and D. G. Shand, Physiological approach to hepatic drug clearance, *Clin. Pharmacol. Ther. 18*:377 (1975).
59. M. E. Morris, V. Yuen, B. K. Tang, and K. S. Pang, Competing pathways in drug metabolism. I. Effect of input concentration on the conjugation of gentisamide in the once-through in situ perfused rat liver preparation, *J. Pharmacol. Exp. Ther. 245*:614 (1988).
60. M. Kukan, Š. Bezek, W. F. Pool, and T. F. Woolf, Metabolic disposition of tacrine in primary suspensions of rat hepatocytes and in single-pass perfused liver: In vitro/in vivo comparisons, *Xenobiotica 24*:1107 (1994).
61. A. Rane, G. R. Wilkinson, and D. G. Shand, Prediction of hepatic extraction ratio from in vitro measurement of intrinsic clearance, *J. Pharmacol. Exp. Ther. 200*:420 (1977).
62. T. Trnovec, M. Ďurišová, P. Burdátš, L. Šoltés, Z. Kállay, Š. Bezek, and L. B. Piotrovskiy, Dose-dependent pharmacokinetics of a xanthine-related nootropic drug, ethimizol in rats, *Pharmacology 34*:149 (1987).
63. Š. Bezek, M. Kukan, Z. Kállay, T. Trnovec, M. Štefek, and L. B. Piotrovskiy, Disposition of ethimizol, a xanthine-related nootropic drug, in perfused rat liver and isolated hepatocytes, *Drug Metab. Dispos. 18*:88 (1990).
64. M. Kukan, Š. Bezek, T. Trnovec, and L. B. Piotrovskiy, The effect of primary metabolites of the xanthine-related nootropic drug ethimizol on its hepatic extraction ratio, *Drug Metab. Dispos. 18*:383 (1990).
65. R. A. Weisiger, Dissociation from albumin: a potentially rate-limiting step in the clearance of substances by the liver, *Proc. Natl. Acad. Sci. USA 82*:1563 (1985).
66. R. A. Weisiger, J. Gollan, and R. Ockner, Receptor for albumin on the liver cell surface may mediate uptake of fatty acids and other albumin-bound substances, *Science 211*:1048 (1981).
67. M. Kukan, Š. Bezek, M. Štefek, T. Trnovec, M. Ďurišová, and L. B. Piotrovskiy, The effect of enzyme induction and inhibition on the disposition of the xanthine-related nootropic drug ethimizol in perfused liver and hepatocytes of rats, *Drug Metab. Dispos. 18*:96 (1990).
68. Š. Bezek, M. Kukan, W. F. Pool, and T. F. Woolf, The effect of cytochrome P4501A induction and inhibition on the disposition of the cognition activator tacrine in rat hepatic preparations, *Xenobiotica 26*:935 (1996).
69. M. Kukan, M. Hricovíny, and Š. Bezek, Biosynthesis and isolation of an ether glucuronide of azidothymidine from rat bile by thin-layer chromatography, *Chromatographia 35*:685 (1993).
70. K. S. Pang, Acinar factors in drug processing: Protein binding, futile cycling, and cosubstrate, *Drug Metab. Rev. 27*:325 (1995).

17
Metabolic Models of Cytotoxicity

Munir Pirmohamed and B. Kevin Park
The University of Liverpool, Liverpool, United Kingdom

I. INTRODUCTION

Cell death can be defined as the irreversible loss of essential cellular functions and structures. It has long been used as an endpoint to assess both drug efficacy and drug toxicity in vitro. For the former, assessment of cytotoxicity of cancer cell lines is used as a measure of the possible efficacy of chemotherapeutic drugs towards human cancers (1). For most anticancer agents, toxicity toward the cell is due to the parent compound itself. However, several agents, such as mitomycin, cyclophosphamide, and ifosfamide, undergo intracellular metabolic activation before causing cell death (1). In terms of drug toxicity, cytotoxicity assays have been used to assess the formation of chemically reactive intermediates from drugs that have been implicated in idiosyncratic drug reactions (2); (Fig. 1).

The aim of this chapter is to review the role of cytotoxicity assays in assessing drug efficacy and drug toxicity. However, we will limit ourselves only to examples for which metabolism is essential for either efficacy or toxicity. A large amount of literature exists

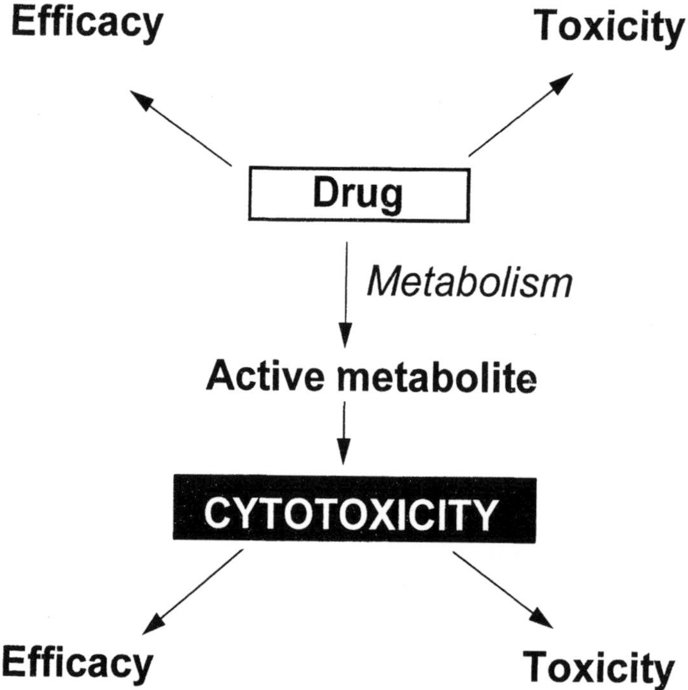

Fig. 1 The role of metabolism in mediating drug (metabolite) cytotoxicity, which may, in turn, be important for efficacy or toxicity.

on the *direct* effects of drugs on cell viability, particularly for anticancer agents; this will not be covered in the chapter.

II. METHODS FOR THE ASSESSMENT OF CELL VIABILITY

Many methods for the assessment of cell viability have been described (Table 1); (1,3–25). These can be divided into direct and indirect methods (3). The direct assays include clonogenic assays and assays that back extrapolate cell densities in exponentially proliferating cultures; they are sensitive and directly measure surviving fractions of cells, but suffer from the major disadvantages of being labor-intensive, technically difficult, and require several weeks to obtain results. In view of these limitations, various indirect, more rapid assays have been developed. These include measurement of

- Inhibition of incorporation of labeled RNA or DNA precursors, amino acids, or other metabolites
- Exclusion of dyes or enzymes
- Decrease of intracellular ATP content
- Formation of a colored or fluorescent product from a precursor
- Release of a labeled compound or an enzyme from damaged cells
- Total cell biomass

Table 1 Some of the Methods Used to Assess Cell Death

Assay	Type	Comments
Trypan blue dye exclusion	Exclusion of dye	Test of membrane integrity, simple, inexpensive, but labor-intensive and subjective. May under- or overestimate true level of cell death.
Lactate dehydrogenase (LDH) release assay	Release of intracellular enzyme	Cytosolic enzyme is released on cell damage. Nonradioactive, colorimetric assay, sensitive, can be automated. May not apply to cells with low LDH levels.
MTT assay (tetrazolium salt reduction)	Formation of colored product	Yellow tetrazolium compound is reduced by succinate dehydrogenase in active mitochondria to form an insoluble blue formazan product. Reduction is dependent on concentration of cells, MTT, and pH, which may lead to underestimating degree of cell death. Low cost but laborious. Activating systems (i.e., microsomes) may also reduce MTT. Performs best with adherent cells.
BCECF proliferative assay	Formation of fluorescent product from precursor	Nonfluorescent acetoxymethyl ester of BCECF (2′,7′-bis-(2-carboxyethyl)-5(6)-carboxyfluorescein) diffuses across membrane and is hydrolyzed inside the cell to the charged fluorescent compound BCECF, which becomes trapped within the cell. Method is accurate, not labor-intensive, does not require radioisotopes, and is amenable to automation. Results take 3–4 days. Hydroxylamines inhibit the esterase and, therefore, prevent the generation of BCECF from its precursor.
Tritiated thymidine (^{3}H-TdR) release assay	Release of labeled compound	^{3}H-TdR incorporates into DNA that is released on cell lysis; low background and sensitive; Disadvantage of using radioisotopes including the use of radioactive disposal facilities. Not suitable for resting cells.
^{51}Chromium (Cr) release assay	Release of radiolabeled compound	^{51}Cr is taken up into cells, becomes covalently bound to basic amino acids of cytoplasmic proteins, which are released from damaged cells. Suffers from disadvantage of spontaneous loss and, therefore, high background. May lack sensitivity in nondividing cells. Expensive, radioisotope requiring special handling and disposal facilities, short half-life, and emits gamma rays.
Europium (Eu^{3+}) release assay	Release of fluorescent compound	Cells labeled with europium chelated to diethylene triaminopentaacetic acid (EuDTPA), which on release from cell, can be quantified after mixing with an enhancing solution. Faster than ^{51}Cr release assay, nonisotopic method, cheap and nonhazardous. Requires a fluorimeter for analysis, and there is high rate of spontaneous loss with some cells. Compound not widely available.
ATP bioluminescence	Release of ATP from damaged cell	Released ATP from cell converts luciferin into a form that is oxidized by luciferase in a high quantum yield chemiluminescent reaction that can be measured by a luminometer. Avoids the use of radioisotopes; is sensitive, quick, simple, and reproducible.

This is not an exhaustive list of the many different assays that have been used to assess cell death. Many assays have not been mentioned, including those utilizing flow cytometry, and the many systems used for determining programmed cell death.

All of these different methods have their advantages and disadvantages (see Table 1), and the choice of method should be tailored to the aims of the study, the cell type, and compounds being studied. However, remember that all these assays are carried out in vitro in a milieu that only approximates that found in a physiological system (1). In particular, the assays lack the pharmacokinetic and pharmacodynamic variables that affect drug action in vivo. Additionally, the use of cell death as an endpoint does not take into account other subcytotoxic damage or biochemical changes that may have been occurring in the cell before its death, which may have been as important, if not more so. Nevertheless, when used in conjunction with parallel in vivo studies and other measures of cell damage, cytotoxicity assays can be illuminating. Their undoubted advantage is that one particular aspect of a drug's toxic or therapeutic effect can be studied in isolation; for example, genetic variation in response may be studied by using cells from different individuals that have been characterized at a molecular level for mutation(s) in the gene of interest (2).

Almost any cell can be used as a target in cytotoxicity assays. For testing anticancer agents, cell lines derived from the tumor type to be treated by the drug are often used (1). When toxicity of an agent to a particular organ is being assessed, ideally human cells derived from that organ should be used as targets because they will reflect the content of activating and detoxifying enzymes in that organ. However, this is often not possible, for example, in liver or kidney toxicity, for which cells from animals, such as rodents, are often substituted. The obvious disadvantage is that it may not be possible to directly extrapolate the data obtained from the animal to humans. An alternative is to use readily accessible human cells, such as peripheral blood cells (PBC) as targets (24). Here, it is important to ensure that the cell used in the assay is a true surrogate for the cells in the organ involved in toxicity, although often this is not possible.

More recently, with the advent of cDNA expression technology, it has become possible to transfect cell lines with metabolizing enzymes, in particular the cytochrome P450 (CYP450) isoforms, to delineate the precise enzymes involved in either the activation of a compound to its cytotoxic metabolite or the enzymes involved in its detoxification (26). Various heterologous expression systems, both prokaryotic and eukaryotic, have been used for the expression of drug-metabolizing enzymes (27). The prokaryotic systems, although having a high level of expression, suffer from the disadvantage of producing inactive protein because of lack of glycosylation and incorrect folding (28). The eukaryotic systems, on the other hand, have lower levels of expression and, in general, tend to be more difficult to culture than the prokaryotic systems (27). The lymphoblastoid cell lines have been the most widely used for cytotoxicity studies (29). In some studies, the cDNAs coding for both the P450 enzymes and detoxification enzymes such as microsomal epoxide hydrolase have been coexpressed to determine the pathways involved in both the activation and detoxification of specific compounds (30,31).

III. MODE OF CELL DEATH: APOPTOSIS OR NECROSIS?

There are two fundamental types of cell death: apoptosis and necrosis (32–41). They can be differentiated as indicated in Table 2. *Necrosis* occurs when the cell is exposed to severe and sudden injury, such as physical and chemical trauma. Because cell injury is extensive, there is often an inflammatory response that is required to clear the debris. In contrast, *apoptosis* (''to fall away'') is more subtle, and it is the most important form of programmed cell death. On exposure to the noxious substance, the cell undergoes a com-

Table 2 Morphological Distinction Between Apoptosis and Necrosis

Apoptosis	Necrosis
Death of isolated cells	Death of contiguous patches or areas of tissue
Chromatin condensation, nuclear shrinkage, and cell shrinkage	Nuclear and organelle swelling, and whole-cell swelling
Budding of plasma membrane	Blebbing of plasma membrane
Late loss of membrane integrity	Early loss of membrane integrity
No inflammatory infiltrate	Inflammatory infiltrate present
Phagocytosis of dying cells by neighboring cells	Phagocytosis of dying cells by professional phagocytes
DNA laddering on gel electrophoresis	Nonspecific DNA degradation as a late event

plex series of molecular changes (which are not fully understood) that ultimately result in its death in the absence of an inflammatory response. However, in the late stages of apoptosis, changes similar to necrosis may be seen within the cell; hence, simple dye exclusion assays cannot be used to differentiate the two forms of cell death (33).

Apoptosis is an energy-dependent process that can run a very fast course (34 min from the onset of budding to complete break-up has been reported; (33)). A biochemical feature of most forms of apoptosis is DNA fragmentation. Initially, DNA fragmentation occurs at 300- or 50-kilobase (Kb) pair intervals; this is followed by cleavage into 180 to 200-bp internucleosomal-sized fragments, which can be visualized on agarose gels as DNA laddering (32,33,39,42). This is caused by activation of calcium–magnesium-sensitive nucleases, which have not yet been fully characterized. The complex series of molecular events underlying apoptosis are not fully understood; several genes appear to be involved (Table 3); (34).

Necrosis in contrast to apoptosis is an energy-independent process, and is characterized by a progressive reduction in the cellular ATP content (33,42). Membrane damage seems to be the key event in the pathogenesis of necrosis; as the injury becomes irreversible, there is a progression from subtle changes in the membrane ionic pumps to nonspecific increases in membrane permeability and, ultimately, to physical membrane disruption.

To date, most of the cytotoxicity assays used in pharmacology and toxicology have used cell death as the endpoint, without differentiating between apoptosis and necrosis. For future studies, consideration should be given to differentiating between apoptosis and

Table 3 Gene Products Influencing Apoptosis

Enhancers of apoptosis	Inhibitors of apoptosis
Bcl-x_S	*Bcl*-2
Bax	*Bcl*-x_L
Bak	*Bcl*-w
Bad	*Mcl*-1
Nbk	p53
Bik 1	Colony-stimulating factors
TNF-α	
Fas/Apo1/CD95	
Interleukin-1β-converting enzyme (ICE)	
c-*myc*	

necrosis for two fundamental reasons: first, it may be possible to design drugs that target a specific gene (product) in apoptosis, induce cell death, and thereby, prevent the progression of the tumor. Second, identification of the pathway of cell death in the pathogenesis of the toxicity of a drug may allow the identification of susceptible individuals or the development of new analogues that do not affect the same pathway.

IV. THE ROLE OF METABOLISM IN DRUG CYTOTOXICITY

In general, drug metabolism can be considered a detoxification process, in that it converts therapeutically active compounds to inactive metabolites that can then be excreted harmlessly from the body. This process may require one or more than one drug-metabolizing enzyme that may be a phase I or phase II enzyme (43; Fig. 2). A drug may undergo sequential phase I and Phase II metabolism, or alternatively, it may undergo only phase I or phase II metabolism (44).

The pharmaceutical industry has also made use of the body's drug-metabolizing enzymes by developing prodrugs that are metabolized (by phase I, phase II, or both, enzymes), either within the liver or in the target tissue, to their active components that then mediate the drug's therapeutic action. Several anticancer agents fall into this category; for example, ifosfamide (45), cyclophosphamide (45), and mitomycin (46); for these drugs, activation to the active component is essential for the anticancer actions and can be seen in vitro as cytotoxicity.

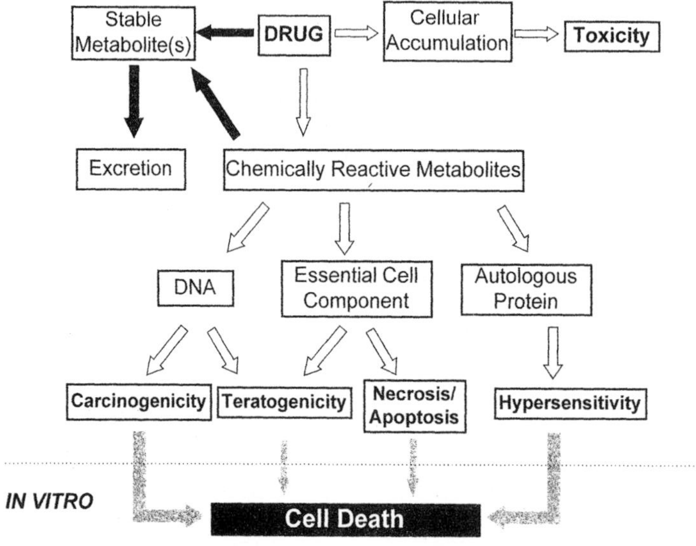

Fig. 2 The role of metabolism in drug toxicity: Metabolism can lead to the formation of chemically reactive intermediates that, if not adequately detoxified, can bind to various essential cellular macromolecules in vivo and, thereby, result in different forms of toxicity. In vitro the formation of chemically reactive metabolites will often be manifested as cellular death, irrespective of the nature of the macromolecule to which the toxic metabolite binds.

In certain circumstances, the drug-metabolizing enzymes can convert a therapeutically active drug to a toxic, chemically reactive metabolite, a process termed *bioactivation* (47; see Fig. 2). Their formation has been implicated in the pathogenesis of idiosyncratic drug reactions. The body is equipped with formidable defense mechanisms, and usually the chemically reactive metabolite will be detoxified (a process that can be termed *bioinactivation*) before it can initiate tissue damage. Indeed, it is possible that most therapeutically used drugs undergo some degree of bioactivation, but do not cause toxicity because the amount of toxic metabolite formed is below a ''toxic'' threshold, or it is promptly detoxified (48). However, given that bioactivation may represent less than 1% of the overall metabolism of a drug, techniques that are sensitive enough to detect this are currently unavailable. In the future, improvements in techniques, such as mass spectrometry and immunochemistry, may allow the detection of such minor metabolites. Both phase I and phase II enzymes can cause drug bioactivation, but usually, it is the former (i.e., the cytochrome P450 enzymes) that are responsible (47).

Inadequate detoxification of a chemically reactive metabolite formed as a result of drug bioactivation is often the first step in the initiation of idiosyncratic drug toxicity (2,47). This may occur if there is an imbalance between drug bioactivation and bioinactivation pathways. Tissue-specific expression of enzymes involved in drug bioactivation and drug detoxification may lead to a selective imbalance in that tissue, resulting in tissue-selective toxicity (49). An imbalance may be the consequence of a genetically determined deficiency of an enzyme or groups of enzymes, or alternatively, it may be acquired as a result of environmental factors, such as infection, diet, or concomitant drug intake. It is important that inadequate detoxification of a chemically reactive metabolite, although an important first step in the occurrence of toxicity, is not necessarily the ultimate step (2). Other factors, such as tissue repair enzymes, immune responsiveness, and the biochemical processes that modulate tissue injury, all may serve as factors determining not only whether idiosyncratic toxicity occurs, but also its severity.

In vivo, an inadequately detoxified chemically reactive metabolite can combine with or damage cellular macromolecules, such as proteins and nucleic acids, and result in various forms of toxicity, including teratogenicity, carcinogenicity, cellular necrosis, and hypersensitivity (2; see Fig. 2). Binding of a chemically reactive metabolite to nucleic acid may result in teratogenicity or carcinogenicity (see Fig. 2), whereas binding to cellular macromolecules may result in either direct or immune-mediated toxicity (47; see Fig. 2). With direct toxicity, binding of the chemically reactive metabolite to a protein will interfere with its normal physiological function, leading to cellular necrosis. Alternatively, the chemically reactive metabolite (CRM) can act as a hapten and initiate an immune reaction that may be due to a specific humoral (antibody) response, a cellular response (T lymphocytes), or a combination of both (50,51). The factors that determine what type of toxicity is mediated by a CRM are poorly understood, but are likely to include the following (50,52,53):

- The relative stability of the CRM, and thus, its reactivity
- The half-life of any drug–protein adducts formed and their concentration within the cell
- The epitope density (i.e., the number of groups of the CRM that are covalently bound to a protein molecule)
- The nature, physiological function, and subcellular site of the carrier protein to which the chemically reactive metabolite binds

An in vitro cytotoxicity assay clearly does not differentiate between the different forms of toxicity occurring in vivo. The demonstration of cytotoxicity in vitro merely serves to provide evidence of the formation of a potentially toxic metabolite (see Fig. 2), but does provide an early alert for appropriate in vivo studies.

V. THE USE OF CYTOTOXICITY AS AN INDICATOR OF DRUG EFFICACY

The chemotherapy regimen used for treatment of a particular cancer depends on its site and histology, rather than the sensitivity of the tumor to a particular agent (1). This has to be contrasted with antimicrobial therapy in which the choice of drug is dictated by the sensitivity of the organism responsible for the infection. In vitro chemosensitivity testing has as yet never gained widespread acceptance because the sensitivity of the tumor may not always be correctly predicted for various reasons, including (1) increased enzymatic activity, leading to drug detoxification; (2) decreased activity of drug-activating enzymes; and (3) tumor heterogeneity. Additionally, tumor growth rate in vitro may be significantly faster than in vivo and therefore, may provide a false-positive indication of tumor sensitivity to the chemotherapeutic agent. Nevertheless, in vitro cytotoxicity assays are useful to investigate the potential sensitivity of different tumors to anticancer agents, which can then be included (or not) in standard chemotherapy regimens. Cytotoxicity assays are also useful in (1) identifying patterns of cross-resistance and cross-sensitivity and evaluating their mechanisms, and (2) initial screening of new agents. Most anticancer agents show direct cytotoxicity, with a few being prodrugs that require activation to the active metabolites. Where a prodrug is being tested for its cytotoxic potential, the in vitro test should include an appropriate activating system; for example, microsomes or, alternatively, the active metabolite should be used. Metabolism and, in particular, induction of metabolizing enzymes, such as the glutathione transferases, plays an important role in determining chemoresistance to many anticancer agents; this is considered in the following.

All anticancer agents induce apoptosis in sensitive cells by targeting DNA, DNA repair enzymes, or metabolic processes, the functions of which are thought to be more critical for the viability of rapidly dividing tumor cells than for their normal untransformed counterparts (35,40). The ability of a malignant cell to resist drug-induced apoptosis, therefore, is one mechanism of chemotherapy resistance (54). Of particular importance here seems to be *BCL-2* and *BCL-X*$_L$ expression, which delays the onset of apoptosis induced by almost all classes of cytotoxic agents. The overexpression of these genes allows a malignant cell to survive, despite the presence of a considerable degree of chemotherapy-induced DNA damage and, importantly, allows time for repair mutations and chromosomal damage.

An additional factor determining the sensitivity of a tumor cell to drug-induced apoptosis is the glutathione (GSH) status of the cell (55). GSH and its associated enzymes act as detoxifying systems within the cell and protect it from various oxidants (56). GSH will conjugate to various electrophiles, either spontaneously or the reaction may be catalyzed by the various glutathione transferases (GST) that enhance both the rate and extent of the reaction (57). Many drug-resistant cancer cell lines have increased expression or activity of GSH or its enzymes, which results in more rapid detoxification and, thereby, diminishes the effectiveness of the treatment (55). Resistance to alkylating agents has been particularly associated with increased GST activity; many of these agents, including nitrogen mustard, chlorambucil, and melphalan, are substrates for the different forms of GST in vitro (55). Transfection experiments have also confirmed the role of GSH and GSTs in mediating resistance; for example, transfection of *GSTA*2 into NIH 3T3 mouse fibroblasts conferred 5.8- and 10.8-fold increased resistance to chlorambucil and mechlorethamine, respectively (58). The glutathione systems not only act as a resistance mechanism for drugs known to be substrates for the GST enzymes, but also for other drugs, such as doxorubicin and cisplatin, which are not GST substrates (55). With doxorubicin, resistance is thought to occur indirectly by detoxifying free radicals that are produced from the metabolism of doxorubicin. With cisplatin, the mechanism of resistance is thought to involve sequestration of the free drug or its metabolites by the enzyme. These mechanisms of resistance have been delineated using in vitro cell systems, and they provide useful indicators of possible patient response in vivo. Similarly, methods of overcoming resistance, and thus improving tumor response, have initially been tested in vitro using modulators of GSH levels and GST enzyme activity before their use in vivo in patients (55). Such agents include buthionine sulfoximine (which inhibits γ-glutamylcysteine synthetase), sulfasalazine (inhibits GST), and ethacrynic acid (inhibits GST); these drugs should still be considered to be experimental, and their widespread use to overcome resistance awaits results of clinical studies. From the perspective of the mechanism of cell death, in vitro depletion of GSH increases the cytotoxicity of many anticancer agents, and with alkylating agents (but not with actinomycin-D or etoposide), switches the mode of cell death from apoptosis to necrosis (59).

A. Ifosfamide

Ifosfamide is an isomer of cyclophosphamide (45). Both compounds are prodrugs that require P450-mediated activation to exert their cytotoxic effects. In vitro studies have shown that ifosfamide is converted by human CYP2B6, CYP3A4, and several CYP2C enzymes (and rat CYP2B1, CYP2C6, CYP2C11, CYP3A1) to a 4-hydroxylated metabolite (60–63); (Fig. 3), which then undergoes spontaneous β-elimination to yield the therapeutically active DNA-alkylating ifosphoramide mustard and the nonalkylating acrolein. Transfection of CYP2B1 into the breast cancer cell line MCF-7 (which has a low P450 level) sensitized the cells to cytotoxicity by ifosfamide and cyclophosphamide; this was prevented by the use of metyrapone, which inhibits CYP2B1 (64). Cytotoxicity was also observed in cocultured nontransfected cells despite there being no cell–cell contact, indicating that the activated metabolites were able to diffuse out of the P450-transfected cells. Other cell lines transfected with the CYP2B1 gene were also rendered susceptible to the drugs. The findings of these in vitro studies were confirmed by in vivo studies whereby intratumoral expression of CYP2B1 increased the therapeutic efficacy of the drugs by 15 to 20-fold, when treating MCF-7 tumors in nude mice, without any increase in toxicity (64). These novel experiments confirm the importance of P450-activation in determining the efficacy of ifosfamide and cyclophosphamide. Because the antitumor effect of the drugs was independent of liver cell activation, the authors suggest that P450 gene transfer–ifosfamide therapy could be combined with an inhibitor of the liver P450-catalyzed drug activation, thereby markedly improving the benefit to risk profile of these

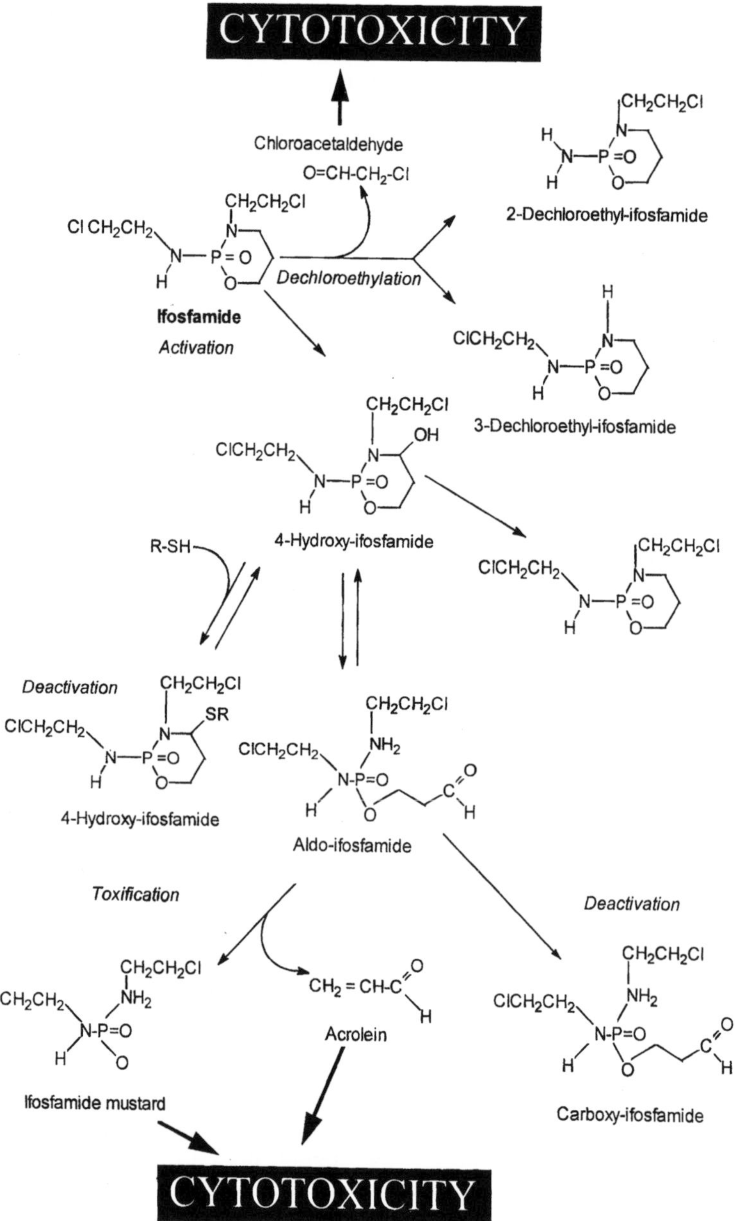

Fig. 3 The metabolism of ifosfamide to active metabolites that are important for both the efficacy and toxicity of the compound. Ifosfamide mustard, acrolein, and chloroacetaldehyde have been implicated as the cytotoxic metabolites. (From Ref. 45.)

drugs. Clearly, the difficulty with this approach will be the identification of a selective inhibitor (i.e., one that inhibits liver P450 without affecting the P450 enzyme in tumor cells).

As with efficacy, the toxicity of ifosfamide is also mediated by its active metabolites (45). The metabolism of ifosfamide is complex (see Fig. 3), and several metabolites, including acrolein, 4-hydroxyifosfamide, and chloroethanol, have been implicated in causing urotoxicity. However, in vivo studies have shown that the level of acrolein in human urine is only 0.001% of the concentration required to produce hemorrhagic cystitis in rats, whereas the concentration of the 4-hydroxy metabolite in urine is 50-fold higher than in plasma, suggesting that the latter, rather than the former, metabolite is responsible for the bladder toxicity (65). The hemorrhagic cystitis can be prevented by administration of thiols, such as high-dose *N*-acetylcysteine; however, this also attenuates the therapeutic efficacy of ifosfamide (66). To prevent cystitis without affecting efficacy, the current practice is to administer mesna, which provides selective detoxification of the urinary tract (47). Mesna circulates in the blood as the inactive disulfide metabolite (dimesna), which cannot be taken up by hepatocytes (67), but is taken up by renal tubular cells in which it is reduced to the active compound mesna by interaction with glutathione, thereby protecting against the urotoxic metabolites (68). Several of the metabolites formed from ifosfamide react with thiols, both inside and outside cells (45). Indeed, administration of ifosfamide depletes cysteine, glutathione, and homocysteine in plasma, and intracellular glutathione, which will, on the one hand, increase efficacy of the drug, but, on the other hand, also increase toxicity (69).

Unlike cyclophosphamide, ifosfamide also causes nephrotoxicity and neurotoxicity (45). The chloroacetaldehyde metabolite, which is not formed from cyclophosphamide, has been implicated as the toxic intermediate (70). Up to 30% of the ifosfamide dose is converted to chloroacetaldehyde (70), and this metabolite is toxic to renal tubular cells in vitro (71). However, a recent study in children, who are more susceptible to nephrotoxicity than adults, did not show an association with chloroacetaldehyde production and, in fact, nephrotoxicity seemed to be greater in those patients showing a decrease in dechlorethylation (72). Local activation of ifosfamide within the kidney itself by CYP3A4 and CYP3A5 may play a role in the pathogenesis of the nephrotoxicity, although the nature of the toxic intermediate remains unclear. This needs to be investigated in vitro using renal cell lines expressing the relevant P450 isoforms.

Chloroacetaldehyde, which is structurally similar to acetaldehyde, has also been implicated in the neurotoxicity (45). Recent studies in mice have suggested that intracerebral glutathione may be important in protecting against neurotoxicity (73). The occurrence of neurotoxicity in these mice was enhanced by the CYP3A inducers phenobarbital (phenobarbitone) and dexamethasone, and accompanied by depletion of glutathione. Administration of chloroethanol, a precursor of chloroacetaldehyde, caused both glutathione depletion and neurotoxicity. Prior depletion of glutathione enhanced neurotoxicity, whereas concomitant administration of *N*-acetylcysteine prevented toxicity. Inhibition of CYP2E1 by administration of either ethanol or 4-methylpyrazole also prevented neurotoxicity. Taken together, these results suggest that ifosfamide requires activation by CYP2E1 within the brain to chloroacetaldehyde, which causes glutathione depletion, and results in neurotoxicity, possibly by interfering with mitochondrial function (73). The relevance of this animal model to neurotoxicity in humans needs further investigation, probably initially by the use of in vitro cell systems.

B. Mitomycin

Mitomycin is a quinone antibiotic that has been found to possess some efficacy in the treatment of solid epithelial tumors, such as breast, lung, and bladder (46). It is a prodrug that is activated by reduction through one- or two-electron pathways to reactive semiquinone and hydroquinone intermediates that subsequently cross-link DNA. In vitro studies have shown that several enzymes can catalyze the activation of mitomycin, including cytochrome P450 reductase, xanthine oxidase, and cytochrome b5 reductase (acting as one-electron reductases), and DT diaphorase (also called naphthoquinone reductase 1 or NQO1) and xanthine dehydrogenase (acting as two-electron reductases) (46). Different enzymes seem to prevail under different physiological conditions, although the enzyme that is present in the largest amount is likely to carry out most bioreduction. Thus, if DT-diaphorase is present in high levels (as it is in many tumors; see later discussion), then it acts as the main-metabolizing enzyme (under both aerobic and anaerobic conditions). If DT-diaphorase is present at low levels, it still predominates in aerobic cells (for one-electron reductases are inhibited by oxygen), whereas under anaerobic conditions, the one-electron reductases will also participate to a significant extent.

The metabolism of mitomycin is complex, and several intermediates have been suggested using in vitro methods (Fig. 4), although the key intermediate is thought to be the quinone methide (46,74). As would be expected for a prodrug, the antitumor activity of mitomycin does not correlate with the levels of the parent compound. In vitro studies have identified a correlation between 2,7-diaminomitosene levels (a metabolite of the quinone methide) and cytotoxicity in human colon cancer cells (75); this has also been confirmed in in vivo studies (76).

There is no doubt that DT-diaphorase is the key enzyme involved in determining the efficacy of mitomycin (46). In in vitro cell models, a good correlation has been obtained between cellular enzyme activity and cytotoxicity. Thus, tumors that contain high levels of DT-diaphorase, such as liver, colon, breast, and non–small-cell lung cancer, are the malignancies most likely to respond to mitomycin (77–79).

DT-diaphorase is polymorphically expressed, being absent in 4% of the human population (80). Recent studies have shown that the mutation responsible for the polymorphism is at position 609 of the cDNA and confers a proline to serine substitution (81,82). In a human non–small-cell lung cancer cell line, this mutation resulted in an elevated mRNA level, but the enzyme activity was 2% of the wild-type, and protein could not be detected by Western blotting, suggesting that the mutation results in the formation of an unstable mRNA species that is degraded (82). Comparative studies showed that the frequency of the homozygous mutant was 7% and the heterozygote was 42% in both tumor and nontumor tissue from the same individuals, suggest that this is a true polymorphism. The polymorphism may act as a genetic predisposing factor for lung cancer, for one study has also shown an elevated frequency of the mutant allele (0.13) in affected individuals when compared with controls (83); however, this needs to be repeated in larger studies using patients of different ethnic backgrounds. The identification of the polymorphism in DT-diaphorase and its functional consequences affords an opportunity to prospectively individualize therapy with mitomycin in patients with solid tumors (even when tumor tissue is unavailable), and thus avoid the use of the drug in patients unable to activate the compound. Similarly, the availability of cell lines with different levels of expression of DT-diaphorase also means that analogues of mitomycin can be screened for their activity using cytotoxicity assays (81).

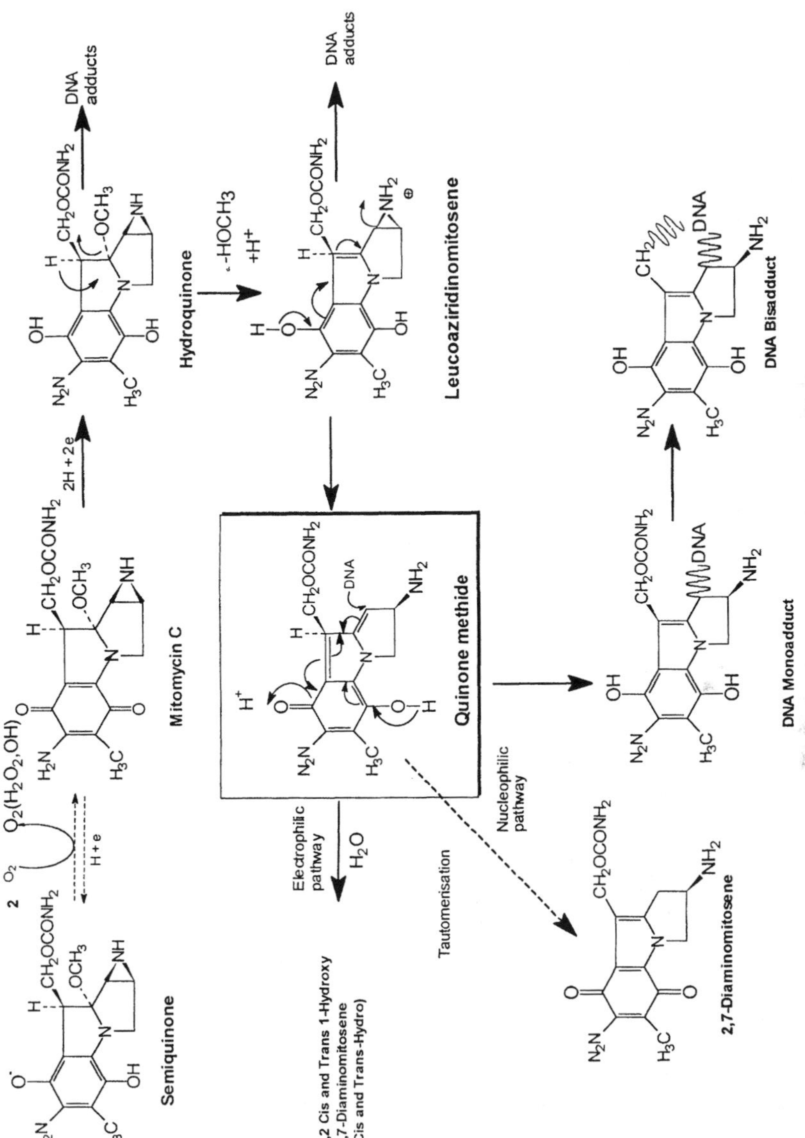

Fig. 4 The pathways for metabolic activation of mitomycin: The key intermediate in the whole scheme is the quinone methide (boxed); this is thought to be the most important pathway determining the efficacy of mitomycin. Other pathways shown on the scheme may also contribute to activation of mitomycin, and thus its efficacy. (From Ref. 46).

VI. THE USE OF CYTOTOXICITY AS AN INDICATOR OF DRUG TOXICITY

By definition, a *chemically reactive metabolite* is unstable and, therefore, it may not be possible to characterize it directly by routine analytical methods, such as high-performance liquid chromatography (HPLC) and mass spectrometry (MS). To assess the formation of these metabolites and their possible role in idiosyncratic drug toxicity, various methods have been used as indirect markers for their formation (84); these include the following:

- Covalent binding to proteins
- Immunological detection of drug protein conjugates
- Trapping of the reactive metabolites as thiol adducts and subsequent characterization by nuclear magnetic resonance (NMR) and mass spectroscopy
- Cytotoxicity assays

A two-stage in vitro cytotoxicity assay (Fig. 5) was devised by Spielberg (24) in an attempt to recreate the in vivo generation of chemically reactive drug metabolites, by the incorporation of a drug–metabolite-generating system (microsomes and NADPH) and mononuclear leukocytes (MNL) as a readily available target cell. The degree of bioactivation of the drug to a chemically reactive metabolite is then assessed by the determination of the viability of the leukocytes. The use of a functional assay is complementary to the chemical and immunochemical methods described in the foregoing. The importance of the assay lies in the fact that several variables can be altered independently or in combination to answer different questions. First, cells from patients with and without adverse

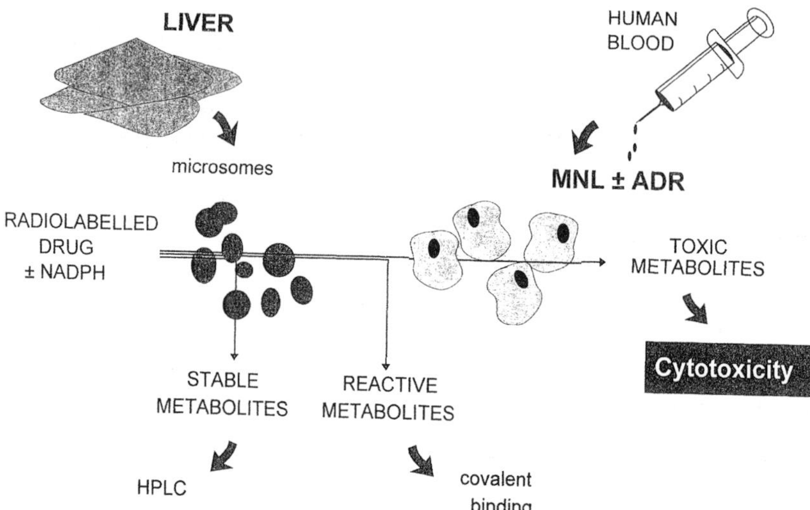

Fig. 5 A schematic representation of the in vitro cytotoxicity assay: The drug is incubated with a metabolizing system comprising liver microsomes and NADPH, and incubated with peripheral blood mononuclear leukocytes (MNL) taken from patients with and without idiosyncratic adverse reactions. Cytotoxicity can be determined by various methods, the most common being trypan blue dye exclusion. Metabolism to stable and protein-reactive metabolites within the system can be assessed by using HPLC and determining covalent binding to the microsomal protein, respectively.

drug reactions can be used to determine interindividual variation in cellular detoxification processes (85). Thus, MNL from patients hypersensitive to phenytoin (86–89), phenobarbital (89), carbamazepine (89–91), sorbinil (92), sulfonamides (91,93), amineptine (94), and cefaclor (95), all exhibit higher in vitro sensitivity when coincubated with the respective drug, than appropriate controls, suggestive of a defect in detoxification. Secondly, by keeping the cells constant, the ability of a drug to undergo bioactivation to a reactive cytotoxic metabolite can be assessed (25). This can be used to determine the functional group within the molecule that is responsible for the toxicity. Furthermore, the biochemical basis of bioactivation can be assessed further by using microsomes induced with an appropriate enzyme inducer, or by coincubating with an inhibitor of the enzyme responsible for bioactivation (discussed in greater detail later).

In this assay, MNL are used as target cells (see Fig. 5). This is clearly not ideal because for most forms of toxicity, the target cell is not the lymphocyte. The use of lymphocytes as surrogate markers, however, can be justified as follows (24,85):

- They can be obtained easily by venepuncture and are robust, requiring simple culture conditions.
- They usually cannot, by themselves, bioactivate drugs because they contain low amounts of P450 enzymes.
- They contain the relevant detoxification mechanisms, such as glutathione and epoxide hydrolase.
- Any heritable deficiency in detoxification mechanisms will be expressed phenotypically in lymphocytes.

In most of the studies, either trypan blue dye exclusion or the MTT assay has been used to estimate cell viability. The MTT assay can be used only if a microsomal system is not present in the incubations because microsomes themselves may metabolize MTT. Thus, the MTT assay is suitable only for use when the toxic metabolite itself can be synthesized and then incubated with the cells, as in sulfamethoxazole hydroxylamine (93). Trypan blue dye exclusion correlates with other indices of cell damage, including lactate dehydrogenase (LDH) release from the cells and loss of ability to respond to concanavalin A blastogenesis (24).

The major problem with the in vitro cytotoxicity assay is that it lacks sensitivity in that any changes observed are small (90), and thus, a negative result may not exclude the formation of a toxic metabolite from the coincubated drug. In addition, a chemically reactive metabolite that is not cytotoxic, but may bind irreversibly to protein, may be formed. Therefore, when a radiolabeled compound is available, it may be more useful to combine the determination of both the cytotoxic and protein-reactive metabolites (as well as the stable metabolites) within the same system (see Fig. 5).

The advantages and disadvantages of using the in vitro cytotoxicity assay in elucidating the pathogenesis of idiosyncratic toxicity is illustrated with reference to several compounds that have been associated with different forms of toxicity.

A. Carbamazepine

Carbamazepine belongs to the aromatic group of anticonvulsants (which also includes phenytoin and phenobarbital). Carbamazepine is the most widely used anticonvulsant in the United Kingdom. In most patients it is well tolerated, causing no more than mild dose-dependent reactions (96). On occasions, however, it can lead to a hypersensitivity reaction characterized by rash, fever, arthralgia, eosinophilia, and sometimes involvement of other organ systems, such as liver and bone marrow (89,90,97,98). The clinical picture seen with carbamazepine is similar to that described for both phenytoin and phenobarbital, suggesting a common pathogenesis and, thus, has been termed the anticonvulsant hypersensitivity syndrome (89). A major advance in understanding the hypersensitivity reaction came about with the development of the in vitro cytotoxicity (24,86; described in the foregoing), which suggested that predisposition to toxicity was due to a defect in detoxification.

In our studies (90,97–100) investigating the mechanism of anticonvulsant hypersensitivity, we have used carbamazepine (CBZ) as the model compound. In initial studies, ten patients with a clinical diagnosis of carbamazepine hypersensitivity were identified and investigated using an in vitro cytotoxicity assay that comprised lymphocytes and hepatic microsomes prepared from mice pretreated with phenobarbital (90). Lymphocytes from hypersensitive patients were more susceptible to metabolites of carbamazepine generated in situ than in controls, which comprised normal healthy volunteers and patients receiving CBZ without adverse effects (Fig. 6). No in vitro chemical cross-sensitivity was observed when the patients lymphocytes were exposed to phenytoin, which suggests that the system was highly specific, for five of the patients were receiving long-term phenytoin therapy without adverse effects (90). Furthermore, there was no difference in the in vitro cellular

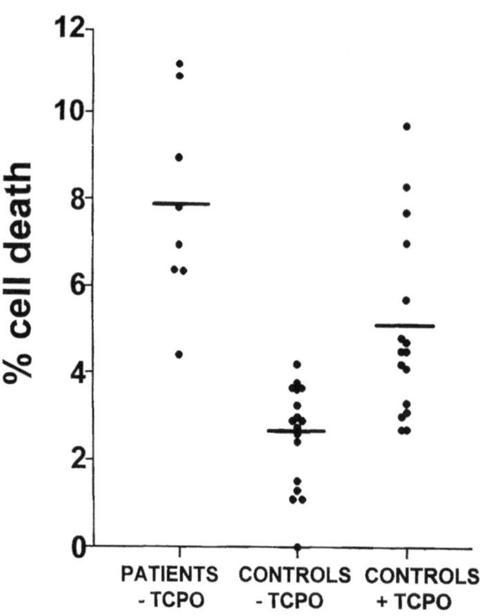

Fig. 6 The metabolism-dependent sensitivity of mononuclear leukocytes (MNL) to carbamazepine in the presence and absence of trichloropropene oxide (TCPO), an inhibitor of microsomal epoxide hydrolase, as assessed using the in vitro cytotoxicity assay depicted in Fig. 5. MNL were taken from patients with carbamazepine hypersensitivity and from controls (comprising healthy volunteers and patients taking carbamazepine without adverse effects), and cytotoxicity was assessed by trypan blue dye exclusion. The horizontal bars represent the mean cytotoxicity values for each group. There was no difference in mean cell death when MNL from hypersensitive patients were exposed to TCPO (data not shown), whereas with control cells, there was a significant increase in cell death ($p < 0.05$) on exposure to TCPO.

sensitivity when cells were exposed to chemically unrelated compounds, such as dapsone hydroxylamine and amodiaquine quinoneimine. The specific nature of the sensitivity of cells taken from the anticonvulsant-hypersensitive patients has been confirmed by a more recent study that showed that the cells exhibit increased cell death when exposed to reactive metabolites derived from the implicated anticonvulsant, but do not show a nonspecific increase in sensitivity when exposed to menadione and formaldehyde (91). In further studies, using microsomes prepared from ten histologically normal livers, we were able to demonstrate that all human livers are capable of bioactivating carbamazepine to a cytotoxic metabolite, further reinforcing that the predominant abnormality in these patients was one of enhanced sensitivity to the metabolite, rather than interindividual variation in bioactivation (99).

The nature of the cytotoxic metabolite was investigated further by the addition of chemical modifiers and enzyme inhibitors. The metabolism-dependent cytotoxicity of carbamazepine in control cells, but not in patient cells, was enhanced by coincubation with trichloropropene oxide (TCPO; (90); see Fig. 6), an inhibitor of microsomal epoxide hydrolase (101). Furthermore, purified microsomal epoxide hydrolase, but not cytosolic epoxide hydrolase, protected against cytotoxicity (99). Covalent binding of carbamazepine was also reduced by microsomal epoxide hydrolase, suggesting that the cytotoxic and protein-reactive metabolites were the same metabolite. Taken collectively, these results suggested that the reactive metabolite was an arene oxide, and that the affected patients may have a deficiency of microsomal epoxide hydrolase (Fig. 7). More recently, we have detected detoxification products from the postulated arene oxide in rat bile (102), and their presence has also been suggested in human urine (103).

In line with the initial studies undertaken by Spielberg (24,86,89), we (90,97,98)

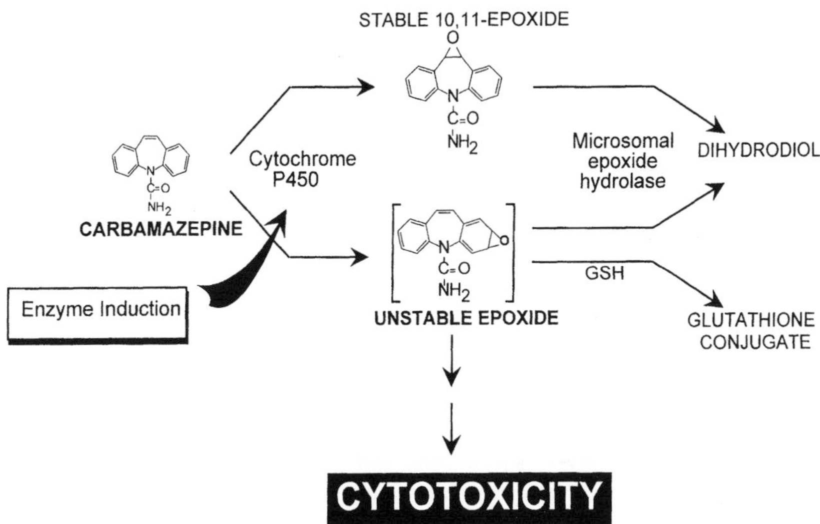

Fig. 7 A schematic representation of the role of metabolism in mediating carbamazepine idiosyncratic toxicity: In vitro studies have shown that bioactivation of carbamazepine is catalyzed by P450 enzymes. Whereas detoxification is performed by microsomal epoxide hydrolase and glutathione (GSH). The unstable epoxide (or its further metabolites) are thought to be responsible for causing cytotoxicity.

and others (91,94) used induced animal microsomes in the in vitro cytotoxicity assay to provide a consistent level of bioactivation and thereby, overcome the problem of interindividual variability in expression of the drug-metabolizing enzymes, which is well recognized in humans. To define more precisely the enzymatic activities associated with this activating system, naphthalene was used as an in vitro probe for bioactivation to epoxides and their bioinactivation to dihydrodiols (104). In accordance with the fact that mouse liver contains less microsomal epoxide hydrolase than human liver (105), and phenobarbital induction has a greater effect on drug oxidation than epoxide hydrolysis (106), the net balance between bioactivation and bioinactivation in phenobarbital-induced mouse microsomes provides a quantitative increase in reactive metabolite formation.

To further investigate the nature of the detoxification deficiency suggested by the in vitro cytotoxicity assay in patients with carbamazepine hypersensitivity, we developed a sensitive assay (using high-specific activity *cis*-stilbene oxide as a substrate) to measure the kinetics of microsomal epoxide hydrolase in lymphocytes (107). No difference in K_m and V_{max} for lymphocyte microsomal epoxide hydrolase was seen between the hypersensitive patients and controls, although the hypersensitive patients had greater variability in activity of the enzyme. This is in accordance with studies using radioimmunoassay that again showed no deficiency of microsomal epoxide hydrolase in patients with carbamazepine hypersensitivity (108). Taken together, these studies showed that carbamazepine-hypersensitive patients do not have either a qualitative or quantitative deficiency of microsomal epoxide hydrolase. Genetic analysis of the coding regions of the microsomal epoxide hydrolase gene has confirmed the findings of the functional studies, in that no specific mutations or pattern of mutations were identified in carbamazepine-hypersensitive patients (109,110).

In the investigation of carbamazepine and anticonvulsant hypersensitivity, the in vitro cytotoxicity assay has been valuable in providing possible evidence of a deficiency of detoxification in hypersensitive patients. However, initial evidence of a deficiency of microsomal epoxide hydrolase has not been confirmed by later studies, which used both functional and genetic analysis of the enzyme. The studies, however, still do not completely exclude a deficiency of microsomal epoxide hydrolase, for there is evidence of tissue-specific expression and inducibility of the enzyme (111). Thus, the use of the lymphocyte as a surrogate marker of deficiency of microsomal epoxide hydrolase may not have accurately reflected a deficiency in other organs. Alternatively, the arene oxide may be only the proximate toxic metabolite, because it may undergo further metabolism to the ultimate toxins, such as catechols and quinones; in fact, recent studies in mice have suggested that carbamazepine is bioactivated to chemically reactive quinones and catechols (112). With naphthalene, which forms an unstable 1,2-epoxide, the ultimate cytotoxic metabolite(s) are the quinones, rather than epoxides (113). However, genetic analysis of polymorphisms in quinone reductase and catechol O-methyltransferase has failed to show any linkage to carbamazepine hypersensitivity (114). Glutathione may serve to detoxify all these different chemically reactive species and, indeed, in vitro, it affords protection to cells (99). However, no association has yet been demonstrated between polymorphisms in glutathione transferase mu and theta isoforms and carbamazepine hypersensitivity (115,116). The final possibility is that the increase in in vitro cell death in carbamazepine hypersensitive patients is secondary to antigen recall, and the secretion of cytotoxic cytokines, such as tumor necrosis factor-α (TNF-α). In support of this, current (mainly clinical) evidence suggests that carbamazepine idiosyncratic toxicity is immune-mediated

(97,98,117,118). Again, by extension of the existing experimental in vitro model, this hypothesis could be explored.

B. Tacrine

Tacrine is an anticholinesterase agent used in the treatment of Alzheimer's disease (119,120). It causes dose-dependent and reversible elevations in serum transaminase levels (termed tacrine transaminitis) in 40–50% of patients undergoing therapy, although frank hepatotoxicity, such as hepatitis and jaundice, is rare (119–122). The mechanism of tacrine transaminitis is unknown. It has been suggested that the parent compound may be responsible by inducing ribosomal dysfunction (123). However, tacrine undergoes rapid and extensive metabolism (124); Fig. 8), and thus, the possibility that the transaminitis may be related to the formation of chemically reactive metabolites has also been raised.

To investigate the latter possibility, by using MNL as target cells and human liver microsomes as the metabolite-generating system, NADPH-dependent formation of cytotoxic metabolites from tacrine has been demonstrated (125). Similarly, in the presence of both NADPH and liver microsomes, tacrine was metabolized to a species that remained irreversibly bound to the microsomal protein, indicative of the generation of chemically reactive species (125).

By the addition of chemical modifiers to the in vitro system, it was possible to gain further information on the nature of the metabolites being generated. Incorporation of reduced glutathione and ascorbic acid inhibited both the metabolism-dependent cytotoxicity and covalent binding of tacrine. Furthermore, sulfhydryl, but not amine nucleophiles inhibited covalent binding. Coupled with the lack of inhibition seen in the presence of the epoxide hydrolase inhibitor cyclohexene oxide we were able to conclude that the reactive metabolites generated from tacrine were electrophilic and that they were unlikely to be epoxides (125).

The enzymes involved in the generation of the cytotoxic and protein-reactive metabolites were assessed using a range of isozyme-specific P450 inhibitors. Bioactivation and overall metabolism of tacrine was inhibited only by the CYP1A2 inhibitors enoxacin and furafylline (125,126). The role of CYP1A2 in chemically reactive metabolite formation was confirmed by a correlation of the extent of bioactivation with the levels of CYP1A2 apoprotein in microsomes from 16 individual livers (127).

As a result of their inherent chemical instability, chemically reactive metabolites are often difficult to identify. However, a stable thioether conjugate of the reactive species was generated when tacrine (or its 7-hydroxy metabolite) was coincubated with microsomes and mercaptoethanol. In this system, the chemically reactive metabolite undergoes reduction to generate a 1:1 thioether adduct of the 7-hydroxy metabolite and mercaptoethanol (128). Thus, the bioactivation of tacrine appears to be a two-step process that involves CYP1A2-dependent hydroxylation at the 7-position, followed by a further CYP1A2-dependent oxidation to yield a reactive quinone methide (128–130); see Fig. 3). This quinone methide can be reduced by interaction with cellular thiols, such as glutathione, or it may interact with essential cellular macromolecules, such as protein or DNA, to initiate cell damage.

Reactive quinone methides are generated from a variety of anticancer drugs that are known to cause redox cycling and generation of oxygen radicals. In fact the cardiotoxicity associated with anthracyclines is thought to result from the inability of myocytes to defend

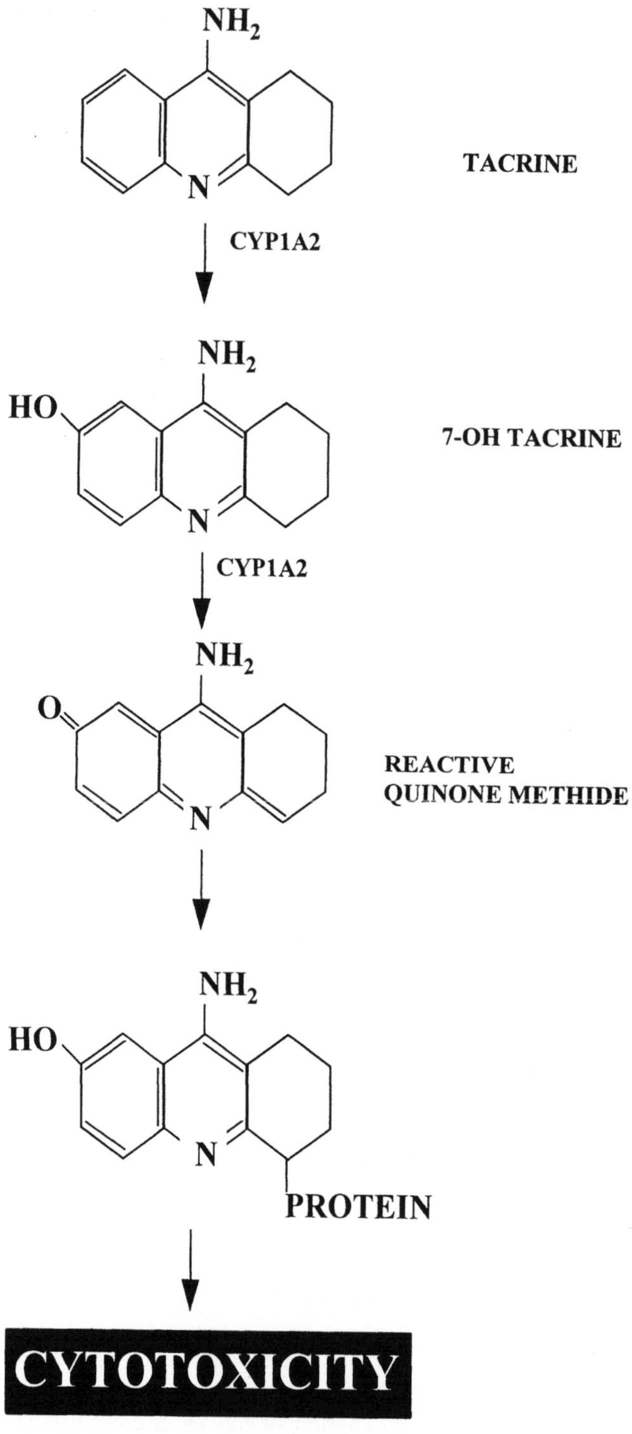

NH_2

N

TACRINE

CYP1A2

NH_2

HO

N

7-OH TACRINE

CYP1A2

NH_2

O

N

REACTIVE
QUINONE METHIDE

NH_2

HO

N

PROTEIN

CYTOTOXICITY

Fig. 8 The metabolic activation of tacrine to its postulated cytotoxic metabolite, the quinone methide.

themselves against the redox cycling (131). One measure of redox cycling is a depletion in the levels of cellular glutathione. Dogterom et al. (132) have demonstrated cytotoxicity and glutathione depletion in rat hepatocytes incubated with tacrine. Thus, quinone–methide-induced redox cycling may also play a role in tacrine-associated toxicity.

C. Dapsone

Dapsone is used to treat a wide variety of diseases including leprosy, inflammatory disorders involving polymorphonuclear leukocyte infiltration, rheumatoid arthritis, dermatitis herpetiformis, and *Pneumocystis carinii* pneumonia (133). The use of dapsone is associated with the dose-dependent toxicity methemoglobinemia. Additionally, dapsone also causes idiosyncratic toxicities, such as agranulocytosis, skin rash, and pneumonitis (134).

The toxic effects of dapsone are thought to be mediated by its hydroxylamine metabolite, formed either by the cytochrome P450- or myeloperoxidase-catalyzed oxidation of the drug (135,136). Initial studies using the in vitro cytotoxicity assay in the presence of human liver microsomes showed that dapsone caused NADPH-dependent MNL cytotoxicity (137). The addition of subphysiological concentrations of ascorbic acid, glutathione, and *N*-acetylcysteine all prevented cellular death. There was a correlation between metabolism-dependent cytotoxicity and covalent binding. Dapsone was metabolized in this in vitro system to a hydroxylamine; this metabolite was synthesized and showed concentration-dependent cytotoxicity in the absence of a metabolizing system, indicating its likely role in dapsone toxicity.

To investigate this further, a two-compartment system (termed the dianorm; Fig. 9) was developed in which human target cells, either MNL or erythrocytes, are separated from a drug-metabolizing system by a semipermeable membrane (138). By using this system, it was possible to demonstrate the ability of human liver microsomes to generate dapsone hydroxylamine in one compartment, with toxicity toward human target cells in a second compartment. These series of experiments also indicated that the hydroxylamine

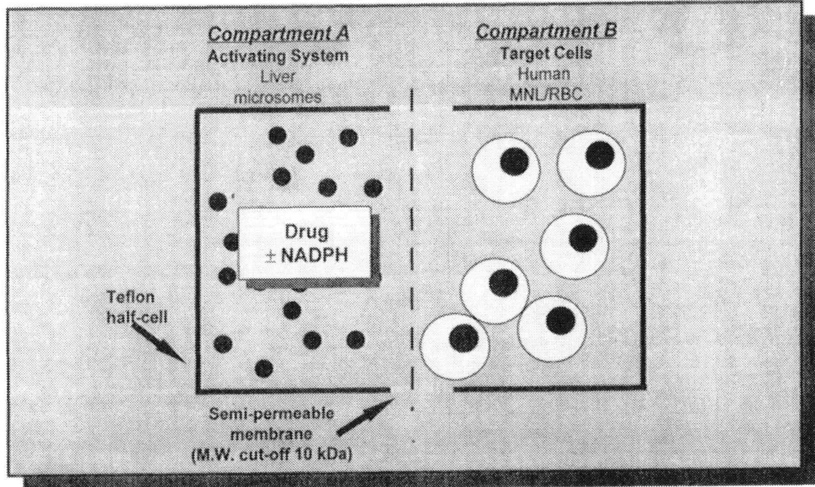

Fig. 9 Schematic representation of the two-compartment system (dianorm) used to assess the metabolism-dependent toxicity (cytotoxicity or methemoglobinemia) of dapsone.

metabolite was sufficiently stable to be formed in the liver and travel through the bloodstream to initiate toxicity in other target organs, such as skin and bone marrow.

In vitro studies showed that the formation of dapsone hydroxylamine and, thus, its toxicity, could be reduced by two P450 inhibitors: ketoconazole and cimetidine (139,140). Subsequent in vitro metabolism studies using enzyme inhibitors, inhibitory antibodies, enzyme expression systems, and correlations with levels of individual isozymes have identified CYP2C, CYP2E1, and CYP3A4 (141–143) as the major forms involved in the bioactivation of dapsone. The involvement of P450 enzymes in the bioactivation of dapsone in vivo has been confirmed by the use of an enzyme inhibitor. Cimetidine, which is a nonspecific inhibitor of P450 (144), blocks the *N*-hydroxylation and hemotoxicity of dapsone in both healthy volunteers and patients (145–147).

Dapsone causes rare and idiosyncratic white cell toxicities (incidence less than 1: 2000), while resulting in dose-dependent red cell toxicity to some extent in all individuals who take the drug (133,134). To investigate this observation, a three-compartment model (''trianorm''; Fig. 10) was developed (148). In this system, the drug-metabolizing enzymes are contained within the central compartment and separated from both white and red blood cells by semipermeable membranes. This allowed a simultaneous exploration of the cell-selective toxicity of dapsone metabolites and what effect the presence of one cell type had on the toxicity observed on the other cells. Dapsone was metabolized to a species capable of diffusing out of microsomes, crossing a semipermeable membrane, and causing

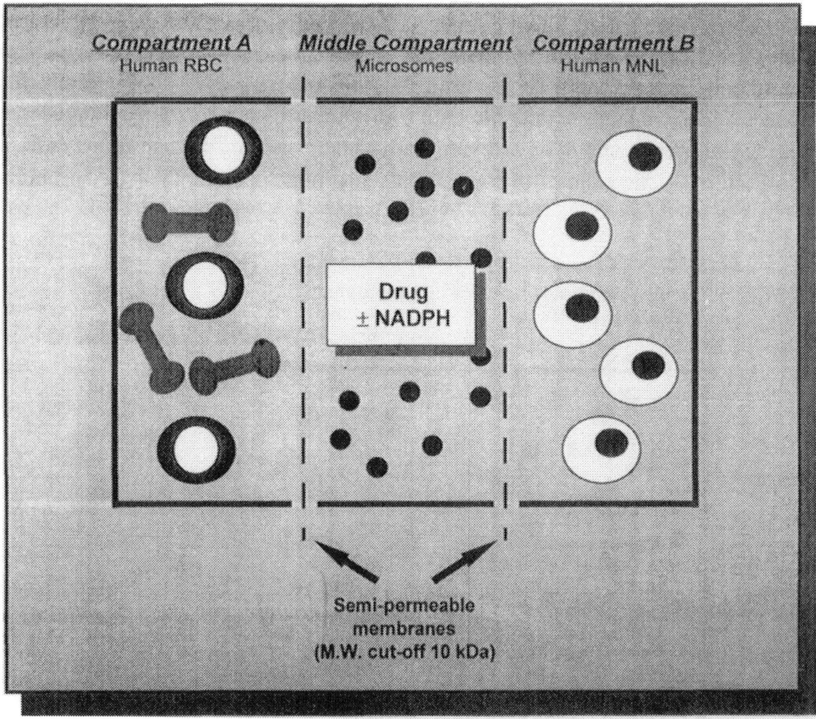

Fig. 10 Schematic representation of the three-compartment system (trianorm) used to simultaneously assess the metabolism-dependent cytotoxicity of dapsone, and the effect of erythrocytes on this toxicity (explained in the text).

toxicity, in the form of methemoglobinemia, to red cells. However, when red cells were present in the three-compartment model, no significant toxicity toward white cells was observed. In contrast, when the red cells were replaced by buffer, significant white cell toxicity was observed, which suggests that red cells protect white cells from the effects of the toxic metabolite. This may be explained by the uptake and accumulation of dapsone hydroxylamine into the red cell, where it is cooxidized with hemoglobin to form nitrosodapsone and methemoglobin (Fig. 11). The nitroso intermediate may be reduced back to the hydroxylamine by intracellular glutathione, whereas the methemoglobin is reduced back to hemoglobin by both NADH- and NADPH-dependent methemoglobin reductase

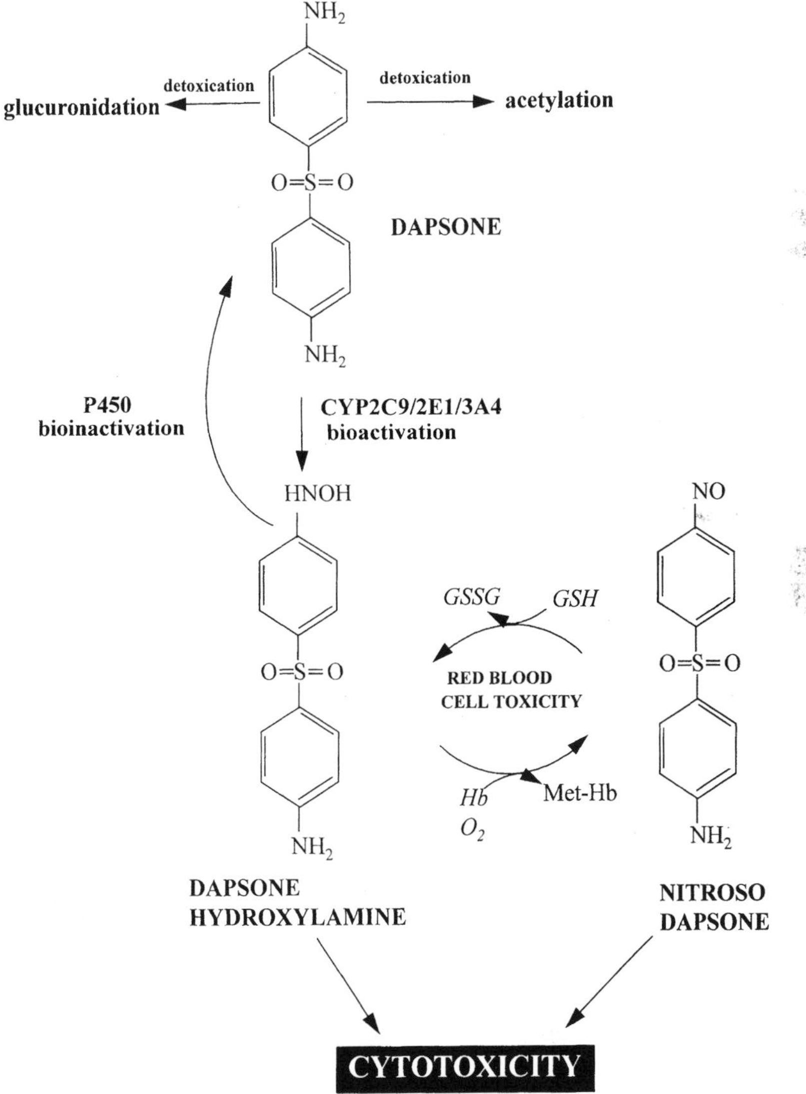

Fig. 11 Metabolic activation of dapsone to its cytotoxic metabolites, dapsone hydroxylamine and nitrosodapsone.

enzymes so that a futile cycle exists (149). This cycle will continue until the dapsone metabolites become irreversibly bound to hemoglobin in a manner similar to 4,4′-methylene dianiline, a close structural analogue (150). Moreover, the ability of low concentrations of red cells (12.5% hemtocrit) to protect white cells suggests that there is a large reserve capacity within the red cells for the uptake and detoxification of dapsone hydroxylamine.

D. Clozapine

Clozapine is an atypical neuroleptic that not only is more efficacious than conventional antipsychotic drugs, but also affects the negative symptoms of schizophrenia, an effect not seen with the conventional drugs (151–153). Its greater efficacy, however, has not been translated into increased clinical use, largely because of a high frequency of agranulocytosis (estimated to be 0.8%) associated with its use (154,155).

The mechanism of the agranulocytosis is unclear, but both peripheral blood neutrophils and bone marrow precursor cells are depleted, suggesting that the bone marrow is the target organ (156). The clinical features of the disease, in particular experience with rechallenge, are not consistent with an immune-mediated mechanism (157,158).

Clozapine is a highly lipophilic compound that undergoes extensive metabolism (Fig. 12), with only 2–5% of the drug being excreted unchanged (159). Its major stable metabolites are desmethyl clozapine and clozapine *N*-oxide, although their serum levels show a great deal of interindividual variability (160). Demethylation of clozapine is dependent on the P450 isoform CYP1A2 (161,162), whereas *N*-oxidation may be due to either the P450 enzymes or flavin monooxygenases (161).

In vitro with human liver microsomes clozapine is converted to a reactive metabolite that binds covalently to microsomal protein (161). This biotransformation was catalyzed

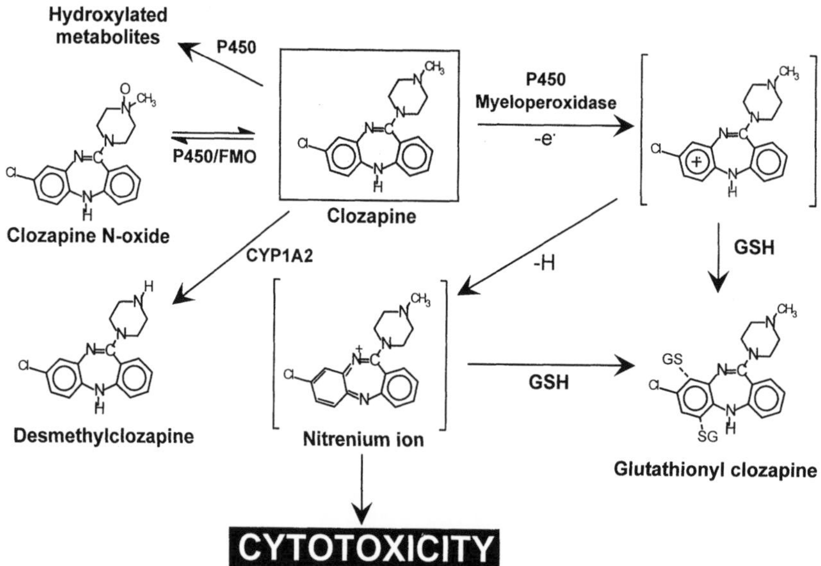

Fig. 12 Metabolism of clozapine by P450 and myeloperoxidase to the postulated nitrenium ion, which is thought to be responsible for cell death.

by several P450 isoforms, but not CYP1A2. The major adduct formed in the presence of glutathione was C-6 glutathionyl clozapine (see Fig. 12). In vivo studies in mice and rats have also demonstrated bioactivation of clozapine to the same adduct (163).

Clozapine is also bioactivated to a chemically reactive metabolite by incubating with myeloperoxidase (164; the major enzyme present in neutrophils), isolated neutrophils (163,165), and their bone marrow precursors (163). With the cellular system, bioactivation is observed only when the cells are activated by phorbol myristate acetate (PMA). Importantly, the adduct formed by neutrophils in the presence of glutathione was the same as that observed with microsomes (163). The toxic metabolite has been tentatively identified as a nitrenium ion (165).

The functional effects of generating this chemically reactive intermediate have been shown using an in vitro assay in which clozapine was coincubated with lymphocytes and human, mouse or rat liver microsomes (166,167); clozapine by itself was nontoxic, but was transformed to a cytotoxic metabolite by the addition of NADPH, a cofactor for P450-mediated metabolism. However, the target cell was the lymphocyte, rather than the neutrophil. This assay cannot be used to assess toxicity to neutrophils because binding to the cell surface of the neutrophil results in its activation (unpublished data). Additionally, although the chemically reactive metabolite can be synthesized, it has a short half-life, estimated to be less than 1 min (165), making it difficult to use in direct cytotoxicity assays.

To assess functional toxicity to neutrophils, a novel assay in which horseradish peroxidase (HRP) and hydrogen peroxide (H_2O_2) were used to generate the reactive metabolite from clozapine in vitro, was developed. Both HRP and myeloperoxidase can metabolize clozapine to chemically reactive species (164). Furthermore, parallel HPLC–MS analysis showed that the metabolites produced by HRP were identical with those produced by both neutrophils and neutrophil precursors (163).

With this assay, we were able to show that the nitrenium ion was cytotoxic to both neutrophils and MNL and, indeed, cytotoxicity was seen at therapeutic concentrations (3 μM) of clozapine (168). Cytotoxicity was accompanied by glutathione depletion and was ameliorated by the exogenous glutathione, both of which suggest that glutathione serves as a detoxification mechanism. Chemical analysis showed that bioactivation of clozapine was accompanied by the formation of the same glutathione conjugates as seen with liver microsomal incubation and when the drug was bioactivated by neutrophils and neutrophil precursors. Demethylclozapine, but not clozapine *N*-oxide, also exhibited cytotoxicity in this system, although the cell death seen with demethylclozapine was less than that observed with equivalent concentrations of clozapine. Preliminary studies in our laboratory have also shown that clozapine may be enhancing apoptosis of neutrophils (unpublished data). The relevance of these in vitro findings to the in vivo situation is unclear and needs further investigation. Of particular importance with clozapine agranulocytosis is identification of the factor precipitating or predisposing to neutrophil depletion, which will allow the development of preventive strategies and the use of the drug in a wider group of patients.

VII. CONCLUSIONS

The aim of this chapter was to demonstrate the use of in vitro model systems to understand and define the role of drug metabolism in drug response in humans in vivo. Such model systems are of particular use in situations for which there is no well-defined animal model available to determine drug efficacy or drug toxicity. We have taken two examples to illustrate general points: anticancer drugs and drugs that cause idiosyncratic toxicities, for bioactivation is often *assumed* to play a pivotal role in such biological responses. A limitation of in vitro models is that they do not represent the dynamic pharmacokinetic situation that occurs in vivo. Nevertheless, careful chemical analysis of drug and both stable and unstable metabolites can facilitate the interpretation of the relevance of in vitro observations to the clinical situation. Furthermore, the inclusion of human cells or tissues provides a powerful means with which to determine the particular role of cells as either initiators of or targets for cell-selective toxicity. Additionally, accurate genotyping and phenotyping of patients (tissue or cells) can begin to provide some insight into the marked interpatient response seen in cancer chemotherapy and with idiosyncratic drug reactions.

REFERENCES

1. W. T. Bellamy, Prediction of response to drug therapy of cancer. A review of in vitro assays, *Drugs 44*:690–708 (1992).
2. B. K. Park, M. Pirmohamed, and N. R. Kitteringham, Idiosyncratic drug reactions: A mechanistic evaluation of risk factors, *Br. J. Clin. Pharmacol. 34*:377–395 (1992).
3. J. R. Sellers, S. Cook, and V. S. Goldmacher, A cytotoxicity assay utilizing a fluorescent dye that determines accurate surviving fractions of cells, *J. Immunol. Methods 172*:255–264 (1994).
4. J. L. Merlin, S. Azzi, D. Lignon, C. Ramacci, N. Zeghari, and F. Guillemin, MTT assays allow quick and reliable measurement of the response of human tumor-cells to photodynamic therapy, *Eur. J. Cancer 28A*:1452–1458 (1992).
5. N. G. Papadopoulos, G. V. Z. Dedoussis, G. Spanakos, A. D. Gritzapis, C. N. Baxevanis, and M. Papamichail, An improved fluorescence assay for the determination of lymphocyte-mediated cytotoxicity using flow-cytometry, *J. Immunol. Methods 177*:101–111 (1994).
6. T. Mosmann, Rapid colorimetric assay for cellular growth and survival: Application to proliferation and cytotoxicity assays, *J. Immunol. Methods 65*:55–63 (1983).
7. R. Arnould, J. Dubois, F. Abikhalil, A. Libert, G. Ghanem, G. Atassi, M. Hanocq, and F. J. Lejeune, Comparison of 2 cytotoxicity assays—tetrazolium derivative reduction (MTT) and tritiated-thymidine uptake on 3 malignant mouse-cell lines using chemotherapeutic-agents and investigational drugs, *Anticancer Res. 10*:145–154 (1990).
8. U. Keilholz, R. Dummer, H. Welters, B. Brado, F. Galm, P. Matheiowetz, and W. Hunstein, A modified cytotoxicity assay with high-sensitivity, *Scand. J. Clin. Lab. Invest. 50*:879–884 (1990).
9. J. Carmichael, W. G. DeGraff, A. F. Gazdar, J. D. Minna, and J. B. Mitchell, Evaluation of a tetrazolium-based semiautomated colorimetric assay: Assessment of chemosensitivity testing, *Cancer Res. 47*:936–942 (1987).
10. L. Bordenave, R. Bareille, F. Lefebvre, and C. Baquey, A comparison between chromium-51 release and LDH release to measure cell-membrane integrity—interest for cytocompatibility studies with biomaterials, *J. Appl. Biomater. 4*:309–315 (1993).
11. G. Basha, P. Yap, and F. Penninckx, Comparative-study of classical, colorimetric and immunological staining methods for the assessment of tumor-cell viability, *Tumor Biol. 17*:354–361 (1996).
12. R. F. Hussain, A. M. E. Nouri, and R. T. D. Oliver, A new approach for measurement of cytotoxicity using colorimetric assay, *J. Immunol. Methods 160*:89–96 (1993).
13. B. E. Loveland, T. G. Johns, I. R. Mackay, F. Vaillant, Z. X. Wang, and P. J. Hertzog, Validation of the MTT dye assay for enumeration of cells in proliferative and antiproliferative assays, *Biochem. Int. 27*:501–510 (1992).

14. J. A. Plumb, R. Milroy, and S. B. Kaye, Effect of the pH dependence of 3-(4,5-dimethylthiazol-2-yl)-2,5-diphenyl-tetrazolium bromide-formazan absorption on chemosensitivity determined by a novel tetrazolium-based assay, *Cancer Res. 49*:4435–4440 (1989).
15. S. P. M. Crouch, R. Kozlowski, K. J. Slater, and J. Fletcher, The use of ATP bioluminescence as a measure of cell proliferation and cytotoxicity, *J. Immunol. Methods 160*:81–88 (1993).
16. P. von Zons, P. Crowley-Nowick, D. Friberg, M. Bell, U. Koldovsky, and T. L. Whiteside, Comparison of europium and chromium release assays: Cytotoxicity in healthy individuals and patients with cervical carcinoma, *Clin. Diagn. Lab. Immunol. 4*:202–207 (1997).
17. A. K. Patel and P. N. Boyd, An improved assay for antibody-dependent cellular cytotoxicity based on time-resolved fluorometry, *J. Immunol. Methods 184*:29–38 (1995).
18. I. P. Beletsky and S. R. Umansky, A new assay for cell death, *J. Immunol. Methods 134*: 201–205 (1990).
19. P. Borella, A. Bargellini, S. Salvioli, C. I. Medici, and A. Cossarizza, The use of nonradioactive chromium as an alternative to Cr-51 in NK assay, *J. Immunol. Methods 186*:101–110 (1995).
20. M. C. Elia, R. D. Storer, L. S. Harmon, A. R. Kraynak, T. W. McKelvey, P. R. Hertzog, K. P. Keenan, J. G. Deluca, and W. W. Nichols, Cytotoxicity as measured by trypan blue as a potentially confounding variable in the in vitro alkaline elution rat hepatocyte assay, *Mutat. Res. 291*:193–205 (1993).
21. H. Y. Jiao, Y. Soejima, Y. Ohe, and N. Saijo, A new MTT assay for examining the cytotoxicity of activated macrophages towards the nonadherent P388-leukemia cell-line, *J. Immunol. Methods 153*:265–266 (1992).
22. F. A. Barile, S. Arjun, and D. Hopkinson, In vitro cytotoxicity testing—biological and statistical significance, *Toxicol. In Vitro 7*:111–116 (1993).
23. T. Husoy, T. Syversen, and J. Jenssen, Comparisons of 4 in vitro cytotoxicity tests—the MTT assay, NR assay, uridine incorporation and protein measurements, *Toxicol. In Vitro 7*: 149–154 (1993).
24. S. P. Spielberg, Acetaminophen toxicity in human lymphocytes in vitro, *J. Pharmacol. Exp. Ther. 213*:395–398 (1980).
25. R. J. Riley, C. Lambert, J. L. Maggs, N. R. Kitteringham, and B. K. Park, An in vitro study of the microsomal metabolism and cellular toxicity of phenytoin, sorbinil, and mianserin, *Br. J. Clin. Pharmacol. 26*:577–588 (1988).
26. F. J. Gonzalez, C. L. Crespi, and H. V. Gelboin, cDNA-expressed human cytochrome P450s: A new age of molecular toxicology and human risk assessment, *Mutat. Res. 247*:113–127 (1991).
27. D. F. Grant, J. F. Greene, F. Pinot, B. Borhan, M. F. Moghaddam, B. D. Hammock, B. McCutchen, H. Ohkawa, G. Luo, and T. M. Guenthner, Development of an in-situ toxicity assay system using recombinant baculoviruses, *Biochem. Pharmacol. 51*:503–515 (1996).
28. T. D. Porter and J. R. Larson, Expression of mammalian P450s in *Escherichia coli, Methods Enzymol. 206*:108–116 (1991).
29. C. L. Crespi, R. Langenbach, and B. W. Penman, Human cell-lines, derived from AHH-1 $TK^{+/-}$ human lymphoblasts, genetically-engineered for expression of cytochromes-P450, *Toxicology 82*:89–104 (1993).
30. C. L. Crespi, F. J. Gonzalez, D. T. Steimel, T. R. Turner, H. V. Gelboin, B. W. Penman, and R. Langenbach, A metabolically competent human cell line expressing five cDNAs encoding procarcinogen-activating enzymes: Application to mutagenicity testing, *Chem. Res. Toxicol. 4*:566–572 (1991).
31. J. A. Styles, A. Davies, C. K. Lim, F. Dematteis, L. A. Stanley, I. N. H. White, Z. X. Yuan, and L. L. Smith, Genotoxicity of tamoxifen, tamoxifen epoxide and toremifene in human lymphoblastoid-cells containing human cytochrome p450s, *Carcinogenesis 15*:5–9 (1994).
32. M. R. Alison and C. E. Sarraf, Apoptosis: Regulation and relevance to toxicology, *Hum. Exp. Toxicol. 14*:234–247 (1995).

33. G. Majno and I. Joris, Apoptosis, oncosis, and necrosis. An overview of cell death [see comments], *Am. J. Pathol. 146*:3–15 (1995).
34. E. White, Life, death and pursuit of apoptosis, *Genes Dev. 10*:1–15 (1996).
35. C. Dive and J. A. Hickman, Drug-target interactions: Only the first step in the commitment to a programmed cell death? *Br. J. Cancer 64*:192–196 (1991).
36. D. L. Vaux, Toward an understanding of the molecular mechanisms of physiological cell death, *Proc. Natl. Acad. Sci. USA 90*:786–789 (1993).
37. L. M. Buja, M. L. Eigenbrodt, and E. H. Eigenbrodt, Apoptosis and necrosis—basic types and mechanisms of cell-death, *Arch. Pathol. Lab. Med. 117*:1208–1214 (1993).
38. A. H. Wyllie, Apoptosis (The 1992 Frank Rose Memorial lecture), *Br. J. Cancer 67*:205–208 (1993).
39. F. G. Que and G. J. Gores, Cell death by apoptosis—basic concepts and disease relevance for the gastroenterologist, *Gastroenterology 110*:1238–1243 (1996).
40. S. Orrenius, Apoptosis: Molecular mechanisms and implications for human disease, *J. Intern. Med. 237*:529–536 (1995).
41. A. R. Haake and R. R. Polakowska, Cell-death by apoptosis in epidermal biology, *J. Invest. Dermatol. 101*:107–112 (1993).
42. G. B. Corcoran, L. Fix, D. P. Jones, M. T. Moslen, P. Nicotera, F. A. Oberhammer, and R. Buttyan, Apoptosis: Molecular control point in toxicity, *Toxicol. Appl. Pharmacol. 128*:169–181 (1994).
43. T. F. Woolf and R. A. Jordan, Basic concepts in drug metabolism: Part 1, *J. Clin. Pharmacol. 27*:15–17 (1987).
44. T. R. Tephly and B. Burchell, UDP-glucuronyl transferases: A family of detoxifying enzymes, *Trends Pharmacol. Sci. 11*:276–229 (1990).
45. T. Wagner, Ifosfamide clinical pharmacokinetics, *Clin. Pharmacokinet. 26*:439–456 (1994).
46. J. Cummings, V. J. Spanswick, and J. F. Smyth, Reevaluation of the molecular pharmacology of mitomycin-C, *Eur. J. Cancer 31A*:1928–1933 (1995).
47. M. Pirmohamed, N. R. Kitteringham, and B. K. Park, The role of active metabolites in drug toxicity, *Drug Safety 11*:114–144 (1994).
48. M. Pirmohamed, S. Madden, and B. K. Park, Idiosyncratic drug reactions: Metabolic bioactivation as a pathogenic mechanism, *Clin. Pharmacokinet. 31*:215–230 (1996).
49. B. K. Park, M. Pirmohamed, and N. R. Kitteringham, The role of cytochrome P450 enzymes in hepatic and extrahepatic human drug toxicity, *Pharmacol. Ther. 68*:385–424 (1995).
50. B. K. Park, J. W. Coleman, and N. R. Kitteringham, Drug disposition and drug hypersensitivity, *Biochem. Pharmacol. 36*:581–590 (1987).
51. L. R. Pohl, H. Satoh, D. D. Christ, and J. G. Kenna, Immunologic and metabolic basis of drug hypersensitivities, *Annu. Rev. Pharmacol. 28*:367–387 (1988).
52. J. R. Gillette, S. S. Lau, and T. J. Monks, Intra- and extra-cellular formation of metabolites from chemically reactive species, *Biochem. Soc. Trans. 12*:4–7 (1984).
53. U. A. Boelsterli, Specific targets of covalent drug–protein interactions in hepatocytes and their toxicological significance in drug-induced liver injury, *Drug Metab. Rev. 25*:395–451 (1993).
54. J. A. Hickman, Apoptosis and chemotherapy resistance, *Eur. J. Cancer 32A*:921–926 (1996).
55. M. L. Obrien and K. D. Tew, Glutathione and related enzymes in multidrug-resistance, *Eur. J. Cancer 32A*:967–978 (1996).
56. D. J. Reed, Glutathione: Toxicological implications, *Annu. Rev. Pharmacol. Toxicol. 30*:603–631 (1990).
57. T. D. Boyer, The glutathione-*S*-transferases: An update, *Hepatology 9*:486–496 (1989).
58. M. Greenbaum, S. Letourneau, H. Assar, R. L. Schecter, G. Batist, and D. Cournoyer, Retrovirus-mediated gene transfer of rat glutathione *S*-transferase Yc confers alkylating drug resistance in NIH 3T3 mouse fibroblasts, *Cancer Res. 54*:4442–4447 (1994).

59. R. S. Fernandes and T. G. Cotter, Aopotosis or necrosis: Intracellular levels of glutathione influence mode of cell death, *Biochem. Pharmacol.* *48*:675–681 (1994).
60. L. Clarke and D. J. Waxman, Oxidative-metabolism of cyclophosphamide—identification of the hepatic monooxygenase catalysts of drug activation, *Cancer Res.* *49*:2344–2350 (1989).
61. G. F. Weber and D. J. Waxman, Activation of the anticancer drug ifosphamide by rat-liver microsomal P450 enzymes, *Biochem. Pharmacol.* *45*:1685–1694 (1993).
62. T. K. H. Chang, G. F. Weber, C. L. Crespi, and D. J. Waxman, Differential activation of cyclophosphamide and ifosphamide by cytochrome-P4502B and cytochrome-P4503A in human liver-microsomes, *Cancer Res.* *53*:5629–5637 (1993).
63. D. Walker, J. P. Flinois, S. C. Monkman, C. Beloc, A. V. Boddy, S. Cholerton, A. K. Daly, M. J. Lind, A. D. J. Pearson, P. H. Beaune, and J. R. Idle, Identification of the major human hepatic cytochrome-P450 involved in activation and *N*-dechloroethylation of ifosfamide, *Biochem. Pharmacol.* *47*:1157–1163 (1994).
64. L. Chen, D. J. Waxman, D. S. Chen, and D. W. Kufe, Sensitization of human breast-cancer cells to cyclophosphamide and ifosfamide by transfer of a liver cytochrome-P450 gene, *Cancer Res.* *56*:1331–1340 (1996).
65. T. Wagner, D. Heydrich, T. Jork, G. Voelcker, and H. J. Hohorst, Comparative-study on human pharmacokinetics of activated ifosfamide and cyclophosphamide by a modified fluorometric test, *J. Cancer Res. Clin. Oncol.* *100*:95–104 (1981).
66. T. Wagner, M. Zink, and G. Schwieder, Influence of mesna and cysteine on the systemic toxicity and therapeutic efficacy of activated cyclophosphamide, *J. Cancer Res. Clin. Oncol.* *113*:160–165 (1987).
67. K. Ormstad and N. Uehara, Renal transport and disposition of Na-2-mercaptoethane sulfonate disulfide (dimesna) in the rat, *FEBS Lett.* *150*:354–357 (1982).
68. C. R. DeVries and F. S. Freiha, Hemorrhagic cystitis: A review, *J. Urol.* *143*:1–7 (1990).
69. B. H. Lauterburg, T. Nguyen, B. Hartmann, E. Junker, A. Kupfer, and T. Cerny, Depletion of total cysteine, glutathione, and homocysteine in plasma by ifosfamide mesna therapy, *Cancer Chemother. Pharmacol.* *35*:132–136 (1994).
70. A. V. Boddy, S. M. Yule, R. Wyllie, L. Price, A. D. J. Pearson, and J. R. Idle, Pharmacokinetics and metabolism in children of ifosfamide administered as a continuous infusion, *Cancer Res.* *53*:3758–3764 (1993).
71. M. Mohrmann, A. Pauli, M. Ritzer, B. Schonfeld, B. Seifert, and M. Brandis, Inhibition of sodium-dependent transport systems in LLC-PK1 cells by metabolites of ifosfamide, *Renal Physiol. Biochem.* *15*:289–301 (1992).
72. A. V. Boddy, M. English, A. D. J. Pearson, J. R. Idle, and R. Skinner, Ifosfamide nephrotoxicity—limited influence of metabolism and mode of administration during repeated therapy in pediatrics, *Eur. J. Cancer* *32A*:1179–1184 (1996).
73. C. Sood and P. J. O'Brien, 2-Chloroacetaldehyde-induced cerebral glutathione depletion and neurotoxicity, *Br. J. Cancer* *74*:S287–S293 (1996).
74. R. T. Dorr, New findings in the pharmacokinetic, metabolic and drug-resistance aspects of mitomycin C, *Semin. Oncol.* *15*:32–41 (1988).
75. D. Siegel, N. W. Gibson, P. C. Preusch, and D. Ross, Metabolism of mitomycin C by DT-diaphorase: Role in mitomycin C-induced DNA damage and cytotoxicity in human colon carcinoma cells, *Cancer Res.* *50*:7483–7489 (1990).
76. J. Cummings, L. Allan, and J. F. Smyth, Encapsulation of mitomycin-C in albumin microspheres markedly alters pharmacokinetics, drug quinone reduction in tumor-tissue and antitumor-activity—implications for the drugs in-vivo mechanism of action, *Biochem. Pharmacol.* *47*:1345–1356 (1994).
77. A. M. Malkinson, D. Siegel, G. L. Forrest, A. F. Gazdar, H. K. Oie, D. C. Chan, P. A. Bunn, M. Mabry, D. J. Dykes, S. D. Harrison, and D. Ross, Elevated DT-diaphorase activity and messenger-RNA content in human non–small-cell lung-carcinoma—relationship to the response of lung-tumor xenografts to mitomycin-C, *Cancer Res.* *52*:4752–4757 (1992).

78. M. Nishiyama, S. Saeki, N. Hirabayashi, and T. Toge, Relevance of DT-diarphorase activity to mitomycin-C (MMC) efficacy on human cancer-cells—differences in in vitro and in vivo systems, *Int. J. Cancer 53*:1013–1016 (1993).
79. R. M. Phillips, A. de-la-Cruz, R. D. Traver, and N. W. Gibson, Increased activity and expression of NAD(P)H:quinone acceptor oxidoreductase in confluent cell cultures and within multicellular spheroids, *Cancer Res. 54*:3766–3771 (1994).
80. Y. H. Edwards, J. Potter, and D. A. Hopkinson, Human FAD-dependent NAD(P)H diaphorase, *Biochem. J. 187*:429–436 (1980).
81. D. Ross, H. Beall, R. D. Traver, D. Siegel, R. M. Phillips, and N. W. Gibson, Bioactivation of quinones by DT-diaphorase, molecular, biochemical, and chemical studies, *Oncol. Res. 6*:493–500 (1994).
82. R. D. Traver, D. Siegel, H. D. Beall, R. M. Philips, N. W. Gibson, W. A. Franklin, and D. Ross, Characterization of a polymorphism in NAD(P)H:quinone oxidoreductase (DT-diaphorase), *Br. J. Cancer 75*:69–75 (1997).
83. E. A. Rosvold, K. A. McGlynn, E. D. Lustbader, and K. H. Buetow, Identification of an NAD(P)H-quinone oxidoreductase polymorphism and its association with lung-cancer and smoking, *Pharmacogenetics 5*:199–206 (1995).
84. M. Pirmohamed and B. K. Park, Cytochromes P450 and immunotoxicity, *Cytochromes P450: Metabolic and Toxicological Aspects* (C. loannides, ed.), CRC Press, Boca Raton, FL, 1996, pp. 329–354.
85. S. P. Spielberg, In vitro assessment of pharmacogenetic susceptibility to toxic drug metabolites in humans, *Fed. Proc. 43*:2308–2313 (1984).
86. S. P. Spielberg, G. B. Gordon, D. A. Blake, D. A. Goldstein, and H. F. Herlong, Predisposition to phenytoin hepatotoxicity assessed in vitro, *N. Engl. J. Med. 305*:722–727 (1981).
87. S. P. Spielberg, G. B. Gordon, D. A. Blake, E. D. Mellitis, and D. S. Bross, Anticonvulsant toxicity in vitro: Possible role of arene oxides, *J. Pharmacol. Exp. Ther. 217*:386–389 (1981).
88. W. T. Gerson, D. G. Fine, S. P. Spielberg, and L. L. Sensenbrenner, Anticonvulsant-induced aplastic anaemia: Increased susceptibility to toxic drug metabolites in vitro, *Blood. 61*:889–893 (1983).
89. N. H. Shear, S. P. Spielberg, M. Cannon, and M. Miller; Anticonvulsant hypersensitivity syndrome: In vitro risk assessment, *J. Clin. Invest. 82*:1826–1832 (1988).
90. M. Pirmohamed, A. Graham, P. Roberts, D. Smith, D. Chadwick, A. M. Breckenridge, and B. K. Park, Carbamazepine hypersensitivity: Assessment of clinical and in vitro chemical cross-reactivity with phenytoin and oxcarbazepine, *Br. J. Clin. Pharmacol. 32*:741–749 (1991).
91. P. Wolkenstein, D. Charue, P. Laurent, J. Revuz, J. C. Roujeau, and M. Bagot, Metabolic predisposition to cutaneous adverse drug reactions, *Arch. Dermatol. 131*:544–551 (1995).
92. S. P. Spielberg, N. H. Shear, M. Cannon, J. C. Hutson, and K. Gunderson, In-vitro assessment of a hypersensitivity syndrome associated with sorbinil, *Ann. Intern. Med. 114*:720–724 (1991).
93. M. J. Reider, J. P. Uetrecht, N. H. Shear, M. Cannon, M. Miller, and S. P. Spielberg, Diagnosis of sulfonamide hypersensitivity reactions by in-vitro ''rechallenge'' with hydroxylamine metabolites, *Ann. Intern. Med. 110*:286–289 (1989).
94. D. Larrey, A. Berson, F. Habersetzer, M. Tinel, A. Castot, G. Babany, P. Letteron, E. Freneaux, J. Loeper, P. Dansette, and D. Pessayre, Genetic predisposition to drug hepatotoxicity: Role in hepatitis caused by amineptine, a tricyclic antidepressant, *Hepatology 10*:168–173 (1989).
95. G. L. Kearns, J. G. Wheeler, S. H. Childress, and L. G. Letzig, Serum sickness-like reactions to cefaclor—role of hepatic-metabolism and individual susceptibility, *J. Pediatr. 125*:805–811 (1994).
96. W. E. Crill, Drugs 5 years later: Carbamazepine, *Ann. Intern. Med. 79*:844–847 (1973).
97. M. Pirmohamed, N. R. Kitteringham, A. M. Breckenridge, and B. K. Park, Detection of an

autoantibody directed against a human liver microsomal protein in a patient with carbamazepine hypersensitivity, *Br. J. Clin. Pharmacol.* *33*:183–186 (1992).

98. P. S. Friedmann, I. Strickland, M. Pirmohamed, and B. K. Park, Investigation of mechanisms in toxic epidermal necrolysis induced by carbamazepine, *Arch. Dermatol.* *130*:598–604 (1994).
99. M. Pirmohamed, N. R. Kitteringham, T. M. Guenthner, A. M. Breckenridge, and B. K. Park, An investigation of the formation of cytotoxic, protein-reactive and stable metabolites from carbamazepine in vitro. *Biochem. Pharmacol.* *43*:1675–1682 (1992).
100. M. Pirmohamed, N. R. Kitteringham, A. M. Breckenridge, and B. K. Park, The effect of enzyme induction on the cytochrome P450-mediated bioactivation of carbamazepine by mouse liver microsomes, *Biochem. Pharmacol.* *44*:2307–2314 (1992).
101. K. M. Ivanetich, M. R. Ziman, and J. J. Bradshaw, 1,1,1-Trichloropropene-2,3-oxide: An alternate mechanism for its inhibition of cytochrome P-450, *Res. Commun. Chem. Pathol. Pharmacol.* *35*:111–119 (1982).
102. S. Madden, J. L. Maggs, and B. K. Park, Bioactivation of carbamazepine in the rat in-vivo—evidence for the formation of reactive arene oxide(s), *Drug. Metab. Dispos.* *24*:469–479 (1996).
103. J. L. Maggs, M. Pirmohamed, N. R. Kitteringham, and B. K. Park, Characterization of the metabolites of carbamazepine in patient urine by liquid chromatography/mass spectrometry, *Drug Metab. Dispos.* *25*:275–280 (1997).
104. M. D. Tingle, M. Pirmohamed, E. Templeton, A. S. Wilson, S. Madden, N. R. Kitteringham, and B. K. Park, An investigation of the formation of cytotoxic, genotoxic, protein-reactive and stable metabolites from naphthalene by human liver in vitro, *Biochem. Pharmacol.* *46*: 1529–1538 (1993).
105. T. M. Guenthner, Epoxide hydrolases, *Conjugation Reactions in Drug Metabolism: An Integrated Approach* (G. J. Mulder, ed.), Taylor & Francis, London, 1990, pp. 365–404.
106. J. P. Hardwick, F. J. Gonzalez, and C. B. Kasper, Transcriptional regulation of rat liver epoxide hydratase, NADPH–cytochrome P-450 oxidoreductase, and cytochrome P-450b genes by phenobarbital, *J. Biol. Chem.* *258*:8081–8085 (1983).
107. C. D. Davis, M. Pirmohamed, N. R. Kitteringham, R. L. Allott, D. Smith, and B. K. Park, Kinetic parameters of lymphocyte microsomal epoxide hydrolase in carbamazepine hypersensitive patients: Assessment by radiometric HPLC, *Biochem. Pharmacol.* *50*:1361–1366 (1995).
108. T. M. Guenthner, J. Kuk, M. Nguyen, C. W. Wheeler, M. Pirmohamed, and B. K. Park, Epoxide hydrolases: Immunochemical detection in human tissues, *Human Drug Metabolism: From Molecular Biology to Man* (E. H. Jeffrey, ed.), CRC Press, Boca Raton, FL, 1993, pp. 65–80.
109. A. Gaedigk, S. P. Spielberg, and D. M. Grant, Characterization of the microsomal epoxide hydrolase gene in patients with anticonvulsant adverse drug-reactions, *Pharmacogenetics* *4*: 142–153 (1994).
110. V. J. Green, M. Pirmohamed, N. R. Kitteringham, A. Gaedigk, D. M. Grant, M. Boxer, B. Burchell, and B. K. Park, Genetic analysis of microsomal epoxide hydrolase in patients with carbamazepine hypersensitivity, *Biochem. Pharmacol.* *50*:1353–1359 (1995).
111. W. Honscha, F. Oesch, and T. Friedberg, Tissue-specific expression and differential inducibility of several microsomal epoxide hydrolase mRNAs which are formed by alternative splicing, *Arch. Biochem. Biophys.* *287*:380–385 (1991).
112. J. H. Lillibridge, B. M. Amore, J. T. Slattery, T. F. Kalhorn, S. D. Nelson, R. H. Finnell, and G. D. Bennett, Protein-reactive metabolites of carbamazepine in mouse-liver microsomes, *Drug. Metab. Dispos.* *24*:509–514 (1996).
113. A. S. Wilson, C. D. Davis, D. P. Williams, M. Pirmohamed, A. Buckpitt, and B. K. Park, Naphthalene toxicity is a function of quinone, but not epoxide metabolites, *Hum. Exp. Toxicol.* *15*:685 (1996).

114. A. Gaedigk, M. Pirmohamed, B. K. Park, and J. S. Leeder, Genetic polymorphisms of NAD(P)H:quinone oxido-reductase (NQO1) and catechol-*O*-methyl transferase (COMT) in patients with anticonvulsant hypersensitivity syndrome, Microsomes meeting abstract, Los Angeles, P-198 (1996).
115. V. J. Green, M. Pirmohamed, N. R. Kitteringham, M. J. Brodie, and B. K. Park, Glutathione-*S*-transferase μ genotype (GSTM1*O) in patients with carbamazepine hypersensitivity, *Br. J. Clin. Pharmacol. 39*:555P (1995).
116. R. L. Allott, M. Pirmohamed, N. R. Kitteringham, and B. K. Park, Glutathione *S*-transferase T1 polymorphism and susceptibility to carbamazepine hypersensitivity, *Br. J. Clin. Pharmacol. 41*:463P (1996).
117. J. M. Zakrzewska and L. Ivanyi, In vitro lymphocyte proliferation by carbamazepine, carbamazepine-10-11-epoxide, and oxcarbazepine in the diagnosis of drug-induced hypersensitivity, *J. Allergy Clin. Immunol. 82*:110–115 (1988).
118. C. C. Vittorio and J. J. Muglia, Anticonvulsant hypersensitivity syndrome, *Arch. Intern. Med. 155*:2285–2290 (1995).
119. M. Farlow, S. I. Gracon, L. A. Hershey, K. W. Lewis, C. H. Salowsky, and J. Dolan-Ureno, A controlled trial of tacrine in Alzheimer's disease, *JAMA 268*:2523–2529 (1992).
120. D. R. Forsyth, D. J. Sormon, R. A. Morgan, and G. K. Wilcock, Clinical experience with and side effects of tacrine hydrochloride in Alzheimer's disease: A pilot study, *Age Ageing 18*:223–229 (1989).
121. J. T. O'Brien, S. Eagger, and R. Levy, Effects of tetrahydroaminoacridine on liver function in patients with Alzheimer's disease, *Age Ageing 20*:129–131 (1991).
122. P. B. Watkins, H. J. Zimmerman, M. J. Knapp, S. I. Gracon, and K. W. Lewis Hepatotoxic effects of tacrine administration in patients with Alzheimer's disease, *JAMA 271*:992–998 (1994).
123. M. W. Farris, S. A. Johnsen, L. P. Walton, W. R. Mumaw, and S. D. Ray, Tetrahydroaminoacridine-induced ribosomal changes and inhibition of protein synthesis in rat hepatocyte suspensions, *Hepatology 20*:240–246 (1994).
124. W. F. Pool, S. M. Bjorge, T. Chang, and T. F. Woolf, Metabolic disposition of the cognition activator in man: identification of phenol glucuronide metabolites in urine, *ISSX Proc. 2*:164 (1992).
125. S. Madden, T. F. Woolf, W. F. Pool, and B. K. Park, An investigation into the formation of stable, protein-reactive and cytotoxic metabolites from tacrine in vitro: Studies with human and rat liver microsomes, *Biochem. Pharmacol. 46*:13–20 (1993).
126. V. Spaldin, S. Madden, W. F. Pool, T. F. Woolf, and B. K. Park, The effect of enzyme-inhibition on the metabolism and activation of tacrine by human liver-microsomes, *Br. J. Clin. Pharmacol. 38*:15–22 (1994).
127. V. Spaldin, S. Madden, D. A. Adams, R. J. Edwards, D. S. Davies and B. K. Park, Determination of human hepatic cytochrome P4501A2 activity in vitro: Use of tacrine as an isozyme specific probe, *Drug Metab. Dispos. 23*:929–934 (1995).
128. S. Madden, V. Spaldin, R. N. Hayes, T. F. Woolf, W. F. Pool, and B. K. Park, Species variation in the bioactivation of tacrine by hepatic microsomes, *Xenobiotica 25*:103–116 (1995).
129. S. Madden, V. Spaldin, and B. K. Park, Clinical pharmacokinetics of tacrine, *Clin. Pharmacokinet. 28*:449–457 (1995).
130. B. K. Park, S. Madden, V. Spaldin, T. F. Woolf, and W. F. Pool, Tacrine transaminitis—potential mechanisms, *Alzheimer Dis. Assoc. Disord. 8*:S39–S49 (1994).
131. J. H. Doroshow, G. Y. Locker, and C. E. Myers, Enzymatic defences of the mouse heart against reactive oxygen metabolites: Alterations produced by doxorubicin, *J. Clin. Invest. 65*:128–135 (1980).
132. P. Dogterom, J. F. Nagelkerke, and G. J. Mulder, Hepatotoxicity of tetrahydroaminoacridine in isolated rat hepatocytes: Effects of glutathione and vitamin E, *Biochem. Pharmacol. 37*: 2311–2313 (1988).

133. M. D. Coleman, Dapsone toxicity—some current perspectives, *Gen. Pharmacol. 26*:1461–1467 (1995).
134. J. Zuidema, E. S. M. Hilbers-Moddermann, and F. W. H. M. Merkus, Clinical pharmacokinetics of dapsone, *Clin. Pharmacokinet. 11*:299–315 (1986).
135. B. F. Glader and M. E. Conrad, Haemolysis by diphenylsulfones: Comparative effects of DDS and hydroxylamine-DDS, *J. Lab. Clin. Med. 81*:267–272 (1973).
136. J. Uetrecht, N. Zahid, N. H. Shear, and W. D. Biggar, Metabolism of dapsone to a hydroxylamine by human neutrophils and mononuclear cells, *J. Pharmacol. Exp. Ther. 245*:274–279 (1988).
137. M. D. Coleman, A. M. Breckenridge, and B. K. Park, Bioactivation of dapsone to a cytotoxic metabolite by human hepatic microsomal enzymes, *Br. J. Clin. Pharmacol. 28*:389–395 (1989).
138. R. J. Riley, P. Roberts, M. D. Coleman, N. R. Kitteringham, and B. K. Park, Bioactivation of dapsone to a cytotoxic metabolite: In vitro use of a novel two compartment system which contains human tissues, *Br. J. Clin. Pharmacol. 30*:417–426 (1990).
139. M. D. Tingle, M. D. Coleman, and B. K. Park, The effect of preincubation with cimetidine on the *N*-hydroxylation of dapsone by human liver microsomes, *Br. J. Clin. Pharmacol. 32*: 120–123 (1991).
140. M. D. Tingle, M. D. Coleman, and B. K. Park, An investigation of the role of metabolism in dapsone-induced methaemoglobinaemia using a two compartment in vitro test system, *Br. J. Clin. Pharmacol. 30*:829–838 (1990).
141. H. J. Gill, M. D. Tingle, and B. K. Park, *N*-Hydroxylation of dapsone by multiple enzymes of cytochrome P450: Implications for inhibition of haemotoxicity, *Br. J. Clin. Pharmacol. 40*:531–539 (1995).
142. C. M. Fleming, R. A. Branch, G. R. Wilkinson, and F. P. Guengerich, Human liver microsomal *N*-hydroxylation of dapsone by cytochrome P-4503A4, *Mol. Pharmacol. 41*:975–980 (1992).
143. A. K. Mitra, K. E. Thummel, T. F. Kalhorn, E. D. Kharasch, J. D. Unadkat, and J. T. Slattery, Metabolism of dapsone to its hydroxylamine by CYP2E1 in vitro and in vivo, *Clin. Pharmacol. Ther. 58*:556–566 (1995).
144. M. Nazario, The hepatic and renal mechanisms of drug interactions with cimetidine, *Drug Intell. Clin. Pharm. 20*:342–348 (1986).
145. M. D. Coleman, A. K. Scott, A. M. Breckenridge, and B. K. Park, The use of cimetidine as a selective inhibitor of dapsone *N*-hydroxylation in man, *Br. J. Clin. Pharmacol. 30*:761–767 (1990).
146. M. D. Coleman, L. E. Rhodes, A. K. Scott, J. L. Verbov, P. S. Friedmann, A. M. Breakenridge, and B. K. Park, The use of cimetidine to reduce dose-dependent methaemoglobinaemia in dermatitis hepertiformis patients, *Br. J. Clin. Pharmacol. 34*:244–249 (1992).
147. L. E. Rhodes, M. D. Tingle, B. K. Park, P. Chu, J. L. Verbov, and P. S. Friedmann, Cimetidine improves the therapeutic toxic ratio of dapsone in patients on chronic dapsone therapy, *Br. J. Dermatol. 132*:257–262 (1995).
148. M. D. Tingle and B. K. Park, The use of a three compartment in vitro model to investigate the role of hepatic drug metabolism in drug-induced blood dyscrasias, *Br. J. Clin. Pharmacol. 36*:31–39 (1993).
149. P. A. Kramer, B. E. Glader, and T. K. Li, Mechanism of methaemoglobin formation by diphenylsulfones. Effect of 4-amino-4′-hydroxy-aminodiphenylsulfone and other *p*-substituted derivatives, *Biochem. Pharmacol. 21*:1265–1274 (1972).
150. E. Bailey, A. G. Brooks, I. Bird, P. B. Farmer, and B. Street, Monitoring exposure to 4,4′methylenedianiline by the gas chromatography–mass spectrometry determination of adducts to haemoglobin, *Anal, Biochem. 190*:175–181 (1990).
151. J. Kane, G. Honigfeld, J. Singer, and H. Meltzer, Clozapine for the treatment-resistant schizophrenic: A double-blind comparison with chlorpromazine, *Arch. Gen. Psychiatry. 45*:789–796 (1988).

152. J. A. Lieberman and A. Z. Safferman, Clinical profile of clozapine: Adverse reactions and agranulocytosis, *Psychiatr. Q. 63*:51–70 (1992).
153. L. H. Lindstrom, A retrospective study on the long-term efficacy of clozapine in 96 schizophrenic and schizoaffective patients during a 13-year period, *Psychopharmacology 99*:S84–S86 (1989).
154. J. M. J. Alvir, J. A. Lieberman, A. Z. Safferman, J. L Schwimmer, and J. A. Schaaf, Clozapine-induced agranulocytosis. Incidence and risk factors in the United States, *N. Engl. J. Med. 329*:162–167 (1993).
155. K. Atkin, F. Kendall, D. Gould, H. Freeman, J. Lieberman, and D. O'Sullivan, The incidence of neutropenia and agranulocytosis in patients treated with clozapine in the UK and Ireland, *Br. J. Psychiatry 169*:483–488 (1996).
156. S. L. Gerson and H. Meltzer, Mechanisms of clozapine-induced agranulocytosis, *Drug Safety 7 (suppl 1:)*17–25 (1992).
157. A. Z. Safferman, J. A. Lieberman, J. M. Alvir, and A. Howard, Rechallenge in clozapine-induced agranulocytosis [letter], *Lancet 339*:1296–1297 (1992).
158. M. Pirmohamed and B. K. Park, Mechanism of action of clozapine-induced agranulocytosis: Current status of research and implications for drug development, *CNS Drugs 7*:139–158 (1997).
159. M. W. Jann, S. R. Grimsley, E. C. Gray, and W. H. Chang, Pharmacokinetics and pharmacodynamics of clozapine, *Clin. Pharmacokinet. 24*:161–176 (1993).
160. S. K. Lin, W. H. Chang, M. C. Chung, Y. W. F. Lam, and M. W. Jann, Disposition of clozapine and desmethylclozapine in schizophrenic-patients, *J. Clin. Pharmacol. 34*:318–324 (1994).
161. M. Pirmohamed, D. Williams, S. Madden, E. Templeton, and B. K. Park, Metabolism and bioactivation of clozapine by human liver in vitro, *J. Pharmacol. Exp. Ther. 272*:984–990 (1995).
162. M. Jerling, L. Lindstrom, U. Bondesson, and L. Bertilsson, Fluvoxamine inhibition and carbamazepine induction of the metabolism of clozapine—evidence from a therapeutic drug-monitoring service, *Ther. Drug Monit. 16*:368–374 (1994).
163. J. L. Maggs, D. Williams, M. Primohamed, and B. K. Park, The metabolic formation of reactive intermediates from clozapine: A drug associated with agranulocytosis in man, *J. Pharmacol. Exp. Ther. 275*:1463–1475 (1995).
164. V. Fischer, J. A. Haar, L. Greiner, R. V. Lloyd, and R. P. Mason, Possible role of free radical formation in clozapine (Clozaril)-induced agranulocytosis, *Mol. Pharmacol. 40*:846–853 (1991).
165. Z. C. Liu and J. P. Uetrecht, Clozapine is oxidized by activated human neutrophils to a reactive nitrenium ion that irreversibly binds to cells, *J. Pharmacol. Exp. Ther. 275*:1476–1483 (1995).
166. M. Pirmohamed, E. Templeton, A. M. Breckenridge, and B. K. Park, In vitro bioactivation of clozapine to a chemically reactive metabolite by human and mouse liver microsomes, *Br. J. Clin. Pharmacol. 34*:168P (1992).
167. A. C. Tschen, M. J. Rieder, K. Qyewumi, and D. J. Freeman, In-vitro toxicity of clozapine metabolites, *Clin. Pharmacol. Ther. 59*:PII12 (1996).
168. D. P. Williams, M. Primohamed, D. J. Naisbitt, J. L. Maggs, and B. K. Park, Neutrophil cytotoxicity of the chemically reactive metabolite(s) of clozapine: Possible role in agranulocytosis, *J. Pharmacol. Exp. Ther. 283*:1375–1382 (1997).

18

In Vitro Metabolism: FMO and Related Oxygenations

John R. Cashman
Human BioMolecular Research Institute, San Diego, California

I. INTRODUCTION TO IN VITRO STUDIES

This chapter will focus on the in vitro drug metabolism of chemicals and drugs by the flavin-containing monooxygenase (FMO) and other closely related monooxygenases. Although the presentation will be as comprehensive as possible, the examples will be largely taken from those of the author's laboratory and many excellent studies from the literature will not be included for the sake of brevity. In addition, when possible, the examples of in vitro and in vivo metabolic studies will concentrate mainly on studies relevant to human drug metabolism and emphasize developments since the last review (1). Throughout this chapter a distinction will be made as to sequential one-electron *oxidation* or two-electron *oxygenation* that characterizes the action of the cytochromes P-450 (CYP) and FMOs, respectively.

As the drug development process becomes more expensive and time-consuming, methods to expedite the drug discovery and drug development process are emerging. With the advent of combinational chemistry approaches, an increasing number of drug discovery candidates are coming through the drug discovery pipeline. In the face of an ever-increasing number of candidates, possibly the most critical point in the drug development process is the decision of which entity is to be taken forward to a more sophisticated stage for clinical testing in humans (2). Because clinical testing of new drug candidates can take 6 years or more, it is essential to increase the ''hit rate'' when deciding on candidates for clinical evaluation. One way to achieve a higher percentage of successful drug candidates in the clinical pipeline would be to introduce human drug metabolism studies earlier in the drug discovery process than is currently done (3). The critical strategy would be to discard the drug candidates that are likely to be unsuitable and to focus the drug development resources on agents that are most likely to lead to successful drug candidates.

The use of in vitro metabolic studies, in principle, could provide a path to evaluate the likelihood that a drug candidate would not pose a problem from a metabolic perspective (4) or from the standpoint of drug–drug interactions (5). The procurement of human in vitro metabolic information could also aid in establishing the animal species most likely to be appropriate for toxicological testing and safety evaluation (3). Finally, in vitro–in vivo correlations of human drug metabolism studies could provide a method for selecting among a short list of equally promising drug candidates the one to be taken forward to more expensive human clinical trials.

II. MAMMALIAN FLAVIN-CONTAINING MONOOXYGENASES

Among the flavoproteins existing in nature that activate molecular oxygen and participate in substrate oxidation, the FMO is distinct in its ability to form a relatively stable hydroperoxyflavin intermediate (6–10). Other flavoprotein oxidases react quite rapidly with molecular oxygen to yield hydrogen peroxide and the oxidized flavin of the flavoprotein (6,11). As seen in other enzyme systems, a hydroperoxyflavoprotein intermediate is not indefinitely stable (12). Apparently, the protein environment of the FMO protects the hydroperoxy species from decomposing, reacting with apoprotein, or disproportionating into reactive oxygen species (13). However, if the required cofactor NADPH is not present to serve as a ''gatekeeper'' then the peroxyflavin can decompose to hydrogen peroxide, and significant enzyme activity can be lost (14,15). The mechanism whereby FMO stabilizes the peroxyflavin is not understood, but clearly, this is an important feature in preventing the wasteful oxidation of NADPH (Scheme 1).

Several predictions concerning FMO enzyme function stem from the fundamental information about the flavoprotein peroxyflavin moiety. First, it is clear that in the absence of NADPH (and especially at elevated temperatures that often accompany hepatic tissue at the time of death or postmortem), FMO activity is quite labile and is rapidly lost (16). This will be discussed in greater detail later. Second, the chemistry of the FMO as a two-electron hydroperoxide-oxygenating agent is distinct from those of substrate oxidation by

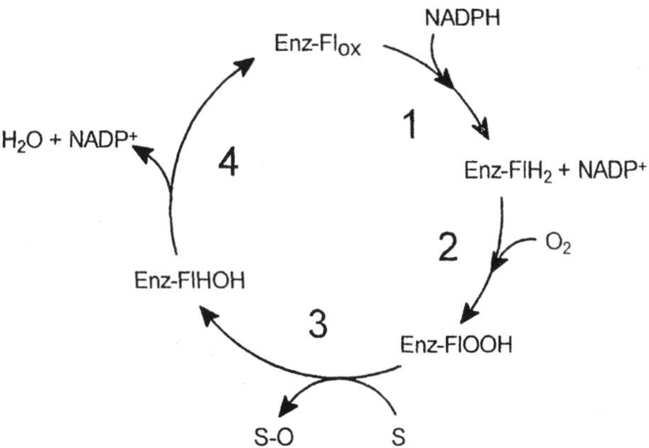

Scheme 1 Catalytic cycle of FMO: S and S-O are the substrate and the oxygenated substrate, respectively. (From Ref. 1)

cytochromes P450, for example, that employ sequential one-electron transfer chemistry. The prediction is that, barring steric constraints, chemicals that are readily oxygenated by alkyl hydroperoxides, peracids or other similar-oxidizing agents should be readily oxygenated by FMO (17,18). This view has been largely supported by the experimental data. Third, although a few examples of nucleophilic oxygenation have been reported previously (18,19), the vast majority of examples in the literature show that FMO acts as an electrophilic-oxygenating catalyst (1,16). This is another feature that distinguishes FMO from other flavoprotein oxidases and other monooxygenases. The important point is that the chemistry of the hydroperoxyflavin determines the types of FMO products formed, and this observation permits relatively easy chemical synthesis and evaluation of FMO activity for a wide variety of substrates. Thus, in most reactions examined, the identity of a putative metabolite arising from FMO can be predicted to a great degree of certainty based on the oxidation product produced from the treatment of a substrate with a peroxide or peracid. In conclusion, it is apparent that the molecular architecture of the substrate-binding domain surrounding the FMO hydroperoxyflavin moiety plays a pivotal role in controlling not only the type of products formed, but also the longevity of the FMO oxygenating species.

A. Flavin-Containing Monooxygenase Enzyme Nomenclature

Evidence now exists for five forms of mammalian FMO that have deduced amino acid sequences, ranging between 52 and 57% identical with that found in rabbit and between 50 and 58% identical across species lines (20,21). The primary methods by which FMO sequences have been ascertained have come from cDNA oligonucleotide sequencing (1), although automated Edman degradation sequencing (22) and mass spectral sequencing of FMO peptides (23) have also provided data. Given the peptide sequence or the amino acid sequence deduced from the cDNA data, FMO forms possessing greater than about 40% amino acid identity can be assigned to one of the five FMO families. FMO orthologues of more than about 80% amino acid identity can thus be grouped in one of the five enzyme families.

A systematic nomenclature has replaced the trivial names used previously for FMO

(20,21). Thus, FMO1 corresponds to pig liver FMO and its orthologues in rabbit and other species. FMO2 is the name given to the major form present in rabbit lung. FMO3 was previously named form 2 from rabbit liver and the human orthologue was named HLFMO II. FMO4 replaces the name human FMO2 as well as the rabbit orthologue named FMO 1E1. Finally, FMO5 is the name given to rabbit FMO 1C1. Comparative studies have shown that flavoproteins with some identity, such as the bacterial equivalent cyclohexanone monooxygenase from *Acinetobacter* (i.e., approximately 25% amino acid sequence homology) and other flavoproteins do not belong to the FMO family of proteins because they are only distantly related in amino acid sequence (24). This is true despite the fact that distantly related flavoproteins in some cases employ the same mechanism and give the same enzymatic products (25). In the future, undoubtedly additional examples of distantly related flavoproteins will be observed in other organisms (e.g., yeast), and there may be cause for reappraisal of the evolutionary relation thus far established.

B. Oxygenation of Nitrogen- and Sulfur-Containing Compounds

Nitrogen- and sulfur-containing functional groups are present in many important drugs, pesticides, and herbicides (14,26,27). Most pharmacologically active compounds contain nitrogen- and sulfur-containing functionalities because of the desirable pharmacological and pharmaceutical chemical properties that these heteroatoms bestow. Nitrogen- and sulfur-containing compounds participate in numerous biotransformations, at least partly because of the nucleophilicity and oxidizability of the heteroatom and also because of the ability of the heteroatom to activate relatively difficult-to-metabolize centers (e.g., methylene functionalities) adjacent to the heteroatom. In general, metabolism of heteroatom-containing drugs and chemicals accomplishes what is desirable in most biotransformations: namely, the change of a relatively lipophilic entity into a more polar, readily excreted metabolite.

Most examples of human liver drug and chemical metabolism have implicated the cytochrome P450 family of monoxygenases, even for oxidations that could potentially be recognized as arising from FMO (28,29). The overlap in FMO and cytochrome P450 type of metabolite has been particularly true for compounds of the nitrogen- and sulfur-containing class of xenobiotics. The paucity of reported human FMO-dependent oxygenation products may be partly due to the similar types of chemistry that cytochromes P450 and FMO can accomplish. Sometimes, however, the lack of apparent involvement of FMO in a particular heteroatom oxygenation may be due to a lack of viability of FMO in the human liver preparation that was used to investigate the substrate. Alternatively, it is possible that the bioanalytical method employed to investigate a role of human FMO in a particular substate oxygenation was less than adequate. This section will provide a detailed description of some of the reported successes and pitfalls associated with studying the role of FMO in a drug or chemical's metabolism. The essential bioorganic features that many laboratories have relied on to examine a role of FMO in metabolism include (1) the establishment of synthetic chemical approaches to the anticipated metabolite, (2) the development of a robust bioanalytical method to examine for the formation of the metabolite in the biological matrix, and (3) the use of an appropriate biological or enzymatic preparation suitable to answer the FMO-related metabolism question posed.

1. S-Oxygenation by the Flavin-Containing Monooxygenase

A variety of studies have shown that human liver microsomes are capable of tertiary amine *N*-oxygenation and sulfide *S*-oxygenation (3,30–40). Generally, *N*-oxygenation or *S*-oxygenation of a xenobiotic represents a detoxication process in that the metabolite is considerably more polar than the starting material and is usually efficiently excreted in the urine (41). The physiological role of the FMO is unknown, but there are some data to suggest that FMO evolved to rid the organism of potentially deleterious lipophilic heteroatom-containing compounds present in the diet (26). The FMOs may have evolved to detoxicate nucleophilic compounds present in dietary foodstuffs. On the other hand, it is possible that FMO has some fundamentally important role in cellular homeostasis, because endogenous nucleophilic nitrogen- and sulfur-containing chemicals are certainly present in the cellular milieu, and some are substrates for FMO. Apparently, the substrate-binding channel of FMO has evolved to preferentially accept nonphysiological nitrogen- and sulfur-containing compounds because endogenous nucleophiles, such as cysteine or glutathione, are not substrates for the FMOs (18). This is important, because if an endogenous sulfur-containing nucleophile, for example, was a substrate for FMO, then it is possible that efficient oxygenation would produce copious quantities of oxygenated product, consume NADPH, and possibly deplete glutathione. Such a situation could potentially place the cell at risk. Sulfur-containing substrates that have a full negative charge, such as an amino acid, generally are very poor substrates for FMO, if they are substrates at all (42). Conversely, some physiologically important nitrogen-containing compounds are *N*-oxygenated by at least some forms of FMO and, thus, it is possible that FMO may serve in a cleanup capacity to prevent the accumulation of unwanted elevated levels of biological amines (e.g., trimethylamine (43)); but, this will be discussed in great detail later.

A significant number of nucleophilic sulfur compounds are present in foodstuffs that are efficiently converted to *S*-oxygenated products (44). One postulate for the evolution of FMO efficiency in the detoxication of chemicals is to prevent other cellular macromolecules from being damaged from reactive metabolites generated by other enzymes (e.g., sulfines, sulfenes, and sulfenic acids; 1,18). For example, thione-containing compounds are sequentially *S*-oxygenated (45–53; Fig. 1). Even though some of these metabolites are quite electrophilic, in all cases examined, FMO is not impaired, whereas occasionally,

$$R\text{-}C(=S)\text{-}NH_2 \longrightarrow R\text{-}C(=S^{+}\text{-}O^{-})\text{-}NH_2 \longrightarrow R\text{-}C(\text{-}S\text{-}OH)=NH$$

$$\downarrow$$

$$R\text{-}C(=O)\text{-}NH_2 \longleftarrow R\text{-}C(\text{-}OH)=NH \longleftarrow R\text{-}C(\text{-}S^{+}(O^{-})\text{-}OH)=NH$$

Fig. 1 Schematic metabolism of a thione-containing compound by FMO (if R = benzene, the compound is thiobenzamide; if R = NHR the compound is a thiourea; if R = OR, the compound is a thiocarbamate). The scheme shows the sulfine (or sulfenic acid) metabolite (i.e., R—S—OH) and the sulfene or sulfinic acid metabolite [i.e., $R—S(OH)_2$]. The electrophilic sulfene is trapped by water to form the ultimate stable metabolite [i.e., $RC(O)\cdot NH_2$].

neighboring endoplasmic reticulum proteins, such as cytochromes P450, can be covalently modified (54–57). Many thiols, thioamides, thioureas, thiocarbamides, and thiocarbamates are efficiently *S*-oxygenated by FMO to form relatively electrophilic metabolic intermediates that inactivate cytochromes P450 (55–58). In the absence of other proteins, FMO converts thiols to disulfides, and thiones to fully oxidized species, such as sulfate.

It is when other proteins intercept the electrophilic metabolites that are substrates for further FMO-dependent oxygenation and detoxication that cellular damage can result. Depending on the neighboring substituents—for example, mercaptoimidazoles or thioureas are excellent substrates for the FMO—sequential *S*-oxygenation affords relatively electrophilic sulfine metabolites that can either be hydrolyzed to nontoxic materials or undergo acylation of nearby biological macromolecules (see Fig. 1). It is possible that covalent binding of macromolecules exerts some toxic cellular consequence. Thus, the relative rate of hydrolysis to covalent binding of an oxidized sulfur-containing compound (i.e., sulfines) probably determines the potential for thiones to participate in drug interactions or toxic cellular sequelae. Other metabolic processes can also intercede and further complicate the disposition and toxicity of a chemical. Intermediate sulfenic acid metabolites can be trapped by cellular nucleophiles, such as glutathione, to form disulfides (59) that can serve as subsequent sources for disulfide exchange and net thiol (e.g., glutathione) oxidation and substrate regeneration (60). Such a futile cycle could deplete the cell of NADPH (a required cofactor for the catalytic cycle of FMO) as well as deplete the cell of glutathione (54). It is possible that this scenario could render the cell susceptible to the toxic properties of other electrophilic metabolites (61–63). In addition, there are a few examples of FMO-catalyzed *S*-oxygenation in which the products are unstable and the metabolite eliminates or rearranges to a new material that possesses some unusual or unanticipated toxicity (64,65). In summary, owing to the enhanced nucleophilicty of the heteroatom, the sulfur atom is a preferred site of FMO oxygenation (66; i.e., *S*-oxygenation is favored over *N*-oxygenation), but this is a site of oxygenation that can lead to complex metabolic and toxicological scenarios.

2. *N-Oxygenation by the Flavin-Containing Monooxygenase*

An increasing number of naturally occurring and xenobiotic nitrogen-containing chemicals *N*-oxygenated by the FMOs have been reported (67–69). Several plant alkaloids including (*S*)-nicotine (35,41,70) and the pyrrolizidine alkaloid plant toxins senecionine, retrorsine, and monocrotaline, are detoxicated by *N*-oxygenation at the tertiary amine center (71–74). Interestingly, in species with a relatively high level of *N*-oxygenating FMO and low pyrrole-forming cytochrome P450 activity, the animals are relatively immune to the toxic properties of pyrrolizidine alkaloids (71,74). In a species that is quite susceptible to the toxic properties of pyrrolizidine alkaloids, such as the rat, apparently the pyrrole-forming cytochrome P450 activity is present to a greater extent than the detoxicated *N*-oxide-forming pathway. Tertiary amine *N*-oxygenation of (*S*)-nicotine is another example whereby an alkaloid is shunted away from the cytochrome P450 metabolic pathway by formation of (*S*)-nicotine *N*-1′-oxide (75–77). (*S*)-Nicotine *N*-1′-oxide has only a fraction of the biological activity of the parent compound (78) and is rapidly excreted unchanged in humans (41). Whereas (*S*)-nicotine *N*-oxygenation represents a true detoxication pathway in humans, in contrast with animals, the contribution to the overall metabolism of (*S*)-nicotine (i.e., 4–5%) is small and is minor in comparison with the cytochrome P4502A6-catalyzed oxidation that ultimately leads to the major (*S*)-cotinine metabolic pathway products (35).

Another example of a protoxin that is metabolized in different species to metabolites that exert either greater or lesser toxic potential comes from the neurotoxin 1-methyl-4-phenyl-1,2,3,6-tetrahydropyridine (MPTP). MPTP was discovered as a contaminant in an illicit synthetic street narcotic that showed neurotoxic properties (i.e., parkinsonism) in humans. In human preparations, a significant amount of MPTP is converted to the neurotoxic metabolites 1-methyl-4-phenyl-2,3-dihydropyridinium ion ($MPDP^+$) and *N*-methyl-4-phenylpyridinium ion (MPP^+) by the sequential action of monoamine oxidase. In contrast, MPTP is metabolized by FMO in an apparent detoxication reaction (43). In a species such as the mouse, MPTP *N*-oxygenation represents a major detoxication pathway (80,81). It is possible that the relative balance of metabolic activation to $MPDP^+$ and MPP^+ versus detoxication to MPTP *N*-oxide contributes to the relative susceptibility of an animal to the neurotoxicity of MPTP. On the other hand, species-dependent differences in the type of neuronal targets that allow an animal to be more or less sensitive to the toxic properties of MPTP may be more important in the overall susceptibility to a disease state such as parkinsonism (82).

Trimethylaminuria is an autosomal recessive human disorder affecting a small subpopulation as an inherited polymorphism (83,84). Activity of human FMO3 has been proposed to be deficient in the metabolism of trimethylamine (TMA) associated with trimethylaminuria (83). Individuals diagnosed with trimethylaminuria excrete relatively large amounts of TMA in their urine, sweat, and breath, and this results in a fishy odor characteristic of TMA, hence, trimethylaminuria individuals have been designated as suffering from the ''fish odor'' syndrome (84). In humans, TMA is exclusively eliminated by metabolism to trimethylamine *N*-oxide (TMA *N*-oxide; 85; (Fig. 2). Urinary TMA *N*-oxide formation is thus a convenient and sensitive biomarker of human FMO activity. It is possible that abnormalities of the gene that encodes human FMO3 may play a substantial role in trimethylaminuria as well as other human diseases that are related to abnormal amine metabolism. For examples, individuals suffering from trimethylaminuria also have an impaired ability to metabolize (*S*)-nicotine (86).

3. *Metabolic Pathways of Other Nitrogen-Containing Compound*

In addition to the initial *N*-oxygenation by FMO, there may be other metabolic pathways that determine the overall metabolic disposition of an amine (87–90). For example, it is possible that tertiary amine *N*-oxides undergo reduction to the parent tertiary amine. In principle, this type of metabolic futile cycling could result in depletion of NADPH and could also provide a reservoir of parent amine to exert prolonged or sustained pharmacological action. For (*S*)-nicotine *N*-1′-oxide, it was observed that after intravenous injection,

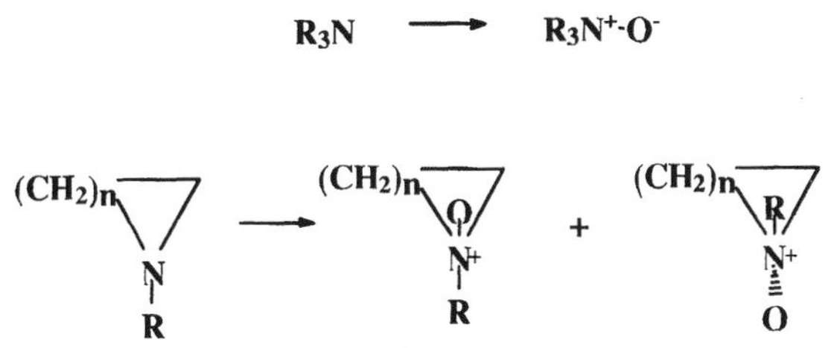

$$R_3N \longrightarrow R_3N^+\text{-}O^-$$

$(CH_2)_n$ ring N–R ⟶ $(CH_2)_n$ ring (O) N^+–R + $(CH_2)_n$ ring (R) N^+–O

Fig. 2 Schematic representation of the *N*-oxygenation of a tertiary amine (i.e., trimethylamine) or a cyclic tertiary amine to form a tertiary amine *N*-oxide.

this metabolite was excreted unchanged in the urine and was not reduced or further metabolized to other metabolites in either humans (41) or animals (78). This is in contrast to what was observed when animals were administered (*S*)-nicotine *N*-1′-oxide by the oral route (91,92). Apparently, extensive gastrointestinal bacterial reduction of (*S*)-nicotine *N*-1′-oxide was observed when the *N*-oxide was administered orally (77). It is unclear if this observation about the recalcitrance to (*S*)-nicotine *N*-1′-oxide reduction will be extended to other tertiary amine *N*-oxides, or if this represents an exception. Regardless, the possible metabolic cycling of *N*-oxides back and forth to amines is an important consideration in their overall disposition.

Another possible determinant in the overall metabolism of an amine to a relatively nontoxic metabolite, such as a tertiary amine *N*-oxide or metabolic bioactivation to a potentially more toxic material, may also depend on the propensity of the amine to undergo *N*-methylation versus *N*-oxygenation (93). For example, secondary arylamines, in principle, may be *N*-oxygenated to hydroxylamines or may undergo *N*-methylation to form the tertiary amine. For the hydroxylamine, further metabolism to form reactive esters has been implicated in the carcinogenic properties of some amines (94–96). On the other hand, formation of the tertiary amine may provide a means for efficient tertiary amine *N*-oxygenation formation, detoxication, and elimination in the urine. Another metabolic possibility is also that *N*-oxygenation of a secondary hydroxylamine to a nitrone may result in a hydrolytically unstable metabolite that efficiently decomposes to an aldehyde or a ketone, depending on the structure of the secondary amine (97,98; Fig. 3). Along these same lines, another nonenzymatic process that may occasionally be possible is for tertiary amine *N*-oxides to undergo Cope-type elimination reactions (99). Cope-type elimination reactions have been reported for a few unstable tertiary amine *N*-oxides formed by the action of FMO (100–102). It is possible that additional examples have occurred in the past, but have led to unanticipated or undetected metabolites (e.g., unrecognized rearrangement products). It is possible that the unexpected amine *N*-oxide rearrangements products (e.g., olefins and hydroxylamines) may possess unusual pharmacological or toxicological properties.

The study of the metabolism of primary amines by FMO has established that FMO2,

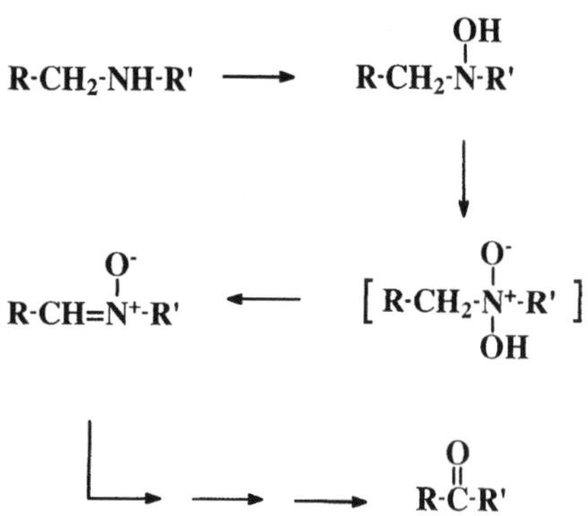

Fig. 3 Schematic representation of the *N*-oxygenation of a secondary amine to initially form a hydroxylamine that is *N*-oxygenated a second time to produce a nitrone after the elimination of water. Depending on the stability of the nitrone, hydrolysis can occur to form an aldehyde or ketone, depending on the nature of the initial secondary amine starting material.

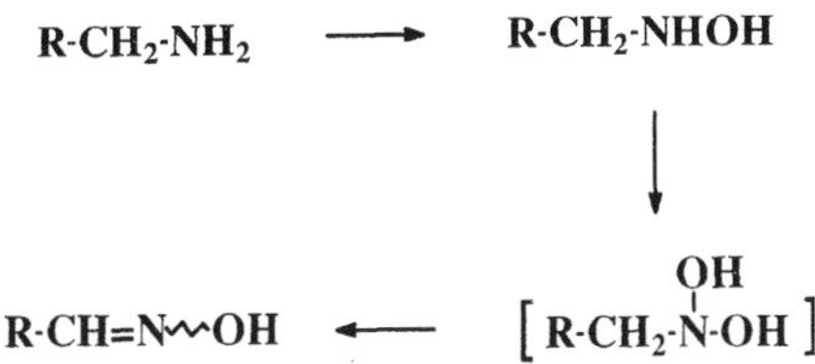

Fig. 4 Schematic representation of the *N*-oxygenation of a primary amine to form a hydroxylamine that, in turn, is sequentially *N*-oxygenated to form the putative bis-*N*,*N*-dioxygenated species that eliminates the element of water to form the relatively stable oxime (*cis*- and *trans*-isomers).

FMO3, and FMO5 are competent to catalyze the *N*-oxygenation of aliphatic primary amines. It is likely that FMO3 is most efficient and sequentially *N*-oxygenates primary amines to form the corresponding hydroxylamine. Another *N*-oxygenation of the hydroxylamine forms a symmetrical bis-*N*,*N*-dioxy intermediate that apparently decomposes with loss of water to form an oxime (90; Fig. 4). Surprisingly, FMO-mediated oxime formation often possesses very great stereoselectivity. Whether the *cis*- or the *trans*-oxime isomer forms is apparently dependent on the chemical character of the substrate. Because of the widespread importance of primary amines in biological systems, this FMO-mediated transformation may have significance from a pharmacological perspective.

III. EXPERIMENTAL MEANS OF DISTINGUISHING THE METABOLISM OF A TERTIARY AMINE OR SULFIDE BY CYTOCHROME P450 FROM THAT OF THE FLAVIN-CONTAINING MONOOXYGENASES

There are a variety of approaches at many levels to distinguish cytochrome P450 activity from that of FMO. At the organismal level, one could select an animal and procure tissue such that it would favor or disfavor the presence of FMO (1,103). The use of different animal models to assess toxicity can readily provide information about the relative role of FMO or cytochrome P450 in the metabolic bioactivation of a protoxin. In addition, induction regimens can be used to increase the amount of a particular cytochrome P450 present in the tissue of an animal without altering FMO activity (3). In general, typical cytochrome P450 induction does not markedly change the specific activity of the microsomal FMO of the animal from which it was induced. In many animals, hormonal modulation can influence the activity of FMO (104,105). Rat liver FMO levels are upregulated by testosterone and repressed by estradiol (106). Rabbit FMO2 activity correlates with progesterone and corticosterone plasma concentrations (107). In addition, gender-related differences for FMO are apparent, especially in animals (103). Thus, mouse FMO activity is very distinct for female and male animals (108,109). Another way to experimentally alter FMO activity in animals is by use of dietary modulation (110). For example, treatment of rats for 7 days with total parenteral nutrition was reported to decrease the activity of microsomal FMO isolated from these same treated animals. Similarly, control of the diet alters the pharmacokinetics of trimethylamine *N*-oxygenation and ethyl methyl sulfide *S*-oxygenation, two chemicals proposed to be selective substrates for FMO (111,112). It is not clear from these studies whether this is a direct or an indirect effect on FMO, because it was unclear whether the animals received appropriate levels of choline, an important dietary constituent that, when absent, causes widespread hepatic dysfunction (113). It is possible that dietary manipulations perturb critical cellular functions, thereby depriving the cell of NADPH and compromising the integrity of FMO. Further studies of the effect of diet on FMO activity are warranted.

IV. STUDYING FLAVIN-CONTAINING MONOOXYGENASE ENZYME ACTIVITY

Aside from treatments of animals to alter cytochromes P450 and FMO, it is likely that most of the differences in the activity of FMO from tissue preparations is due to postmortem deactivation of the original enzyme activity. Because hepatic tissue temperature can rise rapidly after death and because FMO is quite thermally labile, it is critical to immediately cool the tissue to 0–4°C on exsanquination (1,14). Even perfusion of the liver with cold Ringer solution immediatedly after removal from the animal, a procedure commonly used to remove hemoproteins in the preparation of liver tissue for evaluation of cytochrome P450 activity, significantly reduces the amount of FMO activity present. During the first few seconds after an animal has died, NADPH levels drop precipitously and because FMO activity is rapidly lost in the absence of NADPH, it is likely that the greatest percentage of FMO activity that is lost in preparations of microsomes is lost in the first few minutes of tissue procurement (14). For retention of optimal FMO activity from tissue preparations, the best way to preserve activity is to rapidly excise the tissue from the animal and immediately wrap the tissue in tin foil, plunge in liquid nitrogen to completely freeze the tissue as soon as possible, and then store at −80°C.

Another step in the microsomal preparation during which FMO activity can be lost is in disruption of the tissue. Using a blender to break open hepatocytes is generally too harsh on the tissue and results in a large loss of FMO enzymatic activity. A mild way to disrupt the tissue, while at the same time preserving its enzymatic activity, and still work at very cold temperature, is to grind the tissue into small portions with a ceramic mortar and pestle in the presence of liquid nitrogen. When the tissue is fully ground into a fine powder, the mass can be transfered to a glass handheld homogenization device and the ground tissue homogenized in the presence of ice-cold sucrose, buffered with phosphate buffer. The homogenate is thus kept at a cold temperature throughout all the steps and the tissue is gently disrupted. After homogenization of the ground-up tissue to a fine consistency, conventional fractionation by centrifugation should afford microsomal preparations with maximal FMO activity. Once the microsomal pellet is in hand, rehomogenization and recentrifugation of the initial pellet in the presence of KCl affords a microsomal pellet considerably devoid of hemoproteins arising from the blood present in the initially procurred tissue.

Even if one takes great effort to preserve FMO activity in microsomal preparations, inappropriate storage of these preparations can lead to significant loss of enzyme activity. Even storage at −80°C of microsomal fractions diluted in buffer can result in perceptible proteolysis of FMO (and cytochromes P450, as well), especially over prolonged periods. To circumvent this possibility, one useful procedure is to prepare a microsomal pellet and store directly with a blanket of frozen glycerol over the protein.

Once substantial *N*- or *S*-oxygenase activity is in hand from a particular microsomal preparation, various methods can be used to distinguish the contribution of FMO from that of the cytochromes P450 in the overall oxygenation activity. Advantage can be taken of the relative thermal lability of the FMO, compared with cytochromes P450 to distinguish which enzyme is responsible for the metabolism of a particular drug or chemical. With the exception of FMO2, those microsomal FMOs that have been examined are unstable to thermal inactivation in the presence of temperatures in excess of 45°C for 1 min or more in the absence of NADPH. Generally, placement of microsomes in an argon environment at 55°C for 1 min in the absence of NADPH decreases FMO activity 85% or more, while at the same time, reducing cytochrome P450 activity only 15% or less (34). It is important to quickly cool the microsomal sample immediately after thermal inactivation so that cytochrome P450 activity is not further impaired. Because heating microsomes generates considerable oxygen radical species, it is wise to add catalase to decompose any lipid hydroperoxides or other peroxides that may have formed during the course of the heat treatment. Alkylperoxide-mediated autooxidation of *N*- or *S*-containing compounds can be a significant confounding effect that ruins the selective inactivation efforts accomplished by heating the microsome sample. Addition of NADPH to the heat-inactivated microsomes is required to allow the normal resumption of the cytochrome P450 activity to occur.

Organic solvents commonly used in extractive workup can also contribute to autooxidation of FMO substrates. For example, the trace oxidizing contaminants in isopropanol can cause *S*-oxidation of sulfides, such as cimetidine. This example should make the investigator aware of peroxides arising from various potential autooxidation sources as a possible confounder to substrate oxidation (37). On a general note, it is better to initiate the FMO-dependent reactions by addition of substrate to a previously preequilibrated microsomal preparation containing NADPH. For cytochrome P450, often reactions have traditionally been initiated by addition of NADPH. Historically, this may have led to loss of FMO activity and confounded the picture relative to the contribution or the identity of the monooxygenases in *N*-and *S*-containing compound metabolism.

Another way to distinguish FMO from the cytochrome P450s' activity is by using a variety of microsomal incubation conditions. For example, the pH optimum for animal (14) and human (105,106) FMO-mediated *N*-and *S*-oxygenation is generally equal to or above pH 9.0 (14). On the other hand, cytochromes P450 activity is maximal at pH 7.4. In addition, solubilization of microsomes with detergent to abrogate cytochromes P450 activity at elevated pH is another approach to selectively examine FMO activity (36). Thus, treatment of microsomes with 0.2% Lubrol at pH 8.5 largely removed cytochrome P450s' activity. It is likely that not all FMOs from all species and tissues are unaffected by detergent treatments (e.g., rabbit FMO2 is probably affected differentially), but this technique apparently works for the major form of FMO present in adult human liver microsomes. Use of selective antibodies to cytochrome P450 reductase is another effective method to decrease cytochrome P450s' activity, while at the same time preserving FMO activity for selective enzyme-mediated metabolism investigations (103,116).

The use of selective substrates or inhibitors of FMO to obviate the involvement of FMO in the metabolism of a specific chemical has been confounded by the imperviousness of the enzyme to inactivation. The lack of inhibition probably stems from the nature of the FMO substrate-binding site, alluded to earlier, as well as the fundamental two-electron mechanism of the enzyme. Recently, however, several inhibitors of FMO activity have been observed, although the generality of the activity to all of the forms of FMO must be established. Nevertheless, it appears that a class of *N,N*-dimethylaminostilbene carboxylates are selective inhibitors of pig FMO1 (117) and do not operate as anionic detergent inhibitors of FMO (118). Sometimes, apparent inhibition of an FMO substrate is observed, but the mode of action is actually as an alternative competitive substrate for the FMO. The notable feature of the aminostibene carboxylates is that they appear to interfere with the NADPH-binding domain in an uncompetitive fashion or, alternatively, the aminostilbene derivatives may bind at a different site and a conformational change that is induced may prevent NADPH binding.

Another potential method to distinguish the involvement of FMO and cytochromes P450 in the *N*- or *S*-oxygenation of a chemical or a drug is to examine the stereochemistry of the product. This is possible only in the enzyme-mediated formation of a product with a stable center of chirality (119). For metabolic products of cyclic tertiary amines and aliphatic sulfides, however, this has been observed in several examples. From the data, a general, although not ironclad, rule has emerged to suggest that the stereochemistry of FMO-and cytochrome P450-mediated *N*- and *S*-oxygenations may be distinct. This is occasionally useful to distinguish the contribution of the two classes of monoxygenases. However, care must be exercised in these measurements because of the possibility that both monoxygenases could give products with the same stereochemistry.

A. Bioanalytical Approaches to the Study of Flavin-Containing Monooxygenase-Mediated Reactions

For the study of FMO-catalyzed oxygenations, it cannot be stressed too much that the synthesis, characterization, and bioanalysis of putative metabolites is key to the direct quantification of enzymatic action. There are several indirect FMO assays available (128,129), but for answering the question of whether or not a particular compound is a substrate of FMO, it is essential to have a facile synthetic operation to produce a sufficient amount of metabolite. Fortunately, the synthesis of *N*-or *S*-oxides of potential substrates of FMO is relatively straightforward. Table 1 lists several examples of the synthetic methods useful for the procurement of metabolites of FMO. In some compounds more than one site of potential chemical oxidation is present in the molecule under investigation and the use of selective-oxidizing agents is essential. Often, use of a preparation of FMO itself is an efficient way to obtain a desired metabolite when the chemical oxidation gives a

Table 1 Some Examples of Synthetic Methods to Produce FMO-Mediated Metabolites

Product	Reagent	Ref.
S-Oxide	H_2O_2, $NaIO_4$	157
S-Oxide	$NaIO_4$	58
S-Oxide	*t*BuOOH	158
S-Oxide	$NaIO_4$/BSA	120
S-Oxide	$Ti(OiPr)_4$/DET/*t*BuOOH/H_2O	122
S-Oxide	*m*CPBA, FlHOOH[a]	159
S,*S*-Dioxide	$NaIO_4$	124
RNH_2[b]	$LiAlH_4$	90
RNHOH[c]	$NaBH_3CN$	90
$R_3N^+(O^-)$	*m*CPBA	101

[a] FlHOOH, 4α-hydroperoxy-5-ethyl-3-methyl lumiflavin.
[b] Reduction of the RNHOH with $LiAlH_4$.
[c] Reduction of the oxime with $NaBH_3CN$.

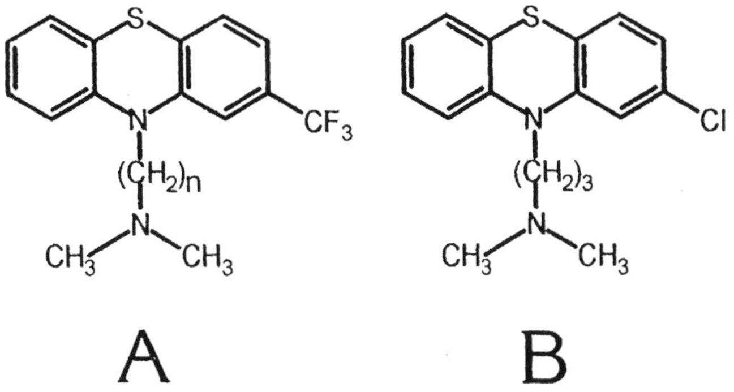

Fig. 5 Chemical structures of (A) 10-[(*N*,*N*-dimethylamino)-alkyl]-2-(trifluoromethyl)phenothiazine and (B) chlorpromazine. (From Ref. 1.)

mixture of difficult-to-separate *N*- or *S*-oxides (130). For example, the 10-([*N*,*N*-dimethylamino]alkyl)-2-(trifluoromethyl)phenothiazine substrate (Fig. 5) that has been used to functionally characterize human FMO3 and other FMOs has three potential sites of mono-oxidation, and chemical oxidation provides mixture of all three as well as some of the dioxide compounds. By employing pig liver microsomes or pig FMO1 in preparative runs, the side chain tertiary amine *N*-oxide can be readily obtained (114,115). This type of approach can be extended to the biosynthesis and purification by preparative high-performance liquid chromatography (HPLC) of any number of desired metabolites and thus provides a route to obtain sufficient material to develop an HPLC bioanalytical system for use in examining other tissues or FMOs for substrate turnover.

In addition to helping with the bioanalytical aspects of the project, synthesis of putative FMO *N*- or *S*-oxide metabolites can also provide valuable information about the aqueous stability and shelf life of a metabolite and can provide material for biological testing. The *N*- and *S*-oxide metabolites of FMO can undergo rearrangement and elimination reactions. Knowledge of the chemical stability of an FMO metabolite can be very useful in deconvoluting complicated metabolic pathways, especially if some of the materials arise from degradation or decomposition of the primary metabolite. Another area for which synthesis of the putative FMO metabolite comes in handy is in the examination of the stereoselectivity of the FMO-catalyzed reaction. Generally, chemical synthesis of *N*- and *S*-oxidation products gives a racemic mixture. Because FMO-catalyzed reactions can give exclusive stereoselectivity, procurement of the racemic material by chemical synthesis is quite useful when establishing chiral HPLC systems to separate the isomers. What is also needed is a highly efficient method for producing one or other of the stereoisomers of the *N*- or *S*-oxidized metabolites, but with recent advances in oxidative asymmetrical synthetic chemistry, this is more and more readily accomplished. Table 2 provides several examples of the use of chiral catalysts in the preparation of sufficient amounts of oxidation products of nitrogen- and sulfur-containing compounds of relevance to the study of FMO.

B. Approaches to the Study of the Stereoselectivity of the Flavin-Containing Monooxygenase

The stereoselective *N*- or *S*-oxygenation by FMO or other mammalian monooxygenases is a hallmark of enzyme action. When the substituents of a sulfide or a tertiary amine are

Table 2 Some Example of Separations of Compounds Possessing a Center of Chirality

Product	Catalyst	Separation	Ref.
(*S*)-Arylalkyl *S*-oxides	Cyclohexanone Monooxygenase, FMO5	HPLC	125,160
(*R*)-Arylalkyl *S*-oxides	FMO 1 and 2	HPLC	125,126
(*R*)-Arylakyl *S*-oxides	Chloroperoxidase, using *t*BuOOH	pTLC	158
(+) Phorate *S*-oxide	Mouse CYP-B2, CYP-PB	TLC	161
(−) Phorate *S*-oxide	Female mouse liver FMO	TLC	161
trans-2-Aryl-1,3-dithiolane *S*-oxides	Hog FMO1; rat or mouse CYP IIB	HPLC	124
trans-2-Aryl-1,3-oxathiolane *S*-oxides	Rat, hog FMO	HPLC	157
Methionine-*d*-*S*-oxide	Rat FMO3	HPLC	162
R-Flosequinan *S*-oxide	Rat FMO1	HPLC	163
cis-2-methyl-1,3-benzodithiol *S*-oxide	Rabbit FMO2; *P. putida; A. foetidus*	HPLC	120
3,5-Dimethyl-2-(3-pyridyl) thiazolidin-4-one *S*-oxide	Rat FMO1 bioreactor	HPLC	164
trans-(*S*)-Nicotine *N*-1′-oxide	Rabbit FMO2; human FMO3	HPLC	41
(+) Cimetidine *S*-oxide	Humans: human FMO3	HPLC	37,39

dissimilar, then the lone pairs of the nitrogen or sulfur atom can exist as prochiral or stereoheterotopic centers and can be oxidatively converted to *N*- or *S*-oxides that may exist as enantiomers or diastereomers. The pyramidal nature of the *N*- or *S*-oxide metabolites that are formed enzymatically may be sufficiently stable to spontaneous racemization and allow the determination of optical activity during FMO-catalyzed reactions. Generally, if the *N*- or *S*-oxide is stable and if a center of chirality has been generated, then the determination of enzyme-mediated stereoselectivity boils down to an exercise in bioanalytical problem-solving and determining the optical purity of a minute amount of an enzyme reaction.

A key step in determining FMO enzyme stereoselectivity is to determine the true optical purity of the enzyme reaction in the face of confounding contributions from nonenzymatic oxidizing agents. This is especially true when examining the stereoselectivity of crude preparations, such as hepatic microsomes, but it is also true for doing product stereoselectivity determinations from other complex biological matrices such as urine. Nonenzymatic oxidation of readily oxidizable *N*-and *S*-containing compounds is a continual problem in the investigation of the stereoselectivity of this class of compounds. Probably the most significant problem in this general area of investigation, however, is the contribution of multiple monooxygenases that are oftentimes present in multiple enzyme forms, affording mixed stereoselectivity and permitting ambiguity in the determination of FMO enzyme stereoselectivity. Care must be exercised in the interpretation of stereoselectivity results and it is useful to determine stereoselectivity under a variety of conditions.

1. Determination of Optical Purity in Flavin-Containing Monooxygenase-Catalyzed Reactions

In principle, several methods are available for the determination of the optical purity of FMO-catalyzed reactions. Stereoselectivity can be measured by ^{1}H NMR spectroscopy of a chiral metabolite in the presence of a chiral auxiliary. Gas–liquid chromatography (GLC) using a chiral column has been used in the separation of lipophilic *S*-oxides, but this is not a routine method of choice for the determination of optical purity. Another method for the determination of absolute configuration is the use of optical rotatory dispersion and circular dichroism; the optical purity of several metabolites has been determined by this means. The most unambiguous method to determine absolute configuration of *N*- or *S*-oxides is by X-ray crystallographic means. However, from a practical point of view, obtaining sufficient amounts of metabolite of a high enough quality makes the use of this approach rare. Currently, the most common method to determine the absolute configuration of an FMO-mediated *N*- or *S*-oxidation is by chiral-phase HPLC (119).

Chiral stationary-phase HPLC is the method of choice for the rapid and relatively inexpensive determination of the absolute configuration of metabolites of FMO. Because only submicrogram amounts of metabolite are generally required for chiral-phase HPLC, this represents about a 1 to 10 thousandfold increase in sensitivity, over other techniques, such as NMR, to determine and quantify centers of chirality. By integrating the relative peak areas of the HPLC chromatograms, a minimum amount of material can be evaluated from an enzymatic incubation after an extractive workup, to provide the relative stereoselectivity information. Currently, three major commercial firms supply most of the chiral stationary-phase materials: the Pirkle columns (Regis Technologies, Morton Grove, IL): the Chiralcel columns (Chiral Technologies, Exton, PA), and the macrocycle-based columns (Advanced Separation Technologies, Whippany, NJ). All do a reasonable job at separation of sulfoxides that contain a center of chirality. Table 2 lists several examples of separations of FMO metabolites by chiral-phase HPLC that have been useful in determining enzyme stereoselectivity. Sometimes, chiral separations can be limited owing to the nature of the solvents required to chromatograph polar *S*-oxide metabolites, for example. For some solvents, it is possible that the fundamental chiral recognition mechanism normally used for diastereomeric complexation, is disrupted, and chiral separation appears to be impossible. However, with the development and introduction of new chiral matrices, progress in the determination of the stereoselectivity of FMO-related reactions should increase in the future.

The determination of stereoselectivity for monooxygenase-related reactions is important because it provides a simple means of segregating the enzyme-dependent processes from the nonenzymatic oxidations that can be quite significant for *N*-, *S*-, or other heteroatom-containing compounds. Depending on the tissue and the way FMO activity was prepared, significant loss of product stereoselectivity can be observed owing to nonenzymatic processes. For example, hydrogen peroxide or molecular oxygen-derived free radical species often efficiently convert amines, sulfides, and phosphorus-containing compounds to racemic *N*-, *S*-, or *P*-oxides, respectively. Sulindac sulfide is stereoselectively converted to its *S*-oxide in the presence of FMO (125,131), but free radical processes also can contribute to *S*-oxidation. Because generation of peroxides from uncoupling monooxygenases is at least partly dependent on NADPH, indirect assays of substrate turnover by monitoring NADPH consumption or molecular oxygen uptake may indicate only a portion of the overall enzyme-dependent process. For tissue preparations that produce substantial quantities of active oxygen species, it is important to rule out the involvement of hydrogen peroxide, alkyl hydroperoxides, hydroxyl radicals, and superoxide by employing catalase, iron chelators, and superoxide dismutase, respectively, during the metabolic incubation. The use of diethylenetriaminepentaacetic acid (DETAPAC) to chelate metals and prevent autooxidative processes in metabolic reactions has recently found widespread use.

Generally, in compounds that possess both nucleophilic nitrogen and sulfur atoms, it is the sulfur atom that is the first site of FMO-mediated oxygenation. Likewise, sulfur-containing compounds are more prone to autooxidation that nitrogen-containing compounds and this can be observed in stereoselectivity studies; for example, in humans, (*S*)-nicotine is not readily subject to autooxidation and *N*-oxygenation is absolutely stereoselective, forming 100% of the *trans*-isomer whereas cimetidine is relatively easily autooxidized and the major metabolite in humans does not give exclusively one *S*-oxide enantiomer. In summary, because of the numerous enzymatic and nonenzymatic pathways that can contribute to the oxidation of *S*-, *N*-, *P*-, and other heteroatom-containing substrates, use of multiple approaches to the elucidation of the enzyme-dependent oxidation should be employed including stereoselectivity studies.

V. RECENT IN VITRO MICROSOME STUDIES

One of the important reasons to obtain a full picture of the in vitro metabolism of a chemical or drug in animals or humans is to use this information to make predictions about in vivo pharmacokinetics. With a clear understanding of the molecular basis for the metabolic clearance and pharmacokinetic parameters, extrapolation to in vivo conditions of humans is possible (2,3). In addition to providing information about the disposition of a chemical, in vitro information is useful to determine the potential toxicities and predict possible drug–drug interactions. In addition to providing insight into possible toxicity, in vitro–in vivo correlative data can be used to predict the animal model that would be most appropriate to use in drug development studies (134). Interspecies scaling can be used to extrapolate from animals to humans, and this may reduce the number of animals as well as humans used in drug development paradigms (135). The general approach to in vitro–in vivo correlations is outlined in the following:

For in vitro–in vivo correlations, hepatic clearance is an important parameter for predicting in vivo clearance from in vitro data. Assuming the instantaneous and complete mixing of the chemical and the blood of the animal, hepatic clearance (CL_h) can be expressed as

$$CL_h = \frac{Q_h f_u CL_i}{Q_h + (f_u CL_i)} \tag{1}$$

where Q_h is the hepatic bloodflow, f_u is the fraction of the unbound drug, and CL_i is the intrinsic ability of the organ to clear the chemical. Because CL_h is the product of hepatic bloodflow and the hepatic extraction ratio ER_h, the hepatic extraction ratio can be written as

$$ER_h = \frac{f_u CL_i}{Q_h + (f_u CL_i)} \tag{2}$$

For chemicals such as sulfides with a high hepatic extraction ratio (i.e., a drug that is efficiently *S*-oxygenated), clearance can approach hepatic bloodflow. Because the intrinsic ability of an organ to *S*-oxygenate a sulfide, for example, may be related to the activity of the FMO enzymes, Eq. (2) can be related to the Michaelis–Menten equation:

$$\text{Rate of metabolism} = \frac{V_{max} C}{K_{mapp} + C} \tag{3}$$

If it is assumed that V_{max} is the maximum rate at which the drug is oxidized and K_{mapp} is the apparent Michaelis–Menten constant, dividing both sides of Eq. (3) by the systemic concentration of the drug C gives Eq. (4).

$$\frac{\text{Rate of metabolism}}{C} = \text{CL}_{\text{m}} = \frac{V_{\text{max}}}{K_{\text{mapp}} + C} \tag{4}$$

Equation five shows the classic relation between the Michaelis–Menten equation and pharmacokinetics:

$$f_{\text{u}}\text{CL}_{\text{i}} = \frac{V_{\text{max}}}{K_{\text{mapp}} + C} \tag{5}$$

Thus, determination of the K_{mapp} for an FMO-catalyzed reaction for drugs with a low extraction ratio from an in vitro experiment can provide a means to predict in vivo clearance parameters. FMO is a three-substrate enzyme that requires NADPH and molecular oxygen in addition to the substrate; therefore the K_m value is only an apparent one. FMO-mediated oxygenation may not be the only metabolic path for formation of the observed product in vitro and hepatic clearance may not be the only process leading to formation of the metabolite in vivo (e.g., renal, pulmonary, intestinal, or other organs may also contribute to elimination). However, the relatively inexpensive determination of in vitro kinetic parameters can lead to a very powerful method for estimating relatively costly in vivo pharmacokinetic data.

Sometimes, it is possible that the in vitro data (e.g., Lineweaver Burk kinetic analysis) suggest that more than one pathway is contributing to the formation of the metabolite (136). If that is so, then using other kinetic treatments (e.g., Eadie Hofstee kinetic analysis) can help quantify the different contributions. In addition, the use of selective functional inhibitors of monooxygenase activity to largely abrogate one enzyme activity often allows the investigator to identify other enzyme activities contributing to product formation. The critical aspect of any in vitro correlation is to have on hand sufficient amounts of microsomes that have been carefully phenotyped for selective functional monooxygenase activity as well as fully characterized for enzyme immunoreactivity. By using regression analysis, one can compare the profile of metabolism of a compound in a library of microsome preparations that have been characterized for FMO or cytochromes P450, correlate monooxygenase activity and immunoreactivity, and ascertain whether one or more enzymatic activity correlates with the product formation of the compound of interest (2,3). Correlation analysis has been performed in the presence of a few human liver microsome libraries examining several FMO-dependent *N*- and *S*-oxygenations, with some success (3,35,36,39,75,137). As additional human liver microsome libraries are characterized for FMO activity and immunoreactivity, it is likely that more and more compound oxygenations will be identified as dependent on human FMO.

VI. SPECIES COMPARISON FOR FMO ACTIVITY

By using different species as a source of tissue from which to prepare FMO activity, it is possible to enrich the microsomal preparation with one or another FMO form (138). From inspection of Table 3, it is clear that tissues of some species have a high titer of

Table 3 Animal and Human Tissue with Prominent Forms of FMO Enzymes Present[a]

Animal	Liver	Lung	Kidney	Intestine
Mouse	FMO3	FOM2	FMO1 and 3	?[b]
Rat	FMO1	FMO2	FMO1 and 3	?
Rabbit	FMO1	FMO2	FMO1 and 3	FMO1
Monkey	FMO3	FMO2	FMO1 and 3	?
Human	FMO3	NP[c]	FMO1 and 3	FMO1

[a] Estimates from references cited in (1).
[b] A question mark indicates that no data are available or the presence of an FMO form is in doubt.
[c] NP, apparently not present; FMO2 is almost completely absent in the Caucasian human lung tissue examined.

largely one form of FMO. For example, pig liver appears to have only FMO1 present. There are a number of anecdotal stories about the type of pig that may give the highest FMO activity (e.g., an animal in estrous or one without red hair), but as described earlier, the most important aspect of preserving FMO activity is to prevent the tissue from warming up—especially in the absence of NADPH. There are other examples of particular tissues enriched with a particular type of FMO activity. Rabbit lung microsomes prepared from pregnant rabbits have elevated amounts of rabbit FMO2 present (139). Female mice apparently have significantly greater amounts of hepatic FMO3 than male mice. Rat liver does not appear to have as high a level of FMO3 as humans, but rats may have another form of FMO present in addition to FMO1 that is similar to FMO1. Because humans (39,140) and nonhuman primates (141,142) have FMO3 as the major form of FMO present, the choice of an appropriate small animal model of human drug metabolism should be made with great care (143).

A. FMO and Cytochrome P450 Enzymology: Drug Interactions

The fundamental mechanisms of FMO (i.e., two-electron oxygenation) and cytochromes 450 (i.e., sequential one-electron oxidation) are quite distinct. Often this mechanistic imperative is manifested in different products being formed from the same substrate. Sometimes, the same product is formed by the two enzymes. The significance of two enzymes making the same product may have consequences for drug interactions (4,5).

The term *drug interactions* means different things to the clinician, the basic researcher, and the patient. Drug interactions have been implicated in various situations resulting in serious human conditions, including seizures, hemorrhage, anaphlaxis, and even death. Most cases reported involving drug metabolism-dependent drug interactions have involved cytochromes P450. An increasing number of drug interactions may involve the UDP-glucuronosyltransferases and other conjugative enzymes (133). Recently, a few reports have emerged to implicate the FMOs in drug interactions.

The formation by FMO3 of human metabolites that constitute a major fraction of the excreted drug administered has been shown in only a few cases. It is likely that as the metabolic involvement of human FMO3 is more vigorously examined, additional examples of major participation of human FMO3 in human drug metabolism will become more apparent. In those instances for which human FMO3 has been implicated in drug metabolism [e.g., cimetidine, (*S*)-nicotine, tamoxifen, and olanzapine] the use of other drugs in vivo that are also good substrates for FMO3 may slow the overall rate of drug metabolism. This may be true if the drug concentration exceeds the K_m for the FMO3 substrate. Significant drug–drug interaction caused by the inhibition of FMO3-mediated metabolism with concomitant use of another drug may become more important as more examples of human FMO3-dependent metabolism are observed.

Epidemiological studies have suggested that consumption of cruciferous vegetables may result in a decreased cancer risk (144,145). The effect of consumption of glucosinolate-containing brussels sprouts or indole-3-carbinol (a major plant alkaloid abundant in cruciferous vegetables) on various metabolic enzymes has been examined, and it has been observed that ingestion of brussels sprouts is anticarcinogenic and increases the activity of glutathione transferase in animals and in humans (146). Indole-3-carbinol (or more likely, the acid condensation products of indole-3-carbinol) apparently binds to the Ah receptor and induces cytochromes P450 1A1, 1A2 (148), and 2B1/2 and 3A (149). In addition, dietary ingestion of indole-3-carbinol produces a dose- and time-dependent repression of both selective functional FMO activity and FMO1 protein levels in the liver and intestine of male Fischer 344 rats. If the same phenomenon is occurring in humans, which results in concurrent induction of cytochrome P450 and repression of FMO, it is possible that the metabolism, pharmacokinetics, and hence, pharmacological activity and potential toxicity of a number of drugs and xenobiotics could be dramatically altered. The influence of diet in human FMO activity may provide interesting insight into the physiological role of FMO.

VII. FUTURE DIRECTIONS

There has been significant progress reported in the literature in the past 6 years that has provided insight into the molecular and structural biology of FMO. It is likely that FMO will receive increased evaluation as its role in the metabolism and disposition of chemicals, drugs, and endogenous materials becomes more apparent. The continued use of in vitro–in vivo correlations will also likely increase and lead to a fuller understanding of a role of FMO in human drug and chemical metabolism. The recognition of a more complete understanding of FMO in human drug metabolism will ultimately provide advances in the development of human therapeutics. This fundamental information will undoubtedly lead to a better understanding of drug–drug interactions and presumably this knowledge may lead to the development of safer human drugs.

Although significant progress has been made toward the elucidation of the structure and regulation of FMO, it is likely that additional advances will continue to emerge. For example, the structure and underlying mechanism of regulation of the *FMO* gene is just beginning to emerge. From the elegant studies of Hines and co-workers, the results suggest that it is likely that *FMO* gene will be quite large and regulation may be quite distinct for FMOs from different tissues. Nontraditional tissue-specific expression mechanisms may be in operation for different FMOs. Regulation may involve the utilization of tissue-specific transcription factors. Regardless, knowledge of a more detailed structure of the *FMO* gene may provide important information on the tissue-specific regulation of metabolism of xenobiotics or endogenous materials by FMO.

Another area of research that is certain to see extensive developments concerns the role of extrahepatic FMO in human and animal drug and chemical metabolism. For example, several recent reports have provided information about the nature of FMO in the brain (150–154), intestine (107), skin (155), nasal (107), and other tissues. With the observation that FMO metabolizes several endogenous materials, including primary amines (90) and sulfur-containing compounds (1), the role of brain FMO in endogenous substrate metabolism may become quite important. Evidence for a distinct type of FMO has been observed in rat brain (150–153). In rabbit brain, the presence of FMO4 has been reported (154). If brain FMO metabolizes biogenic amines efficiently, it is possible that FMO may contribute to the pharmacological activity of this important class of endogenous compound.

Preparations of cDNA-expressed FMO from lower organisms have played an important role in determining the contribution of FMO to substrate specificity and other metabolic processes. The use of recombinant FMO technology has provided enzyme in substantially pure form, especially in species for which tissues are thermally labile or difficult to procure, and the use of cDNA-expressed FMO is certain to increase in the future. In addition to providing sufficient quantities of FMO for substrate specificity studies, it is likely that recombinant FMO may provide enzyme to examine the metabolism of unusual substrates and novel mechanisms of action. For example, the recent report that FMO catalyzes the oxidation of aryl aldehydes by Baeyer-Villiger type chemistry (156) suggests that future studies of FMO may show other novel and selective catalytic properties. cDNA-expressed FMO and variant forms of FMO may provide useful catalysts that could find applications in the chemical or biotechnology industry.

Finally, the role of FMO in human disease conditions is certain to be studied in increasing detail in the future. The abrogation of human FMO3 has been associated with trimethylaminuria, but it is likely that variant forms of FMO are associated with other abnormal metabolic conditions because trimethylaminuria patients appear to suffer from a number of other disease states as well. With the emergence of new and sensitive selective molecular probes of human FMO, it is possible that a role of FMO in human disease conditions will be clarified.

ACKNOWLEDGMENT

I acknowledge the many collaborators and co-workers in my laboratory and in the laboratories of collaborating facilities that have made valuable contributions to the work discussed. The financial support of the National Institutes of Health (GM36426) is gratefully acknowledged.

REFERENCES

1. J. R. Cashman, Structural and catalytic properties of the mammalian flavin-containing monooxygenase, *Chem. Res. Toxicol. 8*:165 (1995).
2. J. R. Cashman, Drug discovery and drug metabolism, *Drug Discov. Today 1*:209 (1996).
3. S. A. Wrighton, M. Vandendranden, J. C. Stevens, L. A. Shipley, B. J. Ring, A. E. Rettie,

and J. R. Cashman. In vitro methods for assessing human hepatic drug metabolism: Their use in drug development, *Drug Metab. Rev.* *25*:453 (1993).
4. C. C. Peck, R. Temple, and J. M. Collins, Understanding consequences of concurrent therapies, *JAMA.* *269*:1550 (1993).
5. G. T. Tucker, The rational selection of drug interaction studies: Implication of recent advances in drug metabolism, *Int. J. Clin. Pharmacol. Ther.* *30*:550 (1992)
6. J. R. Cashman, Monoamine oxidase and flavin-containing monooxygenases, *Comprehensive Toxicology*, vol. 3, pp. 69–96 (F. P. Guengerich, ed.), Elsevier, New York, 1997, Chap. 6.
7. C. Kemal, T. W. Chan, and T. C. Bruice, Reaction of $^3O_2$ with dihydroflavins. 1. $N^{3,5}$-Dimethyl-1,5-dihydrolumiflavin and 1,5-dihydroisoalloxazines, *J. Amr. Chem. Soc.* *99*:7272 (1977).
8. K. C. Jones and D. P. Ballou, Reactions of the 4a-hydroperoxide of liver microsomal flavin-containing monooxygenase with nucleophilic and electrophilic substrates, *J. Biol. Chem.* *261*: 2553 (1986).
9. N. B. Beaty and D. P. Ballou, The oxidative half-reaction of liver microsomal FAD-containing monooxygenase, *J. Biol. Chem.* *256*:4619 (1981)
10. L. L. Poulsen and D. M. Ziegler, The liver microsomal FAD-containing monooxygenases. Spectral characterization and kinetic studies, *J. Biol. Chem.* *254*:6449 (1979)
11. C. Walsh, Flavin coenzymes: At the crossroads of biological redox chemistry, *Acc. Chem. Res.* *13*:148 (1980).
12. J. W. Hastings, C. Balny, C. LePeuch, and P. Douzou, Spectral properties of an oxygenated luciferase–flavin intermediate isolated by low temperature chromatography, *Proc. Natl. Acad. Sci. USA.* *70*:3468 (1973).
13. L. L. Poulsen and D. M. Ziegler, Multisubstrate flavin-containing monooxygenases: Application of mechanism to specificity, *Chem. Bio. Interact.* *96*:57 (1995).
14. D. M. Ziegler, Microsomal flavin-containing monooxygenase: Oxygenation of nucleophilic nitrogen and sulfur compounds, *Enzymatic Basis of Detoxication* (W. B. Jakoby, ed.), Academic Press, New York, 1980, p. 201.
15. N. B. Beaty and D. P. Ballou, The reductive half-reaction of liver microsomal FAD-containing monooxygenase, *J. B. Chem.* *256*:4611 (1981).
16. D. M. Ziegler, Flavin-containing monooxygenases: Catalytic mechanism and substrate specificities, *Drug Metab. Rev.* *19*:1 (1988).
17. T. C. Bruice, J. B. Noar, S. S. Ball, and U. V. Venkataram, Monooxygen donation potential of 4-a-hydroperoxyflavins as compared with those of percarboxylic acid and other hydroperoxides. Monooxygen donation to olefin, tertiary amine, alkyl sulfide, and iodide ion, *J. Am. Chem. Soc.* *105*:2452 (1983).
18. D. M. Ziegler, Flavin-containing monooxygenases: Enzymes adapted for multisubstrate specificity, *Trends Pharmcol. Sci.* 2:321 (1990).
19. T. C. Bruice, Mechanism of flavin catalysis, *Acc. Chem. Res.* *13*:256 (1980).
20. R. N. Hines, J. R. Cashman, R. M. Philpot, D. E. Williams, and D. M. Ziegler, The mammalian flavin-containing monooxygenases: Molecular characterization and regulation of expression, *Toxicol. Appl. Pharmacol.* *125*:1 (1994).
21. M. P. Lawton, J. R. Cashman, T. Cresteil, C. Dolphin, A. Elfarra, R. N. Hines, E. Hodgson, T. Kimura, J. Ozols, I. Phillips, R. M. Philpot, L. L. Poulsen, A. E. Rettie, D. E. Williams, and D. M. Ziegler, A nomenclature for the mammalian flavin-containing monooxygenase gene family based on amino acid sequence identities, *Arch. Biochem. Biophys.* *308*:254 (1994).
22. J. Ozols, Isolation and structure of a third form of liver micosomal flavin-monooxgenase, *Biochemistry.* *33*:3751 (1994).
23. S. Guan, A. M. Falick, D. E. Williams, and J. R. Cashman, Evidence for complex formation between rabbit lung flavin-containing monooxygenase and calreticulin, *Biochemistry* *30*:9892 (1991).

24. N. A. Donoghue, D. B. Norris, and P. W. Trudgill, The purification and properties of cyclohexanone oxygenase from *Nocardia globerula* CL1 and *Acinetobacter* NCIB 9871, *Eur. J. Pharmacol. 63*:175 (1996).
25. C. C. Ryerson, D. P. Ballou, and C. Walsh, Mechanistic studies on cyclohexanone oxygenase, *Biochemistry 21*:2644 (1982).
26. D. M. Ziegler, Recent studies on the structure and function of multisubstrate flavin-containing monooxygenases, *Annu. Rev. Pharmacol. Toxicol. 33*:179 (1993).
27. E. Hodgson and P. E. Levi, The role of the flavin-containing monooxygenase in the metabolism and mode of action of agricultural chemicals, *Xenobiotica 22*:1175 (1992).
28. F. P. Guengerich, Cytochrome P-450 enzymes and drug metabolism, *Prog. Drug Metab. 10*: 1 (1987).
29. S. A. Wrighton and J. C. Stevens, The human hepatic cytochromes P-450 involved in drug metabolism, *Crit. Rev. Toxicol. 22*:1 (1992).
30. D. M. Ziegler and M. S. Gold, Oxidative metabolism of tertiary amines by human liver tissue, *Xenobiotica 1*:325 (1971).
31. M. S. Gold and D. M. Ziegler, Dimethylaniline *N*-oxidase and aminopyrine *N*-demethylase activities of human liver tissues, *Xenobiotica 3*:179 (1973).
32. M. E. McManus, I. Stupans, W. Burgess, J. A. Koenig, P. Dela, M. Hall, and D. J. Birkett, Flavin-containing monooxygenase activity in human liver microsomes, *Drug Metab. Dispos. 15*:256 (1987).
33. A. Lemoine, M. Johann, and T. G. Cresteil, Evidence for the presence of distinct flavin-containing monooxygenases in human tissue, *Arch. Biochem. Biophys. 276*:336 (1990).
34. J. R. Cashman, A.-C. Yang, L. Yang, and S. Wrighton, Stereo- and regioselective *N*- and *S*-oxidation of tertiary amines and sulfides in the presence of adult human liver microsomes, *Drug Metab. Dispos. 25*:453 (1993).
35. J. R. Cashman, S. B. Park, Z.-C. Yang, S. A. Wrighton, P. Jacob III, and N. L. Benowitz, Metabolism of nicotine by human liver microsomes: Stereoselective formation of *trans* nicotine *N′*-oxide, *Chem. Res. Toxicol. 5*:639 (1992).
36. A. J. M. Sadeque, A. C. Eddy, G. P. Meier, and A. E. Rettie, Stereoselective sulfoxidation by human flavin-containing monooxygenase, *Drug Metab. Dispos. 20*:832 (1992).
37. J. R. Cashman, S. B. Park, Z.-C. Yang, C. B. Washington, D. Y. Gomez, K. M. Giacomini, and C. M. Brett, Chemical, enzymatic and human enantioselective *S*-oxygenation of cimetidine, *Drug Metab. Dispos. 21*:587 (1993).
38. J. C. Stevens, L. A. Shipley, J. R. Cashman, M. Vandenbranden, and S. A. Wrighton, Comparison of human and rhesus monkey in vitro phase I and phase II hepatic drug metabolism, *Drug Metab. Dispos. 21*:753 (1993).
39. J. R. Cashman, S. B. Park, C. E. Berkman, and L. E. Cashman, Role of hepatic flavin-containing monooxygenase 3 in drug and chemical metabolism in adult humans, *Chem. Biol. Interact. 96*:33 (1995).
40. J. R. Cashman, S. B. Park, A.-C. Yang, S. A. Wrighton, P. Jacob III, and N. L. Benowitz, Metabolism of nicotine by human liver microsomes: Stereoselective formation of *trans* nicotine *N′*-oxide, *Chem. Res. Toxicol. 5*:639 (1992).
41. S. B. Park, P. Jacob III, N. L. Benowitz, and J. R. Cashman, Stereoselective metabolism of (*S*)-(−)-nicotine in humans: Formation of *trans* (*S*)-(−)-nicotine *N*-1′-oxide, *Chem. Res. Toxical. 6*:880 (1993).
42. D. M. Ziegler, *S*-Oxygenases I—chemistry and biochemistry, *Sulfur-Containing Drugs and Related Organic Compounds* (L. A. Damani, ed.), Ellis Horwood, Chichester, 1989, p. 53.
43. R. Ayesh, S. C. Mitchell, A. Zhang, and R. L. Smith, The fish odour syndrome: Biochemical, familial, and clinical aspects, *Br. Med. J. 307*:655 (1993).
44. L. L. Poulsen, The multisubstrate FAD-containing monooxygenase, *Chemistry and Biochemistry of Flavoenzymes* (F. Miller, ed.), CRC Press, Boca Raton, FL, 1991, p. 87.

45. C. J. Decker and D. R. Doerge, Rat hepatic microsomal metabolism of ethylene thiourea. Contributions of the flavin-containing monooxygenase and cytochrome P-450 isozymes, *Chem. Res. Toxicol. 4*:482 (1991).
46. C. J. Decker, D. R. Doerge, and J. R. Cashman, Metabolism of benzimidazoline-2-thiones by rat hepatic microsomes and hog liver flavin-containing monooxygenase, *Chem. Res. Toxicol. 5*:726 (1992).
47. L. L. Poulsen, R. M. Hyslop, and D. M. Ziegler, *S*-Oxygenation of *N*-substituted thioureas catalyzed by the pig liver microsomal FAD-containing monooxygenase, *Arch. Biochem. Biophys. 198*:78 (1979).
48. R. P. Hanzlik and J. R. Cashman, Oxidation and other reactions of thiobenzamide derivatives of relevance to their hepatotoxicity, *J. Org. Chem. 47*:4645 (1982).
49. C. J. Decker and D. R. Doerge, Covalent binding of ^{14}C and ^{35}S-labeled thiocarbamides in rat hepatic microsomes, *Biochem. Pharmacol. 43*:881 (1992).
50. M. C. Dryoff and R. A. Neal, Identification of the major protein adduct formed in rat liver after thioacetamide administration, *Cancer Res. 41*:3430 (1981).
51. A. E. Miller, J. J. Bischoff, and K. Pae, Chemistry of aminoiminomethanesulfinic and sulfonic acids related to the toxicity of thioureas. *Chem. Res. Toxicol. 1*:169 (1988).
52. R. A. Neal, Microsomal metabolism of thiono-sulfur compounds: Mechanisms and toxicological significance, *Rev. Biochem. Toxicol. 2*:131 (1980).
53. E. Chiele and G. Malvaldi, Role of the microsomal FAD-containing monooxygenase in the liver toxicity of thioacetamide *S*-oxide, *Toxicology 31*:41 (1984).
54. C. J. Decker, J. R. Cashman, K. Sugiyama, D. Maltby, and M. A. Correia, Formation of glutathionyl-spironolactone disulfide by rat liver cytochrome P-450 or hog liver monooxygenase: A functional probe of two-electron oxidations of the thiosteroid? *Chem. Res. Toxicol. 4*:669 (1991).
55. G. L. Kedderis and D. E. Rickert, Loss of a liver microsomal cytochrome P-450 during methimazole metabolism. Role of flavin-containing monooxygenase, *Drug Metab. Dispos. 13*:58 (1985).
56. R. P. Hanzlik and J. R. Cashman, Microsomal metabolism of thiobenzamide and thiobenzamide *S*-oxide, *Drug Metab. Dispos. 11*:201 (1983).
57. J. R. Cashman, Thioamides, *Sulfur-Containing Drugs and Related Organic Compounds*, Vol. 1, Part B (L. A. Damani, ed.), Ellis Horwood, Chichester, 1983 p. 35.
58. J. R. Cashman, L. D. Olsen, R. S. Nishioka, E. S. Gray, and H. A. Bern, *S*-Oxygenation of thiobencarb (Bolero) in hepatic preparations from striped bass (*Morone saxatilis*) and mammalian systems, *Chem. Res. Toxicol. 3*:433 (1990).
59. P. A. Kreiter, D. M. Ziegler, K. E. Hill, and R. P. Burk, Increased biliary GSSG efflux from rat livers perfused with thiocarbamide substrates for the flavin-containing monooxygenases, *Mol. Pharmacol. 26*:122 (1984).
60. D. M. Ziegler, Role of reversible oxidation–reduction of enzyme thiol-disulfides in metabolic regulation, *Annu. Rev. Biochem. 54*:305 (1985).
61. R. P. Hanzlik, Chemistry of covalent binding: Studies with bromobenzene and thiobenzamide, *Adv. Exp. Med. Biol. 187*:31 (1986).
62. D. H. Hutson, *S*-Oxygenation in herbicide metabolism in mammals, *Sulfur in Pesticide Action and Metabolism*. ACS Symp. *Ser.* American Chemical Society, Washington, DC, 1981, p. 65.
63. D. M. Ziegler, L. L. Poulsen, and B. M. York, Role of the flavin-containing monoxygenase in maintaining cellular thiol:disulfide balance, *Function of Glutathione: Biochemical, Physiological, Toxicological and Clinical Aspects* (A. Larson, et al., eds.), Raven Press, New York, 1983, p. 297.
64. S. B. Park, W. N. Howald, and J. R. Cashman, *S*-Oxidative cleavage of farnesylcysteine and farnesylcysteine methyl ester by the flavin-containing monooxygenase, *Chem. Res. Toxicol. 7*:191 (1994).

65. S. B. Park, J. D. Osterloh, S. Vamvakas, M. Hashimi, M. W. Anders, and J. R. Cashman, Flavin-containing monooxygenase-dependent stereoselective *S*-oxidation and cytotoxicity of cysteine *S*-conjugates and mercapturates, *Chem. Res. Toxicol. 5*:193 (1992).
66. D. M. Ziegler and L. L. Poulsen, Hepatic microsomal mixed function amine oxidase, *Methods Enzymol.* 52 (part C):142 (1978).
67. D. M. Ziegler, Molecular basis for *N*-oxygenation of *sec*- and *tert*-amines, *Proceedings of the Third International Symposium on the Biological Oxidation of Nitrogen in Organic Molecules* (L. A. Damani and J. W. Gorrod, eds.) Ellis Horwood, Chichester, 1985, p. 40.
68. D. M. Ziegler, Mechanism, multiple forms and substrate specificities of flavin-containing monooxygenases, N-*Oxidation of Drugs* (P. Hlavica and L. A. Damani, eds.), Chapman & Hall, 1991, p. 59.
69. P. Hlavica and I. Golly, On the genetic polymorphism of the flavin-containing monooxygenase, N-*Oxidation of Drugs* (P. Hlavica and L. A. Damani, eds.) 1991, p. 71.
70. L. A. Damani, W. F. Pool, P. A. Crooks, R. K. Kaderlik, and D. M. Ziegler, Stereoselectivity in the *N'*-oxidation of nicotine isomers by flavin-containing monooxygenase, *Mol. Pharmacol. 33*:702 (1988).
71. D. E. Williams, R. L. Reed, B. Kedzierski, F. P. Guengerich, and D. R. Buhler, Bioactivation and detoxication of the pyrrolizidine alkaloid senecionine by cytochrome P-450 isozymes in rat liver, *Drug Metab. Dispos. 17*:387 (1989).
72. D. E. Williams, R. L. Reed, B. Kedzieski, D. M. Ziegler, and D. R. Buhler, The role of flavin-containing monooxygenase in the *N*-oxidation of the pyrrolizidine alkalid senecionine, *Drug Metab. Dispos. 17*:380 (1989).
73. K. Yuno, H. Yamada, K. Oguri, and H. Yoshimura, Substrate specificity of guinea pig liver flavin-containing monooxygenase for morphine, tropane, and strychnos alkaloids, *Biochem. Pharmacol. 40*:2380 (1990).
74. C. L. Mirand, W. Cung, R. E. Reed, X. Zhao, M. C. Henderson, J.-L. Wang, D. E. Williams, and D. R. Buhler, Flavin-containing monooxygenase: A major detoxifying enzyme for the pyrrolizidine alkaloid senecionine in guinea pig tissues, *Biochem. Biophys. Res. Commun. 178*:546 (1991).
75. C. E. Berkman, S. B. Park, S. A. Wrighton, and J. R. Cashman, In vitro–in vivo correlations of human (*S*)-nicotine metabolism, *Biochem. Pharmacol. 50*:565 (1995).
76. P. Jacob III, N. L. Benowitz, and A. J. Shulgin, Recent studies of (*S*)-nicotine metabolism in humans, *Pharmacol. Biochem. Behav. 30*:249 (1988).
77. A. M. Beckett, J. W. Gorrod, and P. Jenner, The analysis of nicotine-1′-oxide in urine, in the presence of nicotine and cotinine and its application to the study of in vivo nicotine metabolism in man, *J. Pharm. Pharmacol. 23*:553 (1971).
78. M. Duan, L. Yu, C. Savanapridei, P. Jacob III, and N. L. Benowitz, Dispositional kinetics and metabolism of (*S*)-nicotine *N*-1′-oxide in rabbits, *Drug Metab. Dispos. 19*:667 (1991).
79. J. R. Cashman and D. M. Ziegler, Contribution of *N*-oxygenation to the metabolism of MPTP (1-methyl-4-phenyl-1,2,3,6-tetrahydropyridine) by the flavin-containing monooxygenase by various liver preparations, *Mol. Pharmacol. 29*:163 (1986).
80. K. Chiba, E. Kubota, T. Miyakawa, Y. Kato, and T. Ishizaki, Characterization of hepatic microsomal metabolism as an in vivo detoxication pathway of 1-methyl-4-phenyl-1,2,3,6-tetrahydropyridine in mice, *J. Pharmacol. Exp. Ther. 246*:1108 (1988).
81. K. Chiba, H. Horii, I. Kubata, T. Ishizaki, and Y. Kato, Effects of *N*-methylmercaptoimidazole on the disposition of MPTP and its metabolites in mice, *Eur. J. Pharmacol. 180*:59 (1990).
82. D. A. Di-Monte, E.Y. Wu, I. Irwin, L. W. Delanney, and J. W. Langston, Biotransformation of 1-methyl-4-phenyl-1,2,3,6-tetrahydropyridine in primary cultures of mouse astrocytes, *J. Pharmacol. Exp. Ther. 258*:594 (1991).
83. J. R. Humbert, K. B. Hammond, W. E. Hathaway, J. G. Marcoux, and D. O'Brien, Trimethyaminuria—the fish odor syndrome, *Lancet*:770 (1970).

84. R. Ayesh, S. C. Mitchell, A. Zhang, and R. L. Smith, The fish odor syndrome: Biochemical, familial and clinical aspects, *Br. Med. J. 301*:655 (1993).
85. M. Al-Waiz, S. C. Mitchell, J. R. Idle, and R. L. Smith, The metabolism of ^{14}C-labled trimethylamine and its *N′*-oxide in man, *Xenobiotica 17*:551 (1987).
86. R. Ayesh, M. Al-Waiz, and N. J. Crother, Deficient nicotine *N*-oxidation in two sisters with trimethylaminuria, *Br. J. Clin. Pharmacol. 25*:664 (1988).
87. P. R. Ortiz de Montellano, *Cytochromes P-450, Structure Reduction Mechanism and Biochemistry* (P. R. Ortiz de Montellano, ed.), Plenum Press, New York, 1986.
88. F. F. Kadlubar, E. M. McKee, and D. M. Ziegler, Reduced pyridine nucleotide-dependent *N*-hydroxy amine oxidase and reductase activities in hepatic microsomes, *Arch. Biochem. Biophys. 156*:46 (1973).
89. B. Clement, M. G. Schultze-Mosqau, P. H. Richten, and A. Besch. Cytochrome P-450-dependent *N*-hydroxylation of an aminoquanidine (aminohydrazone) and microsomal retroreduction of the *N*-hydroxylated product, *Xenobiotica 24*:671 (1994).
90. J. Lin, C. E. Berkman, and J. R. Cashman, *N*-Oxidation of primary amines and hydroxylamines and retroreduction of hydroxylamines by adult human liver microsomes and adult human flavin-containing monooxygenase 3, *Chem. Res. Toxicol. 9*:1183 (1996).
91. A. H. Beckett, J. W. Gorrod, and P. Jenner, Absorption of (−)-nicotine-1′-*N*-oxide in man and its reduction in the gastrointestinal tract, *J. Pharm. Pharmacol. 22*:722 (1970).
92. P. Jenner, J. W. Gorrod, and A. M. Beckett, Species variation in the metabolism of (*R*)-(+)-, and (*S*)-(−)-nicotine by α-*C*- and *N*-oxidation in vitro, *Xenobiotica 3*:563 (1973).
93. D. M. Ziegler, S. S. Ansher, T. Nagata, F. F. Kadlubar, and W. B. Jakoby, *N*-Methylation: Potential mechanism for metabolic activation of carcinogenic primary arylamines, *Proc. Natl. Acad. Sci. USA 85*:2514 (1988).
94. K. W. Bock, UDP-glucuronosyl transferanses and their role in metabolism and disposition of carcinogens, *Adv. Pharmacol. 27*:367 (1994).
95. A. Miller and E. C. Miller, Electrophilic sulfuric acid ester metabolites as ultimate carcinogens, *Adv. Exp. Med. Biol. 197*:583 (1986).
96. F. F. Kadlubar, J. A. Miller, and E. C. Miller, Hepatic microsomal *N*-glucuronidation and nucleic acid binding of *N*-hydroxy arylamines in relation to urinary bladder carcinogenesis, *Cancer Res. 37*:805 (1977).
97. J. R. Cashman, Z.-C. Yang, and T. Hogberg, Oxidation of *N*-hydroxynorzimeldine to a stable nitrone by hepatic monooxygenases, *Chem. Res. Toxicol. 3*:428 (1990).
98. B. Clement, K. L. Lustig, and D. M. Zielger, Oxidation of desmethylpromethazine catalyzed by pig liver flavin-containing monooxygenase. Number and nature of metabolites, *Drug Metab. Dispos. 21*:24 (1993).
99. J. R. Cashman, Metabolism of tertiary amines by rat and hog liver microsomes: Role of enzymatic Cope-elimination to a *N*-dealkylated product, *Prog. Pharmacol. Clin. Pharmacol. 8*:117 (1991).
100. J. R. Cashman, J. Proudfoot, D. W. Pate, and T. Hogberg, Steroselective *N*-oxygenation of zimeldine and homozimeldine by the flavin-containing monooxygenase, *Drug Metab. Dispos. 16*:616 (1988).
101. J. R. Cashman, Enantioselective *N*-oxygenation of verapamil by the hepatic flavin-containing monooxygenase, *Mol. Pharmacol. 36*:497 (1989).
102. C. Mani and D. Kupfer, Cytochrome P-450-mediated activation and irreversible binding of the antiestrogen tamoxifen to proteins in rat and human liver: Possible involvement of flavin-containing monooxygenases in tamoxifen activation, *Cancer Res. 51*:6052 (1991).
103. R. E. Tynes and E. Hodgson, The measurement of FAD-containing monooxygenase activity in microsomes containing cytochrome P-450, *Xenobiotica 14*:515 (1984).
104. M. W. Duffel, J. M. Graham, and D. M. Ziegler, Changes in dimethylaniline *N*-oxdase activity of mouse liver and kidney induced by steroid sex hormones, *Mol. Pharmacol. 19*:134 (1981).

105. A. Lemoine, D. E. Williams, T. Cresteil, and J. P. Leroux, Hormonal regulation of microsomal flavin-containing monooxygenase: Tissue dependent expression and substrate specificity, *Mol. Pharmacol. 40*:211 (1991).
106. G. A. Dannan, F. P. Guengerich, and D. J. Waxman, Hormonal regulation of rat liver microsomal enzymes. Role of gonadal steroids in programming maintenance and suppression of Δ^4-steroid 5-α-reductase, flavin-containing monooxygenase, and sex-specific cytochromes P-450, *J. Biol. Chem. 261*:10728 (1986).
107. S. E. Shehin-Johnson, D. E. Williams, S. Larsen-Su, D. M. Stresser, and R. N. Hines, Tissue-specific expression of flavin-containing monooxygenase (FMO) forms 1 and 2 in the rabbit, *J. Pharmacol. Exp. Ther. 272*:1293 (1995).
108. E. Hodgson, R. L. Rose, D.-Y. Ryu, G. Falls, B. L. Blake, and P. E. Levi, Pesticide-metabolizing enzymes, *Toxicol. Lett. 82/83*:73 (1995).
109. P. J. Wirth and S. S. Thorgeirsson, Amine oxidase in mice—sex differences and developmental aspects, *Biochem. Pharmacol. 27*:601 (1981).
110. R. F. Kaderlik, E. Weser, and D. M. Ziegler, Selective loss of liver flavin-containing monooxygenases in rats on chemically defined diets, *Prog. Pharmacol. Clin. Pharmacol. 3*:95 (1991).
111. I. P. Nnane and L. A. Damani, Ethylmethyl sulfide: A potential pharmacokinetic probe for monitoring the activity of the flavin-containing monooxygenase in vivo in experimental animals, *Br. J. Clin. Pharmacol. 34*:160p (1992).
112. I. P. Nnane and L. A. Damani, Pharmacokinetics of trimethylamine: An alternative approach for monitoring the activity of the falvin-containing monooxygenase in vivo, *J. Pharm. Pharmacol. 44*:1060 (1992).
113. A. L. Buchman, M. D. Dubin, A. A. Moukazel, D. J. Jenden, M. Roch, K. M. Rice, J. Gornbien, and M. E. Ament, Choline deficiency: A cause of hepatic steatosis during parenteral nutrition that can be reversed with intravenous choline supplementation, *Hepatology 22*:1399 (1995).
114. N. Lomri, Z.-C. Yang, and J. R. Cashman, Expression in *Escherichia coli* of the flavin-containing monooxygenase from adult human liver. Determination of a distinct tertiary amine substrate specificity, *Chem. Res. Toxicol. 6*:425 (1993).
115. N. Lomri, Z.-C. Yang, and J. R. Cashman, Regio- and stereoselective oxygenations by adult human liver flavin-containing monooxygenase 3. Comparison with forms 1 and 2. *Chem. Res. Toxicol. 6*:800 (1993).
116. P. J. Sabourin, B. P. Smyser, and E. Hodgson, Purification of the flavin-containing monooxygenase from mouse and pig liver microsomes, *Int. J. Biochem. 16*:713 (1984).
117. B. Clement, M. Weide, and D. M. Ziegler, Inhibition of purified and membrane-bound flavin-containing monooxygenase 1 by (*N,N*-dimethylanimo)stilbene carboxylates. *Chem. Res. Toxicol. 9*:599 (1996).
118. K. G. Taylor and D. M. Ziegler, Studies on substarte specificity of the hog liver flavin-containing monooxygenase: Anionic organic sulfur compounds, *Biochem. Pharmacol. 36*: 141 (1987).
119. C. R. Boyd, C. T. Walsh, and T.-C. J. Chen, *S*-Oxygenases, II: Chirality of sulphoxidation reactions, *Sulfur-Containing Drugs and Related Organic Compounds* Vol. 2, part A (L. S. Damani, ed.), Ellis Horwood, Chichester, 1989, p 67.
120. J. R. Cashman, L. D. Olsen, D. R. Boyd, R. Austin, R. A. S. McMordie, R. Dunlop, and H. Dalton, Stereoselectivity of enzymatic and chemical oxygenation of sulfur atoms in 2-methy-1,3-benzodithiol, *J. Am. Chem. Soc. 114*:8772 (1992).
121. J. R. Cashman and D. E. Williams, Enantioselective *S*-oxygenation of 2-aryl-1,3-dithiolanes by rabbit lung enzyme preparations, *Mol. Pharmacol. 37*:333 (1990).
122. J. R. Cashman, L. D. Olsen, and L. M. Bornheim, Oxygenation of dialkyl sulfides by a modified Sharpless reagent: A model system for the flavin-containing monooxygenase, *J. Am. Chem. Soc. 112*:3191 (1990).

123. J. R. Cashman, L. D. Olsen, and L. M. Bornheim, Enantioselective *S*-oxygenation by flavin-containing and cytochrome P-450 monooxygenases, *Chem. Res. Toxicol. 3*:344 (1990).
124. J. R. Cashman and L. D. Olsen, Stereoselective *S*-oxygenation of 2-aryl-1,3-dithiolanes by the flavin-containing and cytochrome P-450 monooxygenases, *Mol. Pharmacol. 38*:573 (1990).
125. D. R. Light, D. J. Waxman, and D. T. Walsh, Studies on the chirality of sufoxidation catalyzed by bacterial flavoenzyme cyclohexanone monooxygenase and hog liver flavin adenine dinucleotide containing monooxygenase, *Biochemistry 21*:2490 (1982).
126. A. E. Rettie, B. D. Bogucki, I. Lim, and G. P. Meier, Sterioselective sulfoxidation of a series of alkyl *p*-tolyl sulfides by microsomal and purified flavin-containing monooxygenase, *Mol. Pharmacol. 37*:643 (1990).
127. M. B. Fisher, M. P. Lawton, E. Atta-Asafo-Adjei, R. M. Philpot, and A. E. Rettie, Selectivity of flavin-containing monooxygenase 5 for the (*S*)-sulfoxidation of short-chain aralkyl sulfides, *Drug Metab. Dispos. 23*:1431 (1995).
128. J. R. Cashman and R. P. Hanzlik, Microsomal oxidation of thiobenzamide: A photometric assay for the flavin-containing monooxygenase. *Biochem. Biophys. Res. Commun. 98*:147 (1981).
129. W. X. Guo and D. M. Ziegler, Estimation of flavin-containing monooxygenase activities in crude tissue preparations by thiourea-dependent oxidation of thiocholine, *Anal. Biochem. 198*:143 (1991).
130. S. Sofer and D. M. Ziegler, Microsomal mixed-function amine oxidase: Oxidation products of piperazine-substituted phenothiazines drugs, *Drug Metab. Dispos. 6*:232 (1978).
131. L-O. Eriksson and H. Bostrom, Deactivation of sulindac-sulfides by human renal microsomes, *Pharmacol. Toxicol. 62*:177 (1988).
132. D. E. Duggan, Sulindac: Therapeutic implications of the prodrug/pharmacophore equilibrium, *Drug Metab Rev. 12*:325 (1981).
133. J. R. Cashman, B. Y. T. Perotti, C. E. Berkman, and Jing Lin, Pharmacokinetics and molecular detoxication, *Environ. Health Perspect. 104*:23 (1996).
134. J. Mordenti, Man versus beast: Pharmacokinetic scaling in mammals, *J. Pharm. Sci. 75*:1028 (1986).
135. H. Boxenbaum, Interspecies scaling, allometry, physiological time, and the ground plan of pharmacokinetics. *J. Pharmacokinet. Biopharm. 10*:201 (1982).
136. J. B. Houston, Utility of in vitro drug metabolism data in predicting in vivo metabolic clearance, *Biochem. Pharmacol. 47*:1469 (1988).
137. B. J. Ring, J. Catlow, T. J. Lindsay, T. Gillespie, L. K. Roskos, B. F. Cermele, S. P. Swanson, M. A. Hamman, and S. A. Wrighton, Identification of the human cytochromes P-450 responsible for the in vitro formation of the major oxidative metabolites of the antipsychotic agent olanzapine, *J. Pharmacol. Exp. Ther. 276*:658 (1996).
138. R. E. Tynes and E. Hodgson, Catalytic activity and substrate specificity of the flavin-containing monooxygenase in microsomal systems: Characterization of the hepatic, pulmonary and renal enzymes of the mouse, rabbit and rat, *Arch. Biochem. Biophys. 240*:77 (1985).
139. D. E. Williams, S. E. Hale, A. S. Muerhoff, and B. S. S. Masters, Rabbit lung flavin-containing monooxygenase. Purification, characterization and induction during pregnancy, *Mol. Pharmacol. 28*:381 (1985).
140. C. T. Dolphin, T. E. Cullingford, E. A. Shephard, R. L. Smith, and I. R. Phillips, Differential developmental and tissue-specific regulation of expression of the genes encoding three members of the flavin-containing monooxygenase family of man, FMO1, FMO3 and FMO4, *Eur. J. Biochem. 235*:883 (1996).
141. A. J. M. Sadeque, K. E. Thummel, and, A. E. Rettie, Purification of macaque liver flavin-containing monooxygenase: A form of the enzyme related immunochemically to an isoenzyme expressed selectively in the adult human liver, *Biochim. Biophys. Acta 1162*:127 (1992).

142. A. D. Rodrigues, M. F. Kukula, J. Gerrero, and J. R. Cashman, In vitro hepatic metabolism of ABT-418 in chimpanzee (*Pan troglodytes*), *Drug Metab. Disops. 23*:1143 (1995).
143. R. E. Tynes and R. M. Philpot, Tissue and species-dependent expression of multiple forms of mammalian microsomal flavin-containing monooxygenase, *Mol. Pharmacol. 31*:569 (1987).
144. K. A. Steinmetz and J. D. Potter, Vegetables, fruit and cancer. I. Epidemiology, *Cancer Causes Control 2*:325 (1991).
145. R. McDanell, A. E. M. McLean, A. B. Manley, R. K. Heanhey, and G. R. Fenwick, Chemical and biological properties of indole glucosinolates (glucobrassicins): A review, *Food. Chem. Toxicol. 26*:59 (1988).
146. C. A. Bradfeld and L. F. Bjeldanes. Effect of dietary indole-3-carbinol on intestinal and hepatic monooxygenase, glutathione transferase and epoxide hydrolase activities in the rat, *Food Chem. Toxicol. 22*:977 (1984).
147. W. A. Nijhoff, T. P. J. Mulden, H. Verhagen, G. van Poppel, and W. H. M. Peters, Effect of consumption of brussel sprouts on plasma and urinary glutathione *S*-transferase class-α and -π in humans, *Carcinogenesis 16*:955 (1995).
148. L. F. Bjeldanes, J-Y. Kim, D. R. Grose, J. C. Bartholomew, and C. A. Bradfield, Aromatic hydrocarbon responsiveness—receptor agonists generated from indole-3-carbinol in vitro and in vivo: Comparisons with 2,4,7,8-tetrachlorodibenzo-*p*-dioxin, *Proc. Natl. Acad. Sci. USA. 88*:9543 (1991).
149. C. M. Stresser, D. E. Williams, D. A. Griffin, and G. S. Bailey. Mechanisms of tumor modulation by indole-3-carbinol, *Drug Metab. Dispos. 23*:965 (1995).
150. A. Kawaji, T. Miki, and E. Takabatake, Partial purification and substrate specificity of flavin-containing monooxygenase from rat brain microsomes, *Biol. Pharm. Bull. 18*:1657 (1995).
151. V. Ravindranath, S. Bhamre, S. M. Bhagwat, H. K. Anandatheerthavarada, S. K. Shankar, and P. S. Tirumalai, Xenobiotic metabolism in brain, *Toxicol. Lett. 82/83*:633 (1995).
152. S. V. Bhagwat, S. Bhamre, M. R. Boyd, and V. Ravindranath, Further characterization of rat brain flavin-containing monooxygenase, *Biochem. Pharmacol. 51*:1469 (1996).
153. S. Bhamre, S. V. Bhagwat, S. K. Shankar, D. E. Williams, and V. Ravindranath, Cerebral flavin-containing monooxygenase-mediated metabolism of antidepressants in brain: Immunochemical properties and immunocytochemical localization, *J. Pharmacol. Exp. Ther. 267*: 555 (1993).
154. B. L. Blake, R. M. Philpot, P. E. Levi, and E. Hodgson, Xenobiotic biotransforming enzymes in the central nervous system: An isoform of flavin-containing monooxygenase (FMO4) is expressed in rabbit brain, *Chem. Biol. Interact. 99*:253 (1996).
155. K. Venkatesh, P. E. Levi, A. D. Inman, N. A. Monteiro-Riviere, R. Misra, and E. Hodgson, Enzymatic and immunohistochemical studies on the role of cytochrome P-450 and flavin-containing monooxygenase of mouse skin in the metabolism of pesticides and other xenobiotics, *Pestic. Biochem. Physiol. 43*:53 (1992).
156. G.-P. Chen, L. L. Poulsen, and D. M. Ziegler, Oxidation of aldehydes catalyzed by pig liver flavin-containing monooxygenase, *Drug Metab. Dispos. 23*:1390 (1995).
157. J. R. Cashman, J. Proudfoot, Y.-K. Ho, M. S. Chin, and L. D. Olsen, Chemical and enzymatic oxidation of 2-aryl-1,3-oxathiolanes: Mechanism of the hepatic flavin-containing monooxygenase, *J. Am. Chem. Soc. 111*:4844 (1989).
158. S. Colonna, N. Gaggero, A. Manfredi, L. Casella, and M. Gullotti, Asymmetric oxidation of sulphides, catalyzed by chlorperoxidase, *J. Chem. Soc. Chem. Commun*. p. 1451 (1988).
159. S. Oae, A. Mikami, T. Matsuura, K. Ogawa-Asada, Y. Watanabe, K. Fujimori, and T. Iyanagi, Comparison of sulfide oxygenation mechanism for liver microsomal FAD-containing monooxygenase with the for cytochrome P-450, *Biochem. Biophys. Res. Commun. 131*:567 (1985).
160. M. B. Fisher, M. P. Lawton, E. Atta-Asafo-Adjei, R. M. Philpot, and A. E. Rettie, Selectivity of flavin-containing monooxygenase 5 for the (*S*)-sulfoxidation of short-chain aralkyl sulfides, *Drug Metrab. Dispos. 23*:1431 (1995).

161. P. E. Levi and E. Hodgson, Stereospecificity in the oxidation of phorate and phorate sulfoxide by purified FAD-containing monooxygenase and cytochrome P-450 isozymes, *Xenobiotica 18*:29 (1988).

162. R. J. Krause, S. L. Ripp, P. J. Sausen, L. H. Overby, R. M. Philpot, and A. A. Elfarra, Characterization of the Methionine *S*-oxidase activity of rat liver and kidney microsomes: Immunological and kinetic evidence for FMO3 being the major catalyst, *Arch. Biochem. Biophys. 333*:109 (1996).

163. E. Kashiyama, T. Yokoe, K. Itoh, S. Itoh, M. Odomi, and J. Kamatake, Stereoselective *S*-oxidation of flosequinan sulfide by rat hepatic flavin-containing monooxygenase 1A1 expressed in yeast, *Biochem. Pharmacol. 47*:1357 (1994).

164. K. Nunoya, Y. Yokoi, K. Itoh, S. Itoh, K. Kimura, and T. Kamataki, S-Oxidation of (+)-*cis*-3,5-dimethyl-2-(3-pyridyl)-thiazolin-4-one hydrochloride by rat hepatic flavin-containing monooxygenase 1 expressed in yeast, *Xenobiotica 25*:1283 (1995).

19

Applications of Caco-2 Cells in Drug Discovery and Development

Shiyin Yee and Wesley W. Day
Pfizer Inc., Groton, Connecticut

I. INTRODUCTION

A major challenge of drug design is the selection of candidates that have optimal molecular features for interaction with the receptor and for delivery between the point of administration and the final target site. A primary role for the drug metabolism scientist is to quantify systemic circulation or target site concentrations of drug following oral administration, the major dosing route. Thus, the intestine, becomes a significant variable because it is often the primary barrier for systemic exposure, owing to limited aqueous solubility, poor permeability, or extensive gut wall metabolism.

Many studies have been performed to understand the relationship between the physicochemical properties of a drug and its absorption in vivo. Although some relationships between these variables have been observed, there are enough exceptions to preclude any generalizations. Therefore, to predict and understand the complex process of absorption and to eliminate confounding interspecies differences in absorption, development and the use of tissue culture from man is warranted. A well-characterized and commonly used drug transport model is the Caco-2 monolayers.

Caco-2 cells are derived from human colon cancer cells and spontaneously differentiate to monolayers of polarized cells possessing the differentiated function of intestinal enterocytes when grown on porous membranes. The major advantages of using Caco-2 cell monolayers include the following.

1. The monolayers provide a reliable and versatile human absorption surrogate that can provide mechanistic and quantitative absorption or exsorption predictions at all stages of discovery and development.
2. The ease of the assay and the small requirement of compound permits rapid assessment of transport in a higher throughput format. Thus, Caco-2 may be implemented as a primary drug metabolism screen in early discovery.
3. The model has proven correlation with human absorption data.

This chapter begins with an overview of the major mechanisms of absorption, followed by a brief description of variables that can influence the validation and results from Caco-2 absorption model. Finally, specific examples from the literature will be presented to demonstrate the achievements of this model within a discovery or development program. The goal of this chapter is to leave the reader with a basic understanding of absorption mechanisms and how they relate to Caco-2 data for selection of compounds to advance into the clinic; and how Caco-2 cell may be validated and applied by drug metabolism scientists in their own programs.

II. ABSORPTION AND EXSORPTION TRANSPORT MECHANISMS

There are five pathways of absorption: namely, transcellular, paracellular, carrier-mediated, vesicular or pinocytosis, and lipid transport (Fig. 1). A prerequisite for absorption is that the drug must be present in aqueous solution at the absorption site. This is valid for all mechanisms of absorption, except vesicular or pinocytosis. The last two mechanisms play a minor role in the absorption of drugs ($< 1\%$) and will not be discussed in this chapter. Understanding and isolating the specific mechanism limiting the absorption

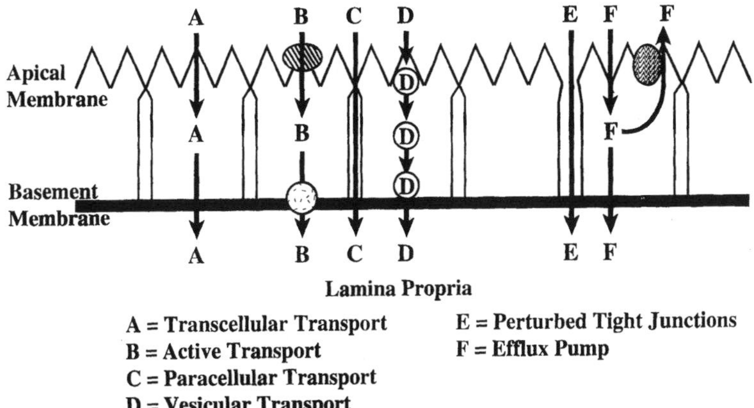

Fig. 1 Pathways across the intestinal barrier. (From Borchardt, 1994.)

of a compound is important in discovery (to accurately predict in vivo absorption in man) and development (to address controlled-release feasibility) mode.

A. Transcellular Transport

Transcellular transport is the most common mode by which drugs are absorbed. Drugs pass through membranes by passive diffusion, and the rates of absorption are determined by the physicochemical properties of the solute (pK_a, log$P$, solubility, hydrogen bonding) and the membrane, and by the concentration gradient. Passive diffusion is usually restricted to the absorption of nonionized lipid-soluble molecules and follows first-order kinetics. This means that the transfer rate is proportional to the concentration of the drug at the absorption site. Because transport by transcellular diffusion is possible along the length of the gastrointestinal (GI) tract, compounds undergoing transcellular absorption are better candidates for a controlled-release formulation. However, controlled-release candidates must not have dissolution or solubility limitations in the colon. Mostly hydrophobic molecules such as propranolol, dexamethasone, and testosterone, are transported by this mechanism.

B. Facilitated and Active Transport

Facilitative or active transport mechanisms of absorption utilize specialized carriers present on the membrane that mediate the transfer of solutes across the cell membrane. These mechanisms are saturable, with the carrier being rate-limiting in the absorption process. The carrier molecule is usually highly selective, and competition for the carrier may occur between drugs of similar structure.

During active transport, chemical energy must be supplied for the transfer of solutes from the lumen into the intracellular compartment. Often the transport occurs against a concentration gradient (e.g., amino acids, small peptides, ACE inhibitors, cephalosporins, and such) in the mucosal-to-serosal direction only. The active site may have structural- and site-specificity. The site-specificity of the active transport process requires that a drug absorbed by this process be fully available when it reaches the specific absorption site.

Drugs that are actively absorbed should not be administered in delayed-release form, because absorption from sites in the gastrointestinal tract other than the ''specific site'' may be extremely limited.

In facilitated diffusion the transfer of substrate takes place along the concentration gradient of the permeating solute (e.g., vitamin B_{12}). No energy expenditure is involved and thus transport is not inhibited by metabolic inhibitors that interfere with energy production.

Some natural substrates and drugs are absorbed simultaneously by active transport and passive diffusion processes. With low drug concentrations at the absorption site, the active transport generally proceeds at a much faster rate than the rate of passive diffusion. The contribution of the passive diffusion process becomes more important at concentrations that are high enough to saturate the active transport process.

C. Paracellular Transport

Two pathways exist for the uptake of water and electrolytes across the intestinal mucosa: transcellular and paracellular. The transcellular pathway allows the passage of hydrophilic molecules of low molecular weight ($<$ 180 Da) and with small molecular radii ($<$ 4 Å) through the water-filled pores in the cell membranes. The paracellular pathway allows access of larger molecules through the junctional complex between the cells. Although the intercellular spaces occupy less than 1% of the surface area of the epithelium, the paracellular pathway accounts for as much as 90% of the transepithelial conductance, and it is by this route that hydrophilic drug molecules are absorbed. The paracellular route of drug transport is important for various hydrophilic drug molecules (i.e., ranitidine, acyclovir, atenolol, and cimetidine).

Negative charges line the transjunctional channels; therefore, the preferential order of absorption via the paracellular pathway is cation $>$ neutral $>$ anion. It has been demonstrated that if the charge/mass ratio is high enough, an anionic solute may be completely impermeant through the paracellular pathway. The three-dimensional structure and flexibility of a molecule is also important in paracellular transport. Artursson et al. (1993) found a much higher permeability of PEG 194 compared with mannitol, which has a similar molecular weight. Furthermore, it has been suggested that the pore size decreases along the GI tract (6.7–8.8 Å, jejunum; $>$ 2.9–3.8 Å, ileum; $>$ 2.3 Å, colon), and the number of pores are also reduced. The larger mean pore radii together with larger water and ion fluxes may result in proportionally higher absorption by solvent drag through the tight junctions in the upper part of the Gl tract (duodenum and jejunum). Regional differences in paracellular transport in the intestine means that drugs utilizing this route would be unfavorable candidates for controlled-release formulations.

Paracellular transport can be studied using Caco-2 model by selectively opening the intercellular spaces with calcium chelators such as EDTA and EGTA (Gan et al., 1993; Artursson and Magnusson, 1990b), low Ca^{2+} (Artursson and Magnusson, 1990b), insulin (McRoberts et al., 1990), various pharmaceutical additives (e.g., sodium lauryl sulfate), and absorption enhancers such as sodium caprate (Anderberg et al., 1993) in the extracellular medium. Attempts to coculture goblet cells that have leakier tight junctions (induced HT29 cells) are in progress (Allen et al., 1991; Wikman et al., 1993). This association may also enhance the predictive capability of the monolayers by maintaining a more physiologically heterogeneous population of cells.

D. Efflux Pump

In addition to the foregoing absorption mechanisms, there is also an exsorption mechanism present in the epithelial cells. P-glycoprotein (Pgp), a 140- to 170-kDa membrane protein, is present on the apical membrane of Caco-2 cell monolayers at high levels and is associated with multiple drug resistance in tumor cells (Gan and Thakker, 1997). In vivo, it is present in several tissues, including the gut, bile duct, kidney, brain, and others. It functions as an ATP-dependent membrane efflux pump that expels drug from the cytoplasm, thereby preventing cellular accumulation and transport. It interacts nonspecifically with a variety of drugs, especially large lipophilic bases (e.g., verapamil, quinidine, vinblastine, digoxin, and erythromycin). Wacher et al. (1995) observed several striking overlaps between the substrates for and the inhibitors of CYP3A and Pgp. Thus, it appears that these enzymes may have complementary roles in the pharmacokinetics of drug absorption and elimination of cancer chemotherapy agents. These types of compounds may not be suitable candidates for controlled-release formulation because limited absorption of slowly released drugs will occur owing to the efflux pump or gut wall metabolism. However, use of an immediate-release dosage form can reduce the extent of the efflux, owing to saturation of the process.

III. CONSIDERATIONS FOR THE VALIDATION OF THE Caco-2 CELL ABSORPTION MODEL

The detailed process of creating a Caco-2 absorption model is beyond the scope of this chapter; however, there are key steps that are important to include when undertaking this task. These steps are listed in Table 1.

It is important to consider that the properties of the monolayer may vary depending on both, culturing and experimental conditions (Wilson et al., 1989; Artursson, 1991b). To obtain reproducible and optimal properties of the cell lines, the cells should be used within a limited number of passages, and the culturing conditions should be optimized. There are specific requirements involved for studying transport across Caco-2 cell monolayers, which are outlined in the following. Figures 2 through 4 graphically depict the culturing and experimental conditions adopted for Caco-2 transport studies in our laboratory.

Table 1 Variables to Characterize When Validating the Caco-2 Cell Absorption Model

Characterization of different routes of absorption (transcellular, paracellular, and carrier-mediated) and exsorption mechanism
Comparison of in vitro kinetic parameters with in vivo data
Establishing a correlation between P_{app} and percentage absorbed in human
Absorption versus metabolism
Thickness of aqueous layer adjacent to the monolayers
Nonspecific binding
Integrity of monolayer
Effect of various organic solvents on monolayer integrity
Effect of culturing days on carrier, enzyme, and Pgp expression
Maintenance of the same cell population for transport studies

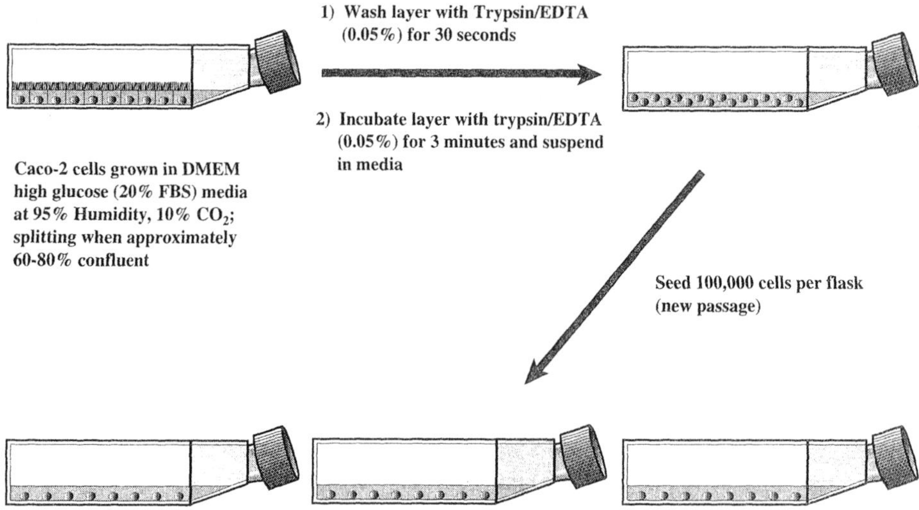

Fig. 2 Culturing of Caco-2 cells in flasks. (From Borchardt, 1994.)

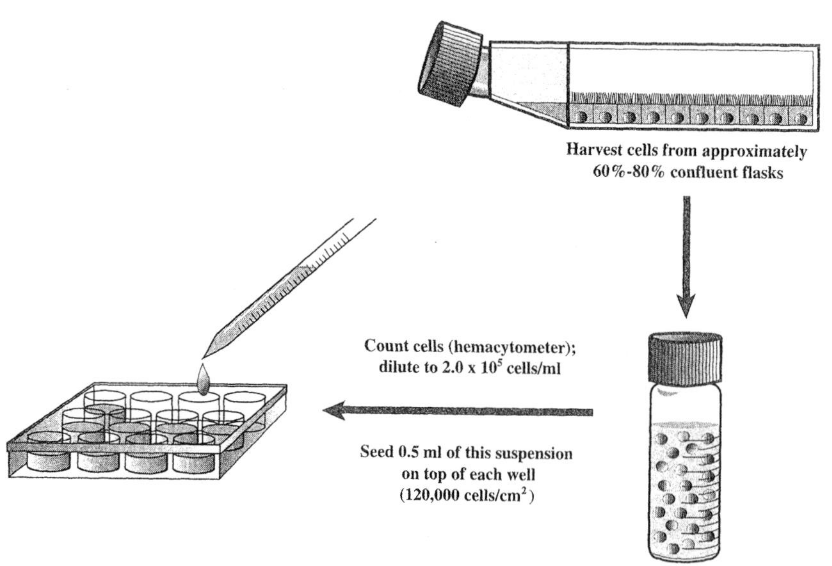

Fig. 3 Seeding of Caco-2 cells onto transwell inserts. (From Borchardt, 1994.)

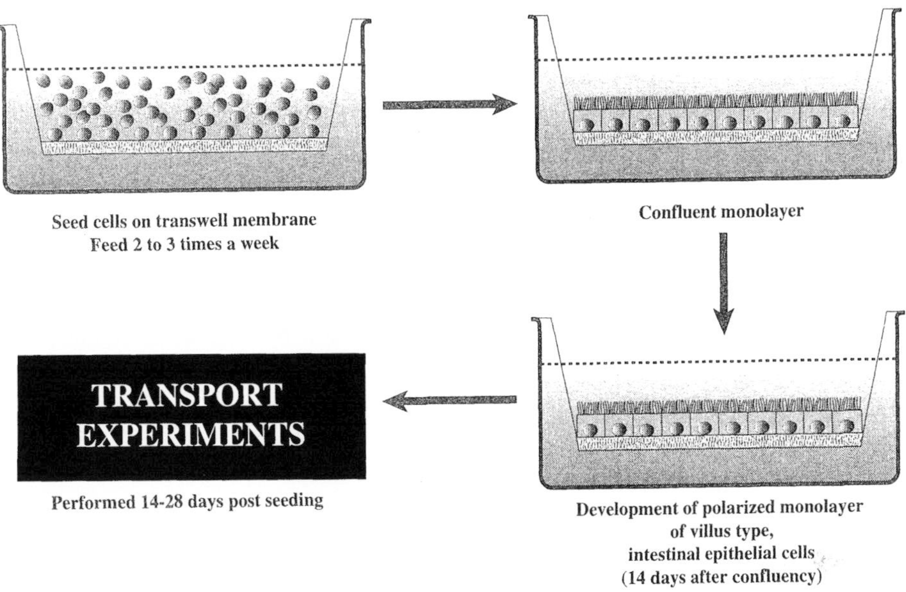

Fig. 4 Culturing of Caco-2 cells on transwell inserts. (From Borchardt, 1994.)

A. Age of Cells and Culturing Conditions

The procedures for Caco-2 cell culturing and experimental procedure have been outlined in several publications (Yee, 1997; Hidalgo et al., 1989; Hidalgo, 1996). It has been observed that dissimilarities in morphological characteristics, leakiness of monolayer, and transport properties between laboratories can arise from the differences in the source of the cells, heterogeneity of cells, and passage number of cells (Hu et al., 1995; Caro et al., 1995; Artursson, 1991b). Monolayers are subcultured when they reach approximately 60–80% confluence (7×10^6 to 10×10^6 cells per flask, 4–5 days postseeding). If cells grow to greater than 80% confluency, then cells should be discarded to avoid differentiation of cells into heterogeneous subpopulations. During the subculturing process, overtrysinization may cause clumping of cells and can be avoided by using lower concentrations of trypsin-EDTA (0.05%:0.02%). Monolayers are usually fed two to three times a week. The choice of filter material is important to minimize adsorption of drugs and proteins to the filter matrix (Cogburn et al., 1991; Artursson, 1990a). A thin filter of polycarbonate or polyethylene terephthalate is generally preferred. Collagen coating of the filter is not necessary for optimal cell attachment (unpublished data). For the sake of increased throughput, cells are grown on filters in a 12- or 24-well format.

B. Monolayer Integrity

The Caco-2 cell line has been extensively investigated by several researchers in terms of similarities of its morphological, biological, and transport properties to normal enterocytes (Hidalgo et al., 1989; Artursson, 1991b). The morphological examination of the Caco-2 cell monolayers can be conducted by electron microscopy and the biological or functional assessment can be made by measurement of marker enzymes (alkaline phosphatase, sucrase–isomaltase, or others). An examination of the barrier properties of the monolayers by use of impermeable markers, such as mannitol, inulin, or PEG4000, will indicate if the cells form tight, intact monolayers. The criteria for mannitol flux should be more stringent for paracellular transport than for transcellular diffusion (≤ 0.5% vs. ≤ 2%/h). This is because paracellular transport is greatly affected by the leakiness of the monolayers, whereas transcellular absorption is not (except for Pgp substrates). Some compounds are cytotoxic and can induce leakiness on the monolayer. It is important to conduct a mannitol check, especially when investigating transport of polar molecules, to give reliable and consistent data. This can be conducted in conjunction with substrate incubation or at the conclusion of transport experiments. Transepithelial electrical resistance (TEER) is not a sensitive measurement of monolayer integrity and should be used with caution (Yee, 1997). Routinely, positive controls representing different types of transport should be run to confirm reproducibility of the cell line.

C. Transport Mechanism

The expression of carriers and enzymes often varies with time and nutrient in cell culture (Hu and Borchardt, 1990). Therefore, it is important to establish the time-dependent expression of an enzyme or transporter to optimize the system and obtain reproducible results (Artursson, 1991b). Because carrier expression is time-dependent, active drug transport studies with Caco-2 monolayers are conducted using 21- to 30-day–cultured cells, compared with transcellular and paracellular absorption and exsorption mechanism studies, which use 15- to 30-day–cultured cells.

In addition to time-dependent expression of carriers, the permeability values for carrier-mediated transport is affected by the time interval between feeding and the transport experiment. Transport has been shown to be optimized with a longer postfeeding time interval (Hu and Borchardt, 1990). Experimental conditions, such as presence or absence of proton gradient, serum, and nutrients such as glucose and amino acids, hydrodynamics of transport system, temperature, and so forth, can also have a profound effect on transport (Hu and Borchadt, 1990; Tsuji, et al., 1994). The thickness of the aqueous boundary layer adjacent to the Caco-2 cell monolayer is an important consideration for compounds transported by the transcellular and carrier-mediated mechanisms, and should be taken into account when comparing data in the literature. Some laboratories have used the gas-lift side-by-side diffusion model to reduce the thickness of the unstirred aqueous layer adjacent to the monolayer. However, this can lead to a significant reduction in pH of the transport medium. In our system, a water bath with shaking speed of 70 cycles/min has produced acceptable permeability values for transcellular and carrier-mediated compounds. Therefore, to use this model successfully, it is very important to characterize the Caco-2 cells using model compounds under a well-defined set of culturing and experimental conditions (Yee, 1997).

The kinetic parameters of some of the carrier-mediated and Pgp substrates are described in Table 2. The criteria for selection of substrate concentration should be such that drug transported across the monolayer is detectable using the developed assay, the drug is not cytotoxic to the cells (mannitol flux ≤ 2%/h), transport kinetics are linear, and drug is in solution. It has been observed that acetonitrile or methanol, up to 1% v/v, can be used to dissolve lipophilic drugs without affecting the monolayer integrity or active transport (unpublished data). Dimethyl sulfoxide (DMSO) affects active transport and monolayer integrity at 0.5% v/v, and its use should be avoided.

Table 2 Summary of Transport Kinetics for P-Glycoprotein and Carrier-Mediated Substrates Across the Caco-2 Cell Monolayers

Substrate	Type	Apical pH/method	V_{max}(nmol/min/cm^2)	K_m(mM)	Ref.
Rifabutin	P-glycoprotein	6.5	1.87	0.93	Yee, unpublished data
Azithromycin	P-glycoprotein	6.5	0.01	0.37	Yee, unpublished data
Digoxin	P-glycoprotein	7.4/net transport	0.77	0.64	Cavet, 1996
Vinblastine	P-glycoprotein	7.4/net transport	0.02	0.02	Hunter, 1993
Ciprofloxacin	P-glycoprotein	7.4/net transport	0.74	0.89	Griffiths, 1993
Glycine	Amino acid	6.5/transport	34.70	8.40	Yee, unpublished data
L-α-methyldopa	Amino acid	6.5/transport	0.05	0.70	Hu, 1990
L-Phe	Amino acid	7.4/transport	0.24	0.56	Hidalgo, 1990
D-Cycloserine	Amino acid	6.0/transport	6.22	15.8	Thwaites, 1995
D-Phe-L-Pro	Dipeptide	6.5/transport	3.40	1.80	Yee, unpublished data
Gly-Sar	Dipeptide	6.0/transport	0.45	2.8	Thwaites, 1993
		7.4/transport	0.28	3.4	Thwaites, 1993
Cefaclor	Peptide drug	6.0/uptake	3.04	7.6	Dantzig, 1992
Cefixime	Peptide drug	6.0/uptake	0.80	17.4	Dantzig, 1994
Cephalexin	Peptide drug	6.0/transport	0.13	4.7	Gochoco, 1994
		6.0/uptake	0.43	2.9	Gochoco, 1994
		6.0/uptake	2.6	7.5	Dantzig, 1994
Cephradine	Peptide drug	6.0/uptake	1.27	8.3	Inui, 1992
		6.0/uptake	0.66	3.17	Hu, 1996
		7.4/uptake	0.09	4.37	Hu, 1996
Loracarbef	Peptide drug	6.0/transport	0.16	0.79	Hu, 1994
		7.4/transport	0.32	8.26	Hu, 1994
		6.0/uptake	2.6, 0.41	8.1, 2.06	Dantzig, 1993; Hu, 1996
Thrombin inhibitor 1	Peptide drug	6.0/transport	0.01	1.67	Walter, 1995
L-Val-acyclovir	Peptide drug	6.0/transport	0.26	0.29	Lee, 1996
Lisinopril	Peptidomimetic	6.5/transport	0.17	3.42	Motheram, 1996
CP-195,543	Monocarboxylic acid	6.5/transport	22.3	2.4	Yee, unpublished data
Benzoic acid	Monocarboxylic acid	6.5/transport	0.05	4.83	Tsuji, 1994
Sodium taurocholate	Bile acid	7.4/transport	0.20	0.06	Chandler, 1993

D. Alternative Systems

Recently, Biocoat Intestinal Epithelium Differentiation Environment (Becton Dickinson) claims to provide a ready-to-use confluent and fully differentiated Caco-2 cell monolayer in 3 days. The Caco-2 cells are seeded on a Biocoat fibrillar collagen-coated insert and cultured in a fully defined, serum-free medium containing butyric acid, hormones, and growth factors (Dando et al., 1996). Our experience using this environment suggests that the 3- to 7-day–cultured monolayers are about four- to sixfold leakier than the 3-week–cultured cells (mannitol flux 2–3% vs. 0.5%/h). The 3-day-old cells gave reasonable permeability values for transcellularly transported compounds; however, permeabilities for compounds transported by paracellular or efflux and carrier-mediated mechanisms are over- and underestimated, respectively. This can lead to problems in terms of rank-ordering compounds, irrespective of transport mechanisms.

IV. APPLICATIONS OF Caco-2 CELL MONOLAYERS IN DISCOVERY AND DEVELOPMENT

A. Rank-Ordering Compounds in the Discovery Mode

Yee (1997) demonstrated that the Caco-2 cell monolayers can be used as a tool to predict absorption in man, independent of the mechanism of transport, under well-defined cell-culturing and experimental conditions. A strong correlation was observed between in vivo human absorption and in vitro P_{app} (apparent permeability coefficient) for a variety of compounds. For compounds that are substrates of Pgp, use of a Pgp inhibitor resulted in a better estimate of absorption in humans. The results of this study suggested that the overall ranking of compounds with P_{app} less than 1×10^{-6} cm/s, between 1×10^{-6} and 10×10^{-6} cm/s, and more than 10×10^{-6} cm/s can be classified as poorly (0–20%), moderately (20–70%), and well (70–100%) absorbed compounds, respectively. A good correlation has also been observed between oral absorption in humans and P_{app} values for compounds absorbed either by the transcellular or paracellular route (Artursson, 1991a; Rubas et al., 1996). With recent advances in automation and mass spectrometry, high-throughput screening of Caco-2 cells in the discovery mode is rapidly achievable (Mandagere et al., 1996).

B. Rapid Assessment of Structure–Transport Relations

Several studies have been conducted using Caco-2 cell monolayers to successfully investigate structure–transport-relationships (STRs) for various therapeutic classes of drugs. Conradi et al. (1992), studied the influence of hydrogen bonding in the amide backbone of peptides on permeability across the Caco-2 cells. Methylation reduced the overall hydrogen bond potential of the peptide and increased passive absorption potential for this class of compounds. Hovgaard et al. (1995) elucidated that the formation of the *O*-cyclopropane carboxylic acid ester derivatives of various β-adrenergic blocking agents increased permeability across the Caco-2 cell monolayers. The enhancement in permeability was more pronounced with a greater increase in lipophilicities, in comparison with the parent compounds. Futhermore, several investigators have attempted to elucidate structural requirements for amino acid, small peptide, and cephalosporins transport using Caco-2 cell lines. Hidalgo et al. (1995) used Caco-2 cell monolayers to investigate the structural require-

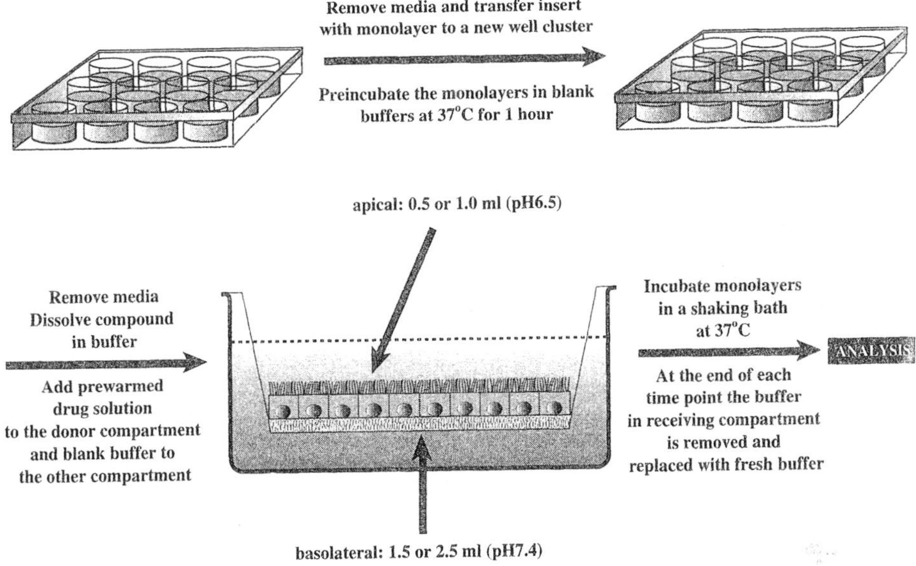

Fig. 5 Transport experiments using Caco-2 cell monolayers. (From Borchardt, 1994.)

ments for interaction with the oligopeptide transporter in Caco-2 cells. They showed that there was no difference in interaction with the oligopeptide transporter by linear dipeptides, regardless of the net charge, that a free carboxy group is needed for interaction, and that acidic amino acids can be included in cyclic peptides. D-Amino acids can be incorporated into dipeptides to increase metabolic stability without compromising their ability to optimally interact with the oligopeptide transporter, as long as one of the amino acids has an L-configuration. Walter et al. (1995) used the Caco-2 cell monolayers to investigate the STR for the peptidomimetic thrombin inhibitors and attempted to rank-order compounds. Tamura et al. (1996) investigated the stereospecificity of the oligopeptide transporter(s) for the stereoisomers of Val-Val-Val.

C. Elucidation of Pathways for Drug Absorption

The relative importance of the transcellular and paracellular routes in drug absorption was studied with β-adrenergic blocking agents of different lipophilicities in the Caco-2 model (Artursson and Magnusson, 1990b). When the paracellular pathways were widened by the sequestration of calcium ions from the tight junctions, the permeabilities of the more hydrophilic and poorly permeable drugs increased severalfold; however, the permeabilities of the more lipophilic and highly permeable drugs remained unchanged. The results indicate that the paracellular pathway is of quantitative importance only for drugs that have a low epithelial permeability. Symmetry in bidirectional transport is observed for compounds transported by a paracellular or transcellular mechanism (except for efflux substrates, for which basolateral B to apical A permeability is far greater than in the A → B direction).

The Caco-2 cell line has been used to study the active transport of a variety of endogenous (amino acids, peptides, bile acids, folic acid, biotin, glucose; see Table 2) and exogenous (benzoic acid, renin inhibitors, antimicrobial agents, and such) compounds, and the endocytic–vesicular pathways of vitamin B_{12} and other macromolecules. For these compounds, permeability in the A → B direction is far greater than in the reverse direction, owing to the presence of carriers on the apical membrane. In addition to the foregoing mechanisms, Caco-2 cells show polarized secretion of lipoproteins, which may be important to study delivery of drug to the lymphatic system (Traber et al., 1987). Recently, Desai et al. (1996) and McClean et al. (1996) investigated the interaction of microparticles with Caco-2 cells. Size-dependent uptake of particles in the order of $0.1 > 1 > 10\ \mu m$ was observed.

D. Assessment of Sustained-Release Potential

Sustained-release (SR) oral formulations are increasingly being used as a means of providing a once-a-day dosing for compounds with shorter half-lives. Therefore, the feasibility of designing an SR formulation for a drug candidate is being examined early on in development. To determine feasibility, a detailed mechanistic study on the transport mechanisms needs to be performed. Because transit time through the small intestine is about 4 h, it is likely that for an SR formulation (tablet or capsule) that is designed to deliver drug over a prolonged period, most of the drug will be delivered to the large intestine. The extent of drug absorption from the colon and the residence time of the formulation in the colon are important considerations in the development of a successful SR dosage form, and compounds that are well absorbed in both the small and large intestine are the best candidates for SR formulations.

An optimum SR formulation is not feasible for a compound absorbed by the carrier or paracellular mode of transport because these mechanisms occur predominantly in the upper GI tract. Transport systems for amino acid, peptide, bile acid, monocarboxylic acid, and such have been shown to be present exclusively in the small intestine, and absent in the colon. For paracellular and carrier-mediated absorption, once-a-day dosing is possible only if the colonic absorption is enhanced or the formulation is retained in the upper GI tract for a longer duration. Conversely, compounds absorbed transcellularly by a diffusive mechanism are good candidates for SR formulations. For these compounds, gut wall metabolism or Pgp efflux may become rate-limiting.

E. Assessment of Formulation and Chemical Strategies to Enhance the Intestinal Permeability of a Drug

The use of permeability enhancers (interferon-gamma, insulin, tumor necrosis factor, palmitoylcarnitine chloride, bile acids), pharmaceutical additives (Tween 80, sodium dodecylsulfate), and prodrugs for oral delivery of polar and high molecular weight drugs have been widely investigated using the Caco-2 cell monolayers (McRoberts, et al., 1990; Mullin and Snock, 1990; Madara and Stafford, 1989; Fix et al., 1990, Artursson, 1991b; Anderberg et al., 1993). In general, the concentration of enhancers required to obtain an effect on the monolayer tight junction is at least ten times lower than that needed in the rat tissues, owing to lack of the protecting mucus layer on the Caco-2 cell monolayers. The Caco-2 cell monolayers were used to evaluate the potential for exploitation of the bile acid carrier and vitamin B_{12} to deliver peptide drugs with poor absorption across intestinal mucosa (Kim et al., 1993; Dix et al., 1990). In addition, Dias and Yatscoff (1996) have used Caco-2 cell monolayers to study the effect of different solubility-enhancing formulations on cyclosporine and rapamycin absorption.

Caco-2 cell monolayers can also be used to assess potential toxic or cytoprotective effects of drug candidates or formulation components on this biological barrier. Schipper et al. (1996) studied the effect of chitosan on epithelial permeability and toxicity using the Caco-2 cell monolayers as a model epithelium. They observed that one out of eight chitosans studied (35% degree of acetylation) had an early onset of permeability enhancement, with very low toxicity and had a flat dose–absorption enhancement response relation. Tang et al. (1993) demonstrated a potential use of Caco-2 cell monolayers to investigate the cytoprotective effects of sucralfate.

F. Assessment of Efflux Potential of Compounds

Transepithelial efflux of cardiac glycosides, fluoroquinolone, vinblastine, peptides, and cyclosporin, have been investigated using Caco-2 cell monolayers (Cavet et al., 1996; Hunter et al., 1993; Burton, et al., 1993; Chiu and Amidon, 1996; Griffiths et al., 1993). The bidirectional transport of Pgp substrates suggests that A → B transport is much slower (at least threefold) than in the B → A direction. Furthermore, the efflux mechanism can be suppressed by use of Pgp inhibitors such as verapamil (0.2 mM), such that the A → B transport of the substrate is similar or closer to the B → A transport.

G. Elucidation of Potential Metabolism Pathways

The use of Caco-2 cell monolayers as a metabolism model is less popular than the use as an absorption model, mainly because of underexpression of metabolizing enzymes. This may be related to the colonic origin of the cells and the time-dependent expression of enzymes in cell cultures. Although the capacity of these cells to metabolize are not as high as normal human enterocytes, it can be used to profile the metabolism pathway in a qualitative sense. Cytochrome P450 1A1 and 1A2, and recently 3A4 and 3A5 have been identified in clones of Caco-2 cells (Hidalgo, 1996). Crespi et al. (1996) were successful in developing Caco-2 cell derivatives that expresses high levels of cDNA-derived cytochrome P450 2A6 and 3A4. Phase II enzymes detected in the Caco-2 cells include glutathione *S*-transferase, phenol sulfotransferases, glucuronidase, and glutathione *S*-transferase (Artursson, 1991b).

V. CONCLUSIONS

Recent advances in genetic engineering and cell-culturing techniques, with a better understanding of transport characteristics of Caco-2 cells versus human epithelium, have enabled exciting applications of Caco-2 cell monolayers as a human absorption and metabolism surrogate model with high throughput in discovery applications.

REFERENCES

Allen, J. D., G. P. Martin, C. Marriott, I. Hassan, and I. Williamson, Drug transport across a novel mucin secreting cell model: Comparison with the Caco-2 cell system, *J. Pharm. Pharmacol. 43*:63P (1991).

Anderberg, K., T. Lindmark, and P. Artursson, Sodium caprate elicits dilatations in human intestinal tight junctions and enhances drug absorption by the paracellular route, *Pharmacol. Res. 10*: 857–864 (1993).

Artursson, P., Epithelial transport of drugs in cell culture. I: A model for studying the passive diffusion of drugs over intestinal absorptive (Caco-2) cells, *J. Pharm. Sci. 79*:476–482 (1990a).

Artursson, P. and C. Magnusson, Epithelial transport of drugs in cell culture. II: Effect of extracellular calcium concentration on the paracellular transport of drugs of different lipophilicities across monolayers of intestinal epithelial (Caco-2) cells, *J. Pharm. Sci. 79*:595–600 (1990b).

Artursson, P. and J. Karlsson, Correlation between oral drug absorption in humans and apparent drug permeability coefficients in human intestinal epithelial (Caco-2) cells, *Biochem. Biophys. Res. Commun. 175*:880–885 (1991a).

Artursson, P., Cell cultures as models for drug absorption across the intestinal mucosa. *Crit. Rev. Ther. Drug Carrier Syst. 8*:305–330 (1991b).

Artursson, P., Selective paracellular permeability in two models of intestinal absorption: Cultured monolayers of human intestinal epithelial cells and rat intestinal segments, *Pharmacol. Res. 10*:1123–1129 (1993).

Borchardt, R., AAPS short course. *Pharmaceutical Applications of Cell and Tissue Culture Systems* (1994).

Burton, P. S., R. A. Conradi, A. R. Hilgers, and N. F. H. Ho, Evidence for a polarized efflux system for peptides in the apical membrane of Caco-2 cells, *Biochem. Biophys. Res. Commun. 190*: 760–766 (1993).

Cavet, M. E., M. West, and N. L. Simmons, Transport and epithelial secretion of the cardiac glycoside, digoxin, by human intestinal epithelial (Caco-2) cells, *Br. J. Pharm. 118*:1389–1396 (1996).

Caro, I., X. Boulenc, M. Rousset, V. Meunier, et al., Characterization of a newly isolated Caco-2 clone (TC-7), as a model of transport processes and biotransformation of drugs, *Int. J. Pharmacol. 116*:147–158, (1995).

Chandler, C. E., L. M. Zaccaro, and J. B. Moberly, Transepithelial transport of cholytaurine by Caco-2 cell monolayers is sodium dependent, *Am. J. Physiol. 264*:G1118–G1125 (1993).

Chiu, Y. and G. L. Amidon, P-Glycoprotein (P-GP) effect to cyclosporin A (CsA) transport in the Caco-2 system, *Pharm. Res. 13*:S239 (1996).

Cogburn, J. N., M. G. Donovan, and C. S. Schasteen, A model of human small intestinal absorptive cells. 1. Transport barrier, *Pharm. Res. 8*:210–216 (1991).

Conradi, R. A., A. R. Hilgers, N. F. H. Ho, and P. S. Burton, The influence of peptide structure on transport across Caco-2 cells, *Pharm. Res. 9*:435–439 (1992).

Crespi, C. L., B. W. Penman, and M. Hu, Development of Caco-2 cells expressing high levels of cDNA-derived cytochrome P4503A4, *Pharm. Res. 13*:1635–1641 (1996).

Dando, S. A., S. Chong, and R. A. Morrison, Evaluation of Biocoat intestinal epithelium differentiation environment (3-day cultured Caco-2 cells) as an absorption screening model with improved productivity, *Pharm. Res. 13*:S240 (1996).

Dantzig, A. H., L. B. Tabas, and L. Bergin, Cefaclor uptake by the proton-dependent dipeptide transport carrier of human intestinal Caco-2 cells and comparison to cephalexin uptake, *Biochim. Biophys. Acta 1112*:167–173 (1992).

Dantzig, A. H., D. C. Duckworth, and L. B. Tabas, Transport mechanisms responsible for the absorption of loracarbef, cefixime, and cefurixime axetil into human intestinal Caco-2 cells, *Biochim. Biophys. Acta 1191*:7–13 (1994).

Desai, M. P., V. Labhasetwar, E. Walter, R. J. Levy, and G. L. Amidon, Comparative uptake of biodegradable microparticles in vitro by Caco-2 cells and in situ by rat intestine, *Pharmacol. Res. 13*:S236 (1996).

Dias, V. C. and R. W. Yatscoff, An in vitro method for predicting in vitro oral bioavailability of novel immunosuppressive drugs, *Clin. Biochem. 29*:43–49 (1996).

Dix, C. J., H. Hassan, H. Y. Obray, R. Stah, and G. Wilson, The transport of vitamin B_{12} through polarized monolayers of Caco-2 cells, *Gastroenterology 98*:1272–1279 (1990).

Fix, J. A., J. Hochman, and E. LeCluyse, Palmitoylcarnitine increases gastrointestinal mucosal cell tight junction permeability, *FASEB J. 4*:A984 (1990).

Gan, L., P. Hsyu, J. F. Pritchard, and D. Thakker, Mechanism of intestinal absorption of ranitidine and ondansetron: Transport across Caco-2 cell monolayers, *Pharm. Res. 10*:1722–1725 (1993).

Gan, L. and D. R. Thakker, Applications of the Caco-2 model in the design and development of orally active drugs: Elucidation of biochemical and physical barriers posed by the intestinal epithelium, *Adv. Drug Deliv. Rev. 23*:77–98 (1997).

Gochoco, C. H., F. M. Ryan, J. Miller, P. L. Smith, and I. J. Hidalgo, Uptake and transepithelial transport of the orally absorbed cephalosporin cephalexin, in the human intestinal cell line, Caco-2, *Int. J. Pharmacol. 104*:187–202 (1994).

Griffiths, N. M., B. H. Hirst, and N. L. Simmons, Active secretion of the fluoroquinolone by human intestinal epithelial Caco-2 cell layers, *Br. J. Pharmacol. 108*:575–576 (1993).

Hidalgo, I. J., T. J. Raub, and R. T. Borchardt, Characterization of the human colon carcinoma cell line (Caco-2) as a model system for intestinal epithelial permeability, *Gastroenterology 96*: 736–749 (1989).

Hidalgo, I. J. and R. T. Borchardt, Transport of a large neutral amino acid (phenylalanine) in a human intestinal epithelial cell line: Caco-2, *Bichim. Biphys. Acta 1028*:25–30 (1990).

Hidalgo, I. J., P. Bhatnagar, C. Lee, J. Miller, G. Cucullino, and P. Smith, Structural requirements for interaction with the oligopeptide transporter in Caco-2 cells, *Pharm. Res. 12*:317–319 (1995).

Hidalgo, I. J., Cultured intestinal epithelial cell models, *Models for Assessing Drug Absorption and Metabolism* (R. T. Borchardt, P. L. Smith, and G. Wilson, eds.), Plenum Press, New York, 1996, pp. 35–50.

Hovgaard, L., H. Brøndsted, A. Buur, and H. Bundgaard, Drug delivery studies in Caco-2 monolayers. Synthesis, hydrolysis, and transport of *O*-clyclopropane carboxylic acid ester prodrugs of various β-blocking agents, *Pharm. Res. 12*:387–392 (1995).

Hu, M. and R. Borchardt, Mechanism of L-α-methyldopa transport through a monolayer of polarized human intestinal epithelial cells (Caco-2), *Pharm. Res. 7*:1313–1319 (1990).

Hu, M., J. Chen, Y. Zhu, A. H. Dantzig, R. E. Stratford, and M. T. Kuhfeld, Mechanism and kinetics of transcellular transport of a new β-lactam antibiotic loracarbef across an intestinal epithelial membrane model system (Caco-2), *Pharm. Res. 11*:1405–1413 (1994).

Hu, M., L. Zheng, J. Chen, L. Liu, Y. Li, A. H. Dantzig, and R. E. Stratford, Jr., Peptide transporter function and prolidase activities in Caco-2 cells: A lack of coordinated expression, *J. Drug Target. 3*:291–300 (1995).

Hu, M., J. Chen, L. Zheng, A. H. Dantzig, and R. E. Stratford, Uptake characteristics of loracarbef and cephalexin in the Caco-2 cell culture model: Effects of the proton gradient and possible presence of a distinctive second component, *J. Pharm. Sci. 85*:767–772 (1996).

Hunter, J., M. A. Jepson, T. Tsuruo, N. L. Simmons, and B. H. Hirst, Functional expression of P-glycoprotein in apical membranes of human intestinal Caco-2 cells, *J. Biol. Chem. 268*: 14991–14997 (1993).

Inui, K. I., M. Yamamoto, and H. Saito, Transepithelial transport of oral cephalosporins by monolayers of intestinal epithelial cell line Caco-2: Specific transport systems in apical and basolateral membranes, *J. Pharmacol. Exp. Ther. 261*:195–201 (1992).

Kim, D. C., A. W. Harrison, M. J. Ruwart, K. F. Wilkinson, K. F. Fisher, I. J. Hidalgo, and R. T. Borchardt. Evaluation of bile acid transporter in enhancing intestinal permeability to renin-inhibitory peptides, *J. Drug Target. 1*:347–359 (1993).

Lee C. P., R. L. A. de Vrueh, and P. L. Smith, Transport of a prodrug of acyclovir, L-val acyclovir, via the oligopeptide transporter, *Proc. Int. Symp. Control. Release Bioact. Mater. 23*:47–48 (1996).

Madara, J. L. and J. Stafford, Interferon-γ directly affects barrier function of cultured intestinal epithelial monolayers. *J. Clin. Invest. 83*:724 (1989).

Mandagere, A. K., M. A. Correll, J. P. Moooney, J. C. Poole, K. Hwang, and L. K. Cheng, Application of automation to Caco-2 drug diffusion studies, *Pharmacol. Res. 13*:S237 (1996).

McClean, S., N. Clarke, Z. Ramtoola, E. Prosser, Interaction of microparticles with Caco-2 cells and mammalian ileum in situ, *Pharm. Res. 13*:S240 (1996).

J. A. McRoberts, R. Aranda, N. Riley, and H. Kang, Insulin regulates the paracellular permeability of cultured intestinal epithelial cell monolayers, *J. Clin. Invest. 85*:1127 (1990).

Motheram, R., D. K. Goins, and A. H. Hikal, Mechanism of transport of an ACE inhibitor (lisinopril) in Caco-2 cell monolayers, *Pharm. Res. 13*:S239 (1996).

Mullin, J. M. and K. V. Snock, Effect of tumor necrosis factor on epithelial tight junctions and transepithelial permeability, *Cancer Res. 50*:2172 (1990).

Rubas, W., M. E. M. Cromwell, Z. Shahrokh, J. Villagran, T. N. Nguyen, M. Wellton, T. H. Nguyen, and R. J. Mrsny, Flux measurements across Caco-2 monolayers may predict transport in human large intestinal tissue, *J. Pharm. Sci. 85*:165–169 (1996).

Schipper, N. G. M., K. M. Varum, and P. Artursson, Chitosans as absorption enhancers for poorly absorbable drugs. 1: Influence of molecular weight and degree of acetylation on drug transport across human intestinal epithelial (Caco-2) cells, *Pharm. Res. 13*:1686–1692 (1996).

Tang, A. S., P. J. Chikhale, P. K. Shah and R. T. Borchardt, Utilization of a human intestinal epithelial cell culture system (Caco-2) for evaluating cytoprotective agents, *Pharm. Res. 10*: 1620–1626 (1993).

Tsuji, A., H. Takanaga, I. Tamai, and T. Terasaki, Transcellular transport of benzoic acid across Caco-2 cells by a pH-dependent and carrier-mediated transport mechanism, *Pharm. Res. 11*: 30–37 (1994).

Tamura, K., C. Lee, P. L. Smith, and R. T. Borchardt, Effect of charge on oligopeptide transporter-mediated permeation of cyclic dipeptides across Caco-2 cell monolayers, *Pharm. Res. 13*: 1752–1754 (1996).

Thwaites, D. T., C. D. A. Brown, B. H. Hirst, and N. L. Simmons, H^+-coupled dipeptide (glycyclsarcosine) transport across apical and basal borders of human intestinal Caco-2 cell monolayers display distinctive characteristics, *Biochim. Biophys. Acta 1151*:237–245 (1993).

Thwaites, D. T., G. Armstrong, B. H. Hirst, and N. L. Simmons, D-Cycloserine transport in human intestinal epithelial (Caco-2) cells: Mediation by a H^+-coupled amino acid transporter, *Br. J. Pharmacol. 115*:761–766 (1995).

Traber, M. G., H. J. Kayden, and M. J. Rindler, Polarized secretion of newly synthesized lipoproteins by the Caco-2 human intestinal cell line, *J. Lipid Res. 28*:1350 (1987).

Wacher, V. J., C. Wu, and L. Z. Benet, Overlapping substrate specificities and tissue distribution of cytochrome P450 3A and P-glycoprotein: Implications for drug delivery and activity in cancer chemotherapy, *Mol. Carcinog. 13*:129–134 (1995).

Walter, E., T. Kissel, M. Reers, G. Dickneite, D. Hoffmann, and W. Stueber, Transepithelial transport of peptidomimetic thrombin inhibitors in monolayers of a human intestinal cell line (Caco-2) and their correlation to in vivo data, *Pharm. Res. 12*:360–365 (1995).

Wilson, G., *Pharmaceutical Applications of Cell and Tissue Culture to Drug Transport* (G. Wilson, S. S. Davis, L. Illum, and A. Zweibaum, eds.), Plenum Press, New York, 1989.

Wikman, A., J. Karlsson, J. Carstedt and P. Artursson, A drug absorption model based on the mucus layer producing human intestinal goblet cell line HT29-H, *Pharm. Res. 10*:843–852 (1993).

Yee, S., In vitro permeability across Caco-2 cells (colonic) can predict in vivo (small intestinal) absorption in man—fact or myth, *Pharm. Res. 14*:763–766 (1997).

20

The Role of Nuclear Magnetic Resonance Spectroscopy in Drug Metabolism

I. D. Wilson
Zeneca Pharmaceuticals, Macclesfield, Cheshire, United Kingdom

J. K. Nicholson and J. C. Lindon
Imperial College School of Medicine, London, United Kingdom

I. INTRODUCTION

The determination of the structure of the metabolites of a new drug or xenobiotic has always been one of the most exciting and technically challenging aspects of any metabolic study. The two techniques most often applied to this process have been mass spectrometry (MS) and nuclear magnetic resonance (NMR) spectroscopy. In the past, the perceived difference in sensitivity between MS and NMR meant that recourse to the latter was only made if MS was unable to provide an unequivocal answer. However, technical advances in NMR, including the introduction of ever-higher field strengths (500 and 600 MHz NMR spectrometers are now relatively common and, at the time of writing, 800 MHz instruments have just become commercially available) have made NMR spectroscopy much more sensitive. In addition, the nondestructive nature of NMR spectroscopy makes it an attractive first option before MS. Indeed, in our laboratories, there is a trend toward performing NMR before any subsequent MS wherever possible to obtain the maximum amount of information on a precious sample. In addition, the ability of the NMR spectrum to give an overall estimate of the purity of the metabolite, which may be a single radiolabeled peak isolated from a chromatogram, but which might contain a myriad of endogenous contaminants, can be very valuable to the mass spectrometrist should problems arise when MS is attempted.

It is certainly true that with a modern high-field NMR spectrometer, operating at 500 or 600 MHz, high-quality one-dimensional (1D) proton NMR spectra of drug metabolites on a few tens of micrograms of a pure drug metabolite can now be obtained in a matter of minutes. The real challenge now is to obtain the material in a suitably pure form for spectroscopy, and this relies on the skills of the metabolism chemist in the area of isolation from the matrix, both from other metabolites and from endogenous compounds. By its nature the isolation of metabolites can be tedious and time-consuming, and it often forms the rate-limiting step in any metabolic study. Sometimes the nature of the problem (i.e., very low doses, or a multiplicity of minor metabolites) makes this isolation and identification strategy the only feasible option for the elucidation of the metabolic fate of a novel xenobiotic. However, over the past decade, there has been an increasing realization of the potential of NMR for the analysis of untreated biofluids (reviewed by Nicholson and Wilson, 1987; Malet-Martino and Martino, 1989). Indeed, it is often possible to detect resonances for the xenobiotic and its metabolites even in the presence of the endogenous compounds and to derive some metabolic information from the NMR spectrum. Once detected the metabolites present in the sample can be identified using a whole range of methods and techniques from the simple to the highly sophisticated, depending on the problem, as described later. Although not described in detail here, the quantitative nature of NMR can be used to obtain quantitative excretion balance data without the need for radiolabeling studies.

Here we present modern approaches to the use of NMR spectroscopy in the study of xenobiotic metabolism, illustrated by examples from our own work, for which a range of NMR techniques have been used for the detection and elucidation of metabolite structures. These methods are centered initially on the use of NMR techniques directly on biological fluids, but are equally applicable to materials partially, or completely, purified from the relevant biomatrix.

II. THE DETECTION OF XENOBIOTICS AND THEIR METABOLITES IN BIOLOGICAL FLUIDS

The detection of xenobiotics and their metabolites in biological fluids depends on being able to pick out the relevant signals among those of the endogenous metabolites normally present. At first sight this might seem a formidable task. However, over the last decade or so, many of the signals from these endogenous substances in biofluids, such as plasma, urine, bile, cerebrospinal fluid (CSF) or other, have been assigned (e.g., see Wilson and Nicholson, 1995). Thus, in Fig. 1 a typical ^{1}H NMR spectrum of human urine, obtained

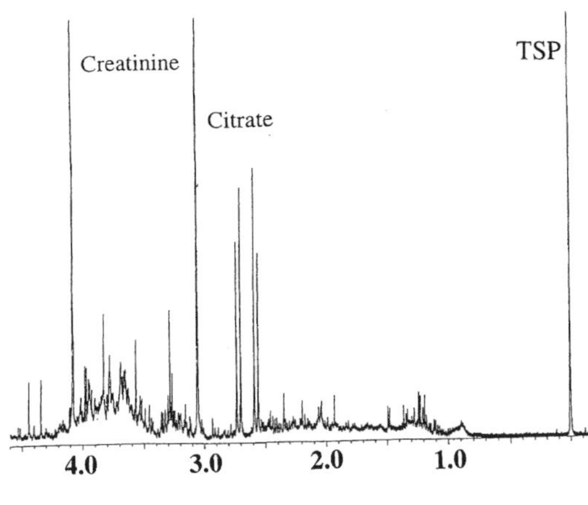

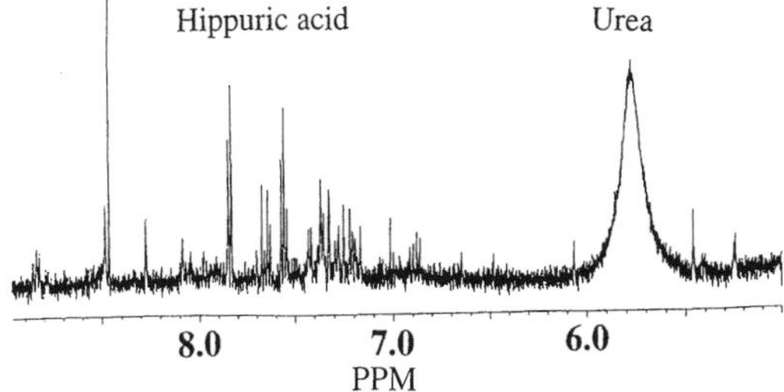

Fig. 1 A 500-MHz ^{1}H NMR spectrum of control human urine with assignment of endogenous metabolites as marked.

at 500 MHz, is shown, with its characteristic chemical shift and spin–spin-coupling patterns for the major endogenous substances, such as creatinine, amino acids, hippurate, and citrate. One of the most remarkable features of such a spectrum is not simply the richness of the data that it provides about the composition of the urine, but that it was obtained on essentially an untreated sample. This is possible because of the enormous advances that have been made in the techniques of suppression of the water resonance in recent years, which means that the potentially interfering water peak can be easily eliminated without detriment to the quality of the resulting spectrum. The spectrum in Fig. 1 was obtained on a control urine to which about 10% v/v of D_2O had been added as a frequency lock for the spectrometer. Water peak suppression can be achieved by several methods, such as presaturation of the signal by separate irradiation at the water frequency or by the use of a nuclear Overhauser enhancement (NOE) scheme that has the effect of improving the suppression in the wings of the broad water resonance. Alternatively, the water can be selectively omitted from the profile of the NMR excitation profile such that no water signal results. This has the added advantage that any NH or OH protons that are in exchange with the water protons can be observed because they are no longer saturated by virtue of exchange with the presaturated water peak.

A detailed examination of the ^1H NMR spectra of such urine samples reveals the presence of thousands of signals, most of which are minor. Indeed the identification of this myriad of minor endogenous metabolites remains a considerable problem, but one that when solved offers the possibility of deeper insights into changes of biochemical, toxicological, or clinical importance. However, the challenge of identifying new signals arising from the presence of xenobiotic metabolites is slightly less daunting as the investigator at least knows the structure of the substance that has been administered and can thus look for resonances characteristic of the administered materials. If present, the signals from drug-related material may be used both for the purposes of quantification, to obtain an excretion balance, and for an attempt at identification.

The detection of xenobiotic metabolites using proton NMR spectroscopy of intact biofluids is very dependent on the composition and, hence, the proton NMR spectrum of the biofluid. The three main biofluids that would be used in a drug metabolism study are urine, bile, and blood plasma, although in special cases others would need to be examined, including seminal fluid or CSF. Urine, as can be seen from Fig. 1, gives rise to a very complex proton NMR spectrum, but because of its essentially isotropic nature, the NMR peaks are generally sharp and well resolved. Thus, if a drug metabolite resonance appears in a region of the spectrum where few endogenous metabolite signals resonate, then it is possible to make some deductions about metabolite structures. One example here is the observation of ester glucuronides of carboxylate drugs that give a resonance from the β-1 proton of the glucuronic acid moiety in a relatively clear region. Blood plasma presents more difficulties on two counts. First, plasma concentrations of drug metabolites tend to be much lower than in excretory fluids, such as urine or bile, and the presence of high protein concentration in plasma means that many metabolites are strongly protein-bound. This results in broad NMR linewidths for the metabolite signals that, therefore, are often indistinguishable from the protein peaks themselves. Detection of metabolites in blood plasma is generally very difficult and is successful only in special cases. The use of NMR spectral-editing techniques, such as the use of spin-echo spectra, can sometimes alleviate the problem. Bile is also a complex fluid in physicochemical terms, with a multiphasic nature containing micellar regions. Thus, drug metabolites can be partitioned between the various compartments and, again, this can lead to broad NMR resonances for the drug metabolites. For blood plasma and bile, it may be preferable to carry out metabolite identification after extraction of the biofluid in an appropriate manner (see later discussion).

III. NMR NUCLEI SUITABLE FOR USE IN METABOLIC STUDIES

A. ^{1}H

Undoubtedly the most useful nucleus with which to obtain structural information on metabolites of xenobiotics is ^{1}H because hydrogen is present in most compounds. The ^{1}H nuclide is essentially 100% abundant and has the highest NMR sensitivity of any nonradioactive nuclide. It has a rather limited chemical shift range (about 12 ppm for organic molecules) and ^{1}H–^{1}H spin–spin coupling causes many spectral complications. However, the only real problem with the use of ^{1}H NMR is the potential for endogenous compounds to interfere, but in an "uncrowded" region of the spectrum for a biofluid, such as urine, as little as 1 μg can be detected at 600 MHz with an acquisition period of 1 h. For this type of study one would typically use approximately a 600-μL sample, to which 100 μL of D_2O would be added (containing an internal chemical shift and quantification standard such as trimethylsilylpropiopionic acid; TSP). If the amount of sample was limited, then the use of a microprobe could reduce the amount required to near 100 μL. A typical example of the results that can be obtained in this way is shown in Fig. 2 for acetamino-

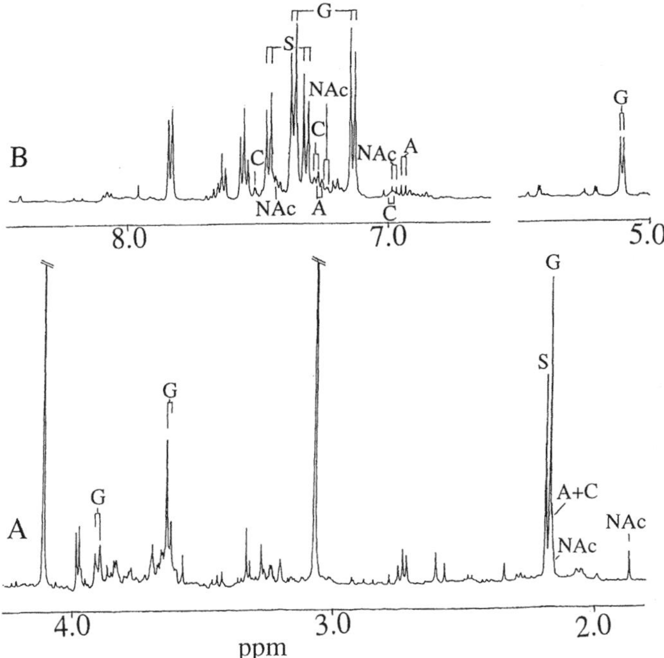

Fig. 2 An ^{1}H NMR spectrum of urine containing metabolites of acetaminophen: (A) aliphatic region and (B) aromatic region. Assignments are as follows: A, acetaminophen; G, acetaminophen glucuronide; S, acetaminophen sulfate; C, acetaminophen-L-cysteinyl derivative; NAc, acetaminophen *N*-acetyl-L-cysteinyl derivative.

phen. As can be seen from even a cursory examination of this spectrum, NMR readily shows the presence of several metabolites, including signals for the anomeric proton of a glucuronide, and aromatic resonances for both the glucuronide and sulfate metabolites. More careful examination of the spectrum reveals the presence of all of the major known metabolites of the drug (Bales et al. 1984).

B. ^{2}H and ^{3}H

In addition to ^{1}H, the other isotopes of hydrogen, deuterium (^{2}H) and tritium (^{3}H) are also detectable by NMR, and providing that suitably labeled compounds are available, there is the potential to use them in metabolic studies. ^{2}H NMR suffers from the fact that the deuteron is a quadrupolar nucleus, and this gives rise to rapid relaxation and, hence, broad NMR lines for anything other than very small molecules. Also the chemical shift range is the same in part per million terms as for ^{1}H NMR, but this is only about 15% in frequency (Hz) terms, giving rather poor spectral resolution. In addition ^{1}H–^{2}H spin-couplings are also only about 15% the value of ^{1}H–^{1}H couplings and are often not resolved. Hence, structural identification based on ^{2}H NMR has to rely largely on chemical shift values. Examples of the use of this strategy include investigations of the metabolic fate of dimethylformamide (DMF) and deuterated acetaminophen and phenacetin (Farrant et al. 1992; Nicholls et al., 1995, 1997a,b). The application of deuterium labeling and NMR spectroscopy to phenacetin and acetaminophen involved the study of futile deacetylation of these acetanilides in an attempt to gain an insight into the role of this process in nephrotoxicity (Nicholls et al. 1995, 1997a,b). In this work, the acetanilide was synthesized with a fully deuterated *N*-acetyl moiety, thereby rendering this group ''invisible'' in the ^{1}H NMR spectrum. However, following administration, it was possible to detect, using ^{1}H NMR spectroscopy, undeuterated acetanilide formed by the metabolic process of deacetylation followed by reacetylation. This process might be important in explaining the nephrotoxicity of such drugs as acetaminophen for the futile deacetylation involves a flux through 4-aminophenol, a known and potent nephrotoxin.

The ^{3}H nuclide has the highest sensitivity of any nuclear spin, but it is radioactive. Its observation frequency at a given magnetic field strength is about 6% higher than for ^{1}H, therefore, chemical shifts and coupling constants are similar to those for ^{1}H NMR. The natural abundance of ^{3}H is negligible; accordingly, ^{3}H NMR studies are carried out on specifically labeled drugs. Several pharmaceutical research organizations have investigated in ^{3}H NMR probes specifically to carry out metabolic studies on ^{3}H-labeled compounds, for these are often synthesized during drug development. Because ^{3}H is radioactive, special precautions have to be taken before introducing the sample into the NMR spectrometer: the practicalities of ^{3}H NMR have been reviewed (Jones, 1993; Desk et al., 1991).

C. ^{13}C

The low natural abundance of ^{13}C in the environment (about 1.1% uniformly distributed), coupled with low inherent sensitivity, makes the use of this nucleus rather less attractive than ^{1}H as a general means of performing metabolic studies. However, providing that a suitably isotopically enriched compound can be obtained, then ^{13}C NMR can be used in the same way that has been described for deuterium-labeled compounds, with similar benefits in specificity. ^{13}C NMR spectra are usually acquired with decoupling of all ^{1}H–^{13}C spin couplings so that each ^{13}C site in a molecule gives a singlet resonance, unless two ^{13}C nuclei have been introduced within about three or four bonds apart. Indeed, we have used ^{13}C-labeled acetyl groups to investigate futile deacetylation of acetanilides in exactly the same manner as described earlier for deuterated materials (Nicholls et al. 1995). Carbon 13-labeling tends to be more widely used for metabolic studies than ^{2}H or ^{3}H because suitable NMR probes are more commonly available for this nucleus compared with deuterium and tritium. With modern NMR instruments it is also possible to detect the ^{13}C NMR spectrum of a compound indirectly by the ^{1}H NMR spectrum using the fact that all CH, CH_2, and CH_3 groups are connected by a one-bond ^{1}H–^{13}C spin coupling. This means that the ^{13}C NMR data are collected at the much superior ^{1}H NMR sensitivity, and this has resulted in various metabolite structures being determined using a combination of ^{1}H and ^{13}C NMR chemical shifts (for example, see Sec. VI.C on the structure of metabolites of the anti-HIV compound BW524W91).

D. ^{19}F

Probably one of the most useful nuclei for NMR studies of the metabolic fate of xenobiotics is ^{19}F, which if present in the compound under study provides a sensitive and specific ''NMR handle'' with which to detect, profile, and quantify metabolites. The ^{19}F nucleus is 100% abundant and has about 83% of the NMR sensitivity of ^{1}H. However, the much larger chemical shift range of ^{19}F NMR, combined with the lack of any endogenous interferences make this technique particularly well suited to this type of work. In addition, as a result of the large chemical shift range, the ^{19}F nucleus provides a sensitive indicator of metabolic changes at positions on the molecule quite some distance away (up to at least eight carbons) from the site of fluorine ''labeling,'' and the presence of ^{19}F–^{1}H spin coupling can also provide useful diagnostic structural information from both one- and two-dimensional (1-D and 2-D) NMR experiments.

E. ^{15}N

The ^{15}N nucleus has both a very low natural abundance and a low inherent NMR sensitivity making it a very difficult nucleus for NMR studies. In addition, the nuclear relaxation times are very long, which means that lengthy relaxation delays need to be introduced into the NMR acquisition, and this makes data collection inefficient. If, as in ^{13}C NMR, the ^{1}H interactions are decoupled it is possible for the ^{15}N signals to become smaller because the NOE is negative (unlike ^{13}C) and for signals to disappear. Thus ^{1}H-decoupling in ^{15}N NMR has to be used with caution. Nevertheless, for ^{15}N-enriched compounds, the possibility of ^{15}N NMR should be explored, for it can give valuable information on parts of the molecules that may have few protons.

F. ^{31}P

Phosphorus NMR has been used extensively in studies of endogenous metabolism, particularly for spectra obtained on whole tissues or in vivo. The ^{31}P nuclide is 100% naturally abundant, has a reasonable inherent sensitivity and, therefore, is a favorable nucleus for NMR spectroscopy. Accordingly, it should find use in metabolism studies of phosphorus-containing drugs or when metabolism includes phosphorylation. We have used this approach to study the rat and human metabolites of the anticancer drug ifosfamide (Foxall et al., 1996).

IV. NMR-BASED TECHNIQUES FOR SIGNAL ASSIGNMENT AND COMPOUND IDENTIFICATION

A. Introduction

After having detected, and possibly quantified, the metabolites present in a biofluid, the next aim must be either identification, or confirmation of identity, if that can be deduced relatively easily from the spectrum of metabolites in the biofluid itself.

Broadly, approaches to metabolite identification can be categorized as follows:

1. Methods based solely on NMR spectroscopy (i.e., using single and multipulse–multidimensional techniques only)
2. Modification (physical, chemical, or enzymatic) monitored by NMR spectroscopy
3. Isolation or removal of interferences by chromatography, solid-phase or solvent extraction, with subsequent identification by NMR and other spectroscopic techniques as appropriate
4. Directly coupled chromatography–NMR-based methods.

The isolation techniques employed may be relatively simple, such as solvent or solid-phase extraction, or they may use more sophisticated methods, such as HPLC with fraction collection. Alternatively, the recent development of directly coupled, or ''hyphenated'' chromatographic systems such as HPLC–NMR may be the most appropriate solution, depending on the problem. In addition, there is the option of isolating the target analyte in the traditional way, without using NMR spectroscopy to monitor the process. Purification can obviously be based on a radiolabel if present, or some other specific means of detection, with structure elucidation by a range of spectroscopic techniques only when the material has been substantially cleaned up.

For clarity each of the four NMR-based methods (see foregoing 1–4) will be considered in turn. However, in practice, they are employed singly or in combination and in any appropriate order, to achieve the required identification.

Several examples of the application of NMR-based techniques to aid the identification of both xenobiotics and their metabolites are given later to illustrate the way in which NMR-based approaches can be used. A particular advantage of NMR-based methods is that signals resulting from the presence of an unknown compound contain structural information that can aid, or enable, identification. This contrasts markedly with the appearance of new peaks in a chromatogram when the information content is low. The chemical shift of the NMR signal allows relatively simple identification according to chemical type (e.g., CH_3—C, CH_3—O, aromatic, or other). If present, spin–spin couplings allow connectivity of the signals present in the spectrum to be determined enabling the structure of the unknown metabolite to emerge.

B. Homonuclear and Heteronuclear Two-Dimensional NMR Techniques as an Aid to Identification of Sample Components

On many occasions simple 1-D NMR spectroscopy will not provide an answer because of overlapping signals from a number of similar metabolites, or interference from endogenous materials (e.g., Bales et al., 1985). In such circumstances the application of 2-D NMR techniques may be useful. The 2-D ^{1}H COSY spectrum illustrated in Fig. 3 shows the aromatic region of the acetaminophen metabolite-containing urine from Fig. 2. In the 1-D spectrum the resonances for the unchanged acetaminophen and its *N*-acetylcysteinyl metabolite cannot easily be detected and differentiated from each other, and certainly, not all of the resonances for these substances can be seen owing to the presence of the sulfate and glucuronide metabolites of acetaminophen and hippurate. In the 2-D spectrum, however, all of these problems are overcome, owing to the separation of their off-diagonal cross-peaks. This example shows how 2-D COSY experiments can detect the presence of known metabolites obscured by the presence of a range of interfering substances. This 2-D technique is one of a suite of NMR methods that can be used to obtain more information about components in a sample. A further example of a 2-D technique that we have found useful for biofluid analysis, although we have not yet applied it to metabolic work, is homonuclear 2-D J-resolved (JRES) ^{1}H NMR spectroscopy, which can be used rapidly to detect peak overlap that might have resulted in poor quantification. One of the most useful 2-D NMR experiments for structural identification is the total correlation or TOCSY experiment. This provides connectivity information along an unbroken chain of proton–proton couplings, such as from H-1 round to H-6 in a glycoside ring. We have used this approach extensively in stop-flow HPLC–NMR studies of drug metabolism. If the drug metabolite has one clearly resolved spin-coupled signal, then a much more rapid 1-D analogue of the 2-D experiment can be carried out to map out the chemical shifts that are coupled along a chain of couplings to the resolved peak. This method has been used to identify signals from endogenous metabolites hidden under large envelopes of NMR resonances in seminal fluid, and others have used it to identify drug glucuronide signals (e.g., Leo and Wu, 1992).

It is also possible to use 2-D techniques for NMR studies of nuclei other than ^{1}H.

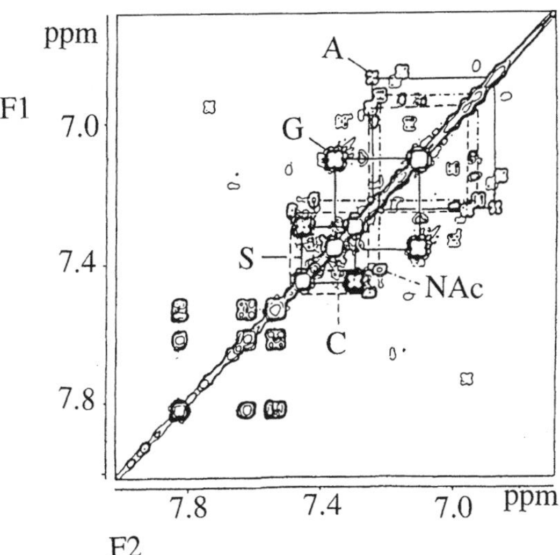

Fig. 3 A two-dimensional COSY ^{1}H NMR spectrum of urine (see Fig. 2) containing metabolites of acetaminophen.

As discussed earlier, ^{19}F NMR provides a useful means of obtaining both the number and proportions of fluorinated metabolites in samples. Further interpretation is limited because 1-D ^{1}H-decoupled ^{19}F NMR spectra are generally featureless, with singlet signals for each isolated fluorine. The information content of such spectra, therefore, is relatively low and, in many ways, similar to a chromatogram. However, the application of 2-D NMR techniques, in which the spin–spin coupling of the fluorine "label" to adjacent protons is exploited, enables more information to be obtained. Thus, 2-D ^{1}H–^{19}F COSY spectra can be used to obtain the chemical shift of any proton coupled to each ^{19}F nucleus. This technique was originally employed in studies on the urinary excretion of flucloxacillin and its metabolites in the urine of rats dosed at 200 mg/kg (Everett et al., 1985). The 2-D ^{1}H–^{19}F COSY spectra obtained revealed a total of four fluorine signals. These were identified as flucloxacillin and three metabolites [(5*R*)-flucloxacillin penicilloic acid, 5′-hydroxymethylflucloxacillin, and (5*S*)-flucloxacillin penicilloic acid]. Each of these four fluorine NMR peaks showed correlations with two coupled protons (corresponding to the protons *ortho* and *meta* to the fluorine nuclei). In the contour plot, overlapping NMR peaks from the *ortho* and *meta* protons were clearly resolved from overlap with one another or the resonances of endogenous compounds.

A similar approach can also be used for compounds in which ^{15}N or ^{13}C are present as a result of synthetic enrichment. For example, in studies on dimethylformamide the 2-D ^{1}H–^{13}C long-range chemical shift correlated (HETCOR) NMR spectrum for a rat urine sample obtained after intraperitoneal dosing with [^{13}C] dimethylformamide (DMF) showed a total of seven metabolites. In the NMR spectrum of urine for the ^{15}N-enriched DMF, nine metabolites were observed (Nicholson et al., unpublished observations). The reason that fewer metabolites were detected after administration of the ^{13}C-labeled DMF was simply a result of the loss of the label during metabolism. As for the flucloxacillin example, this type of experiment enables the analyst to obtain access to those signals in the ^{1}H NMR spectrum that relate to the compound under study. The current approach for this type of work would be to use "inverse" ^{1}H NMR detection, with correlation to the heteronuclei based on one-bond or long-range couplings, rather than direct detection of the ^{13}C or ^{15}N nucleus.

One of the first stages in assigning the ^{1}H signals present in the NMR spectrum of a sample is the identification of CH_3, CH_2, or CH resonances. This process can now be achieved unambiguously through the use of the MAXY NMR approach. This results in 1-D or 2-D NMR spectra being edited into separate subspectra according to whether the resonances are from CH_3, CH_2, or CH groups. Nevertheless, the sensitivity of this approach is less than that for conventional ^{1}H NMR spectroscopy because the method relies on selective detection of ^{1}H nuclei bonded to the 1.1% naturally abundant ^{13}C nuclei.

C. Enzymatic and Chemical Modification as an Aid to Metabolite Identification by NMR

1. *Phase II Metabolites Identified Using Enzymatic Hydrolysis*

Conjugates are among the most widely encountered metabolites and, for many years, have been characterized by specific enzymatic hydrolysis. In particular, the enzymes β-glucuronidase and aryl sulfatase have been widely used in metabolic drug studies for liberating drugs and their metabolites from their conjugates. Although chromatographic analysis before and after enzymatic hydrolysis is normally used, the application of NMR spectroscopy to the analysis of such samples is an obvious alternative, especially if the metabolites have been detected by NMR spectroscopy in the first place. Hydrolysis can be performed in the NMR tube itself, for which often the only necessary change is adjustment of the pH to that required by the enzyme for maximum activity. Thus, with appropriate controls, the identity of glucuronides and sulfates can easily be established (see later) using NMR spectroscopic detection. The major technical difficulty that we have encountered in using enzymes in this way is that, on occasion, the analyte (or liberated aglycone) can become bound to the enzyme, or perhaps some other contaminating protein, and thus its NMR signal is attenuated (or lost). Acidification of the sample can be used to overcome this binding.

A typical example of the use of this approach is given by a study on the urinary metabolites of the model compound 2-trifluoromethylphenol for which, following administration to the rat, a combination of ^{19}F NMR spectroscopy and selective enzyme hydrolysis was used to characterize the metabolites (Bollard et al., 1996). In this instance, 376 MHz ^{19}F NMR spectroscopy was used to obtain the metabolite profile illustrated in Fig. 4a. This spectrum shows the sample obtained for the 0- to 8-h urine following intraperitoneal administration at 10 mg/kg and reveals the absence of the parent phenol and presence of

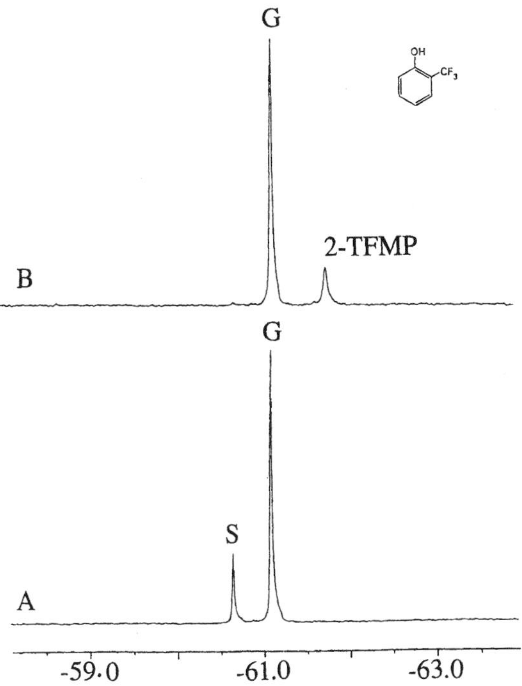

Fig. 4 (A) A 376-MHz ^{19}F NMR spectrum of rat urine containing metabolites of 4-trifluoromethylphenol (TFMP); (B) same urine after incubation with aryl sulfatase.

two metabolites. This example clearly shows how NMR can be used to provide excretion balance data, as quantification of the metabolites in the sample under such circumstances is technically undemanding. Incubation of this sample with β-glucuronidase or aryl sulfatase resulted in the reduction of the signals of glucuronide and sulfate conjugates, respectively, and a corresponding increase in the signal of the aglycone, which in this instance, was a hydroxylated metabolite of the parent phenol. In Fig. 4b the effect of incubation with sulfatase is shown. This is a typical example of what can be done using classic enzymatic hydrolysis, and works equally well for other nuclei.

However, although this example shows the value of enzymes as a way of obtaining further information on the nature and identity of metabolites, the results of enzymatic hydrolysis followed by NMR spectroscopy must be treated as cautiously as if they had been monitored by chromatography. For example, certain glucuronides are resistant to enzymatic hydrolysis with β-glucuronidase because of steric effects or, in those formed from carboxylic acids, transacylation. The failure of an enzyme to cause a change is not conclusive evidence that the particular metabolite is not a conjugate. Furthermore, the disappearance of a signal, and the concomitant appearance of another could easily be due to chemical instability, rather than a result of enzyme action. Lastly, it has to be remembered that not all preparations are pure, and contamination of the enzyme preparation with other hydrolytic enzymes might be the cause of observed changes in the NMR spectrum. Confirmation of the identity of a glucuronide or sulfate still requires either the demonstration of the inhibition of that hydrolysis by, for example, saccharolactone or phosphate, respectively, or unambiguous spectroscopic data.

However, performed with the appropriate control and inhibition studies enzymatic hydrolysis monitored by NMR can provide useful data, and the method development required is often minimal.

2. *Physical and Chemical Modification of Samples as an Aid to NMR Signal Assignment*

As well as using enzymes it is also possible to probe the structures of the metabolites by simple physical modification (such as changes in pH), or by chemical reaction. The reaction types that could be employed include hydrolytic reactions using acid or bases, oxidations and reductions, as well as more specific chemistry targeted at particular functional groups. The use of such reactions, monitored by NMR spectroscopy can result in diagnostic changes in the spectra of the affected compounds.

3. *Effect of pH Changes on NMR Spectra of Biofluids*

Many xenobiotics and metabolites contain acidic or basic functional groups, or both, and will demonstrate pH-dependent chemical shifts. However, probably the most useful effect of acids and bases is their ability to hydrolyze labile groups. This property is particularly useful for the base hydrolysis of ester glucuronides. Thus, although β-glucuronidase can be used to hydrolyze ester and ether glucuronides, in a mixture of both types of glucuronide, or when both types were potential metabolites for a given aglycone, distinguishing between them could be problematic based solely on enzymatic hydrolysis. However, ester glucuronides can be readily hydrolyzed with alkali, in contrast to ether glucuronides, which are relatively stable. Alkaline hydrolysis thus results in the loss of the signals owing to the ester glucuronides and a corresponding increase in the size of the peaks corresponding to the aglycone.

Apart from showing the way in which intentional alkaline hydrolysis can be used to identify ester glucuronides, this example also acts as a reminder of the need to ensure that samples containing alkali-labile groups, such as ester glucuronides, are correctly stored. Urine samples left to stand, unfrozen, will become alkaline (even if not alkaline already), and any ester glucuronides will either slowly hydrolyze or transacylate. Precautions can be taken to reduce or prevent this happening and merely require the acidification of the sample on collection (although little can be done to prevent in vivo hydrolysis or transacylation before collection).

4. *Detection of Free Thiols in Urine using Ellman's Reagent*

Specific chemical reactions for particular functional groups provide another route to identification. Ellman's reagent (5,5′-dithiobis-(2-nitrobenzoic acid), is a thiol-specific reagent that we have used as an aid to NMR signal assignment. This has been used for the analysis of the drug penicillamine (dimethylcysteine) and its metabolites in human urine (Nicholson et al., 1988). The drug is excreted in the urine both as the free thiol, the homodisulfide and as a mixed disulfide with cysteine. On standing, penicillamine, along with other thiols, slowly oxidizes to form disulfides. Measures to stabilize the free thiol by freezing, acidification, or addition of EDTA, were not completely effective in preventing this reaction. When added to urine samples, Ellman's reagent reacts with any free penicillamine present by forming a disufide derivative. This reaction both prevents the oxidation of penicillamine to its internal disulfide and causes a marked change in the chemical shift of the signals for the two nonequivalent methyl groups. This change in chemical shift thus immediately identifies the signals from any free thiols present in the sample.

5. *Addition of Cyclodextrins to Biofluids for the Identification of Enantiomers*

The importance of chirality, and its effects on the disposition and metabolic fate of xenobiotics has resulted in an increased need for methods of analysis that can determine enantiomeric composition. The bulk of such methods has been based on chromatography, but alternative methods for biofluids using NMR spectroscopy can also be developed using cyclodextrins (Taylor et al. 1992; Matthews et al., 1997). Naturally occurring α, β, and γ-cyclodextrins have been supplemented by a range of synthetically derivatized versions (some with greatly superior water solubility compared with β-cyclodextrin), and these offer further opportunities for chiral resolution in NMR spectroscopy (Taylor et al., 1991). Because cyclodextrins are built from D-glucose units (5, 6, and 7, respectively, for α, β, and γ-cyclodextrin) they are optically active. Therefore, drug enantiomers can form transient diastereomeric inclusion complexes with cyclodextrins. The application of cyclodextrins has particular advantages in NMR spectroscopic studies of biofluids because the formation of the inclusion complex works well in aqueous solution. An example of the use of these compounds is illustrated in Fig. 5a and b, in which the ^{19}F NMR spectra of a racemic mixture of Moshers acid (for structure see inset), obtained in a urine sample, in the presence and absence of β-cyclodextrin, are shown. The appearance of the signals for the individual enantiomers produced by the addition of β-cyclodextrin is quite clear, with good separation between the resonances for the *R* and *S* enantiomers of Moshers acid. The results were the same whether the analyte was present in buffer or urine, and were readily applied to the study of the elimination of both Moshers acid and *p*-trifluoromethylmandelic acid in the rat (Mathews, 1997).

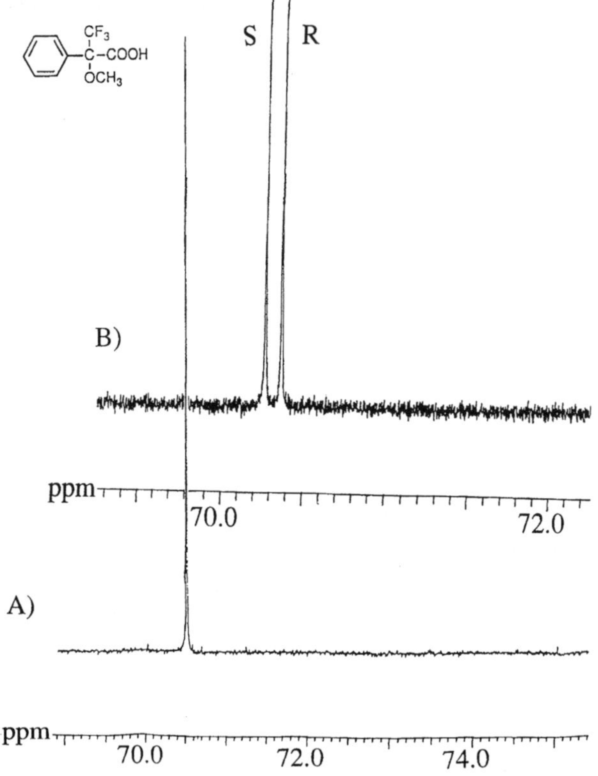

Fig. 5 (A) A 235.36-MHz ^{19}F NMR spectrum of rat urine following administration of Moshers acid; (B) the same sample following the addition of β-cyclodextrin (1:1 molEq).

V. EXTRACTION AND PURIFICATION AS AN AID TO THE IDENTIFICATION OF SAMPLE COMPONENTS

A. Introduction

A great deal of structural information on xenobiotic metabolites present in a biofluid sample can be derived using NMR-based techniques alone, or in combination with enzymatic or chemical reactions. When, as often occurs, this information is insufficient for structure elucidation, the next step is the isolation of the unknown(s) in a suitable form to obtain the requisite structural information, often using a combination of techniques, such as NMR spectroscopy and mass spectrometry to aid identification. Simple extractions may often be enough on their own to remove the endogenous interferences and enable further work to be done on the sample.

B. Liquid–Liquid Extraction

Liquid–liquid extraction is a tried and tested method that can be used to isolate and concentrate compounds of interest from a biomatrix and, if the extraction solvent is deuterated (e.g., deuterochloroform), the organic phase can be analyzed directly by NMR spectroscopy without further pretreatment. Liquid–liquid extraction has been used by us to isolate the *N*-acetylcysteinyl conjugate of *N*-methyl formamide, resonances for which had been observed in the ^{1}H NMR spectra of rat urine (Tulip et al., 1986). Ethyl acetate and acidic conditions (pH 2) were employed for the extraction and isolation of the metabolite, followed by identification with TLC and MS in addition to further NMR spectroscopic studies. Although applicable to some problems, liquid–liquid extraction tends not to be particularly suitable for the extraction and purification of polar, ionic drugs and metabolites, such as drug conjugates. Thus, the need to use very polar solvents for this type of compound to obtain high extraction efficiencies results in the coextraction of the very polar endogenous interferences that the analyst wishes to remove.

In practice, solid-phase extraction (SPE; liquid–solid extraction) is a somewhat more versatile method for preparing drug metabolite samples for NMR.

C. Solid-Phase Extraction

In solid-phase extraction (SPE), the biofluid sample, after suitable pretreatment (such as adjustment of pH) is passed through a sorbent bed, usually contained within the barrel of a polypropylene syringe. Provided that an appropriate sorbent has been selected, any metabolites will be extracted from the aqueous sample and, if suitable extraction conditions have been chosen, many unwanted contaminants will pass through the sorbent bed unretained. Careful selection of the elution solvent used to recover the metabolite can also provide a further cleanup of the analyte by leaving more unwanted contaminants adsorbed to the cartridge.

A typical example of the use of SPE to concentrate, and partially purify, drug metabolites from urine is shown in Fig 6a and b. Thus, Figure 6a shows the ^{1}H NMR spectrum of human urine obtained following administration of 600 mg of acetylsalicylic acid (aspirin). In addition to the normal range of urinary components, the spectrum reveals several new resonances in the aromatic region of the spectrum, arising from drug-related material. To extract the aspirin metabolites (the glycine conjugate salicyhippuric acid), 10 mL of urine was taken, acidified to pH 2 with 0.1 M HCl (to suppress ionization), and then loaded on to a 500 mg, 3 mL C_{18} BondElut cartridge. The cartridge had been "activated" by prewashing with methanol (5 mL) and then 0.1 M HCl (5 mL). Once the sample had been loaded, the cartridge was washed with 1 mL of 0.1 M HCl (to elute nonretained compounds), and the aspirin metabolites were then recovered using 5 mL of methanol. The solvent and residual water were then removed by a combination of evaporation and freeze-drying before NMR spectroscopic measurement in D_2O (see Fig. 6b). As can be seen, this relatively simple procedure has resulted in a sample that contains mainly drug-related material, essentially salicylhippuric acid, together with endogenous hippuric acid. Most of the other urinary components were not retained under the conditions employed for extraction and were eluted to waste. The use of deuterated solvents for the wash and elution steps could enable direct analysis by NMR spectroscopy if speed was an important consideration. Once an metabolite has been extracted, it is quite feasible to redissolve it in a nonaqueous solvent for NMR analysis. Indeed changing the solvent can be of benefit under certain circumstances. Thus, chemical reactions may be more easily performed in organic solvents, and if an aprotic solvent (e.g., DMSO-d_6) is used, it is possible to detect the NMR signals for protons, such as those on NH or OH groups, that would normally be lost by chemical exchange in D_2O.

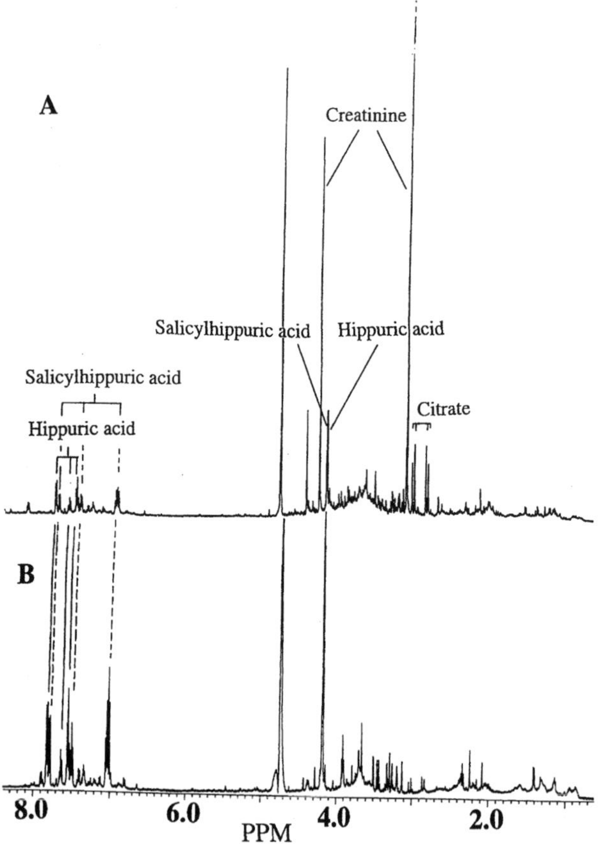

Fig. 6 (A) An ^{1}H NMR spectrum of human urine following administration of 600 mg of aspirin; (B) after solid-phase extraction. Assignments are as marked.

D. Solid-Phase Extraction–Chromatography

As illustrated in the foregoing, SPE can enable the extraction and concentration of metabolites to be performed rapidly and efficiently from a biofluid or from an in vitro incubation. Under favorable circumstances, it is also possible to obtain some degree of purification of the selected analytes, but the procedure still often results in a multicomponent mixture. It is possible to obtain a pure compound by SPE providing that a somewhat more sophisticated method is used.

The technique that we have developed to achieve greater purity of compounds of interest is basically stepwise gradient elution chromatography, using solvent mixtures of increasing eluotropic strength. Once extraction conditions have been developed, the solid-phase extraction–chromatography (SPEC) approach can often be used very successfully and rapidly to fractionate relatively complex mixtures (e.g., see Wilson and Ismail, 1986; Wilson and Nicholson, 1987, 1988).

A generic approach that we have used is sequential elution of the adsorbed metabolites with a gradient of 20, 40, 60, 80, and 100% methanol in 0.1% HCl (v/v). The results

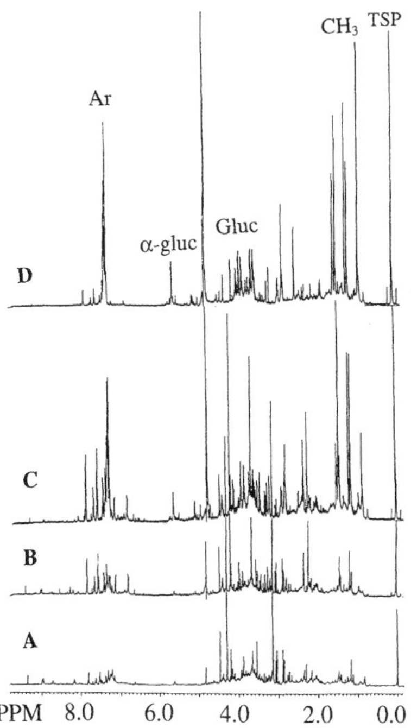

Fig. 7 (A) An ^{1}H NMR spectrum of human urine obtained following administration of ibuprofen. (b–d) The spectra of SPEC extracts showing the results obtained with stepwise elution using methanol–water mixtures of increasing eluotropic strength; (B) 20% methanol, this fraction contains little compound-related material; (C) 40% methanol, this fraction contains a large amount of phase 2 conjugates of phase 1 hydroxylated and carboxylated ibuprofen metabolites; and (d) 60% methanol, this fraction contains mainly ibuprofen glucuronide together with some of the material eluting in the 40% fraction. Key. Ar, aromatic resonances; CH$_3$, methyl resonances, α-gluc, anomeric proton of glucuronide conjugate; Gluc, glucuronide resonances.

of such an experiment for ibuprofen metabolites are shown in Fig. 7a–d. Here the 20% fraction contains hippuric acid (see Fig. 7a), with the 40 and 60% fractions (see Fig. 7b, c) containing only the ibuprofen metabolites. Close examination of the spectrum for the 60% fraction shows it to consist largely of ibuprofen glucuronide, with a small amount of two other metabolites present. The 80 and 100% fractions (not shown) contained no ibuprofen-related material, but did contain most of the coextracted urinary pigments.

By resubjecting each of the relevant fractions to SPEC–NMR again, using a shallower solvent gradient can enable the isolation of the individual components in a pure form, if required. Given the capacity of SPE cartridges, SPEC can be used to isolate milligram quantities of metabolites in a pure form from large volumes of biofluid. There are many examples where SPEC-NMR using C_{18} material for the separation has been successfully applied to the isolation of drug metabolites.

There are many types of phases available for SPE, including ion-exchange materials, various types of reversed-phase material, mixed mode, silica, and polymeric materials, providing a range of selectivities that can be exploited to achieve a purification. One potentially useful material for metabolism studies is immobilized phenylboronic acid, which can be used to selectively extract certain types of glucuronide (Tugnait et al., 1992, 1994) with good specificity.

VI. COUPLING OF HIGH-PERFORMANCE LIQUID CHROMATOGRAPHY TO NMR SPECTROSCOPY

A. Introduction

As illustrated earlier, SPEC–NMR can, under favorable circumstances, provide surprisingly good results with a minimum of effort. However, the limitations of SPEC are obvious, and it simply does not have the resolving power required to separate several similar metabolites that could be present in the same sample. In such cases, it will be necessary to employ a more powerful separation technique such as HPLC, but this can cause problems if the only means of working out when the component of interest elutes involves NMR. The logistical problems of analyzing a large number of HPLC fractions from a gradient run are clear, and many of the benefits of the use of NMR are lost.

The obvious solution to this problem is the direct coupling of HPLC and NMR spectroscopy, and fortunately, recent developments in HPLC–NMR probe design, digital electronics, and improved software have made this possible. Indeed it can now be said that HPLC–NMR can be considered as a routine application. The main technical difficulties that had to be solved before HPLC and NMR could be successfully hyphenated centered on areas such as a loss of chromatographic resolution as a result of long column-to-NMR transfer times and the large volume of flow cells used in NMR probes. A further difficulty was the result of the large dynamic range required by the need to detect signals from low concentrations of metabolites in the presence of much larger signals from the mobile phases. Another consideration was obviously the relatively low sensitivity of NMR in a flowthrough system relative to the concentrations of drug metabolites likely to be present in the HPLC eluates. Such technical problems surrounding HPLC–NMR have been effectively addressed and HPLC–NMR probes are now produced by several manufacturers. Technical aspects of HPLC–NMR, including its applications to drug metabolism have been reviewed in detail elsewhere (Spraul et al., 1993b; Lindon et al., 1995, 1996, 1997), and an illustrative example of the use of these techniques will be given in the following.

For the implementation of HPLC–NMR it is no longer a prerequisite that deuterated organic solvents need to be used for the chromatography. This problem has been solved, even for reversed-phase HPLC systems, because very efficient solvent NMR peak suppression can be performed. This eliminates the need for expensive deuterated solvents, although in practice D_2O, which is relatively inexpensive, is still often used in mobile-phase mixtures to make multiple solvent suppression easier. Methanol and acetonitrile, both commonly used for RP–HPLC, each give rise to singlet resonances in the ^{1}H NMR spectrum that can easily be suppressed. That said, ^{13}C satellite peaks for these solvents are also present following the suppression of the main peak and have the potential to cause problems, for these signals can be much larger than those of the metabolites. The suppression of these unwanted satellite peaks can be achieved in one of two ways.

In the first of these, suppression irradiation is set over the central peak and the two satellite peaks in a cyclic fashion. Alternatively, if an inverse geometry probe is used that includes a ^{13}C coil, decoupling of the ^{13}C–^{1}H splitting is possible which collapses the satellite peaks under the central peak and allows single frequency suppression. A valuable recent innovation enables automatic solvent suppression in gradient elution. Thus, if when a gradient elution method is used to resolve a metabolite mixture with, for example, acetonitrile in the solvent, the resonance frequency of the acetonitrile will change as the composition of the eluant is altered. An NMR software routine can be used that automatically finds ^{13}C satellite peaks for the solvent, interpolates them to find the main signal, and setting the suppression frequency accordingly. Other solutions to the problems associated with gradient elution will also be considered.

B. Modes of Operation of HPLC–NMR

In principle, it is possible to perform HPLC–NMR detection for any of the magnetically active nuclei, but those of most importance for drug metabolism studies are ^{1}H, ^{2}H, ^{19}F, ^{13}C, and ^{31}P. As a direct consequence of the generally low levels of drug metabolites, the most sensitive nuclei, ^{1}H, ^{19}F, and ^{31}P have been used most extensively. Obviously, as in conventional NMR spectroscopy of the biofluid sample, the use of ^{1}H NMR as the detection mode provides a general method for the detection of compounds, including the drug metabolites, eluting from the column. The use of ^{13}C NMR in HPLC–NMR can be facilitated through indirect detection of ^{13}C resonances by the much more sensitive ^{1}H NMR signals of attached protons. Again, as seen with direct biofluid analysis, the use of ^{19}F NMR spectroscopy for detection can provide major benefits in terms of selectivity in situations for which fluorine-containing metabolites are present.

There are currently five main operational modes that can be employed for HPLC–NMR (isocratic or gradient elution) as follows:

1. Continuous-flow
2. Stop-flow
3. Time-sliced stop-flow
4. Peak collection into capillary loops for postchromatographic analysis
5. Automatic detection of chromatographic peaks with triggered NMR acquisition

Clearly, the simplest option is continuous-flow detection. However, continuous-flow methods can usually be performed only with samples containing relatively large amounts of material with either ^{1}H or ^{19}F NMR detection (unless isotopically enriched compounds are used). That said, there are examples of HPLC–NMR drug metabolic studies using ^{2}H and ^{31}P NMR detection (Farrant et al., 1997; Foxall et al., 1996). As considered briefly earlier, in continuous-flow NMR with gradient elution, the NMR resonance positions of the solvent peaks will change with the changing eluant composition during the run. To ensure effective solvent suppression, these changing solvent resonance frequencies must be determined as the chromatographic run proceeds. One method of obtaining these frequencies is to use a ''dummy run.'' The frequencies obtained are stored in the data system, which then uses these values to set the irradiation frequency during the actual separation. An alternative, real-time approach, determines the solvent suppression frequencies with a single exploratory scan as soon as a chromatographic peak is detected and then performs solvent suppression.

If the retention times of the analytes are known, or there is a good method for their detection on-line, such as UV or radioactivity, stop-flow HPLC–NMR becomes a viable option. In the stop-flow option, all of the usual techniques of high-resolution NMR spectroscopy become available to the analyst. In particular, these include valuable 2-D techniques for structure determination, as described earlier. In practice, it is possible to acquire NMR data on a number of peaks in a chromatogram using a series of stops during elution without on-column diffusion causing an unacceptable loss of chromatographic resolution.

There are two further special categories of stop-flow experiment. First, fractions eluting from the column can be stored in capillary loops for later off-line NMR study (''peak picking''). Second, the flow can be halted at short intervals during the passage of the eluting peak through the NMR flow cell (''time-slicing'') in a manner analogous to the use of diode-array UV detector to obtain spectra from various portions of the peak. This allows chromatographic peak purity to be estimated. Time-slicing is most useful when the separation is poor, or when the compounds under study have weak, or no UV chromophores, making it difficult to determine the retention times.

Whichever type of HPLC–NMR experiment is performed, there is no need for compromises to be made on the chromatography used, including, for example, the type of gradient elution. In addition, although HPLC–NMR analyses are generally robust relative to changes in chromatographic conditions, the powerful structural capabilities of NMR spectroscopy ensure that complete chromatographic separation is not necessary for full structural identification of metabolites. Fully automated analysis is also an option wherein the samples are placed in an autosampler and predefined HPLC–NMR experiments are performed. The software allows automatic detection of UV peaks in the chromatogram based on predetermined time windows or peak intensities. Each UV peak triggers the chromatographic computer system to stop the flow, and then data relating to it (intensity, retention time) are transferred to the NMR computer and used to define the automatically acquired NMR spectrum. This automatic NMR operation includes field homogeneity optimization, setting and optimization of all NMR acquisition parameters, and the predefinition of the resultant signal–noise ratio required in the spectrum. The measurement of 2-D NMR spectra can also be performed. With currently available commercial software, the automated run can be halted at any time with reversion to manual control if desired.

To date HPLC–NMR spectroscopy has been applied to the profiling and identification of the metabolites of several drugs and xenobiotics present in biofluids, such as plasma, urine, and bile samples, from rat and humans and in vitro microsomal incubations. In general, the simple stop-flow approach has predominated in these studies. Clearly, each type of biofluid presents its own particular difficulties for chromatography and NMR spectroscopy, and these result from the different endogenous components present. The use of stop-flow HPLC–NMR to extend NMR spectral acquisition times has also been investigated, enabling spectra with high digital resolution to be obtained for relatively minor metabolites present in the samples.

Not all drug metabolism studies will be amenable to the HPLC–NMR approach, and those such as cyclosporine or steroid drugs, will be probably present in a sample at far too low a concentration to be detected. Nevertheless, HPLC–NMR spectroscopy will have a valuable role to play in the determination of the structures of xenobiotic metabolites. This approach is boosted by the widespread use of postacquisition data-processing methods for enhancing the recovery of information from NMR spectra. One technique that we have explored is the use of the maximum entropy method, based on bayesian statistics for which marked improvements in the NMR spectral signal–noise ratio was obtained for the minor *N*-acetylcysteinyl metabolite of acetaminophen, and this allowed detailed analysis of the NMR peak-splitting patterns and gave unequivocal information on the metabolite structure (Seddon et al., 1994).

C. Examples of HPLC–NMR Spectroscopy Applied to Drug Metabolism

The first example of the use of HPLC–NMR to examine a metabolism problem was the detection and identification of ibuprofen metabolites in human urine (Spraul et al., 1992, 1993). Since that time several applications of HPLC–NMR to in vivo and in vitro drug metabolism studies have been described, including acetaminophen (Spraul et al., 1994), antipyrine (Wilson et al., 1993), flurbiprofen (Spraul et al., 1993a), and ifosphamide (Foxall et al., 1996). The metabolism of various drugs currently under development has also been studied using HPLC–NMR spectroscopy. These include BW524W91 an anti-HIV compound (Shockor et al., 1996a); BW1370U87, a monoamine oxidase inhibitor (Shockor et al., 1996); and iloperidone, a candidate antipsychotic agent (Mutlib et al., 1995). In addition, the metabolism of several model compounds has been evaluated using HPLC–NMR spectroscopy. Finally, HPLC–NMR is uniquely suited to the study of the reactivity of ester glucuronide metabolites of various drugs and xenobiotics (e.g., see Lenz et al., 1996; Sidelmann et al., 1995a,b, 1996a–d). These applications have been reviewed in detail (Lindon et al., 1996, 1997).

A good example of the usefulness of HPLC–NMR in metabolic work is the study of the metabolic fate of racemic flurbiprofen ([±]-2-(2-fluoro-4-biphenylyl)propionic acid) in humans which forms a number of phase 1 and 2 metabolites (Spraul et al., 1993) (principally hydroxyflurbiprofen). The flurbiprofen-related material in urine is largely present in the form of glucuronides.

In this instance the presence of fluorine in the drug enabled the use of ^{19}F NMR spectroscopy for selective postchromatographic detection of metabolites. The ^{19}F NMR spectrum (with ^{1}H decoupling), measured at 564 MHz, of a human urine obtained following oral administration (200 mg) is shown in Fig. 8 and, from this, it would appear that

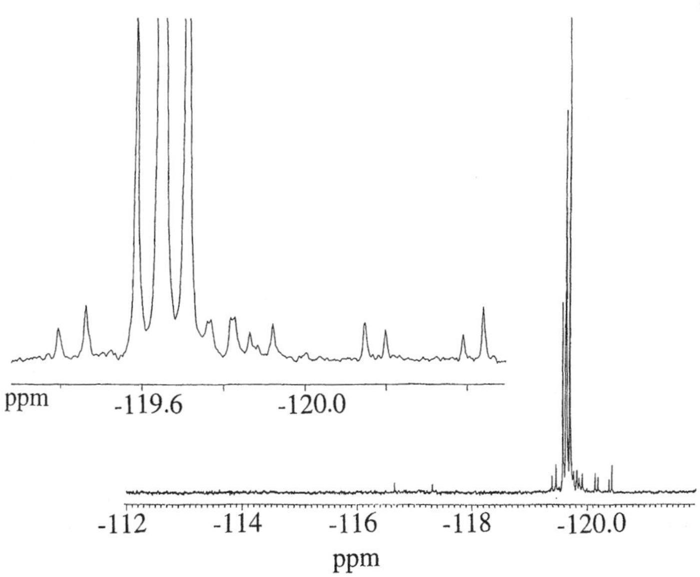

Fig. 8 The ^{19}F NMR spectrum of human urine following administration of 200 mg of flurbiprofen.

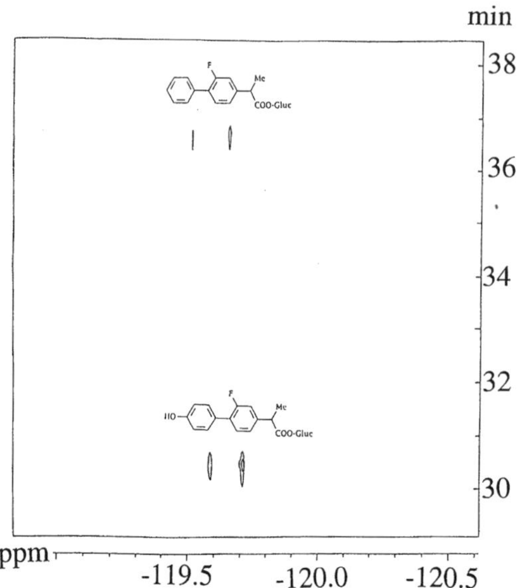

Fig. 9 The on-flow ^{19}F NMR spectrum obtained using HPLC–NMR for the sample shown in Fig. 8: The two pairs of peaks result from the presence of two pairs of diastereoisomers of flurbiprofen glucuronide and 4-hydroxyflurbiprofen glucuronide (see also Fig. 10).

three major (and a considerable number of minor) fluorine-containing species are in the sample. From the chemical shifts of these metabolites it would appear that the major sites of metabolism had occurred at positions distant to the fluorine-containing phenyl ring.

The sample was subjected to HPLC–NMR spectroscopy with ^{19}F detection and Fig. 9 shows the pseudo–2-D contour plot for the continuous-flow experiment. As this reveals, four ^{19}F resonances are present as two pairs of signals. The metabolites eluting at 30.5 min corresponded in chemical shift to the two largest resonances seen in Fig. 9, whereas the second pair corresponded to the remaining major components. The elution of these substances as pairs of peaks was a result of the metabolites being present as diastereoisomeric ester glucuronides. By repeating the separation using stop-flow ^{1}H NMR detection at 600 MHz, ^{1}H spectra of the metabolites were obtained (Fig. 10a,b), which were consistent with the β-D-glucuronic acid conjugate of the hydroxylated metabolite and flurbiprofen glucuronide. For both sets of glucuronides, the β-D-glucuronic acid H-1 proton is present as two pairs of doublets at δ5.49 and δ5.52 but, because the diastereoisomers are not present in equal proportions, the ratios of these two are not 1 : 1. Similarly, the resonances for the methyl resonances of the propionic acid side chain, readily detected as a pair of doublets in the aliphatic portion of the spectrum are not equal in intensity. Using the time-slicing technique, and eluting for a further 30 s, it was possible to obtain an ^{1}H NMR spectrum for only one of the diastereoisomers of flurbiprofen-β-D-glucuronide as a result of the partial chromatographic resolution. Further studies with ^{1}H NMR in stop-flow mode enabled a further, minor, flurbiprofen metabolite to be identified as the free 4′-hydroxyflurbiprofen.

As this example illustrates, the specificity and simplicity of the ^{19}F NMR spectra greatly eased the task of identifying which of the peaks in the chromatogram were drug-related, enabling the analyst to concentrate on identifying the metabolites of the analyte.

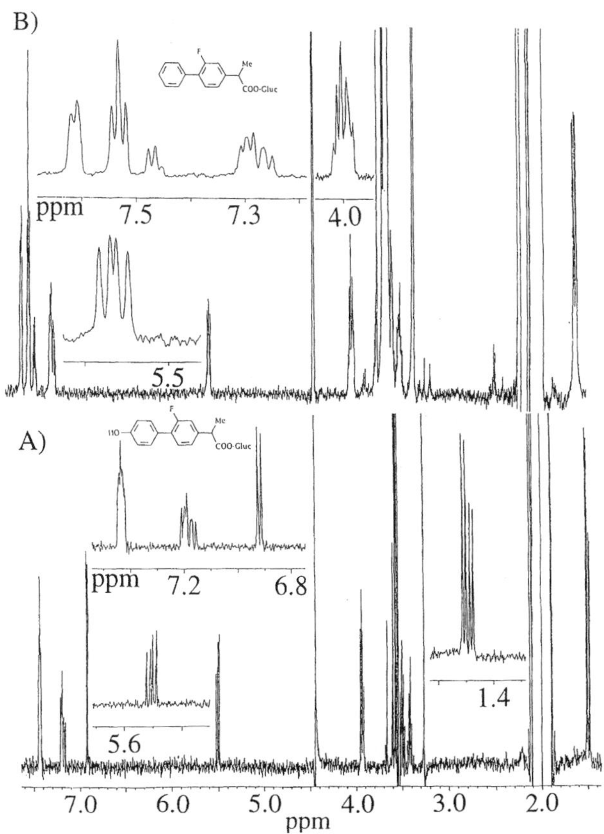

Fig. 10 Stop-flow HPLC–^{1}H NMR spectrum from human urine following administration of 200 mg flurbiprofen: (A) HPLC peak at 30.5 min, corresponding to diastereomeric 4-hydroxyflurbiprofen glucuronides; (B) HPLC peak at 36.6 min, corresponding to diastereomeric flurbiprofen glucuronides.

VII. HPLC–NMR–MASS SPECTROMETRY AS AN AID TO METABOLITE IDENTIFICATION

In the identification of unknown metabolites, NMR spectroscopy and mass spectrometry (MS) are clearly complementary, and data from both is often required to confirm a definite metabolite structure. Mass spectrometry, and its application to metabolism studies is described elsewhere in this volume. However, the recent combination of mass spectrometry and HPLC–NMR, to give the doubly hyphenated technique of HPLC–NMR–MS is noteworthy in that it enables high quality NMR and mass spectra to be obtained from the same HPLC run, at the same time, with enormous benefits for efficient metabolite identification. Problems that have to be addressed include ensuring that the chromatographic system used is compatible with both types of spectrometer in terms of solvents for NMR spectroscopy and ionization conditions for the MS, integration of the data capture, the optimal configuration of the spectrometers, and the need to ensure that the magnetic field of the NMR instrument does not adversely affect the performance of the mass spectrometer.

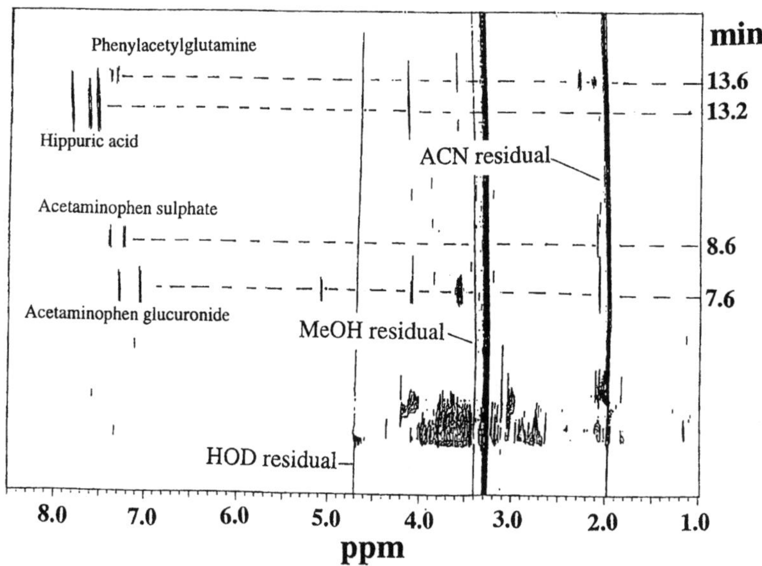

Fig. 11 HPLC–NMR–MS of acetaminophen: The on-flow NMR trace showing the compound-related material. Assignments are as shown.

Certainly, such systems are still evolving; however, the practicability of the approach has been demonstrated. There have now been several applications of LC–NMR–MS to the identification of drug or xenobiotic metabolites in urine (Shockor et al., 1996c; Scarfe et al., 1997; Burton et al., 1997) involving the identification of acetaminophen metabolites in humans and bromotrifluoromethylaniline metabolites in rat.

In the acetaminophen study, conventional RP-HPLC was used with a mobile phase consisting of acetonitrile–water and trifluoroacetic acid. The eluant from the C_{18} column was split, with 95% of the flow directed to the NMR spectrometer, and 5% to the ion-trap MS. This setup enabled the rapid identification of acetaminophen glucuronide, which would have been relatively easy by NMR alone, and acetaminophen sulfate. The confirmation of the identity of the sulfate metabolite was greatly facilitated by having both NMR and MS available, for the sulfate moiety is NMR "invisible." The data generated by this experiment are given in Figs. 11 and 12. Also, in the experiment it was possible to identify an unknown endogenous component that was present in the sample as phenylacetylglutamine, and again, both the NMR and MS data were required for unequivocal identification.

More recently we have studied the metabolic fate of bromotrifluoromethylaniline in the rat by HPLC–NMR–MS (Scarfe et al. 1997) and were able to confirm the identity of the major compound-related material as the sulfate conjugate of a ring-hydroxylated metabolite. The trifluoromethyl group was used to provide an NMR handle with which to monitor the separation, and ^{1}H NMR and MS were obtained on several peaks. Again, both techniques were required to identify the metabolite, with the MS data distinguishing between simple hydroxylation and sulfation, and NMR providing the site of hydroxylation on the ring.

Although it is arguable that in both cases the same results could easily have been obtained on separate systems, there seems no doubt that the double-hyphenation of these

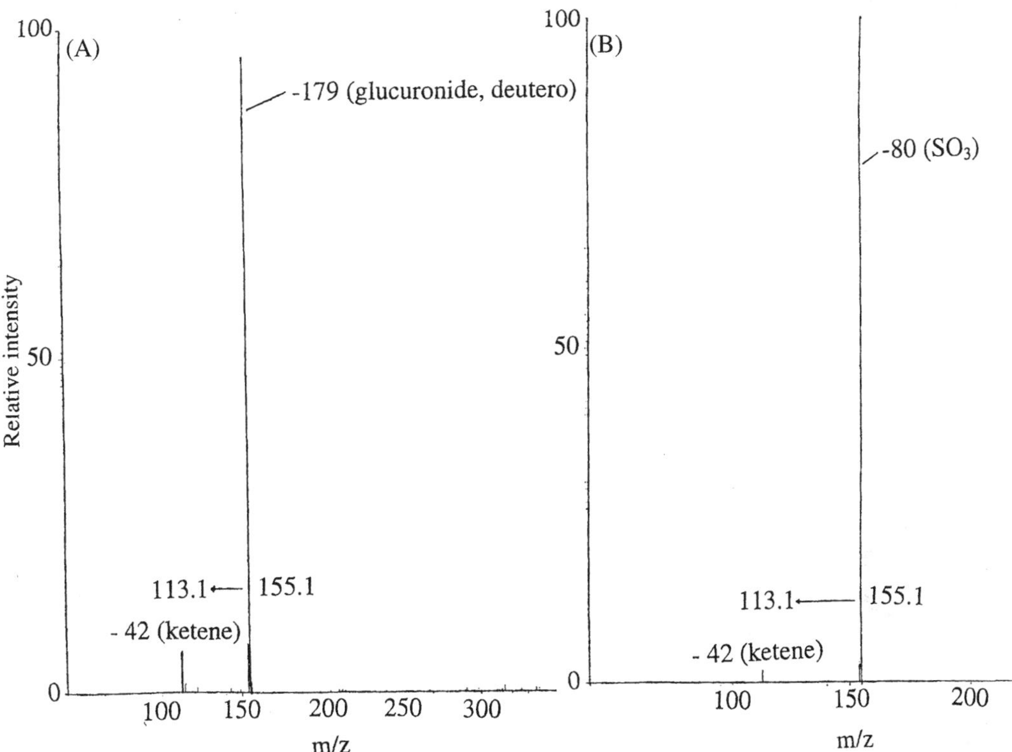

Fig. 12 HPLC–NMR–MS of acetaminophen: Mass spectra of the product ions of (A) acetaminophen glucuronide (m/z 334) and (B) acetaminophen sulfate (m/z 235; see also Fig. 11).

instruments made the whole process more efficient and ensured that the NMR and MS data were acquired on the same chromatographic peak and, hence, on the same material.

VIII. CONCLUSIONS

Nuclear magnetic resonance continues to be an invaluable aid to the identification of xenobiotic metabolites. With the recent increases in technological capabilities, we are sure that the role of NMR spectroscopy will increase. Furthermore, with the recent advent of HPLC–NMR, which is now commercially available, and the latest developments in HPLC–NMR–MS and in capillary HPLC–NMR, the use of NMR-based methods to study drug metabolism will be considered a routine and mainstream approach.

REFERENCES

Bales, J. R., P. J. Sadler, J. K. Nicholson, and J. A. Timbrell, Urinary excretion of acetaminophen and its metabolites as studied by proton NMR spectroscopy, *Clin. Chem. 30*:1631–1636 (1984).

Bales, J. R., J. K. Nicholson, and P. J. Sadler, Two dimensional proton nuclear magnetic resonance ‘‘maps’’ of acetaminophen metabolites in human urine. *Clin. Chem. 31*:757–762 (1985).

Bollard, M., E. Holmes, C. A. Blackledge, J. C. Lindon, I. D. Wilson, and J. K. Nicholson, ^{1}H and ^{19}F-NMR studies on the metabolism and urinary excretion of mono- and disubstituted phenols in the rat, *Xenobiotica 26*:255–273 (1996).

Burton, K. I., J. R. Everett, M. J. Newman, F. S. Pullen, D. S. Richards, and A. G. Swanson, On-line liquid chromatography coupled with high field NMR and mass spectrometry (LC–NMR–MS): A new technique for drug metabolite structure elucidation. *J. Pharm. Biomed. Anal. 15*:1903–1912 (1997).

Everett, J. R., K. Jennings, and G. Woodnut, ^{19}F NMR spectroscopy of the metabolites of flucloxacillin in rat urine, *J. Pharm. Pharmacol. 37*:869–873 (1985).

Foxall, P. J. D., E. M. Lenz, J. C. Lindon, G. H. Neild, I. D. Wilson, and J. K. Nicholson, NMR and HPLC-NMR studies on the toxicity and metabolism of ifosfamide, *Ther. Drug Monitor. 18*:498–505 (1996).

Hesk, D., J. R. Jones, and W. J. S. Lockley, Regiospecific deuteration and tritiation of various drugs using a homogeneous rhodium trichloride catalyst, *J. Pharm. Sci. 80*:887–890 (1991).

Jones, J. R., Tritium-synthesis, NMR and applications. *J. Labeled Compounds Radiopharm. 33*:369 (1993).

Lenz, E. M., D. Greatbanks, I. D. Wilson, M. Spraul, M. Hofmann, J. C. Lindon, J. Troke, and J. K. Nicholson, Direct characterisation of drug glucuronide isomers in human urine by HPLC–NMR spectroscopy: Application to the positional isomers of 6,11-dihydro-11-oxo-dibenz[*b*,*e*]oxepin-2-acetic acid glucuronide, *Anal. Chem. 68*:2832–2837 (1996).

Leo, G. C. and W.-N. Wu, The utility of one dimensional homonuclear Hartman-Hahn spectroscopy (1D HOHHAHA) for identifying the suprofen glucuronides fraction from an HPLC separation, *J. Pharm. Biomed. Anal. 10*:607–613 (1992).

Lindon, J. C., J. K. Nicholson, and I. D. Wilson, The development and application of coupled HPLC–NMR spectroscopy, Adv. Chromatogr. *36*:315–382 (1995).

Lindon, J. C., J. K. Nicholson, and I. D. Wilson, Direct coupling of chromatographic separations to NMR spectroscopy, *Prog. NMR Spectrosc. 29*:1–49 (1996).

Lindon, J. C., J. K. Nicholson, U. G. Sidelmann, and I. D. Wilson, Development of directly-coupled HPLC-NMR and its application to drug metabolism, *Drug Metab. Rev. 29*:705–746 (1997).

Malet-Martino, M. C. and R. Martino, Application of nuclear magnetic resonance spectroscopy to drug metabolism studies, *Xenobiotica 19*:583–607 (1989).

Matthews, D. B., R. H. Hinton, B. Wright, I. D. Wilson, and D. Stevenson, Bioanalysis of *p*-trifluoromethylmandelic acid and Moshers acid by chiral gas chromatography and fluorine nuclear magnetic resonance to study chiral inversion: Application to rat urine samples, *J. Chromatogr. B 696*:279–285 (1977).

Mutlib, A. E., J. T. Strupczewski, and S. M. Chesson, Application of hyphenated LC/NMR and LC/MS techniques in rapid identification of in vitro and in vivo metabolites of loperidone, *Drug Metab. Dispos. 23*:951–964 (1995).

Nicholls, A. W., S. T. Caddick, I. D. Wilson, R. D. Farrant, J. C. Lindon, and J. K. Nicholson, High resolution NMR spectroscopic studies on the metabolism and futile deacetylation of 4-hydroxyacetanilide-d_3 (paracetamol) in the rat, *Biochem. Pharmacol. 49*:1155–1164 (1995).

Nicholls, A. W., R. D. Farrant, J. P. Shockcor, S. E. Unger, I. D. Wilson, J. C. Lindon, and J. K. Nicholson, NMR and HPLC–NMR spectroscopic studies of futile deacetylation in paracetamol metabolites in rat and man, *J. Pharm. Biomed. Anal. 15*:901–910 (1997).

Nicholls, A. W., J. C. Lindon, S. Caddick, R. D. Farrant, I. D. Wilson, and J. K. Nicholson, NMR spectroscopic studies on the metabolism and futile deacetylation of phenacetin in the rat, *Xenobiotica* (in press, 1997).

Nicholson, J. K., J. A. Walshe, and I. D. Wilson, Application of high resolution ^{1}H-NMR spectroscopy to the detection of penicillamine and its metabolites in human urine, *Drug Metab. Drug Interact. 6*:439–446 (1988).

Nicholson, J. K. and I. D. Wilson, High resolution nuclear magnetic resonance spectroscopy of biological fluids as an aid to drug development, *Prog. Drug Res. 31*:427–479 (1985).

Scarfe, G. B., I. D. Wilson, M. Spraul, M. Hofmann, U. Brauman, J. C. Lindon, and J. K. Nicholson, Application of directly-coupled high-performance liquid chromatography–nuclear magnetic resonance–mass spectrometry to the detection and characterisation of the metabolites of 2-bromo-4-trifluoromethylaniline in rat urine, *Anal. Commun. 34*:37–39 (1997).

Seddon, M. J., M. Spraul, I. D. Wilson, J. K. Nicholson, and J. C. Lindon, Improvement in the characterisation of minor drug metabolites from HPLC–NMR studies through the use of quantified maximum entropy processing of NMR spectra, *J. Pharm. Biomed. Anal. 12*:419–424 (1994).

Sidelmann, U., C. Gavaghan, H. A. J. Carless, R. D. Farrant, J. C. Lindon, I. D. Wilson, and J. K. Nicholson, Identification of the novel positional isomers of 2-fluorobenzoic acid 1-*O*-acyl-glucuronide by directly coupled HPLC–NMR, *Anal. Chem. 67*:3401–3404 (1995a).

Sidelmann, U., C. Gavaghan, H. A. J. Carless, M. Spraul, M. Hofmann, J. C. Lindon, I. D. Wilson, and J. K. Nicholson, 750 MHz directly coupled HPLC–^{1}H NMR: Application for sequential characterisation of the positional isomers and anomers of 2-, 3-and 4-fluorobenzoic acid glucuronides in equilibrium mixtures, *Anal. Chem. 67*:4441–4445 (1995b).

Sidelmann, U., E. M. Lenz, P. N. Sanderson, M. Hofmann, M. Spraul, J. C. Lindon, I. D. Wilson, and J. K. Nicholson, 750 MHz HPLC–NMR spectroscopic studies on the separation and characterisation of the positional isomers of the glucuronides of 6,11-dihydro-11-oxobenz [*b,e*]oxepin-2-acetic acid, *Anal. Chem. 68*:106–110 (1996a).

Sidelmann, U. G., A. W. Nicholls, P. Meadows, J. Gilbert, J. C. Lindon, I. D. Wilson, and J. K. Nicholson, Directly-coupled ^{19}F and ^{1}H HPLC–NMR analysis of mixtures of isomeric ester glucuronide conjugates of trifluoromethylbenzoic acids, *J. Chromatogr. A 728*:377–385 (1996b).

Sidelmann, U. G., S. H. Hansen, C. Cavaghan, H. A. J. Carless, J. C. Lindon, R. D. Farrant, I. D. Wilson, and J. K. Nicholson, Measurement of internal acyl migration reaction kinetics using directly-coupled HPLC–NMR: Application for the positional isomers of synthetic 2-fluoro-benzoyl-D-glucopyranuronic acid, *Anal. Chem. 68*:2564–2572 (1996c).

Sidelmann, U. G., S. H. Hansen, C. Gavaghan, A. W. Nicholls, H. A. J. Carless, J. C. Lindon, I. D. Wilson, and J. K. Nicholson, Development of a simple HPLC method for the separation of mixtures of positional isomers and anomers of synthetic 2-, 3- and 4-fluorobenzoic acid glucuronides formed via acyl migration reactions, *J. Chromatogr. Biomed. Appl. B 685*:113–122 (1996d).

Shockcor, J. P., L. S. Silver, R. W. Wurm, P. N. Sanderson, R. D. Farrant, B. C. Sweatman, and J. C. Lindon, Characterisation of in vitro metabolites from human liver microsomes using directly coupled HPLC–NMR: Application to a phenoxathiin monoamine oxidase-A inhibitor, *Xenobiotica 26*:41–48 (1996a).

Shockcor, J. P., R. W. Wurm, L. W. Frick, P. N. Sanderson, R. D. Farrant, B. C. Sweatman, and J. C. Lindon, HPLC–NMR identification of the human urinary metabolites of (−)-*cis*-5-fluoro-1-[2-(hydroxymethyl)-1,3-oxathiolan-5-yl]cytosine, a nucleoside analogue active against human immunodeficiency virus (HIV), *Xenobiotica 26*:189–199 (1996b).

Shockcor, J. P., S. E. Unger, I. D. Wilson, P. J. D. Foxall, J. K. Nicholson, and J. C. Lindon, Combined hyphenation of HPLC, NMR spectroscopy and ion-trap mass spectrometry (HPLC–NMR–MS) with application to the detection and characterisation of xenobiotic and endogenous metabolites in human urine, *Anal. Chem. 68*:431–4435 (1996c).

Spraul, M., M. Hofmann, P. Dvortsak, J. K. Nicholson, and I. D. Wilson, Liquid chromatography coupled with high-field proton NMR for profiling human urine for endogenous compounds and drug metabolites, *J. Pharm. Biomed. Anal. 10*:601–605 (1992).

Spraul, M., M. Hofmann, P. Dvortsak, J. K. Nicholson, and I. D. Wilson, High-performance liquid chromatography coupled to high-field proton nuclear magnetic resonance spectroscopy: Application to the urinary metabolites of ibuprofen, *Anal. Chem. 65*:327–330 (1993).

Spraul, M., M. Hofmann, I. D. Wilson, E. Lenz, J. K. Nicholson, and J. C. Lindon, Coupling of HPLC with ^{19}F- and ^{1}H-NMR spectroscopy to investigate the human urinary excretion of flurbiprofen metabolites, *J. Pharm. Biomed. Anal. 11*:1009–1015 (1993a).

Spraul, M., M. Hofmann, J. C. Lindon, J. K. Nicholson, and I. D. Wilson, Liquid chromatography coupled with high-field proton NMR: Current status and future prospects, *Anal. Proc. 30*: 390–392 (1993b).

Spraul, M., M. Hofmann, J. C. Lindon, J. K. Nicholson, and I. D. Wilson, Evaluation of liquid chromatography coupled with high-field ^{1}H NMR spectroscopy for drug metabolite detection and characterisation: The identification of paracetamol metabolites in urine and bile, *NMR Biomed. 7*:295–303 (1994).

Taylor A., D. A. R. Williams, and I. D. Wilson, Derivatised β-cyclodextrins combined with high field NMR for enantiomer analysis: Application to ICI 185,282 (5(*Z*)-7-([2*RS*,4*RS*,5*SR*]-4-*O*-hydroxyphenyl-2-trifluoromethyl-1,-dioxan-5-yl) heptenoic acid), *J. Pharm. Biomed. Anal. 6*:493–496 (1991).

Taylor, A., C. A. Blackledge, J. K. Nicholson, D. A. R. Williams, and I. D. Wilson, Use of cyclodextrins and their derivatives for chiral analysis by high field nuclear magnetic resonance spectroscopy, *Anal. Proc. 29*:229–231 (1992).

Tugnait M., F. Y. K. Ghauri, I. D. Wilson, and J. K. Nicholson, NMR-monitored solid-phase extraction of phenolphthalein glucuronide on phenylboronic acid and C_{18} bonded phases, *J. Pharm. Biomed. Anal. 9*:895–899 (1992).

Tugnait, M., I. D. Wilson, and J. K. Nicholson, High resolution 1H NMR spectroscopic monitoring of extraction of model glucuronides on phenylboronic acid and C_{18} bonded phases, *Sample Preparation for Biomedical and Pharmaceutical Analysis* (D. Stevenson and I. D. Wilson, eds.), Plenum, New York, 1994, p. 127–138.

Tulip, K., J. A. Timbrell, J. K. Nicholson, I. D. Wilson, and J. Troke, A proton magnetic resonance study of the metabolism of *N*-methylformamide in the rat, *Drug Metab. Dispos. 14*:746–749 (1986).

Wilson, I. D. and I. M. Ismail, A rapid method for the isolation and identification of drug metabolites from human urine using solid phase extraction and proton NMR spectroscopy, *J. Pharm. Biomed. Anal. 4*:663–665 (1986).

Wilson, I. D. and J. K. Nicholson, Solid-phase extraction chromatography and nuclear magnetic resonance spectrometry for the identification and isolation of drug metabolites in urine, *Anal. Chem. 59*:2830–2832 (1987).

Wilson, I. D. and J. K. Nicholson, Solid-phase extraction chromatography and NMR spectroscopy (SPEC–NMR) for the rapid identification of drug metabolites in urine, *J. Pharm. Biomed. Anal. 6*:151–165 (1998).

Wilson, I. D. and J. K. Nicholson, Nuclear magnetic resonance spectroscopy applications: Biological fluids, *Encyclopaedia of Analytical Science*, 1995, 3518–3531.

Wilson, I. D., J. K. Nicholson, M. Hofmann, M. Spraul, and J. C. Lindon, Investigation of the human metabolism of antipyrine using coupled liquid chromatography and NMR spectroscopy of urine, *J. Chromatogr. 617*:324–328 (1993).

21

Preclinical Drug Metabolism Studies and Strategies

Philip B. Inskeep and Wesley W. Day
Pfizer Inc., Groton, Connecticut

I. INTRODUCTION

A major function of preclinical drug metabolism studies has been, and continues to be, to define absorption, distribution, metabolism, and elimination (ADME) of a drug in animal models for extrapolation to humans. If these profiles are similar in animal models and human, results from pharmacological and toxicological studies in animals may be more directly related to the clinical efficacy and safety in humans. When significant differences exist between animals and humans, these studies may define additional experimentation that may be required to fully evaluate the benefits and risks of the new agent. For example, if a major metabolite in humans is not observed in the toxicology species, additional toxicology studies may be needed to evaluate the toxicological profile of that metabolite.

In addition to this historically important mission of preclinical drug metabolism, pharmaceutical research organizations in recent years have placed a greater emphasis on use of similar studies in the candidate discovery process. Early identification of ADME problems in new chemical series may help drive synthetic efforts to identify potent pharmacologically active agents that are more likely to possess an appropriate pharmacokinetic profile in humans.

With the growth in technology and everpresent competition in the health care industry, it is not enough to just provide a potent therapeutic agent. Depending on the therapeutic area, providing a drug with the optimal pharmacokinetic profile may give that product a competitive edge. Moreover, with emphasis on price, many programs require once-a-day administration to reduce cost of goods as well as providing convenience to the patient. This is certainly true for lowering cholesterol, blood sugar, or blood pressure, for which a long $t_{1/2}$ or sustained-release for around-the-clock coverage has become the rule and not the exception. This is exemplified by the success of nifedipine (Procardia XL; Michelson, 1991). This compound is a calcium channel blocker developed to treat hypertension. Originally, the compound was efficacious, but required frequent dosing. After reformulation, the sustained-release form, allowing once-a-day dosing with round-the-clock coverage, rapidly expanded market share for the product. Conversely, for a compound used in a critical care or perioperative setting that is dosed through intravenous administration, a short $t_{1/2}$ may be preferable to facilitate rapid cessation of the pharmacological action beyond the desired period and, thereby, minimize the potential for adverse effects.

With preclinical drug metabolism data, one can apply predictive models using in vivo allometry or other in vitro human metabolism or absorption techniques. Application of these models in early stages may provide insight into any potential weakness in the transport mechanism or metabolism in humans; consequently, a program actively applying these models can have a greater degree of confidence before expending resources required to advance a compound into the clinic. There are many examples of successful application of human microsomal metabolism screening used in programs in which polymorphic isozymes may be involved. Smith et al. (1996) from Pfizer have summarized how knowledge of the reactions catalyzed by cytochrome P450 enzymes are being used to improve the design of drugs. Although numerous studies have been published that use Caco-2 cell monolayers as an in vitro surrogate of human absorption (Artusson et al., 1991), it was only recently demonstrated that this system could be used in discovery, to prospectively predict both the extent and the mechanism of absorption in humans (Yee, 1997). Through combination of human in vitro surrogates for metabolism and absorption with pharmacokinetics in two or more species, accurate predictions of bioavailability, C_{max}, T_{max}, and $t_{1/2}$ can be made prospectively to doses in humans (Obach et al., 1996; Liston et al., 1996).

In this chapter, we will discuss potential drug metabolism strategies for early preclinical support and typical approaches for a preclinical drug metabolism program once a lead candidate is identified. Thus, the discussion is divided into the two phases often used to describe preclinical research: candidate identification and candidate development. In the candidate identification or discovery phase, the pharmacokinetics and metabolism of pharmacologically active agents are determined in animal models as part of the project progression strategy. Compounds clearing the discovery process are then progressed into the candidate development phase, where more detailed characterization of ADME as well as toxicology are completed. Often, the beginning of the development phase is defined as when toxicological evaluation begins. Despite the term ''preclinical'' many of the advanced preclinical studies may be conducted after early clinical studies have started.

II. CANDIDATE IDENTIFICATION

Historically, ADME studies occurred after a compound was selected, based solely on pharmacological endpoints, for advancement into development. At this point, information was generated to establish toxicology doses, to guide the dose selection for phase I and to provide information to regulatory agencies on exposure–activity–toxicity relations. Over the last 10 years, industrial needs for improved efficiency, greater throughput of compounds, with less chance of attrition in the clinic have resulted in drug metabolism scientists participating early in the discovery process of a drug. In a report describing compounds developed by the seven UK-owned pharmaceutical companies through 1985, 39.4% of the 198 compounds discontinued for development had been terminated owing to inappropriate pharmacokinetics (Smith, 1994; Prentis et al., 1988). Therefore, from a dispositional perspective, it is important to design the most efficient screening processes to select pharmacokinetically viable as well as therapeutically viable compounds in the discovery phase.

A. Objectives and Roles of Drug Metabolism Scientist

There are three major areas in which an effective drug metabolism program can have an impact in the discovery phase.

1. Selection of compounds with relevant pharmacokinetic parameters for the therapeutic area.
2. Quantitative prediction of in vivo exposure and $t_{1/2}$ in humans for design of phase I studies.
3. Customization of any ADME characteristics that may differentiate a compound from others in its class. This may be especially important if it is anticipated that the compound will not be first-in-class.

The responsibility of the drug metabolism scientist in the discovery process is to determine and optimize the relation between the chemistry and pharmacology profile through quantitative and qualitative consideration of ADME. Ordinarily, this can be accomplished with some strategic in vivo or in vitro screening to facilitate an understanding of the pharmacokinetic or physiochemical limitations of the chemical series in question. Often these limitations are directly responsible for limiting in vivo activity in the pharmacodynamic model. Therefore, an iterative feedback with the medicinal chemists should be established for simultaneous consideration of both pharmacological (structure–activity relation; SAR) and dispositional characteristics (structure–metabolism relation; SMR, or structure permeability relation; SPR) of the chemical series.

Examples of how the chemistry of drugs can dramatically affect the disposition and activity profile were well summarized in a report by Gumbleton and Sneader (1994). Omeprazole, a proton pump inhibitor used to decrease gastric acid secretion of the parietal cells in the small intestine, is a weak base (p*K*a = 4.0) that is rapidly absorbed in the basic environment of the upper intestine. Once omeprazole enters the parietal cell, the acidic intracellular environment results in complete ionization and entrapment of drug. Subsequently, it is converted to the active moiety, a sulphenamide metabolite, which acts irreversibly on the target enzyme. Omeprazole itself has bioavailability of 40% and a 1-h $t_{1/2}$, but owing to the activation of the drug following intracellular entrapment, a single dose has a 1-day duration of action. Although this example typifies an extreme example of what effective molecule design based on dispositional understanding may yield, as the understanding and application of ADME studies earlier in discovery is increased, the ability to design molecules that retain activity and possess desirable dispositional attributes will improve.

B. Identification of Relevant ADME Characteristics for Predevelopment Compounds

The question now arises as to how one defines the relevant dispositional characteristics for an active discovery program. Campbell (1994) has described two strategies for moving quickly through preclinical discovery. The first involves initiating pharmacology, drug metabolism, and toxicology regulatory nonclinical studies quickly and in parallel (brute force approach). Although this approach provides a lot of data, it involves extensive resources. Consequently, it is costly and risky and may be practical only for global companies. The second strategy involves selection and design of all studies to sequentially obtain information for compound progression and avoid studies that could be delayed until the success of the compound is more assured. In reality, with the growth in technology and understanding of mechanisms regulating disposition, the latter approach can be just as productive as the former, and is much more efficient.

With the analytical resources now available, extensive dispositional data can be generated from the in vivo model concurrent with screening in a timely manner. Furthermore, if ADME information obtained from the pharmacology model is used to develop parallel in vivo or in vitro screening models, improvements in the ADME profile can be made quite rapidly. Often, physiochemical data can explain limitations of in vivo exposure. Listed below are some general physiochemical evaluations that can provide a preliminary understanding of limitations of compounds within a therapeutic area (Ritschel, 1992; Campbell, 1994). Discussion of application of this information to the drug metabolism strategy is available (see Sec. II.D).

pK_a and ionization
Solubility
Particle size
Lipophilicity
Molecular weight

In addition to physiochemical variables, Table 1 outlines the numerous dispositional variables that can affect systemic exposure. The most basic involvement of a drug metabolism scientist is often to resolve discrepancies between in vitro potency data (inhibitory concentration; IC) and the in vivo pharmacological model (effective dose; ED). Often there is little correlation between ED and IC values within a series of compounds because the interactions of many in vivo variables complicate the relation. Although there are many dispositional reasons a compound may not work in vivo, the first step taken to understand why early leads fail should be determination of an exposure–response relation

Table 1 Dispositional Variables Affecting Pharmacokinetic and Pharmacodynamic Relationships

Intestinal			
	Luminal		
		Dissolution	
		Physicochemical properties	
			MW
			Molecular volume
			pK_a—ionization
			Lipophilicity
			Solubility
			Hydrogen bonding
		Microfloral metabolism	
	Gut wall		
		Permeability	
			Transcellular transport
			Paracellular transport
			Active transport
			P-glycoprotein efflux
		Metabolism	
			Oxidation (CYP3A)
			Glucuronidation
Hepatic metabolism			
	Phase I		
		P450	
		Non-P450	
	Phase II		
		Glucuronidation	
		Sulfation	
		Glutathione–amino acid conjugation	
		Methylation	
Extrahepatic metabolism			
	Renal		
	Brain		
	Lung		
Target tissue			
	Penetration or Metabolism		
	Transport mechanisms		
	Nonspecific binding		
Plasma protein binding			

within the animal model. If in vivo activity is not observed and the drug is not quantifiable in the animal, then absorption or metabolism should be examined for limiting in vivo activity. If the compound is present in the plasma without activity, then target tissue levels or protein binding may be issues. In vivo or in vitro studies may be designed to address these issues (see Sec. III.C.1 and III.D.1).

If there is sufficient sensitivity with the analytical assay, knowledge of in vivo pharmacokinetics at therapeutically relevant doses is difficult to match for the amount of information it can supply relative to all aspects of ADME. Initial examination of pharmacokinetics within a series can often isolate weaknesses in metabolism or absorption that, if necessary, can be taken to an in vitro model using animal or human surrogates to increase throughput. In vivo approaches discussed later in this chapter involve cocktail dosing or analysis or, if possible, direct sampling from the animal model when in vivo activity is determined. In vitro screening, with occasional strategic in vivo analysis as SAR evolves, can be very useful if throughput from a dedicated in vivo approach is inadequate.

If an in vivo model is not available for a project, it is particularly relevant to initially determine if the compound has acceptable pharmacokinetics in animals. However, an in vitro approach may be more appropriate, especially when a large number of compounds need examination, when metabolic lability is a known weakness, or when absorption is limiting exposure. The ultimate objective should be to profile strengths and weaknesses of all compounds in the series and identify the structural attributes most favorable for the desired activity. The next section will describe some examples of efficient approach designs for a high throughput discovery program.

C. Strategies for Discovery Drug Metabolism Studies

The ability to elucidate the relations between metabolism, absorption, physiochemical properties (e.g., lipophilicity), pharmacokinetics, and the pharmacological endpoint will enable scientists to discover increasingly potent candidates in vivo that do not possess many of the nonspecific side effects observed with existing medications. The strategy taken to understand the relative importance of the various parameters affecting ADME is very important. Although there are numerous twists to the complexity of the SAR that may convey limitations on a chemical series of interest, a general approach toward disposition issues can be developed. The major limitation will be in the analytical capabilities needed to quantify drug in the samples generated.

1. Analytical Strategies

The predominant rate-limiting step in almost any drug metabolism discovery program is quantitation of drug in the samples. Recent technological advances in drug quantitation using mass spectrometry (MS) has greatly facilitated increases in the throughput within in vitro and in vivo drug metabolism studies. Compared with high-performance liquid chromatography (HPLC) with UV or florescence detection, mass spectral analysis can reduce assay development, sample preparation, and analysis time, while also providing greater sensitivity and metabolite identification (Hiller et al., 1997). The sensitivity and low volumes necessary for MS are ideal for use of 96-well plates from in vitro screens (Tweedie, 1996). Most importantly, simultaneous analysis of multiple compounds is possible because the analysis is based on parent or fragment mass weight. With the ability to profile multiple entities in a single sample, one can use in vivo cocktail dosing to a single animal with 5 to as many as 20 compounds (Halm et al., 1996; Olah et al., 1997; Berman et al., 1997). Conversely, samples obtained from multiple animals each dosed with a different compound can be pooled to reduce the number of samples being analyzed. Finally, as described by Olah et al. (1997) and Hiller et al. (1997), tandem mass spectrometry technology allows reliable, accurate, and precise profiling of parent drug and metabolite in biological matrices at concentrations of 1–1000 ng/mL.

2. Strategies in a Pharmacodynamic-Driven Discovery Program

A standard discovery paradigm involves identification of a chemical series or structural class possessing affinity or activity for an in vitro receptor or enzyme. This in vitro model is subsequently used to base a synthetic program for identification of a compound with sufficient potency and selectivity to advance to in vivo models. The in vivo model should be pursued whenever possible because it can provide essential information for prediction of therapeutic dose in clinical studies and the margin of safety. If an in vivo pharmacology model does not exist or has very low throughput, the discovery program can be driven by drug metabolism using the approaches discussed in the following.

If a validated pharmacology model is available with acceptable throughput, the drug metabolism scientist can work in parallel with biologists using in vivo or in vitro tools to provide information on advantages or liabilities of compounds within the structural series based on metabolism, absorption, or physiochemical properties. In addition, through concurrent analysis of drug exposure and pharmacological activity, unexpected phenomena affecting activity, such as target organ partitioning or differential cellular uptake, can be assessed. Table 2 outlines some of the information that can be derived when pharmacokinetics is examined with the pharmacological endpoints.

If the biological model is reflective of the in vitro activity, the drug will have to be absorbed and delivered to the site of action for activity. Thus, the application of high-throughput in vitro screening for metabolism or absorption can be useful for providing the chemist with data to develop an SMR or an SPR concurrent with synthesis around biological activity (SAR). If the pharmacokinetics and pharmacological activity cannot be examined simultaneously in the pharmacology model, the disposition of selected compounds can be assessed in parallel using a high-throughput cocktail approach, described later. This is especially important if there appears to be a discrepancy between the in vitro activity and efficacy or potency in the in vivo model.

When a compound lacks in vivo activity, but has good in vitro potency and does not appear to be highly metabolized, it is important to determine if the drug is available to the target. If there is acceptable drug exposure without pharmacological effect, it should be determined if a high level of protein binding is limiting the free concentration. If a high level of protein binding is determined to be a limiting variable within a chemical series, it may be useful to correct the total exposure data to free concentration. When a high level of protein binding is occurring, the difference between 98 and 99% bound is actually a doubling of the free concentration that can affect in vivo activity. If protein binding is not an issue, other explanations may be pursued. First, determine if the pharmacology endpoint is relevant for the enzyme or receptor of interest or that there are no

Table 2 Information Derived from Pharmacokinetic–Pharmacodynamic Approaches in Preclinical Screening

Efficacious dose and efficacious concentration derived from dose and exposure response curves
Knowledge of favorable and poor ADME characteristics
Identification of active metabolites
Identification of nonlinearity in animals
Identification of interspecies differences
Better prediction of dose and dose interval for phase I studies
Better assessment of therapeutic index in conjunction with toxicology data

confounding interspecies differences in the enzyme of interest. This can occur when an in vitro program is developed using a human enzyme and the pharmacology model is rat or mouse. One hopes that there is a lead or model compound that possesses the biochemical attributes of interest that can validate the pharmacology model. Once the pharmacology model issues are resolved, compounds with the desired or optimal pharmacokinetics can be profiled for in vivo activity.

It is also possible that activity in vivo is greater than expected, based on in vitro pharmacology data (relative to the corrected free concentration in plasma). This may suggest the formation of an active metabolite in vivo. In vitro metabolism studies can be performed to confirm formation and also to provide adequate amount of the metabolite for structural identification. Here, mass spectroscopy can provide structural data to the chemist to facilitate synthesis and testing of activity of the metabolite(s). Alternatively, the metabolite isolated in HPLC fraction collection can be tested for activity using the in vitro pharmacology model. With the in vivo pharmacology model, metabolic inhibitors can be administered to confirm the relation between metabolism and in vivo activity. For example, if the metabolite was formed by cytochrome P450, administration of 1-aminobenzotriazole (Mugford et al., 1992) before pharmacological testing would be expected to reduce or abolish the pharmacological activity.

3. *Strategies in a Drug-Metabolism-Driven Discovery Program*

The increased need for drug metabolism support being created by the vast numbers of compounds being synthesized with combinatorial chemistry has resulted in the brute force attitude to disposition, during which many compounds are screened until the desirable pharmacokinetic traits are found (White, 1996; Berman et al., 1997; Olah et al., 1997). Often there is acceptable exposure within a series and the primary focus becomes fine tuning the SMR or SPR for optimal $t_{1/2}$, clearance, or bioavailability.

This drug metabolism approach may be used if there is no in vivo pharmacology model, if the in vivo model does not differentiate activity for compounds, or if the model is so labor-intensive that throughput is limiting. For clarity, the methods for creating a drug metabolism-driven discovery program are divided into in vivo or in vitro evaluations, respectively. Although in vitro methods for metabolism or absorption can provide much data in a timely fashion, in vivo data are difficult to match for the multifaceted insight they provide the disposition scientist. In practice, most discovery projects employ both in vitro and in vivo techniques in the relative proportions to meet specific needs. Additionally, because in vivo preclinical studies must use nonhuman species, the in vitro studies using human tissue such as Caco-2 or microsomes can confirm that selection of a compound is not confounded by interspecies differences.

a. In Vivo Drug Metabolism Approach. This method is becoming increasingly popular in several major pharmaceutical companies (Halm et al., 1996; White, 1996; Berman et al., 1997; Olah et al., 1997). The limitation of an in vivo disposition approach is that the throughput may be lower than in vitro-driven approaches; however, with a good analytical program and use of cocktail dosing or sampling, the throughput may be acceptable for most programs.

Depending on the route of administration, this approach can provide information on approximate peak plasma levels, hepatic extraction, half-life, clearance, and fraction absorbed. The best application of this program is when quantitative improvements are desired in an endpoint, such as C_{max} or clearance, and there is a large number of compounds to examine, perhaps derived from combinatorial chemistry. It is more important here to find the best compound, rather than elucidate mechanisms explaining the limitations for the SMR or SPR. After identification of compounds that have superior in vivo exposure, human in vitro surrogates can be used to ensure that interspecies differences are not an issue.

If exposure is low following oral dosing, the initial challenge is to distinguish between high first-pass metabolism and poor oral absorption. In a higher-throughput discovery program, examples such as this take time to address and may require some special studies (Nickerson and Toler, 1996), described in the following to identify the limiting variable. At this time the screening may be halted, or if compound throughput is not too demanding, a limited number of structural representatives can be examined to identify dispositional mechanisms limiting exposure. With this information, chemists can alter synthetic approaches to facilitate the improvement of pharmacokinetic profile.

b. In Vitro Drug Metabolism Approach. An in vitro-screening approach can generally accommodate a large number of compounds for evaluation of metabolism or absorption in animal or human surrogates. Application of in vitro models for screening is relevant if there are known in vivo limitations that can be specifically addressed using the in vitro model. If the synthetic effort is very broad, these models may provide timely feedback on the SMR or SPR. The data from the in vitro studies can be examined in conjunction with physiochemical data, such as $\log P$ or pK_a, and superior compounds can be advanced to in vivo pharmacokinetic and/or pharmacological profiling. The main advantage of this approach is that information derived from these in vitro models can be readily integrated into the ongoing synthetic effort. The applications and fine-tuning of these approaches to best support an ongoing discovery program are expanded in the following section.

D. Lead Optimization

By now it may appear that rational drug design is not important or is in conflict with a high-throughput paradigm during which huge numbers of compounds are synthesized and screened until desired attributes are found. With what is now known about the variables affecting ADME, rapid in vitro screens can be designed to select and define optimal characteristics for a higher number of compounds, with occasional checks against in vivo activity or exposure to verify that the desired SMR or SPR is tracking. Thus, the iterative approach to designing molecules will still occur, just on a much broader scale.

Regardless of whether the project will be driven by pharmacology, metabolism, or even brute force high-throughput, it is important to correctly use physiochemical, in vitro metabolism and absorption data, and in vivo pharmacokinetics to help optimize the in vivo activity of the series. Often, to achieve acceptable in vivo pharmacological activity, some sacrifice in in vitro potency may be necessary to optimize in vivo potency-limiting parameters, such as lipophilicity, absorption, or metabolism. However, if enough data are produced on a variety of compounds, the chances of choosing a successful clinical candidate are increased. From a dispositional perspective, the most important and practical areas in which modifications can be made to improve the PK/PD relation are the physiochemical properties, metabolism, and absorption.

1. Physiochemical Properties

This overview will discuss only how this type of information can be helpful in optimizing exposure and activity within a chemical series. The *solubility* of the compound at various gastrointestinal (GI) pH and environments can be easily determined, and in conjunction to dissolution, is an important determinant of the absorption for the compound. The extent of GI absorption is related to the dissolution of the compound and the permeability across the epithelial membrane. Ironically, in some programs, slow dissolution has been desirable to lower C_{max} and protract T_{max}, thus providing a sustained-release effect that can increase therapeutic duration or decrease side effects linked to peak drug concentrations.

Most drugs are either a weak acid or a weak base and are absorbed by passive diffusion in a nonionized state; therefore, the pK_a of a compound is very important because it defines how the pH of an environment, whether it be plasma, urine, tissue, or the GI tract, may affect absorption. The relation between pH and pK_a is defined by the Henderson–Hasselbach equation. However, even with less than 1% of the compound in the nonionized state, if lipophilicity of the compound is sufficient, there may be adequate membrane permeation. An explanation for this is related to the rapid equilibrium between the ionized and nonionized state. In addition, if there are any permeation processes involved other than passive diffusion, the ionization state may be less important (refer to Chap. 19 for a detailed explanation how physiochemical properties can affect different mechanisms of absorption).

Lipophilicity can be assessed using *M*log*P* or *C*log*P* which are computer generated semiquantitative estimates of octanol:water partitioning that are calculated based on the chemical structure (Stewart et al., 1997). Lipophilicity and larger molecular size are often a consequence of medicinal chemistry efforts to increase the potency of the compound. Unfortunately, this is often at the price of poor in vivo activity, because large, lipid-soluble drugs are less soluble in an aqueous environment, more labile to metabolism, and more susceptible to biliary elimination. All of these factors may contribute to lower bioavailability. In addition, higher lipophilicity in an aqueous environment (blood) may result in higher protein binding levels. In general, if *M*log*P* values are higher than 4, protein binding can be expected to be high ($>$85% bound); whereas values less than 1 may suggest low to moderate protein binding ($\leq$50%). High lipophilicity can also result in a greater volume of distribution, a longer $t_{1/2}$ for the compound, and better endothelial or epithelial cell penetration.

2. Metabolism

Most new chemical entities are of sufficient lipophilicity that they are not excreted unchanged and, as a consequence, metabolism is a major component of total body clearance. Given that there are significant interspecies differences in the major enzymes of xenobiotic metabolism, many companies have established rigorous metabolism screening programs using hepatic microsomes (phase I metabolism), slices, or hepatocytes (phase I and II metabolism) from appropriate species, including humans. This type of screening rapidly assesses the metabolic lability of a series of compounds and facilitates creation of a SMR around the obtained information. Defining the initial sites of metabolism for a molecule or chemical series is important to facilitate understanding of what may regulate the clearance of the compound in vivo (refer to Chapters 4 and 6 for detailed discussion about assessment of phase I and II drug metabolism).

One type of in vitro approach involves the assessment of specific cytochrome P450 (CYP450) isozymes such as CYP2D6 or CYP3A4. This information provided early in the discovery process allows chemists to understand the role of different functional groups on the parent molecule in altering the systemic exposure of a drug. With increasing concerns about drug–drug interactions or the possibility for a narrow therapeutic index, elimination of compounds that induce or inhibit CYP3A4 or are primarily metabolized by polymorphic isozymes (CYP2D6 or CYP2C19) is of paramount importance. Isozymal-specific effects on drug metabolism and pharmacokinetics are discussed in detail in Chapter 7.

A precaution when using an in vitro approach for screening is to verify that metabolic lability observed in vitro has relevance to in vivo models. The relevance of in vitro lability data can be demonstrated experimentally by verifying that compounds with good activity in vivo are slowly metabolized. If the metabolic pathway is understood, the animal model can be treated with a nonspecific inhibitor of metabolism to determine if activity can be increased when the compound is not metabolized. 1-Aminobenzotriazole can be administered to animals before drug treatment to inhibit the P450 enzymes (Mugford et al., 1992). It would be expected that the metabolic clearance would drop or in vivo pharmacological activity would increase. If not, P450-mediated metabolism may not be the limiting variable for clearance or systemic exposure.

Another value of in vitro metabolism screening is to identify and characterize metabolites for potential activity. Active metabolites may be suspected when there is sustained activity of a compound in vivo without the presence of an unchanged parent, or activity in excess of that predicted from in vitro potency. Induced microsomes can be used to form larger amounts of the suspected active moiety for in vitro potency testing.

3. Absorption

Development of a structure–permeability relation within a series is difficult because assessment of absorption can involve metabolism, various transport mechanisms, as well as numerous physiochemical properties. Stewart et al. (1997) have outlined physiochemical and biological approaches for evaluation of the intestinal absorption for discovery level compounds. Selection of the appropriate experimental system is up to the investigator, but it should be timely and provide information relative to the expected limitations in vivo. Chapter 19 will discuss mechanisms and variables regulating GI absorption.

A common physiochemical problem for dose-limited oral exposure is poor aqueous solubility. Although solubility can be measured easily, it is important to determine if the compound stays in solution and for how long. If solubility is determined to be low, it is worthwhile to determine the reason. Low aqueous solubility can be caused by high crystal-lattice energy, which can be approximated by melting point, or it can be due to desolvation and solvation factors. In the latter example, lipid solubility and partition coefficients are useful indicators (Stewart et al., 1997).

For biological approaches, a variety of novel recent methods have been documented for determination of rate and extent of intestinal absorption using in vitro (Yee, 1997) in situ and in vivo methods (Stewart et al., 1997; Hoffman et al., 1995), as well as some computational methods involving pharmacokinetic data and some assumptions (Kwon and Inskeep, 1996).

A common method for estimation of rate and extent of intestinal absorption is the gut loop in situ method in rats (Schurgers et al., 1986; Stewart et al., 1997). This method is labor-intensive, and often overestimates the absorption rates of compounds because it measures disappearance in the lumen, as opposed to measurement of drug in portal blood. Alternatively, an accurate and reproducible method to estimate the rate and extent of intestinal absorption is estimation or measurement of the fraction of a dose entering the portal blood following an oral dose (Hoffman et al., 1995; Kwon and Inskeep, 1996).

Although this method does not separate fraction lost by metabolism or permeability, this can be estimated from in vitro metabolism or absorption models. For a discovery program, a cannulated portal vein animal model may be an effective screening paradigm because it is a dynamic tool encompassing all mechanisms involving intestinal absorption. However, that this model is surgically compromised must also be considered.

The Caco-2 cell monolayers are an excellent in vitro human surrogate that have been used successfully for both prospective and retrospective prediction of absorption in man (see Chap. 19). Because they can predict all mechanisms of absorption quantitatively (Yee, 1997), they are now routinely used as a high-throughput model in the discovery setting at Pfizer. When the preclinical in vivo pharmacology model has absorption problems that do not correlate with Caco-2 study results, the other methods can be employed. However, because of the potential for interspecies differences in absorption, the Caco-2 model should be used to provide confidence that any compound selected from studies with animals can be absorbed in humans.

Discrimination between high first-pass hepatic extraction and poor oral absorption is a crucial task. One effective in vivo method to identify the barrier(s) to oral bioavailability is the use of different routes of administration of the drug(s). Separation between low absorption and hepatic extraction can be accomplished through direct comparison of the area under the plasma concentration–time curves (AUC) following oral and intraperitoneal dosing (Nickerson and Toler, 1996). The limitation to this method is that, owing to rapid entry of the compound into portal circulation, first-pass hepatic extraction may be saturated. Alternatively, with a portal infusion model, with a rate that mimics intestinal absorption, saturation is less likely to occur.

III. CANDIDATE DEVELOPMENT

Many of the activities in developmental research are similar to those in discovery research. However, the tone of the research project changes as the objectives change from progressing lead series and identifying potential candidates to the process of bringing one of them to the marketplace. Whereas discovery efforts often emphasize speed and the use of techniques and procedures that will hasten elimination of unsuitable candidates, development efforts emphasize assembling a data package suitable to fully understand the characteristics of each chosen candidate and support its regulatory filing. Discovery studies often emphasize screening a large number of candidates, using the fewest replicates to distinguish major differences among the candidates, whereas development studies will use sufficient replicates to provide a reliable determination of variability. Similarly, discovery analytical efforts are based on speed, so that a useful assay may possess minimal precision, accuracy, and selectivity while providing for maximum throughput. Assays for development studies, however, are developed and validated to provide sensitive, reproducible results with less concern for very rapid throughput.

Many studies in preclinical developmental research, especially those defining toxicokinetics, are conducted under the auspices of Good Laboratory Procedures (GLPs), which require good recordkeeping in support of scientific procedures used in drug development. All scientists involved in developmental ADME research should fully understand the requirements of GLPs, which can be found in the Code of Federal Regulations (21 CFR § 58). One of the requirements of GLP is that organizations involved in nonclinical laboratory studies of agents intended for use in humans or animals must prepare and follow standard operating procedures (SOPs). Although following SOPs for ADME studies may appear to be overly confining, it is important to note that well-written SOPs provide a convenience, rather than a burden, for laboratory scientists. SOPs are formalizations of how procedures are normally carried out. When the SOP is followed, the only documentation required for the method is a reference to that SOP. If a particular project requires a deviation from that SOP, documentation is provided simply by recording the changes in the procedure in the data file and notifying the study director that a deviation from the SOP occurred. Compliance with the principles of GLP within each organization conducting such studies is monitored by federally mandated quality assurance units (QAU). Effective communication between researchers and the QAU helps ensure that study documentation will meet the requirements of regulatory agencies.

A. Development of Suitable Bioanalytical Techniques for Use in Toxicology and Early Clinical Studies

The first step in any new preclinical program is development and validation of analytical methods for measuring drug or key metabolites in blood (plasma or serum) or another relevant biological matrix. Very often, the assays used in discovery activities are not adequate for preclinical drug metabolism development studies. Although assay sensitivity requirements for toxicology studies are often less demanding than for discovery studies (because of the higher doses used), assay selectivity, precision, accuracy, and reproducibility may be more important for development studies. Drug metabolism departments have SOPs that outline the requirements of analytical methods used to support GLP studies. It is important that good records are maintained for lot numbers of all analytical standards used in the assay validation and verification procedures. Guidance, from the International Conference on Harmonization (ICH), on assay validation has been published (ICH, 1995a).

1. Assay Sensitivity and Selectivity

Sensitivity and selectivity of assays needed for assessment of exposure in toxicology studies often are not issues because the high doses used result in relatively high exposure. Nevertheless, these characteristics need to be established for the expected conditions of use. It is important to use a source of matrix that is similar to actual study matrix. Commercially available pooled plasma lots often have a different chromatographic profile than plasma from on-site control toxicology animals. Selectivity should be demonstrated in individual matrix samples obtained from several individual animals. Pooled lots of plasma are suitable for preparing standard curves and quality control samples, but not for assessing selectivity and sensitivity of an assay.

2. Assay Validation

Procedures for assay validation are varied throughout the industry, but the key factors are selectivity (see foregoing), recovery, stability, and intra-assay precision and accuracy at several different concentrations, spanning the dynamic range of the assay; usually the lower limit of quantification (LLOQ), the upper limit of quantification (ULOQ), and three additional concentrations between the extremes. Whereas, precision and accuracy should achieve criteria established by company SOPs (usually within $\pm$ 15%), criteria for recovery of drug and internal standard are more difficult to define. In general, the higher the recovery, the more sensitive and more precise the assay, but as long as recovery is reproducible and similar for drug and internal standard (IS), relatively low recoveries (50%) may be acceptable.

3. *Assay Verification*

Assay verification (also termed assay performance) may be more important than assay validation, because it demonstrates whether or not the assay performs within specifications whenever it is used. Usually, this is achieved by including quality control (QC) samples, prepared in batch at three or more concentrations spanning the dynamic range of the assay, to be assayed with study samples. QC samples must be prepared independently from standard curve samples. SOPs should establish the criteria that QC samples must achieve to use the results from an assay run. For example, one popular approach is to include replicate QC samples at each of three or more concentrations (one near the LLOQ, one near the midpoint of the dynamic range, and one near the ULOQ). The concentrations of each QC standard determined by the standard curve must fall within a range defined by the SOPs (usually ±15–20% of the intended concentrations). Usually, the SOPs will define a limit on the number of QC samples that may fall outside the acceptable range before a run is disqualified.

B. Characterization of Absorption and Pharmacokinetics to Assist in Dose Selection for Toxicological Evaluation

An early objective of a developmental research program is to evaluate the potential toxicity of the drug candidate. A key factor in designing toxicology studies is the use of a dosing regimen that will provide appropriate systemic exposure in the toxicology animal models of the test agent. Pharmacokinetic studies at the early development level should help define the appropriate formulation, dose levels, and dosing regimen for evaluation of toxicity in the specific species chosen for toxicological evaluation.

Some of the information needed to decide how a drug should be administered to the toxicology species may already have been generated as part of the discovery effort (clearance, volume of distribution, bioavailability), but often the data are collected using pharmacologically active doses that are much lower than will be used in toxicology studies. Thus, a key feature of early developmental studies is to examine or predict the pharmacokinetics of candidates at the higher doses anticipated for use in toxicology studies. Key activities involved in the earliest assessments include selection of a suitable formulation, selection of dose levels and dosing regimen, and determination of an appropriate sampling scheme for the evaluation of toxicokinetics (C_{max}, AUC, dose proportionality, and potential changes in exposure with multiple dosing). Some of these studies may be initiated before a validated assay is available.

1. *Selection of an Appropriate Formulation*

New pharmaceuticals with ever-increasing pharmacological potency and selectivity often have physicochemical properties (solubility, lipophilicity, pK_a) that are not ideal for absorption across the gastrointestinal tract. Even for compounds with acceptable physical characteristics for administration of pharmacologically active dose levels, problems may occur when the drug is administered at the higher doses required for toxicological evaluation. However, drugs with good aqueous solubility often do not need to be extensively tested for their potential for absorption before starting the first toxicology study. In these cases, little risk is taken by assuming increases in exposure (C_{max} or AUC) with increasing dose, and then confirming (or refuting) the assumption by monitoring exposure in the first toxicology studies. Often, a simple solution or suspension of drug in an aqueous vehicle, such as methylcellulose, provides an adequate formulation to obtain dose-dependent exposure in the toxicology species.

For compounds with anticipated absorption problems, pretoxicology formulation studies may be required to avoid wasting resources on a toxicology study that does not achieve increased systemic exposure with increasing dose. Establishment of an appropriate dose formulation for toxicological assessment is often a combined effort of toxicology, pharmaceutical research, analytical chemistry, and drug metabolism groups, occurring at an early stage of development. When simple aqueous formulations are not expected to be adequate to achieve dose-related exposure, several options are available, but their assessment will likely delay the start of definitive toxicology studies. Therefore, it is important to carefully design studies that will rapidly evaluate potential formulations. If the pharmacokinetics of a candidate have been determined in the toxicology model species at lower doses for which formulation was not a problem, these data can be used to design studies to appraise new formulations at the higher doses.

Several approaches may be taken to increase the delivery of an agent to the toxicology species. Improving the physical characteristics of the agent (e.g., particle size or salt form), formulating with absorption enhancers (e.g., oils, surfactants, or polymers), or dosing with food may enhance the rate or extent of oral absorption. Once adequate systemic exposure is achieved for a minimal toxicology dose level, greater increments of exposure may be obtained by using bid, tid, or qid oral-dosing regimens. Occasionally when adequate exposure cannot be achieved with oral dosing, parenteral dosing may be used.

2. Selection of Dose Levels and Dosing Regimen

Selection of doses for early toxicology studies involves several components (most of which involve toxicology issues outside the scope of this chapter). The key contribution of early drug metabolism studies toward dose selection is verification that systemic exposure is likely to increase with increasing dose. However, an important objective of toxicology studies is to demonstrate the potential toxicity of the agent, even if that toxicity is not observed until unrealistically high doses are administered to the animals. When working with such a compound it may be important to determine the dose level beyond which no further systemic exposure can be obtained. Paradoxically, more effort may be spent on developing formulations to increase systemic exposure (or demonstrate that a ceiling has been achieved) for a relatively innocuous agent than for an agent with greater intrinsic toxicity.

Toxicology studies in animals provide information that will assist in the assessment of risk of the agent for humans. The greater the similarities in the disposition of the drug in animals to the disposition in humans, the more confidence can be placed in the risk assessment. However, interspecies differences in ADME are more the rule than the exception. Often, clearance of drug is much more rapid from animals than it is from humans. Although, the disposition of drug in a human is usually unknown at this stage of drug development, the *desired* or predicted pharmacokinetics in humans can assist in selecting an appropriate dosing scheme in animals. For example, if the desired pharmacokinetic profile of the agent in humans is rapid absorption and rapid clearance with SID dosing, a similar regimen should be used in toxicology studies. However, if the agent is expected to maintain fairly consistent plasma levels over the dosing interval in humans with daily dosing, but has a short half-life in animals, bid or tid dosing may be required to achieve a similar profile of overall exposure in the animals.

3. Selection of a Toxicokinetics Sampling Scheme

Assessment of systemic exposure in toxicology model species (toxicokinetics) is an important part of a candidate development program. This assessment has two major functions. The first is to help understand the toxicology models, and the second is to understand the relevance of the toxicology model to potential human toxicity. Measurement of systemic exposure in model species is necessary to understand the effect (or lack of effect) of dose, sex, species, and such on toxicological effects in the animal models. As the drug development program progresses, these data will help estimate the margins of safety for therapeutics. Guidance from the ICH on the assessment of systemic exposure in toxicity studies has been published (ICH, 1995, b).

We use the term *toxicokinetics* to denote measurement of systemic exposure to the test agent in toxicology model species, whether that measurement is made using the animals from the toxicology study, animals from a satellite group included in the study, or a separate study designed to mimic dosing conditions of the toxicology study. In any case, the systemic exposure may be characterized by measurement of parent compound or metabolites at a single timepoint or at several time points. The term toxicokinetics is somewhat of a misnomer, because kinetic parameters, such as elimination rate, are not always determined in these studies. The primary objectives of toxicokinetics are (1) to help understand the relation between dose and effects within the study, (2) to provide an assessment of exposure that can be compared across toxicology studies, and (3) to augment human risk assessment by providing a basis of interspecies exposure comparisons.

A major challenge in determining an effective toxicokinetic sampling strategy is maintaining a balance between obtaining sufficient exposure data and not compromising the integrity of the main toxicology study through excessive sampling of study animals. The use of satellite animals or separate exposure studies avoids oversampling of toxicology study animals, but precludes the direct correlation of toxicology findings with exposure. Thus, obtaining a complete toxicokinetic analysis from any one toxicology study that will meet all the needs for that study is extremely difficult, especially for smaller species with limited blood volumes. However, because several toxicology studies will be required for any successful drug candidate, it is not essential to obtain complete toxicokinetics from any one study. In fact, a structured approach across several studies, with each new study building on previous ones, provides a means to fully meet the objectives of toxicokinetic analysis without jeopardizing any of the toxicology endpoints of the individual studies.

The approach outlined here is based on a toxicological progression that is fairly standard throughout the industry, where less-expensive studies that can provide definite go/no-go decision endpoints are initiated in early development and then followed by longer term, more expensive studies. For each type of study, exposure measurements fulfill different objectives. Thus, the sampling design for each study needs to consider the specific study objectives as well as the overall toxicokinetics package that will evolve as the candidate progresses through developmental research.

a. Indices of Exposure. The major roles of toxicokinetics, regardless of the type of study, are to verify and quantify the systemic exposure of the test animals to the test agent or its representative metabolite. The maximum plasma concentration (C_{max}) and AUC are the two primary indices of systemic exposure. Depending on the pharmacology of the test agent, for some studies (e.g., genotoxicity and reproductive toxicology), C_{max} may be more meaningful for correlating effects with exposure, whereas for others (e.g., chronic toxicity), AUC may be more meaningful. Neither of these parameters can be determined accurately without sampling at multiple time points. For some studies, multiple sampling with four to six time points will provide both C_{max} and AUC. However, it is not practical to sample extensively for all toxicology studies. For example, for extensive sampling in rodents when only one time point can be sampled from each animal (because of blood volume limitations), a complete toxicokinetic design using satellite animals could require more animals for kinetics than for toxicology endpoints. For such studies, systemic exposure may be assessed at a single time point. However, choosing that time point may be difficult, because the time when C_{max} is first observed (T_{max}) often varies for different dose levels.

Often, an approximate C_{max}, or C_x, is defined in toxicokinetic protocols as the concentration obtained at a time expected to be near the T_{max}. This approach is useful when sufficient pharmacokinetic data exists to support selection of the sampling time point. However, even if T_{max} does not shift as the dose level is increased, it is possible that C_{max} will not increase with dose, even though AUC does, because of saturation of absorption. For this reason, the concentration at a later time point C_t is sometimes used as a single time point measurement of exposure. Often C_t is a minimum concentration, C_{min}, taken at 24 h after dosing or just before the next daily dose. An advantage of C_{min} is that it may provide a reliable assessment of total exposure when T_{max} shifts with dose level. However, a disadvantage of measuring C_{min} instead of C_{max} is the requirement for greater assay sensitivity. In addition, some toxicities are better correlated with C_{max} than with other measures of exposure (AUC and C_{min}).

b. Genotoxicity Studies. Usually, one of the earliest studies requiring toxicokinetics is in vivo genotoxicity in rodents. Because a positive outcome in a genotoxicity screen is normally an unequivocal kill signal for a project, it is important to verify that a negative outcome (that is, no genotoxicity in the model) is not a reflection of poor systemic exposure to the agent in the model species. Normally, single-time–point plasma concentrations provide a suitable measure, although tissue concentrations may provide a stronger assessment of exposure for specific targets (e.g., liver for unscheduled DNA synthesis studies, bone or bone marrow for micronucleus studies). Toxicokinetic sampling times in these studies normally will be based on the preliminary pharmacokinetics, as described earlier, and on the timing of the genotoxicity endpoints.

c. Exploratory Range-Finding Toxicology. The next studies requiring toxicokinetics are usually exploratory range-finding studies, often using a rodent species and a large animal species (usually dog or monkey). In these studies, four to five plasma samples on the first or last day of dosing can provide key information: assessment of steady state C_{max} and AUC for each dose level, the potential for plasma concentrations to increase or decrease with multiple dosing, and the potential for gender differences in exposure. These data are easily collected from large animals (usually two to three animals per sex per dose level), because blood volumes of the animals are large enough to support four or five samples of sufficient volume for standard analytical methods. However, obtaining toxicokinetic data from the rodent species may be limited by the smaller blood volumes of these animals.

If the assay is sensitive enough to quantify drug in small volumes of rodent plasma (<100 μL), multiple sampling of each animal may be possible, especially if a satellite group of animals is included exclusively for toxicokinetics. If multiple bleeding is not possible or desired, single time points from each animal may be obtained. Two strategies are possible. In the first strategy, blood at a single time point, selected as described earlier, is obtained from main study animals on day 1 and the last day of the study. This approach provides for a direct correlation of observed toxicities and systemic exposure, but may not be able to demonstrate dose-related exposure under some circumstances (e.g., dose-related saturation of absorption or metabolism). A second strategy is to collect blood from individual animals at different sampling times at the end of the study (van Bree et al., 1994; Nedelman et al., 1995). For example, if five animals per sex per dose level are used, blood would be collected from each animal at a different time point. This provides two samples (one from a male and one from a female) at each of five time points and can provide an estimation of AUC for each dose level. If blood volumes required for analysis are small enough to justify bleeding each animal at two or three time points instead of one, more confidence in the estimated exposure (both AUC and C_{max}) can be achieved.

At this point, it is important to emphasize that the goal of this approach to exploratory toxicology studies is not to provide comprehensive toxicokinetic information, but to provide sufficient information to plan appropriate sampling schemes in more definitive GLP toxicology studies that will follow. Whereas using one or two animals per time point may not be sufficient for definitive toxicokinetics, such limited sampling usually provides an adequate database without consuming excessive analytical resources at an early stage of development. As the toxicology program progresses, there will be several opportunities to obtain more complete toxicokinetic data.

d. Definitive Subchronic Toxicology. Once exploratory dose range-finding studies have been completed, definitive (GLP) subchronic toxicology studies are started. Typically, studies of 2–4 weeks duration in two species are needed before phase I studies in humans can be initiated. For these studies, a strategy similar to that outlined for exploratory studies can be used, sampling on day 1 and at the end of the study. The information obtained from the exploratory studies should be used to select optimum sampling times. The toxicokinetics obtained from these studies will be derived from a larger number of samples per time point (because of the larger number of animal replicates); thus, they will provide a more definitive assessment of exposure. In general, sampling for toxicokinetics near the end of these studies (just before necropsy) is recommended so that exposure can be closely correlated with pathological findings. However, ADME can be altered by certain toxicities, so it may be difficult from results in a single study to fully understand the relation between exposure and toxicity.

Many phase II programs will require toxicology studies of 3–6 months duration. As long as dose levels and formulations are not significantly altered, there is no need to duplicate sampling days used in previous studies. If this approach is followed across several toxicology studies, exposure assessments on day 1, at 2 weeks, at 1 month, and at 3 or 6 months can be compiled.

e. Carcinogenicity Studies. Toxicology studies to assess carcinogenicity in rodents are often conducted by dosing in feed, whereas many of the other studies described in the foregoing are conducted by dosing orally (e.g., gavage, capsules, or tablets). When this occurs, extensive sampling for toxicokinetics of in-feed range-finding studies is needed. Usually, AUC at steady state for animals dosed in-feed will be similar to AUC at steady state for animals dosed by gavage, but C_{max} and C_{min} can be substantially different. Because rodents feed throughout the day (although most food consumption occurs in the dark cycle), systemic exposure for animals dosed in-feed can be relatively constant, even for compounds with short half-lives. Therefore, four to five time points over a 24-h period near the end of the study are usually sufficient to assess steady-state C_{max}, C_{min}, and AUC.

It is important to obtain a reliable assessment of steady-state exposure in range-finding studies, so that toxicokinetic sampling of main study animals in the 2-year carcinogenicity studies can be kept to a minimum. Sampling of main study or satellite animals at a single time point at 6 months is normally adequate to verify exposure.

f. Reproductive Toxicology Studies. In reproductive function studies involving species not studied in earlier toxicology studies, complete sampling to provide C_{max} and AUC may be needed. Fertility studies are usually conducted with rodents. Because a substantial database will have been accumulated throughout the chronic toxicology program, extensive toxicokinetic sampling for fertility studies is not necessary. Single-time point sampling for female rats approximately 1 week before mating and for male rats 1 week after mating should suffice for most studies. Occasionally, concentrations of drug in milk may be assessed.

Because sampling for toxicokinetics may interfere with teratology study endpoints, satellite groups dosed in parallel with main study animals are often required. These studies are often conducted with rats or rabbits, and there are two key objectives for toxicokinetics: (1) verify systemic exposure in pregnant animals; and (2) determine the extent of fetal exposure to drug. Use of radiolabeled drug on the day before sampling may assist in monitoring fetal exposure to drug and metabolites.

g. Other Toxicology Studies. Other types of toxicology studies, such as injection site toleration and ocular or dermal irritation, are often conducted to meet specific needs for a particular project. Assessments of systemic exposure may or may not be necessary, and the decision to monitor exposure should be made considering the integrated toxicokinetics program (see following section).

h. Integrated Toxicokinetics Program. The goal of this approach, building on the toxicokinetics database as the toxicology program develops, is to provide relevant exposure data for the program while using a realistic amount of bioanalytical resources. The amount of effort required for each toxicokinetics program will vary with each program. For example, if the formulation and dose levels remain unchanged, with subsequent studies less sampling may be required. If observed toxicities in earlier studies appear to be correlated with systemic exposure, more extensive monitoring (increased time points or increased sampling days) may be needed to fully understand the relation. In this way, the design of toxicokinetic sampling is customized for each new toxicology study so that necessary information will be obtained without excessive sampling of study animals.

C. Evaluation of Drug Distribution and Metabolic Profiles in Toxicology Species and Humans

Metabolic profiles in human as well as the toxicology species are needed to confirm the relevance of toxicology data to human safety (de Sousa et al., 1995). In general, the toxicology model species are considered adequate as long as all major metabolites observed in humans are also observed in animals, even if more metabolites are observed in animals. When, a major human metabolite is not formed by the toxicology species, these studies may identify a more relevant toxicology model or lead to additional toxicology studies with the major metabolites that are unique to humans. Although new approaches using stable isotopes are emerging (Abramson et al., 1996), nearly all comparative metabolism studies for new drug applications are completed using ^{3}H or ^{14}C.

1. Tissue Distribution Studies

There are two major roles for preclinical tissue distribution studies. They provide the database for dosimetry calculations that define an acceptable dosage of radioactivity to be used in definitive human biotransformation studies. In addition, these studies may assist in understanding efficacy and toxicity issues that arise in a development or discovery program. In these instances, evidence can be obtained about the intensity and duration of exposure of a specific target organ to drug or metabolites, confirming that the project is worth pursuing further or providing direction for seeking backup candidates.

Tissue distribution studies are usually conducted in a rodent species with radiolabeled drug. Either ^{3}H or ^{14}C can be used for these studies, but ^{14}C offers several advantages over ^{3}H (e.g., greater metabolic stability and higher disintegration energy) and should be used if possible. The radioisotope should be located in a metabolically stable position so label will not be lost to volatile metabolites (e.g., $^{14}CO_2$, ^{3}HOH, formic acid). Requirements for future biotransformation studies (see following section) should be considered at this time.

Radiochemical stability of the isotope can be an issue, especially when specific activity of the stock material is high. Therefore, storage of radioactive compounds for tissue distribution and other drug metabolism studies must be carefully considered. Unfortunately, there are no tried and true storage guidelines that can be guaranteed to work with every radiolabeled compound. Nevertheless, the following principles apply (Evans, 1982).

1. Obtain small quantities of radiolabeled compound. Although the cost of synthesizing or purchasing a large amount of radiolabeled compound is only marginally higher than the cost for a small amount, the expense of holding unused portions under a site license or trying to dispose of excess or radiodecomposed radiolabel will usually exceed the savings on synthesis. Many manufacturers of radiolabeled compounds will store excess amounts of custom-synthesized compounds for a nominal charge.
2. Store radiolabeled materials as a dilute solution (ethanol, toluene, or water are preferred solvents), preferably with a small amount of a free radical scavenger such as ethanol (1–10%). If no radiostability information is available, it is advisable to divide the stock material into several sublots, each stored under different conditions (e.g., dry, in solution, frozen, refrigerated, reduced specific activity). Samples from each sublot should be examined at various times to determine which storage conditions are most appropriate.
3. Radiochemical purity should be determined using at least two separate methods of analysis (e.g., reverse-phase HPLC and normal-phase HPLC; not two different mobile phases with the same chromatographic column). Radiochemical purity should be reestablished, with at least one of the methods, immediately before each study.
4. Initiate radiostability studies immediately on receipt of radiolabeled material. Some radiolabeled materials will degrade very rapidly under some storage conditions. If these studies are delayed and there is a significant degradation prob-

lem, the entire lot may be lost before appropriate storage conditions can be identified. Even if the material is stable, documentation of its stability will be needed.

5. Radioisotopes with very high specific activity are especially prone to instability. If a range of specific activities is needed for a program, the dilutions of the high specific activity stock should be made immediately on receipt.
6. Ideally, all the radiolabeled isotope needed for a preclinical and clinical drug metabolism program can be prepared in a single synthesis. However, compliance with GMPs, required for clinical studies, complicates this approach. Therefore, consultation with a member of pharmaceutical process and development or manufacturing groups with GMP experience is recommended before synthesis.

There are two approaches for the conduct of tissue distribution studies. In both approaches, animals are dosed with radiolabeled drug and killed at suitable times after dosing for assessment of radioactivity levels in the tissues as a function of time. However, different methods for quantification of radioactivity in tissues distinguish the two approaches. Tissues from the animals can be dissected and the radioactivity in the tissue or an aliquot from that tissue can be measured by liquid scintillation analysis (LSA) in an approach referred to as dissect and quantify (D&Q). Aliquots of tissues obtained from the D&Q technique are either combusted in a sample oxidizer so that $^{14}CO_2$ or ^{3}HOH are quantitatively recovered and quantified by LSA, or digested in alkali, neutralized, or bleached, and quantified by LSA. Alternatively, the animals can be frozen in a carboxymethyl cellulose block and sagittally sectioned. The thin-body sections can be applied to x-ray film for classic whole-body autoradiography (WBA) or to storage phosphor imaging screens for whole-body autoradioluminography (WBAL).

Regardless of the technique used to assess tissue distribution of radioactivity, there are several tissues defined by regulations (21 CFR § 361.1) that are particularly important: gonads, lens of the eye, and active blood-forming organs. In addition, all of the major organs are usually sampled or analyzed for total radioactivity to understand the general distribution of drug throughout the body. Finally, other tissues of particular interest because of the mode of action of the drug or potential specific toxicities may be studied. In most cases, tissue distribution is determined only after administration of a single dose. However, additional repeated-dose tissue distribution studies may be considered when tissue half-lives of radioactivity greatly exceed plasma half-life; when steady-state plasma concentrations of drug are markedly higher than those predicted from single dose pharmacokinetics; when toxicity that may affect distribution is observed only after long-term dosing; or when the agent is being developed for site-specific–targeted delivery (ICH, 1995c).

The D&Q method offers the advantage of greater sensitivity because larger sample sizes are available for quantification. However, WBA and WBAL offer the advantage of comprehensive measurement of radioactivity throughout the body. For example, in the D&Q method, radioactivity in large tissues, such as the liver, can be readily detected, even when present at low concentrations because large aliquots of tissue can be analyzed. However, radioactivity in tissues that are poorly defined (e.g., some regions of the brain) or too small (e.g., rodent adrenal gland) for reproducible dissection cannot be detected by D&Q. In WBA or WBAL, however, radioactivity may be difficult to detect in some tissues because of the constraints of the small mass of material that can be exposed to the developing medium. However, all tissues are potentially available for detection, provided a sufficient number of tissue slices are analyzed. Until recently, D&Q was considered to be more quantitatively reliable than WBA, but recent advances in WBAL have eliminated this concern (Potchoiba et al., 1995).

Regardless of the methods used to quantify radioactivity in animal tissues, dosimetry calculations will be required to estimate the total body exposure and potential for specific tissue accumulation of radioactivity in humans during biotransformation studies with radiolabeled drug (Dain et al., 1994). The basic assumption for the validity of this approach is that the tissue distribution of radioactivity in humans will be linearly related to the tissue distribution of radioactivity in the animal model. The purpose of dosimetry is to estimate "the amount of radioactive material to be administered [to human subjects] . . . such that the subject receives the smallest radiation dose with which it is practical to perform the study without jeopardizing the benefits to be obtained from the study" (21 CFR 361.1). In any event, the radiation dose for a single administration may not exceed 3 rem for whole-body, active blood-forming organs, lens of the eye, and gonads, or 5 rem for other tissues. The following approach is one method to calculate the estimated human radiation dose (expressed in rem) from tissue distribution data in animals. (Note: For the purposes of this discussion 1 rem is equivalent to 1 rad.)

Tissue concentrations of radioactivity in tissue distribution studies are normally expressed in units of microcuries per gram (μCi/g) tissue. In a study with four to five time points (over an interval covering at least three plasma half-lives or excretion of at least 75% of the dose), AUC(0–∞) can be calculated (units of μCi h/g) for each tissue of interest. The radiation dose for each tissue can be calculated using an adaptation of the method of Loevinger and Berman (1968).

$$D = \mathrm{AUC}(0\text{–}\infty)\Delta_i \mathrm{DCF}$$

where D is the absorbed radiation dose in rem, Δ_i is the equilibrium radiation dose constant (0.105 g rem/μCi h for ^{14}C or 0.021 g rem/μCi h for ^{3}H), and DCF is the dose conversion factor to extrapolate from the animal model to man ($\mathrm{dose}_{man}/\mathrm{dose}_{animal}$, where doses are expressed in μCi). This function can be applied to each tissue of interest for the proposed clinical dose (perhaps 50 μCi). If the limiting tissue were to result in a calculated absorbed dose of 1 rem, the clinical dose could be increased by as much as threefold, thereby increasing the sensitivity of the study without exceeding the exposure limits set in 21 CFR 361.1. However, if the study design involves multiple administrations of radiation (including sources such as X rays), all sources of radiation should be summed so that total exposure does not exceed the limits specified in the regulation.

2. *Biotransformation Studies*

The major goal of studying the biotransformation of a drug in animals is to help establish the relevance of the animal models in evaluating the potential human toxicity of the compound. This is more important for the animal models used to evaluate reproductive toxicology and carcinogenicity than for the models used to evaluate acute toxicity and, in fact, phase I clinical trials usually are started in the absence of in vivo biotransformation data. The key objective of biotransformation studies is to verify that the major human metabolites formed are also formed in the toxicological model species. Normally, studies will be conducted to quantify the major metabolites, and to identify, but not necessarily quantify the minor metabolites. Typically, major metabolites are those accounting for more than 5–10% of total excreted radioactivity, and metabolites present at less than 1% of total dose are generally considered to be insignificant. However, the precise assessment of what constitutes a major (or insignificant) metabolite will be case-dependent.

When major metabolites present in humans are not observed in toxicology model species, additional work may be required to verify that these metabolites are not toxic. Because it is normally impractical to repeat an entire toxicology program (especially, long-term chronic exposure studies) with synthetically prepared metabolite (or with the parent drug in a species that does form the metabolite), in vitro approaches (i.e., genotoxicology) or short-term in vivo studies may be sufficient to address safety concerns about the metabolite.

a. Radiolabeled Material. A key concern in preparing radiolabeled compounds for use in biotransformation studies is the metabolic stability of the position containing the radioisotope. Compounds containing ^{14}C are generally preferred to compounds containing ^{3}H. The radiolabel should be positioned at a portion of the structure where metabolism is unlikely to displace the label. Methyl groups, carboxylic acids, and other functional groups that could be easily displaced metabolically should not be considered. Frequently, the radiolabeled material synthesized for tissue distribution studies also can be used for animal and human biotransformation studies. However, GMP procedures are generally required for the preparation of the dosing formulation for clinical biotransformation studies.

b. Toxicology Animal Models. Biotransformation studies for toxicology species should be initiated as soon as a decision is made to progress a compound to phase I investigation, and certainly before long-term toxicology studies are started. Biotransformation in the species used for carcinogenicity studies normally will be the focus.

c. In Vivo–In Vitro Correlations (Predictions for Humans). In vitro biotransformation studies, including enzyme and cell preparations from animals and humans, should be initiated as early in a candidate discovery or development program as possible. These studies can help identify appropriate species to use for toxicology studies when significant interspecies differences in metabolism exist. The proper choice of toxicology model species early in a development program can avert problems when metabolites observed in clinical studies (conducted, by definition, after initial toxicology studies have been completed) are not seen in the toxicology species.

d. Clinical Studies. Clinical biotransformation studies should be scheduled after tissue distribution studies and before the start of phase II studies and long-term toxicology studies. Ideally, these studies should be carried out with radiolabeled drug in a formulation similar to the anticipated commercial form. However, this objective can rarely be met in practice. Usually, a bioavailable formulation (solution or suspension) of radiolabeled drug that can be prepared through GMP procedures is used.

Selection of the dose for these studies is often difficult, because the clinically efficacious dose(s) usually are unknown. Usually, a well-tolerated dose near the high end of phase I dose titration studies is used. The specific activity of the dose should be high enough to characterize the important metabolites, but not expose the subjects in the study to excessive amounts of radioactivity. Although arbitrary limits for the total radioactive dose (such as 100 μCi for administration of ^{14}C) are often used, the specific activity of the dose should be determined using dosimetry techniques based on the results of animal tissue distribution studies (see Sec. III.C.1).

The radiolabeled drug is administered to approximately four to eight volunteers housed in a clinical research facility. Urine and feces are quantitatively collected over intervals spanning the time anticipated to recover 95–100% of the administered radioactivity. Plasma samples should also be collected over the duration of the study, even if it is known from other phase I studies that the drug cannot be quantified at the latter time points. For drugs that may be expired directly or metabolized to volatile compounds, volunteers in the study may be required to breathe into a gas-trapping apparatus to quantify the expired radioactivity over defined intervals. Because these latter studies are much more difficult to conduct and to obtain satisfactory total radioactivity recoveries, careful thought should go into the synthesis of the radiolabeled compound (see foregoing).

Because of concern to minimize human exposure to radioactivity, the requirements for extended housing of study subjects, the expense of the studies, and the importance of quantitative recovery, the conduct of these studies must be carefully controlled. Key features of the study design to minimize the likelihood of having to repeat them include the following: (1) using appropriately positioned radiolabel, of the appropriate specific activity, in the test drug; (2) strict adherence to the sample collection scheme by on-site housing of the study subjects throughout the duration of the study; (3) use of a simple sample collecting system to ensure that urine and feces samples are quantitatively recovered; (4) use of a fail-safe–labeling system to ensure that study samples can be collected, stored on site, and shipped to analysis laboratories without error, and (5) carefully controlled access of study subjects to food and drink throughout the study.

e. In Vitro Metabolic Studies. In vitro metabolic studies during the development phase are geared toward identification of enzymes or isoforms involved in the metabolism of the drug candidate. This knowledge may assist in predicting whether subpopulations may exist that would have reduced or exaggerated systemic exposure to the drug (i.e., polymorphism) and whether drug–drug interactions are likely. The use of in vitro studies in drug development is discussed in more detail in Part 3: Technologies.

D. Additional Studies in Candidate Identification and Development

Several other studies may provide additional data that will help understand the disposition of drug in animals and human and interspecies relations.

1. *Protein Binding*

Many drugs are bound noncovalently to plasma proteins (e.g., albumin, α-acid glycoprotein, and lipoproteins). It is generally recognized that only the free fraction of drug is available for transport into cells or to exert pharmacological effects. Thus, comparisons of systemic exposure across species often take into account species differences in protein binding. Drugs that are highly bound to plasma proteins may also interact with other highly bound drugs, although examples in which such an interaction has had a clinically relevant effect are rare.

2. *Red Blood Cell Partitioning*

Most pharmacokinetic studies involve monitoring plasma or serum drug concentrations because of the ease of working with these matrices. Theoretically, blood concentrations of drug may be more meaningful in understanding a drug's pharmacokinetics, especially when the drug partitions into red blood cells. However, since therapeutic drug monitoring usually involves measurement of drug in plasma or serum (the standard for most hematological measurements), the extra effort required to develop blood assays does not seem to be justified. Instead, in vitro studies to determine the partitioning of drug into red blood cells provides a data base to estimate whole blood concentrations from plasma concentrations. However, if red blood cell partitioning is extensive or highly dependent on total drug concentration, development and utilization of a whole-blood assay for drug may be justified.

3. *Tissue Penetration*

Pharmacokinetics describe the behavior of drugs in the plasma compartment of animals and man, but can only imply whether the drug actually reaches its target tissue. Measuring tissue concentrations of drugs in animals often offers an analytical challenge, and measuring concentrations of drugs in tissues from humans is hindered by obvious sampling limitations. Nevertheless, determinations of tissue levels of drugs in animals can offer vital insight for interpreting the degrees of potency or toxicity in animals and their human relevance.

REFERENCES

Abramson, F. P., Y. Teffera, J. Kusmierz, R. C. Steenwyk, and P. G. Pearson, Replacing ^{14}C with stable isotopes in drug metabolism studies, *Drug Metab. Dispos. 24*:679–701 (1996).

Artursson, P. and J. Karlsson, Correlation between oral drug absorption in humans and apparent drug permeability coefficients in human intestinal epithelial (Caco-2) cells, *Biochem. Biophys. Res. Commun. 175*:880–885 (1991).

Berman, J., K. Halm, K. Adkison, and J. Shaffer, Simultaneous pharmacokinetic screening of a mixture of compounds in the dog using API LC/MS/MS analysis for increased throughput, *Med. Chem. 40*:3–5 (1997).

Campbell, D. B., Are we doing too many animal biodisposition investigations before phase I studies in man? A reevaluation of the timing and extent of ADME studies, *Eur. J. Drug Metab. Pharmacokinet. 19*:283–93 (1994).

Dain, J. G., J. M. Collins, and W. T. Robinson, A regulatory and industrial perspective of the use of carbon-14 and tritium isotopes in human ADME studies, *Pharm. Res. 11*:925–928 (1994).

de Sousa, G., N. Florence, B. Valles, P. Coassolo, and R. Rahmani, Relationships between in vitro and in vivo biotransformation of drugs in humans and animals: Pharmaco-toxicological consequences, *Cell Biol. Toxicol. 11*:147–153 (1995).

Evans, E. A., Self-decomposition of radiochemicals: Principles, control, observations and effects. Review 16, Amersham Corp., Arlington Heights, IL, 1982.

Gumbleton, M. and W. Sneader, Pharmacokinetic considerations in rational drug design. *Clin. Pharmacokinet. 26*:161–168 (1994).

Halm, K. A., K. Adkison, J. Berman, and J. E. Shaffer, N-in-one dosing in the dog: LC/MS as a tool for higher throughput in vivo pharmacokinetic screening of drug discovery lead candidate mixtures [Abstr] *Mol. Divers. Comb. Chem.* (1996).

Hiller, D. L., T. J. Zuzel, J. A. Williams, and R. O. Cole, Rapid scanning technique for the determination of optimal tandem mass spectrometric conditions for quantitative analysis, *Rapid Commun. Mass Spectrosc. 11*:593–597 (1997).

Hoffman, D. J., T. Seifert, A. Borre, and H. N. Nellans, Method to estimate the rate and extent of intestinal absorption in conscious rats using an absorption probe and portal blood sampling, *Pharm. Res. 12*:889–894 (1995).

ICH. International Conference on Harmonisation; Guideline on Validation of Analytical Procedures: Definitions and Terminology; Availability, *Fed. Reg. 60*:11260–2, March 1 (1995a).

ICH. International Conference on Harmonisation; Guideline on the Assessment of Systemic Exposure in Toxicity Studies; Availability, *Fed. Reg. 60*:11264–8, March 1 (1995b).

ICH. International Conference on Harmonisation; Guideline on Repeated Dose Tissue Distribution Studies; Availability, *Fed. Reg. 60*:11274–5, March 1 (1995c).

Kwon, Y. and P. B. Inskeep, Theoretical considerations on two equations for estimating the extent of absorption after oral administration of drugs, *Pharm. Res. 13*:566–569 (1996).

Liston, T. E., J. G. Baxter, B. C. Jones, F. MacIntyre, R. S. Obach, D. J. Rance, B. M. Silber, and P. Wastall, Predicting in vivo clearance and bioavailability in animals and man from in vitro metabolism data: A comparison of seven methods, *ISSX Abstr.* p. 297 (1996).

Loevinger, R. and M. Berman, Medical Internal Radiation Dose Committee (MIRD) Pamphlet No. 1—A schema for absorbed-dose calculations for biologically-distributed radionuclides, *J. Nucl. Med. Suppl. 1*:7 (1968).

Michelson, E. L., Calcium antagonists in cardiology: Update on sustained-release drug delivery systems, *Clin. Cardiol. 14*:947–950 (1991).

Mugford, C. A., M. Mortillo, B. A. Mico, and J. B. Tarloff, 1-Aminobenzotriazole-induced destruction of hepatic and renal cytochromes P450 in male Sprague-Dawley rats, *Fundam. Appl. Toxicol. 19*:43–49 (1992).

Nedelman, J. R., E. Gibiansky, and D. T. W. Lau, Applying Bailer's method for AUC confidence intervals to sparse sampling, *Pharm. Res 12*:124–128 (1995).

Nickerson, D. F. and S. M. Toler, Intraperitoneal and intraportal administration of droloxifene to the Sprague-Dawley rat: assessing the first-pass effect. *Xenobiotica 27*:627–632 (1997).

Obach, R. S., J. G. Baxter, B. C. Jones, T. E. Liston, F. MacIntyre, D. J. Rance, B. M. Silber, and P. Wastall, A comprehensive retrospective analysis of the prediction of human pharmacokinetic parameters from preclinical data using allometric scaling, *ISSX Abstr.* p. 264 (1996).

Olah, T. V., D. A. McLoughlin, and J. D. Gilbert, The simultaneous determination of mixtures of drug candidates by LC/API/MS/MS as an in vivo drug screening procedure, *Rapid Commun. Mass Spectrom. 11*:17–23 (1997).

Potchoiba, M. J., T. G. Tensfeldt, M. R. Nocerini, and B. M. Silber, A novel quantitative method for determining the biodistribution of radiolabeled xenobiotics using whole-body cryosectioning and autoradioluminography, *J. Pharmacol. Exp. Ther. 272*:953–962 (1995).

Prentis, R. A., Y. Lis, and S. R. Walker, Pharmaceutical innovation by the seven UK owned pharmaceutical companies (1964–85). *Br. J. Clin. Pharmacol. 25*:387–396 (1988).

Ritschel, W. A., *Handbook of Basic Pharmacokinetics*, 2nd ed., Drug Intelligence Publications, 1982.

Schurgers, N., J. Bijdendijk, J. J. Tukker, and D. J. A. Crommelin, Comparison of four techniques for studying drug absorption kinetics in the anesthetized rat in situ, *J. Pharm. Sci. 75*:117–119 (1986).

Smith, D. A., Design of drugs through a consideration of drug metabolism and pharmacokinetics, *Eur. J. Pharmacokinet. 3*:193–199 (1994).

Smith, D. A., B. C. Jones, and D. K. Walker, Design of drugs involving the concepts and theories of drug metabolism and pharmacokinetics. *Med. Res. Rev. 16*:243–266 (1996).

Stewart, B. H., O. H. Chan, N. Jezky, and D. Fleisher, Discrimination between drug candidates using models for evaluation of intestinal absorption, *Adv. Drug Deliv. Rev. 23*:27–45 (1997).

Tweedie, D. J., Higher throughput metabolic screening in drug discovery—more better, faster? *ISSX Abstr.* 1996.

van Bree, J., J. Nedelman, J.-L. Steimer, F. Tse, W. Robinson, and W. Niederberger, Application of sparse sampling approaches in rodent toxicokinetics: A prospective view. *Drug Inform. J. 28*:263–279 (1994).

White, R. E., Evolving Strategies of Drug Metabolism Support of Drug Discovery, *ISSX Abstr.* 1996.

Yee, S. In vitro permeability across Caco-2 cells (colonic) can predict in vivo (small intestinal) absorption in man—fact or myth, *Pharm. Res. 14*:763–766 (1997).

22

Clinical Drug Metabolism Studies

William F. Pool

Parke-Davis Pharmaceutical Research Division, Warner-Lambert Company, Ann Arbor, Michigan

I. INTRODUCTION

Human drug metabolism studies are perhaps the single most important metabolism study conducted in support of drug development. Human metabolism studies provide information related to major routes of drug disposition, recovery of dose, absorption, intersubject variability, as well as formation of active or toxic metabolites. Given the importance of this information to drug development, it may seem somewhat surprising that this information historically has often been generated in the later stages of clinical development. Late availability of information often meant that metabolism data would not be available to assist in design of the clinical development program. Realization of the benefits of early clinical metabolic information to the drug development plan is leading many investigators to incorporate a nonradiolabeled metabolic profiling component early in phase I clinical trials. Here, liquid chromatography–mass spectrometry (LC/MS/MS) is used for metabolic profiling with metabolite identification being assisted by information gained from the prior preclinical metabolism investigations. Clinical absorption, distribution, metabolism, and elimination (ADME) studies, targeted for late phase I to early phase II, employ radiolabeled drug, generally carbon 14 (^{14}C), which allows the identification of recovery, excretion routes, metabolic profile, as well as the identification of drug-derived metabolites.

The metabolic-profiling information obtained from clinical drug metabolism studies validates preclinical metabolism and toxicology studies relative to exposure, and thus the appropriateness of species chosen as well as support for pharmacology studies. This information also provides support for the validity of in vitro metabolism studies utilizing human tissue preparations.

Because of the rapid advancement of in vitro technologies, significant information on human metabolism of a new chemical entity (NCE) can be obtained before the initiation of clinical metabolism studies. This is partly due to the increasing availability of human tissue, as well as the advent of various expressed human metabolizing enzyme systems. In vitro information is generated in cellular subfractions (e.g., cytosol, microsomes, and such), whole cells (e.g., hepatocytes), and in some studies organ perfusions. The use of expressed enzyme systems assists in the identification of specific cytochrome P450 (CYP450) enzymes responsible for NCE metabolism, which when investigated along with preclinical induction–inhibition studies, can provide guidance for development of a clinical drug–drug interaction program.

The doses used in clinical metabolism studies are commonly much lower than those administered to the preclinical species and thus, can present significant challenges for the metabolism scientist. These challenges encompass metabolic profiling, metabolite isolation, and metabolite identification. The purpose of this chapter is to provide practical approaches to clinical drug metabolism studies along with some specific examples.

II. REGULATIONS AND GUIDELINES FOR CLINICAL METABOLISM STUDIES

Regulations and guidelines for the use of radiolabeled drugs in human, can be found in the current Code of Federal Regulations 21 CFR § 361.1 (1). Contained within these regulations are administrative requirements for the establishment of a Radioactive Drug Research Committee (RDRC) whose approval, as well as that of an appropriate Institutional Review Board (IRB), are necessary to conduct radiolabeled studies in human subjects. Also defined are regulations and guidelines pertaining to study populations (age, gender, size), magnitude and range of pharmacological dose, dosimetry (based on animal tissue distribution studies), maximum yearly exposure, and limits on radiation dose. It is this last point which provides guidance to the sponsor for the amount of radioactivity (μCi/subject) that may be administered to humans. 21 CFR § 361.1 (b) (3) states "The amount of radioactive material to be administered shall be such that the subject receives the smallest radiation dose with which it is practical to perform the study without jeopardizing the benefits to be obtained from the study." Thus, it is the objectives of the study that determine the amount of radioactivity administered, as long as exposure obtained from dosimetry calculations does not exceed the maximum yearly exposure limits. For single-dose administration, the maximum yearly exposure for whole-body, active blood-forming organs, lens of the eye, and gonads is 3 rem, whereas other organs may not exceed 5 rem.

A. Dose Selection

Clinical metabolism studies are often conducted before the identification of a clinically efficacious dosage. When efficacy data is unavailable, information obtained from phase

I tolerance studies are used to identify a well-tolerated, high-end dose. If efficacy information is available, the highest dose to be marketed is generally used in clinical metabolism studies.

The objectives of the clinical metabolism study determine the amount of radioactivity to be used, as long as exposure remains compliant with the regulations. Several issues relative to the radiolabeled dose need to be addressed while planning a clinical metabolism study.

1. Selection and Regiochemical Placement of Nuclide

Tritium (^3H) and carbon 14 (^{14}C) are the nuclides generally used in clinical metabolism studies. Issues involving ^3H exchange with protons (^1H) and possible isotope effects relative to metabolism, limit the usefulness of ^3H in metabolic studies. Carbon 14 does not have these liabilities and, thus, is the nuclide of choice for most metabolic studies. The regiochemical position of the nuclide in the NCE has far-reaching ramifications on the outcome and quality of the data obtained from these studies. The position of the label must be such that it is both chemically and metabolically stable.

2. Amount of Radioactivity and Exposure

In accordance with the spirit of 21 CFR § 361.1 (b) (3) (see Ref. 1), the amount of radioactivity administered and safety (exposure) are determined by the study objectives and dosimetry calculations, respectively. The results of a survey of pharmaceutical houses indicated that the amounts of radioactivity administered in clinical metabolism studies were variable (2). According to this study, approximately 50% of the respondents administered 100 μCi or more per subject (range, 10–300 μCi) of ^{14}C and 70% administered 200 μCi or more per subject (range, 50–1000 μCi) of ^3H. The results also showed that, although approximately 75% of the respondents used historical information in selecting the amount of radioactivity administered, safety was assured in all cases through dosimetry calculations with the rat identified as the most common species used. Generally, radiochemical instability (radiolysis) is not an issue in clinical metabolism studies; however, if the dose is low and thus the specific activity high, radiolysis may become a problem. If a high specific activity dose is planned, the possibility of radiochemical instability should be investigated before study initiation.

3. Dose Formulation

Ideally, clinical metabolism studies should use the targeted formulation; however, as with any clinical study, clinical metabolism studies require Good Manufacturing Practices (GMP) compliance. Equipment used to formulate clinical supplies often is not suitable for formulating radioactive doses, for a variety of reasons, including waste and equipment contamination; therefore, a solution or suspension dose is oftentimes administered. If a suspension dose is chosen care must be taken to ensure uniform distribution of radioactivity within the suspension.

4. Dose Regimen

Clinical ADME studies are generally conducted using a single-dose design; however, if substantially long half-lives of radioactivity are observed in the single-dose study, it may be appropriate to conduct a multiple dose ADME study. This study would provide multiple dose pharmacokinetic data for the parent drug and total radioactivity as well as allow identification of accumulating metabolites not detectable in a single-dose study. Study design must ensure steady-state conditions for both parent and total radioactivity and must be conducted using multiple doses of the same specific activity. Under these circumstances, however, mass balance data would be of questionable value. Recovery of radiolabel would be greater than 100% owing to accumulation from previous doses.

III. OBJECTIVES OF SINGLE-DOSE CLINICAL METABOLISM STUDY

Clinical metabolism studies are generally performed in four to eight healthy normal male volunteers in a controlled research facility after an overnight fast (2,3). Quantitative collection of urine and feces is performed over predetermined time intervals, projected to result in 95–100% recovery of administered dose. Serial blood samples are also taken for metabolic profiling as well as pharmacokinetic assessments of parent versus total radioactivity (parent and metabolites). Objectives of clinical metabolism studies include the following:

1. Obtain estimates of the rate and extent of absorption, as well as determination of the rate and amount of elimination from blood and excretory pathways (mass balance).
2. Provide biotransformation data in humans: This information can then be used to assess the relevance of preclinical studies (exposure; validation of appropriateness of toxicology species) and human in vitro data (validation of in vitro prediction studies).
3. Generate pharmacokinetic parameters for both parent and total ^{14}C (parent and metabolite; i.e., if $t_{1/2}$ of ^{14}C is longer than that of parent then a longer-lived metabolite may be present).

IV. SINGLE-DOSE CLINICAL METABOLISM STUDIES WITH [^{14}C]TACRINE

Tacrine (1,2,3,4-tetrahydro-9-acridinamine monohydrochloride monohydrate) is an acetylcholinesterase inhibitor developed for the treatment of senile dementia of the Alzheimer type. Oral bioavailability of tacrine is low and highly variable, even though it is rapidly absorbed (4–6). Tacrine also undergoes extensive first-pass metabolism, which is the most likely explanation for the low bioavailability observed. During the course of clinical development, reversible increases in serum alanine transaminase (ALT) were observed that were not seen in preclinical toxicology species.

Single-dose clinical metabolism studies with [^{14}C]tacrine were initially conducted in a nonblinded, sequential-dose, crossover design in six healthy normal male volunteers, ages 50–62 (7). Volunteers received a single 10-mg dose of [^{14}C]tacrine (100 μCi), followed by a 40-mg dose (100 μCi) 28 days later. The position of the ^{14}C label is shown in Fig. 1. This position is both chemically and metabolically stable, as evidenced by complete recovery of administered radioactivity in preclinical animal studies. Blood was collected over 48 h, and urine and feces were collected over 96 h postdose. Unchanged drug and several metabolites in plasma were analyzed by a validated high-performance liquid chromatography (HPLC) method, using fluorescence detection.

Figure 2 shows mean blood levels following a single 10-mg dose of [^{14}C]tacrine, and Fig. 3 shows mean cumulative excretion. Qualitatively similar results were obtained

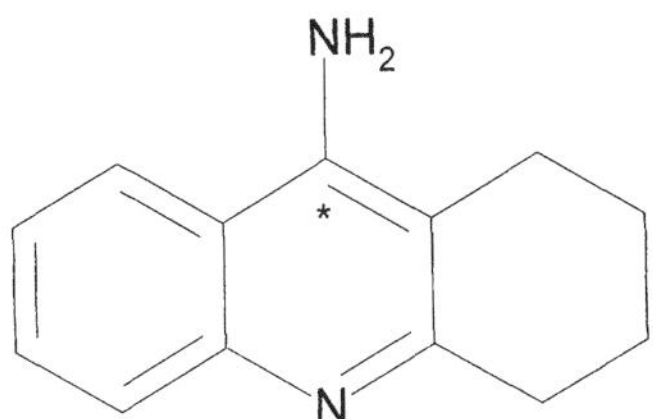

Fig. 1 Structure of tacrine: asterisk denotes position of ^{14}C label.

for the 40-mg dose. Radioactivity recoveries were similar following 10- and 40-mg doses (79 and 75%, respectively), with incomplete recovery observed. Careful analysis of the excretion data suggested a small, but significant, amount of radioactivity was recovered at the 96-h time point; however, data obtained from a repeated study at 40 mg (100 μCi) in which collection periods were extended to 2 weeks failed to significantly improve recovery. This lack of mass balance is suggestive of sequestration of tacrine-derived material.

Gradient HPLC radioactivity profiling of plasma and urine showed tacrine to be rapidly and extensively metabolized to a variety of mono- and dihydroxylated metabolites including dihydrodiol metabolites and phenol glucuronides, with profiles that were qualitatively similar to those obtained in rat and dog preclinical metabolic studies (7). Quantitative differences, however, were noted across the three species investigated and may play a role in the observed lack of toxicity seen in the toxicology species. This study confirmed the existence of a high first-pass effect previously seen in rat and dog.

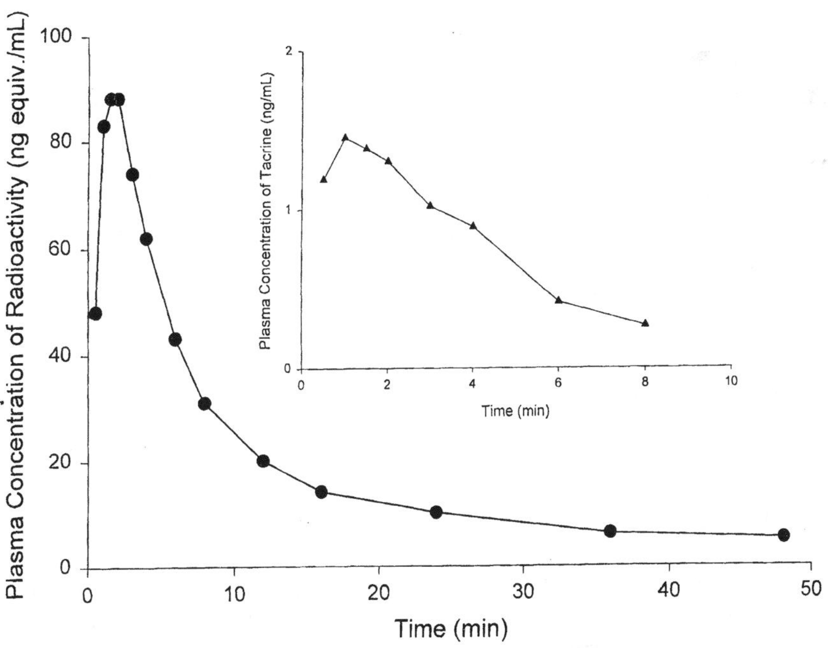

Fig. 2 Mean plasma levels following a single 10-mg oral solution dose of [^{14}C]tacrine to six healthy normal male volunteers.

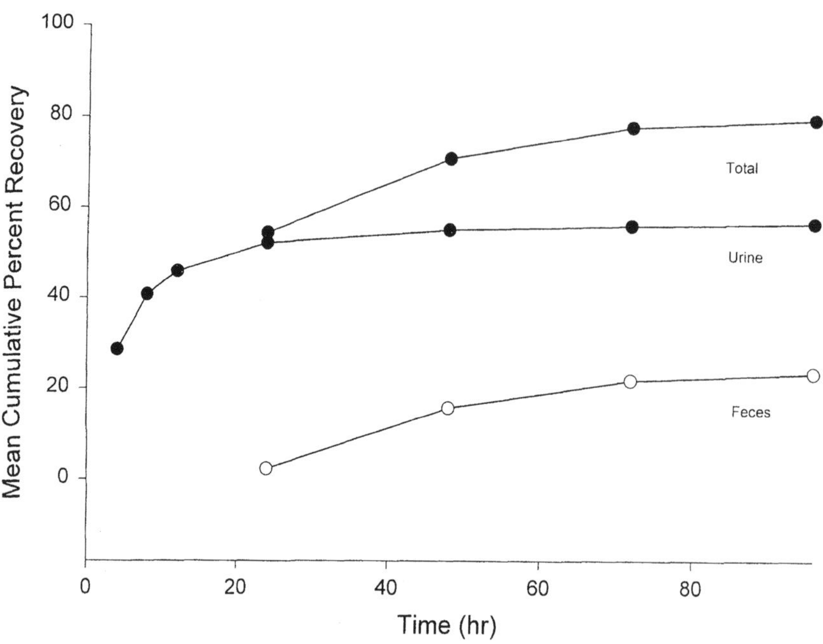

Fig. 3 Mean cumulative excretion of radioactivity following a single 10-mg oral solution dose of [^{14}C]tacrine to six healthy normal male volunteers.

In vitro metabolism studies of tacrine in rat, dog, and human liver microsomes showed NADPH-dependent formation of stable, protein-reactive, and cytotoxic metabolites (8–10). During the course of these investigations the authors showed similar microsomal tacrine metabolic profiles between the species studied, with differences observed attributed to quantitative differences in the rates of metabolism. In addition, in vitro studies conducted with heterologous-expressed human P450 enzymes and induction–inhibition studies in hepatic microsomes identified CYP1A2 as the major enzyme system responsible for tacrine metabolism and bioactivation (11). This enzyme is characterized as a ''high-affinity–low-capacity'' system and thus, saturable. CYP1A2 is inducible by cigarette smoking (12,13) and, indeed, studies have shown enhanced tacrine clearance in subjects who smoke (14). Significant intersubject variability in the expression of CYP1A2 has been reported (15,16) as well as ethnic and gender differences (17). Clinical studies with tacrine have reported wide interindividual variation in pharamacokinetic parameters consistent with the variable nature of CYP1A2 (4–6).

V. SINGLE-DOSE CLINICAL METABOLISM STUDY WITH [^{14}C]MILAMELINE

Milameline [(*E*)-1,2,5,6-tetrahydro-1-methyl-3-pyridinecarboxaldehyde-*O*-methyloxime monohydrochloride] is a nonspecific muscarinic agonist undergoing clinical trials for the treatment of age-related cognitive disorders. Previous metabolism studies in rat have identified urine as the primary excretion route following milameline administration. Metabolic profiling of rat urine showed that milameline is extensively metabolized, with the *N*-oxide derivative identified as a major metabolite.

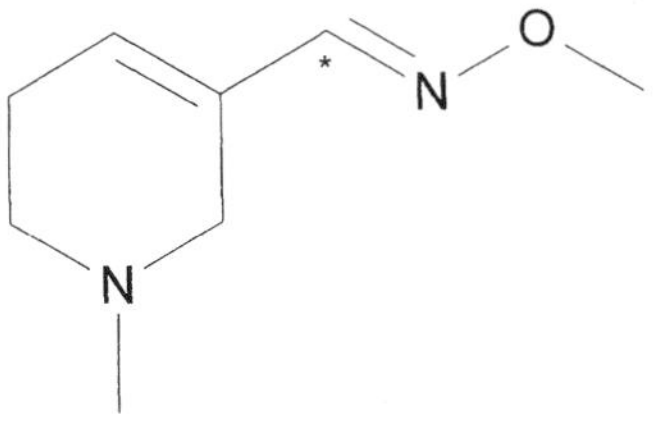

Fig. 4 Structure of milameline: asterisk denotes position of ^{14}C label.

A single-dose clinical metabolism study with [^{14}C]milameline was conducted in eight healthy normal male volunteers, ages 20–38. Volunteers received a single 2-mg–solution dose of [^{14}C]milameline (106 μCi). The position of the label is shown in Fig. 4. This position is both chemically and metabolically stable, as demonstrated by complete recovery in rat studies following [^{14}C]milameline administration. Urine and feces were collected for 96 h postdose. Recovery of radiolabel was essentially complete, with approximately 95% of dose recovered in urine.

Metabolic profiling of the initial 24-h urine collection showed that milameline is extensively metabolized to several metabolites. The major metabolite was identified as the *N*-oxide derivative by LC/MS/MS comparisons to reference standard. Identification of the other metabolites required the concentration and semipreparative HPLC purification of several hundred milliliters of urine before nuclear magnetic resonance (NMR) or LC/MS/MS analyses owing to the low dose administered.

VI. EXCRETION AND IDENTIFICATION OF METABOLITES IN HUMAN BILE FOLLOWING SINGLE-DOSE ADMINISTRATION OF [^{14}C]ATORVASTATIN

Atorvastatin ([*R*-(*R**, *R**)]-2-(4-fluorophenyl)-β,δ-dihydroxy-5-(1-methylethyl)-3-phenyl-4-[(phenylamino) carbonyl]-1*H*-pyrrole-1-heptanoic acid calcium salt; 2:1), is a recently approved HMG-CoA reductase inhibitor marketed for the treatment of hypercholesterolemia. The structure and position of ^{14}C label is shown in Fig. 5. Preclinical disposition studies showed that atorvastatin undergoes extensive first-pass metabolism, with biliary excretion of metabolites. A [^{14}C]atorvastatin clinical ADME study showed essentially complete recovery of radioactivity in feces, suggestive of a significant biliary contribution to high fecal recovery. To address this issue, a study was designed to measure biliary

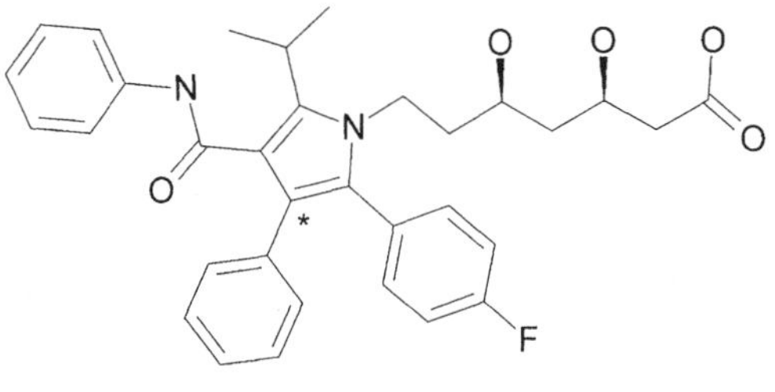

Fig. 5 Structure of atorvastatin: asterisk denotes position of ^{14}C label.

excretion and to profile and identify drug-derived metabolites in bile from postcholecystectomy patients fitted with a T-tube. Two subjects received a single 40-mg (50-μCi) dose of [^{14}C] atorvastatin. Urine, bile, and fecal samples were collected serially for 120 h postdose. In both subjects, most of the radioactivity was excreted in bile, with peak excretion occurring in the 12- to 24-h interval. Biliary excretion of radioactivity was prolonged with detectable amounts present through 120 h. Gradient HPLC radioactivity profiling of bile samples collected over the first 24 h showed the presence of eight components accounting for approximately 65% of biliary radioactivity. LC/MS analysis of these components showed the presence of hydroxylated metabolites, conjugates and β-oxidation products.

VII. FUTURE DIRECTIONS IN CLINICAL METABOLISM STUDIES

Realization of the importance of human metabolism information early in drug development will require the metabolism scientist to provide data as soon as possible. To conduct clinical ADME studies before late phase I increases the risk of needlessly exposing subjects to radioactivity, not to mention the significant expense associated with these studies should development be discontinued for safety and tolerability reasons. Metabolism information can be gleaned from early nonradiolabeled phase I studies by judicious use of LC/MS/MS analyses of plasma and urine. An obvious caveat, however, is that without the "tag" afforded by radiolabeled material, metabolites may be missed. This issue can be minimized if careful attention is paid to the design, conduct, and interpretation of preclinical in vitro and in vivo metabolic studies. The rapid advancement of in vitro technologies, particularly relative to increased availability of human tissue and heterologous expression systems, will permit timely availability of human metabolism information. It is possible that this could result in less preclinical in vivo information available at the start of clinical development. Steps must be taken to validate in vitro methods to ensure the usefulness of in vitro data.

Increased awareness of issues, such as drug–drug interactions, effects on metabolism, induction effects, and environmental and genetic factors, will influence protocol design. For example, the realization that polymorphism in metabolism may significantly affect the metabolism of a drug would suggest that studies be conducted in the respective polymorph populations.

In addition, with the reintroduction of combinatorial techniques in chemistry, and the development of in vitro mass-screening technologies, as well as in vivo cassette dosing (*N* in 1), the number of clinical metabolism studies will surely increase. This coupled with the ever increasing knowledge obtained from such projects as the Human Genome Project is affording new targets and allowing the identification of more selective and potent drugs. Also, the development of peptide and peptoid NCE's, or more appropriately NME (new molecular entities), are surely going to challenge metabolism scientists both in clinical and preclinical development. These new developments are going to require a constant reevaluation of how studies are conducted to maximize resource and time utilization.

The use of ^{14}C as a tag on NCEs has long been the mainstay of metabolism studies. However, the cost associated with handling and disposal of low-level β-emitters, as well as potential risk to human subjects from exposure to ionizing radiation have always plagued this technique. An advance in the area of LC interfaces with mass spectrometry has resulted in the development of LC/CRIMS (chemical reaction interface mass spectrometry; 18). This method permits the selective detection of stable isotopes such as ^{13}C

and ^{15}N. The interface involves thermal decomposition and oxidation of the analyte to such molecular species as $^{13}CO_2$ and ^{15}NO in a microwave-induced plasma (MIP) oven using an oxidizing gas, such as SO_2. The oxidized products are then detected by quadrupole or magnetic sector mass spectrometry using selected ion monitoring (SIM). This method has been used recently with favorable results compared with radioactivity detection (19,20).

The sensitivity and selectivity of the mass spectrometer coupled with LC-compatible interfaces, such as electrospray and atmospheric pressure chemical ionization (APCI), have made this the detector of choice for most LC applications. The selectivity of LC/MS and LC/MS/MS allows minimal chromatography and very short run times. Mass spectrometry has always played a valuable role in metabolite identification. With the increased use of isotopically stable labeled NCEs in metabolism studies, LC/MS/MS techniques, such as isotope dilution, will be useful for profiling. This will be especially true in matrices, such as plasma, in which often low levels of metabolites and thus radioactivity complicate the use of conventional radioactivity flow detection. Isotope dilution (cluster) is an MS/MS technique that profiles both the labeled analyte and unlabeled analyte, the ratio of which is determined by the enrichment of the administered isotopically stable labeled NCE (21). Another useful technique for LC/MS/MS profiling is metabolite mapping, which monitors compound class-selective fragmentation (constant neutral loss). Thus, the mass spectrometer screens for metabolites by looking for the loss of ''key'' fragment ions.

High-field nuclear magnetic resonance (NMR) spectroscopy can play an important role in the identification of drug-derived metabolites present in biological fluids (see Chap. 20). In the standard one-dimensional ^{1}H NMR experiment, however, resonances from drug-derived metabolites are often overlapped with those of endogenous compounds, thereby necessitating the use of spectral-editing methods for structural elucidation, or more often, time- and resource-consuming isolation and purification techniques. Recent technical advances have allowed the separation and NMR analysis of complex mixtures of drug-derived metabolites in biological fluids by directly coupling HPLC separation systems with ultrahigh-field NMR spectrometers (HPLC–NMR); the NMR being a novel type of LC detector delivering high structural content information (22–24). A field strength of 11.74 tesla (500 MHz for protons) is considered the minimum, whereas the increased sensitivity afforded by higher-field strengths expand the usefulness of this technology. The system comprises a fully automated HPLC system that is capable of diverting analytes of interest directly to the spectrometer in continuous-flow or stop-flow modes, or to a removable multiloop fraction collector. Current ultrahigh-field HPLC–NMR technology allows the detection of compounds at a low nanogram range, with the capability of performing high-level multidimensional experiments on-line. This exciting technology is complementary to mass spectrometry and will, no doubt, play an increasing role in metabolite identification by providing structural information earlier in the drug development process.

REFERENCES

1. 21 CFR 361.1 Revised April 1, 1996.
2. J. G. Dain, J. M. Collins, and W. T. Robinson, A regulatory and industrial perspective of the use of carbon-14 and tritium in human ADME studies, *Pharm. Res. 11*:925 (1994).
3. H. F. Schran and J. M. Jaffee, Pharmacokinetics in drug discovery and development: Clinical

studies, *Drugs and the Pharmaceutical Sciences*, Vol. 67, *Pharmacokinetics Regulatory, Industrial, Academic Perspectives* (P. G. Welling and F. L. S. Tse, eds.), Marcel Dekker, New York, 1995, p. 335.

4. D. R. Forsyth, G. K. Wilcock, R. A. Morgan, C. A. Truman, J. M. Ford, and C. J. C. Roberts, Pharmacokinetics of tacrine hydrochloride in Alzheimer's disease, *Clin. Pharmacol. Ther. 46*: 634 (1989).
5. A. Ahlin, A. Adem, T. Junthe, G. Ohman, and H. Nyback, Pharmacokinetics of tetrahydroaminoacridine: Relation to clinical and biochemical effects in Alzheimer's patients, *Int. Clin. Psychopharmacol. 7*:29 (1992).
6. N. R. Cutler, A. J. Sedman, P. Prior, B. A. Underwood, A. Selen, L. Balough, A. W. Kinkel, S. I. Gracon, and E. R. Gamzu, Steady state pharmacokinetics of tacrine in patients with Alzheimer's disease, *Psychopharmacol. Bull. 28*:231 (1990).
7. W. F. Pool, M. D. Reily, S. M. Bjorge, and T. F. Woolf, Metabolic disposition of the cognition activator tacrine in rats, dogs, and humans: Species comparisons, *Drug Metab. Dispos. 25*: 590 (1997).
8. S. Madden, V. Spaldin, R. N. Hayes, T. F. Woolf, W. F. Pool, and B. K. Park, Species variation in the bioactivation of tacrine by hepatic microsomes, *Xenobiotica 25*:103 (1995).
9. S. Madden, T. F. Woolf, W. F. Pool, and B. K. Park, An investigation into the formation of stable, protein-reactive, and cytotoxic metabolism from tacrine in vitro, *Biochem. Pharmacol. 46*:13 (1993).
10. T. F. Woolf, W. F. Pool, S. M. Bjorge, T. Chang, O. P. Goel, C. F. Purchase II, M. C. Schroder, K. L. Kunze, and W. F. Trager, Bioactivation and irreversible binding of the cognition activator tacrine using human and rat liver microsomal preparations: Species differences, *Drug Metab. Dispos. 21*:874 (1993).
11. V. Spaldin, S. Madden, W. F. Pool, T. F. Woolf, and B. K. Park, The effect of enzyme inhibition on the metabolism and activation of tacrine by human liver microsomes, *Br. J. Clin. Pharmacol. 38*:15 (1994).
12. W. Kalow and B.-K. Tang, Caffeine as a metabolic probe: Exploration of the enzyme-inducing effect of cigarette smoking, *Clin. Pharmacol. Ther. 49*:44 (1991).
13. S. Wanwimolruk, S. M. Wong, P. F. Coville, S. Viriyayudhakorn, and S. Thitiarchakul, Cigarette smoking enhances the elimination of quinine, *Br. J. Clin. Pharmacol. 36*:610 (1993).
14. D. Welty, W. Pool, E. Posvar, and A. Sedman, The effect of cigarette smoking on the pharmacokinetics and metabolism of Cognex in healthy volunteers, *Pharm. Res. 10 (Suppl.)*:S334 (1993).
15. M. A. Butler, N. P. Lang, J. F. Young, N. E. Caporaso, P. Vineis, R. B. Hayes, C. H. Teitel, J. P. Massengill, M. F. Lawsen, and F. F. Kadlubar, Determination of CYP1A2 and NAT2 phenotypes in human populations by analysis of caffeine urinary metabolites, *Pharmacogenetics 2*:116 (1992).
16. F. F. Kadlubar, Biochemical individuality and its implications for drug and carcinogen metabolism: Recent insights from acetyltransferase and cytochrome P4501A2 phenotyping and genotyping in humans, *Drug Metab. Rev. 26*:37 (1994).
17. M. V. Relling, J.-S. Lin, G. D. Ayers, and W. E. Evans, Racial and gender differences in *N*-acetyltransferase, xanthene oxidase, and CYP1A2 activities, *Clin. Pharmacol. Ther. 52*:643 (1992).
18. F. P. Abramson, CRIMS: Chemical reaction interface mass spectrometry, *Mass Spectrom. Rev. 13*:341 (1994).
19. F. P. Abramson, Y. Teffera, J. Kusmierz, R. C. Steenwyk, and P. G. Pearson, Replacing ^{14}C with stable isotopes in metabolism studies, *Drug Metab. Dispos. 24*:697 (1996).
20. C. A. Goldthwaite, Jr., F.-Y. Hsieh, S. W. Womble, B. J. Nobes, I. A. Blair, L. J. Klunk, and R. F. Mayol, Liquid chromatography/chemical reaction interface mass spectrometry as an alternate to radioisotopes for quantitative drug metabolism studies, *Anal. Chem. 68*:2996 (1996).

21. R. N. Hayes, W. F. Pool, M. W. Sinz, and T. F. Woolf, Recent developments in drug metabolism methodology, *Drugs and the Pharmaceutical Sciences*, Vol. 67, *Pharmacokinetics Regulatory, Industrial, Academic Perspectives* (P. G. Welling and F. L. S. Tse, eds.), Marcel Dekker, New York, 1995, p. 202.
22. J. C. Lindon, J. K. Nicholson, and I. D. Wilson, The development and application of coupled HPLC–NMR spectroscopy, *Adv. Chromatogr. 36*:315 (1996).
23. J. C. Lindon, J. K. Nicholson, I. D. Wilson, Direct coupling of chromatographic separations to NMR spectroscopy, *Prog. NMR Spectrosc. 29*:1 (1996).
24. S. A. Korhammer and A. Bernreuther, Hyphenation of high-performance liquid chromatography (HPLC) and other chromatographic techniques (SFC, GPC, GC, CE) with nuclear magnetic resonance (NMR): A review, *Fresenius J. Anal. Chem. 345*:131 (1996).

Index